EDITED BY
Paolo Fornasiero and **Mauro Graziani**

RENEWABLE RESOURCES *and* RENEWABLE ENERGY

A GLOBAL CHALLENGE

SECOND EDITION

CRC Press
Taylor & Francis Group
Boca Raton London New York

CRC Press is an imprint of the
Taylor & Francis Group, an **informa** business

CRC Press
Taylor & Francis Group
6000 Broken Sound Parkway NW, Suite 300
Boca Raton, FL 33487-2742

First issued in paperback 2016

© 2012 by Taylor & Francis Group, LLC
CRC Press is an imprint of Taylor & Francis Group, an Informa business

No claim to original U.S. Government works

Version Date: 20110803

ISBN 13: 978-1-138-19852-4 (pbk)
ISBN 13: 978-1-4398-4018-4 (hbk)

Library of Congress Cataloging-in-Publication Data

Renewable resources and renewable energy : a global challenge / editors Paolo Fornasiero and Mauro Graziani. -- 2nd ed.
    p. cm.
Includes bibliographical references and index.
ISBN 978-1-4398-4018-4 (alk. paper)
    1. Renewable energy sources. I. Fornasiero, Paolo, 1968- II. Graziani, Mauro.

TJ808.R458 2011
621.042--dc23   2011027261

Visit the Taylor & Francis Web site at
http://www.taylorandfrancis.com

and the CRC Press Web site at
http://www.crcpress.com

Contents

PART I   Technologies for Application and Utilization of Renewable Resources

PART II   Plastics and Materials from Renewable Resources

PART III   Technologies for Renewable Energy

PART IV   Trends, Needs, and Opportunities in Selected
Biomass-Rich Countries

Preface to the Second Edition

In the five years since the publication of the first edition of *Renewable Resources and Renewable Energy: A Global Challenge*, there has been increasing interest in the design of new eco-efficient chemical processes. This has prompted us to review the current work in these areas. In Part I, we have updated the chapters on bioplastics and bio-based chemical feedstocks and have added new chapters on aqueous-phase catalytic processing and glycerol upgrading. In Part II, we have included revised chapters on environmentally degradable plastics and the performance of bio-based materials and have added new chapters on the use of fish gelatin and its applications and on polymer blends based on rice straw and bagasse. In Part III, we have added chapters on the uses of bioethanol and on solid oxide fuel cells and electrolyzers and have revised the chapters on gasification technologies, on the uses of H_2 in transportation, and on molten carbonate fuel cells. In Part IV, we have added chapters dealing with developments in India, Brazil, and the African continent and have updated the work being carried out in Argentina.

Although it cannot be exhaustive, this book is meant to serve as a reference on renewable resources. General principles are critically discussed together with practical applications, considering especially industrial viewpoints. Particular attention has been devoted to discussing the situation of some representative transitional countries, whose richness in renewable resources offers tremendous opportunities.

We wish to thank all contributors to this second edition.

Preface to the First Edition

The continuous increase of world population, especially in developing countries, is exponentially enhancing the requirements of energy and raw materials. These goods represent the fundamental needs to ensure a correct human development. The limited availability of resources makes urgent the adoption of suitable strategies in the raw materials and energy sectors, in order to prevent an economic and social emergency that, in the absence of adequate strategies, even not considering pessimistic scenarios, will arrive sooner or later.

The fulfillment of the basic human needs, such as food, health, and acceptable environment quality, requires enormous amount of energy. It is therefore urgent to act immediately, before it is too late and the gap between developed and developing countries cannot be any longer filled. The lack of adequate actions can lead to social worldwide perturbations that can be easily foreseen on the bases of today's situation.

The large-scale use of renewable resources is becoming an urgent must. It is generally accepted that the renewable energy produced nowadays is well below the world potentiality and it contributes only marginally to the human needs. This is due not only to economic reasons, related to the cost of traditional–fossil resources and to political choices, but also to technological limitations. The real problem is to know when science and technology will be able to answer to the open questions regarding renewable materials and energy, allowing the adoption on global scale of well-considered decision before the impoverishment of traditional energy sources becomes an unsolvable problem. Science and technological development are the key tools for the achievement of these objectives. Therefore we must become conscious that the progresses of scientific research and of its applications represent the primary way to solve the major human problems. This "scientific optimism" cannot solve the problem by itself, especially in the developing countries. In fact, in those countries the lack of infrastructures and human resources with adequate scientific knowledge often precludes the possibility of carrying out innovative research as well as the adoption of technologies derived from the developed world.

Many are the proposed sustainable solutions in the energy sector, but so far, only few are competitive with the use of fossil hydrocarbons. These, besides being our primary energy source, are important building block for the synthesis of most of the chemical products that we commonly use every day. Therefore, the end of oil will lead not only to energy crisis, but it will affect also the availability of most of the products that satisfy our elementary needs. In fact, 90% of the organic substances derive, through chemical transformations/reactions, from seven oil derivates: ethylene, propylene, butenes, benzene, toluene, xylene, and methane. Approximately 16% of oil is transformed into chemicals. Notable is the fact that the extraction of oil is getting more and more expensive. In 1920, the energy of 1 barrel of oil was sufficient for the extraction and the refinement of 50 barrels of oil. Nowadays, with the same energy input it is possible to get only 5 barrels.

The replacement of oil with new energy/raw feedstocks should not be our primary target. Our reasonable objective must be the maximization of sources diversification. Some general aspects must be constantly kept in mind. First, each geographic area has its own characteristic and resources. Then, the distance between energy/raw materials sources and the cities, where the electricity and the products are used, requires huge investments for transportation and distribution.

Beside the investments for the production, we have always to consider also those related to the exploitation. The discontinuous availability of some energy/feedstock sources can be also a serious problem.

However, the richness in biodiversity, even considering the sensitivity to climate changes, and the exploitation of agro-overproduction represent a great opportunity for some developing countries. The possible use of agro-food waste for countries lacking in fossil resources, the needs of diversification of resources, and the development of national/local capacities is an additional big challenge.

Great attention is dedicated nowadays to the so-called "hydrogen economy." It must be immediately clarified that hydrogen is not an energy source, but it is an energy vector and it can be considered renewable. Its utilization in fuel cells leads to water as final product, and from water it can be produced using solar energy. However, besides some promising experimental successes, today hydrogen is still mainly produced from methane. New, clean, and efficient methods for hydrogen production are necessary to really move to the "hydrogen economy." In any case, science and technology, with their limitations, are the only way to the solution of the problem.

The importance of the topics related to the exploitation of renewable resources and renewable energy has been reflected in several international institutions and programs on global level. European Union and United States have been coordinating their strategic plans setting up the working groups and cooperative programs. Need of promotion of these issues, especially in developing countries, is underlined also within UNIDO programs on renewable energy.

Particular attention to the technologies of renewable feedstocks exploitation has been paid by International Centre for Science and High Technology of UNIDO. Series of awareness and capacity building programs together with pilot projects promotion in developing countries have been organized by ICS-UNIDO. In fact, their effort has inspired the present initiative resulting in the preparation of a comprehensive survey in this field. We appreciate the collaboration of Professor Stanislav Miertus in this initiative.

Editors

Paolo Fornasiero received his PhD in heterogeneous catalysis in 1997. A postdoctoral fellow at the Catalysis Research Center of the University of Reading (United Kingdom), he became an assistant professor of inorganic chemistry at the University of Trieste in 1998, and subsequently an associate professor in 2006. His scientific interests are in the technological application of material science and heterogeneous catalysis to the solution of environmental problems, such as the design of innovative materials for catalytic converters, the development of catalysts for the reduction of nitrogen oxides under oxidizing conditions, the photo-catalytic degradation of pollutants, and the design of new catalysts for the production and purification of hydrogen to be used in fuel cells. He is the coauthor of more than 150 publications in international journals and books, 3 patents, and a number of communications to national and internationals meetings, in many cases as invited lecturer. He was awarded the Stampacchia Prize in 1994 for his first publication, and he received the Nasini Gold Medal in 2005, awarded by the Italian Chemical Society, for his contribution to research in the field of inorganic chemistry.

Mauro Graziani has been full professor of inorganic chemistry at the University of Trieste since 1975. His scientific interests range from organometallic chemistry, to homogeneous catalysis, to catalyst heterogeneization on various types of supports, and, finally, to heterogeneous catalysis using transition metals supported on different oxides. In particular, with regard to the last-mentioned system, hydrogen production, water–gas shift, and NO + CO reactions have been studied. Graziani is the coauthor of more than 200 publications, 4 patents, and has been invited to present lectures at the most prestigious congresses in the field. He was a visiting research associate at The Ohio State University and MIT; visiting professor at the Universities of Cambridge, Campinas, and Zaragoza; UNESCO scientific advisor; pro-rector at the University of Trieste; dean of the Faculty of Science of the University of Trieste; vice president of Elettra Synchrotron; and vice president of the Area Science Park of Trieste. He is also an associate fellow of the Third World Academy of Science.

Contributors

Adelaide Maria de Souza Antunes
Brazilian Institute of Industrial Property
and
School of Chemistry
Federal University of Rio de Janeiro
Rio de Janeiro, Brazil

Carlos R. Apesteguía
Consejo Nacional de Investigaciones Científicas
   y Técnicas
Instituto de Investigaciones en Catálisis y
   Petroquímica
Universidad Nacional del Litoral
Santa Fe, Argentina

Donato Alexandre Gomes Aranda
Department of Chemical Engineering
Federal University of Rio de Janeiro
Rio de Janeiro, Brazil

Roberto J. Avena-Bustillos
Department of Biological and Agricultural
   Engineering
University of California, Davis
Davis, California

and

Agricultural Research Service
Western Regional Research Center
United States Department of Agriculture
Albany, California

Arianna Barghini
Department of Chemistry and Industrial
   Chemistry
University of Pisa
Pisa, Italy

Francesco Basile
Dipartimento di Chimica Industriale e dei
   Materiali
Alma Mater Studiorum
Università di Bologna
Bologna, Italy

Peter J. Bechtel
Agricultural Research Service
Subarctic Agricultural Research Unit
United States Department of Agriculture
Kodiak, Alaska

Drew J. Braden
Department of Chemical and Biological
   Engineering
University of Wisconsin-Madison
Madison, Wisconsin

Miguel Brandão
Institute for Environment and Sustainability
Joint Research Centre
European Commission
Ispra, Italy

Michael Carus
Nova-Institut GmbH
Hürth, Germany

Francesco Cherchi
Chemtex Italia Srl
Gruppo Mossi & Ghisolfi
Tortona, Italy

Emo Chiellini
Department of Chemistry and Industrial
   Chemistry
University of Pisa
Pisa, Italy

Bor-Sen Chiou
Agricultural Research Service
Western Regional Research Center
United States Department of Agriculture
Albany, California

Stefania Cometa
Department of Chemistry and Industrial
   Chemistry
University of Pisa
Pisa, Italy

Andrea Corti
Department of Chemistry and Industrial
    Chemistry
University of Pisa
Pisa, Italy

Loredana De Rogatis
Consiglio Nazionale delle Ricerche
Istituto di Chimica dei Composti
    OrganoMetallici
Firenze, Italy

Renzo Di Felice
Dipartimento di Ingegneria Chimica e di
    Processo "G.B. Bonino"
Università degli Studi di Genova
Genova, Italy

Tommaso Di Felice
Chemtex Italia Srl
Gruppo Mossi & Ghisolfi
Tortona, Italy

James A. Dumesic
Department of Chemical and Biological
    Engineering
University of Wisconsin-Madison
Madison, Wisconsin

Mohamed El-Newehy
International Centre for Science and High
    Technology
United Nations Industrial Development
    Organization
Trieste, Italy

and

Department of Chemistry
Tanta University
Tanta, Egypt

Paolo Fornasiero
Department of Chemical and Pharmaceutical
    Sciences
and
Consiglio Nazionale delle Ricerche
Istituto di Chimica dei Composti
    OrganoMetallici
University of Trieste
Trieste, Italy

Greg M. Glenn
Agricultural Research Service
Western Regional Research Center
United States Department of Agriculture
Albany, California

Raymond J. Gorte
Department of Chemical and Biomolecular
    Engineering
University of Pennsylvania
Philadelphia, Pennsylvania

Michael D. Gross
Department of Chemical Engineering
Bucknell University
Lewisburg, Pennsylvania

Elif I. Gürbüz
Department of Chemical and Biological
    Engineering
University of Wisconsin-Madison
Madison, Wisconsin

Juliane Haufe
Department of Chemistry
Utrecht University
Utrecht, the Netherlands

Peter Heidebrecht
Process System Engineering Group
Max Planck Institute for Dynamics of Complex
    Technical Systems
Magdeburg, Germany

Barbara Hermann
PepsiCo International
Leicester, United Kingdom

Syed H. Imam
Agricultural Research Service
Western Regional Research Center
United States Department of Agriculture
Albany, California

Arvind Lali
DBT-ICT Centre for Energy Biosciences
Institute of Chemical Technology
Mumbai, India

Tara H. McHugh
Agricultural Research Service
Western Regional Research Center
United States Department of Agriculture
Albany, California

Marco Merlo
Dipartimento di Ingegneria Chimica e di
    Processo "G.B. Bonino"
Università degli Studi di Genova
Genova, Italy

Stanislav Miertus
International Centre for Applied Research
    and Sustainable Technology
Bratislava, Slovakia

Ramani Narayan
Department of Chemical Engineering
    and Materials Science
Michigan State University
East Lansing, Michigan

Ademola Olufolahan Olaniran
School of Biochemistry, Genetics, and
    Microbiology
University of KwaZulu-Natal
Durban, Republic of South Africa

William J. Orts
Agricultural Research Service
Western Regional Research Center
United States Department of Agriculture
Albany, California

Piero Ottonello
Chemtex Italia Srl
Gruppo Mossi & Ghisolfi
Tortona, Italy

Mario Pagliaro
Consiglio Nazionale delle Ricerche
Istituto per lo Studio dei Materiali
    Nanostrutturati
Palermo, Italy

Martin K. Patel
Department of Chemistry
Utrecht University
Utrecht, the Netherlands

Nei Pereira, Jr.
Department of Biochemical Engineering
Federal University of Rio de Janeiro
Rio de Janeiro, Brazil

Dorsamy (Gansen) Pillay
School of Biochemistry, Genetics, and
    Microbiology
University of KwaZulu-Natal
and
Department of Biotechnology
Durban University of Technology
Durban, Republic of South Africa

and

National Research Foundation of South Africa
Pretoria, Republic of South Africa

Michele Rossi
Dipartimento di Chimica Inorganica,
    Metallorganica e Analitica
Università degli Studi di Milano
Milano, Italy

Eduardo Falabella Sousa-Aguiar
Centro de Pesquisas Leopoldo Américo
    Miguez de Mello
Petrobras Research Centre
and
Department of Organic Processes
School of Chemistry
Federal University of Rio de Janeiro
Rio de Janeiro, Brazil

Kai Sundmacher
Process Systems Engineering
Otto von Guericke University
and
Process System Engineering Group
Max-Planck-Institute for Dynamics of Complex
    Technical Systems
Magdeburg, Germany

Paolo Torre
Chemtex Italia Srl
Gruppo Mossi & Ghisolfi
Tortona, Italy

Ferruccio Trifirò
Dipartimento di Chimica Industriale e dei
  Materiali
Alma Mater Studiorum
Università di Bologna
Bologna, Italy

Herman van Bekkum
Department of Chemical Engineering
Delft University of Technology
Delft, the Netherlands

Martin Weiss
Department of Chemistry
Utrecht University
Utrecht, the Netherlands

Part I

Technologies for Application and
Utilization of Renewable Resources

1 Bioplastics
Principles, Concepts, and Technology

Ramani Narayan

CONTENTS

1.1 INTRODUCTION

Bio-based and biodegradable plastics based on renewable biomass carbon feedstock offers the intrinsic value proposition of a reduced carbon footprint depending on the amount of renewable carbon in the product and in harmony with the rates and timescales of natural biological carbon cycle. The use of biomass feedstocks contributes to a country's economic growth, especially developing countries, all of which have abundant biomass resources, and the potential for achieving self-sufficiency in materials. Bio-based plastics must be organic and contain in whole or part biogenic carbon (carbon from biological sources) to be classified as bio-based ("bio"). Identification and quantification of bio-based content is based on the radioactive C-14 signature associated with (new) biocarbon and is measured as the percent weight of biocarbon to the total organic carbon present in the product. Using experimentally determined bio (renewable) carbon content values, one can calculate the intrinsic CO_2 emissions reduction achieved by substituting petro carbon with biocarbon—the material carbon footprint value proposition. It is equally important to calculate the carbon footprint arising from all operations involved in converting feedstock to product—the process carbon footprint. In addition, environmental impacts other than carbon for all operations need to be calculated. The process carbon and environmental footprint is calculated using life cycle assessment (LCA) methodology. However, process and product improvements and end-of-life options with respect to energy use, and environmental emissions are occurring at a rapid pace. Therefore static LCA's based on old or outdated data or end-of-life scenarios are misleading and provide the wrong picture.

Biodegradability in concert with disposal options like composting (compostable plastic), offers a viable, end-of-life option to completely remove single use, short-life disposable products like packaging and consumer articles from the environmental compartment via microbial assimilation. However, not all bio-based polymers are biodegradable, and not all biodegradable polymers are bio-based. Most importantly, complete biodegradability (complete utilization of the polymer by the microorganisms present in the disposal environment) is necessary as per ASTM and ISO standards otherwise there will be serious health and environmental consequences.

Bio-based products technology platform is discussed with examples of emerging bio-based products.

1.2 BIO-BASED PLASTICS/PRODUCTS

1.2.1 BIOMASS/RENEWABLE CARBON DRIVERS

There is an abundance of natural, renewable biomass resources as illustrated by the fact that the primary production of biomass estimated in energy equivalents is 6.9×1017 kcal/year [1]. Mankind utilizes only 7% of this amount, that is, 4.7×10^{16} kcal/year. In terms of mass units, the net photosynthetic productivity of the biosphere is estimated to be 155 billion ton/year [2] or over 30 ton per capita, and this is the case under the current conditions of nonintensive cultivation of biomass. Forests and croplands contribute 42% and 6%, respectively, of that 155 billion ton/year. The world's plant biomass is about 2×10^{12} ton and the renewable resources amount to about 10^{11} ton/year of carbon, of which starch provided by grains exceeds 10^9 ton (half of which comes from wheat and rice) and sucrose accounts for about 10^8 ton. Another estimate of the net productivity of the dry biomass gives 172 billion ton/year, of which 117.5 and 55 billion ton/year are obtained from terrestrial and aquatic sources, respectively [3].

Fortunately, we are growing trees faster than they are being consumed, although sometimes the quality of the harvested trees is superior to those being planted. Agriculture represents only a small fraction of these vast biomass resources, and this acreage does not include idle croplands and pastures. Again, these figures clearly illustrate the potential for biomass utilization.

It is estimated that U.S. agriculture accounts directly and indirectly for about 20% of the GNP by contributing $750 billion to the economy through the production of foods and fiber, the manufacture of farm equipment, the transportation of agricultural products, etc. It is also interesting that while agricultural products contribute to the U.S. economy with $40 billion of exports, and each billion of export dollars creates 31,600 jobs (1982 figures), foreign oil imports drains the economy and makes up 23% of the U.S. trade deficit (U.S. Department of Commerce 1987 estimate). Given these scenarios of abundance of biomass feedstocks, the value added to a country's economy, it seems logical to pursue the use of agricultural and biomass feedstocks for production of materials, chemicals, and fuels [4].

Biomass-derived materials are being produced at substantial levels. For example, paper and paperboard production from forest products was around 139 billion lb in 1988 [5], and biomass-derived textiles production around 2.4 billion pounds [6]. About 3.5 billion pounds of starch from corn is used in paper and paperboard applications, primarily as adhesives [7]. However, biomass use in production of plastics, coatings, resins, composites, and other articles of commerce is negligible. These areas are dominated by synthetics derived from oil and represent the industrial materials of today.

1.2.2 ENVIRONMENTAL DRIVERS

Global warming/climate change issues are front and center on the agenda of nations, companies, and the public at large. Carbon footprint reductions related to products, processes, and energy are being increasingly researched by companies and academe. Replacement of petro/fossil carbon in products with bio/renewable carbon offers a reduced carbon footprint as detailed later.

End-of-life issues like what happens to a product after use when it enters the waste stream is becoming increasingly important. Recycling, waste-to-energy, and biodegradability in targeted biological disposal systems like composting and anaerobic digesters are becoming important design criteria for products. This has opened up new market opportunities for developing bio-based and biodegradable/compostable products as the next generation of sustainable materials that meets ecological and economic requirements—ecoefficient products [8–11].

1.2.3 CARBON VALUE PROPOSITION FOR BIOPLASTICS

Carbon is the major basic element that is the building block of polymeric materials and fuels—bio-based products, petroleum-based products, biotechnology products, fuels, and even life itself. Therefore, discussions on sustainability, sustainable development, and environmental responsibility center on the issue of managing carbon (carbon-based materials) in a sustainable and environmentally responsible manner. Indeed, the burning issue of today is concerns over increasing man-made CO_2 emissions with no offsetting fixation and removal of the released CO_2. Reducing our carbon footprint is a major issue facing us today. The use of annually renewable biofeedstocks for manufacture of plastics and products offers an intrinsic zero or neutral carbon footprint value proposition.

The intrinsic "zero carbon" value proposition is best explained by reviewing and understanding nature's biological carbon cycle. Nature cycles carbon through various environmental compartments with specific rates and timescales (see Figure 1.1).

Carbon is present in the atmosphere as inorganic carbon in the form of CO_2. The current levels of CO_2 in the atmosphere are around 380 ppm (parts per million). This life-sustaining heat-trapping value of CO_2 in the atmosphere (maintains the earth's temperature) is changing to life threatening because of increasing man-made carbon (CO_2) and other heat-trapping gas emissions to the atmosphere. While one may debate the severity of effects associated with this or any other target level of CO_2, there can be no disagreement that uncontrolled, continued increase in levels of CO_2 in the atmosphere will result in a slow perceptible rise of the earth's temperature, global warming, and with it associated severity of effects affecting life on this planet as we know. It is therefore prudent and necessary to try and maintain current levels—the "zero carbon" approach. This can best be done by using renewable biomass crops as feedstocks to manufacture our carbon-based products, so that the CO_2 released at the end of life of the product is captured by planting new crops in the next season. Specifically, the rate of CO_2 release to the environment at end of life equals the rate of photosynthetic CO_2 fixation by the next-generation crops planted—a "zero carbon" footprint. In the case of fossil feedstocks, the rate of carbon fixation is in millions of years while the end-of-life release rate into the environment is in 1–10 years—the math is simple, this is not sustainable and results in more CO_2 release than fixation, resulting in an increased carbon footprint, and with it the attendant global warming and climate change problems [12].

Based on the earlier carbon-cycle discussions and basic stoichiometrics, for every 100 kg of polyolefin (polyethylene [PE] and polypropylene) or polyester manufactured from a fossil feedstock, there is a net 314 kg CO_2 (85.7% fossil carbon) or 229 kg of CO_2 (62.5% fossil carbon) released into

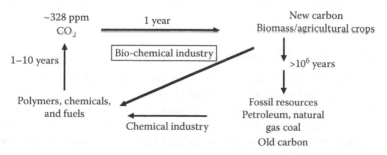

FIGURE 1.1 Global carbon cycle showing rates and timescales.

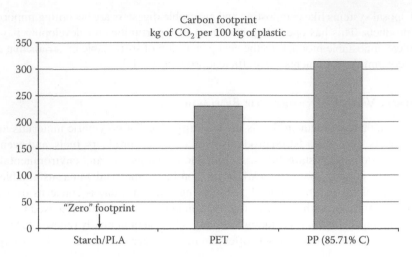

FIGURE 1.2 Intrinsic material's carbon footprint.

the environment, respectively, at end of life. However, if the polyester or polyolefin is manufactured from a biofeedstock, the net release of CO_2 into the environment is zero because the CO_2 released is fixed immediately by the next crop cycle (Figure 1.2).

This is the fundamental intrinsic value proposition for using a bio/renewable feedstock, and is typically ignored during life-cycle assessment (LCA) presentations. Incorporating bio-content into plastic resins and products would have a positive impact—reducing the carbon footprint by the amount of biocarbon incorporated, for example, incorporating 29% biocarbon content, using, say, cellulose or starch into a fossil-based polyolefin resin offers an intrinsic CO_2 emissions reduction of 42%—the material carbon footprint reductions. These are significant environmental benefits that accrue for using bio-based plastics.

However, another important consideration that must be taken into account is the CO_2 emissions arising from the conversion of the feedstock to product, CO_2 emissions during product use, and ultimate disposal. The major contributory component in this step is the fossil carbon energy usage. Currently, in the conversion of biofeedstocks to product, for example, corn to poly(lactic acid) (PLA) resin, fossil carbon energy is used. PLA is a bio-based and biodegradable plastic that has found major commercial applications. The CO_2 released per 100 kg of plastic during the conversion process for biofeedstocks as compared to fossil feedstock is in many cases higher, as in the case of PLA. However, in the PLA case [13], the total (net) CO_2 released to the environment taking into account the intrinsic carbon footprint as discussed earlier is lower, and will continue to get even better, as process efficiencies are incorporated and renewable energy is substituted for fossil energy (Figure 1.3).

For PLA and other bio-based products, it is important to calculate the conversion "carbon costs" using LCA tools, and ensure that the intrinsic "neutral or zero carbon" footprint is not negated by the conversion "carbon costs" and the net value is lower than the product being replaced from feedstock to product or resin manufacture.

1.2.4 Biocarbon Content Determination

In order to calculate the intrinsic CO_2 reductions from incorporating biocarbon content, one has to identify and quantify the bio-based carbon content [12,15].

As shown in Figure 1.4, ^{14}C signature forms the basis for identifying and quantifying bio-based content. The CO_2 in the atmosphere is in equilibrium with radioactive $^{14}CO_2$. Radioactive carbon is formed in the upper atmosphere through the effect of cosmic ray neutrons on ^{14}N. It is rapidly

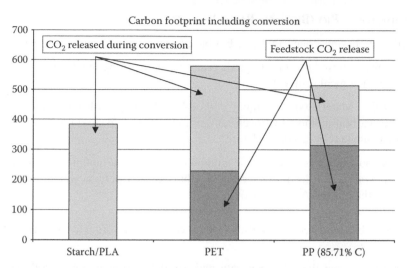

FIGURE 1.3 Material and process carbon footprint. (From Narayan, R., *ACS Symp. Ser.*, 939, Chapter 18, 282, 2006. With permission.)

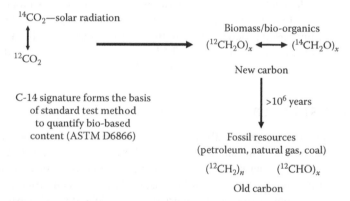

FIGURE 1.4 Carbon-14 methodology to determine bio content—ADTM D6866. (From Narayan, R., *ACS Symp. Ser.*, 939, Chapter 18, 282, 2006. With permission.)

oxidized to radioactive $^{14}CO_2$, and enters the Earth's plant and animal life ways through photosynthesis and the food chain. Plants and animals that utilize carbon in biological food chains take up ^{14}C during their lifetimes. They exist in equilibrium with the ^{14}C concentration of the atmosphere, that is, the numbers of C-14 atoms and nonradioactive carbon atoms stay approximately the same over time. As soon as a plant or animal dies, they cease the metabolic function of carbon uptake; there is no replenishment of radioactive carbon, only decay. Since the half-life of carbon is around 5730 years, the fossil feedstocks formed over millions of years will have no ^{14}C signature. Thus, by using this methodology, one can identify and quantify bio-based content. ASTM subcommittee D20.96 [14] has codified this methodology into a test method (D6866) to quantify bio-based content. D6866 test method [15] involves combusting the test material in the presence of oxygen to produce carbon dioxide (CO_2) gas. The gas is analyzed to provide a measure of the products. $^{14}C/^{12}C$ content is determined relative to the modern carbon-based oxalic acid radiocarbon standard reference material (SRM) 4990c (referred to as HOxII).

1.2.5 Terminology: Bio (Biomass)-Based Plastics

From the earlier discussions, one can define bio-based or biomass-based plastic as follows:

Bio-based or biomass-based plastics: Organic material/s containing in whole or part biogenic carbon (carbon from biological sources).

Organic material/s: Material(s) containing carbon-based compound(s) in which the carbon is attached to other carbon atom(s), hydrogen, oxygen, or other elements in a chain, ring, or three-dimensional structures (IUPAC nomenclature).

Bio content: The bio content is based on the amount of biogenic carbon present, and defined as the amount of *biocarbon* in the plastic as fraction weight (mass) or percent weight (mass) of the total organic carbon in the plastic (ASTM D6866).

% bio or bio-based content = Bio(organic) carbon/total (organic carbon) × 100

The U.S. Congress passed the Farm Security and Rural Investment Act of 2002 (P.O. 107–171) and expanded by the Food, Conservation, and Energy Act of 2008 (2008 Farm Bill) to increase the purchase and use of bio-based products. The U.S. Department of Agriculture (USDA) was charged with developing guidelines for designating bio-based products and to publish a list of designated bio-based product classes for mandated federal purchase—the USDA biopreferred program. In addition, the biopreferred program developed a voluntary labeling program for the broad-scale consumer marketing of bio-based products. The methodology described earlier for identifying and quantifying bio-based content and the use of ASTM D6866 to establish bio-based content of products forms the basis for the USDA biopreferred program [16].

1.3 END-OF-LIFE STRATEGY: COMPOSTING

Currently, most products are designed with limited consideration to its ecological footprint especially as it relates to its ultimate disposability. Of particular concern are plastics used in single-use disposable packaging and consumer goods. Designing these materials to be completely biodegradable and ensuring that they end up in an appropriate disposal system is environmentally and ecologically sound. For example, composting our biodegradable plastic and paper waste along with other "organic" compostable materials like yard, food, and agricultural wastes can generate much-needed carbon-rich compost (humic material). Compost-amended soil has beneficial effects by increasing soil organic carbon, increasing water and nutrient retention, reducing chemical inputs, and suppressing plant disease. Composting is increasingly a critical element for maintaining the sustainability of our agriculture system [17,18].

There exists confusion between the terms bio-based and biodegradability, and these are erroneously used interchangeably. Not all bio-based products are biodegradable and not all biodegradable products are bio-based. Bio-based refers to the origins of the carbon in the plastic or product, and its value proposition is for reducing the carbon footprint. Biodegradability is an end-of-life option with disposal systems like composting and anaerobic digestion. It is a functional property attribute to be designed and engineered into a product when needed or necessary. Thus, single-use, short-life, disposable packaging and consumer goods lend themselves to biodegradability design in concert with disposal systems like composting, anaerobic digestors, soil, and marine environment.

Unfortunately, there is much confusion, misunderstanding, misinformation, and even misleading claims made on terms like biodegradability, compostability, anaerobic digestion, landfill biodegradation, and marine biodegradation. Claims are made without substantiation with hard scientific data or the data provided has little or nothing to do with substantiating biodegradation.

The rationale as to why and how biodegradation of plastics is good for the environment has been lost. Making the plastic polymer to break down into small fragments, even making them so small that they are invisible to the naked eye by chemical (hydrolytic, oxidative, or photo) or biological means is not good for the environment and could have serious negative environmental consequences (as shown later in this chapter). In other words, "degradation" or "partial biodegradation" is not an acceptable option. Environmental biodegradability is good if and only if the degraded fragments are completely consumed by the microorganism present in the disposal environment—that is removed from the environment and safely enters the food chain of the microorganisms. Many papers in the literature report on how to design and engineer polymers to break down in the environment and manufacturers offer products that are designed to break down in the environment. However, little or no evidence is offered that these fragments are completely consumed by the microorganisms present in the disposal environment in a reasonable defined time period, or evidence presented shows partial consumption of the degraded fragments [19,20].

1.3.1 Measurement of Biodegradability

Microorganisms use the carbon substrates to extract chemical energy for driving their life processes by aerobic oxidation of glucose and other readily utilizable C-substrates as shown by the equation in Figure 1.5.

Thus, a measure of the rate and amount of CO_2 evolved in the process is a direct measure of the amount and rate of microbial utilization (biodegradation) of the C-polymer. This forms the basis for ASTM and International Standards for measuring biodegradability or microbial utilization of the test polymer/plastics. Thus, one can measure the rate and extent of biodegradation or microbial utilization of the test plastic material by using it as the sole carbon source in a test system containing a microbially rich matrix-like compost in the presence of air and under optimal temperature conditions (preferably at 58°C—representing the thermophilic phase). Figure 1.6 shows a typical graphical output that would be obtained if one were to plot the percent carbon converted to CO_2

$$\text{Glucose/C-bioplastic} + 6O_2 \longrightarrow 6CO_2 + 6H_2O; \quad \Delta G^{0'} = -686 \text{ kcal/mol}$$

FIGURE 1.5 Basic equation for utilization of carbon substrate by microorganisms.

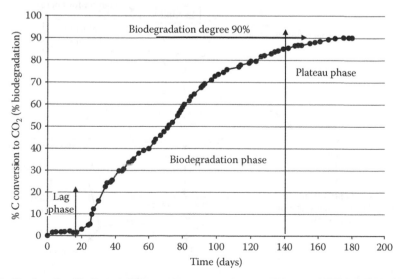

FIGURE 1.6 Testing for biodegradability under composting conditions—ASTM D5338, D6400. (From Narayan, R., *ACS Symp. Ser.*, 939, Chapter 18, 282, 2006. With permission.)

as a function of time in days. First, a lag phase during which the microbial population adapts to the available test C-substrate. Then, the biodegradation/utilization phase during which the adapted microbial population begins to utilize the carbon substrate for its cellular life processes, as measured by the conversion of the carbon in the test material to CO_2. Finally, the output reaches a plateau when all of the substrate is completely utilized.

1.3.2 Integration of Biodegradable Plastics with Disposal Infrastructure

Making or calling a product biodegradable or recyclable has no meaning whatsoever if the product after use by the customer does not end up in a disposal infrastructure that utilizes the biodegradability or recyclability features. Recycling makes sense if the recyclable product can be easily collected and sent to a recycling facility to be transformed into the same or new product. Biodegradable plastics would make sense if the product after use ends up in a disposal infrastructure that utilizes biodegradation. Composting, wastewater/sewage treatment facilities, and managed, biologically active landfills (methane/landfill gas for energy) or anaerobic digestors are established biodegradation infrastructures. Therefore, producing biodegradable plastics using annually renewable biomass feedstocks that generally end up in biodegradation infrastructures like composting is ecologically sound and promotes sustainability. Materials that cannot be recycled or biodegraded can be incinerated with recovery of energy (waste to energy). Landfills are a poor choice as a repository of plastic and organic waste. Today's sanitary landfills are plastic-lined tombs that retard biodegradation because of little or no moisture and negligible microbial activity. Organic waste such as lawn and yard waste, paper, food, biodegradable plastics, and other inert materials should not be entombed in such landfills. Figure 1.7 illustrates the integration of biodegradable plastics with disposal infrastructures that utilize the biodegradable function of the plastic product.

Among disposal options, composting is an environmentally sound approach to transfer biodegradable waste, including the new biodegradable plastics, into useful soil amendment products. Composting is the accelerated degradation of heterogeneous organic matter by a mixed microbial population in a moist, warm, and aerobic environment under controlled conditions. Biodegradation of such natural materials will produce valuable compost as the major product, along with water and

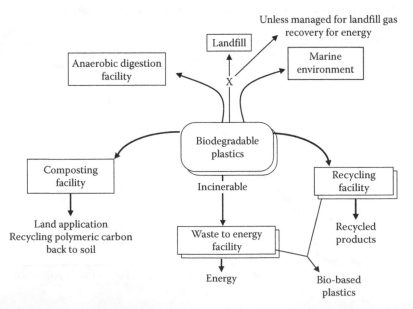

FIGURE 1.7 End-of-life options for biodegradable plastics—integration with disposal systems. (From Narayan, R., *ACS Symp. Ser.*, 939, Chapter 18, 282, 2006. With permission.)

carbon dioxide. The CO_2 produced does not contribute to an increase in greenhouse gases, because it is already part of the biological carbon cycle as discussed earlier. Composting our biowastes not only provides ecologically sound waste disposal but also provides much-needed compost to maintain the productivity of our soil and sustainable agriculture. Figure 1.7 shows disposal infrastructures that can receive biodegradable plastics.

Composting is an important disposal infrastructure because greater than 50% of the municipal solid waste (MSW) stream is biowastes like yard trimmings, food, and nonrecyclable paper products [17–19].

1.3.3 Degradable vs. Biodegradable Plastics: An Issue

Designing products to be degradable or partially biodegradable causes irreparable harm to the environment. Degraded products may be invisible to the naked eye. However, out of sight will not make the problem disappear. One must ensure complete biodegradability in a short defined time frame (determined by the disposal infrastructure). Typical time frames would be up to one growing season or 1 year. As discussed earlier, the disposal environments are composting, anaerobic digestion, marine/ocean, and soil.

Unfortunately, there are products in the marketplace that are designed to be degradable, that is, they fragment into smaller pieces and may even degrade to residues invisible to the naked eye. However, there is no data presented to document complete biodegradability within the one growing season/1-year time period. It is assumed that the breakdown products will eventually biodegrade. In the meanwhile, these degraded, hydrophobic, high-surface-area plastic residues migrate into the water table and other compartments of the ecosystem causing irreparable harm to the environment. In a recent *Science* article [21], researchers report that plastic debris around the globe can erode (degrade) away and end up as microscopic granular or fiber-like fragments, and that these fragments have been steadily accumulating in the oceans. Their experiments show that marine animals consume microscopic bits of plastic, as seen in the digestive tract of an amphipod. The Algalita Marine Research Foundation [22] reports that degraded plastic residues can attract and hold hydrophobic toxic elements like PCB and DDT up to one million times background levels. The PCBs and DDTs are at background levels in soil, and diluted out so as to not pose significant health risk. However, degradable plastic residues with these high-surface areas concentrate these highly toxic chemicals, resulting in a toxic time bomb, a poison pill floating in the environment posing serious risks.

Recently, Japanese researchers confirmed these findings. They reported that PCBs, DDE, and nonylphenols (NPs) were detected in high concentrations in degraded polypropylene (PP) resin pellets collected from four Japanese coasts. The paper documents that plastic residues function as a transport medium for toxic chemicals in the marine environment [23].

Therefore, designing hydrophobic polyolefin plastics, like PE to be degradable, without ensuing that the degraded fragments are completely assimilated by the microbial populations in the disposal infrastructure in a very short time period poses more harm to the environment than if it was not made degradable.

ASTM committee D20.96 [14,15] has developed a specification standard for products claiming to be biodegradable under composting conditions or compostable plastic. The specification standard ASTM D6400 identifies the following three criteria:

- Conversion to CO_2, water, and biomass via microbial assimilation of the test polymer material in powder, film, or granule form
- 60% carbon conversion of the test polymer to CO_2 for homopolymer and 90% carbon conversion to CO_2 for copolymers, polymer blends, and addition of low MW additives or plasticizers
- Same rate of biodegradation as natural materials—leaves, paper, grass, and food scraps
- Time—180 days or less; if radiolabeled polymer is used 365 days or less

Disintegration

- <10% of test material on 2 mm sieve using the test polymer material in the shape and thickness identical to the product's final intended use—see ISO 16929 and ISO 20200.

Safety

- The resultant compost should have no impacts on plants, using OECD Guide 208, Terrestrial Plants, Growth Test.
- Regulated (heavy) metals content in the polymer material should be less than 50% of EPA (USA, Canada) prescribed threshold.

The aforementioned specification standard is in harmony with standards in Europe, Japan, Korea, China, and Taiwan, for example, EN13432 titled "Requirements for Packaging Recoverable through Composting and Biodegradation—Test Scheme and Evaluation Criteria for the Final Acceptance of Packaging" is the European standard (norm) and similar to D6400. At the International level, the International Standards Organization (ISO) has published ISO 17088, "Specification for Compostable Plastics," which is in harmony with ASTM D 6400 and the European norms.

1.4   BIO-BASED MATERIALS TECHNOLOGY

Polymer materials based on annually renewable agricultural and biomass feedstocks can form the basis for a portfolio of sustainable, environmentally preferable alternatives to current materials based exclusively on petroleum feedstocks. Two basic routes are possible. Direct extraction from biomass yields a series of natural polymer materials (cellulose, starch, and proteins), fibers, and vegetable oils that can form the platform on which polymer materials and products can be developed as shown in Figure 1.8 (the bolded items in the figure represent our work in this area).

Alternatively, the biomass feedstock (annually renewable resources) can be converted to biomonomers by fermentation or hydrolysis. The biomonomers can be further modified by a biological or chemical route. As shown in Figure 1.9, the biomonomers can be fermented to give succinic acid, adipic acid, 1,3-propane diol—precursor chemicals for the manufacture of polyesters. An example of this is DuPont's Sorona polyester made from a bio 1,3-propanediol. Biomonomers can be fermented to lactic acid, which is then converted into PLA—currently being commercialized by NatureWorks LLC with a 300 mm lb manufacturing plant in Blair, Nebraska [24]. They can also be microbially transformed to biopolymers like the polyhydroxyalkanoates (PHAs), which are being commercialized by an ADM–Metabolix joint venture. A chapter relating to PHA polymers can be found in this book. Braskem in Brazil has announced a 200,000 ton bio-polyethylene from ethanol to ethylene to polyethylene, and 30 kton of bio-polypropylene.

Instead of microbial fermentative processes, chemical conversion of biomonomers yields intermediate chemicals like ethylene and propylene glycols. Vegetable oils offer a platform to make a portfolio of polyols, lubricants, polyesters, and polyamides. We have reported on a new ozone-mediated transformation of vegetable oils to polyols, urethane foams, polyesters, and polyamides, including biodiesel fuel [25–27].

Surfactants, detergents, adhesives, and water-soluble polymers can be engineered from biomass feedstocks. As discussed earlier, bio-based materials targeted for short-life, single-use, disposable packaging materials, and consumer products can and should be engineered to retain inherent biodegradability properties, thereby offering an environmentally responsible disposal option for such products.

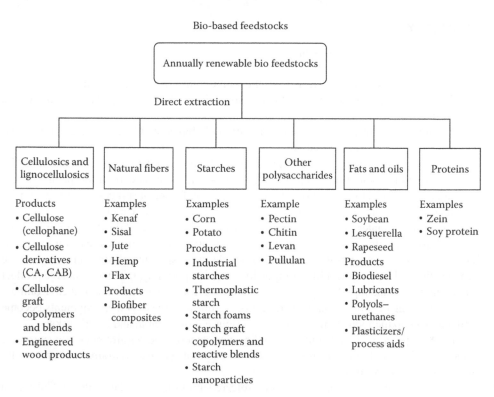

FIGURE 1.8 Direct extraction of biomass to provide biopolymers for use in manufacture of bio-based products.

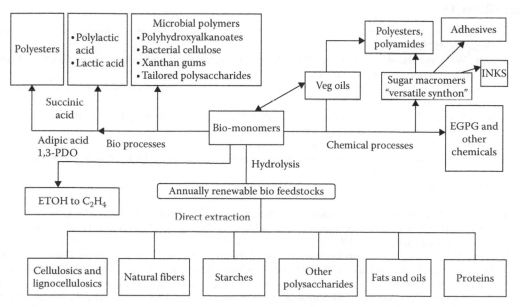

FIGURE 1.9 Conversion to biomonomers, chemicals, and polymers.

1.5 SUMMARY

The following conclusions can be drawn:

- The use of biomass/renewable feedstock for manufacture of plastics, chemicals, and fuels (bioplastics or bioproducts) offers a "value proposition" for reducing the carbon and the environmental footprint.
- Methodology for identification and quantification of "bio content" has been developed and codified into ASTM Standard D6866.
- Using bio content calculations, one can calculate the intrinsic CO_2 reductions achieved by incorporating bio content into a plastic or product—the material carbon footprint.
- It is equally important to report on the process carbon footprint (process energy footprint arising from the conversion of feedstock to product) using LCA methodology to ensure that the intrinsic material carbon value proposition is not negated during the conversion, use, and disposal life-cycle phases of the plastic or product—the process carbon footprint.
- Biodegradability is an end-of-life option for single-use disposable plastics and needs to be tied to a disposal environment like composting. More importantly, if a biodegradable product is not completely and rapidly (1 year or less) removed from the defined disposal environment (like composting or soil, or anaerobic digestion, or marine), the degraded fragments become toxin carriers up the food chain resulting in serious environmental and health risks!
- ASTM, European, and ISO standards define and specify the requirements for complete biodegradability and must be strictly adhered to so that serious environmental and health consequences can be averted.
- U.S. federal government through its biopreferred program and the State of California strictly enforce the standards and principles outlined in this chapter and providing the market pull for these bioplastics.
- A number of bio-based and biodegradable plastic products are in the marketplace and increasing.

REFERENCES

1. Lieth, H. and H. R. Whittaker (Eds.), *Primary Productivity of the Biosphere*, Springer Verlag, Berlin, Germany, 1975.
2. Jack, W. White and Wilma McGrew, Institute of Gas Technology, *Symposium Papers on Clean Fuels from Biomass and Wastes*, Orlando, FL, 1977.
3. Szmant, H. H., *Industrial Utilization of Renewable Resources*, Technomic Publishing Co., Lancaster, Basel, U.K., 1986.
4. Rowell, R. M., T. P. Schultz, and R. Narayan (Eds.), *Emerging Technologies for Materials and Chemicals from Biomass*, ACS Symposium Series 476, Washington, DC, 1991.
5. Cavaney, R., News article, Pulp Pap. Int., 31, July, 37, 1989.
6. Chum, H. L. and A. J. Power, Opportunities for the cost-effective production of biobased materials, *ACS Symp. Ser.*, 476, 28–41, 1991.
7. Doane, W. M., C. L. Swanson, and G. F. Fanta, Emerging polymeric materials based on starch, *ACS Symp. Ser.*, 476, 197–230, 1991.
8. Narayan, R., Environmentally degradable plastics, *Kunststoffe*, 79(10), 1022–1026, 1989.
9. Narayan, R., Biomass (renewable) resources for production of materials, chemicals, and fuels—A paradigm shift, *ACS Symp. Ser.*, 476, 1, 1992.
10. Narayan, R., Polymeric materials from agricultural feedstocks, in *Polymers from Agricultural Coproducts*, Eds. M. L. Fishman, R. B. Friedman, and S. J. Huang, *ACS Symposium Series* 575, Washington, DC, p. 2, 1994.
11. Narayan, R., Commercialization technology: A case study of starch based biodegradable plastics, in *Paradigm for Successful Utilization of Renewable Resources*, Eds. D. J. Sessa and J. L. Willett, AOCS Press, Champaign, IL, p. 78, 1998.
12. Narayan, R., Biobased and biodegradable polymer materials: Rationale, drivers, and technology exemplars, *ACS Symp. Ser.*, 939, Chapter 18, 282, 2006.

13. Vink, E. T. H. et al., The ecoprofiles for current and near future nature works polylactide (PLA) production, *J. Ind. Biotechnol.*, 3(1), Spring, 58–81, 2007.
14. ASTM International, *Annual Book of Standards; Standards D6866; D6400, D6868, D7021*, ASTM International, Philadelphia, PA, Vol. 8.03, 2007.
15. ASTM International, Committee D20 on plastics, Subcommittee D20.96 on biobased and environmentally degradable plastics, www.astm.org, accessed January 2011.
16. United States Department of Agriculture (USDA) Biopreferred Program, www.biopreferred.gov, accessed 2011.
17. Narayan, R., Biodegradation of polymeric materials (anthropogenic macromolecules) during composting, in *Science and Engineering of Composting: Design, Environmental, Microbiological and Utilization Aspects*, Eds. H. A. J. Hoitink and H. M. Keener, Renaissance Publications, Worthington, OH, p. 339, 1993.
18. Narayan, R., Impact of governmental policies, regulations, and standards activities on an emerging biodegradable plastics industry, in *Biodegradable Plastics and Polymers*, Eds. Y. Doi and K. Fukuda, Elsevier, New York, p. 261, 1994.
19. Song, J. H., R. J. Murphy, R. Narayan, and G. B. H. Davies, Biodegradable and compostable alternatives to conventional plastics, *Phil. Trans. R. Soc. B*, 364, 2127–2139, 2009.
20. Narayan, R., Misleading claims and misuse of standards continues to proliferate the bioplastics industry space, *BioPlastics Magazine*, Issue 1, January 2010, http://bioplastics-cms.de/bioplastics/download/downloads.php, Retrieved January 2011.
21. Thompson, R. C., Y. Olsen, R. P. Mitchell, A. Davis, S. J. Rowland, A. W. G. John, D. McGonigle, and A. E. Russell, Lost at sea: Where is all the plastic? *Science*, 304, 838, 2004.
22. Algalita Marine Research Foundation, www.algalita.org/pelagic_plastic.html, updated February 2008.
23. Mato, Y., T. Isobe, H. Takada, H. Kahnehiro, C. Ohtake, and T. Kaminuma, Plastic resin pellets as a transport medium for toxic chemicals in the marine environment, *Environ. Sci. Technol.*, 35, 318–324, 2001.
24. Nature Works LLC., www.natureworksllc.com, Retrieved January 2011.
25. Narayan, R. and D. Graiver, Value-added chemicals from catalytic oyanation of vegetable oils, *Lipid Technol.*, 16, 2, 2006.
26. Tran, P., D. Graiver, and R. Narayan, Ozone-mediated polyol synthesis from soybean oil, *J. Am. Oil Chem. Soc.*, 82(9), 653–659, 2005.
27. Baber, T., D. Graiver, C. Lira, and R. Narayan, Application of catalytic ozone chemistry for improving biodiesel product performance, *Biomacromolecules*, 6(3), 1334–1344, 2005.

2 Bio-Based Key Molecules as Chemical Feedstocks

Herman van Bekkum

CONTENTS

2.1 INTRODUCTION

World biomass production amounts to 120,000 million ton per annum [1], of which some 20% is being cultivated, harvested, and used (food, feed, and nonfood). Human consumption ranges from 6% (South America) to 80% (South Central Asia). A recent joint effort of 18 groups worldwide arrives at 123 billion ton of fixed carbon annually [1a]. Tropical forest and savannahs account for 60% of this amount.

Bio-based chemicals and materials may be approached in various ways:

Nature already produces the desired structures [2], and isolation of these components requires only physical methods, for example, polysaccharides (cellulose, starch, alginate, pectin, guar gum, chitin, inulin, etc.), disaccharides (sucrose and lactose [animal origin]), triglycerides, lecithin, natural rubber, gelatin (animal origin), various flavors and fragrances, and quinine (flavor as well as pharmaceutical). Some present-day production volumes are sucrose, 160×10^6 metric ton/year, triglycerides 180×10^6 ton/year, and natural rubber 6.5×10^6 ton/year. Cotton, the natural cellulose fiber, is produced in a volume of over 20×10^6 ton/year.

One-step (bio)chemical modification of naturally produced structures, for examples, cellulose and starch derivatives, glucose and fructose, glycerol, and fatty acids. Nature offers various starting materials for pharmaceuticals. Thus, morphine is converted by one methylation step into the antitussive and analgesic codeine (200 ton/year), whereas one (generally illegal) acetylation step leads to heroin. Fermentation is a one-step process in which the enzymes of a microorganism catalyze the conversions in a multistep synthesis without isolation of intermediates. Bulk chemical examples include ethanol $85.9 \times 10^6 \, m^3$/year (2010 estimate), citric acid 10^6 ton/year, sodium glutamate 2.2×10^6 ton/year, and lactic acid 300×10^3 ton/year, starting from starch or sucrose.

Multistep derivation of organic chemicals and organic materials from natural products, for examples, vitamin C in several steps from glucose; (S)-β-hydroxybutyrolactone in two steps from lactose; the fragrance linalool in four steps from α-pinene, (–)-menthol in six steps from β-pinene [3]; fatty alcohols and amines from triglycerides; and alkyl polyglucosides from glucose and fatty alcohols, poly-L-lactate in two steps from L-lactic acid.

2.2 BIOREFINERY

Biorefinery, also indicated as total crop use—in which coproduction of bulk and fine chemicals takes place, while waste organic materials are liquefied or gasified or are fermented to methane (biogas)—is coming to the fore. Direct burning is also an option which is executed worldwide with bagasse, delivering the energy required in sugarcane refineries. A concise definition is "biorefinery is the sustainable processing of biomass into a spectrum of marketable products and energy" [3a].

As an example, we mention the soybean crop, which is first split into beans and waste biomass. The beans are rolled and divided, by hydrocarbon extraction, into soybean meal (goes mainly to animal feed but also provides an isoflavones concentrate, a health-enhancing food supplement) and crude soybean oil. The crude oil undergoes several purification steps in which, in the modern technology, valuable side products like lecithin and steroids (e.g., β-sitosterol) are isolated and refined soybean oil is obtained.

The steroids can serve as starting compounds for various steroid-type pharmaceuticals [4] and, after transesterification with sunflower oil, find application in cholesterol level–lowering margarines. Tall oil, a by-product of wood pulp manufacture, is a source of related steroids, serving the same health purpose [5].

Sanders et al. designed biorefineries with grass and several leaf materials as feedstocks [5a]. Main products are proteins and fibers. A grass-based biorefinery is under construction in the Netherlands.

In a recent book, entitled *The Biobased Economy. Materials and Chemicals in the Post-oil Era* [5b], several sections are devoted to biorefineries, including their strong and weak sides.

It is often the case that waste material volumes are substantial and sometimes much larger (e.g., sugarcane) than the volumes of the primary target compounds or materials. It has been estimated [6] that the caloric value of agricultural waste streams is more than 50% of the world's annual oil consumption.

2.3 BIOMASS TO TRANSPORTATION FUELS

Processes that convert biomass to (green) transportation fuels or to fuel boosters (see Figure 2.1) are receiving much attention nowadays. Two classics are ethanol and 1-butanol. On a large and ever-increasing scale, hydrolysis of the polysaccharide starch and the disaccharide sucrose to the monosaccharides is carried out, followed by fermentation to ethanol. Two ethanol molecules and two CO_2 molecules are formed from one C_6-monosaccharide (glucose). By far, the largest ethanol producers are Brazil [7,8] and the United States [9,9a]. Zeolites play a role (adsorption or membrane techniques) in the dewatering of ethanol, which is required for mixing with gasoline. Forthcoming are processes in which lignocellulosics, as present in agricultural waste streams (wheat straw, cornstalks, and bagasse), are also hydrolyzed (enzymatically) and converted to so-called second-generation ethanol [10] by microorganisms that convert C_6-monosaccharides, glucose, as well as C_5-monosaccharides (from hemicellulose) to ethanol.

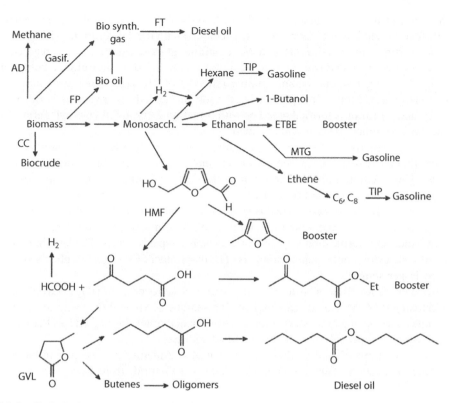

FIGURE 2.1 Carbohydrates to transportation fuels. AD, anaerobic digestion; FP, fast pyrolysis; FT, Fischer–Tropsch; TIP, total isomerization process; MTG, methanol-to-gasoline; ETBE, ethyl *t*-butyl ether; HMF, hydroxymethylfurfural; CC, catalytic cracking; GVL, gamma-valerolactone.

When the ethanol is not blended but used as such as fuel, hydrous ethanol (e.g., 95%) can be applied. In several other chemical conversions of ethanol, there is no need to remove all the water. For example, in a cascade-type continuous setup [11], a partial evaporation of the fermentation liquid was carried out and the ethanol–water mixture was passed over a H-ZSM5 zeolite catalyst (350°C, 1 atm). As in the methanol-to-gasoline (MTG) [12] process, a mixture of alkanes and aromatics was obtained. In this approach, the ethanol–water separation is avoided, as the hydrocarbons and water are nonmiscible and separated by gravity. Though its octane number is good, there is no future in MTG because of the trend to lower the aromatics content.

When passing the ethanol–water mixture over H-ZSM5 at lower temperature (200°C) or over zeolite H-Y, only dehydration occurs and ethene is obtained [13]. The track to gasoline is then oligomerization to C_6–C_8 alkene and hydrogenation/isomerization, for example, over the TIP-catalyst Pt-H-mordenite.

Recently, Dumesic et al. [14,14a] disclosed a route from glucose to *n*-hexane (together with some pentane and butane) consisting of hydrogenation to sorbitol followed by stepwise hydrogenolysis over a Pt catalyst in acid medium (pH 2). The hydrogen required is obtained [15] by aqueous-phase reforming of sorbitol or glycerol over a relatively inexpensive Raney NiSn catalyst. Roughly 1.6 mol of glucose are required per mol hexane.

By gasification of biomass, and with supplementation of hydrogen, the Fischer–Tropsch synthesis can be entered toward clean diesel oil.

Figure 2.1 also shows routes via hydroxymethylfurfural (HMF), a compound that can be selectively made from fructose using a dealuminated H-mordenite or a niobium-based catalyst. HMF is becoming a new key chemical; its chemistry (and that of furfural) has been reviewed by Moreau et al. [16] and by Chheda et al. [16a]. When preparing HMF, it is advantageous to apply a cascade reaction by using a fructose precursor. Thus, the hydrolysis of the fructan inulin (glucose [fructose]$_n$) or of the glucose–fructose combination sucrose is coupled with the dehydration to HMF. In the case of sucrose as starting compound, HMF and the remaining glucose can be easily separated.

HMF may be hydrogenated over an RuCu catalyst [17] to 2,5-dimethylfuran, a compound with the very high blending research octane number (BRON) of 215 [18]. When executing the HMF hydrogenation over a Pd catalyst in 1-propanol as the solvent [18a], the propyl ether 5-hydroxymethyl-2-(propyloxymethyl) furan is formed together with 5-methyl-2-(propyloxymethyl) furan. Here, the dipropyl acetal is assumed to be initially formed.

Due to the high reactivity of HMF, its formation may be coupled to a consecutive reaction, for example, to ethyl ether formation when adding ethanol to the solvent, or to an ester, when acetic acid is added [18b]. Avantium researchers coined the term "furanics" for the family of furan and tetrahydrofuran derivates obtained from HMF.

An interesting HMF-derived compound is levulinic acid formed together with formic acid by solid acid catalysis. A one-pot cascade route from, for example, inulin seems feasible. Fagan et al. [19] give examples in which an esterification step with an alkene is coupled as well. The levulinic esters, such as ethyl levulinate, exhibit good octane numbers. The by-product formic acid might also be esterified or used as hydrogen source.

By hydrogenating levulinic acid to gamma valerolactone (over a Pt/TiO$_2$ catalyst) and then to valeric acid (using Pt/ZSM-5 as the catalyst), Shell researchers explored the "valeric biofuels." This family includes various valerate esters, such as pentyl valerate (cf. Figure 2.1) and ethylene glycol bis-valerate, which compounds are suitable as diesel additives.

Finally, it may be mentioned in this section that also 2-methyltetrahydrofuran has been proposed for fuel application. This compound is made, via furfural, from C$_5$ sugars, as present in hemicellulose.

2.4   ETHANOL AND METHANOL AS KEY CHEMICALS

Some further conversions of ethanol are shown in Figure 2.2. We have already mentioned preparation of today's number one organic chemical, ethene. The decrease of mass in the conversion of carbohydrates via ethanol to ethene is large (65%) and the annual production of ethene exceeds 110×10^6 ton, so the total world sugar production (160×10^6 ton) would not nearly be enough to cover that amount. Nevertheless, the construction of a polyethylene plant in Brazil, based on

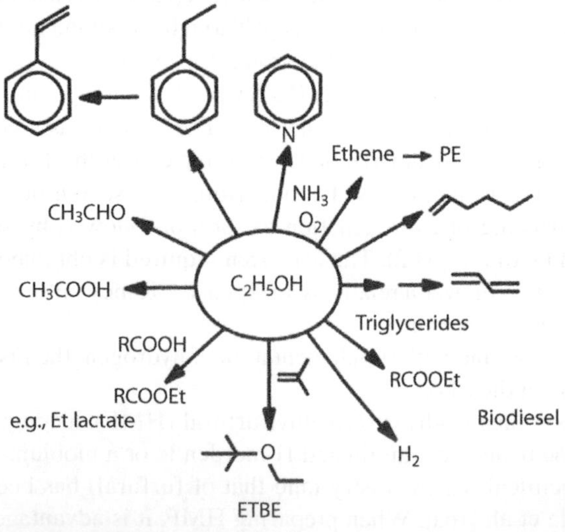

FIGURE 2.2   Ethanol as key chemical.

bioethanol is under consideration. In India, where cheap sugar streams are available, over 400,000 ton of ethanol were used in 1997 [20] to make "alco-chemicals" as the main product. In China and India, aqueous ethanol is directly applied in aromatic ethylation (ethylbenzene, 1,4-diethylbenzene, and 4-ethyltoluene).

The reaction of ethanol with isobutene affords ethyl *t*-butyl ether (ETBE), a partly green compound, which is a successor of methyl *t*-butyl ether (MTBE) as a gasoline booster. MTBE is under environmental pressure and has been banned in several regions. Biodiesel becomes fully renewable based when ethanol is applied instead of methanol in the transesterification of triglycerides.

The oxidative dehydrocyclization of ethanol and ammonia toward pyridine and 4-methylpyridine was studied in the author's group [21]. The medium-pore zeolites H-ZSM-5 and H-TON are the catalysts of choice.

Finally, aqueous ethanol can be considered as a future hydrogen carrier. Up to 6 mol of hydrogen can be obtained from 1 mol of ethanol.

Methanol is a bulk chemical that is mainly made from natural gas, via syngas. A challenge is the direct air oxidation of methane to methanol. There is some progress in achieving this goal [21a,b]. Syngas can also be made from coal or from biomass by gasification. Recently, a methanol plant that uses cheap technical glycerol as feedstock came into operation in the Netherlands.

The chemical network around methanol is shown in Figure 2.3. As oil will be depleted faster than gas and coal, it is expected that methanol is to play an important role in the supply of ethene and propene. In the methanol-to-olefins (MTO) processes [12], methanol is converted over, for example, zeolite SAPO-34 (a microporous silicoaluminophosphate with chabazite structure) toward ethene and propene (with tunable ratio).

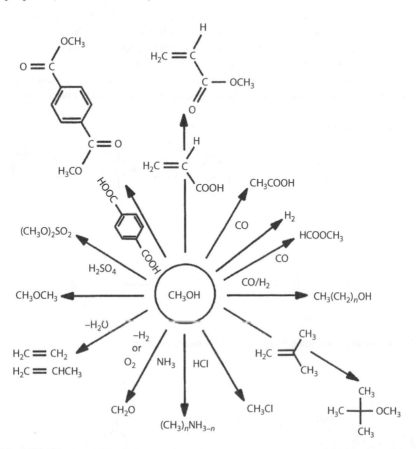

FIGURE 2.3 Methanol as key chemical.

2.5 MONOTERPENES

Limonene and α- and β-pinene are natural key molecules for the monoterpene (C_{10}) sector of the fragrance industry.

Limonene is a component of orange and lemon peels (different enantiomers present) and is a cheap by-product of the citrus industry. The two pinenes are major components of crude sulfate turpentine (CST), a by-product of wood processing, and are relatively inexpensive starting compounds.

Catalytic conversions in the monoterpene field have recently been reviewed [22]. For instance, limonene is commercially converted to alkoxylated systems by solid acid–catalyzed addition of lower alcohols [23]. Quite another limonene conversion is dehydrogenation toward "green" aromatic, p-cymene, which can be converted by oxidation to the hydroperoxide and rearranged to p-cresol.

The networks around α- and β-pinene are versatile. By way of example, Figure 2.4 shows how the important fragrance linalool is approached industrially from both α- and β-pinene. These semisynthetic linalool products are in competition with synthetic linalool, made by building up the C_{10} system starting from isobutene.

Another example is the synthesis of campholenic aldehyde by epoxidation of α-pinene followed by isomerization [24]. The conversion has been executed as a one-pot procedure [24a] using a mesoporous Ti-silicate and t-butyl hydroperoxide as the oxidant. Campholenic aldehyde is the starting compound for several sandalwood fragrances.

2.6 TRIGLYCERIDES, GLYCEROL

Triglycerides (oils and fats) production keeps growing fast, in fact faster than the world population. Figure 2.5 shows the world growth curve as well as the growth of the major individual triglycerides [24a]. For a long time, soybean oil was the number one as to volume produced, but in 2008, palm oil took the lead. An interesting newcomer in the triglyceride field is jatropha oil.

Most triglycerides serve primarily food applications. Some 15% of the annual production of 180×10^6 ton (2010 estimate) goes to oleochemicals, such as fatty acids and soaps, fatty amines,

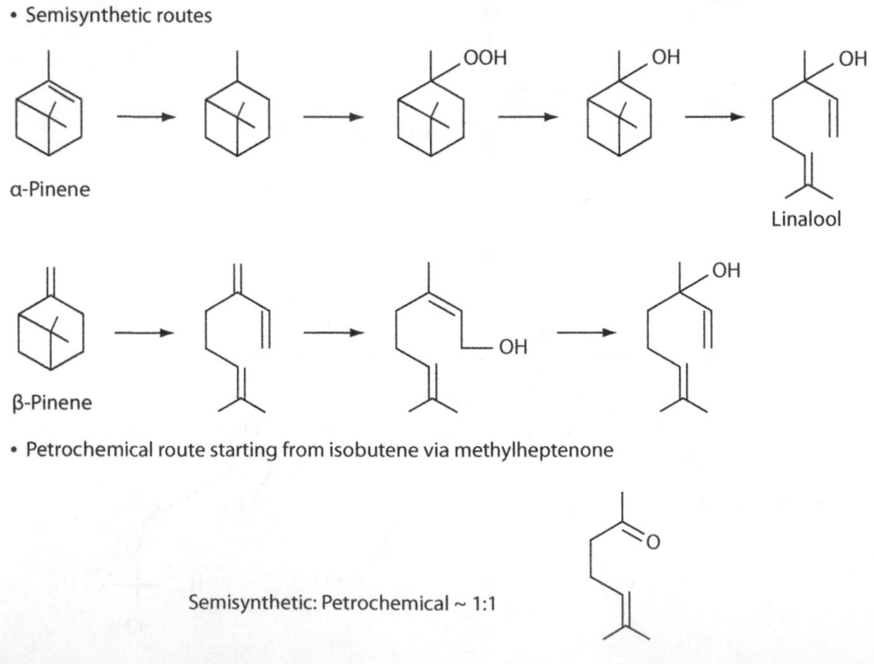

FIGURE 2.4 Linalool syntheses.

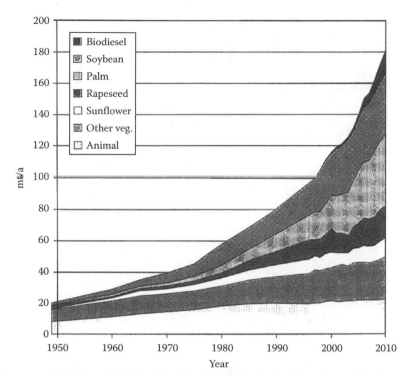

FIGURE 2.5   Global oils and fats production.

long-chain alcohols, and biodiesel (fatty acid methyl esters). In almost all cases, glycerol is a by-product. An exception is the Neste oil biodiesel process in which, by catalytic hydrogenolysis, complete deoxygenation takes place. Here, the glycerol moiety ends as propane. Especially because of the growth of biodiesel (2010 estimate 21×10^6 ton, with Germany as leading producer [25]), world glycerol production is now approaching 2 million ton/year. Less than 5% of that amount is made by the conventional petrochemical process starting with chlorination of propene. The natural way is a clear winner here in terms of both economy and eco-friendliness.

Glycerol is applied as such in pharmaceutical and cosmetic formulations as well as in food, tobacco, and cellophane. Some chemical conversions are given in Figure 2.6.

A major reaction application of glycerol is as triol in alkyd resin formulations. Moreover, it is used to grow polyether polyols to be used in polyurethanes. Only a few percent of the glycerol production is used for manufacturing glycerol trinitrate which, adsorbed on silica to give dynamite, was the first commercial derivative (1866) of glycerol. New opportunities are selective oxidation over Au catalysts [26] toward glyceric acid, synthesis of glycidol [27] via the carbonate, and partial hydrogenolysis toward 1,2-propanediol over a NiNaX zeolite catalyst [27a].

In particular, a new process to make epichlorohydrin is expected to absorb a substantial amount of glycerol. A recent booklet [27b] is devoted to new uses for glycerol.

2.7 TRIGLYCERIDES, FATTY ACIDS

The fatty acid composition of triglycerides depends strongly on the crop that provides them. Thus, coconut oil and palm-kernel oil contain mostly relatively short (C_{10}, C_{12}, C_{14}, and C_{16}) saturated fatty acids, whereas soybean, rapeseed, and sunflower oil contain mainly the C_{18} unsaturated acids oleic and linoleic acid, with one and two double bonds, respectively. In palm oil, with the largest production volume, the dominant fatty acid components are palmitic acid (C_{16} saturated) and oleic acid (C_{18} monounsaturated).

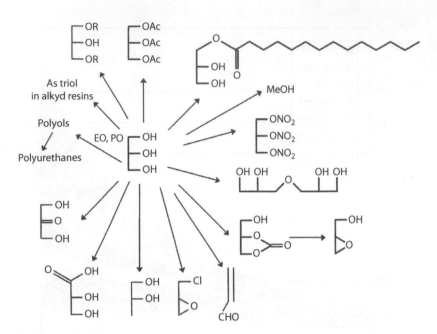

FIGURE 2.6   Glycerol conversions.

Insight into health aspects has been subject to fluctuation. Linolenic acid (C_{18}, three double bonds), belonging to the so-called omega-3 acids, was earlier removed from margarine triglycerides by selective hydrogenation, nowadays it is treasured. Moreover, it is agreed now that trans double bonds, formed upon partial hydrogenation of natural cis double bonds (catalytic hardening) are unhealthy.

Commercial products obtained from saturated fatty acids include a broad spectrum of esters, linear alcohols, primary and secondary linear amines, amides, and various metal salts. Relatively new are the direct hydrogenation of carboxylic acids to aldehydes over chromia [28] and the coupling of fatty acids (by an amide bond) to amino acids leading to a new class of surfactants of natural origin (Ajinomoto). Biocatalytic terminal- and α-hydroxylation of fatty acids is a challenge.

In the unsaturated oleic acid molecule, the double bond is an extra reaction site. As shown in Figure 2.7, metathesis [29] especially opens up interesting conversions, although these are not yet operated industrially. Thus, self-metathesis of oleic acid (as methyl ester) gives a C_{18} dicarboxylic acid (together with a C_{18} alkene), whereas metathesis with ethene leads to 1-decene and 9-decenoic acid, a precursor of nylon-10. By oxidative ozonization, oleic acid is industrially converted in the C_9 dicarboxylic acid azelaic acid and nonanoic acid.

The C_{10} dicarboxylic acid, sebacic acid, is made from ricinoleic acid, which is the major carboxylic acid constituent of castor oil. Ricinoleic acid is a C_{18} fatty acid containing a double bond at C_9 and a hydroxyl group at C_{12}. There is renewed interest in Nylon-610 (Rhodia, BASF, and DSM) for use in fuel lines.

Another special fatty acid is erucic acid, the major component of high erucic rapeseed oil and of crambe oil. Erucic acid is a C_{22} compound with a double bond at C_{13}. Its amide is a slip agent for plastics.

Calendula oil and Chinese wood oil (tung oil) contain up to 60% and 80%, respectively, C_{18} acids with three conjugated double bonds. Accordingly, these oils find use in fast-drying paints.

Some recent patents deal with isomerization (branching) of oleic and stearic acid, with the aim of lowering the melting point. For instance, oleic acid is isomerized [30] at 250°C over an acidic zeolite, followed by hydrogenation over Pd. The product obtained is similar to the "isostearic acid" found as by-product in the dimerization of oleic acid over an acid clay.

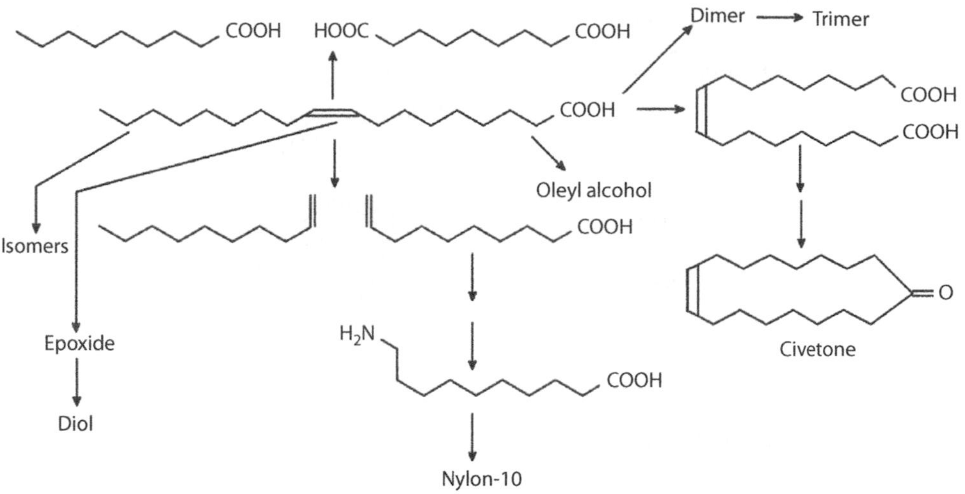

FIGURE 2.7 Oleic acid as key chemical.

Recently, it has been found [30a] that saturatic fatty acids can be deoxygenated over Pd nanoparticles supported on mesoporous SBA-15 material. The reaction is carried out at 300°C under 17 bar of 5 vol% H_2 in Ar using dodecane as solvent. Stearic acid gives heptadecane with 98% selectivity at 96% conversion. Oxygen is removed as CO and CO_2. Heptadecanes are observed as intermediates, so isomerization could be coupled.

Finally, some attention will be paid in this section to triglyceride transesterification as applied in the manufacture of biodiesel, and also in the redistribution of acyl groups over the glycerol moieties in palm oil and palm oil fractions in order to alter their melting properties.

Biodiesel production processes are conventionally based on homogeneous alkaline catalysis by sodium or potassium methoxides or hydroxides. Though these catalysts are very active, there are several drawbacks such as soap formation due to the presence of free fatty acids in crude oils (e.g., some 5 wt% in crude palm oil).

Alternatives are the use of enzymes (lipases) or of heterogeneous chemocatalysts. A recent special issue of Topics in Catalysis [30b] is fully devoted to the application of heterogeneous catalysts in biodiesel production.

Basic as well as acid catalysts can be used in transesterification [30c]. Solid acids require higher reaction temperatures than solid bases. Catalysts should be tolerant to free fatty acid and to water and should not be subject to leaching into the reaction mixture. As basic catalysts have been studied, for instance, CaO and MgO nanocrystals, on various supports, several heterogeneous or heterogenized sulfonic acids have been tested as catalysts in the transesterification of triglycerides.

Some 7000 ton of the emulsifier sorbitan oleate were used by BP in 2010 to disperse the leaked oil in the Gulf of Mexico.

2.8 CARBOHYDRATES, GENERAL

Carbohydrates are by far the most abundant class of renewables. The big three among the carbohydrates (Figure 2.8) are the glucose polymers (glucans), cellulose and starch, and the disaccharide sucrose. Chitin is also widespread, but its actual production—from waste material of the seafood industry—is small, an estimation is 5000 ton/year. Its monomer, glucosamine, is receiving much attention as a health supplement.

Another widespread polysaccharide is hemicellulose, a copolymer of C_5 sugars (xylose and arabinose) and C_6 sugars (mannose and galactose). Hemicellulose forms, together with cellulose and lignin, the main constituents of lignocellulosic materials such as wood or bagasse.

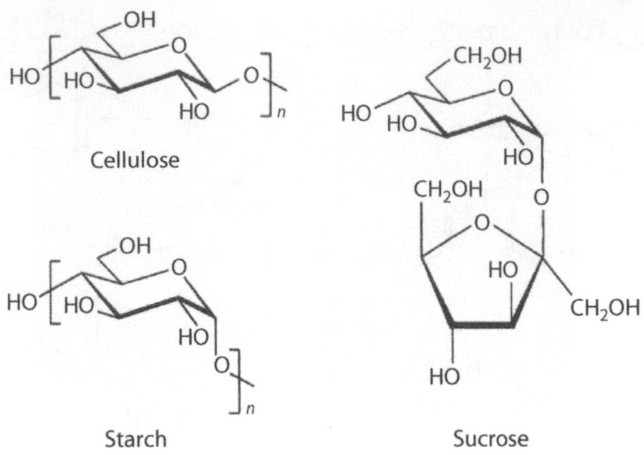

FIGURE 2.8 Carbohydrates, the big three.

Several natural polysaccharides find use [31] as thickening and stabilizing agents (hydrocolloids) [32] in food and beverages. Sources can be seaweed (agar, alginate, and carrageenan), seeds (guar gum and locust bean gum), fruits (pectin), or bacteria (xanthan gum). A relative newcomer in the polysaccharide field is the fructan inulin.

2.9 CELLULOSE

Wood harvested annually for energy, construction, paper and cardboard, and hygiene products is estimated to amount to over 4 billion m^3 [33]. In many countries, forestry is in a sustainable balance, that is, harvesting is fully compensated by replanting; however, this is not the case in all countries. Certification is an instrument here.

The world production of paper, cardboard, and pulp was 380×10^6 ton in 2008 [33a]. The top 4 producing countries were the United States, China, Japan, and Germany, with production volumes of 80, 80, 31, and 23×10^6 ton, respectively. The numbers 5, 6, and 7 are the forestry countries Canada, Sweden, and Finland, respectively.

Recycle streams contribute substantially to the feedstock demand. The worldwide paper and cardboard recycle stream was 210×10^6 ton in 2008. Regenerated cellulose includes fibers (rayon, mainly used in tires) and films (cellophane was once the leading clear packaging film). Classic cellulose solvents used in regeneration are carbon disulfide/sodium hydroxide and an ammoniacal copper solution. More recent solvents are N-methylmorpholine-N-oxide and phosphoric acid. The most recently developed cellulose solvents are ionic liquids. For example, 1-butyl-3-methyl-imidazolium chloride dissolves 100 g/L of cellulose at 100°C [34]. Improved solubility of carbohydrates in ionic liquids is observed when dicyanamide is applied as the anion [35].

Another ionic liquid, 1-ethyl-3-methyl-imidazolium acetate, was found to be able to dissolve hard as well as soft wood samples completely [35a]. Moreover, upon adding the proper reconstitution solvents (e.g., acetone/water 1:1 v/v), carbohydrate-free lignin and low-lignin cellulose/hemicellulose materials could be obtained.

Recently, the Moulijn group used zinc chloride hydrate as solvent and catalyst to convert cellulose to glucose [35b]. By coupling with catalytic hydrogenation, the relatively stable compound sorbitol was obtained, which was converted in two steps into isorbide (dianhydrosorbitol) (Figure 2.9). Altogether, a fine example of a four-step cascade reaction toward the stable diol isosorbide that can be used among others as polyester comonomer.

A target in the processing of waste wood and other lignocellulosic rest materials is to arrive at a satisfactory separation of hemicellulose (as such or as its monosaccharides), cellulose, and lignin. The aromatic biopolymer lignin is mainly used as fuel, but application in phenolic resins,

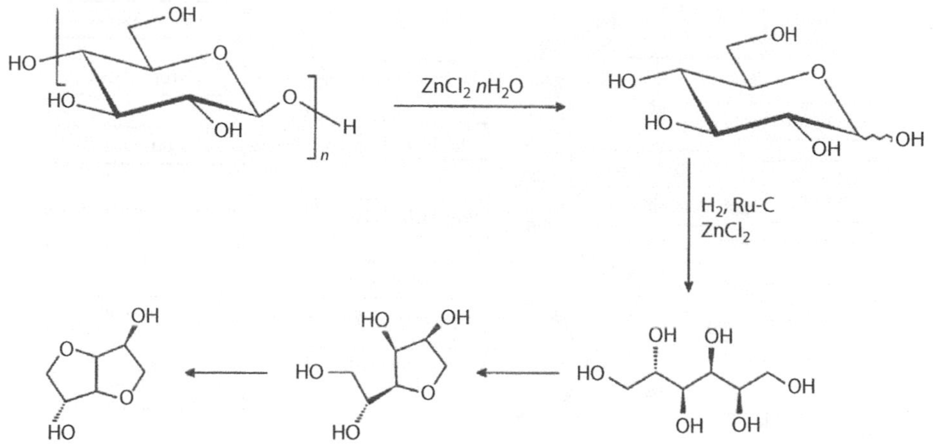

FIGURE 2.9 Cascade from cellulose to isosorbide.

in chipboard, or in plywood might be considered too [35c]. A metabolic of lignin, 2-pyrone-4, 6-dicarboxylic acid might be of interest as dicarboxylic acid component of polyesters [35d].

Cellulose derivatives [36] can be divided into nonionic and anionic materials. The degree of substitution—up to three—is always an important variable. The nonionic derivatives comprise ethers (hydroxyethyl and hydroxypropyl cellulose and methyl cellulose) and esters (cellulose acetate and cellulose nitrate). The anionic carboxymethyl cellulose is produced in the largest volume. The cellulose derivatives serve a broad spectrum of applications. For instance, cellulose acetate is applied in cigarette filters, as membranes, as fibers, etc. The anionic class may be extended by 6-carboxycellulose [37] and 2,3-dicarboxycellulose; both materials show promise but are still in the research and development stage. A challenge is to arrive at cellulose-based superabsorbing materials.

2.10 STARCH AND GLUCOSE

Starch is a mixture of a linear α-1,4-glucan (amylose, see Figure 2.8) and a branched glucan (amylopectin), containing also 1,4,6-bonded glucose units. Generally, the weight ratio for amylopectin: amylose is about 75:25, but high-amylopectin starches can be obtained by genetic modification of corn or potato.

The big starch-containing grains, wheat, rice, and corn are each annually produced at amounts of over $650 \cdot 10^6$ ton. Some $60 \cdot 10^6$ ton of starch is industrially isolated. The dominant raw material (almost 80%) is corn.

The starch serves food and nonfood applications. In Europe, the ratio is about 1:1. The largest nonfood application in Europe as well as in the United States is in the paper and board area. Both native and modified starches are applied here. Of the starch derivatives used in paper making, cationized starch is of particular importance. Here, starch is equipped with C_3 chains carrying a quaternary ammonium group. The reagent is made by reacting epichlorohydrine with trimethylamine.

Figure 2.10 lists the major starch-derived chemicals and materials. The conversion pathways are marked with either a circle or a square, indicating whether the conversion step is industrially biocatalyzed or chemocatalyzed, respectively. For instance, the starch-to-vitamin C route involves consecutively enzymatic steps (hydrolysis), a metal-catalyzed hydrogenation, a biocatalytic bacterial regioselective oxidation to L-sorbose, and chemocatalytic protection/oxidation/deprotection/ ring-closure steps. North China Pharmaceutical Corp. possesses fermentation technology for the direct conversion of glucose to vitamin C. With an annual production of over 100,000 ton, vitamin C is becoming a bulk chemical.

Glucose-to-fructose isomerization is an important conversion process in the sweetener field. On a weight-base, fructose is 1.5 times sweeter than sucrose. Using the enzyme glucose isomerase,

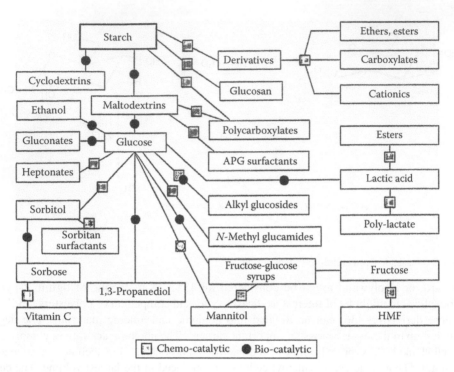

FIGURE 2.10   Starch and glucose network.

the isomerization is executed on a 10×10^6 ton/year scale in the United States, leading to a roughly 1:1 mixture of glucose and fructose. By chromatographic techniques, this mixture can be further enriched in fructose.

In a recent paper of the Davis group [38a], zeolite-Sn-beta is disclosed as a heterogeneous Lewis-acid catalyst for the isomerization of glucose toward a mixture of glucose, fructose, and mannose. Tin-beta is acid stable enabling to couple isomerization with acid-catalyzed reactions as starch hydrolysis and fructose to HMF conversion.

Fermentative processes for the polyester monomer 1,3-propanediol and for the diabetic-friendly sweetener mannitol have been developed by Du Pont and zuChem [38], respectively. A newcomer among chemicals obtained by fermentation is succinic acid. This C4-dicarboxylic acid is obtained in a CO_2-binding process from sucrose or glucose. There is European succinic acid cooperation between PURAC and BASF and between DSM and Roquette Fr. [38b]. Succinic acid has the potential to become a key chemical. In chemocatalysis, sometimes steps can be combined; thus, starch can be directly converted to sorbitol by applying a bifunctional Ru-HUSY zeolitic catalyst [39]. The outer surface of the zeolite provides the Bronsted acidity required for the starch hydrolysis. The Ru component catalyzes the hydrogenation of glucose. This process has recently come into industrial practice.

Note that there are also some commercial green surfactants shown in Figure 2.10: the sorbitan esters, known for a long time and the more recently developed (Henkel) alkyl polyglucosides (APG surfactants) and the N-methyl glucamides (Procter and Gamble/Hoechst). Major classes of green surfactants are triglycerides derived: soaps, long-chain alcohol sulfates, and the nonionic linear alcohol ethene oxide adducts. Together with dicarboxylate-polysaccharides as Ca-complexing materials, peracetylated sugar polyols as peracetate precursors, and carboxymethyl cellulose as anti-redeposition agent, it is not difficult to imagine production of fully green detergent formulations.

Anionic starches can be obtained by sulfatation, by carboxymethylation, or by oxidation. Starch oxidation skills have improved substantially. In particular, the TEMPO-catalyzed oxidation [40] displays an amazing selectivity. Indeed, potato starch has been oxidized to 6-carboxy starch with a selectivity of >98% at 98% conversion. Salt-free enzymatic TEMPO oxidations (O_2/laccase/TEMPO) have also been patented recently. Here, 6-carboxylate as well as 6-aldehyde groups are introduced. Gallezot et al. reported recently [41] on the use of iron tetrasulfonatophthalocyanine as catalyst and hydrogen peroxide as oxidant in the oxidation of several starches. Carboxyl as well as carbonyl functions are introduced leading to hydrophilic materials.

For further starch derivatives, the reader is referred to the book of Gotlieb and Capelle [42].

2.11 SUCROSE

With its present-day world market price of about 30 U.S. cents/kg, the disaccharide sucrose [43] is probably the cheapest chiral compound. The four largest producers are Brazil, India (sugarcane), the EU (sugar beet), and China (cane as well as beet). The cane: beet ratio is presently 80:20. The top five exporters are Brazil >> Thailand > Australia > SADC > Guatemala [43a].

Some industrial sucrose conversions are shown in Figure 2.11. Part of the conversions, for example, alcohol manufacture, can be executed with molasses, the mother liquor of the sugar crystallization. Potential side products of beet sugar manufacture are pectin (from the pulp) and betaine and raffinose (from the molasses).

Sucrose can be transesterified with fatty acid methyl esters toward mono- and diesters, applied as emulsifiers or to the sucrose polyesters (SPEs) which materials have been proposed as fat replacers. The fully esterified sucrose octaacetate is known for its bitterness.

Full use of the hydroxyl groups of sucrose is also made in the reaction with ethene oxide or propene oxide leading to polyether polyols, which are used in polyurethane manufacture.

In the high-intensity sweetener sucralose, three hydroxyl groups of sucrose have been replaced by chlorine. This pertains to the 1- and 6-position in the fructose part and to the 4-position in the glucose part of sucrose. Sucralose has a similar taste profile to sucrose [44]; it is nontoxic, non-nutritive, and 60 times more stable to acid hydrolysis than sucrose. Sucralose is 650 times as sweet as sucrose.

The main producer of sucralose is Tate & Lyle. There is still some discussion on the biodegradability of sucralose.

Sucrose is biocatalytically isomerized (Südzucker) whereby the $1 \rightarrow 2$ bond between glucose and fructose changes to a $1 \rightarrow 6$ bond. The disaccharide obtained is named Palatinose, which upon hydrogenation gives a 1:1 mixture of two C_{12} systems under the name Palatinit.

Both Palatinose and Palatinit are commercial sweetening compounds. Due to the low rate of hydrolysis compared with sucrose, both systems are suitable for diabetics and are mild to the teeth.

By fermentation, several organic acids are made from sucrose or glucose. Besides the carboxylic acids listed in Figure 2.11, we mention fumaric acid, malic acid, and oxalic acid. Moreover, the amino

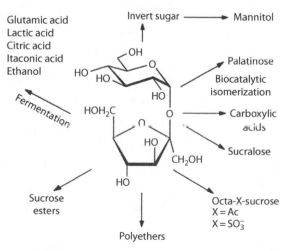

FIGURE 2.11 Sucrose as key chemical.

acids lysine, threonine, and tryptophan are industrially made. The amino acid glutamic acid, as the sodium salt, is made on a large scale. It is a flavoring compound, umami, the fifth taste.

2.12 INULIN AND FRUCTOSE

A relative newcomer in the carbohydrate field is inulin, a fructan (Figure 2.12) consisting of β-(2 → 1)-linked fructofuranose units with an α-glucopyranose unit at the reducing end. Inulin (GF_n) is obtained in Belgium and the Netherlands from the roots of chicory. The average degree of polymerization is relatively low, $n = {\sim}12$ [45]. Small amounts of oligofructose without a terminal glucose unit are also present in inulin. Another interesting inulin-containing crop is Jerusalem Artichoke [46]. Though the tuber yield (~45 ton/ha) and the inulin content (15–18 wt%) are about the same as for chicory, the average degree of polymerization (6) is lower.

Inulin is applied as such, as health additive, in food applications; it is claimed to improve the intestinal bacterial flora. Moreover, it is a direct source (hydrolysis) of a high-fructose syrup that can be applied as a sweetener. Fructose is also the precursor of the versatile compound HMF from which several furan-2,5-disubstituted "biomonomers" (diol, dialdehyde, and dicarboxylic acid) can be derived. Furthermore, HMF is a precursor of levulinic acid (cf. Figure 2.1).

Worldwide, derivatization studies on inulin are ongoing. For a review, see Stevens et al. [47]. A first successful example has been carboxymethylation leading to the new material carboxy-methyl inulin (CMI). In the reaction with chloroacetate (Figure 2.12), position 4 of the fructose units turned out to be the most reactive [48] but positions 3 and 6 contribute too. CMI appeared to be an excellent low-viscosity inhibitor of calcium carbonate crystallization [49] and is industrially manufactured now.

Another new commercial inulin derivative is inulin lauryl carbamate [50], an efficient stabilizer of oil droplets and hydrophobic particles against coalescence and/or flocculation. The material is used in personal-care applications.

Other promising inulin derivatives are 3,4-dicarboxyinulin, an excellent calcium-complexing material [51] and alkoxylated inulins [52].

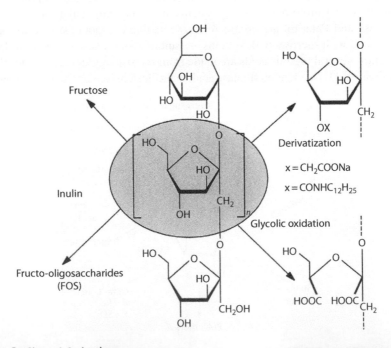

FIGURE 2.12 Inulin and derivatives.

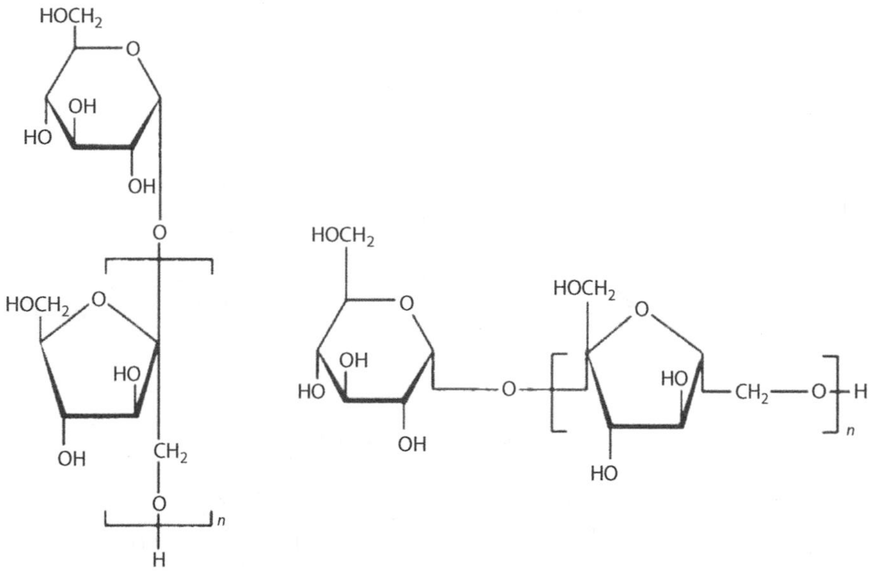

FIGURE 2.13 Inulin (left) and levan (right).

Another fructan with a terminal nonreducing glucose unit is levan [52a]. Here, the β-D-fructofuranose units are 2,6-linked, so the fructofuranose rings are part of the polymeric chain, whereas in inulin the five-membered rings can be seen as side groups of the $(C_2 C_1 O)_n$ backbone (Figure 2.13).

Levan is a constituent of various grasses. The average degree of polymerization is about 25. Some grass levans contain a single branch point.

A variety of bacterial strains produce levans when grown on sucrose in a suitable nutrient medium. These bacterial levans are of high molecular weight and contain 1,2,6-branching points though the majority of the linkages is of the 2,6 type.

2.13 LACTOSE

The major source of the disaccharide lactose (a galactose–glucose combination) is cheese whey, which contains 4.4 wt% lactose (together with 0.8% of protein and 0.8% of minerals). Lactose can be crystallized in its α- or β-form as monohydrate or anhydrous, respectively. Common grades are edible and pharmaceutical [53]. Recently, an inhalation grade lactose was added. Altogether, it is admirable that the processes of collecting, transporting, concentrating, demineralization, and crystallization allow such a modest price for lactose.

Lactose is used as such in the food industry, in infant nutrition, and in the pharmaceutical industry (matrix material of pills). In view of the growing chemical network around it (Figure 2.14), lactose is developing to a key molecule. Industrial conversions include hydrogenation to the low-caloric sweetener lactitol, isomerization to the laxative lactulose (galactose–fructose combination), oligomerization to health supplements, and oxidation to the metal-complexing lactobionic acid. Moreover, lactose serves to make chiral compounds such as (S)-3-hydroxy-γ-butyrolactone. Part of the lactose molecule is sacrificed in this synthesis. Via its monomers, new vitamin C-type antioxidants can be made from lactose [54].

2.14 LACTIC ACID

Lactic acid was the first chiral compound of which the two enantiomers were isolated (Liebig, 1847). In 1874, Van't Hoff explained the phenomenon by the concept of the tetrahedral carbon bonds (Figure 2.15, left L-(ı)-, right D-(−)-lactic acid).

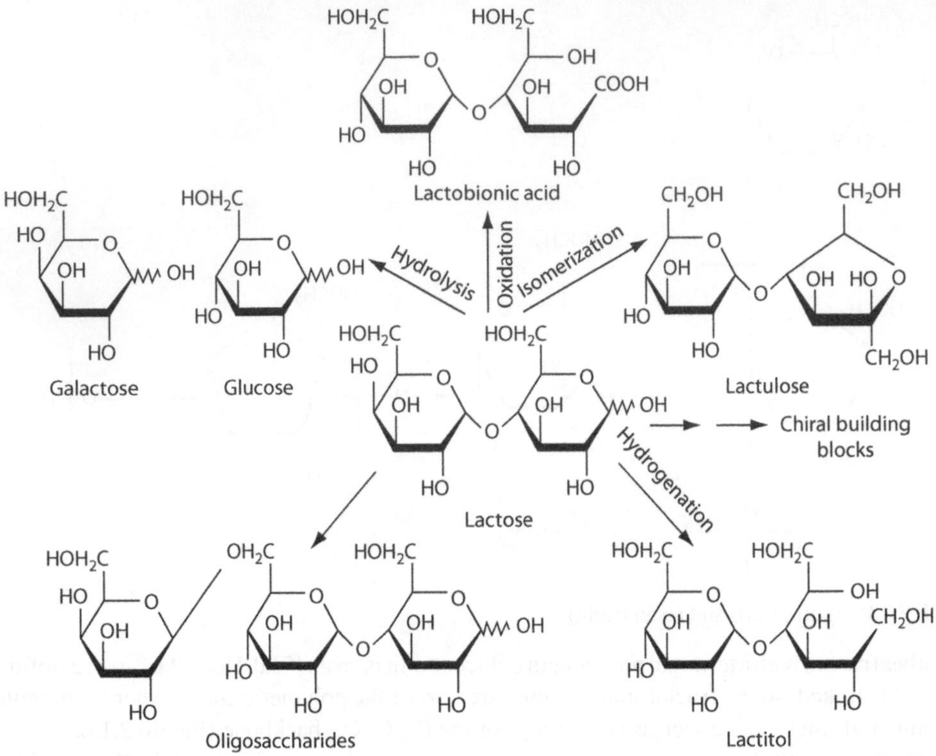

FIGURE 2.14   Lactose as key chemical.

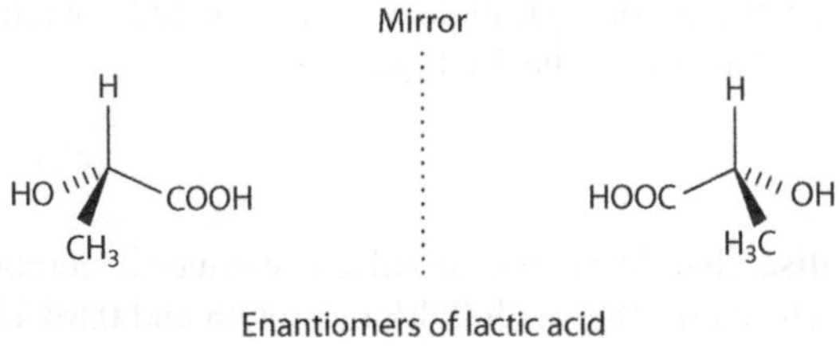

FIGURE 2.15   The two lactic acids.

L-(+)-Lactic acid is the natural form existing in the human body, where it is produced and metabolized in an amount of some 125 g/day. Some bacteria make exclusively the L-(+)-isomer, while others produce selectively the D-(−)-isomer. Most lactic acid made by fermentation of sucrose or glucose is the L-isomer. A description of the production process can be found in Ref. [55].

Lactic acid is used traditionally in meat preservation, but it is increasingly developing as a key chemical (see Figure 2.16). Major outlets are in esters (biodegradable solvents and bread improvers), in the use as chiral building blocks in syntheses and especially in the biodegradable polyester poly-L-lactate.

Poly-lactate manufacture is preferably carried out by making first the cyclic dimeric lactide and then applying a catalytic ring–opening polymerization [56]. So far, mainly poly-L-lactate has been produced—via the LL-lactide—but after the discovery [57] that a 1:1 mixture of poly-L- and poly-D-lactate has largely improved properties (distinctly higher melting temperature), interest in the D-compounds increased. Apparently, a stereocomplex is formed by the two enantiomeric poly-lactates. Furthermore, it turned out to be possible to make the stereocomplex starting from racemic lactic acid [58]. The poly-L/poly-D lactate mixture is called second-generation PLA.

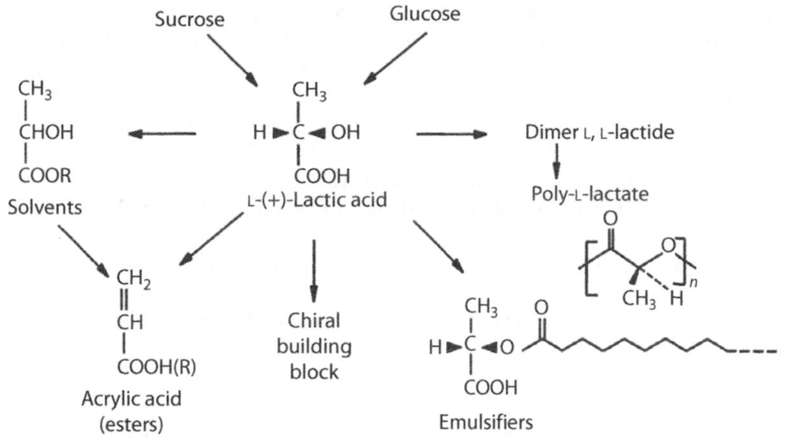

FIGURE 2.16 Lactic acid as key chemical.

Also a mixed lactide–glycolide cyclic lactone is industrially made, giving a biodegradable lactate–glycolate polymer. The world market leader in lactic acids and lactides is PURAC, a division of CSM.

The classic chemical synthesis of (racemic) lactic acid involves cyanohydrin formation from acetaldehyde, followed by hydrolysis of the lactonitrile. On a modest scale, this process is still executed.

Meanwhile, several other approaches have been published. Thus, Exxon researchers designed a five-step lactic acid synthesis from propanal [59]. In most steps zeolites are applied as catalysts and high selectivities are obtained. Another access to lactic acid is Au-catalyzed oxidation of 1,2-propanediol.

Recently, it was reported that the C_3 sugars, dihydroxyacetone and glyceraldehyde, can be converted into methyl or ethyl lactate over zeolite USY [60,61] or tin-beta [62] in methanol or ethanol, respectively. Medium to high selectivities were attained. Over tin-beta (Si/Sn 125), the yield of methyl lactate from dihydroxyacetone and methanol was >99%. In this way, a route from cheap glycerol to alkyl lactates is available. A plausible reaction mechanism—via pyruvic aldehyde—was advanced.

Though a somewhat higher temperature is required, also the hexoses glucose and fructose and the disaccharide sucrose can be converted to methyl lactate over Sn-beta in methanol [63]. Here, a retro-aldol reaction is assumed to occur from C_6 to $2C_3$, glyceraldehyde and dihydroxyacetone. In the sucrose conversion, the yield of methyl lactate was 68%. Also, the Ti-beta and Zr-beta were active as catalysts with methyl lactate yields from sucrose of 44% and 40%, respectively.

2.15 CONCLUSION

In conclusion, it can be stated that renewables offer a broad spectrum of structures that are applied as such or form starting points for further upgrading. The networks around the renewable key molecules are steadily expanding. In future, we will rely more and more on green feedstocks.

REFERENCES

1. Imhoff, M.L. et al., Global patterns in human consumption of net primary production, *Nature*, 429, 870, 2004. (a) Beer, C. et al., Terrestrial gross carbon dioxide uptake: Global distribution and covariation with climate, *Science*, 329, 834, 2010.
2. Steglich, E., Fugmann, B., and Lang-Fugmann, S. (Eds.), *Römpp Encyclopedia of Natural Products*, Thieme, Stuttgart, Germany, 2000.

3. Sheldon, R.A., *Chirotechnology*, Marcel Dekker, New York, p. 304, 1993. (a) Cherubini, F. et al., Toward a common classification approach for biorefinery systems, *Biofuels, Bioprod. Bioref.*, 3, 534, 2009.
4. Kleemann, A. and Engel, J., *Pharmaceutical Substances*, 3rd edn., Thieme, Stuttgart, Germany, 1999.
5. De Guzman, D., Sterol investments on the rise, *Chem. Mark. Rep.*, 25, April 11, 2005. (a) Sanders, J. et al., Bio-refinery as the bio-inspired process to bulk chemicals, *Macromol. Biosci.*, 7, 105, 2007. (b) Langeveld, H., Sanders, J., and Meeusen, M. (Eds.), *The Biobased Economy: Biofuels, Materials and Chemicals in the Post-Oil Era*, Earthscan, London, U.K., 2010.
6. Groeneveld, M.J., Research for a sustainable shell, in *Conference Report Gratama Workshop*, Osaka, Japan, p. 44, 2000.
7. Knight, P., No sign of a halt to Brazil's sugar output—Madness of foresight? *Int. Sugar J.*, 106, 474, 2004.
8. Schmitz, A., Seale, J.L., and Schmitz, T.G., Determinants of Brazil's ethanol sugar blend ratios, *Int. Sugar J.*, 106, 586, 2004.
9. 2009 US ethanol production exceeds 10.75 bn gallons, *Sugar Ind.*, 135, 167, 2010. (a) Global ethanol production to reach 85.9 bn L in 2010, *Sugar Ind.*, 135, 209, 2010.
10. Iogen Energy Canada, Demo-plant near Ottawa since 2004, 2010 production 509 m^3 of cellulosic ethanol. Shell participates.
11. De Boks, P.A. et al., Process of the continuous conversion of glucose to hydrocarbons, *Biotechnol. Lett.*, 4, 447, 1982.
12. Haw, J.F. and Marcus, D.M., Methanol to hydrocarbon catalysis, in *Handbook of Zeolite Science and Technology*, Auerbach, S.M., Carrado, K.A., and Dutta, P.K., Eds., Marcel Dekker, New York, p. 833, 2003.
13. Oudejans, J.C., van den Oosterkamp, P.F., and van Bekkum, H., Conversion of ethanol over zeolite H-ZSM-5 in the presence of water, *Appl. Catal.*, 3, 109, 1982.
14. Dumesic, J.A., Aqueous-phase catalytic processes for the production of hydrogen and alkanes from biomass-derived compounds over metal catalysts, Presented at the *6th Netherlands Catalysis and Chemistry Conference*, Noordwijkerhout, the Netherlands, March 7–9, 2005. (a) Huber, G.W. et al., Production of liquid alkanes for transportation fuel from biomass-derived carbohydrates, *Science*, 308, 1446, 2005.
15. Huber, G.W., Slabaker, J.W., and Dumesic, J.A., Raney Ni-Sn catalyst for H_2 production from biomass-derived hydrocarbons, *Science*, 300, 2075, 2003.
16. Moreau, C., Belgacem, M.N., and Gandini, A., Recent catalytic advances in the chemistry of substituted furans from carbohydrates, *Top. Catal.*, 27, 11, 2004. (a) Chheda, J.N., Huber, G.W., and Dumesic, J.A., Katalytische Flüssigphasenumwandlung oxygenierter Kohlenwasserstoffe aus Biomasse zu Treibstoffen und Rohstoffen für die Chemiewirtschaft, *Angew. Chem.*, 119, 38, 7298, 2007.
17. Roman-Leshkov, Y. et al., Production of dimethylfuran for liquid fuels from biomass-derived carbohydrates, *Nature*, 447, 982, 2007.
18. Papachristos, M.J. et al., The effect of the molecular structure of antiknock additives on engine performances, *J. Inst. Energy*, 64, 113, 1991. (a) Luyckx, G.C.A. et al., Ether formation in the hydrogenolysis of hydroxymethylfurfural over palladium catalysts in alcoholic solution, *Heterocycles*, 77, 1037, 2009. (b) Gruter, G.J.M. and de Jong, E., Furaria, novel fuel options from carbohydrates, *Biofuel Technol.*, 1, 11, 2009.
19. Fagan, P.J. et al., Preparation of levulinic acid esters and formic acid esters from biomass and olefins, PCT Int. Appl. WO 03 85071, Du Pont, 2003. (a) de Lange, J.-P. et al., Valeric biofuels: A platform of cellulosic transportation fuels, *Angew. Chem. Int. Ed.*, 49, 4479, 2010.
20. Kadakia, A.M., Alcohol-based industry marching into the 21st century, *Chem. Weekly*, August 5, p. 57, 1997.
21. le Febre, R.A., Hoefnagel, A.J., and van Bekkum, H., The reaction of ammonia and ethanol or related compounds towards pyridines over high-silica medium pore zeolites, *Recl. Trav. Chim. Pays-Bays*, 115, 511, 1996. (a) Schoonheydt, R.A. et al., A $[Cu_2O]^{2+}$ core in Cu-ZSM-5, the active site in the oxidation of methane to methanol, *PNAS*, 106, 18908, 2009. (b) Beznis, N.V., Weckhuijsen, B.M., and Bitter, J.H., Partial oxidation of methane over Co-ZSM-5: Tuning the oxygenate selectivity by altering the preparation route, *Catal. Lett.*, 136, 52, 2010.
22. Swift, K.A.D., Catalytic transformations of the major terpene feedstocks, *Top Catal.*, 27, 143, 2004; Ravasio, N. et al., Mono- and bifunctional heterogeneous catalytic transformation of terpenes and terpenoids, *Top Catal.*, 27, 157, 2004; Monteiro, J.L.F. and Velosos, C.O., Catalytic conversions of terpenes into fine chemicals, *Top. Catal.*, 27, 169, 2004.
23. Hölderich, W.F. and Laufer, M.C., Zeolites and nonzeolitic molecular sieves in the synthesis of fragrances and flavours, in *Zeolites for Cleaner Technologies*, Guisnet, M. and Gilson, J.-P., Eds., Imperial College Press, London, U.K., 2002, p. 301.

24. Kunkeler, P.J. et al., Application of zeolite titanium beta in the rearrangement of α-pinene oxide to campholenic aldehyde, *Catal. Lett.*, 53, 135, 1998. (a) Suh, Y.W. et al., Redox mesoporous molecular sieve as a bifunctional catalyst for the one-pot synthesis of campholenic aldehyde from α-pinene, *J. Mol. Catal. A*, 174, 249, 2001. (b) Courtesy Dr. G. van Duijn, Unilever Research Laboratory, Vlaardingen, the Netherlands.
25. Knoth, G., Biodiesel: Current trends and properties, *Top. Catal.*, 53, 714, 2010.
26. Hutchings, G.J. et al., New directions in gold catalysis, *Gold Bull.*, 37, 1, 2004; Oxidation of glycerol using supported gold catalysts, *Top. Catal.*, 27, 131, 2004.
27. Yoo, J.-W. et al., Process for the preparation of glycidol, US Pat. 6.316.641, 2001. (a) Zhao, J. et al., Ni/NaX: A bifunctional efficient catalyst for selective hydrogenolysis of glycerol, *Catal. Lett.*, 134, 184, 2010. (b) Pagliaro, M. and Rossi, M., *The Future of Glycerol: New Usages for a Versatile Raw Material*, RSC Green Chemistry Book Series, Cambridge, U.K., 2008.
28. Yokoyama, T. and Setoyama, T., Carboxylic acids and derivatives, in *Fine Chemicals through Heterogeneous Catalysis*, Sheldon, R.A. and van Bekkum, H., Eds., Wiley-VCH, Germany, 2001, p. 370.
29. Mol, J.C., Catalytic metathesis of unsaturated fatty acid esters and oils, *Top. Catal.*, 27, 97, 2004.
30. Zhang, S., Zonchao, S., and Steichen, D., Skeletal isomerization of fatty acids with a zeolite catalyst, Pat. Appl. WO 03 082464, 2003. (a) Lestari, S. et al., Diesel-like hydrocarbons from catalytic deoxygenation of stearic acid over supported Pd nanoparticles on SBA-15 catalysts, *Catal. Lett.*, 134, 250, 2010. (b) Heterogeneous catalysis for biodiesel production, special issue, Lee, A.F. and Wilson, K., Eds., *Top. Catal.*, 53, 11–12, 713–829, 2010. (c) Yan, S. et al., Advancements in heterogeneous catalysis for biodiesel synthesis, *Top. Catal.*, 53, 721, 2010.
31. Mirasol, F., Focus on nutrition gives boost to hydrocolloid sector as supply tightens, *Chem. Mark. Rep.*, 32, March 28, 2005.
32. Nishinari, K. (Ed.), *Hydrocolloids*, Elsevier, Amsterdam, the Netherlands, 2000.
33. Abramovitz, J.N., Mattoon, A.T., Reorienting the forest products economy, in *State of the World 1999*, Strake, L., Ed., Earthscan Publ. London, p. 62, 1999.
34. Swatloski, R.P. et al., Dissolution of cellulose with ionic liquids, *J. Am. Chem. Soc.*, 124, 4974, 2002.
35. Janssen, M.H.A. et al., Room temperature ionic liquids that dissolve carbohydrates in high concentrations, *Green Chem.*, 7, 39, 2005. (a) Sun, N. et al., Complete dissolution and partial delignification of wood in the ionic liquid 1-ethyl-3-methylimidazolium acetate, *Green Chem.*, 11, 646, 2009. (b) De Almeida, R.M. et al., Cellulose conversion to isosorbide in molten salt hydrate media, *ChemSusChem*, 3, 325, 2010. (c) Müller, A., Ein Wald voller Möglichkeiten, *Nachr. Chem.*, 58, 748, 2010. (d) Michinobu, T. et al., Polyesters of 2-pyrone-4,6-dicarboxylic acid (PDC) obtained from a metabolic intermediate of lignin, *Polym. J.*, 40, 68, 2008.
36. Kamide, K., *Cellulose and Cellulose Derivatives: Molecular Characterization and its Applications*, Elsevier, Amsterdam, the Netherlands, p. 180, 2005.
37. Isogai, A. and Kato, Y., Preparation of polyuronic acid form cellulose by TEMPO-mediated oxidation, *Cellulose*, 5, 153, 1998.
38. De Guzman, D., Bio-based mannitol closer to market, *Chem. Mark. Rep.*, 40, April 18, 2005. (a) Moliner, M., Roman-Leshkov, Y., and Davis, M.E., Tin-containing zeolites are highly active catalysts for the isomerization of glucose in water, *PNAS*, 107, 6164, 2010. (b) DSM NV and Roquette establish biobased succinic acid joint venture, *Sugar Int.*, 135, 415, 2010.
39. Jacobs, P.A. and Hinnekens, H., Single-step catalytic process for the direct conversion of polysaccharides to polyhydric alcohols by simultaneous hydrolysis and hydrogenation, Eur. Pat. Appl. 329 923, 1989.
40. Bragd, P.L., van Bekkum, H., and Besemer, A.C., TEMPO-mediated oxidation of polysaccharides: Survey of methods and applications, *Top. Catal.*, 27, 49, 2004.
41. Kachkarova-Sorokina, S.L., Gallezot, P., and Sorokin, A.B., A novel clean catalytic method for waste-free modification of polysaccharides by oxidation, *Chem. Commun.*, 2844, 2004.
42. Gotlieb, K.F. and Capelle, A., *Starch Derivatization, Fascinating and Unique Industrial Opportunities*, Wageningen Academic Publishers, Wageningen, the Netherlands, 2005.
43. Van der Poel, P.W., Schiweck, H., and Schwartz, T. (Eds.), *Sugar Technology*, Verlag Bartens KG, Berlin, Germany, 1998. (a) www.illovo.co.za/world of sugar/sugar statistics int
44. Hough, L., Applications of the chemistry of sucrose, in *Carbohydrates as Organic Raw Materials*, I, Lichtenthaler, F.W., Ed., VCH, Weinheim, New York, 1991, p. 33.
45. de Leenheer, L., Production and use of inulin: Industrial reality with a promising future, in *Carbohydrates as Organic Raw Materials*, III, VCH/CRF, The Hague, the Netherlands, 1996, p. 67.
46. Stolzenburg, K., Topinambur—Rohstoff für die Inulin- und Fructosegewinnung, *Zuckerind.*, 130, 193, 2005.
47. Stevens, C.V., Meriggi, A., and Booten, K., Chemical modification of a valuable renewable resource, inulin and its industrial applications, *Biomacromolecules*, 2, 1, 2001.

48. Verraest, D.L. et al., Distribution of substituents in O-carboxymethyl and O-cyanoethyl ethers of inulin, *Carbohydr. Res.*, 302, 203, 1997.
49. Verraest, D.L. et al., Carboxymethyl inulin: A new inhibitor for calcium carbonate precipitation, *J. Am. Oil Chem. Soc.*, 73, 55, 1996.
50. Booten, K. and Levecke, B., Polymeric carbohydrate-based surfactants and their use in personal care applications, *SOFW J.*, 130(8), 10, 2004.
51. Besemer, A.C. and van Bekkum, H., The relation between calcium sequestering capacity and oxidation degree of dicarboxy-starch and dicarboxy-inulin, *Starch*, 46, 419, 1994.
52. Rogge, T.M. et al., Improved synthesis and physico chemical properties of alkoxylated inulin, *Top. Catal.*, 27, 39, 2004. (a) Collins, P.M. (Ed.), *Dictionary of Carbohydrates*, 2nd edn., Chapman & Hall/ CRC, London, U.K., 2006, p. 681.
53. Timmermans, E., Lactose: Its manufacture and physico-chemical properties, in *Carbohydrates as Organic Raw Materials*, III, VCH/CRF, The Hague, the Netherlands, 1996, p. 3.
54. Abbadi, A. et al., New food antioxidant additive based on hydrolysis products of lactose, *Green Chem.*, 5, 74, 2003.
55. Van Velthuijsen, J.A., Lactic acid production and utilization, in *Carbohydrates as Organic Raw Materials*, III, Van Bekkum, H., Röper, H., and Voragen, A.G.J., Eds., VCH Weinheim, Germany, 1995, p. 129.
56. Drumright, R.E. et al., Polyactic acid technology, *Adv. Mater.*, 12, 1841, 2000.
57. Hada, Y. et al., Stereocomplex formation between enantiomeric poly (lactides), *Macromolecules*, 20, 904, 1987.
58. Radano, C.P. et al., Direct preparation of the polylactic acid stereocomplex from racemic lactic acid, *J. Am. Chem. Soc.*, 122, 1552, 2000.
59. Dakka, J. and Goris, H., Clean process for propanal oxidation to lactic acid, *Catal. Today*, 117, 265, 2006.
60. West, R.M. et al., Zeolite H-USY for the production of lactic acid and methyl lactate from C_3-sugars, *J. Catal.*, 269, 122, 2010.
61. Pescarmona, P.P. et al., Zeolite-catalysed conversion of C_3 sugars to alkyl lactates, *Green Chem.*, 12, 1083, 2010.
62. Taarning, E. et al., Zeolite-catalyzed isomerisation of triose sugars, *ChemSusChem*, 2, 625, 2009.
63. Holm, M.S. et al., Conversion of sugars to lactic acid derivatives using heterogeneous zeotype catalysts, *Science*, 328, 602, 2010.

3 Aqueous-Phase Catalytic Processing in Biomass Valorization to H₂ and Liquid Fuels

Elif I. Gürbüz, Drew J. Braden, and James A. Dumesic

CONTENTS

3.1 INTRODUCTION

3.1.1 CURRENT ENERGY OUTLOOK

Fossil fuels are the primary source of energy in the world. For example, in 2008, the global energy consumption was 474 exajoules (1 EJ = 10^{18} J), of which 88% was derived from fossil fuel resources. As shown in Figure 3.1, renewable resources currently account for only about 7% of total energy consumption (both globally and in the United States). Primary sources of renewable energy include

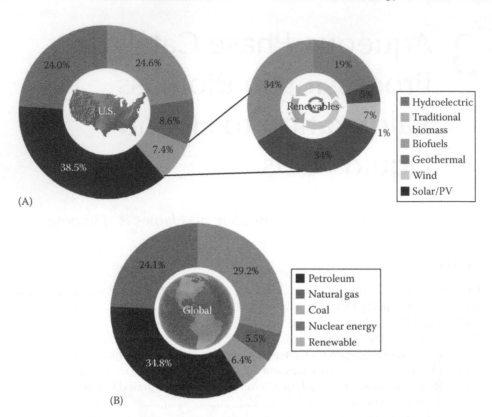

FIGURE 3.1 Distribution of energy usage in (A) the United States and (B) globally by resource. (For U.S. energy usage distribution, refer to the legend for (B). Both legends follow a clockwise order starting from the bottom of the pie-charts.) (From BP Statistical review of world energy, 2009; U.S. Department of Energy Information Administration, State and U.S. historical energy data, 2009; International Energy Agency, Key world energy statistics, 2009.)

hydroelectricity (34% of renewables) and traditional biomass (i.e., open combustion; 34% of renewables). Solar, wind, and geothermal energies along with biofuels account for 1%, 7%, 5%, and 19% of the renewable energy used, respectively (BP 2009; Department of Energy Information Administration 2009; International Energy Agency 2009). Residential and commercial energy consumption is primarily in the form of electrical power, and these two sectors consume primarily coal and natural gas. In contrast, transportation energy is derived almost exclusively from oil (96%). Thus, in general, current global energy systems rely heavily on coal for electrical power generation and oil for transportation energy. However, depletion of these finite fossil fuel resources seems unavoidable in the future since these resources are expended faster than can be regenerated. Recent analyses of proven reserves for oil, coal, and natural gas show that these fuels have lifetimes of approximately 40, 60, and 200 years, respectively, at current usage rates (Boyle 2004; Worldwatch Institute Center for American Progress 2006; Graziani and Fornasiero 2007; Kreith and Goswami 2007). In particular, oil is the resource with the shortest expected lifetime, and importantly, oil is the predominant source of transportation energy, one of the largest and fastest growing global energy sectors. Additionally, the consumption of fossil fuels leads to the emission of CO_2, causing detrimental effects on the environment such as accelerated rates of global warming.

Analyses show that renewable energy resources such as solar, wind, geothermal, and hydroelectric are best suited for the generation of electricity and heat (Boyle 2004; Graziani and Fornasiero 2007; Kreith and Goswami 2007), which makes them (and nuclear energy) ideal for stationary power applications. The transportation industry requires clean burning fuels with high energy densities for efficient storage at ambient conditions. Currently, these criteria are best met by liquid hydrocarbons derived from crude oil. Therefore, a renewable option would be to utilize resources such as biomass to produce liquid fuels that meet the aforementioned criteria.

3.1.2 LIQUID BIOFUELS

The liquid biofuels used most widely today for transportation are ethanol and biodiesel. Ethanol is the predominant biomass-derived fuel at the present time, accounting for 90% of total biofuel usage. Fermentation produces a dilute aqueous solution of ethanol and necessitates an energy-intensive distillation step to completely remove water from the mixture. Low concentration (5%–10%) blends of ethanol with gasoline (i.e., E5–E10) can be employed in current spark ignition engines while ethanol-rich mixtures (E85) require additional engine upgrades (U.S. Department of Energy and Environmental Protection Agency 2010).

Biodiesel is the second most abundant renewable liquid fuel. It can be used in current injection engines in a wide range of blends with petrol–diesel or as a pure fuel (without petrol–diesel). Esterification of fatty acids or transesterification of oils (triglycerides) with alcohols (normally methanol and ethanol), using a basic (Kim et al. 2004; Di Serio et al. 2006; Granados et al. 2007) or acidic catalyst (Lotero et al. 2005; Melero et al. 2009), results in the production of a first generation biodiesel (Ma and Hanna 1999; Fukuda et al. 2001). The primary side product of this reaction is glycerol which is obtained in high concentrations in water. Catalytic routes for the upgrading of this waste stream to fuels and chemicals are being developed and an example is presented later in this chapter. Hydrotreating of fatty acids and triglycerides is also possible (Huber et al. 2007) and can be carried out in petroleum refineries by coprocessing vegetable oils with petroleum-derived feedstocks (Huber and Corma 2007; Li et al. 2010). The main disadvantage of oil-based processes is the unavailability of inexpensive feedstocks. Normally palm, sunflower, canola, rapeseed, and soybean oils are used, but they are expensive and can otherwise be used as food sources. Various nonedible oils have been proposed as appropriate feedstocks (e.g., cynara [Fernández et al. 2006], jatropha, karanja [Patil and Deng 2009]). Additionally, researchers are focusing on the production of third generation biodiesel, which is produced from algae-derived triglycerides (Chisti 2007).

Both of the liquid biofuels presented earlier are oxygenated fuels with molecular compositions that differ from the petroleum-derived fuels used today. However, these biofuels do not meet the criteria required for liquid transportation fuels, namely, to burn cleanly and have high energy densities for efficient storage at ambient conditions. Therefore, in the short term, it is desirable to utilize biomass to generate liquid fuels that meet the physical property requirements of today's liquid fuels and are compatible with modern internal combustion engines. When compared with ethanol, the production of hydrocarbon fuels from biomass has important advantages. First, because it is possible to produce renewable hydrocarbon fuels that are chemically identical to those currently obtained from petroleum, it would not be necessary to modify the existing infrastructure for their implementation and distribution in the transportation sector. Additionally, the processes for the production of hydrocarbon biofuels could be coupled with the fuel production systems of existing and well-developed petroleum refineries. Moreover, the increase in oxygen content of a fuel negatively impacts its heating value (i.e., the heat released when a known quantity of fuel is burned under specific conditions). For instance, ethanol has only 66% of the heating value of gasoline, and thus, cars running on ethanol rich mixtures like E85 get 30% lower gas mileage (U.S. Department of Energy and Environmental Protection Agency 2010). Biomass-based hydrocarbon fuels, in contrast, offer an equivalent energy content and gas mileage performance to petroleum-derived fuels. Also, the addition of oxygenates to gasoline increases the hydrophilicity of the mixture. In the case of ethanol/gasoline blends, water contamination can result in a phase separation of both components, which is an important concern especially in cooler climates. Finally, an expensive energy-consuming distillation step to purify ethanol can be eliminated as hydrocarbons are able to separate spontaneously from the water solvent during the biofuel production process.

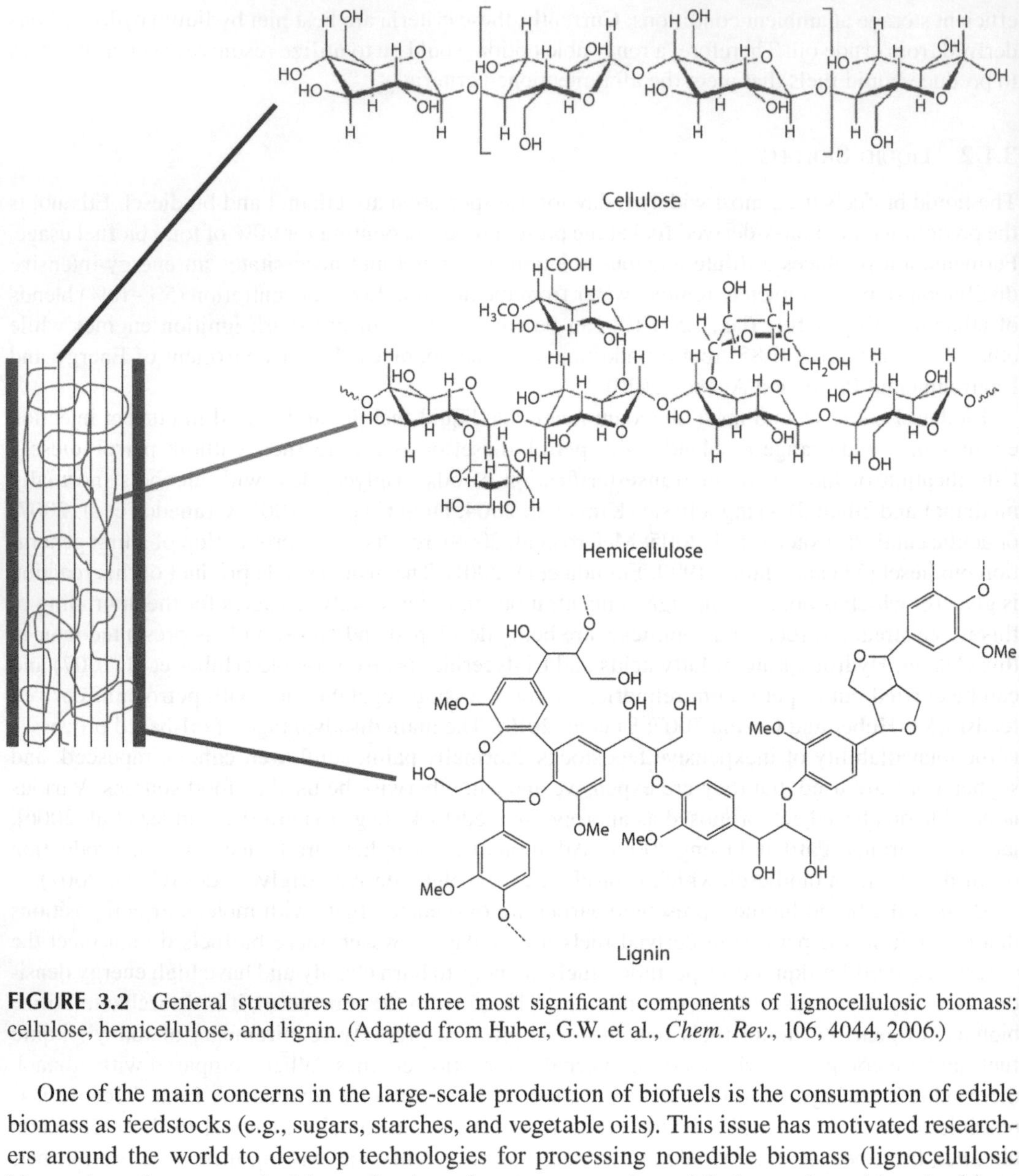

FIGURE 3.2 General structures for the three most significant components of lignocellulosic biomass: cellulose, hemicellulose, and lignin. (Adapted from Huber, G.W. et al., *Chem. Rev.*, 106, 4044, 2006.)

One of the main concerns in the large-scale production of biofuels is the consumption of edible biomass as feedstocks (e.g., sugars, starches, and vegetable oils). This issue has motivated researchers around the world to develop technologies for processing nonedible biomass (lignocellulosic biomass) so that a sustainable production of a new generation of fuels can be achieved without affecting food supplies. In this respect, lignocellulosic biomass is attractive since it is abundant (Perlack et al. 2005) and can be grown faster and more economically than food crops (Klass 2004). Lignocellulosic biomass consists of three major components (Figure 3.2): cellulose (40%–50%), hemicellulose (25%–35%), and lignin (15%–20%) (Lange 2007; Stocker 2008). In the following section, a more detailed description of the major lignocellulose components is presented.

3.1.3 Lignocellulosic Biomass

Lignin is an amorphous polymer composed of methoxylated phenylpropane structures such as coniferyl alcohol, sinapyl alcohol, and coumaryl alcohol (Chakar and Ragauskas 2004; Huber et al. 2006). This material provides plants with strength and structural rigidity as well as a hydrophobic vascular system for the transportation of water and solutes (Vanholme et al. 2008).

The hemicellulose and cellulose fractions are surrounded by lignin, which must first be depolymerized by a pretreatment step so that the cellulose and hemicellulose portions can be accessed for further upgrading (Mosier et al. 2005; Huber et al. 2006). Although lignin can be isolated, it is not readily amenable to upgrading strategies. One option for lignin utilization is to burn it directly to produce heat and electricity. Process heat and power obtained from burning lignin and other residual solids are more than enough to drive the biofuel production process (National Renewable Energy Laboratory 2002; Aden and Foust 2009). In addition, lignin can serve as a feedstock in the production of phenolic resins (Gosselink et al. 2004). Pyrolysis strategies for lignin have been reported for the production of bio-oils (Evans et al. 1986) and aromatics (Carlson et al. 2010).

Hemicellulose is an amorphous polymer generally comprising five different sugar monomers: D-xylose (the most abundant), L-arabinose, D-galactose, D-glucose, and D-mannose (Lynd et al. 1991). To increase the effectiveness of the cellulose hydrolysis steps in the production of glucose, the hemicellulose fraction of biomass is typically removed during pretreatment. The pretreatment process aims to preserve the xylose obtained from hemicellulose and inhibits the formation of degradation and dehydration products (Mosier et al. 2005). Compared to hydrolysis of crystalline cellulose, hemicellulose extraction/hydrolysis is an easier process and allows for high yields of sugar. For example, hemicellulose is readily depolymerized to produce xylose monomers through dilute acid hydrolysis with H$_2$SO$_4$ being the most commonly used acid (Carvalheiro et al. 2008; Mamman 2008). Xylose monomers are used as feedstocks for ethanol production via fermentation (Zaldivar et al. 2001; Saha 2003) or for preparation of furfural via dehydration (Mamman 2008).

Cellulose is a polymer composed of glucose units linked via β-glycosidic bonds, providing the structure with a rigid crystallinity that inhibits hydrolysis (Lynd et al. 1991). Cellulose is largely inaccessible to hydrolysis in untreated biomass before the removal of lignin and hemicellulose (Mosier et al. 2005; Huber et al. 2006). High yields of glucose (>90% of theoretical maximum) can be achieved by enzymatic hydrolysis of cellulose after biomass pretreatment (Lloyld and Wyman 2005; Wyman et al. 2005a). Harsher conditions using solutions of mineral acids (H$_2$SO$_4$) at elevated temperatures can be applied to hydrolyze cellulose; however, these conditions lead to the formation of degradation products such as hydroxymethylfurfural (HMF), levulinic acid, and insoluble humins (Wyman et al. 2005b; Rinaldi and Schuth 2009).

The cost of cellulosic ethanol is approximately twice that of corn ethanol due to the complexity in isolating sugars from lignocellulosic biomass (Regalbuto 2009). Accordingly, research is being conducted worldwide with the aim of simplifying the sugar isolation process so that cost-competitive technologies may be developed for the generation of liquid fuels from nonedible sources. Experts have identified this challenge as the key bottleneck for the large-scale implementation of a lignocellulose-derived biofuel industry (Bozell 2008).

3.1.4 Processes Developed for Conversion of Lignocellullosic Biomass to Fuels

Current strategies for the conversion of lignocellulosic biomass into liquid hydrocarbon fuels involve three major primary routes: gasification, pyrolysis, and aqueous-phase hydrolysis (Figure 3.3). The first approach involves the conversion of biomass to synthesis gas, a mixture of CO and H$_2$, which serves as a precursor of liquid hydrocarbon fuels. The second strategy converts solid biomass into a liquid fraction known as bio-oil, which can be further upgraded to gasoline and diesel fuel components. Finally, the third route involves the aqueous-phase hydrolysis of biomass to produce sugars and valuable intermediates that can be catalytically processed in the aqueous phase to the full range of liquid hydrocarbon fuels including gasoline, diesel, and jet fuels.

Gasification of biomass requires high temperatures (e.g., >1100 K), necessary for the endothermic formation of synthesis gas (Lange 2007). The gasification process takes place with partial combustion of the biomass by cofeeding an oxidizing agent (e.g., oxygen, air) in the gasifier. The synthesis gas obtained can be converted to a distribution of alkanes through Fischer–Tropsch synthesis (Caldwell 1980; Iglesia et al. 1993; Dry 2002) or can be used to produce methanol by methanol synthesis (Klier 1982;

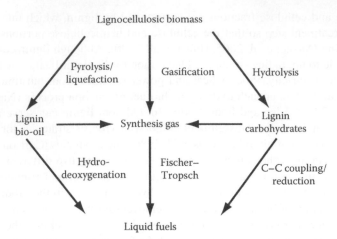

FIGURE 3.3 General processing strategies for the conversion of lignocellulosic biomass conversion into liquid transportation fuels. (Adapted from National Science Foundation, Breaking the chemical and engineering barriers to lignocellulosic biofuels: Next generation hydrocarbon biorefineries, Paper read at University of Massachusetts Amherst, National Science Foundation, Chemical, Bioengineering, Environmental, and Transport Systems Division, Washington, DC, http://www.ecs.umass.edu/biofuels/Images/Roadmap2-08.pdf, 2008.)

Lange 2001). Moreover, the H_2/CO ratio of the synthesis gas can be adjusted by using a water gas shift (WGS) catalyst (Kunkes et al. 2009b) or by in situ CO_2 adsorption (Koppatz et al. 2009) to produce H_2-enriched streams. The low purity of the synthesis gas stream (Soares et al. 2006) and the negative impact on the thermal efficiency by the high moisture content of the biomass (Huber et al. 2006) are challenges for the development of this route.

Pyrolysis and liquefaction involve the thermal decomposition of biomass under an inert atmosphere at lower temperatures (573–973 K) (Mohan et al. 2006; Lange 2007), leading to the formation of bio-oil. This organic liquid consists of a complex mixture of more than 300 different highly oxygenated compounds, polymeric carbohydrates, and lignin fragments (Mohan et al. 2006; National Science Foundation 2008) with a high water content (~25 wt%). The high oxygen content of the molecules present in bio-oils leads to a liquid with a low energy content. Pyrolysis oil requires further deoxygenation to be used as a transportation fuel. Typically, bio-oil processing approaches involve the consumption of H_2 from an external supply (Elliott 2007; Kersten et al. 2007) or additional upgrading reactions over zeolites to produce aromatic fuels (Carlson et al. 2008).

Compared to the temperatures used in gasification and pyrolysis, aqueous-phase acid and enzymatic hydrolysis reactions are effective at relatively low temperatures and can achieve separation of the carbohydrate and lignin fractions of lignocellulosic feeds. The lignin fraction of these feeds must be depolymerized and removed by appropriate pretreatment steps to enable the acids or enzymes to more easily access and hydrolyze the cellulose and hemicellulose fractions of biomass. Lignin can be attacked at mild temperatures in the presence of acids, and hydrolysis of hemicellulose can also occur under the same conditions. Many pretreatment steps involving chemical, physical, and biological approaches have been developed to depolymerize lignocellulosic materials (Carroll and Somerville 2009; Kumar et al. 2009a). Wyman et al. (Kumar et al. 2009b) have recently studied the effects of these pretreatments on the morphology and structure of biomass. The goal of these processes is to obtain an aqueous–sugar monomer solution from lignocellulose at the lowest possible cost, and it has been estimated that sugar solutions can be produced from corn stover at around 10 ¢/kg of sugar (National Renewable Energy Laboratory 2002). Future improvements in pretreatment technologies could decrease the cost of sugar production to as low as 5 ¢/kg (Lynd et al. 1999).

In this chapter, we focus on more recently developed processes that utilize aqueous-phase compounds (e.g., sugars, polyols, lactic acid, and levulinic acid) obtained by lignocellulosic biomass treatments. First, we discuss aqueous-phase reforming (APR), in which aqueous solutions of sugars or polyols are converted to H_2 and alkanes (mainly pentane and hexane), and gas phase reforming of aqueous solutions of glycerol to obtain synthesis gas, which can subsequently be used for the production of liquid hydrocarbon fuels and/or chemicals by means of Fischer–Tropsch and methanol syntheses, respectively. Then, we address another APR process, in which sugars and polyols having a C:O ratio of 1:1 are converted by a combination of C–C and C–O cleavage reactions to produce a mixture of monofunctional organic compounds such as alcohols, ketones, carboxylic acids, and heterocyclics. These monofunctional intermediates can undergo catalytic upgrading reactions to produce hydrocarbon fuels by various routes, such as dehydration, aromatization, and alkylation over acid zeolite catalysts; aldol condensation of alcohols and ketones over bifunctional catalysts containing metal and basic sites; and ketonization of carboxylic acids over basic oxides, to control C–C coupling and oxygen removal. We next consider an approach in which aqueous solutions of sugars formed from lignocellulosic biomass undergo catalytic dehydration to produce furan compounds, such as furfural and 5-hydroxymethylfurfural (HMF). These furanic aldehydes can then undergo aldol condensation reactions over basic catalysts to produce hydrocarbons suitable for diesel fuel applications. We also present the lactic acid platform, in which bifunctional catalysts containing metal and acid sites are used to perform dehydration/hydrogenation, C–C and C–O cleavage reactions, and C–C coupling reactions such as aldol condensation and ketonization, to obtain an organic liquid composed of propanoic acid and ketones in the C_4–C_7 range. Finally, we consider the levulinic acid platform, in which lignocellulosic biomass undergoes acid treatment to produce levulinic acid in a single step. The aqueous solution of levulinic acid (in the presence of formic acid) then undergoes catalytic reduction to form γ-valerolactone (GVL). GVL can then be used to produce nonane for diesel fuel or branched alkanes with molecular weights appropriate for jet fuel. Both of these processes are initiated with the ring opening of GVL to produce pentenoic acid isomers. For the production of the diesel fuel, pentenoic acid isomers are hydrogenated to obtain pentanoic acid, which can be ketonized to obtain 5-nonanone. Finally, 5-nonanone is converted to nonane by hydrodeoxygenation. For the production of jet fuel, pentenoic acid isomers go through decarboxylation to produce butene and CO_2, combined with butene oligomerization to form C_8–C_{20} alkenes.

3.1.5 Conversion of Carbohydrates and Polyols to Fuels: Thermodynamic Issues

We first explore some of the issues involved in the conversion of an oxygenated hydrocarbon to an alkane, and we then address issues involved in producing an alkane containing a larger number of carbon atoms than present in the sugar reactant. To illustrate the concepts, we use ethylene glycol as the reactant because this compound is the smallest molecule containing a C–C bond and has a C:O stoichiometry equal to 1:1, thereby representing a sugar alcohol. We use butane as the alkane product because the production of this molecule would involve a C–C bond formation step that increases the length of the alkane chain.

The conversion of ethylene glycol to butane involves the following stoichiometric equation:

$$C_2O_2H_6 \rightarrow \frac{5}{13}C_4H_{10} + \frac{6}{13}CO_2 + \frac{14}{13}H_2O \tag{3.1}$$

This conversion takes place by a combination of two processes involving the consumption and production of H_2:

$$C_2O_2H_6 + \frac{3}{2}H_2 \rightarrow \frac{1}{2}C_4H_{10} + 2H_2O \tag{3.2}$$

$$\frac{3}{10}(C_2O_2H_6 + 2H_2O \rightarrow 2CO_2 + 5H_2) \tag{3.3}$$

where the H_2 required for the reduction of ethylene glycol is supplied by reforming ethylene glycol with water to form H_2 and CO_2. The exothermic nature of the reaction in Equation 3.1 (–24 kcal/mol) dictates that the energy content of the alkane product, as expressed by the heat of combustion to form CO_2 and water vapor, is lower than that of the ethylene glycol reactant. Fortunately, the decrease in enthalpy for this conversion is sufficiently small such that 91% of the energy content of the ethylene glycol reactant is preserved in the alkane product. Importantly, the alkane product contains approximately 36% of the mass of the reactant, which leads to an increase in the energy density.

It is now important to address the energy changes involved in the reactions employed to increase the length of the alkane chain. These reactions are necessary for the production of alkanes for transportation fuels. In particular, the reduction processes considered in the aforementioned examples would lead to the conversion of glucose to hexane, whereas larger alkanes are required for gasoline, jet, and diesel fuel applications.

The conversion of ethane to butane by the following stoichiometric reaction is an endothermic process (10 kcal/mol):

$$2C_2H_6 \rightarrow C_4H_{10} + H_2 \tag{3.4}$$

Accordingly, the equilibrium constant for this reaction would be low and it would not be possible to achieve high conversions in a single-pass reactor. In contrast, various coupling reactions involving partially oxygenated reactants have favorable equilibrium constants for conversion. For example, the coupling between ethane and ethanol to form butane and water is an exothermic reaction (–12 kcal/mol):

$$C_2H_6 + C_2OH_6 \rightarrow C_4H_{10} + H_2O \tag{3.5}$$

In a similar fashion, the coupling between two ethanol molecules leads to butanol and water in another exothermic reaction (–16 kcal mol⁻¹):

$$2C_2OH_6 \rightarrow C_4OH_{10} + H_2O \tag{3.6}$$

This butanol product can then undergo dehydration to form butane, which can be accomplished with high conversion in view of the positive value for the entropy change of the dehydration process.

Various catalytic scenarios can be imagined to achieve C–C coupling between ethanol and ethane molecules and these reactions will be addressed later in this chapter. For example, two molecules of ethanol can undergo dehydrogenation to acetaldehyde, followed by aldol condensation, dehydration, and hydrogenation over bifunctional catalysts containing metal and basic sites (Chheda and Dumesic 2007). Another option would be to convert ethanol to ethylene by dehydration, coupled with ethylene dimerization, followed by a hydride transfer with ethane over acid catalysts (Chang and Silvestri 1977). While the development of catalysts for these C–C coupling reactions involving biomass-derived compounds is in its infancy, the important conclusion from the previous analysis is that partially oxygenated compounds offer new routes for C–C coupling that are unavailable for C–C coupling reactions between alkanes. Thus, the strategy that emerges for converting a biomass-derived compound to an alkane containing a larger number of carbon atoms than present in the sugar reactant is to utilize the oxygen-containing moieties in the reactant and/or the reaction intermediates to form C–C bonds prior to the formation of the final alkane product.

3.1.6 REACTION CLASSES FOR CATALYTIC CONVERSION OF CARBOHYDRATE-DERIVED FEEDSTOCKS

The development of strategies to process carbohydrates requires an understanding of the fundamental reactions that can take place for these highly oxygenated molecules as outlined in this section. Figure 3.4 represents some chemistries involved in the production of fuels and chemicals from carbohydrate-based feedstocks in a biorefinery.

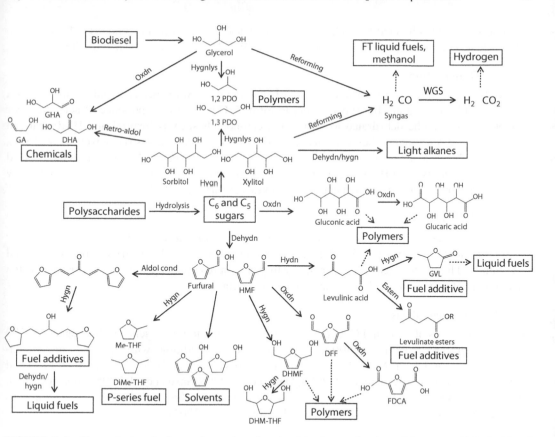

FIGURE 3.4 General chemical transformations involved in the production of transportation fuels and chemicals from carbohydrate-based feedstocks in a biorefinery. Furfural is derived from C₅ sugars and HMF from C₆ sugars. Abbreviations for reactions: Hygn, hydrogenation; Oxdn, oxidation; Hygnlys, hydrogenolysis; Estern, esterification; Dehydn, dehydration; Aldol cond, Aldol condensation. Abbreviations for chemical compounds: GA, glycol-aldehyde; GHA, glyceraldehyde; DHA, dihydroxyacetone; PDO, propanediol; HMF, hydroxymethylfurfural; DFF, diformylfuran; FDCA, 2,5-furan-dicarboxylic acid; DHMF, di(hydroxymethyl)furan; DHM-THF, di(hydroxymethyl)-tetrahydrofuran; Me-THF, methyl tetrahydrofuran; GVL, γ-valerolactone. (Adapted from Chheda, J.N., and Dumesic, J.A., *Catal. Today*, 123(1–4), 59, 2007.)

3.1.6.1 Hydrolysis

Hydrolysis is one of the major reactions for processing lignocellulosic biomass and is used to cleave the glycosidic bonds between the sugar units to obtain simple sugars. Hydrolysis reactions can be carried out using acid or base catalysts at temperatures ranging from 370 to 570 K, depending on the structure and nature of the polysaccharides. Acid hydrolysis is more common because base hydrolysis results in more side reactions, and thus, lower yields (Bobleter 2005). Acid hydrolysis proceeds by C–O–C bond cleavage at the intermediate oxygen atom between two sugar molecules (Bobleter 2005). Cellulose, the most abundant polysaccharide with β-glycosidic linkages, is the most difficult polysaccharide to hydrolyze because of its high crystallinity. Both mineral acids and enzymes can be used for cellulose hydrolysis, but enzymatic hydrolysis is more selective compared to acid hydrolysis (Wyman et al. 2005b). Using enzymatic catalysts, glucose can be produced from cellulose at yields close to 100%, whereas the highest yields obtained for the case of concentrated mineral acid hydrolysis are typically lower than 70% (Huber et al. 2006). Hemicelluose can be hydrolyzed at more modest temperatures and dilute acid concentrations, thereby minimizing degradation of the simple sugars obtained (Wyman et al. 2005b). Soluble starch (a polyglucan with α-glycosidic linkages obtained from corn and rice) and inulin (a polyfructan obtained from chicory) can be hydrolyzed at modest conditions (340–420 K) to form glucose and fructose, respectively (Moreau et al. 1997; Nagamori and Funazukuri 2004).

3.1.6.2   Dehydration

Dehydration reactions of carbohydrates and carbohydrate-derived molecules result in the loss of a water molecule from an alcohol functionality to form C=C bonds in the presence of acid sites. For example, sugars can be dehydrated to form furan compounds such as furfural and HMF that can subsequently be converted to diesel fuel additives (Huber et al. 2005), industrial solvents (e.g., furan, tetrahydrofurfuryl alcohol, and furfuryl alcohol) (Lichtenthaler and Peters 2004), various bio-derived polymers (by conversion of HMF to 2,5-furandicarboxylic acid [FDCA]) (U.S. Department of Energy 2004), and P-series fuels (by subsequent hydrogenolysis of furfural) (Paul 2001).

3.1.6.3   Hydrogenation

Hydrogenation reactions are carried out in the presence of a metal catalyst such as Pd, Pt, Ni, or Ru at moderate temperatures (370–420 K) and moderate pressures (10–30 bar) to saturate C=C and C=O bonds. The product selectivity for the hydrogenation reaction depends on factors such as the solvent, partial pressure of hydrogen, and the nature of catalyst.

3.1.6.4   Hydrogenolysis

The hydrogenolysis of C–C and C–O bonds in polyols occurs in the presence of hydrogen (14–300 bar) at temperatures from 400 to 500 K usually under basic conditions with supported metal catalysts including Ru, Pd, Pt, Ni, and Cu (Zartman and Adkins 1933; Tronconi et al. 1992; Lahr and Shanks 2003, 2005; Chaminand et al. 2004; Dasari et al. 2005; Saxena et al. 2005; Miyazawa et al. 2006). The general objective of hydrogenolysis is to selectively break C–C and/or C–O bonds to produce more valuable polyols and/or diols (Chaminand et al. 2004).

3.1.6.5   Reforming Reactions

Reforming reactions involve the cleavage of C–C bonds of the carbohydrate backbone to yield a mixture of CO and hydrogen. In the presence of water, the water gas shift reaction converts CO to CO_2 and hydrogen. Equation 3.7 shows a steam reforming reaction starting with an alkane, whereas Equation 3.8 shows the reforming of a carbohydrate, and Equation 3.9 shows the water gas shift reaction:

$$C_nH_{2n+2} + nH_2O \leftrightarrow nCO + (2n+1)H_2 \tag{3.7}$$

$$C_nH_{2y}O_n \leftrightarrow nCO + yH_2 \tag{3.8}$$

$$CO + H_2O \leftrightarrow CO_2 + H_2 \tag{3.9}$$

It has recently been shown that APR can be used to convert aqueous solutions of sugars and sugar alcohols to produce H_2 and CO_2 at temperatures near 500 K over metal catalysts (Cortright et al. 2002). Importantly, the selectivity toward H_2 can be controlled by altering the nature of the metal sites (e.g., Pt) and metal alloy (e.g., Ni-Sn) (Huber et al. 2003) components, and by choice of the catalyst support (Shabaker et al. 2003).

3.1.6.6   C–C Coupling Reactions

3.1.6.6.1   Fischer–Tropsch Synthesis

The catalytic hydrogenation of carbon monoxide to produce alkanes and alkenes (together with some oxygenated compounds) is known as Fischer–Tropsch synthesis. The alkane production reaction can be represented as Equation 3.10 (Caldwell 1980):

$$nCO + (2n+1)H_2 \rightarrow C_nH_{2n+2} + nH_2O \tag{3.10}$$

Since Fischer–Tropsch synthesis is a polymerization reaction, it produces a wide molecular-weight range of compounds whose distribution depends on temperature, feed gas composition, pressure, catalyst type, and promoters. A variety of catalysts can be used for the Fischer–Tropsch process, the most common ones containing cobalt, iron, or ruthenium. Currently, there are two Fischer–Tropsch operating modes. The high-temperature process (570–620 K) uses iron-based catalysts to produce gasoline and linear low molecular mass alkenes, whereas the low-temperature process (470–510 K) utilizes either iron or cobalt catalysts to generate high molecular mass linear waxes (Dry 2002). The production of liquid alkanes from biomass takes place by gasification of biomass to produce synthesis gas (H$_2$/CO) followed by Fischer–Tropsch synthesis.

3.1.6.6.2 Ketonization

The synthesis of ketones from carboxylic acids is called ketonization. Various studies on the ketonization of carboxylic acids have been presented using several metal-oxides as catalysts such as Cr$_2$O$_3$, Al$_2$O$_3$, PbO$_2$, TiO$_2$, ZrO$_2$, CeO$_2$, iron oxide, and manganese oxide, as well as Mg/Al hydrotalcites. Two molecules of carboxylic acids combine to form a ketone via ketonization, producing carbon dioxide and water. The stoichiometry of the reaction is as follows:

$$R' \overset{O}{\diagup}\text{OH} \; + \; R \overset{O}{\diagup}\text{OH} \longrightarrow R' \overset{O}{\diagup} R \; + \; H_2O + CO_2 \qquad (3.11)$$

In this reaction, RCOOH reacts with R'COOH via cross ketonization; alternatively, two symmetric ketones of RCOR and R'COR' can be formed via homo ketonization. As a result, two different carboxylic acids can produce three different ketones (Nagashima et al. 2005).

3.1.6.6.3 Aldol Condensation

The base- or acid-catalyzed aldol condensation reaction, a type of C–C coupling reaction, is used for synthesis of higher molecular weight ketones and aldehydes from more abundant low molecular weight homologues (Tichit et al. 2003). In addition to fine chemical synthesis, aldol condensation reaction followed by oxygen removal steps can be used to obtain branched or linear, long carbon chain alkanes suitable for gasoline, diesel, and jet fuel from lower molecular weight ketones and aldehydes such as 2-hexanone, acetone, furfural, and HMF (Kunkes et al. 2008b; West et al. 2008). Aldol condensation requires at least one carbonyl compound having an α-hydrogen atom. The condensation mechanism on basic sites is initiated by the abstraction of the proton in the alpha position to the carbonyl group to form a carbanion intermediate which is stabilized by the enolate resonance isomer. To form the aldol product, the carbanion acts as a nucleophilic reagent and the α-carbon of this donor attacks the carbonyl carbon of the other molecule which acts as the acceptor (Di Cosimo et al. 1996; Tichit et al. 2003; Diez et al. 2008). The aldol condensation between two 2-ketones is shown in the following reaction:

$$R_1 \overset{O}{\diagup} \; + \; R_2 \overset{O}{\diagup} \longrightarrow R_1 \underset{OH}{\diagup} \overset{O}{\diagup} R_2 \qquad (3.12)$$

The aldol adduct can further undergo dehydration to form an unsaturated aldehyde or ketone. The selectivity of the process toward heavier compounds is controlled by the reaction temperature, solvent, reactant molar ratio, structure of reactant molecules, and the nature of the catalyst (Barrett et al. 2006).

3.1.6.6.4   *Oligomerization*

The oligomerization of light alkenes is applied for the production of petrochemicals (nonene and dodecene) and fuels (polymeric gasoline and middle distillate). The general reaction of alkene oligomerization can be expressed by the following equation:

$$nC_mH_{2m} \longrightarrow C_{nm}H_{2nm} \tag{3.13}$$

where

    m defines the number of carbon atoms in the mono-alkene substrate

    n defines the number of monomer molecules reacting to form the product oligomer

Alkenes serve as reactive intermediates for subsequent C–C coupling reactions and are desirable intermediates in biomass processing technologies to obtain fuel-grade compounds (Knifton et al. 1994). Alkene oligomerization reactions can be carried out with high conversion at moderate temperatures (e.g., 470 K) at which the favorable enthalpy change compensates for the unfavorable entropy change caused by the decrease in the number of moles in the reaction.

3.2   AQUEOUS-PHASE REFORMING AND DERIVATIVE TECHNOLOGIES

Initially developed as a strategy to produce hydrogen from carbohydrates (Cortright et al. 2002), APR has become an important process in strategies for the production of light alkanes (Huber et al. 2004), synthesis gas (Soares et al. 2006), and monofunctional compounds (Kunkes et al. 2008b) (discussed in a separate section). These strategies are summarized in Figure 3.5.

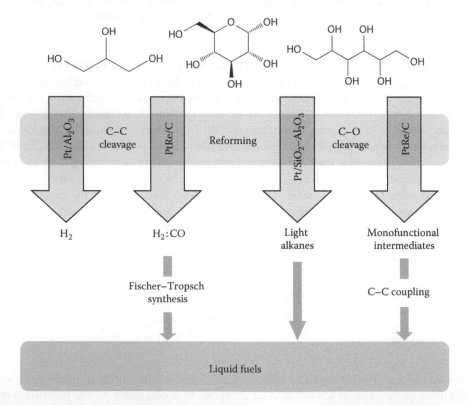

FIGURE 3.5   Catalytic conversion strategies for the production of H_2 and liquid transportation fuels from oxygenated biomass-derived species.

3.2.1 HYDROGEN PRODUCTION FROM AQUEOUS-PHASE REFORMING

Carbohydrates (e.g., glucose) and polyols (e.g., ethylene glycol, glycerol, and sorbitol) can be converted to primarily H$_2$ and CO$_2$ via reforming with water in the aqueous phase over appropriate heterogeneous catalysts at temperatures near 500 K. The resulting hydrogen can be purified, if necessary, and utilized as

1. A chemical reagent for hydrogenation reactions (i.e., hydrogenation of carbohydrates to produce glycols)
2. A chemical feedstock for the production of ammonia and fertilizers
3. A renewable fuel source for PEM fuel cells
4. A hydrogen-rich gas stream that augments the gas stream from biomass gasification to be used for the production of liquid fuels via the Fischer–Tropsch process

Hydrogen production using APR of carbohydrates has several advantages over steam reforming. Firstly, neither water nor the oxygenated hydrocarbon needs to be vaporized for the APR process, reducing the overall energy requirement. Secondly, temperatures and pressures that are favorable for APR are also favorable for the water gas shift reaction. These similarities make it possible to produce hydrogen with low amounts of CO in a single reactor. Thirdly, APR takes place at pressures around 15–50 bar, where hydrogen purifying strategies such as pressure-swing adsorption or membrane separation are effective. Overall, H$_2$ and CO$_2$ can be generated from aqueous solutions of carbohydrates at low temperatures in a single reactor as opposed to steam reforming which requires higher temperatures and a multi-reactor system.

Reaction selectivity issues must be addressed for the APR reaction to produce H$_2$ from biomass resources since the H$_2$ and CO$_2$ produced at low temperatures are thermodynamically unstable with respect to alkanes and water. In this respect, alkanes (especially CH$_4$) can be produced by methanation and Fischer–Tropsch reactions (Vannice 1977; Kellner and Bell 1981; Dixit and Tavlarides 1983; Iglesia et al. 1992). Both of the reactions are initiated by the C–O cleavage reactions within CO/CO$_2$, by H$_2$. Therefore, hydrogen production via APR of oxygenated hydrocarbons requires a catalyst that promotes reforming reactions (C–C scission combined with water gas shift) and inhibits alkane–formation reactions (C–O scission followed by hydrogenation). A high activity for the water gas shift reaction is also important in terms of removing CO from the metal surface at low reforming temperatures. Parallel and series reaction pathways that affect the conversion of oxygenated hydrocarbons to H$_2$ are shown in Figure 3.6.

The APR process is typically carried out in the presence of supported Pt catalysts (Cortright et al. 2002). The nature of both the metal and support has an important effect on the product selectivity of APR reactions (Shabaker et al. 2003). Pt black and Pt supported on Al$_2$O$_3$, TiO$_2$, and ZrO$_2$ (Davda et al. 2005) have been demonstrated to be active and selective for the APR of methanol and ethylene glycol to produce hydrogen. Pd supported catalysts have shown similar selectivity, although with a lower activity compared to Pt-based materials. It has also been reported that Rh, Ru, and Ni favor the production of alkanes from polyols over hydrogen (Davda et al. 2003) due to the fact that C–O bond cleavage is favored more than C–C bond cleavage over these metals. Acidity, either introduced by the use of solid acid supports (i.e., SiO$_2$/Al$_2$O$_3$) or by the addition of mineral acids to the feed (HCl), also has an important influence on product selectivity. The selectivity to alkanes increases (Davda et al. 2005) due to the increased rates of the catalytic cleavage of C–O bonds by dehydration and hydrogenation pathways compared to hydrogenolysis and reforming reactions. The addition of specific promoters can also improve the catalytic activity and selectivity for hydrogen production. For example, nickel supported on SiO$_2$ or Al$_2$O$_3$ shows a low selectivity for hydrogen due to the high rates of light alkane formation. However, the addition of Sn as a promoter to Raney-Ni-based catalysts improves the production rate of hydrogen from ethylene glycol, glycerol, and sorbitol (Huber et al. 2003).

Experimental results for the APR of glucose, sorbitol, glycerol, ethylene glycol, and methanol over a Pt/Al$_2$O$_3$ catalyst at 498 and 538 K show that the selectivity for H$_2$ production improves in the order glucose < sorbitol < glycerol < ethylene glycol < methanol. Although methanol, ethylene glycol, glycerol,

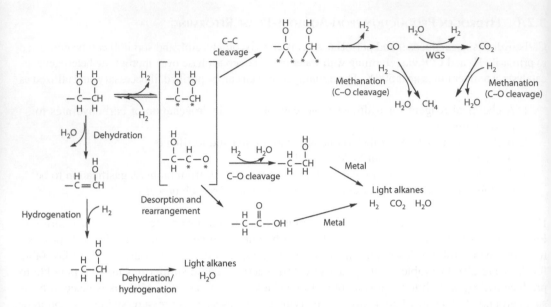

FIGURE 3.6 Reaction pathways involved in the catalytic conversion of biomass-derived oxygenated compounds to H_2 and alkanes over supported metal catalysts. (Adapted from Cortright, R.D. et al., *Nature*, 418, 964, 2002.)

and sorbitol can be derived from renewable feed stocks (Narayan et al. 1984; Tronconi et al. 1992; Blanc et al. 2000; Li et al. 2000), the reforming of less reduced and more immediately available compounds such as glucose would be highly desirable. Finally, lower temperatures result in higher H_2 selectivities, although this trend is in part due to the lower conversions obtained at lower temperatures. The trend for the selectivity for alkane production follows the opposite of that exhibited by the H_2 selectivity.

3.2.2 Production of Heavier Alkanes from Aqueous-Phase Reforming

APR of sorbitol can be tailored to produce a clean stream of heavier alkanes consisting primarily of butane, pentane, and hexane (see Figure 3.6). The conversion of sorbitol to alkanes plus CO_2 and water is an exothermic process that retains approximately 95% of the heating value and only 30% of the mass of the sorbitol feed. This reaction takes place by a bifunctional pathway using both metal and acid sites. Hydrogen and CO_2 are formed on the metal catalyst (such as Pt) and the sorbitol can be dehydrated on a solid acid catalyst (such as silica–alumina) or with a mineral acid. When these steps are balanced properly, the hydrogen produced in the first step is fully consumed by hydrogenation of dehydrated reaction intermediates, leading to the overall conversion of sorbitol to alkanes plus CO_2 and water. The alkanes formed are straight-chain compounds with only minor amounts of branched isomers (less than 5%). The selectivities toward alkanes can be adjusted by modifying the catalyst composition, pH of the feed, reaction conditions, and the reactor design (Huber et al. 2004). For instance, the selectivity toward heavier alkanes can be increased when a solid acid catalyst (SiO_2/Al_2O_3) is added to Pt/Al_2O_3, whereas the H_2 selectivity decreases from 43% to 11%. This behavior indicates that the majority of the H_2 produced by the reforming reaction is consumed by the production of alkanes when the catalyst contains a sufficient number of acid sites.

3.2.3 Glycerol Reforming to Synthesis Gas Coupled with Fischer–Tropsch Synthesis

The production of synthesis gas from oxygenated hydrocarbons over a supported metal catalyst containing Pt may be limited at low temperatures (e.g., lower than 570 K) by CO desorption from the metal surface. The heat of CO adsorption on Pt is high (e.g., 31–43 kcal/mol, depending on the CO coverage), leading to a high CO coverage on the Pt surface. The high CO coverage decreases when CO is converted to CO_2 by the water gas shift reaction, an undesirable side reaction for synthesis gas production. Thus, an active catalyst for the production of synthesis gas from oxygenated hydrocarbons at low temperatures should not only suppress the water gas shift reaction, but also weaken the bonding of CO to the metal surface. The binding energy of CO can be reduced by using appropriate metal alloys. For example, it has been predicted that a surface overlayer of Pt on Ru (or Re) decreases the CO binding strength compared to the bonding of CO on a surface of Pt metal (Christoffersen et al. 2001). Guided by these calculations, it has been found that PtRu and PtRe alloys supported on carbon are excellent catalysts for the conversion of aqueous solutions of glycerol to synthesis gas at temperatures below 570 K, at which temperatures Pt/C catalysts exhibit low catalytic activity (Soares et al. 2006).

As discussed previously, an important characteristic of APR is that it operates at conditions that favor the water gas shift reaction. As a result, the APR process typically generates $H_2:CO_2$ gas mixtures containing low levels of CO (~100 ppm). While this low level of CO is desirable for producing H_2, it is a disadvantage if the goal is to produce $H_2:CO$ gas mixtures for synthesis gas utilization steps, such as Fischer–Tropsch synthesis.

In contrast to the APR process, vapor phase reforming of oxygenated hydrocarbons is best achieved with volatile reactants such as methanol, ethylene glycol, and glycerol. Under vapor phase reforming conditions, it is possible to control the extent of the water gas shift reaction and subsequently, the $H_2:CO$ ratio as well as the amount of CO_2 produced. While it is still possible to generate $H_2:CO_2$ gas mixtures containing small amounts of CO (e.g., 1%) by incorporating water gas shift promoters (e.g., ceria, copper) into the catalyst, it is also possible to use catalysts that are not active for the water gas shift reaction. This way, by using catalysts that do not promote the water gas shift reaction, $H_2:CO$ gas mixtures can be produced and used for synthesis gas utilization steps. As shown in Figure 3.7, one important application of vapor phase reforming of oxygenated hydrocarbons is the coupling of the reforming process (to produce synthesis gas) with Fischer–Tropsch synthesis (to utilize $H_2:CO$), thereby obtaining liquid alkanes from oxygenated hydrocarbon feeds.

The vapor phase reforming of glycerol is especially interesting for potential coupling with Fischer–Tropsch synthesis because glycerol can be produced from renewable resources such as low value waste stream from plant oil transesterification (Klass 1998; Encinar et al. 2005) and animal fat (Klass 1998) in the production of biodiesel. This waste contains glycerol in water at high concentrations (e.g., 80%) (Gerpen 2005). Selling the waste glycerol solution can lower the production costs of biodiesel by 6% (Haas et al. 2006). Glycerol can also be produced by fermentation of sugars

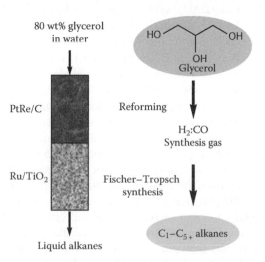

FIGURE 3.7 Schematic representation of the dual-bed catalytic reactor used for the combined glycerol reforming and Fischer–Tropsch synthesis to produce liquid alkanes.

like glucose either directly (Gong et al. 2006) or as a by-product of the industrial conversion of lignocellulose into ethanol (Institute of Local Self-Reliance 2000).

The production of synthesis gas through gas phase reforming of glycerol is described by the following stoichiometric equation (Soares et al. 2006):

$$C_3O_3H_8 \rightarrow 3CO + 4H_2 \tag{3.14}$$

This reaction is typically carried out over platinum catalysts that are active for C–C breaking reactions (leading to CO, H_2, and CO_2) as opposed to C–O breaking reactions (leading to light hydrocarbons) (Cortright et al. 2002; Alcala et al. 2003). As discussed previously, the water gas shift reaction must be inhibited to selectively produce synthesis gas by using inert materials (e.g., carbon) as a catalyst support instead of using oxide supports that can activate water (Soares et al. 2006). The glycerol-derived synthesis gas can subsequently be used for the production of liquid hydrocarbon fuels and/or chemicals by means of Fischer–Tropsch and methanol syntheses, respectively. This processing route has several attractive characteristics. First, glycerol reforming is carried out at moderate temperatures (498–620 K) as opposed to the high temperatures (1100–1300 K) needed for biomass gasification. In addition, glycerol reforming temperatures are also suitable for Fischer–Tropsch synthesis, making it possible to efficiently combine the two processes in a single reactor (Simonetti et al. 2007). Secondly, the synthesis gas obtained from glycerol reforming is not diluted and is free of impurities. Because of the purity of this stream, the capital costs associated with gas-cleaning units are decreased, making the process cost competitive at a small scale to process distributed biomass resources. Finally, coupling the endothermic glycerol reforming with the exothermic Fischer–Tropsch synthesis results in an energy-efficient process for the production of liquid transportation fuels from a renewable biomass resource.

Equation 3.14 shows the conversion of glycerol to CO and H_2. The endothermic enthalpy change of this reaction (84 kcal/mol) corresponds to about 24% of the heating value of the glycerol (354 kcal/mol). The heat generated by Fischer–Tropsch conversion of CO and H_2 to liquid alkanes such as octane (98 kcal/mol) corresponds to about 28% of the heating value of glycerol. Combining these two reactions results in the exothermic process shown in Equation 3.15, with an enthalpy change (15 kcal/mol) that is about 4% of the heating value of glycerol:

$$C_3O_3H_8 \rightarrow \frac{7}{25}C_8H_{18} + \frac{19}{25}CO_2 + \frac{37}{25}H_2O \tag{3.15}$$

As presented by Simonetti et al. (2007), synthesis gas can be produced at high rates and selectivities suitable for Fischer–Tropsch synthesis (H_2/CO between 1.0 and 1.6) from concentrated aqueous glycerol feed solutions at low temperatures (548 K) and high pressures (1–17 bar) over a 10 wt% Pt–Re/C catalyst with an atomic Pt:Re ratio of 1:1. Fischer–Tropsch synthesis was then coupled with the glycerol reforming reaction in a dual-bed single reactor system consisting of a Pt–Re/C catalyst bed (glycerol reforming) followed by a Ru/TiO_2 catalyst bed (Fischer–Tropsch reaction.) This integrated process produced a liquid alkane stream with selectivity for C_{5+} hydrocarbons between 63% and 75% at 548 K and pressures between 5 and 17 bar, with more than 40% of the carbon in the products containing the organic liquid phase at 17 bar. The aqueous liquid effluent from the combined process contained between 5 and 15 wt% methanol, ethanol, and acetone. The aqueous product stream can either be recycled to obtain a gaseous product or the chemicals can be separated from the water by distillation to be used in the chemical industry. This integrated process has the potential to improve the economics of "green" Fischer–Tropsch synthesis by reducing capital costs and increasing thermal efficiency. By coupling the glycerol reforming to synthesis gas and Fischer–Tropsch synthesis, the need to condense water and oxygenated hydrocarbon by-products between the two reactions can be eliminated. Moreover, the highly endothermic and exothermic steps that would result from the separate operations of the two reaction steps can be avoided.

3.3 PRODUCTION OF LIQUID ALKANES FROM CARBOHYDRATES VIA FORMATION OF HMF AND FURFURAL

The dehydration of carbohydrates leads to the formation of furan compounds like furfural and HMF. While furfural is produced industrially using the Quaker oats process (Zeitsch 2000), a large-scale cost-effective production method for HMF does not exist (Kuster 1990). HMF production from fructose and glucose has been carried out in water, organic solvents (DMSO) (Musau and Munavu 1987) and biphasic systems (water-MIBK) (Moreau et al. 1996) using acid catalysts such as mineral acids (HCl, H$_2$SO$_4$) (Asghari and Yoshida 2006), solid acids (zeolites) (Moreau et al. 1996) and ion-exchange resins (Mercadier et al. 1981), and salts (LaCl$_3$) (Seri et al. 2001). Dehydration of fructose to HMF in water is generally not selective. Even though high yields (>90%) are achieved in the presence of aprotic solvents such as DMSO (Szmant and Chundury 1981; Brown et al. 1982; van Dam et al. 1986; Musau and Munavu 1987), product recovery from these solvents leads to thermal degradation of the HMF product. Glucose dehydration to HMF has lower reaction rates and selectivities compared to fructose conversion, such that in pure DMSO the maximum HMF yield obtained is 42% at low concentrations of glucose (3 wt%) (Szmant and Chundury 1981). Other carbonyl compounds can also be formed, such as dihydroxyacetone and glyceraldehydes, from retro-aldol condensation of glucose (Collins and Ferrier 1995). Acetone can be produced from the fermentation of glucose (Klass 1998).

The conversion of furfural and HMF to higher molecular weight alkanes for transportation fuels starts with an aldol condensation in the presence of a base catalyst. Some examples for basic aldol condensation catalysts include alkali and alkaline earth oxides (Shigemasa et al. 1994; Di Cosimo et al. 1996; Zhang et al. 2003; Wang et al. 2004), phosphates (Zeng et al. 2005), MCM41 (Choudary et al. 1999), hydrotalcites (Roelofs et al. 2001; Climent et al. 2004), magnesia–zirconia, and magensia–titania (Aramendia et al. 2004a,b). Cross-condensation of furfural with acetone has been carried out using an amino-functionalized mesoporous base (Choudary et al. 1999) as well as chiral L-Proline (Liu et al. 2003) catalysts in organic solvents. However, not many solid basic catalysts have been reported that are active in aqueous-phase reactions due to leaching of catalyst components into the water phase and poor hydrothermal stability. Even so, homogeneous mineral bases such as NaOH (Gutsche et al. 1967; Shigemasa et al. 1994) have been reported to be active for the aldol condensation of carbohydrate-derived molecules in an aqueous environment.

For the production of fuels from furan compounds, aldol condensation is followed by the hydrogenation of the aldol adduct to increase its solubility in the aqueous phase. It is important to note that selective hydrogenation of the furan ring in HMF and furfural can lead to carbonyl-containing compounds that can undergo self aldol condensation to form heavier alkanes. In particular, thermodynamic considerations favor hydrogenation of the C=C bond over the C=O bond for reactions involving unsaturated aldehydes (Zanella et al. 2004). Hydrogenation of the C=C bond is also kinetically favored over the C=O bond for small molecules (Marinelli et al. 1995; Englisch et al. 1997). However, the C=C bonds of furfural are less reactive than the C=O bond, probably due to steric effects (Marinelli et al. 1993), making the production of tetrahydrofurfural (THF2A) by furfural hydrogenation difficult. Compared with other metals commonly used for hydrogenation, Pd has been reported to have a low rate for C=O bond hydrogenation (Marinelli et al. 1995). Thus, in literature, aldol condensation and subsequent C=C bond hydrogenation have been coupled using a bifunctional metal–base catalyst for the organic phase for producing MIBK (Nikolopoulos et al. 2005) using Pd on hydrotalcites.

The 5-hydroxymethyl-tetrahydrofurfural (HMTHFA) formed by the hydrogenation of HMF can undergo self-condensation to form C$_{12}$ species that would subsequently be hydrogenated further to form water-soluble organic species. According to this approach, the biomass-to-liquid alkane fuel conversion steps involve changing the functionality of the sugars through a series of selective reactions. A change in the molecular weight is then introduced through C–C coupling reactions (aldol condensation), after which large water-soluble compounds are formed by hydrogenating the adducts. Figure 3.8 shows the reaction pathways involved in the production of liquid alkanes from

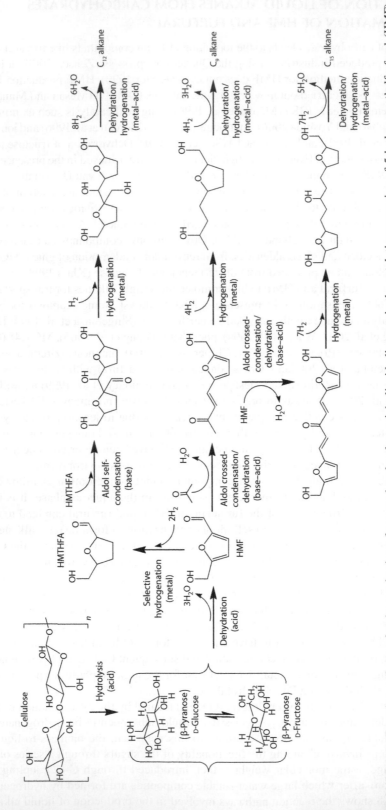

FIGURE 3.8 Reaction pathways for the production of targeted molecular weight liquid alkanes via the catalytic conversion of 5-hydroxymethylfurfural (HMF). (Adapted from Huber, G.W. et al., *Science*, 308, 1446, 2005.)

polysaccharides, based on the formation of furan compounds, followed by aldol condensation and dehydration/hydrogenation. The initial hydrolysis conversion is typically carried out at high temperatures in the presence of mineral acid catalysts and involves breaking C–O–C linkages to form simpler carbohydrate molecules. These carbohydrates can further undergo dehydration by loss of three water molecules to form furan compounds such as HMF and furfural. Aldol condensation involves C–C coupling between two compounds containing carbonyl groups to form larger organic molecules. This reaction is carried out in polar solvents, for example, water or water–methanol, in the presence of base catalysts such as mixed Mg–Al–oxides or MgO–ZrO$_2$ at low temperatures. As indicated in Figure 3.8, a ketone like acetone first reacts with HMF to form a C$_9$ species, which can subsequently react with a second HMF molecule to form a C$_{15}$ molecule. The aldol products have a limited solubility in water and precipitate out of the aqueous phase. The subsequent hydrogenation step saturates the C=C and C=O bonds of the aldol adducts over a metal catalyst (Pd/Al$_2$O$_3$), thereby producing large water-soluble organic compounds. These water-soluble organic molecules can undergo repeated dehydration/hydrogenation reactions in the presence of bifunctional catalysts containing acid and metal sites (Pt/SiO$_2$–Al$_2$O$_3$) to form liquid alkanes. This process occurs in a four-phase dehydration/hydrogenation reactor system consisting of (1) an aqueous inlet stream containing large water-soluble molecules, (2) a hexadecane alkane sweep inlet stream, (3) an H$_2$ inlet gas stream, and (4) a solid catalyst (Pt/SiO$_2$–Al$_2$O$_3$) (Huber et al. 2005). As dehydration/hydrogenation takes place, the organic reactants in the aqueous phase become more hydrophobic and the hexadecane alkane stream serves to remove the hydrophobic species from the catalyst before they react further to form coke.

3.4 REFORMING/DEOXYGENATION OF SUGARS AND POLYOLS OVER PtRe CATALYSTS

As previously discussed, the production of liquid hydrocarbon fuels from biomass-derived carbohydrates is typically initiated by oxygen removal reactions (e.g., C–O hydrogenolysis, dehydration, hydrogenation) and should be combined with C–C bond formation reactions (e.g., aldol condensation, ketonization, oligomerization) to produce compounds with increased molecular weight. It is also economically desirable to carry out these processes with minimal use of external hydrogen and a limited number of operation steps (e.g., reaction and purification/separation steps). Regarding the use of external hydrogen, it is possible to generate the hydrogen necessary for the oxygen-removal reactions in situ (Cortright et al. 2002) by APR of a fraction of the sugar feedstock. The following equation shows the APR reaction of glucose to obtain hydrogen:

$$C_6O_6H_{12} + 6H_2O \rightarrow 6CO_2 + 12H_2 \tag{3.16}$$

Multifunctional catalysts capable of carrying out different reactions in the same reactor can be utilized to decrease the number of steps involved in biomass processing (Simonetti and Dumesic 2009).

Recently, an approach that combines in situ hydrogen generation and multifunctional catalyst usage has been developed to convert aqueous solutions of sugars and sugar alcohols into liquid hydrocarbon fuels by means of a simple, two-step process (Figure 3.9) (Kunkes et al. 2008b). The initial phase of this process converts sugars and polyols into monofunctional chemical intermediates in a single reactor over a carbon-supported PtRe catalyst. In this step, more than 80% of the initial oxygen content of the sugars and polyols is removed by controlling C–C cleavage (leading to CO$_2$ and H$_2$) and C–O cleavage (leading to alkanes) rates. This process operates at moderate pressures (20–30 bar) and temperatures (283–523 K) with highly concentrated aqueous feeds (40%–60%) of sorbitol or glucose. It also yields a spontaneously separating organic phase consisting of a mixture of monofunctional intermediates in the C$_4$–C$_6$ range (including acids, alcohols, ketones, and

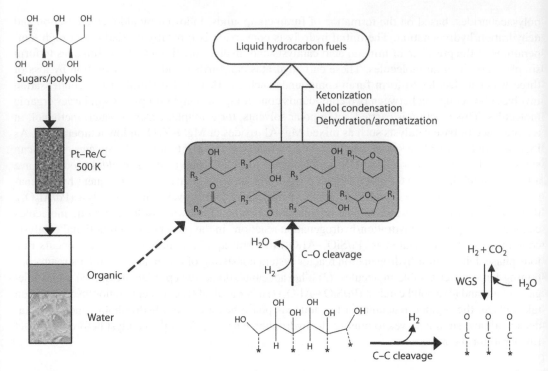

FIGURE 3.9 Schematic representation of the reforming/reduction of sugars and polyols over a PtRe/C catalyst to produce an organic phase containing monofunctional species. The hydrophobic intermediates can be upgraded to liquid fuels through C–C coupling reactions. (Adapted from Kunkes, E.L. et al., *J. Catal.*, 260(1), 164, 2008.)

heterocycles). The exothermic deoxygenation reactions are balanced with endothermic reforming reactions in the same reactor, such that the overall conversion is mildly exothermic and more than 90% of the energy content of the polyol or sugar feed is stored in the reaction products. Cleavage of C–O bonds takes place by hydrogenolysis and is promoted by Re (Pallassana and Neurock 2002; Kunkes et al. 2008a). In this way, the progressive removal of oxygen from the intermediates weakens the binding between the oxygenated molecules and the catalyst's surface. This weakened binding facilitates desorption and results in the production of the monofunctional species, such as acids, alcohols, ketones, and heterocycles (Figure 3.9), before alkanes are generated. Compared to bio-oil obtained from pyrolysis of biomass, the monofunctional liquid contains a well-defined mixture of hydrophobic compounds free of water.

The functionalities contained in these product compounds allow for the use of subsequent upgrading processes like C–C coupling reactions. Partially deoxygenating the feed and subsequently upgrading the resulting intermediates leads to a better control of reactivity. Later in this chapter, this strategy is applied to other important biomass derivatives such as lactic acid and levulinic acid to obtain chemicals and fuels (Serrano-Ruiz et al. 2010b). The organic stream of monofunctional compounds can be converted, through C–C coupling reactions, into targeted liquid hydrocarbon fuels of different classes (Figure 3.9). Different C–C coupling reactions (e.g., aldol condensation, oligomerization, and ketonization) can be applied to upgrade different types of monofunctional compounds in the organic stream (e.g., ketones, alcohols, acids). For example, the hydrophobic effluent derived from the carbohydrate feed can be hydrogenated to obtain alcohols which can subsequently be converted to aromatic compounds (gasoline components) at atmospheric pressure over an acidic catalyst such as H-ZSM5. Alternatively, oligomerization reactions can be used to increase the molecular weight of alkenes that can be obtained by dehydration of these alcohol compounds. The ketones can undergo aldol condensation over a bifunctional catalyst with acid/base and metal sites (i.e., $Cu/Mg_{10}Al_7O_x$) to form larger molecular weight compounds with minimal branching, as required for diesel applications (Kunkes et al. 2009a). Alcohols can also undergo aldol condensation via the formation of ketones by a dehydrogenation reaction over the metal sites available in the bifunctional catalyst (Gurbuz et al. 2010a). Finally, carboxylic acids can be combined through a ketonization reaction to obtain higher molecular weight ketones. This reaction is of particular importance when the organic stream is rich in carboxylic acids as is the case when the monofunctional compounds are produced from glucose (Kunkes et al. 2008b).

Kunkes et al. (2008b) developed a cascade process where monofunctional oxygenates are upgraded to alkanes suitable for diesel fuel by a combination of C–C coupling reactions (i.e., aldol condensation and ketonization reactions). The authors first show that the ketones in the monofunctional effluent can be condensed directly over $Cu/Mg_{10}Al_7O_x$ (where the mixed oxide achieves aldol condensation and Cu hydrogenates the unsaturated aldol adduct) to produce $C_8–C_{12}$ ketones, accounting for 45% of the carbon in the monofunctional mixture. However, the basic $Cu/Mg_{10}Al_7O_x$ catalysts deactivate with time on stream due to the poisoning of basic sites required for aldol condensation by organic acids present in the feed stream containing the monofunctional mixture. To avoid poisoning the aldol condensation catalyst, ketonization of these acids can first be carried out to convert the carboxylic acids to heavier ketones prior to the aldol condensation step. Following this strategy, the monofunctional effluent derived from the sugar or polyol feed would first go through a fixed bed reactor that operates at 573 K and 20 bar and contains a $Ce_1Zr_1O_x$ catalyst. Under these conditions, the acids present in the feed are completely converted to heavier ketones ($C_7–C_{11}$), CO_2, and water. The $C_3–C_6$ ketones present in the feed remain inert over $Ce_1Zr_1O_x$. By operating at a high pressure and low temperature, product vaporization can be minimized and the ketonization effluent goes through a second fixed bed reactor containing a bifunctional aldol condensation catalyst. $Pd/Ce_1Zr_1O_x$ is an appropriate catalyst as it shows a good condensation activity and an improved resistance to organic acids. Using a combination of ketonization and aldol condensation, the authors report a 63% yield of C_{7+} oxygenates from a monofunctional stream obtained from glucose over the PtRe/C catalyst. The production of liquid alkane fuels through similar strategies that are based on a monofunctional platform has also been reported elsewhere by Blommel et al. (2008).

It was explained earlier that the ketonization of carboxylic acids must be performed prior to the aldol condensation/hydrogenation for the catalytic upgrading of monofunctional intermediates to transportation fuels. This order is necessary because the active basic sites in the aldol condensation catalyst are poisoned by the presence of the carboxylic acids (Kunkes et al. 2008b; West et al. 2009b). The ketonization step can be carried out over a $Ce_1Zr_1O_x$ mixed oxide catalyst (Kunkes et al. 2008b) with almost a 100% yield. Aldol condensation/hydrogenation can be carried out over $Pd/Ce_1Zr_1O_x$ with a H_2 cofeed. Since both reactions are performed under similar conditions, it is possible to integrate the ketonization and aldol condensation/hydrogenation reactions in a single reactor with a dual-bed system. This would consist of $Ce_1Zr_1O_x$, for ketonization, followed by a downstream bed of 0.25 wt% $Pd/Ce_1Zr_1O_x$, for aldol condensation/hydrogenation. Integrating the aldol condensation and ketonization steps into a single reactor system is desirable because it would streamline the overall C–C coupling process. Therefore, the possibility of this coupling strategy between ketonization and aldol condensation/hydrogenation was explored by studying the effects of the ketonization byproducts, CO_2 and water, on the aldol condensation reaction of representative ketones (2-hexanone and 2-butanone) over $Pd/Ce_1Zr_1O_x$ (Gurbuz et al. 2010b). Cofeeding a mixture of 10 mol% CO_2 in H_2 resulted in an almost complete loss of self-condensation activity over 0.25 wt% $Pd/Ce_1Zr_1O_x$, decreasing the conversion of 2-hexanone from 60% to 5%. Decreasing the space velocity of the ketone feed was not effective in overcoming the inhibition caused by CO_2. Based on these results, it was concluded that it is necessary to remove CO_2 and water prior to the aldol condensation/hydrogenation step over $Pd/Ce_1Zr_1O_x$ and that it was not feasible to integrate the two C–C coupling steps.

It has been shown in literature that the nature of the interaction of CO_2 with the oxide surface can be modified by changing the composition of mixed oxides (Di Cosimo et al. 1998; Diez et al. 2008). Accordingly, a study was performed to investigate how the composition of ceria–zirconia, mixed with an oxide catalyst, can be modified to formulate an aldol condensation catalyst that is resistant to CO_2 inhibition and will permit the integration of the ketonization and aldol condensation reactions in a single reactor. The catalysts studied include 0.25 wt% Pd/CeO_x, $Pd/Ce_5Zr_2O_x$, $Pd/Ce_1Zr_1O_x$, $Pd/Ce_2Zr_5O_x$, and Pd/ZrO_2. Prior to studying the effects of CO_2 and water on the aldol condensation activity, the 2-hexanone self-condensation activity was measured for all catalysts. It was reported that the conversion of 2-hexanone increases with the increasing zirconia content. In comparison with the 58% conversion obtained with the original catalyst $Pd/Ce_1Zr_1O_x$, pure ZrO_2 has the highest conversion rate with 90% conversion. All ceria-containing catalysts displayed a significantly lower activity with the introduction of CO_2, showing yields of less than 20% to C–C coupling products. Alternatively, Pd/ZrO_2 showed significant resistance to CO_2 poisoning (20% decrease in condensation activity). The acid and base properties of ceria/zirconia mixed oxide catalysts were studied using temperature programmed desorption (TPD) of ammonia and CO_2. It was shown that as the ceria content of the mixed oxide catalyst is increased, the number of total CO_2 binding sites, and specifically, the strong CO_2 binding sites are increased. Alternatively, catalysts with greater proportions of zirconia adsorb larger amounts of ammonia.

In addition to CO_2, another by-product of ketonization is water. A 40% decrease in activity for aldol condensation of 2-butanone was observed for $Pd/Ce_1Zr_1O_x$ when a mixture containing 12 wt% water in 2-butanone was used as the feed (Kunkes et al. 2009a). The water's effect on aldol condensation activity was studied over Pd/ZrO_2 by cofeeding water with 2-butanone. Compared to the 40% decrease in the catalytic activity of $Pd/Ce_1Zr_1O_x$, only a 10% decrease in catalytic activity was observed for the Pd/ZrO_2. The diminished inhibition of aldol condensation reaction by water and CO_2 over Pd/ZrO_2 makes this catalyst ideal for the integration of ketonization and aldol condensation/hydrogenation.

Based on these results, a detailed comparison of the two catalytic upgrading processes (i.e., cascade mode versus dual-bed system) was carried out using a mixture of monofunctional intermediates obtained by converting an aqueous solution of 60 wt% sorbitol over a Pt–Re/C catalyst (Gurbuz et al. 2010a). The cascade mode consists of two separate reactors employed for ketonization over $Ce_1Zr_1O_x$ followed by aldol condensation/hydrogenation over 0.25 wt% Pd/ZrO_2, with the removal of CO_2 and water between reactors. In the dual-bed system, these two catalysts are combined in a single reactor (Gurbuz et al. 2010a). Following either of these strategies, a dehydration/hydrogenation step can be applied using a bifunctional metal/acid catalyst such as Pt/SiO_2-Al_2O_3 to convert the heavier ketones to an alkane stream. Figure 3.10 shows the reaction pathways involved in ketonization, aldol condensation/hydrogenation, and dehydration/hydrogenation for representative C_5 species. Acids are coupled in ketonization reactions to form linear ketones, CO_2, and water. Ketones and alcohols are combined in aldol condensation/hydrogenation reactions to form branched ketones and water. Single methyl-branched ketones are obtained when 2-ketones and/or secondary alcohols condense. Single ethyl-branched ketones are formed when a 3-ketone/alcohol condenses with a 2-ketone/alcohol. Finally, extensive branching is present in the product ketone when 3-ketones or tertiary alcohols undergo self-condensation. Ring opening reactions of heterocyclics can also take place over Pd/ZrO_2 to form aldehydes or ketones, depending on the structure of the heterocyclic molecule and the location of the C–C cleavage in the molecule (see Figure 3.10). Aldehydes can condense with 2-ketones to form linear ketones. In addition to coupling reactions, monofunctional species are also partially converted into alkanes over Pd/ZrO_2. Comparing the overall conversion of monofunctional species and yield of high molecular weight C–C coupling products, it was shown that the two processing strategies are essentially equally active. Therefore, the dual-bed, single reactor system is the preferred mode of catalytic upgrading because two factors can be eliminated, namely, the energy consumption as well as the reactor infrastructure associated with cooling and reheating the products obtained from ketonization prior to the aldol condensation/ hydrogenation step (Gurbuz et al. 2010a).

When the effluent from the upgrading reactions is analyzed in detail, it is observed that among the branched ketones or alkanes, only small amounts of single, ethyl-branched species are formed compared to single, methyl-branched species. This behavior shows that the

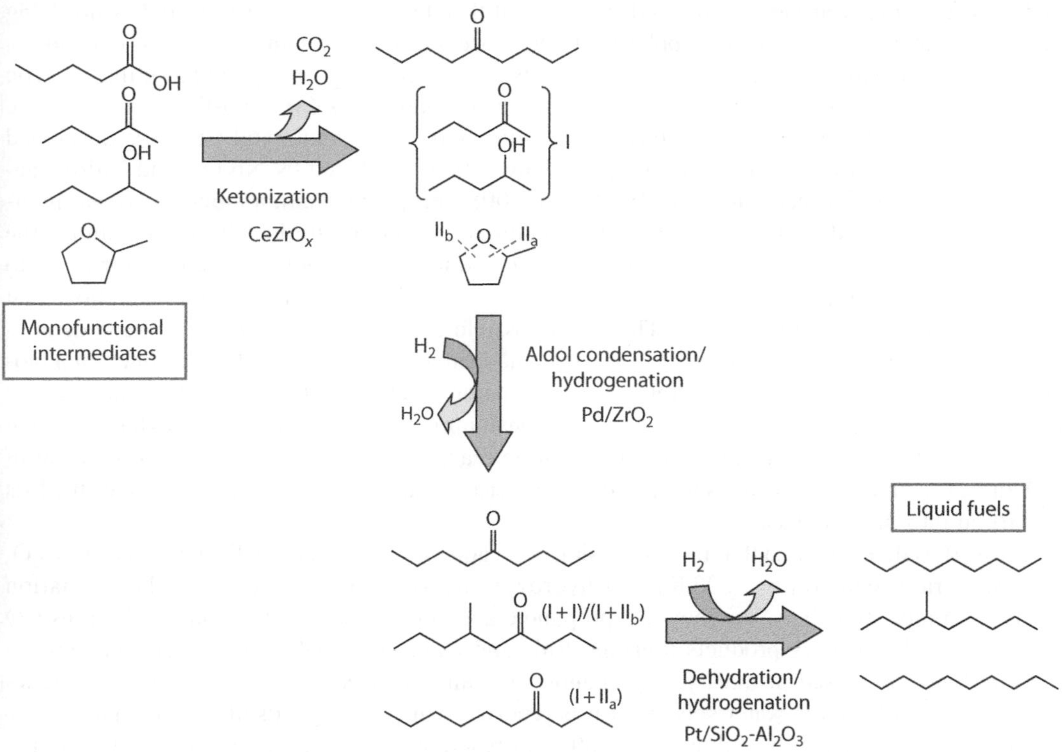

FIGURE 3.10 Reaction pathways for the ketonization, aldol condensation/hydrogenation, and dehydration/ hydrogenation of representative C_5 species derived from sugars and polyols. Species II_a and II_b are formed by ring opening as denoted in the figure. (Adapted from Gurbuz, E.I. et al., *Appl. Catal. B*, 94, 134, 2010.)

self-condensation of 2-ketones or alcohols is more favorable than the cross-condensation of 3-ketones/alcohols with 2-ketones/alcohols. Finally, the absence of species containing two branches indicates that coupling among 3-ketones/alcohols is not significant. This analysis shows that the extent of branching in the effluent stream of these two consecutive C–C coupling reactions (i.e., ketonization and aldol condensation) can be controlled and minimized as opposed to the acid-catalyzed coupling reactions like alkene oligomerization. (O'Connor and Kojima 1990). This characteristic makes ketonization and aldol condensation unique routes to obtaining high molecular weight alkanes valuable for diesel fuel.

3.5 LACTIC ACID PLATFORM

Lactic acid (2-hydroxypropanoic acid) is the most naturally abundant carboxylic acid (Datta 2004). While its traditional uses have involved food-related applications, novel applications for the industrial production of polymers and chemicals have expanded the market for lactic acid usage since the 1990s. The bifunctionality of lactic acid allows it to be converted through several processing strategies to produce a variety of different compounds such as acetaldehyde (Mok et al. 1989) (via decarbonylation/ decarboxylation), acrylic acid (Holmen 1958; Mok et al. 1989; Wadley et al. 1997; Sawicki 1998) (via dehydration), propanoic acid (Serrano-Ruiz and Dumesic 2009a) (via reduction), 2,3-pentanedione (Gunter et al. 1994) (via condensation), and polylactic acid (PLA) (Lipinsky and Sinclair 1986; Lunt 1998) (via self-esterification to dilactide and subsequent polymerization). The presence of two adjacent functional groups (–OH and –COOH) in lactic acid, a small molecule of only three carbon atoms, leads to a high reactivity and an increased potential for thermal decomposition (Fisher and Filachione 1950). From a fundamental perspective, lactic acid can serve as a model compound for more complex and over-functionalized biomass-derived species, and therefore, catalytic reaction studies involving lactic acid can provide information applicable to the processing of more complex biomass derivatives.

Recently, an active site coupling strategy has been applied to the catalytic upgrading of lactic acid in a single reactor for the production of targeted compounds (Serrano-Ruiz and Dumesic 2009a). A bifunctional Pt/Nb_2O_5 catalyst containing both acidic and metal sites was developed to remove oxygen from lactic acid through a dehydration (catalyzed by Nb_2O_5) and hydrogenation (catalyzed by Pt) reaction cycle (West et al. 2008). Propanoic acid, a monofunctional intermediate, was formed. The acidity of Nb_2O_5 catalyzes, in addition to dehydration reactions, the decarbonylation of lactic acid (Mok et al. 1989). For example, lactic acid can be transformed into the monofunctional species acetaldehyde through decarbonylation/decarboxylation reactions that also produce gaseous CO and CO_2. Thus, two possible conversion strategies have been demonstrated that achieve deoxygenation of lactic acid and lead to the formation of monofunctional biomass derivates (1) through a dehydration/hydrogenation reaction cycle that involves the removal of oxygen in the form of water, but requires the consumption of hydrogen from an external source and (2) through a decarbonylation/decarboxylation reaction that removes oxygen in the form of CO and CO_2, and therefore, does not require an external source of hydrogen, but involves the loss of carbon into the gas phase.

It was thus demonstrated that the conversion of an aqueous solution of lactic acid over Pt/Nb_2O_5 at a moderate temperature (523 K) and hydrogen pressure (57 bar) leads to the formation of the two monofunctional compounds, propanoic acid and acetaldehyde, along with gases CO and CO_2, as the primary products (Serrano-Ruiz and Dumesic 2009b). At high conversions, an organic layer composed primarily of propanoic acid and ketones in the C_4–C_7 range spontaneously separates from the water solvent. It was reported that carbon yields of approximately 50% in the organic layer can be achieved at optimum reaction conditions when using concentrated solutions of lactic acid. As a control, a Pd catalyst supported on inert carbon black (Vulcan) was utilized at the same conditions. An organic layer was not observed in the reactor effluent. Alternatively, it was determined that in the absence of the acid support, the majority of the carbon in lactic acid is converted into the gas phase species CO, CO_2, and methane. The conversion indicates that the acidic Nb_2O_5 plays an integral role in the production of the hydrophobic species comprising the organic phase. Figure 3.11 depicts the proposed chemistry involved in the formation of hydrophobic C_4–C_7 ketones from the intermediates acetaldehyde and propanoic acid. The production of 3-pentanone from propanoic acid is achieved by the ketonization of two molecules of acid over the niobia catalyst, thus producing the larger symmetric ketone. A second catalyst bed (following Pt/Nb_2O_5) containing ceria–zirconia can be used at the same reaction pressure and temperature to achieve a more complete conversion of propanoic acid to pentanone (Serrano-Ruiz and Dumesic 2009a). The remaining C_4–C_7 ketones observed in the organic phase are formed via aldol condensation reactions in which acetaldehyde, acting as a building block, is coupled together through the formation of C–C bonds as shown in Figure 3.11. Niobia plays a critical role in directing the synthesis toward the formation of the larger ketones. It is also a suitable material for the deoxygenation of biomass derivatives because of its strong acidity and stability under high water environments (West et al. 2008, 2009a). Additionally, this example demonstrates how the use of a bifunctional catalyst (metal and acid sites) can lead to the desired product through a series of reaction steps (dehydration/hydrogenation and C–C coupling) that can be carried out in a single reactor, thereby reducing the potential operational and capital costs of the process.

Several applications can be anticipated for the catalytic processing of lactic acid over bifunctional Pt/Nb_2O_5 catalysts. For example, the ketone components comprising the organic phase produced from lactic acid can be separated and utilized as a source of valuable chemicals currently obtained exclusively from petroleum. Additionally, the ketones in the organic phase can be converted over a Ru/C catalyst in the presence of hydrogen to the corresponding alcohols. This liquid mixture of alcohols in the C_4–C_7 range can be used as a high-energy-density liquid fuel (Serrano-Ruiz and Dumesic 2009a).

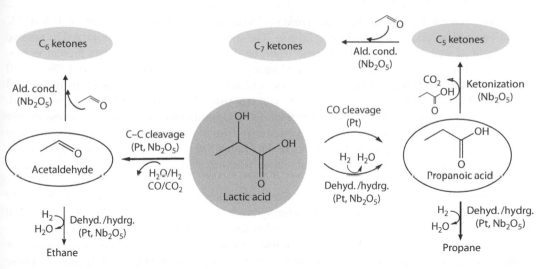

FIGURE 3.11   Reaction pathways for the catalytic conversion of lactic acid over Pt/Nb_2O_5 to produce light alkanes and higher molecular weight ketones. (Adapted from Serrano-Ruiz, J.C. and Dumesic, J.A., *Green Chem.*, 11(8), 1101, 2009.)

3.6   LEVULINIC ACID PLATFORM

Levulinic acid (4-ketopentanoic acid) is another organic acid that has received attention as a potential platform chemical that is readily derived from lignocellusic biomass and agricultural wastes. Several studies in literature demonstrate that aqueous solutions of C_6 sugars (e.g., glucose, fructose) as well as slurries containing insoluble cellulose or lignocellulosic biomass can be converted to levulinic acid and formic acid using strong soluble acid catalysts (Girisuta et al. 2006, 2007). Two common approaches for the levulinic acid production via the acid-catalyzed hydrolysis of cellulose have been applied. In the first approach, a high concentration of a mineral acid (e.g., up to 70 wt% H_2SO_4 or 16 N HCl) is used as a catalyst to convert cellulose at low reaction temperatures (e.g., 320 K). However, the high acid concentrations raise the operating costs as both the hydrolysis reactor and the acid recovery system are constructed of expensive materials. Alternatively, the second approach commonly used involves the use of dilute mineral acid solutions to convert cellulose at higher operating temperatures (420–510 K). The dilute acid approach is typically favored because of the associated lower acid concentrations. The most effective acid catalysts are H_2SO_4, HCl, and trifluoroacetic acid (TFA) at acid concentrations ranging from 0.1 to 2 mol/L (Thomas and Schuette 1931; Asghari and Yoshida 2007; Heeres et al. 2009). Studies suggest that the sugars are first converted to HMF which is rapidly converted to levulinic and formic acids. Proposed reaction mechanisms in literature suggest that glucose is first converted to HMF through an acid-catalyzed dehydration reaction. HMF, a highly reactive intermediate then undergoes an acid-catalyzed rehydration reaction that produces an equal molar mixture of levulinic and formic acids. In addition to the production of levulinic acid, insoluble carbonaceous solids known as humins are irreversibly produced from the degradation/polymerization of glucose and HMF in acidified solutions. The production of humins decreases the final yield of levulinic acid. However, because of its high carbon content (e.g., >65 wt% carbon), humins can serve as a dense energy source that can be burned to produce heat or electricity.

The biofine process is one variation of the acid-catalyzed conversion of lignocelluloses into levulinic and formic acid (Fitzpatrick 1997). In this process, the biomass feedstock is mixed with sulfuric acid (1.5–3 wt%) and is first hydrolyzed and dehydrated to HMF in a plug-flow reactor at 483–493 K and 25 bar with a short residence time (12 s) to minimize the formation of degradation products.

In a second reactor, the intermediates are converted to levulinic acid and formic acid at 463–473 K and 14 bar with a residence time close to 20 min. These conditions have been optimized to remove formic acid as well as the furfural that arise from the dehydration of the C_5 sugars present in biomass. Also, this process has been demonstrated on a pilot plant scale (Biometrics Inc. 1996; Bozell et al. 2000; Hayes 2009). Final yields of levulinic acid are between 70% and 80% of the theoretical maximum. Humins, produced by the degradation of reactive intermediates and lignin (Horvat et al. 1985), account for approximately 30% of the final mass. Optimization of the biofine process for use with inexpensive raw materials has been suggested to potentially decrease the price of levulinic acid to 8–20 ¢/kg (Bozell et al. 2000). Pilot plants in the United States and Italy have used paper waste, agricultural residues, and organic municipal wastes (Biometrics Inc. 1996; Hayes 2009) with successful results.

Levulinic acid has been identified by the United States Department of Energy as one of the top 12 valuable platform chemicals upon which biorefining processes may be established (U.S. Department of Energy 2004). In this respect, catalytic approaches for the conversion of levulinic acid into liquid fuel components have been developed. For example, levulinic acid can be reacted with methanol or ethanol over an acid catalyst at room temperature to form methyl- and ethyl-esters, respectively. These esters can then be blended with diesel fuel up to 20%. Another levulinic acid conversion process is the production of methyltetrahydrofuran (MTHF), a fuel extender, which can be blended with up to 70% gasoline without necessitating engine modifications. Although the direct conversion of levulinic acid to MTHF is possible, an indirect route that involves the production of γ-valerolactone (GVL) can lead to high MTHF yields (85%) when using bimetallic Pd–Re/C catalysts (Elliot and Fry Jr. 1999). Accordingly, the catalytic production of GVL from levulinic acid, and more specifically, from levulinic acid solutions derived from lignocellulosic biomass, has recently gained attention.

Levulinic acid can be converted to GVL at near quantitative yields over heterogeneous and homogeneous metal catalysts at relatively modest reaction temperatures (373–543 K) and pressures (5–150 bar). The hydrogen required for the reduction reaction can be provided externally as molecular hydrogen or can be generated in situ from other molecular sources (e.g., formic acid). As shown in Figure 3.12, levulinic acid can first be converted to 4-hydroxypentanoic acid through a metal-catalyzed hydrogenation reaction. The 4-hydroxypentanoic acid then undergoes a ring closing reaction that forms a cyclic ester bond and liberates one mole of water.

Alternatively, levulinic acid can first self-condense, losing one mole of water, to form α-angelicalactone. The α-angelicalactone is then hydrogenated over a metal catalyst to form GVL (see Figure 3.12). While this reduction is typically performed using an external H_2 supply, it is possible to generate H_2 in situ by decomposing the formic acid, which is produced as a by-product in the production of levulinic acid. The use of formic acid as a hydrogen donor could lower GVL production costs by reducing the need for external H_2 in the production of GVL and eliminating costly levulinic acid purification strategies. There are two potential mechanisms by which the hydrogen transfer could occur. One possible mechanism is that a metal-formate is first formed and then decomposed into CO_2 and a metal-hydride that reacts with levulinic acid to form 4-hydroxypentanoic acid. A second mechanism involves the decomposition of formic acid into CO_2 and molecular H_2 which is then used in the hydrogenation step (Deng et al. 2009).

GVL is a versatile molecule that has many applications for use as a liquid fuel. For example, GVL can be used directly as a fuel additive, but only to a limited extent due to its blending limits, high water solubility, and low energy density (Horvath et al. 2008). Alternatively, GVL can be converted into an array of other fuel-grade compounds or specialty chemicals. Since GVL is typically derived from levulinic acid in aqueous media as described earlier, the direct use of GVL as a fuel requires costly separation/purification steps and distillation/extraction methods to remove water. The increased hydrophobic nature of GVL enables the use of solvents, such as ethyl acetate (Mehdi et al. 2008) or supercritical CO_2 (Bourne et al. 2007), although they can be difficult to operate on large scales. Another alternative is to directly process the aqueous solutions of GVL to produce hydrophobic liquid alkanes with the appropriate molecular weight to be used as liquid fuels.

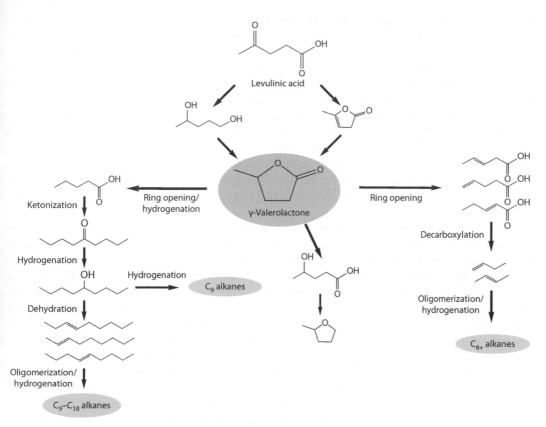

FIGURE 3.12 Reaction pathways for production of targeted molecular weight alkanes via the catalytic upgrading of γ-valerolactone (GVL) derived from levulinic acid. (Adapted from Alonso, D.M. et al., *Green Chem.*, 12, 1493, 2010.)

Recently, two novel approaches have been developed (Bond et al. 2010; Serrano-Ruiz et al. 2010a) for the conversion of GVL into hydrophobic fuel components of targeted molecular weights that are ideal for use in gasoline, diesel, and jet fuels (shown in Figure 3.12). Both approaches utilize heterogeneous acid catalysts to facilitate a high temperature ring opening reaction to convert GVL into a mixture of pentenoic acids. A metal functionality can be added to the acid catalyst and, in the presence of H$_2$, the chemistry can be directed toward the production of pentanoic acid. This ring opening and hydrogenation reaction to form pentanoic acid is also an exothermic reaction and is thermodynamically favorable ($\Delta G = -14$ kcal/mol, $\Delta H = -20$ kcal/mol). Alternatively, continued exposure of the pentenoic acids to a solid acid catalyst causes the decarboxylation of the carboxylic acid group and production of a mixture of butene isomers and CO$_2$. The overall exothermic decarboxylation reaction is thermodynamically favored ($\Delta G = -18$ kcal/mol, $\Delta H = -5$ kcal/mol). These two approaches lead to different molecular weight liquid hydrocarbons with varying physical properties for use in liquid fuels.

The pentanoic acid–based approach was demonstrated by Serrano–Ruiz et al. (2010a) to convert aqueous solutions of GVL (50 wt%) into C$_9$ hydrocarbons. As previously mentioned, GVL can be converted to pentenoic acids by a ring opening reaction (Ayoub and Lange 2008), and this mixture of pentenoic acids can subsequently be hydrogenated to produce pentanoic acid. It was demonstrated that a Pd/Nb$_2$O$_5$ catalyst was stable for use with aqueous solutions and active for the ring opening and hydrogenation reactions. High yields of pentanoic acid (92%) were achieved with a 0.1% Pd at 598 K and 35 bar (50% H$_2$, 50% He). It was determined that the metal loading of the catalyst and the partial pressure of H$_2$ controlled the pentanoic acid yield. For example, high metal loadings and high partial pressures of H_2 favor the formation of by-products, such as butane and CO_x species, which form through pentanoic acid decarboxylation/decarbonylation, and pentane, which forms through hydrogenation/dehydration of pentanoic acid (Serrano-Ruiz et al. 2010a).

The pentanoic acid formed from the ring opening and reduction reaction can be upgraded to 5-nonanone by ketonization (Renz 2005) over $CeZrO_x$ at 698 K and pressures from 1 to 20 bar. In the ketonization reaction, two pentanoic acid molecules react together to form a C_9 ketone molecule (5-nonanone) as well as CO_2 and water. Ketonization of pentanoic acid is thermodynamically favorable ($\Delta G = -16$ kcal/mol) at 523 K (Serrano-Ruiz et al. 2009a). It was demonstrated that a 5-nonanone yield of 60% could be achieved at 623 K. Other ketones of lower molecular weight (2-hexanone and 3-heptanone) were observed and the authors attributed their formation to α-and β-scission of 5-nonanone. It has been previously demonstrated that ceria–zirconia is an active catalyst for organic acid ketonization (Serrano-Ruiz and Dumesic 2009a). In this respect, it was shown for the first time that the conversion of GVL to pentanoic acid and the subsequent conversion of pentanoic acid to 5-nonanone could be combined in a single flow reactor containing a dual catalyst bed operating under similar reaction conditions (698 K and 20 bar). An overall 5-nonanone yield of 84% was achieved with a minor yield (6%) of lower ketone species.

5-Nonanone can be converted to nonane via hydrogenation/dehydration over Pt/Nb_2O_5 at 528–568 K and 60 bar (West et al. 2008). The lower ketones present in the stream are converted to C_6–C_7 alkanes and can be easily separated as a vapor from nonane, which remains a liquid, and can be used as a blender in diesel fuels. Another alternative use of the 5-nonanone is to convert it to nonanol via hydrogenation over Ru/C at 423 K and 50 bar. In return, nonanol can be subsequently dehydrated over an acid catalyst, such as Amberlyst 70 (423 K), to produce nonene which can be coupled by acid-catalyzed oligomerization (Alonso et al. 2010). In this approach, smaller ketones would also be converted to alkenes that would undergo oligomerization along with nonene to produce C_6–C_{27} alkenes that can be hydrogenated over Pt/Nb_2O_5 to liquid alkanes. The molecular weight range for the final alkanes can be modified by varying reaction conditions of the temperature, pressure, or weight hour space velocity (WHSV) such that the final mixture could be used in jet fuel or as a diesel blender (Alonso et al. 2010).

The butene-based approach for GVL conversion was demonstrated by Bond et al. (2010). In this study, a dual reactor processing approach was applied to convert an aqueous solution of GVL (30–80 wt%) to liquid hydrocarbon fuels ranging in carbon length from C_8 to C_{24}. The first reactor contains a solid acid catalyst over which GVL is ring opened into a mixture of pentenoic acids which are subsequently converted via a decarboxylation reaction to a mixture of butene and an equal molar amount of CO_2. High yields of butene (>95%) can be achieved over a SiO_2/Al_2O_3 catalyst operating at 648 K and 36 bar. Two potential mechanisms have been proposed for the decarboxylation. It is suggested that GVL can either react through an acid-catalyzed ring opening to pentenoic acid and subsequent acid-catalyzed decarboxylation, or GVL can undergo a hydride shift and decarboxylation to form butene and CO_2 without passing through pentenoic acid. It was determined that for a given temperature the ring opening of GVL is approximately constant at pressures ranging from 1 to 36 bar, but that the decarboxylation of pentenoic acid decreases with increasing pressure.

The second reactor in the dual reactor approach converts the butene mixture into higher molecular weight alkenes through an oligomerization reaction over a solid acid catalyst. High yields of higher molecular oligomers have been demonstrated for the oligomerization of butene, in the presence of CO_2, over an Amberlyst 70 acid resin catalyst. In the integrated two reactor approach developed by Bond et al., it was determined that both catalytic reactors in series could operate at the same system pressure. The presence of CO_2 in the butene stream fed to the oligomerization reactor reversibly decreased the oligomerization rate only marginally; the decrease in activity is attributed to a dilution of the reactant, butene. However, it was determined that the presence of water in the feed for the butene oligomerization reaction significantly inhibited the catalyst activity; no reaction rate was observed for an equal molar feed of butene, CO_2, and water. For this reason, the integrated conversion process was designed with a liquid–vapor separator connected in series between the two catalytic reactors.

The effluent from the first reactor passes through the separator which is operated at the system pressure (36 bar). The temperature of the separator is controlled between 373 and 393 K to achieve the condensation and removal of most of the water solvent (>98%) while maintaining a high vaporization of butene. The butene and CO_2 are then passed into a second reactor connected in series that contains a solid acid catalyst designed for the alkene oligomerization. Using a SiO_2/Al_2O_3 catalyst for the GVL decarboxylation and an Amberlyst 70 catalyst for alkene oligomerization, an overall yield of approximately 75% can be achieved for the production of C_{8+} liquid alkenes from GVL. The effluent from the butene oligomerization reactor then passes into a vapor–liquid separator connected in series and operated at the system pressure and ambient temperature. In the final separator, the alkene oligomers are condensed and separated from the relatively pure CO_2 vapor phase.

The butene-based approach for the conversion of GVL into liquid fuels has several advantages. The conversion of GVL into alkene oligomers requires no external hydrogen. Moreover, if the GVL is produced via the formic acid hydrogen transfer strategy outlined previously, there is no net hydrogen requirement for the conversion of cellulose into alkene oligomers. The alkene oligomers produced from the oligomerization of butene are ideal for use in jet fuel and gasoline. The addition of alkenes into fuels helps to prevent auto ignition and it would not be necessary to convert the alkenes into fully saturated alkanes. However, if alkanes are desired, then only one mole of hydrogen would be required to convert one alkene oligomer into its corresponding alkane. The resulting oligomer liquid fuel retains approximately 95% of the energy content of the glucose molecules that comprise cellulose. In addition to the liquid oligomers, a relatively pure CO_2 stream can be recovered from the final vapor–liquid separator at high pressure (36 bar), enabling efficient CO_2 sequestration strategies or use as a reagent.

Recently, several works report processes for the production of GVL that integrate hydrolysis/dehydration of the carbohydrates to form levulinic acid and the subsequent hydrogenation to GVL in a single vessel (Deng et al. 2009; Heeres et al. 2009). It has been reported that sulfuric acid (commonly used in cellulose hydrolysis) poisons Ru catalysts (used in levulinic acid hydrogenation). Thus, to allow the integration of both reactions in the same vessel, other acids for cellulose hydrolysis have been investigated, such as trifluoroacetic acid. Heeres et al. (2009) used Ru/C as a catalyst and reported that the overall reaction rate is limited by the hydrogenation of levulinic acid to GVL. Using fructose as a raw material, a final yield of 52% for levulinic acid and 11% for GVL can be achieved at 453 K and 50 bar if external formic acid is added at the beginning of the reaction. These numbers are close to the maximum yield of levulinic acid from fructose reported in the literature, indicating that the reaction rate of fructose hydrogenation to sorbitol/mannitol is slower than the rate of dehydration. The presence of H_2 at the beginning of the reaction does not increase by-product formation. When comparing different feeds using formic acid and external hydrogen to increase reaction rates, they observed that the maximum yield is reached using fructose (62% GVL, 4% levulinic acid) and sucrose (52% GVL, 9% levulinic acid). Yields are lower when using glucose (38% GVL, 4% levulinic acid) and cellulose (29% GVL, 6% levulinic acid). Promising results have recently been reported using only formic acid as a hydrogen donor, HCl for sugar dehydration, and a mixture of $RuCl_3 \cdot 3H_2O$ and PPh_3 as the hydrogenation catalyst (Deng et al. 2009). The authors observed that using a stronger base as a ligand increases the reaction rate at 423 K, and they suggest that specific concentrations of water and CO_2 have a positive effect on the final yield to GVL. For example, they observed that with less than 25% water, the conversion of levulinic acid during hydrogenation increased from 78% to 100% when CO_2 was added at 4 MPa.

3.7 CONCLUDING REMARKS

Biomass is a renewable carbon source that can be processed in an integrated biorefinery, in a manner similar to petroleum in conventional refineries, to produce fuels and chemicals. The conversion of biomass to hydrocarbon biofuels faces two general chemical challenges: increasing the energy density of renewable feedstocks by reducing their high oxygen content, and facilitating the formation of C–C bonds such that parent monomers (generally limited to 5–6 carbon atoms) can be coupled to form hydrocarbons of appropriate molecular weight and volatility for use as transportation fuels.

The process of biorefining lignocellulosic feedstocks to liquid transportation fuels generally comprises two distinct processing components. First, biomass is deconstructed to produce upgradeable gaseous or liquid platforms. This step can be carried out through thermochemical pathways to produce synthesis gas (by gasification) or bio-oils (by pyrolysis or liquefaction), or by APR pathways to produce soluble intermediates, such as sugars or levulinic acid, that are then deoxygenated to form upgradeable intermediates. The functional intermediates from each platform are subsequently upgraded by C–C coupling reactions and finally reduced (if necessary) to form the desired liquid hydrocarbon fuel. The primary focus of this chapter is on aqueous-phase processing strategies for the production of bio-fuels. In this respect, APR of sugars/polyols can be used to produce H_2 by selective cleavage of C–C bonds, and with the addition of C–O bond cleavage, the reforming can be adjusted to form monofunctional compounds that can then be upgraded by C–C bond forming strategies, such as aldol condensation or ketonization to form targeted alkanes. Alternatively, sugars can be dehydrated to form furfurals that can be coupled using aldol condensation strategies to form larger oxygenates that yield linear C_9–C_{15} alkanes upon dehydration/hydrogenation. In addition, sugars can undergo fermentation to produce lactic acid that can be catalytically converted to hydrophobic ketones (for chemical applications) or hydrophobic alcohols (as fuel additives). As another option, cellulose can be converted to levulinic acid and subsequently to GVL, the latter of which can be upgraded by ring opening/ketonization or decarboxylation/oligomerization into targeted alkenes.

In general, we have tried to illustrate in this chapter that it is often possible to integrate multiple reactions in a single reactor by the use of multifunctional catalysts in a single catalyst bed, and/or by the use of dual catalyst beds. Furthermore, it is possible to achieve process simplification by the use of several reactors in a cascade mode, where the effluent from one reactor is simply used as the inlet for a downstream reactor, with minimal separation employed between reactors (e.g., passing through a gas–liquid separator). This coupling between reactions, for example, the integration of reactions that consume hydrogen with those that produce hydrogen, affords new strategies by which lignocellulosic biomass can be utilized efficiently for the production of fuels, chemicals, hydrogen, and any necessary process heat and power.

ACKNOWLEDGMENTS

The results from the University of Wisconsin reported here were supported in part by the U.S. Department of Energy Office of Basic Energy Sciences, by the National Science Foundation Chemical and Transport Systems Division of the Directorate for Engineering, by the DOE Great Lakes Bioenergy Research Center (www.greatlakesbioenergy.org), and by the Defense Advanced Research Projects Agency (Surf-cat: Catalysts for Production of JP-8 range molecules from Lignocellulosic Biomass).

REFERENCES

Aden, A. and T. Foust. 2009. Technoeconomic analysis of the dilute sulfuric acid and enzymatic hydrolysis process for the conversion of corn stover to ethanol. *Cellulose* 16 (4):535–545.

Alcala, R., M. Mavrikakis, and J. A. Dumesic. 2003. DFT studies for cleavage of C–C and C–O bonds in surface species derived from ethanol on Pt(111). *J. Catal.* 218:178–190.

Alonso, D. M., J. Q. Bond, J. C. Serrano-Ruiz, and J. A. Dumesic. 2010. Catalytic conversion of biomass to biofuels. *Green Chem.* 12:1493–1513.

Aramendia, M. A., B. Victoriano, J. Cesar, M. Alberto, M. M. Jose, A. N. Jose, R. R. Jose, and J. U. Francisco. 2004a. Synthesis and textural–structural characterization of magnesia, magnesia-titania and magnesia-zirconia catalysts. *Colloids Surf. A* 234 (1–3):17–25.

Aramendia, M. A., B. Victoriano, J. Cesar, M. Alberto, M. M. Jose, R. R. Jose, and J. U. Francisco. 2004b. Magnesium-containing mixed oxides as basic catalysts: Base characterization by carbon dioxide TPD-MS and test reactions. *J. Mol. Catal. A: Chem.* 218 (1):81–90.

Asghari, F. S. and H. Yoshida. 2006. Acid-catalyzed production of 5-hydroxymethyl furfural from D-fructose in subcritical water. *Ind. Eng. Chem. Res.* 45:2163–2173.

Asghari, F. S. and H. Yoshida. 2007. Kinetics of the decomposition of fructose catalyzed by hydrochloric acid in subcritical water: Formation of 5-hydroxymethylfurfural, levulinic, and formic acids. *Ind. Eng. Chem. Res.* 46:7703–7710.

Ayoub, P. M. and J. P. Lange, Processes for converting levulinic acid into pentanoic acid. 2008. WO 2008/142127 to Shell Internationale. Accessable via: https://data.epo.org/publication-server/getpdf.jsp?pn=2170797&ki=BI&CC=EP.

Barrett, C. J., J. N. Chheda, G. W. Huber, and J. A. Dumesic. 2006. Single-reactor process for sequential aldol-condensation and hydrogenation of biomass-derived compounds in water. *Appl. Catal. B* 66 (1–2):111–118.

Biometrics Inc. 1996. Municipal solid waste conversion project, Final Report, Contract No:4204-ERTER-ER-96.

Blanc, B., A. Bourrel, P. Gallezot, T. Haas, and P. Taylor. 2000. Starch-derived polyols for polymer technologies: Preparation by hydrogenolysis on metal catalysts. *Green Chem.* 2:89–91.

Blommel, P. G., G. R. Keenan, R. T. Rozmiarek, and R. D. Cortrigth. 2008. Catalytic conversion of sugar into conventional gasoline, diesel, jet fuel, and other hydrocarbons. *Int. Sugar J.* 110 (1319):672.

Bobleter, O. 2005. In *Polysaccharides*, ed. S. Dumitriu. New York: Marcel Dekker.

Bond, J. Q., D. M. Alonso, D. Wang, R. M. West, and J. A. Dumesic. 2010. Integrated catalytic conversion of gamma-Valerolactone to liquid alkenes for transportation fuels. *Science* 327 (5969):1110–1114.

Bourne, R. A., J. G. Stevens, J. Ke, and M. Poliakoff. 2007. Maximizing opportunities in supercritical chemistry: The continuous conversion of levulinic acid to gamma-valerolactone in CO_2. *Chem. Commun.* (44):4632–4634.

Boyle, G. 2004. *Renewable Energy: Power for a Sustainable Future*. New York: Oxford University Press.

Bozell, J. J. 2008. Feedstocks for the future–biorefinery production of chemicals from renewable carbon. *Clean* 36:641–647.

Bozell, J. J., L. Moens, D. C. Elliot, Y. Wang, G. G. Neuenschwander, S. W. Fitzpatrick, R. J. Bilski, and J. L. Jarnefeld. 2000. Production of levulinic acid and use as a platform chemical for derived products. *Resour. Conserv. Recy.* 28:227–239.

BP. 2009. BP Statistical Review of World Energy, June 2009. http://www.bp.com/liveassets/bp_internet/globalbp/globalbp_uk_english/reports_and_publications/statistical_energy_review_2008/STAGING/local_assets/2009_downloads/statistical_review_of_world_energy_full_report_2009.pdf (Accessed on 12 May 2010).

Brown, D. W., A. J. Floyd, R. G. Kinsman, and Y. Roshan-Ali. 1982. Dehydration reactions of fructose in non-aqueous media. *J. Chem. Technol. Biotechnol.* 32:920–924.

Caldwell, L. Council for Scientific and Industrial Research. 1980. Selectivity in Fischer–Tropsch synthesis: Review and recommendations for further work. http://www.fischer-tropsch.org/DOE/DOE_reports/81223596/pb81223596.pdf (Accessed on 10 June 2010).

Carlson, T. R., J. Jae, Y.-C. Lin, G. A. Tompsett, and G. W. Huber. 2010. Catalytic fast pyrolysis of glucose with HZSM-5: The combined homogeneous and heterogeneous reactions. *J. Catal.* 270 (1):110–124.

Carlson, T. R., T. P. Vispute, and G. W. Huber. 2008. Green gasoline by catalytic fast pyrolysis of solid biomass derived compounds. *ChemSusChem* 1:397–400.

Carroll, A. and C. Somerville. 2009. Cellulosic biofuels. *Annu. Rev. Plant Biol.* 60:165–182.

Carvalheiro, F., L. C. Duarte, and F. M. Girio. 2008. Hemicellulose biorefineries: A review on biomass pretreatments. *J. Sci. Ind. Res.* 67:849–864.

Chakar, F. S. and A. J. Ragauskas. 2004. Review of current and future softwood kraft lignin process chemistry. *Ind. Crop. Prod.* 20:131–141.

Chaminand, J., L. Djakovitch, P. Gallezot, P. Marion, C. Pinel, and C. Rosier. 2004. Glycerol hydrogenolysis on heterogeneous catalysts. *Green Chem.* 6:359–361.

Chang, C. D. and A. J. Silvestri. 1977. Conversion of methanol and other O-compounds to hydrocarbons over zeolite catalysts. *J. Catal.* 47 (2):249–259.

Chheda, J. N. and J. A. Dumesic. 2007. An overview of dehydration, aldol-condensation and hydrogenation processes for production of liquid alkanes from biomass-derived carbohydrates. *Catal. Today* 123 (1–4):59–70.

Chisti, Y. 2007. Biodiesel from microalgae. *Biotech. Adv.* 25 (3):294–306.

Choudary, B. M., M. L. Kantam, P. Sreekanth, T. Bandopadhyay, F. Figueras, and A. Tuel. 1999. Knoevenagel and aldol condensations catalysed by a new diamino-functionalised mesoporous material. *J. Mol. Catal. A: Chem.* 142:361–365.

Christoffersen, E., P. Liu, A. Ruban, H. L. Skriver, and J. K. Norskov. 2001. Anode materials for low-temperature fuel cells: A density functional theory study. *J. Catal.* 199:123–131.

Climent, M. J., A. Corma, S. Iborra, and A. Velty. 2004. Activated hydrotalcites as catalysts for the synthesis of chalcones of pharmaceutical interest. *J. Catal.* 221:474–482.

Collins, P. and R. Ferrier. 1995. *Monosaccharides*. West Sussex, England, U.K.: Wiley.

Cortright, R. D., R. R. Davda, and J. A. Dumesic. 2002. Hydrogen from catalytic reforming of biomass-derived hydrocarbons in liquid water. *Nature* 418:964–967.

Dasari, M. A., P.-P. Kiatsimkul, W. R. Sutterlin, and G. J. Suppes. 2005. Low-pressure hydrogenolysis of glycerol to propylene glycol. *Appl. Catal. A* 281:225–231.

Datta, R. 2004. Hydrocarboxylic acids. In *Kirk-Othmer Encyclopedia of Chemical Technology*, Vol. 14, 5th edn. Hoboken, NJ: Wiley.

Davda, R. R., J. W. Shabaker, G. W. Huber, R. D. Cortright, and J. A. Dumesic. 2003. Aqueous-phase reforming of ethylene glycol on silica-supported metal catalysts. *Appl. Catal. B* 43 (1):13–26.

Davda, R., J. Shabaker, G. Huber, R. Cortright, and J. Dumesic. 2005. A review of catalytic issues and process conditions for renewable hydrogen and alkanes by aqueous-phase reforming of oxygenated hydrocarbons over supported metal catalysts. *Appl. Catal. B* 56 (1–2):171–186.

Deng, L., J. Li, D. M. Lai, Y. Fu, and Q. X. Guo. 2009. Catalytic conversion of biomass-derived carbohydrates into gamma-valerolactone without using an external H-2 supply. *Angew. Chem. Int. Ed.* 48 (35):6529–6532.

Di Cosimo, J. I., V. K. Diez, and C. R. Apesteguia. 1996. Base catalysis for the synthesis of a,b-unsaturated ketones from the vapor-phase aldol condensation of acetone. *Appl. Catal. A* 137 (1):149–166.

Di Cosimo, J. J., V. K. Diez, M. Xu, E. Iglesia, and C. R. Apesteguia. 1998. Structure and surface and catalytic properties of Mg-Al basic oxides. *J. Catal.* 178:499–510.

Di Serio, M., M. Ledda, M. Cozzolino, G. Minutillo, R. Tesser, and E. Santacesaria. 2006. Transesterification of soybean oil to biodiesel by using heterogeneous basic catalysts. *Ind. Eng. Chem. Res.* 45 (9):3009–3014.

Diez, V. K., J. I. Di Cosimo, and C. R. Apesteguia. 2008. Study of the citral/acetone reaction on Mg_yAlO_x oxides: Effect of the chemical composition on catalyst activity, selectivity and stability. *Appl. Catal. A* 345:143–151.

Dixit, R. S. and L. L. Tavlarides. 1983. Kinetics of the Fischer–Tropsch synthesis. *Ind. Eng. Chem. Proc. Des. Dev.* 22:1–9.

Dry, M. E. 2002. The Fisher-Tropsch process: 1950–2000. *Catal. Today* 71:227.

Elliott, D. C. 2007. Historical developments in hydroprocessing bio-oils. *Energy Fuels* 21 (3):1792–1815.

Elliot, D. C. and J. G. Fry Jr. 1999. Hydrogenated 5-carbon compound and method of making. Richland, WA: Battelle Memorial Institute.

Encinar, J. M., J. F. Gonzales, and A. Rodriguez-Reinares. 2005. Biodiesel from used frying oil. Variables affecting the yields and characteristics of the biodiesel. *Ind. Eng. Chem. Res.* 44:5491–5499.

Englisch, M., A. Jentys, and J. A. Lercher. 1997. Structure sensitivity of the hydrogenation of crotonaldehyde over Pt/SiO_2 and Pt/TiO_2. *J. Catal.* 166 (1):25–35.

Evans, R. J., T. A. Milne, and M. N. Soltys. 1986. Direct mass-spectrometric studies of the pyrolysis of carbonaceous fuels. *J. Anal. Appl. Pyrolysis* 9:207–236.

Fernández, J., M. D. Curt, and P. L. Aguado. 2006. Industrial applications of *Cynara cardunculus* L. for energy and other uses. *Ind. Crop. Prod.* 24 (3):222–229.

Fisher, C. H. and E. M. Filachione. 1950. Properties and Reactions of Lactic Acid—A Review (AIC-279). Bureau of Agricultural and Industrial Chemistry, United States Department of Agriculture, Philadelphia.

Fitzpatrick, S. W. 1997. Production of levulinic acid from carbohydrate-containing materials. U.S. Patent No. 5,608,105.

Fukuda, H., A. Kondo, and H. Noda. 2001. Biodiesel fuel production by transesterification of oils. *J. Biosci. Bioeng.* 92 (5):405–416.

Gerpen, J. V. 2005. Biodiesel processing and production. *Fuel Process. Technol.* 86:1097–1107.

Girisuta, B., L. P. B. M. Janssen, and H. J. Heeres. 2006. A kinetic study on the conversion of glucose to levulinic acid. *Chem. Eng. Res. Design* 84:339–349.

Girisuta, B., L. P. B. M. Janssen, and H. J. Heeres. 2007. Kinetic study on the acid-catalyzed hydrolysis of cellulose to levulinic acid. *Ind. Eng. Chem. Res.* 46:1696–1708.

Gong, C. S., J. X. Du, N. J. Cao, and G. T. Tsao. 2006. Coproduction of ethanol and glycerol. *Appl. Biochem. Biotechnol.* 84:543–559.

Gosselink, R. J. A., E. de Jong, B. Guran, and A. Abacherli. 2004. Co-ordination network for lignin—Standardisation, production and applications adapted to market requirements (EUROLIGNIN). *Ind. Crop. Prod.* 20 (2):121–129.

Granados, M. L., M. D. Z. Poves, D. M. Alonso, R. Mariscal, F. C. Galisteo, R. Moreno-Tost, J. Santamaria, and J. L. G. Fierro. 2007. Biodiesel from sunflower oil by using activated calcium oxide. *Appl. Catal. B* 73 (3–4):317–326.

Graziani, M. and P. Fornasiero. 2007. *Renewable Resources and Renewable Energy: A Global Challenge*. Boca Raton, FL: CRC Press, Taylor & Francis Group.

Gunter, G. C., D. J. Miller, and J. E. Jackson. 1994. Formation of 2,3-pentanedione from lactic acid over supported phosphate catalysts. *J. Catal.* 148:252–260.

Gurbuz, E. I., E. L. Kunkes, and J. A. Dumesic. 2010a. Dual-bed catalyst system for C–C coupling of biomass-derived oxygenated hydrocarbons to fuel-grade compounds. *Green Chem.* 12 (2):223–227.

Gurbuz, E. I., E. L. Kunkes, and J. A. Dumesic. 2010b. Integration of C–C coupling reactions of biomass-derived oxygenates to fuel-grade compounds. *Appl. Catal. B* 94:134–141.

Gutsche, C., D. Redmore, R. Buriks, K. Nowotny, H. Grassner, and C. Armbruster. 1967. Base-catalyzed triose condensations. *J. Am. Chem. Soc.* 89:1235–1245.

Haas, M. J., A. J. McAloon, W. C. Yee, and T. A. Foglia. 2006. A process model to estimate biodiesel production costs. *Bioresour. Technol.* 97:671–678.

Hayes, D. J. 2009. An examination of biorefining processes, catalyst and challenges. *Catal. Today* 145:138–151.

Heeres, H., R. Handana, D. Chunai, C. B. Rasrendra, B. Girisuta, and H. J. Heeres. 2009. Combined dehydration/(transfer)-hydrogenation of C6-sugars (D-glucose and D-fructose) to gamma-valerolactone using ruthenium catalysts. *Green Chem.* 11 (8):1247–1255.

Holmen, R. E. 1958. Production of Acrylates by Catalytic Dehytration of Lactic Acid and Alkyl Lactates. U.S. Patent No. 2,859,240.

Horvat, J., B. Klaic, B. Metelko, and V. Sunjic. 1985. Mechanism of levulinic acid formation. *Tetrahedron Lett.* 26:2111–2114.

Horvath, I. T., H. Mehdi, V. Fabos, L. Boda, and L. T. Mika. 2008. Gamma-Valerolactone—A sustainable liquid for energy and carbon-based chemicals. *Green Chem.* 10 (2):238–242.

Huber, G. W., J. N. Chheda, C. J. Barrett, and J. A. Dumesic. 2005. Production of liquid alkanes by aqueous-phase processing of biomass-derived carbohydrates. *Science* 308:1446–1450.

Huber, G. W., P. O'Connor, and A. Corma. 2007. Processing biomass in conventional oil refineries: Production of high quality diesel by hydrotreating vegetable oils in heavy vacuum oil mixtures. *Appl. Catal. A* 329:120–129.

Huber, G. W. and A. Corma. 2007. Synergies between bio- and oil refineries for the production of fuels from biomass. *Angew. Chem. Int. Ed.* 46:7184–7201.

Huber, G. W., R. D. Cortright, and J. A. Dumesic. 2004. Renewable alkanes by aqueous-phase reforming of biomass-derived oxygenates. *Angew. Chem. Int. Ed.* 43 (12):1549–1551.

Huber, G. W., S. Iborra, and A. Corma. 2006. Synthesis of transportation fuels from biomass: Chemistry, catalysts, and engineering. *Chem. Rev.* 106:4044–4098.

Huber, G. W., J. W. Shabaker, and J. A. Dumesic. 2003. Raney Ni-Sn catalyst for H₂ production from biomass-derived hydrocarbons. *Science* 300:2075–2078.

International Energy Agency. 2009. Key world energy statistics 2009. http://www.iea.org/textbase/nppdf/free/2009/key_stats_2009.pdf (Accessed on 12 May 2010).

Iglesia, E., S. C. Reyes, R. J. Madon, and S. L. Soled. 1993. Selectivity control and catalyst design in the Fischer–Tropsch synthesis sites, pellets, and reactors. *Adv. Catal.* 39:221.

Iglesia, E., S. L. Soled, and R. A. Fiato. 1992. Fischer–Tropsch synthesis on cobalt and ruthenium. Metal dispersion and support effects on reaction rate and selectivity. *J. Catal.* 137:212–224.

Institute of Local Self-Reliance. 2000. *Carbohydrate Economy Bulletin*, to be found under http://www.carbohydrateeconomy.org/library/admin/uploadedfiles/Carbohydrate_Economy_Bulletin_Volume_1_Numb_3.htm (Accessed on 8 Feb 2009).

Jae, J., A. T. Geoffrey, L. Yu-Chuan, R. C. Torren, S. Jiacheng, Z. Taiying, Y. Bin, F. W. Charles, W. Curtis Conner, and W. H. George. 2010. Depolymerization of lignocellulosic biomass to fuel precursors: Maximizing carbon efficiency by combining hydrolysis with pyrolysis. *Energy Environ. Sci.* 3:358–365.

Kellner, C. S. and A. T. Bell. 1981. The kinetics and mechanism of carbon monoxide hydrogenation over alumina-supported ruthenium. *J. Catal.* 70:418–432.

Kersten, S. R. A., W. P. M. van Swaaij, L. Lefferts, and K. Seshan. 2007. Option for catalysis in the thermochemical conversion of biomass into fuels. In *Catalysis for Renewables: From Feedstock to Energy Production*, eds. G. Centi and R. A. V. Santen. Weinheim, Germany: Wiley-VCH.

Kim, H. J., B. S. Kang, M. J. Kim, Y. M. Park, D. K. Kim, J. S. Lee, and K. Y. Lee. 2004. Transesterification of vegetable oil to biodiesel using heterogeneous base catalyst. *Catal. Today* 93–95:315–320.

Klass, D. L. 1998. *Biomass for Renewable Energy, Fuels and Chemicals. Biomass for Renewable Energy, Fuels and Chemicals*. San Diego, CA: Academic Press.

Klass, D. L. 2004. Biomass for the renewable energy and fuels. In *Encyclopedia of Energy*, ed. E. J. Cleveland. London, U.K.: Elsevier.

Klier, K. 1982. Methanol Synthesis. *Adv. Catal.* 31:243–313.

Knifton, J. F., J. R. Sanderson, and P. E. Dai. 1994. Olefin oligomerization via zeolite catalysis. *Catal. Lett.* 28:223–234.

Koppatz, S., C. Pfeifer, R. Rauch, H. Hofbauer, T. Marquard-Moellensted, and M. Specht. 2009. H_2 rich product gas by steam gasification of biomass with in situ CO_2 absorption in a dual fluidized bed system of 8 MW fuel input. *Fuel Process. Technol.* 90:914–921.

Kreith, F. and D.Y. Goswami. 2007. *Handbook of Energy Efficiency and Renewable Energy*. Boca Raton, FL: CRC Press, Taylor & Francis Group.

Kumar, P., D. M. Barrett, M. J. Delwiche, and P. Stroeve. 2009a. Methods for pretreatment of lignocellulosic biomass for efficient hydrolysis and biofuel production. *Ind. Eng. Chem. Res.* 48 (8):3713–3729.

Kumar, R., G. Mogo, V. Balan, and C. E. Wyman. 2009b. Physical and chemical characterizations of corn stover and poplar solids resulting from leading pretreatment technologies. *Bioresour. Technol.* 100:3948–3962.

Kunkes, E. L., E. I. Gurbuz, and J. A. Dumesic. 2009a. Vapour-phase C–C coupling reactions of biomass-derived oxygenates over Pd/CeZrOx catalysts. *J. Catal.* 266 (2):236–249.

Kunkes, E. L., D. A. Simonetti, J. A. Dumesic, W. D. Pyrz, L. E. Murillo, J. G. Chen, and D. J. Buttrey. 2008a. The role of rhenium in the conversion of glycerol to synthesis gas over carbon supported platinum–rhenium catalysts. *J. Catal.* 260 (1):164–177.

Kunkes, E. L., D. A. Simonetti, R. M. West, J. C. Serrano-Ruiz, C. A. Gaertner, and J. A. Dumesic. 2008b. Catalytic conversion of biomass to monofunctional hydrocarbons and targeted liquid-fuel classes. *Science* 322:417–421.

Kunkes, E. L., R. R. Soares, D. A. Simonetti, and J. A. Dumesic. 2009b. An integrated catalytic approach for the production of hydrogen by glycerol reforming coupled with water-gas shift. *Appl. Catal. B* 90:693–698.

Kuster, B. M. F. 1990. 5-Hydroxymethylfurfural (HMF). A review focussing on its manufacture. *Starch* 42:314–321.

Lahr, D. G. and B. H. Shanks. 2003. Kinetic analysis of the hydrogenolysis of lower polyhydric alcohols: Glycerol to glycols. *Ind. Eng. Chem. Res.* 42:5467.

Lahr, D. G. and B. H. Shanks. 2005. Effect of sulfur and temperature on ruthenium-catalyzed glycerol hydrogenolysis to glycols. *J. Catal.* 232:386–394.

Lange, J. P. 2001. Methanol synthesis: A short review of technology improvements. *Catal. Today* 64:3–8.

Lange, J. P. 2007. Lignocellulose conversion: An introduction to chemistry, process and economics. *Biofuels Bioprod. Bioref.* 1:39–48.

Li, L. X., E. Coppola, J. Rine, J. L. Miller, and D. Walker. 2010. Catalytic hydrothermal conversion of triglycerides to non-ester biofuels. *Energy Fuels* 24:1305–1315.

Li, H., W. Wang, and J. F. Deng. 2000. Glucose hydrogenation to sorbitol over a skeletal Ni-P amorphous alloy catalyst (Raney Ni-P). *J. Catal.* 191:257–260.

Lichtenthaler, F. W. and S. Peters. 2004. Carbohydrates as green raw materials for the chemical industry. *C. R. Chim.* 7:65–90.

Lipinsky, E. S. and R. G. Sinclair. 1986. Is lactic acid a commodity chemical? *Chem. Eng. Prog.* 82:26–32.

Liu, H., L. Peng, T. Zhang, and Y. Li. 2003. L-Proline catalyzed asymmetric aldol reactions of protected hydroxyacetone. *New J. Chem.* 27:1159–1160.

Lloyld, T. A. and C. E. Wyman. 2005. Combined sugar yields for dilute sulfuric acid pretreatment of corn stover followed by enzymatic hydrolysis of the remaining solids. *Bioresour. Technol.* 96:1967–1977.

Lotero, E., Y. J. Liu, D. E. Lopez, K. Suwannakarn, D. A. Bruce, and J. G. Goodwin. 2005. Synthesis of biodiesel via acid catalysis. *Ind. Eng. Chem. Res.* 44 (14):5353–5363.

Lunt, L. 1998. Large-scale production, properties and commercial applications of polylactic acid polymers. *Polym. Degrad. Stab.* 59:145–152.

Lynd, L. R., J. H. Cushman, R. J. Nichols, and C. E. Wyman. 1991. Fuel ethanol from cellulosic biomass. *Science* 251 (4999):1318–1323.

Lynd, L. R., C. E. Wyman, and T. U. Grengross. 1999. Biocommodity engineering. *Biotechnol. Prog.* 15:777–793.

Ma, F. R. and M. A. Hanna. 1999. Biodiesel production: A review. *Bioresour. Technol.* 70 (1):1–15.

Mamman, A. S. 2008. Furfural: Hemicellulose/xylose derived biochemical. *Biofuels Bioprod. Bioref.* 2 (5):438–454.

Marinelli, T. B. L. W., S. Nabuurs, and V. Ponec. 1995. Activity and selectivity in the reactions of substituted a,b-unsaturated aldehydes. *J. Catal.* 151 (2):431–438.

Marinelli, T. B. L. W., J. H. Vleeming, and V. Ponec. 1993. Reactions of multifunctional organic compounds— Hydrogenation of acrolein on modified platinum catalysts. *Stud. Surf. Sci. Catal.* 75 (New Frontiers in Catalysis, Pt. B):1211–1222.

Mehdi, H., V. Fabos, R. Tuba, A. Bodor, L. T. Mika, and I. T. Horvath. 2008. Integration of homogeneous and heterogeneous catalytic processes for a multi-step conversion of biomass: From sucrose to levulinic acid, gamma-valerolactone, 1,4-pentanediol, 2-methyl-tetrahydrofuran, and alkanes. *Top. Catal.* 48 (1–4):49–54.

Melero, J. A., J. Iglesias, and G. Morales. 2009. Heterogeneous acid catalysts for biodiesel production: Current status and future challenges. *Green Chem.* 11 (9):1285–1308.

Mercadier, D., L. Rigal, A. Gaset, and J. P. Gorrichon. 1981. Synthesis of 5-hydroxymethyl-2-furancarboxaldehyde catalysed by cationic exchange resins. Part 1. Choice of the catalyst and the characteristics of the reaction medium. *J. Chem. Technol. Biotechnol.* 31:489–496.

Miyazawa, T., Y. Kusunoki, K. Kunimori, and K. Tomishige. 2006. Glycerol conversion in the aqueous solution under hydrogen over Ru/C + an ion-exchange resin and its reaction mechanism. *J. Catal.* 240:213–221.

Mohan, D., C. U. Pittman, and P. H. Steele. 2006. Pyrolysis of wood/biomass for bio-oil: A critical review. *Energy Fuels* 20:848–889.

Mok, W. S., M. J. Antal, and M. Jones. 1989. Formation of acrylic acid from lactic acid in supercritical water. *J. Org. Chem.* 54:4596–4602.

Moreau, C., R. Durand, J. Duhamet, and P. Rivalier. 1997. Hydrolysis of fructose and glucose precursors in the presence of H-form zeolites. *J. Carbohydr. Chem.* 16:709–714.

Moreau, C., D. Robert, R. Sylvie, D. Jean, F. Pierre, R. Patrick, R. Pierre, and A. Gerard. 1996. Dehydration of fructose to 5-hydroxymethylfurfural over H-mordenites. *Appl. Catal. A* 145 (1–2):211–224.

Mosier, N., C. E. Wyman, B. E. Dale, R. T. Elander, Y. Y. Lee, M. Holtzapple, and M. R. Ladisch. 2005. Features of promising technologies for pretreatment of lignocellulosic biomass. *Bioresour. Technol.* 96:673–686.

Musau, R. M. and R. M. Munavu. 1987. The preparation of 5-hydroxymethyl-2-furaldehyde (HMF) from D-fructose in the presence of DMSO. *Biomass* 13:67–74.

Nagamori, M. and T. Funazukuri. 2004. Glucose production by hydrolysis of starch under hydrothermal conditions. *J. Chem. Technol. Biotechnol.* 79:229–233.

Nagashima, O., S. Sato, R. Takahashi, and T. Sodesawa. 2005. Ketonization of carboxylic acids over CeO₂-based composite oxides. *J. Mol. Catal. A* 227:231.

Narayan, R., G. Durrence, and G. T. Tsao. 1984. Ethylene glycol and other monomeric polyols from biomass. *Biotechnol. Bioeng. Symp.* 14:563–571.

National Renewable Energy Laboratory. 2002. Lignocellulosic biomass to ethanol process design and economics utilizing co-current dilute acid prehydrolysis and enzymatic hydrolysis for corn stover. www.nrel.gov/docs/fy02osti/32438.pdf (Accessed on 12 June 2010).

National Science Foundation. 2008. Breaking the chemical and engineering barriers to lignocellulosic biofuels: Next generation hydrocarbon biorefineries. Paper read at University of Massachusetts Amherst. National Science Foundation. Chemical, Bioengineering, Environmental, and Transport Systems Division, Washington, DC. http://www.ecs.umass.edu/biofuels/Images/Roadmap2-08.pdf (Accessed on 12 June 2010).

Nikolopoulos, A. A., B. W.-L. Jang, and J. J. Spivey. 2005. Acetone condensation and selective hydrogenation to MIBK on Pd and Pt hydrotalcite-derived Mg single bond Al mixed oxide catalysts. *Appl. Catal. A* 296:128–136.

O'Connor, C. T. and M. Kojima. 1990. Alkene oligomerization. *Catal. Today* 6 (3):329–349.

Pallassana, V. and M. Neurock. 2002. Reaction paths in the hydrogenolysis of acetic acid to ethanol over Pd(111), Re(0001), and PdRe alloys. *J. Catal.* 209:289–305.

Patil, P. D. and S. Deng. 2009. Optimization of biodiesel production from edible and non-edible vegetable oils. *Fuel* 88 (7):1302–1306.

Paul, S. F. 2001. Alternative fuel. U.S. Patent. 6,309,430.

Perlack, R. D., L. L. Wright, A. F. Turhollow, R. L. Graham, B. J. Stokes, and D. C. Erbach. 2005. Biomass as feedstock for a bioenergy and bioproducts industry: the technical feasibility of a billion-ton annual supply. DOE/GO-102005-2135, Oak Ridge National Laboratory. http://feedstockreview.ornl.gov/pdf/billion_ton_vision.pdf

Regalbuto, J. R. 2009. Cellulosic biofuels—Got gasoline? *Science* 325:822–824.

Renz, M. 2005. Ketonization of carboxylic acids by decarboxylation: Mechanism and scope. *Eur. J. Org. Chem.* (6):979–988.

Rinaldi, R. and F. Schuth. 2009. Acid hydrolysis of cellulose as the entry point into biorefinery schemes. *ChemSusChem* 2 (12):1096–1107.

Roelofs, J. C. A. A., D. J. Lensveld, A. J. van Dillen, and K. P. de Jong. 2001. On the structure of activated hydrotalcites as solid base catalysts for liquid-phase aldol condensation. *J. Catal.* 203:184–191.

Saha, B. C. 2003. Hemicellulose bioconversion. *J. Ind. Microbiol. Biotechnol.* 30 (5):279–291.

Sawicki, R. A. 1998. Catalyst for Dehydration of Lactic Acid to Acrylic Acid. U.S. Patent No. 4,729,978.

Saxena, U., N. Dwivedi, and S. R. Vidyarthi. 2005. Effect of catalyst constituents on (Ni, Mo, and Cu)/Kieselguhr-catalyzed sucrose hydrogenolysis. *Ind. Eng. Chem. Res.* 44:1466–1473.

Seri, K., Y. Inoue, and H. Ishida. 2001. Catalytic activity of lanthanide(III) ions for the dehydration of hexose to 5-hydroxymethyl-2-furaldehyde in water. *Bull. Chem. Soc. Jpn.* 74:1145–1150.

Serrano-Ruiz, J. C. and J. A. Dumesic. 2009a. Catalytic upgrading of lactic acid to fuels and chemicals by dehydration/hydrogenation and C–C coupling reactions. *Green Chem.* 11 (8):1101–1104.

Serrano-Ruiz, J. C. and J. A. Dumesic. 2009b. Catalytic processing of lactic acid over Pt/Nb$_2$O$_5$. *ChemSusChem* 2:581–586.

Serrano-Ruiz, J. C., D. Wang, and J. A. Dumesic. 2010a. Catalytic upgrading of levulinic acid to 5-nonanone. *Green Chem.* 12 (4):574–577.

Serrano-Ruiz, J. C., R. M. West, and J. A. Dumesic. 2010b. Catalytic conversion of renewable biomass resources to fuels and chemicals. *Annu. Rev. Chem. Biomol. Eng.* 1:79–100.

Shabaker, J. W., R. R. Davda, G. W. Huber, R. D. Cortright, and J. A. Dumesic. 2003. Aqueous-phase reforming of methanol and ethylene glycol over alumina-supported platinum catalysts. *J. Catal.* 215 (2):344–352.

Shabaker, J. W., G. W. Huber, R. R. Davda, R. D. Cortright, and J. A. Dumesic. 2003. Aqueous-phase reforming of ethylene glycol over supported platinum catalysts. *Catal. Lett.* 88:1–8.

Shigemasa, Y., K. Yokoyama, H. Sashiwa, and H. Saimoto. 1994. Synthesis of threo- and erythro-3-pentulose by aldol type reaction in water. *Tetrahedron Lett.* 35:1263–1266.

Simonetti, D. A. and J. A. Dumesic. 2009. Catalytic production of liquid fuels from biomass-derived oxygenated hydrocarbons: Catalytic coupling at multiple length scales. *Catal. Rev.* 51:441–484.

Simonetti, D. A., J. Rass-Hansen, E. L. Kunkes, R. R. Soares, and J. A. Dumesic. 2007. Coupling of glycerol processing with Fischer–Tropsch synthesis for production of liquid fuels. *Green Chem.* 9 (10):1073–1083.

Soares, R. R., D. A. Simonetti, and J. A. Dumesic. 2006. Glycerol as a source for fuels and chemicals by low-temperature catalytic processing. *Angew. Chem. Int. Ed.* 45 (24):3982–3985.

Stocker, M. 2008. Biofuels and biomass-to-liquid fuels in the biorefinery: Catalytic conversion of lignocellulosic biomass using porous materials. *Angew. Chem. Int. Ed.* 47:9200–9211.

Szmant, H. H. and D. D. Chundury. 1981. The preparation of 5-hydroxymethylfurfuraldehyde from high fructose corn syrup and other carbohydrates. *J. Chem. Technol. Biotechnol.* 31:135–145.

Thomas, R. W. and H. A. Schuette. 1931. Studies on levulinic acid. I. Its preparation from carbohydrates by digestion with hydrochloric acid under pressure. *J. Am. Chem. Soc.* 53:2324–2328.

Tichit, D., D. Lutic, B. Coq, R. Durand, and R. Teissier. 2003. The aldol condensation of acetaldehyde and heptanal on hydrotalcite-type catalysts. *J. Catal.* 219:167–175.

Tronconi, E., N. Ferlazzo, P. Forzatti, I. Pasquon, B. Casale, and L. Marini. 1992. A mathematical model for the catalytic hydrogenolysis of carbohydrates. *Chem. Eng. Sci.* 47:2451.

U.S. Department of Energy (DOE). 2004. Top value added chemicals from biomass volume 1—Results of screening for potential candidates from sugars and synthesis gas.

U.S. DOE Energy Information Administration. 2009. State and U.S. Historical Energy Data.

U.S. Department of Energy and Environmental Protection Agency. 2010. http://fueleconomy.gov/feg/flextech.shtml

van Dam, H. E., A. P. G. Kieboom, and H. van Bekkum. 1986. The conversion of fructose and glucose in acidic media: Formation of hydroxymethylfurfural. *Starch* 38:95–101.

Vanholme, R, K. Morreel, J. Ralph, and W. Boerjan. 2008. Lignin engineering. *Curr. Opin. Plant Biol.* 11:278–285.

Vannice, M. A. 1977. The catalytic synthesis of hydrocarbons from H$_2$/CO mixtures over the Group VIII metals: V. The catalytic behavior of silica-supported metals. *J. Catal.* 50:228–236.

Wadley, D. C., M. S. Tam, P. B. Kokitkar, J. E. Jackson, and D. J. Miller. 1997. Lactic acid conversion to 2,3-pentanedione and acrylic acid over silica-supported sodium nitrate: Reaction optimization and identification of sodium lactate as the active catalyst. *J. Catal.* 165:162–171.

Wang, G., Z. Zhang, and Y.-W. Dong. 2004. Environmentally friendly and efficient process for the preparation of β-hydroxyl ketones. *Org. Process. Res. Dev.* 8:18–21.

West, R. M., D. J. Braden, and J. A. Dumesic. 2009a. Dehydration of butanol to butene over solid acid catalysts in high water environments. *J. Catal.* 262 (1):134–143.

West, R. M., E. L. Kunkes, D. A. Simonetti, and J. A. Dumesic. 2009b. Catalytic conversion of biomass-derived carbohydrates to fuels and chemicals by formation and upgrading of mono-functional hydrocarbon intermediates. *Catal. Today* 147(2):115–125.

West, R. M., Z. Y. Liu, M. Peter, and J. A. Dumesic. 2008. Liquid alkanes with targeted molecular weights from biomass-derived carbohydrates. *ChemSusChem* 1:417–424.

Worldwatch Institute Center for American Progress. 2006. American energy: The renewable path to energy security. http://www.americanprogress.org/issues/2006/09/american_energy.html/AmericanEnergy.pdf (Accessed on 18 May 2010).

Wyman, C. E., B. E. Dale, R. T. Elander, M. Holtzapple, M. R. Ladisch, and Y. Y. Lee. 2005a. Comparative sugar recovery data from laboratory scale application of leading pretreatment technologies to corn stover. *Bioresour. Technol.* 96:2026–2032.

Wyman, C. E., S. R. Decker, M. E. Himmel, J. W. Brady, C. E. Skopec, and L. Viikari. 2005b. Hydrolysis of cellulose and hemicellulose. In *Polysaccharides: Structural Diversity and Functional Versatility*, ed. S. Dumitriu, 2nd edn. New York: Marcel Dekker Inc., pp. 995–1033.

Zaldivar, J., J. Nielsen, and L. Olsson. 2001. Fuel ethanol production from lignocellulose: A challenge for metabolic engineering and process integration. *Appl. Microbiol. Biotechnol.* 56 (1–2):17–34.

Zanella, R., L. Catherine, G. Suzanne, and T. Raymonde. 2004. Crotonaldehyde hydrogenation by gold supported on TiO₂: Structure sensitivity and mechanism. *J. Catal.* 223 (2):328–339.

Zartman, W. H. and H. Adkins. 1933. Hydrogenolysis of sugars. *J. Am. Chem. Soc.* 55:4559–4563.

Zeitsch, K. J. 2000. *The Chemistry and Technology of Furfural and Its Many By-Products, Sugar Series*, Vol. 13, 1st edn. Amsterdam, the Netherlands: Elsevier, pp. 34–69.

Zeng, R., X. Fu, C. Gong, Y. Sui, X. Ma, and X. Yang. 2005. Preparation and catalytic property of the solid base supported on the mixed zirconium phosphate phosphonate for Knoevenagel condensation. *J. Mol. Catal. A Chem.* 229:1–5.

Zhang, Z., Y.-W. Dong, and G.-W. Wang. 2003. Efficient and clean aldol condensation catalyzed by sodium carbonate in water. *Chem. Lett.* 32:966–967.

4 Valorization of Bio-Glycerol

Mario Pagliaro and Michele Rossi

CONTENTS

4.1 INTRODUCTION

Conversion to epichlorohydrin (estimated consumption: 100,000 ton per year) at France's Tavaux site of Solvay; hydrogenation (hydrogenolysis) to 1,2 propylene glycol (estimated use: 1000 ton) in the United States by Senergy Chemical; bio-glycerol direct employment in new concrete additives (estimated use: 1000 ton) by Grace; and methanol production in the Netherlands by the consortium Biomethanol[1] with a production capacity of 250 million liters are the first large-scale chemical applications using bio-glycerol—glycerol stemming from biodiesel manufacturing—as raw material. At least two new major conversions are currently being scaled up: acrolein production in France by Arkema,[2] and glycerol tertiary butyl ether as fuel oxygenate additive in Spain by Repsol.[3]

Here, we focus on new processes for crude glycerol refinement and on the economic aspects of the bio-glycerol issue. We refer the readers interested in aspects related to the chemoselective catalytic conversion of glycerol into valued chemicals to our recent specialized book.[4]

4.2 BACKGROUND AND CONTEXT

Glycerol (1,2,3-propanetriol, $C_3H_8O_3$) is a valued liquid chemical with more than 1500 end uses due to its remarkable physical and chemical properties. Most frequently called glycerin(e), the polyol glycerol is a hygroscopic, colorless, odorless, viscous, sweet tasting, low boiling, nontoxic, emollient, good solvent, and easily biodegradable.[4] Together, the top three uses for refined glycerol—food, personal, and oral care products—account for 64% of refined glycerol global consumption (Figure 4.1). The remainder is used for alkyd resins, cellophanes, explosives, and other miscellaneous uses throughout industry. The glycerol market, therefore, is intrinsically fragmented with a large number of companies using relatively small amounts of glycerol.[5]

Glycerol occurs as molecular backbone in triglycerides, which are the main constituents of all vegetable and animal fats and oils. Chemical processing of the latter natural products by the soap (fatty acids) and oleochemical (fatty alcohols) industries for the production of soap and oleochemicals, which involves the liberation of glycerol at a level of 10% in weight of the oil or fat (Scheme 4.1), has therefore been the chief source of glycerol for most of the twentieth century.[6]

Glycerol, of course, is the raw material for manufacturing the leading explosive nitroglycerol. Hence, when glycerol demand due to wartimes exceeded the supply from the soap industry, reasons

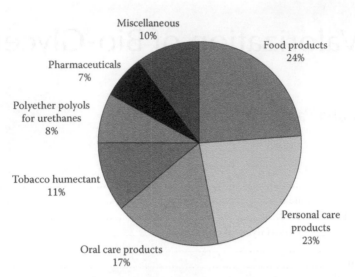

FIGURE 4.1 End use of refined glycerol. (Reproduced from ABG Inc. for the U.S. Soybean Export Council, Glycerol market analysis, Adayana Agribusiness Group, 2007. With permission.)

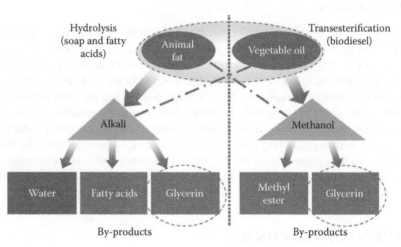

SCHEME 4.1 Both saponification and transesterification reaction yield 10 wt.% glycerol as by-product. (Reproduced from ABG Inc. for the U.S. Soybean Export Council, Glycerol market analysis, Adayana Agribusiness Group, 2007.)

of military security led to the first synthetic plants for glycerol manufacture both in Europe and in the United States: first, during World War I, for glycerol production through microbial sugar fermentation,[7] and then, in 1943, from petroleum feedstock, once the high-temperature chlorination of propene to allyl chloride could be controlled properly[8] (by I. G. Farben in Germany in Oppau and Heydebreck) followed after the end of World War II by Shell in Texas' Houston in 1948. Synthetic glycerol made from propylene via epichlorohydrin is today carried out by only one producer, Dow Chemical in the United States, and is not regarded as influencing the glycerol market anymore. Actually, the high and volatile price of propylene along with good demand for epichlorohydrin have caused reversal of this process, with bio-glycerol from biodiesel manufacturing used by Solvay as raw material for epichlorohydrin (used to synthesize epoxy resins that afford a better profit margin).

In 2003, generous tax credits to biodiesel manufacturing were granted in the largest EU nations, Germany, France, and Italy, followed by similar incentives launched in the United States 2 years later. This initiated a boom in the global production of this biofuel leading to large surplus of glycerol obtained as by-product. In practice, all the increase in the amount of glycerol in the market, doubled to an estimated 1,600 kton in 2010 from 600,000 kg in 1995, is primarily the result of the increase in fatty acids methyl mono-alkyl esters (FAME) production first in Europe and the United States and, more recently, in Southern America and in Asia. In its turn, even if to a lesser extent, also the glycerol quantity obtained from the oleochemical industry has continuously increased. As a result of this glycerol glut, traditional market usages for refined glycerol in the United States, Asia, Latin America, and Europe—for pharmaceuticals, food and beverages, personal and oral care products, and alkyd resins (Figure 4.1)—are not capable anymore to absorb this enormous surplus, because crude glycerol is contaminated with toxic methanol (see the following text) and thus not suitable for use in all main traditional applications.

4.3   REFINING CRUDE GLYCEROL

There are three basic types of glycerol—tallow-based, vegetable-based, and synthetic—based upon derivation, purity, and potential end uses. Vegetable glycerol is derived from natural oils such as coconut, palm-kernel oil, and palm stearin. Most of the glycerol in the market today is manufactured to meet the requirements of the United States Pharmacopeia (USP) and the Food Chemicals Codex (FCC). Crude glycerol is at least 80% pure and is almost always refined to further points of purity up to 99% as glycerol is generally sold as 99.5% or 99.7% pure. Refined glycerol can be classified into six categories (Table 4.1) and three main classes:

- *Technical grade*—used as a building block in chemicals, not used for food or drug formulation
- *USP—glycerol* from animal fat or plant oil sources, suitable for food products and pharmaceuticals
- *Kosher—glycerol* from plant oil sources, suitable for use in kosher foods

Lower grades of tallow-based glycerol are usually designated for polyols and alkyd resins markets, while higher grades—with purities up to 95.5% and 99% pure—compete with vegetable and synthetic glycerol.

Crude glycerol obtained from traditional biodiesel manufacturing has a high salt and free fatty acid content and a substantial color (yellow to dark brown, Figure 4.2). Consequently, crude glycerol has few direct uses, and its value is marginal.

Most biodiesel plants employ obsolete homogeneous transesterification reaction between fat or vegetable oil and methanol catalyzed by NaOH or KOH, using a 6 to 1 molar ratio of methanol to oil in order to drive the equilibrium reaction to completion.[9] Most of the excess alcohol ends up in the bio-glycerol layer.

Classified as crude glycerol, such bio-glycerol stream normally contains approximately 85% glycerol and 10% methanol. The rest is wastewater (a mixture of water, leftover methyl esters and lipids, inorganic salts (catalyst residues), free fatty acids, unreacted mono-, di-, and triglycerides,

TABLE 4.1
Purification Quality of Glycerol Can Be
Identified by Its Grade

Refined Glycerol	Purity (%)
Technical grade	99.5
USP	96, vegetable based
USP	99.5, tallow based
USP/FCC-kosher	99.5
USP/FCC-kosher	99.7

FIGURE 4.2   Upon transesterification, the crude glycerol layer sits at the bottom of biodiesel layer. (Photo courtesy of Biodieselcommunity.org).

and other "matter organic non-glycerol" (MONG, in varying amounts). The typical range of composition is reported in Table 4.2.[10]

If refined to a chemically pure condition, glycerol would be a valuable by-product (Table 4.3) as glycerol, in general, competes with other polyols, such as sorbitol and oil-derived pentaerythritol and trimethylol propane. Glycerol's properties are generally superior, but glycerol often is not employed due to its high cost.

For example, as late as 2007, the Food and Drug Administration blocked all shipments of toothpaste from China to the United States, following reports of contaminated toothpaste and syrups entering *via* Panama. The syrup produced in Panama contained toxic diethylene glycol (DEG) and killed at least 100 people. The poison, falsely labeled as glycerol, had in 2006 been mistakenly mixed into medicines in Panama, resulting in the fatal poisonings. DEG had originated from a Chinese factory that had deliberately falsified records in order to export DEG in place of the more expensive glycerol.[11] Eventually, a large batch of toothpaste contaminated with DEG also reached Europe, and a number of poisoning cases being reported in Italy and southern Europe.

TABLE 4.2
The Typical Range of Bio-Glycerol Composition

Material	wt.%
Glycerol	88%–95%
Ash	4%–6%
Ether extract	0.1%–0.4%
Water	1%–3%
Sodium	0.1%–4%
Potassium	0.1%–5%
Iron	7–11 mg/kg
Phosphorous	60–110 mg/kg

Source: Adapted from Thompson, J. C. and He, B., *Appl. Eng. Agric.*, 22, 261, 2006.

TABLE 4.3

Main Usages of Glycerol in Industry

Medical and pharmaceutical uses: Pure glycerol is widely used in pharmaceutical formulations. Due to its
properties, it is mainly used for improving smoothness, providing lubrication, and as a humectant found in
many cosmetic products where moisturization is desired, such as in moisturizing hair conditioners. It is used
in cough syrups, elixirs, expectorants, ointments, plasticizers for medicine capsules, ear infection medicines,
anesthetics, lozenges, gargles, and as a carrier for antibiotics and antiseptics.

Cosmetics and toiletries: Glycerol also can be used as an emollient in toothpaste, skin creams and lotions,
shaving preparations, deodorants, and makeup.

Production of several types of polymers: Glycerol is used as a component to produce polymers. For example,
polyglycerol ester is used in biodegradable surfactants and lubricants; polyglycerol and polyglycerol
methacrylates are used as wood treatments to improve its stability; and polyester polyols are used to produce
polyurethanes, which are applied as coatings, foams, and sprays.

Purifying crude glycerol is costly—a refinery can cost upward of $20 million—and generally not feasible for small-to-medium-scale biodiesel plants, which recover most of the expensive methanol from the glycerol layer, and are left with salt-grade bio-glycerol. On a global scale, a large amount of bio-glycerol thus is not upgraded and is usually burned. Perhaps not surprisingly, then, a number of grease and bio-glycerol leakages from biodiesel plant have been recently reported in the United States and in Europe, such as in the case of the Alabama river contaminated with 450 times higher than permit levels for two miles downstream in 2007.[12] A recent insight provided by an oleochemical industry insider sheds light on the problem[13]:

The problem with glycerol from biodiesel production is that it has heavy contamination from methanol. This makes it unsuitable to process for the glycerol consumer market. A few years ago the world glycerol market suffered a massive price slump as all of the biodiesel glycerol was coming on to the market. As it was starting to be used, it was discovered that it was unsuitable for most glycerol markets.

As a consequence of this, traditional glycerol is now undergoing a massive price correction due to global shortages. It is obviously a major focus of biodiesel manufacturers to produce a pharmaceutical grade glycerol. Unfortunately, high temperature low pressure distillation is the only way this can currently be done and any raw material that has been in contact with methanol is unsuitable for that type of process.

Each new generation of biodiesel plant is claiming that they have developed the technology for pharmaceutical glycerol production to encourage enthusiastic investors. To this day though, it isn't working. For that reason, I believe that a large portion of biodiesel glycerol goes into animal feed stock.

Traditional purification of raw glycerol employs high-temperature, vacuum-distillation technique as glycerol boils at 290°C. Basically, the purification process starts with neutralization by an acid, and then water and alcohol are removed to produce 80%–88% pure glycerol that is ready to be sold as crude glycerol.

The glycerol phase is typically (Figure 4.3) neutralized with hydrochloric acid, and the cationic component of the catalyst is incorporated as sodium chloride. The salt content in crude glycerol, stemming from the use of homogeneous alkaline catalysts, often ranges from 5% to 7% in weight. The process is energy intensive, and thus a cost-efficient refinery using distillation requires large-scale operation such as in the case of the new Cargill refining plant in Frankfurt, Germany, adjacent to its biodiesel production in the Höchst Industrial Park[14] with which this company has expanded its refining capacity to serve European refined glycerol demand and cope with excessive crude glycerol supplies coming from the booming European biodiesel industry.

New low-cost strategies for the purification of crude glycerol streams have been developed in the last 3 years, which do not require the vaporization of glycerol, and are either based on ion exchange

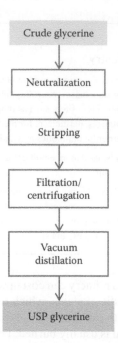

FIGURE 4.3   Common steps to obtain pure glycerol. (Adapted from www.eetcorp.com).

techniques tailored to the high salt content of bio-glycerol, or on a combination of electrodialysis and nanofiltration affording a colorless liquid with low salt content, equivalent to technical-grade purity.

Developed by Rohm and Haas and Novasep Process, the first method (*Ambersep BD50 process*) consumes small quantities of water for the separation of the salt fraction from glycerol and has a recommended minimum capacity of 5000 ton of pure glycerin per year.[15] Depending on the required purity of the final glycerol, it is possible to produce a purified glycerol product with 99.5% purity or to add a polishing step that enables the end user to produce a high-quality glycerol product with 5–10 ppm salt content.

Crude glycerol from storage is first heated to 90°C on a heat exchanger using energy recovered from the purified glycerol stream plus steam. After a safety filtration (to protect the downstream processing steps from fouling by suspended materials), the hot and clean crude glycerol is degassed before entering a chromatographic separator using a chromatographic separation resin to purify crude glycerol with a high salt composition. The raffinate stream contains the salts, other organic impurities (dyes and free fatty acids), and the minor fraction of glycerin not separated by the chromatography unit. The raffinate is thus further processed in an evaporator/crystallizer unit, affording the recovery of the salts and a "secondary glycerin" having a similar composition to the crude glycerin input, thereby avoiding production of glycerol effluents from the plant.

Due to the recycling of the condensates produced during the reconcentration of the purified glycerol fraction, no source of freshwater is necessary for the operation of the chromatographic separator. Water is used for the chromatographic separation, and therefore, the energy consumption is essentially limited to the removal of water from the purified glycerin after purification.

Commercialized by EET Corporation, a small company[16] based in Tennessee, another purification new process offering an alternative to vacuum distillation is the membrane-based technology (High Efficiency Electro-Pressure Membrane, *HEEPM*).[17] Suitable for refinement up to USP grade (99.7% purity), the process has several competitive advantages over distillation technologies (Table 4.4) as it avoids many of the issues associated with evaporation and distillation such as foaming, carryover of contaminants, corrosion, limited recovery, and high capital, energy, and operating costs.

Glycerol purification process begins with pretreatment of the glycerol to remove any solids and fouling organics and partially remove color-causing organics, affording a colorless liquid with low salt content (Figure 4.4).

TABLE 4.4

Main Features of Vacuum Distillation and Membrane-Based Nanofiltration Purification Techniques

Vacuum Distillation	EET Process
Well established technology for glycerol	Production proven technology for glycerols/glycols
Low recovery/yield	95% + recovery
High capital cost	Lower capital cost
Energy intensive	Low energy use
Only suited to operations >30	Scalable process
Tons/day	Low maintenance
High maintenance	Tolerates feed stream variations
Sensitive to feed stream variations	

Source: Adapted from www.eetcorp.com

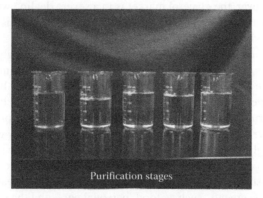

Purification stages

FIGURE 4.4 Purification stages of the HEEPM process. (Adapted from www.eetcorp.com).

The process takes advantage of the respective differences and advantages of electrodialysis and nanofiltration (or reverse osmosis) by combining both processes to optimize the separation characteristics of each. Use of the integrated *HEEPM* approach overcomes inherent limitations of both membrane-based technologies, allowing these to operate optimally and economically, achieving improved efficiencies and product recovery for varying feed chemistries. In addition to chloride and sulfate containing streams, other biodiesel-derived glycerol feeds containing phosphates (potassium or sodium), mixtures of sulfates and chlorides, and mixtures of organic acids (e.g., acetic acid) and inorganic acids (chlorides) have been successfully purified to meet USP 99.7 specifications.

Increased contaminant loading in the feed will only affect production rate and operating cost. Crude glycerol streams with high FFA and MONG content (such as from biodiesel plants using feedstocks from rendering plants and grease traps) are more difficult to pretreat compared to those from plants processing soybean or rapeseed oil, and require further pretreatment to allow optimum separations.

4.4 ECONOMIC ASPECTS OF BIO-GLYCEROL

As mentioned earlier, for decades, the glycerol market was mainly devoted to *nonchemical*, direct and highly profitable commercial applications, with only two major chemical usages, namely, in the manufacture of nitroglycerine and in alkyd resin production. Hence, prior to the biodiesel boom, the economic value of glycerol was an important part of the profitability of the soap and oleochemical

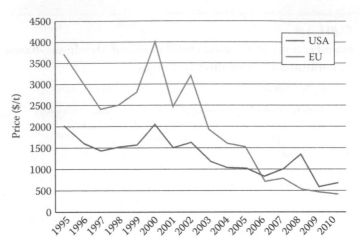

FIGURE 4.5 The change of the market price of 99.5% kosher-grade glycerol in the United States and Europe, 1995–2010. (Realized by the authors using data from ICIS pricing, 2010.)

industry. Since 2003, however, the glycerol market is in a severe decline with most key players having already pulled out of the synthetic glycerol market. In Japan, for instance, the main glycerol production factories ceased operations in October 2005.

In practice, oversupply of bio-glycerol has caused collapse of crude glycerol price almost to $0 per ton (Figure 4.5), that is, it has become a waste product. At the same time, oversupply has caused a dramatic fall also in the price of pure glycerol (99.5% kosher grade). In detail (Figure 4.6), the price for refined glycerol decreased from about €4000/ton in year 2000 to less than €450/ton in early 2010, when the price of crude glycerol was below $50/ton.[18] Clearly, the prolonged high level of the price of crude oil ($80/barrel despite the recession) coupled with very low glycerol prices is opening up new opportunities for glycerol consumption by substitution, especially in China and in Southeast Asia that are emerging as an important market for glycerol suppliers worldwide due to economic growth and rising living standards of Asian consumers, leading to increased demand from different key end-user industry segments, such as food, personal care, and the pharmaceutical industry.

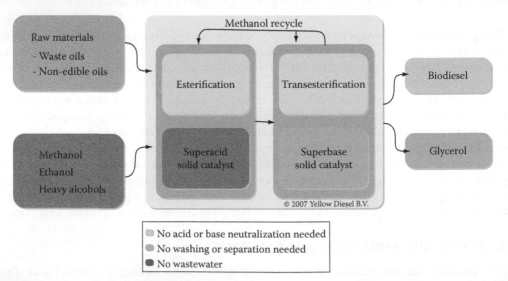

FIGURE 4.6 Structure of a biodiesel-manufacturing process over solid catalyst yielding superior quality biodiesel and pure glycerol, ready for value-added application. (Reproduced from Yellow Diesel B.V., Amsterdam, the Netherlands, www.yellowdiesel.com. With permission.)

TABLE 4.5

**Most Important Factors That Contribute
to the Cost of Biodiesel Production**

Voice of Cost	Percentage of Plant Operation Costs
Feedstock and meal prices	64
Capital costs	20
Electricity	33–50

Source: Adapted from Zhang, Y. et al., *Bioresour. Technol.*, 90, 229, 2003.

Results from a recent cost-benefit analysis clearly indicate that a biodiesel plant requires a glycerol purification unit to be economically viable,[19] as glycerol recovery leads to increased profitability by decreasing the cost of wastewater treatment and by revenues generation due to sale of refined glycerol. Yet, the influence of the oil price on the biodiesel price is larger than the influence of the glycerol price. In detail, an increase of $0.01/kg in waste oil price would cause an increase of $0.02/kg in the biodiesel break-even price, whereas a reduction of $0.01/kg in the biodiesel break-even price requires an increase of $0.06/kg in the price of crude glycerol (85 wt.%).[20]

Seemingly large fluctuation in glycerol price has a relatively small impact on the biodiesel break-even price. The majority of biodiesel production input costs indeed come from feedstock followed by the cost of electricity used to extract the oil from oilseeds (Table 4.5).

This means that the biodiesel industry will be more focused on low feedstock prices than in finding alternative usages for glycerol. Cities such as Kyoto in Japan or Graz in Austria, for example, use biodiesel to fuel garbage trucks and city buses, in which the feedstock is waste vegetable oil separately collected. Similarly, in the United States, even if in 2009 the federal tax credit subsidizing biodiesel was withdrawn, one producer in the State of Washington continued to flourish since it uses grease (waste cooking oil) rather than virgin oil, focusing on energy efficiency.[21] In the words of the company's head the company focuses on "recycling more, having no by-product waste and recovering all the heat and all the chemicals and resusing it over and over again." This demonstrates that by reducing its costs, the biodiesel industry can survive without subsidies, even if a 2003 study[22] reviewing 12 economic feasibility studies concluded that, compared to diesel, biodiesel was not yet economically viable. With pre-tax diesel priced at US$0.18 per liter in the United States and US$0.20–0.24 per liter in some European countries, projected costs of biodiesel from vegetable oil and waste grease in 2003 were, respectively, US$0.54–0.62 and US$0.34–0.42 per liter.

What is required to achieve consistent cost reduction is the employment of new biodiesel-manufacturing technologies based on *heterogeneous* catalysis capable to reduce chemical and energy input costs and consumption, lower the cost of fuel washing, and at the same time improve glycerol profits.[23]

By using a solid acid or base catalyst, for example, the Dutch company Yellow Diesel is able to make high-quality biodiesel and glycerol of almost refined quality that can be used for the cosmetics and food industry (Figure 4.6).[24] The solid catalyst does not leach and the reactant mixture is efficiently converted into FAME meeting the European biodiesel standard (EN 14214). No expensive neutralizations, washing, and separation steps are required, and the process can be conducted continuously in smaller plants of the same efficiency. Overall, the new process allows 40% reduction in the investment cost in biodiesel production installations and 30% decrease in the operational cost. Similarly, one of us has had the good fortune to review the 2011 study[25] where high-yield homogeneous production of biodiesel at 20°C (instead than at 60°C) and using only a 4:1 methanol: oil excess (instead than 6:1) and facilitated biodiesel/bio-glycerol separation is described.

We argue here that generous tax credit and subsidies to the biodiesel (and to agriculture) companies in practice have *delayed* innovation allowing industry to postpone costs reduction and use cumbersome, homogeneous chemical conversion to manufacture a fuel whose advantages are clear. Biodiesel in fact is an excellent alternative source of energy replacing diesel fuel key role in transportation, and in agriculture whose production provides good revenues to farmers and whose use decreases air pollution emissions benefiting society while it does not require engine upgrade, thanks to its excellent fuel characteristics. Further, it is biodegradable, nontoxic, and essentially free of sulfur and aromatics; it has a relatively high flash point (150°C), which makes it less volatile and safer to transport and handle than petroleum diesel. Its lubricating properties can reduce engine wear and extend engine life.

The results of a study,[26] which includes the energy costs of farm machinery and processing facilities, together with recent advances in crop yields and biofuel production efficiency, refute previous assertions that biofuel requires more energy to produce than it yields. Hence, relative to ethanol, biodiesel releases just 1.0%, 8.3%, and 13% as much agricultural nitrogen, phosphorus, and pesticide pollutants, respectively, per net energy gain. Biodiesel also releases less air pollutant per net energy gain than does ethanol. Thus, relative to the fossil fuels they displace, greenhouse gas emissions are reduced 41% by the production and use of biodiesel, compared with 12% for ethanol. The advantage of biodiesel over ethanol arises from its lower agricultural input and the more efficient conversion of its feedstocks to fuel. In some countries, filling stations sell biodiesel for less than conventional diesel (Figure 4.7).

Governments in Europe and the United States provide a subsidy in the form of tax exemption for each liter of biodiesel produced. Furthermore, in the EU, biodiesel producers also benefit from crop subsidies, which give lower soybean prices. The main rationale in subsidizing the biodiesel industry is to support the farmer, reduce unemployment, prevent neglect of the countryside, and, in democracies, to win the votes of the beneficiaries. Starting in 2008, some of these credits have been cancelled and in general reduced. Yet, biodiesel manufacturing has continued to flourish, despite a 50% reduction in U.S. production. Biodiesel production increased 9% in 2009 to 16.6 billion liters globally.[27]

FIGURE 4.7   In countries such as Germany where biodiesel receives generous tax credits, biodiesel at filling stations sells for less than conventional diesel. (Photo courtesy of Wikipedia.)

This compares to a 5-year average (end-2004 through 2009) growth of 51%. The EU represents nearly 50% of total output in 2009, when production increased less than 6%, down from 65% growth in 2005. According to detailed analysis of REN21, at least half of the EU and U.S. existing plants remained idle during 2009. Increasing production by 34% during 2009 to surpass Germany as both the European and world leader, France produced more than 2.6 billion liters, or 16% of global biodiesel. For details, see Ref. [27]. It is remarkable, however, that biodiesel production is far less concentrated than ethanol, with the top 10 countries accounting for only 76% of total production in 2009. Significant expansion occurred in Argentina, Colombia, Indonesia, Brazil, China, Malaysia, and Thailand that are now among in the top 15 of biodiesel producers. India, which ranked 16th in 2009, increased production more than 100-fold to over 130 million liters.

In this context, despite repeated claims that the downward trend of the glycerol market price is "now over," this wishful scenario thus far has not happened, and—we argue in this analysis—it is not likely to happen in the near or medium term, even if demand for crude glycerol will expand following new chemical productions using bio-glycerol as raw material as well as to increasing glycerol demand in China, India, and Southern America where growth in the living standards is causing increased demand.

For example, according to chemical market intelligence service ICIS, U.S. prices for refined vegetable glycerol were "ready for takeoff" being assessed at 26–32 cents/lb ($573–705/ton), and at 29–34 cents/lb for tallow glycerol.[28] For comparison, in 2007, the latter price was from 29 to 35 cents a pound.[29] It is instructive to report the claims of the "glycerol sellers" such as the following taken from the ICIS Web site to learn how this market is dominated by scarcity of information:

> American glycerol sellers expect price increases of 10 dollar cents/lb (157 euro/ton) in the next few months and this will attract imports from Asia and Europe where glycerol is priced lower.[27]

Actually, the crude glycerol production almost stopped in the United States in 2009 and yet the price of crude glycerol in January 2010 was 5 ct/lb! Such exceedingly low price of crude glycerol does not justify anymore transport costs (roughly $80–$100/ton from Europe to the United States) for exporting the crude. Accordingly, Figure 4.8 clearly shows that imports of refined glycerol in the United States from Germany suddenly stopped in 2007.

To summarize, the fundamental restraint on the glycerol market is and will remain oversupply. From estimates by the European Biodiesel Board, 4 billion liter of biodiesel will be produced in Europe only in 2010, with a 10% increase over 2009 despite the serious economic recession started in 2008. This equates to some 0.4 Gl of crude glycerol, namely, a glut that requires either glycerol recovery and refining in biodiesel plants or new value-added chemical conversions of bio-glycerol.

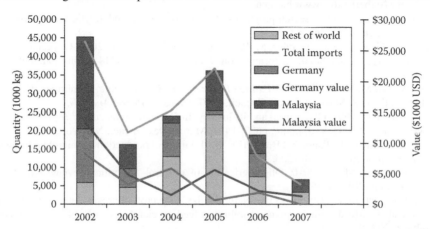

FIGURE 4.8 Crude glycerol imports by country while the overall imports of refined glycerol generally between 45 and 25 kton a year almost ceased between 2005 and 2007. (Reproduced from ABG Inc. for the U.S. Soybean Export Council, Glycerol market analysis, Adayana Agribusiness Group, 2007. With permission.)

4.5 CONCLUSIONS

In 2006, the top three global glycerol suppliers were Procter and Gamble, Cognis, and Uniqema that, combined, had more than one-third of the market share.[30] Currently, biodiesel companies are the leading glycerol producers, with Europe and Asia hosting the majority of the overall production. As of today (December 2010), the price of crude glycerol is 5 cents/lb in the United States (and €50/ton in Europe), namely, low enough to justify switching from petrochemical to bio-based raw materials for a number of chemical products. Along with consistently low prices, the other determining factor for the chemical industry is availability, which is now guaranteed.

Thurmond in 2008 forecasted that by year 2010, more than 200 nations would become biodiesel-producing nations and suppliers.[31] The earnest economic recession started in September 2008 falsified this prophecy. Yet, the number of nations producing biodiesel has certainly increased and now includes large countries such as Argentina and many new African and Asian producing nations. Hence, chemical companies can now safely and profitably start massive production of chemicals using glycerol as raw material. ADM, for example, recently announced the start of production of propylene glycol (PG) through glycerol hydrogenolysis in a U.S. plant capable to process 125,000 ton/year of crude glycerol.[32] Similarly, production of epichlorohydrin in China by Dow is expected to come online in 2012.[33]

Until now, the impact of commercial processes actually using bio-glycerol as raw material was surprisingly modest compared to both the global surplus of glycerol as well as the number and scope of the scientific advances dealing with glycerol new conversions of the last decade.[4] Now, however, the structural economic reasons that caused little practical innovation are over. Many new bio-refineries, plants in which bio-glycerol is used as a primary building block, replacing raw materials of petrochemical will rapidly emerge in many of the countries where older biodiesel plants using homogeneous manufacturing process exist. In this report, we have provided economic and technical arguments that support this scenario.

ACKNOWLEDGMENTS

We gratefully acknowledge the contributions of our coworkers Rosaria Ciriminna and Cristina Della Pina.

REFERENCES

1. BioMcN (the Netherlands). www.biomcn.eu
2. Successful Arkema and hte research project in glycerol to acrolein and acrylic acid conversion», press release from Arkema, December 3, 2009.
3. Melero, J. A., G. Vicente, G. Morales, M. Paniagua, and J. Bustamante. 2010. Oxygenated compounds derived from glycerol for biodiesel formulation: Influence on EN 14214 quality parameters. *Fuel* 89, 2011–2018.
4. Pagliaro, M. and M. Rossi. 2010. *The Future of Glycerol*. RSC Publishing, Cambridge, U.K.
5. ABG Inc. for the U.S. Soybean Export Council. 2007. *Glycerol Market Analysis*.
6. Neumann, W. H. C. 1991. Glycerin(e) and its history. In *Cosmetic Science and Technology Series*, eds. E. Jungermann and N. O. V. Sonntag, pp. 7–1411. Marcel Dekker Inc., New York.
7. Wang, Z. X., J. Zhuge, H. Fang, and B. A. Prior. 2001. Glycerol production by microbial fermentation: A review. *Biotechnol. Adv.* 19: 201–223.
8. Richter, M. and A. Martin. 2010. Oligomerization of glycerol—A critical review. *Eur. J. Lipid. Sci. Technol.* DOI: ejlt/201000386.
9. AllAboutFeed.Net (Reed Business, the Netherlands). Glycerol improves feed efficiency in cattle. http://www.allaboutfeed.net/news/glycerol-improves-feed-efficiency-in-cattle-1439.html (accessed on December 8, 2007).
10. Thompson, J. C. and B. He. 2006. Characterization of crude glycerol from biodiesel production from multiple feedstocks. *Appl. Engineer. Agric.* 22: 261–265.

11. Bogdanich, W. and J. Hooker. May 6, 2007. From China to Panama, a trail of poisoned medicine. *The New York Times*.
12. Goodman, B. 2008. «Pollution is called a byproduct of a 'clean' fuel», *The New York Times*, March 11, 2008.
13. The quotation is taken from a conversation of Luke Hallam with a blog reader who works in the oil industry in Australia: http://envirofuel.com.au/2008/01/23/the-impact-of-biodiesel-on-the-glycerole-and-tallow-markets (last time accessed October 14, 2008).
14. D. de Guzman, Green news from Cargill. www.icis.com/blogs/green-chemicals/2009/06/green-news-from-cargill.html (accessed on June 8, 2009).
15. Lancrenon, X. and J. Fedders. 2008. «An innovation in glycerin purification», *Biodiesel Magazine*, June 2008. See also at http://www.amberlyst.com/glycerol.htm (accessed on August 8, 2011).
16. EET corporation. www.eetcorp.com
17. Sdrula, N. 2010. A study using classical or membrane separation in the biodiesel process. *Desalination* 250: 1070–1072.
18. ICIS news: US glycerine prices set to skyrocket on low biodiesel production. http://www.icispricing.com/il_shared/Samples/SubPage99.asp (accessed on January 6, 2010).
19. Singhabhandhu, A. and T. Tezuka. 2010. A perspective on incorporation of glycerin purification process in biodiesel plants using waste cooking oil as feedstock. *Energy* 35: 2493–2504.
20. Zhang, Y., M. A. Dubé, D. D. McLean, and M. Kates. 2003. Biodiesel production from waste cooking oil: 2. Economic assessment and sensitivity analysis. *Bioresour. Technol.* 90: 229–240.
21. See Baskin, M., «Small Seattle biodiesel producer plans to expand. Secret? Fast-food deep fryers», *Seattle Post Globe*, June 1, 2010.
22. Demirbas, A. 2003. Fuel conversional aspects of palm oil and sunflower oil. *Energ. Source* 25: 457.
23. Kotrba, R. 2010. The dreaded 'what if' question. *Biodiesel Magazine*, July.
24. Yellow Diesel, B. V. www.yellowdiesel.com.
25. Maeda, Y., L. T. Thanh, K. Imamura, K. Izutani, K. Okitsu, L. Van Boi, P. N. Lan, N. C. Tuan, Y. E. Yoo, and N. Takenaka. 2011. New Technology for the production of biodiesel fuel, *Green Chem.* 13: 1124–1128.
26. Hill, J., E. Nelson, D. Tilman, S. Polasky, and D. Tiffany. 2003. Environmental, economic, and energetic costs and benefits of biodiesel and ethanol biofuels. *Proc. Natl. Acad. Sci. USA* 103: 11206–11210.
27. Martinot, E. (lead author) 2010. Renewables 2010. Global status report, REN 21 Renewable energy policy network for the 21st century. Last time accessed December 15, 2010.
28. Lefebvre, B. and J. Taylor. 2010. «Glycerol prices ready to rocket, fatty alcohols and acids see upward pricing pressure». *ICIS Chemical Business*, January 18, 2010. www.icis.com/Articles/2010/01/25/9326793/us-glycerol-prices-ready-to-rocket-fatty-alcohols-and-acids-see-upward-pricing.html
29. Taylor, J. 2007. «US glycerol makers seek to use market changes», *ICIS News*, March 7, 2007. www.icis.com/Articles/2007/03/07/9011588/us-glycerole-makers-seek-to-use-market-changes.html
30. Frost & Sullivan. 2006. What is the global market outlook for glycerol in 2006. http://www.frost.com/prod/servlet/market-insight-top.pag? docid=70283330 (accessed on August 8, 2011).
31. Thurmond, W. 2008. *Biodiesel 2020: A Global Market Survey*, 2nd edn. Emerging Markets Online (Houston, TX) www.emerging-markets.com/biodiesel
32. Archer Daniels Midland Company, ADM: Evolution Chemicals. Propylene Glycol, http://www.adm.com/en-US/products/evolution/Propylene-Glycol/Pages/default.aspx (accessed on July 27, 2010).
33. Kee, W. F. (Frost & Sullivan). 2010. «Market and technology foresights for the oleochemical industry and market outlook for the Southeast Asia», June 1, 2010. www.slideshare.net/FrostandSullivan/market-and-technology-foresights-for-the-oleochemical-industry-and-market-outlook-for-the-sea-region

Part II

Plastics and Materials from Renewable Resources

5 Developments and Future Trends for Environmentally Degradable Plastics

Emo Chiellini and Andrea Corti

CONTENTS

5.1 INTRODUCTION

5.1.1 PETROPOLYMERS AND BIOPOLYMERS: REMARKS ON ENVIRONMENTAL IMPACT

The worldwide consumption of polymeric materials and plastics rises annually by around 7%–10% with approximately 300 million ton in the year 2010, which corresponds nearly to 30 kg per capita with an average of 80–100 kg in industrialized countries and 2–20 kg in emerging countries and countries in transition (Van Os 2001).

More than 99.5% of plastics are based on fossil feedstock (crude oil), the reserve of which is predicted to last for only approximately 80 more years (EC-Funded Project "Packtech" 2002).

Public concern about the environmental consequences bound to production and consumption of various materials and products is increasing. These effects occur at every stage in a product's life cycle—from the extraction of the raw materials through the processing, manufacturing, and transportation phases, ending with use and disposal or recycling (Various web sites 2006).

At the occasion of World Conference on Ecology 1992 in Rio de Janeiro, the United Nations Framework Convention on Climatic Change was signed. With the Kyoto Protocol in 1997, industrialized nations undertook the initiative of reducing greenhouse gas (GHG) emission at least 5% below the amount of 1990. The EU committed itself to cutting GHG emissions to 8% below the level quoted by the year 2008 (Ellerman and Buchner 2007).

In 2007, the consumption of polymeric materials for plastic applications in Western Europe has been above 65 million ton. Many of plastic applications involve a service time of less than 1 year, after that the vast majority of these plastics are discarded to become a waste. In case fossil-based plastics would be replaced by starch-based plastics, a reduction of 0.8–3.2 ton CO_2 per ton material is expected (Sperling and Carraher 1988). It is, however, to remark that either the conversion of starch directly to plastic items or to monomeric building blocks convertible to polymers and hence plastics will interfere with food chain. Nowadays, with the increasing worldwide population by 250,000 individuals/day and hence with the increasing of food and feed, ethical reasons should lead to an halt in the utilization of food chain resources for the production of fuel and plastics.

Recent EU legislation increasingly requires the recovery of plastic waste through recycling, composting, or energy recovery. While the plastic industry mostly recycles its own in-plant scraps, since commercial-scale plastics processing began many years ago, post-consumer plastics recycling is still very limited. Moreover, for recycling to offer true environmental benefits, whatever the material involved, a number of factors should be taken into account. Most broadly, the manpower and economic resources used to collect, sort, and recycle must be less, or at least comparable, to those used to produce virgin materials. Sufficient demand from end-market users for the recycled material is also a vital prerequisite, and a key factor for marketing these products is that the use of recycled materials does not compromise product safety or performance (Van Os 2001).

Organic recycling is a specific recycling option for biodegradable waste, for example, for biodegradable or compostable materials. It diverts biodegradable waste from landfills, preventing emissions of methane that represents a very powerful GHG generated in anaerobic conditions by landfills. Composting technologies, used for the disposal of food and yard waste accounting for 25%–30% of total municipal solid waste, should be suitable for the disposal of biodegradable/compostable plastics from renewable resources as well as plastics from fossil feedstock, together with soiled or food-contaminated paper. Indeed, it has to be taken into account that the composting processes is claimed as "microbial combustion" (Hellmann et al. 1997), thus eventually providing a fast release of CO_2 and other GHG with respect to the timescale of natural carbon cycle.

The term "biopolymers," and hence the converted plastic items (*bioplastics*), refers to natural products that are polymeric in character as grown or directly convertible to plastic items without any significant chemical modification. To the best of our knowledge, this may occur only in the case of microbial polyesters (Bozell 1991). Thus, under that heading one can include *natural polymers* (proteins, polysaccharides, and microbial polyesters). Natural polymers that are chemically or enzymatically modified including also polymers obtained by conventional polymerization procedures of monomeric precursors derived from the natural polymers can be classified as *artificial polymers* (Van Wyk 2001). These last would be better classified as *bio-based polymeric materials* and the relevant manufactured items as *bio-based plastics* (Scheme 5.1) (Young 1997, Chiellini et al. 2004a).

Total demand for biodegradable polymers in the United States, Europe, and Japan was 20 kton in 1998 valued at $95 millions. The European market for bio-based plastics, resulting from the fruits of 20 years of technological development, is growing at very slow pace, not commensurable with practical impact and industrial efforts spent for their takeoff. According to the industrial association IBAW, the bio-based plastics usage in 2004 amounted to 50 kton. Compostable rubbish bags and starch-based loose packing material constitute the major part.

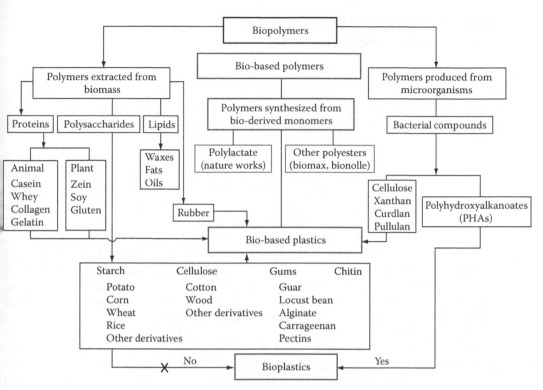

SCHEME 5.1 Main options for the production of bio-based plastics from biopolymers and bio-based polymers. A single option for bioplastics.

The IBAW estimates that 1/10th of all plastics applications in Europe could be catered for by modern bio-based plastics, a very optimistic figure that corresponds to approximately 5 million ton of polymers. However, 2006 worldwide production capacity only amounts to 150 kton (Various web sites 1995), and it is expected, by the very optimistic projectionists, to grow up to 2 million ton by 2011 that is less than 1% of the overall present production of plastics (Tavares and Marreiros, 2009).

Traditional natural polymeric materials, as reported in Scheme 5.1, are represented by polysaccharides (cellulose, starch, chitin, alginic acid, ulvans, xanthanes, and guar gum), proteins (fibroin, keratin, collagen, polynucleotides, RNA, and DNA), natural rubber, lignins, and vegetable oils binders. Fibrous material derived from renewable crops, by-products, or their industrially processed wastes can be considered a good source for the formulation of polymeric blends and composites based either on synthetic and natural components (*hybrids*) and relevant manufactured items will be identified as *bio-based plastics* for quick and meaningful identification (Scheme 5.1) (Chiellini et al. 2004a).

Herewith, plastic items identified as *bio-based plastics* are usually in an ambiguous manner classified as "*bioplastics*," and we again stress that *bioplastics* can be identified only as plastic items obtained by direct conversion of virgin natural polymers. If we exclude the potential conversion of microbial polyhydroxy alkanoates (PHAs) to relevant plastic items, the so-called bioplastics remain as chimeric items floating in the ocean of petrochemical-based plastic items.

Biopolymers, not chemically modified, without producing any net increase in atmospheric CO_2 balance as part of a sustainable material cycle. It is, however, to remark that the relevant bio-based plastic items may result in an unpair final CO_2 balance connected to the chemical modification done on these items either prior or during their processing to plastic items.

5.2 PETROPOLYMERS AND BIOPOLYMERS: ECONOMICAL CONSIDERATIONS

5.2.1 BIO-BASED POLYMERS AND PLASTICS

In contrast with the terminology used in our chapter of the first edition of the book *Renewable Resources and Renewable Energy—A Global Challenge*, in the present contribution we have banned the ambiguous and misleading term "bioplastics" and substituted with the right appropriate "bio-based" plastics.

In industrial production, sustainability must be achieved, keeping in mind that business will fail unless a minimum margin profit is guaranteed. New bio-based plastics are claimed as more appropriate options where the degradation constitutes a plus in specific applications by defraying the cost inherent in the management of the disposal of post-consumer items (Doane 1994, Heyde 1998). Plastic articles that are used once and then disposed off are targeted as the primary market areas such as food packaging films, foams and bags, food service items (containers of milk, water, and soft drinks), mulching films, and transplanting (Mayer and Kaplan 1994). It is, however, taken for granted that infrastructures have to be available to allow for their biorecycling as a final stage of their recovery. Bio-based polymeric materials and relevant plastic items have been claimed as achievable at a fairly low cost in any country, in contrast to petroleum-based materials, and hence improperly economically sustainable. A normative strategy has been proposed for resource choice and recycling to meet the criteria of sustainability (Reijnders 2000). The use of agricultural and forestry biowastes as resources not interfering with food chain could add value to bio-based plastic productions that has been proposed because of their high content of cellulose, hemicellulose, and lignin that can be converted to monomeric building blocks or be used as fibrous fillers in bio-based composites (van Wyk 2001).

A wide range of biodegradable bio-based plastics are under development (Weber and Denmark 1995), and if they are disposed in anaerobic digestion or incinerated in combustion plants, their "natural stored energy" can be utilized for producing heat and electricity.

For example, the contents of energy in important agricultural plants are reported in Table 5.1 (IFEU/BIFA 2003).

By the way, the availability of raw materials from renewable resources, as already stated, should not interfere with food chain productions (Scott 2000). Only 4% of the global production 7.3 Gton of oil is used for plastics accounting for more than 250 Mton annually (Roper 2002). The limited amount of crude oil resources consumed for plastic production when compared with the amount utilized as combustible for energy generation (93% of total fuel feedstock) contributes in an extremely relatively modest extent to the GHG effect. So even though one may imagine to replace all the fossil fuel feedstock–based plastics with those derived from renewable resources, the effect on GHG reduction would be minimal and eventually disputable depending upon the life cycle assessment (LCA) balance that might be heavily affected by the conversion of the biopolymers or biowaste to building blocks and by the further processing to polymers and plastic items.

TABLE 5.1
Energy Content in Agricultural Plants

Plant	Energy[a] (MJ/kg)
Wheat	15.9
Barley	15.9
Potato	14.4
Maize-corn	15.3

[a] Referred to dry material.

However, the crude oil consumption devoted to plastic production represents indeed a small amount when compared to the available biomass that could be used for bio-based plastic items production. The estimated annual production of biomass by biosynthesis is 170 Gton, only about 6 Gton (3.5%) are used by man worldwide (Roper 2002). The global production of crop by-products was, for instance, 2.7 Gton in the year 2000, and after a decade we can assume that to be above 3.0 Gton. The production of specific, pure biopolymers is significantly lower, with the exception of cellulosic fibers and starch. The total potential mass of starch from cereals is about 1.2 Gton annually. Considering that cereal carbohydrates are currently used for food and feed, the amount available for material application should be increasingly reduced and finally banned. But even a small fraction of the total production biowaste yields considerable amounts, which even though could not compete with commodity petropolymers, they could however compete with specialty polymers, provided adequate performances might be exhibited by the derived plastic items. Nowadays, this represents one of the major limiting factor for a wide practical and competitive spreading of bio-based plastics.

In this respect, increasing attention and efforts by academic and industrial researchers are nowadays devoted to a possible use of renewable feedstocks both as energy source (Patel 1999, El Bassam 2001) and as raw materials for economically competitive production processes of building blocks and hence relevant polymeric materials and plastics (Rowell et al. 1997, Warwel et al. 2001).

The acceptance for nonfood use differs widely, with the EU having a surplus of crops and the developing countries needing all edible biopolymers primarily for food and feed. Starch costs about €0.3–0.4/kg and might be available in large amounts in pure form, and their use as raw materials for production of plastic items as those reported in Table 5.2 should be banned.

Despite the relatively low raw material price for starch, there is no starch-based plastic available on the market for less than €3–4/kg. The high conversion cost for starch is remarkable and this fact combined with the fact that the starch-based products are indeed blends with synthetic petro-based polymers aliphatic like polycaprolactone or bionolle or aliphatic-aromatic polyesters such as Ecoflex and Ecovio (BASF), Eastarbio (Eastman Chemical), and Biomax (Dupont). Other edible biopolymers such as gelatin, alginate, and xanthan all cost €3–10/kg (Narayan 1992).

Several corporations are also initiating large-scale production of polylactic acid (PLA). In Table 5.3 are collected the companies declaring a capacity for PLA production higher than 50 kton. The PLA is produced by a combination of biotech and chemical process, and the derived plastic items can be identified as *bio-based plastics*. From starch or sugar is attained L-lactic acid that is afterward submitted to polymerization, according to conventional procedures, leading to PLA with an estimated price of €1–2/kg when it is produced on a large-scale plant with a capacity higher than 50 kton. According to a recent report (Patel 1999) on the production of PLA, there are various companies all over the world that declare high production capacity (Table 5.3), but so far the PLA availability in the market

TABLE 5.2

Examples of Starch-Based Biopolymers: Producers and Brand Names

Company	Brand Name
Biotec GmbH	Bioplast
VTT Chemical Technology	COHPOL
Groen Granulaat	Ecoplast
Japan Corn Starch Ltd. and Grand River Technology	Evercom
Novamont SpA	MaterBi
Starch Tech.	ReNEW
Supol GmbH	Supol
Novon international	Novon

TABLE 5.3

Major Producers of Lactic Acid and Poly(Lactic) Acid

Company Name	Raw Material	Products	Trade Names	Capacity 2009 (kton)
NatureWorks LLC (USA)	Maize, cassava, rice, lignocellulosics from corn stover	PLA polymer	NatureWorks Ingeo™	150
PURAC (Gorinchem, NL)	Tapioca starch, sugarcane	Lactic acid, lactates, lactides monomers	PURALACT	200
Teijin (Tokyo, Japan)	Corn, sugarcane	Heat-resistant fibers	BIOFRONT™	200
Pyramid Bioplastics (Germany)	Sugar beet, sugar cane	D-PLA and L-PLA polymer	—	60
Tate & Lyle (London, U.K.)	Sugarcane	PLA for biomaterials	—	45
Toyobo Biologics	—	PLA polymer	Bioecol	200
Tong-Jie-Liang Biomaterials Co., Ltd. (China)	Starch	PLA resin	—	100

is disputable. This can be compared with the price for conventional plastics in bulk at €0.5–1/kg (polyethylene, PE; polyvinylchloride, PVC; polystyrene, PS; etc.), and €2–4/kg for poly(vinyl alcohol) (PVA). It remains however a major issue bound to its processability that requires speed accuracy.

Among the candidates for biodegradable plastics, poly(hydroxyalkanoates) (PHAs) have been drawing much attention because they can be produced from renewable resources as well as from petrochemical components, have similar properties to conventional plastics specifically with polypropylene, and are considered completely biodegradable (Mergaert et al. 1992, Steinbüchel et al. 1992, Hocking and Marchessault 1994, Lenz et al. 1994, Lee 1996, Steinbüchel and Füchtenbusch 1998). The fairly high production cost of PHAs makes them substantially more expensive than synthetic commodity plastics (Steinbüchel et al. 1992). The cost of carbon substrate significantly affects the overall economics in large production scale (Yamane 1992, Choi and Lee 1997). Therefore, the production cost can be considerably lowered when different kinds of biowaste (whey permeate (EC-Funded Project "Wheypol-G5RD/CT2001/00591" 2001)) and wastewater from olive oil production (EC-Funded Project "Polyver-COOP/CT2006-032967/" 2006) are used, as highlighted by Reddy et al. (2003) and recently experienced in EU-funded projects. The production costs of PHA by using the natural microbial strain *A. eutrophus* are US$16 (€12) per kg which is 10 times more expensive than polypropylene. With recombinant *E. coli* as PHA producer, price can be reduced to US$4/kg (€3.1), which is close to other biodegradable plastic materials such as PLA and aliphatic polyesters based on aliphatic diols $(C_2–C_4)$ and linear aliphatic diacids $(C_4–C_6)$. In Table 5.4 are collected the producers of PHAs.

The effect of substrate cost and yield on the production cost of poly(3-hydroxybutyrate) (PHB) was reviewed by Madison and Huisman and relevant data are collected in Table 5.5 (Madison and Huisman 1999).

At the 227th ACS National Meeting 2004, it was reported by Bohlmann an estimated PHB production cost of $4/lb (€7.0/kg) for the Zeneca process, as based on glucose, propionic acid mixed feed, with wastewater by enzyme permeabilization (Bohlmann 2004).

Commercialization of plant-derived PHA will require the creation of transgenic crop plants (Snell and Peoples 2002). Production of PHA on an agronomic scale could allow synthesis of biodegradable plastics in the million ton scale compared to fermentation, which produces material in the thousand ton scale. PHA could potentially be produced at a cost of US$0.20–0.50/kg (€0.15–0.40) if they could be synthesized in plants to a level of 20%–40% dry-weight and thus competitive with the petroleum-based plastics. This production even though very much intriguing for its positive economic impact expectations has not taken off. Monsanto company that has

TABLE 5.4
Commercially Interesting PHAs and Recent Commercialization Development

Company	Location	Raw Material	PHAs	Trade Name	Stage/Scale
Tianan	China	Corn sugar	P(3HB-co-3HV)	Enmat	Industrial
Telles	USA	Corn sugar	PHB copolymers	Mirel	Industrial
Kaneka	Japan	Vegetable oil	P(3HB-co-3HHx)	Kaneka	Pilot and industrial
Green Bio/DSM	China	Sugar (unspecified)	P(3HB-co-4HB)	Green Bio	Pilot and industrial
PHB Industrial	Brazil	Cane sugar	P(3HB)	Biocycle	R&D, pilot and industrial
			P(3HB-co-3HV)		
Biomer	Germany	Sugar (sucrose)	P(3HB)	Biomer	R&D, pilot
Mitsubishi Gas Chemical	Japan	Methanol (from natural gas)	P(3HB)	Biogreen	R&D, pilot
Biomatera	Canada	Sugar (unspecified)	P(3HB-co-3HV)	Biomatera	R&D, pilot
Meredian	USA	Corn sugar	P(3HB) copolymers	Nodax [1]	Industrial
			P(3HB-co-4HB)		
Tepha	USA	n/a	P(4HB)	Tephaflex	R&D, pilot
				TephElast	
Tianzhu	China	Bamboo	P(3HB-co-3HHx)	Tianzhu	R&D, pilot
P&G	USA	Sugar	P(3HB-co-3HO)	Nodax	Stopped in 2006
			P(HB-co-3Hod)	Nodax	

TABLE 5.5
Poly(3-Hydroxybutyrate) (PHB) Production Cost–Effect of Substrate and PHB Yield

Substrate	Substrate Price (€/kg)	PHB—Yield[a] (%)	Production Cost (€/kg)
Glucose	0.40	38	1.06
Sucrose	0.23	40	0.59
Methanol	0.14	43	0.34
Acetic acid	0.48	38	1.27
Ethanol	0.41	50	0.81
Cane molasses	0.18	42	0.42
Cheese whey	0.06	33	0.18
Hemicellulose hydrolysate	0.06	20	0.28

Source: Madison, L.L. and Huisman, G.W., *Microbiol. Mol. Biol. Rev.*, 63, 21, 1999.
[a] (PHB wt/substrate wt) · 100.

been engaged since 1996 in pilot trial for the PHA production under the trade name of Biopol from transgenic plant abandoned the project few years back (2001) (Metabolix 2001). By the way, problems such as the use of transgenic plants and the competition between food production and plastic production make this pathway not appealing for European industries.

5.2.2 PETROCHEMICAL POLYMERS

Among the synthetic petro-based polymers, PVA represents one of the most interesting materials to be utilized for several ecocompatible uses in many fields of application.

PVA is composed of repeating units of vinyl alcohol monomer. It cannot be prepared by the direct polymerization of its corresponding vinyl alcohol monomer that is the acetaldehyde enolic form, but is usually produced indirectly by alkaline hydrolysis (alcholysis) of polyvinyl acetate. The physical properties of commercial PVA are related to molecular weight, degree of hydrolysis, degree and type of chain branching, amount of cross-linking between polymers chains, and the type and concentration of additives. By varying these factors, useful properties such as strength, stiffness, gas barrier characteristics, and water solubility can be controlled. Because of water-solubility characteristics, PVA is used in several specific fields, such as colloidal protection of other chemical substances, textile and paper industry, hospital laundry bags, and single-dose packaging of solid and liquid components for laundry and dishwash machines. The Italian company Ecopol SRL has developed a technology, using a base of PVA, that allows to manufacture products, particularly extruded films in melt blown process, with good mechanical properties as well as very good hygroscopic characteristics. The product is marketed under the trade name of Hydrolene®.

Among biodegradable polymers of synthetic origin PVA is a particularly well-suited material for the formulation of blends with natural polymers since it is highly polar and can also be manipulated in water solutions and depending upon the grade in polar nonprotic and protic functional organic solvents (Lahalih et al. 1987, Coffin et al. 1996, Chiellini et al. 2000a, 2001a,b, 2003b, Cinelli et al. 2003, Grillo Fernandes et al. 2004).

Ongoing investigation in our laboratories on the formulation and applicability of mixtures of PVA, as synthetic water-soluble polymeric material, and bio-based "fillers" from agroindustrial and forestry waste, has highlighted the potential of attaining ecocompatible articles meant to experience environmental degradation at the end of their service life eventually programmed.

Ongoing cooperation between a research group formerly at USDA station in Peoria (IL, USA) and presently in Albany (CA, USA) and our research group at the University of Pisa has led to the development of several blends and composites based on PVA and lignocellulosic components derived from waste of agroindustrial diary processing such as sugarcane, fruits, corn, wheat, and wood, as well as proteic derived materials from tannery industry (Chiellini et al. 2002, 2003a, Cinelli et al. 2003, Imam et al. 2005).

In the last decade, attentions of researchers and industries have been devoted to the production of hybrid composites based on synthetic polymeric matrices and natural fillers with the aim to reduce production costs as well as to promote the ecological performance of final products. In these regards, the investigation of the ultimate fate of the synthetic polymeric matrices utilized for the hybrid composite productions represents an important aspect to establish the real ecocompatibility of the innovative kinds of materials and relevant manufacture items.

5.3 POLYVINYLICS BIODEGRADATION

The present global situation in the production of synthetic polymers accounts for almost 85%–90% represented by full carbon backbone macromolecular systems (polyvinylics and polyvinylidenics) (Kunstoff 2004), and for about 35%–45% of usage as one-way use items production (disposables and packaging). Therefore, it is reasonable to envisage a dramatic environmental impact as attributable to the accumulation of plastic litter and waste constituted by full carbon backbone polymers, which basically are recalcitrant to physical chemicals and biological degradation processes.

In opposition to "hydro-biodegradation" process of natural and synthetic polymers containing heteroatoms in the main chain (polysaccharides, proteins, polyesters, polyamides, polyethers, and polyurethanes), the mechanism of biodegradation of full carbon backbone polymers requires an initial oxidation step as mediated or not by enzymes followed by the fragmentation again mediated or not by enzymes with substantial reduction in molecular weight. The functional fragments become then vulnerable to microorganisms present in different environments with production under aerobic conditions of carbon dioxide, water, and cell biomass. In Figure 5.1 are outlined the general features of environmentally degradable polymeric materials and plastics that are classified into the specific

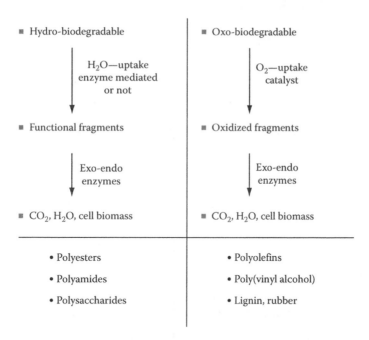

FIGURE 5.1 General classification of environmentally degradable polymers.

sectors of the hydro-biodegradables and oxo-biodegradables. Typical examples of the so-called oxo-biodegradable polymers are represented by PE, PVA, natural rubber (poly-1,4-*cis*-isoprene), and lignin (a natural heteropolymer) (Scott and Wiles 2001). The major biodegradation mechanism of PVA in aqueous media is represented by the oxidative random cleavage of the polymer chains, the initial step being associated to the specific oxidation of methylene carbon bearing the hydroxyl group, as mediated by *oxidase* and *dehydrogenase* type enzymes, to give β-hydroxyketone as well as 1,3-diketone moieties. The latter groups are susceptible of carbon–carbon bond cleavage promoted by specific β-*diketone hydrolase*, leading to the formation of carboxyl and methyl ketone end groups (Suzuki 1979, Sakai et al. 1986).

The ultimate biochemical fate of partially hydrolyzed PVA samples has been recently described by using *Pseudomonas vesicularis* PD strain, a specific PVA-assimilating bacterium (Sakai et al. 1998). This bacterium metabolizes PVA by a secondary alcohol oxidase throughout the oxidation of the hydroxyl groups followed by hydrolysis of the formed β-diketones by a specific hydrolase. Both enzymes are extracellular and the polymer chains are cleaved by repeated enzyme-mediated reactions outside the cells into small fragments, which are further incorporated and assimilated inside the bacterial cytoplasm and metabolized up to carbon dioxide (Figure 5.2) (Sakai et al. 1998).

The initial oxidation step of PVA macromolecules can also be promoted by ligninolytic enzymes [(lignin peroxidase (Lip) and laccase)] produced by white-rot fungal species such as *Phanerochaete crysosporium, Geotrichum fermentans* (Mori et al. 1998, Inés Mejía et al. 1999), and *Pycnoporus cinnabarinus* (Larkin et al. 1999). The monoelectronic enzymatic oxidation reactions lead to free radical formation with the formation of carbonyl groups as well as double bonds, thus increasing the macromolecule unsaturation and hence susceptibility to further oxidation by oxygen uptake (Inés Mejía et al. 1999).

Similarly, the oxidative instability of polyolefins in the environment is due to physical–chemical radical-promoted reactions, enhanced eventually by the presence of sensitizing impurities in the polymer chain. The understanding of the mechanisms of physical aging (e.g., thermal and photolytic degradation) of poly(ethylene) (PE), as well as of the role of heat and light stabilizers in

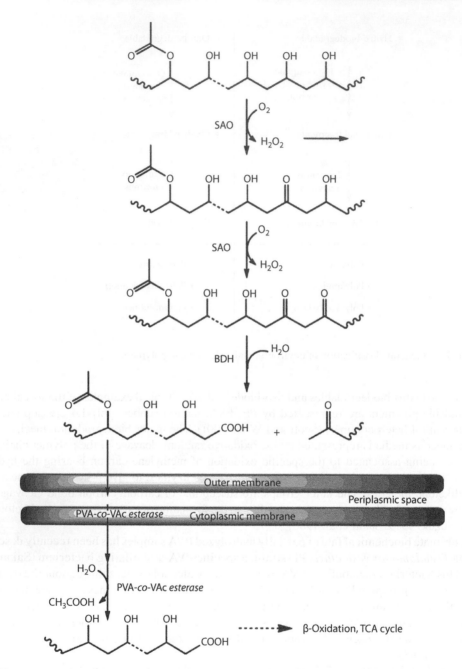

FIGURE 5.2 Biodegradation pathway of partially acetylated PVA.

improving the resistance of PE to the environmental oxidation, has been applied, starting from the 1960s. There is nowadays an opposite strategy aimed at accelerating the degradation rate of PE in order to overcome its intrinsic recalcitrance to any biological attack. In this connection, copolymerization with a small amount of monomers containing carbonyl group (carbon monoxide and methyl vinyl ketone) or the incorporation of transition metal compounds (dithiocarbamates) as photoinitiators or photosensitizers, and prooxidants (fatty acids and salts), constitutes the major strategies for the introduction of functional groups and substances capable of promoting the degradation of macromolecules, when exposed to light and temperature, throughout the generation of free radicals reacting with molecular oxygen to produce peroxides and hydroperoxides (Scott 1994).

Further to the hydroperoxides decomposition in the presence of heat, light, and metallic ions, leading also to the formation of macroalkoxy radicals, the autooxidation of polyethylene proceeds through classical free radical chain reactions (Gugumus 2001, 2002). As a result, chain scission and cross-linking are the major consequences of thermal oxidation of polyolefins (Erlandsson et al. 1997, Burman and Albertsson 2005). In the presence of oxygen, however, chain scission and macromolecules oxidation after thermal degradation (Karlsson et al. 1997) are the predominant reactions (Khabbaz et al. 1999) leading to the significant reduction of molecular weight of poly(ethylene) samples containing a prooxidant, and to the formation of oxidation products including carboxylic acids, ketones, lactones, and low molar mass hydrocarbons (Karlsson et al. 1997).

Rate and extent of radical oxidation of polyolefins are also affected by structural parameters such as chain defects and branching, these latter being representatives of relatively weak links liable to bond cleavage and free radical production for succeeding chain fragmentation. In this connection, it has been stated that the hierarchy in the oxidation susceptibility of polyolefins is iPP > LDPE > LLDPE > HDPE (Winslow 1977, Iring et al. 1986).

By considering the same carbon skeleton and the identical requirement of an initial oxidation step, studies on the environmental fate of EDPs based on PVA and PE may represent an interesting tool for evidencing analogies or differences between the degradation mechanisms and particularly the influence of different environmental parameters in accelerating the biodegradation of these two large-volume carbon backbone synthetic polymers.

5.4 HYBRID POLYMER COMPOSITES

5.4.1 FORMULATIONS FOR MULCHING AND HYDRO-BIOMULCHING PRACTICE

Mulching practice controls radiation, soil temperature and humidity, weed growth, insect infestation, soil compaction, and the degree of carbon dioxide retention. Natural materials, such as straw and leaves, have always been used to provide an insulating layer around the roots of vegetables and protection of soft fruits. In this connection, the fundamental role acquired by plastics in agriculture since the middle of the last century. The definition of "plasticulture" was suggested (Web site: http://en.wikipedia.org/wiki/Plasticulture) as a new practice where the use of plastics for crops production and preservation can be considered as a routine practice capable of increasing unitary yields, inducing earlier harvests, limiting the use and release of herbicides and pesticides, as well as to improve the efficiency of water resources utilization (Lamont 1993). This practice was intended then to contribute significantly to the economic viability of farmers worldwide (Espi et al. 2006). Accordingly, conventional plastic films are therefore extensively used as coverings of greenhouses, low tunnels, and mulching, where low density (LDPE) and linear low density (LLDPE) poly(ethylene) films represent the most employed polymeric materials (Briassoulis 2005) accounting for an estimated worldwide market of 2.0–2.5 million tons per year in the agricultural segments.

The large utilization of these materials in various agricultural regions is posing increasing concerns about their collection and disposal (Italian legislation 2006) after their service life. In addition, visual pollution and detrimental effects on soil management and fertility have also been claimed as problems connected to *plasticulture*. These problems can be solved, specifically for the agricultural plastic wastes, by using in situ degradable/biodegradable materials having unitary costs comparable to those of conventional plastics. In this case, however, the requirement for a controlled service life ranging between several weeks up to 1 year or more depending upon the practical utilizations (i.e., mulching films or tunnels covering) is needed. In keeping, since the second half of the twentieth century, strategies to achieve low cost environmentally degradable polyolefins haven been undertaken (Scott 1965, Guillet 1973a).

5.4.2 Copolymerization Procedures

Copolymerization has been traditionally carried out in order to introduce UV-absorbing groups capable to enhance the photooxidation process. In this connection, carbonyl group can be incorporated into the PE main chains by the copolymerization of ethylene with carbon monoxide (Heskin and Guillet 1968, Harlan and Kmiec 1995), or in the side chain by copolymerization with methyl vinyl ketone giving copolymers commercialized under the trade name Ecolyte® (Guillet 1973b,c). Ketone groups can be rather easily introduced in the side chains by copolymerization with vinyl ketones. However, the structure of the side chains strongly depends on the type of the vinyl ketone used as well as by the polymerization conditions. It has been demonstrated that a great variety of side-chain ketones may rise, when the copolymerization takes places under high pressures, as a result of the so-called backbiting mechanism in polyethylene (Guillet 1980). Moreover, it has also been demonstrated that side-chain carbonyl groups are more photoactive than those contained in the main chains because of the higher quantum yields for the formation of radicals by both Norrish type I and II processes in the solid phase. Additionally, side-chain vinyl ketone copolymers can be used as masterbatches in order to induce photodegradation of pure polyethylenes, thus allowing to control under a certain extent the degradation rates. This strategy is the basis of commercial Ecolyte process (Guillet 1973b,c).

Nevertheless, these materials usually have no induction periods and can be used mainly in short-term applications.

5.4.3 Prodegradant Systems

More recently, an other strategy to make polyethylene susceptible of biodegradation is represented by the use of prooxidant additives. It has been in fact suggested that this alternative may provide a more efficient control of the shelf life, service life, and degradation rate for the polyolefin in several applications (Scott 1981).

Most of these compounds are based on transition metals capable of yielding two metal ions differing in the oxidation number by one unit. Several metal carboxylates and acetylacetonates of Co^{3+}, Fe^{3+}, and Mn^{3+} soluble in polyolefins are very effective photo-prooxidants capable to initiate the degradation process through the metal salts photolysis to give the reduced form of the metal ion and a free radical under UV irradiation FeX, $h\nu > [FeX, X'] > FeX_2 + X'$. The anion radical promotes a fast hydrogen abstraction from the polymer and the relevant formation of hydroperoxide. Afterward, the general radical oxidation mechanisms of the polyeolefins is thought to proceed being enhanced by the usual redox reactions between hydroperoxides and metal ions FeX, $+ ROOH > RO' + FeX_2OH$.

In an other case, the properties of many metal complexes containing sulfur as a ligand play an opposite role acting either as photo or thermal stabilizer and as sensitizer after an induction period. Dithiocarbamates and dithiophosphates (Figure 5.3) are the principal representatives of this class of additives that exert the antioxidant effect by decomposing hydroperoxides by an ionic mechanism (Al-Malaika and Scott 1983, Al-Malaika et al. 1988). After that, however, the ligand is destroyed, thus releasing free transition metal ions which start to behave as a prooxidant component according to the aforementioned mechanism. Hence, antioxidant and photosensitizer properties are both present in the same compound. This evidence constituted the basis of the development of a well-known class of photodegradable polyethylenes having a defined and controlled induction period as started by Scott and further refined in collaboration with Gilead in order to finely control the lifetime before the photooxidation commencement. This has been accomplished by using two component systems in which the length of the induction period is controlled by one metal thiolate and the rate of photooxidation by a second (Scott 1995, Scott and Gilead 1982). The most representative compounds of this class of "delayed action" photosensitizers are the Fe(III) dithiocarbamates and dithiophosphates. The so-called Scott–Gilead (SG) technology led to the commercialization of several photodegradable polyethylenes especially

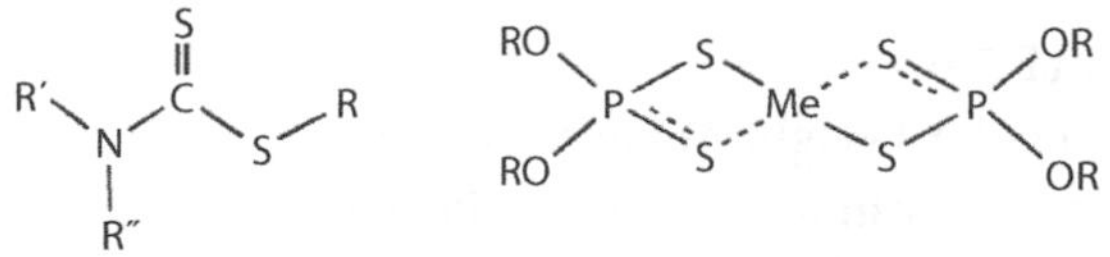

FIGURE 5.3   Structural formula of dithiocarbamates and dithiophospahtes.

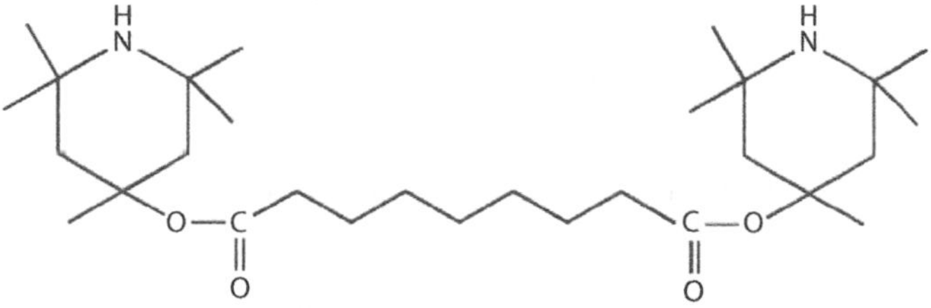

Tinuvin 770 Bis(2,2,6,6-tetramethyl-4-piperidyl)sebacate

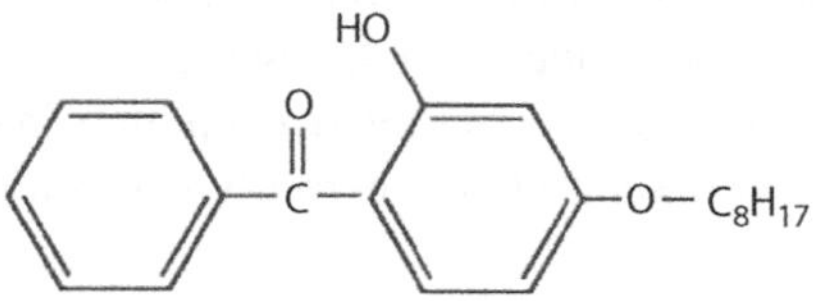

Cyasorb 531 2-hydroxy-4-octyloxy benzophenone

FIGURE 5.4   Structural formulas of two typical commercial photostabilizers.

devoted to applications in agriculture, such as mulching films, which requires a well-defined induction period before the starting of the photodegradation process does occur.

A different system, even constituted by a combination of a photosensitizer and a photoantioxidant, was developed by Allen, by using anthraquinone and Tinuvin 770, respectively (Allen 1980) (Figure 5.4).

Photodegradable PE samples produced according to the SG technology have been repeatedly studied under both photo and thermal exposure, with the aim to establish the mechanism of polymer degradation and the effect of different type of prooxidants and additives.

In a case study, three different photodegradable PE samples containing iron dimethyldithio-carbamate (sample SG1), iron dimethyldithiocarbamate and 0.8% carbon black (sample SG2), and iron dimethyldithiocarbamate and nickel dibuthyldithiocarbamate (sample SG3) were submitted to UV irradiation between 280 and 359 nm, with or without 300 h heat exposure (50°C), in comparison with a not-additivated LDPE films. The oxidative degradation of samples was also studied during 5 weeks of thermal aging at 80°C (Karlsson et al. 1997). Degradation rate by means of molecular weight changes and analysis of the degradation products were therefore assessed.

Depending upon type and combination of photo pro-degradants, different oxidative behaviors were recorded during the UV exposure of the analyzed SG samples. In particular, it was ascertained that molecular weight of samples additivated with iron dimethyldithiocarbamate (SG1 sample) only decreased after a few hours of UV irradiation. Longer induction times were instead required for samples containing both iron and nickel dimethyldithiocarbamates. In addition, molecular weight analysis was also suggesting that the scissions of main chains were the dominant process, whereas cross-linking resulted in a fairly low extent. A complex behavior was also observed by submitting the sample to UV exposure followed by thermal aging. For instance, iron dimethyldithiocarbamate was inducing the drop of molecular weight both during UV irradiation and in the second step of thermal exposure. On the contrary, the same catalyst was promoting a

---

TABLE 5.6
Molecular Weight Analysis of Iron
Dimethyldithiocarbamate LDPE Samples
Submitted to Different Aging Treatment

Sample Aging	$Mw \cdot 10^{-3}$	$Mn \cdot 10^{-3}$	ID
None	190.0	32.7	5.9
UV 100h	52.3	6.0	8.8
UV 100h + 80°C/5 weeks	23.1	4.8	4.8
80°C/5 weeks	202.0	49.8	4.4

Source: Karlsson, S. et al., *Macromolecules*, 30, 7721, 1997.

---

slight increase of the molecular weight when the SG1 LDPE samples were submitted to thermal degradation only (Table 5.6).

Different degradation products profiles were also recorded. Dicarboxylic acids were thus found as the main components in the photodegraded samples, while mono- and dicarboxylic acids, as well as ketones and ketoacids were recorded in higher relative amounts in thermally oxidized specimens with respect to the analogous sample submitted only to UV exposure.

5.4.4 TRANSITION METAL CARBOXYLATES PROOXIDANT ADDITIVES

An other class of prooxidant additives is represented by compounds capable to induce an oxo-degradative process of polyolefins by absorbing energy as heat. Also this class of additives is based on the activity exerted by transition metal ions typically added to the final product in form of fatty acids or acetylacetonate organic ligands. The most employed cations are Mn^{2+} (Jakubowicz 2003) and Co^{2+} (Weiland et al. 1995). Instead of Fe^{3+} complexes that play the main role in photooxidation processes, Mn^{2+} and Co^{2+} are needed to accelerate the radical chain reactions of polyolefin oxidation through the formation and decomposition of hydroperoxides and peroxides as induced by energy (heat or light) absorption.

The mechanism of oxidative degradation of polyethylene catalyzed by transition metal ions has been recognized as a sequence of free radical chain reactions (Osawa et al. 1979, Osawa 1988, Lin 1997). As a typical example, cobalt stearate included in full carbon backbone polymers, when exposed to energy absorption is susceptible to electron transfer in the 3d subshell of cobalt atoms leading to the formation of carboxylic acid free radicals that easily decarboxylates to form alkyl radicals. These latter do react with carbon backbone macromolecules, thus promoting the formation of macroradicals especially in the presence of tertiary carbon atom sites, susceptible to produce hydroperoxides in the presence of oxygen (Scheme 5.2).

Nowadays, several specialties are sold under different trademarks, end products, and master-batches containing either photo or thermal or both prooxidants constituted by organic ligands with transition metal ions.

The effectiveness of cobalt stearate in promoting the accelerated thermal oxidation of LDPE films has been also recently confirmed (Roy et al. 2006a). Indeed, a huge increase of melt flow index (MFI) as a consequence of massive chain scissions was observed after 100h thermal aging at 70°C of LDPE films containing 0.1%–0.3% cobalt stearate. Significant decay of mechanical properties, such as elongation at break, and MFI increase were also recorded when the same films were submitted to UV exposure. In accordance, cobalt stearate can be considered to be effective in promoting both photooxidation and thermal oxidation of LDPE. Nevertheless, largest degree of oxidation was observed during the thermal aging, thus suggesting once more that cobalt organic salts are very effective prooxidants for polyolefins mainly guided by thermal inputs.

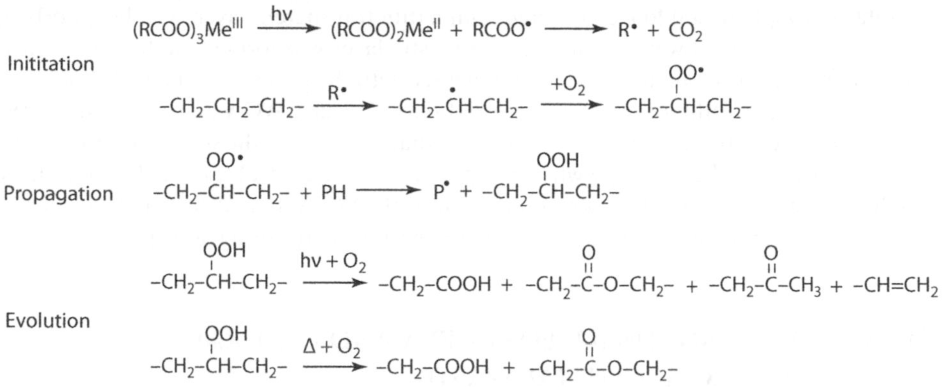

Inititation

$$(RCOO)_3Me^{III} \xrightarrow{hv} (RCOO)_2Me^{II} + RCOO^\bullet \longrightarrow R^\bullet + CO_2$$

$$-CH_2-CH_2-CH_2- \xrightarrow{R^\bullet} -CH_2-\overset{\bullet}{C}H-CH_2- \xrightarrow{+O_2} -CH_2-\overset{\overset{OO^\bullet}{|}}{C}H-CH_2-$$

Propagation

$$-CH_2-\overset{\overset{OO^\bullet}{|}}{C}H-CH_2- + PH \longrightarrow P^\bullet + -CH_2-\overset{\overset{OOH}{|}}{C}H-CH_2-$$

Evolution

$$-CH_2-\overset{\overset{OOH}{|}}{C}H-CH_2- \xrightarrow{hv + O_2} -CH_2-COOH + -CH_2-\overset{\overset{O}{\|}}{C}-O-CH_2- + -CH_2-\overset{\overset{O}{\|}}{C}-CH_3 + -CH=CH_2$$

$$-CH_2-\overset{\overset{OOH}{|}}{C}H-CH_2- \xrightarrow{\Delta + O_2} -CH_2-COOH + -CH_2-\overset{\overset{O}{\|}}{C}-O-CH_2-$$

SCHEME 5.2 Radical chain reactions in polyolefins catalyzed by metal carboxylate salts.

In addition, rate and extent of oxidative degradation were positively correlated with the content of cobalt stearate.

The effectiveness of cobalt organic complexes has been also investigated either as a function of the type of the organic ligand, as well as of the aging conditions (photo and/or thermal exposure). It is known, in fact, that the catalytic activity is depending upon the valence and ionic bonding, as well as upon the miscibility in the blends with polymer chains at molecular level (Osawa et al. 1979).

In other studies, the effect of carboxylate chain lengths (laurate, palmitate, and stearate) as organic ligands for cobalt ions as photosensitizers was investigated. LDPE films containing 0.05%–0.2% of each type of cobalt carboxylates were prepared and tested under UV exposure, by using FT-IR spectroscopy, mechanical testing, and molecular weight determinations as analytical tools (Roy et al. 2006b). It was therefore ascertained that the oxidative degradation increased with the increase of the chain length of the carboxylate residues. Therefore, besides the content of cobalt, during the photo-exposure, higher levels of degradation qualitatively assessed by MFI were obtained with cobalt stearate, followed by palmitate and laurate. This feature can be attributed either to the higher thermal stability of stearate during the LDPE processing, as well as to the better compatibility and hence miscibility of longer carboxylate chain within the LDPE matrix.

Also the techniques "hydro-mulching" or "liquid mulching" provide a conditioning effect on soil structure (Kay et al. 1977, Young et al. 1977). Water-soluble synthetic degradable polymers, such as poly(acrylamide), PVA, carboxymethyl cellulose, and hydrolyzed starch-g-polyacrylonitrile (HSPAN), can be easily sprayed onto the soil alone or in mixture with nutrients or other agronomic valuable fillers (Stefanson 1974, Oades 1976, Painuli and Pagliai 1990a,b, Orts et al. 2000). Polymer solutions can also be applied as tackyfiers, thus helping to hold the mulch in place once applied or aimed at forming a sort of thatch intended to protect seeds and soil against erosion. Polyvinyl alcohol and gelatin (WG) were used as continuous matrices from synthetic and natural source, respectively, for the formulation of hydro-biomulching mixtures with sugarcane bagasse (SCB) and wheat flour (WF) as fillers (Cinelli 1996, Kenawy et al. 1999, Chiellini et al. 2002).

In countries where gelatin and sugarcane juice processing occupy an important position, the combined use of these by-products allows for low cost and in situ formulations of environmentally degradable hydro-biomulching. The addition of SCB fibers to WG hardened the resulting films and darkened film color. Accordingly, films based on WG containing 10%–30% SCB were rather flexible, whereas films containing 40%–50% SCB turned out to be very tough and brittle. A WG/SCB 80/20 ratio (WGSCB20) appeared to be the most interesting composition as far as filler content and mechanical properties are concerned.

Previous investigation on the applications of PVA water solutions to soil outlined that when sprayed onto the soil surface the solutions penetrated about 0.5 cm in the soil. When a PVA solution

5% by weight was applied, soil looked covered with a thin transparent layer, wet, but poorly aggregated. With a 10% by weight water solution a thin plastic layer was formed on the soil surface; soil formed a very fragile crust, about 0.5 cm high, together with the polymer (Cinelli 1996).

The results of sprayed film experiments showed that PVA and WG water suspensions resulted easy to spray, mixture-soil aggregates lasted for more than 3 weeks on the soil, and the soil appeared to be conditioned and in a better shape when compared with the control sample. Formulations containing SCB conferred a marked brown color to the soil. WG, WF, and SCB presence enhances PVA time of permanence on the soil, thus guaranteeing for the resulting soil structuring effect.

5.5 FILMS AND LAMINATES BASED ON PVA AND NATURAL FILLERS FROM AGRICULTURAL WASTE

Hybrid composites were prepared by use of PVA with 96% hydrolysis degree and natural fillers represented by cellulosic materials from three different sources: sugarcane bagasse, orange (OR), and apple (AP) peel, which were the remains of fruit residue after juice extraction (Chiellini et al. 2000a, 2001a, 2003a, Imam et al. 2005).

Unmodified commercial-grade corn starch with approximately 30% amylose and 70% amylopectin (the United States) was added in some formulations to replace as much of PVA as possible without compromising the film properties. Films were prepared by casting of water suspensions, at 10% PVA by weight. Addition of glycerol and urea softened the films as both glycerol and urea are known to act as a plasticizer for PVA in PVA/starch-based films (Lawton and Fanta 1994, Mao et al. 2000, Chiellini et al. 2004a,b).

In PVA/OR blends, starch addition caused significant reduction in elongation at break with somewhat moderate variation in ultimate tensile strength (UTS). Excellent film-forming properties and flexibility were observed in films even when starch concentration in formulations exceeded 25%. In PVA/SCB blends, elongation at break (EB) was mostly unaffected by starch content while UTS increased with starch addition, thus the introduction of starch increased cohesiveness of PVA/SCB films. Whereas in PVA/AP blends, addition of starch increased the presence of defects and small holes were detected in the films, thus indicating a loss in mechanical properties due to increased starch content.

A front of positive results particularly observed for the hybrid composites based on OR fibers prepared by casting, further tests were performed by compression molding. Thus, the substitution of starch with orange waste in hybrid composite based on PVA and glycerol presented a modest variation in mechanical properties.

Composites prepared with AP presented similar EB values (61%), to composites prepared with OR, but higher UTS (9 MPa) and YM (57 MPa). PVA/SCB resulted also harder with a reduced EB (7%) than PVA/OR and PVA/AP composites, but with a relatively high UTS (8 MPa) and YM (171 MPa). This behavior is connected with the fibers composition and type. Both AP and SCB have fibrous shape that confers hardness to the composites when processing allows for orientation as in the Brabender, while in casting films the random distribution of the fibers was a source of fragility.

Composites containing urea presented a higher EB (148%), a decreased UTS (1 MPa) and YM (2 MPa) in comparison with composites based on OR, as expected for the increased percentage of plasticizers additives.

5.6 HYBRID COMPOSITES BY INJECTION MOLDING AND FOAMING

Hybrid composites with PVA and starch as continuous matrices were prepared with corn fiber (CF). CF is an industrial name given to the pericarp fraction of the corn kernel that is a coproduct of ethanol production by wet milling technology. CF contains pericarp, and also starch and protein

TABLE 5.7
Composition of Bio-Based Mixtures Consisting of PVA
and Corn Fibers Processed by Injection Molding

Sample	PVA (%)	Corn Fiber (%)	Starch (%)	Glycerol (%)	Pentaerythritol (%)	PEG (%)
PCF1	42	26	0	21	11	0
PCF2	40	25	0	20	10	5
PCFSt1	36	23	9	18	9	5
PCFSt2	33	21	17	17	8	4
PCFSt3	29	32	14	14	7	4

PVA, Poly(vinylalcohol) grade 5/88; PEG, poly(ethylene glycol) 2000.

from the endosperm. The wet CF (60% moisture) is sold at about $15 a ton, while dried and ground at 10 mesh is sold at about $50 a ton with its main use is in animal feeds. CF had a composition of 1% fat, 14% protein, 25.5% starch, 59% lignocellulosic component, and 0.5% ash. Injection mold specimens were prepared testing, respectively, glycerol, pentaerythritol, and polyethylene glycol 2000 as suitable plasticizers (Cinelli et al. 2003).

The plasticizer proportion as in PCF2 (Table 5.7) was used for the rest of the composites, corn starch was introduced in the formulation and the amount of CF was progressively increased up to a value of 32% in the mixture. As starch and CF volume fraction increased, EB decreased while the modulus generally increased. Interestingly, UTS was not significantly affected by increasing CF. Also, composites made with 32% fibers and only 29% PVA resulted in cohesive extrudes.

For samples PCFSt1, PCFSt2 (Table 5.7) prepared, respectively, with 9% and 17% starch and approximately the same fiber PVA weight ratio as in composite PCF2, EB was slightly increased with storage, and UTS and YM were decreased. These changes are probably an indication that water was absorbed during storage. Composite PCFSt3 containing a higher ratio fibers to PVA had an increase in UTS and YM and a decrease in EB. This indicates that this composite was getting stiffer and less flexible with age. Anyway, composites tested after 1 year of storage had tensile properties similar to composites tested after 7 days of storage.

In consideration of the positive results obtained by injection molding processing of polymeric matrices with CFs, further investigation were done evaluating the effect of CFs introduction into starch-based foamed trays. Foamed food containers contribute in large scale to the amount of plastics in municipal solid waste streams. These are mainly produced by expanded polystyrene (EPS) or coated paperboard. In recent years, several efforts have been devoted to the production of similar items based on polymers from renewable resources such as starch (Shogren et al. 1998, Lawton et al. 1999).

Potato starch, CF, magnesium stearate and PVA having 88% (P88) and 98% (P98) hydrolysis degree, respectively, were first mixed using a kitchen aid mixer provide with a wire whisk attachment. For PVA free batter and with less than 50 weight part of fibers, gum arabic (1% by weight of starch) was added to prevent starch settling (Table 5.8). Water was added to reach the required total solid content (Cinelli et al. 2006).

Foam trays prepared using a lab model–baking machine were submitted to mechanical testing, and the maximum force (MF) and deformation at MF (MFD) were assessed by means of a Universal Testing Machine Instron Mod 4201.

Increase in fiber contents lowered MFD and MF values; thus, fiber's irregular shape was not allowing a strengthening effect, whereas PVA presence improved MFD and MF values especially when grade P98 was used even at 45% fiber content (Shogren et al. 1998).

TABLE 5.8
**Composition of the Bio-Based Batter Used
to Produce Trays by Foaming Technique**

Batter	St[a] (%)	PVA (%)	CF[a] (%)	Arabic Gum (%)	Magnesium Stearate (%)
St	96	0	0	1	3
StCF50	65	0	31	1	3
StCF100	49	0	49	0	2
StCF150	39	0	59	0	2
StP	80	16	0	1	3
StPCF50	58	11	29	0	2
StPCF100	45	8	45	0	2
StPCF150	36	7	55	0	2

[a] St, Starch; CF, corn fibers; PVA, poly(vinyl alcohol) grade 5/88.

5.7　BIODEGRADATION OF FULL CARBON BACKBONE POLYMERS

5.7.1　Biodegradation of Poly(Vinyl Alcohol)

The environmental fate of water-soluble PVA has been primarily investigated due to its large utilization in textile and paper industries that generate considerable amounts of wastewaters contaminated by PVA. Since 1936, it was observed that PVA was susceptible of sustaining ultimate biodegradation when submitted to the action of *Fusarium lini* (Nord 1936). Afterward, the nature of PVA as truly biodegradable synthetic polymer was repeatedly and intensively assessed.

Most of the PVA-degrading microorganisms have been identified as aerobic bacteria belonging to *Pseudomonas*, *Alcaligenes*, and *Bacillus* genus. Some species degrade and assimilate PVA axenically, even though symbiotic association exhibiting complex cross-feeding processes is a rather common feature of PVA biodegradation (Shimao et al. 1983, Matsumura et al. 1994, Mori et al. 1996, Kawagoshi and Fujita 1998, Kawai and Hu 2009).

Nevertheless, the extensive biodegradation of PVA is accomplished almost exclusively by specific degrading microorganisms whose occurrence in the environment appears to be uncommon and in most cases strictly associated with PVA-contaminated environments (Porter and Snider 1976, Solaro et al. 2000). Correspondingly, limited mineralization of PVA-based blown film was recorded during respirometric biodegradation tests in aqueous medium inoculated with municipal sewage sludge (Chiellini et al. 1999b). On the contrary, in the presence of the sewage sludge collected from paper mill wastewater treatment plant, the mineralization extent of PVA and PVA-based blown film was comparable to that of cellulose, even though in a longer incubation time (Figure 5.5) (Chiellini et al. 1999a).

This behavior is demonstrating that the selective pressure exerted by the constant presence of large amount of PVA in wastewater from paper factories is effective in the establishing of microbial consortia able to degrade and assimilate PVA. The enrichment procedure of paper mill sewage sludge, as obtained by repeated sequential transfers of this microbial inoculum in the presence of PVA as sole carbon and energy source, induced a significant acceleration in the degradation rate, with a substantial increase of the extent of PVA mineralization. On the other hand, the acclimation of the microbial strains to PVA led to an almost total abatement of cellulose assimilation (Figure 5.6), thus confirming the high specificity of PVA-degrading microorganisms (Chiellini et al. 1999a).

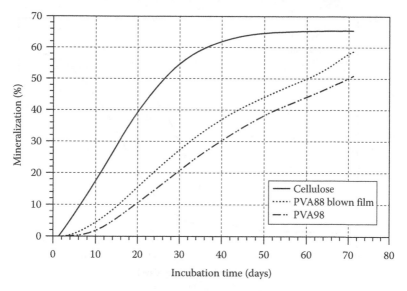

FIGURE 5.5 Time profiles of mineralization of PVA-based blown films, 98% hydrolyzed PVA (PVA98), and cellulose in an aqueous medium in the presence of paper mill sewage sludge.

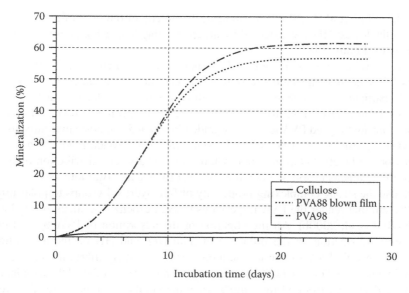

FIGURE 5.6 Time profiles of mineralization of PVA-based blown films, 98% hydrolyzed PVA (PVA98), and cellulose in an aqueous medium in the presence of selected PVA-degrading microorganisms.

Molecular weight, hydrolysis degree (HD) of parent poly(vinyl acetate), at least in the 80%–100% range, and content of head-to-head units do not appear to greatly affect the enzymatic random endocleavage of PVA macromolecules in aqueous media (Suzuki et al. 1973, Watanabe et al. 1976, Solaro et al. 2000). However, earlier investigations are suggesting that some structural characteristics of PVA, such as the hydrophobic character (associated to the residual acetyl group content), may considerably influence the activity of different PVA-degrading enzymes (Matsumura and Toshima 1992, Hatanaka et al. 1995, Chiellini et al. 1999a). Accordingly, no major differences in the extent of mineralization in the presence of an acclimated microbial inoculum have been observed among three commercial PVA samples (PVA72, PVA88, and PVA98) having HD 72%, 88%, and 98%,

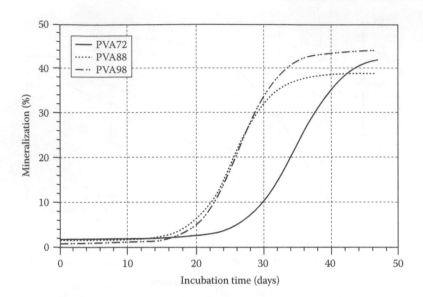

FIGURE 5.7   Time profiles of mineralization of PVA samples having different degree of hydrolysis in an aqueous medium in the presence of acclimated PVA-degrading microorganisms.

respectively (Chiellini et al. 1999b). Nevertheless, the kinetic of the mineralization process of the sample having the lowest HD was affected by an appreciably longer lag phase with respect to the other two samples (Figure 5.7).

By contrast, it has been assessed that both DPn and HD are strongly affecting the biodegradation process of PVA in solid-state incubation tests aimed at reproducing natural environments such as soil or compost. In particular, the preferential assimilation of low molecular weight fractions has been ascertained in soil burial tests (Takasu et al. 1999), whereas a very limited propensity of high molecular weight and fully hydrolyzed PVA to be biodegraded in soil and compost environments was repeatedly observed (Krupp and Jewell 1992, Bloembergen et al. 1994, Sawada 1994, Chen et al. 1997).

This behavior can be attributed in a first instance to the very rare distribution of specific PVA-degrading microorganisms in soil and compost media (Tokiwa and Iwamoto 1992).

However, also the strong complexing propensity of free hydroxyl groups to polar mineral components of the soil that render the PVA impervious to even exoenzyme attack has to be taken into account (Chiellini et al. 2000b). An investigation of the adsorption of PVA by soil components provided several information on the influence of both DPn and HD on the adsorption process. For instance, the specific PVA adsorption on montmorillonite at equilibrium increased from 31 to 37 mg/g with increasing HD from 72.5% to 98%, while in the case of 88% HD sample, the specific PVA adsorption decreased from 34 to 28 mg/g when the number average molecular weight of the sample was increased from 36 to 88 kDa (Chiellini et al. 2000b).

PVA adsorption on whole soil was shown to behave similarly, thus recording also in this case the preferential adsorption of low molecular weight fractions. No detectable amount of PVA was released from montmorillonite samples when suspended in water, suggesting that the adsorption process is almost irreversible (Chiellini et al. 2000b).

The influence of the adsorption by montmorillonite on the biodegradation of PVA was investigated, in the presence of an acclimated microbial inoculum in liquid cultures containing soluble PVA as reference and PVA adsorbed on montmorillonite (Chiellini et al. 1999a). The recorded mineralization profiles clearly showed that PVA in solution was extensively mineralized (34%), whereas when adsorbed on montmorillonite only 4% mineralization was achieved within 1 month of incubation (Figure 5.8), thus indicating that PVA adsorption on inorganic substrates effectively inhibits the biodegradation processes (Chiellini et al. 2000a).

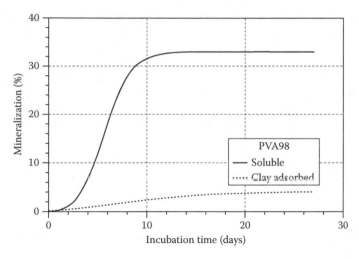

FIGURE 5.8 Time profiles of mineralization of soluble and clay-adsorbed PVA in an aqueous medium in the presence of acclimated PVA-degrading microorganisms.

Based on these information, the dependence of mineralization rate on the HD and DPn has been further evaluated in a soil burial respirometric experiments, by using three different commercial grades PVA having 72%, 88% and 98% HD, respectively, and corresponding Mw of 25.0, 7.2, and 93.5 kDa. Very limited mineralization was observed in the case of the PVA sample having the highest Mw and HD, whereas lower HD and in particular DPn have been shown to promote the biodegradation propensity of PVA in soil (Figure 5.9) (Chiellini et al. 1999a, 2006b).

The information achieved with these studies are evidencing different biodegradation behavior of PVA in aqueous and in solid-state media reproducing different environmental conditions, as well as the role exerted by PVA structural parameters such as molecular weight and degree of hydrolysis in one case or in the other.

To clarify this point, different PVA samples having similar degree of polymerization and noticeably different HDs (ranging between 11% and 75%) have been prepared by controlled acetylation of commercial grade sample (HD = 98%) and submitted to biodegradation respirometric experiments under aqueous, mature compost, as well as soil incubation media (Chiellini et al. 2006b).

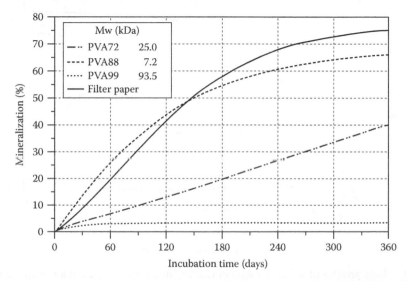

FIGURE 5.9 Time profiles of mineralization of commercial PVA samples in a soil burial respirometric test.

In a soil burial respirometric test, the biodegradation extents of four different PVA (PVA11, PVA30, PVA50, and PVA75) samples in comparison with the biodegradation behavior of PVA98 representing the polymeric substrate utilized in the reacetylation reactions have been assessed.

Reacetylated PVA samples having a molar content of vinyl acetate units ranging between 25% and 75%, PVA75, PVA50, and PVA30, respectively, showed comparable biodegradation profiles, thus reaching fairly high mineralization extents (53%–61%) at the end of the test (Figure 5.10). On the contrary, the almost completely hydrolyzed sample (PVA98), as well as the sample having the highest content of vinyl acetate units (PVA11) (Figure 5.10), did not undergo any significant microbial degradation after approximately 2 years of incubation time (Chiellini et al. 2006b).

A similar behavior, characterized however by lower mineralization extents, was observed in the presence of the same PVA samples when tested in a mature compost respirometric trials (Figure 5.11).

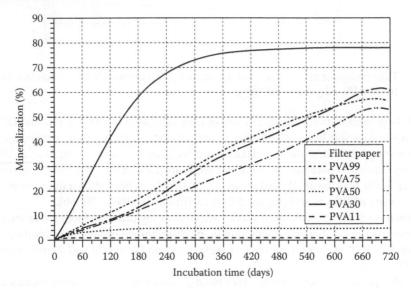

FIGURE 5.10 Time profiles of mineralization of reacetylated PVA samples in a soil burial respirometric test.

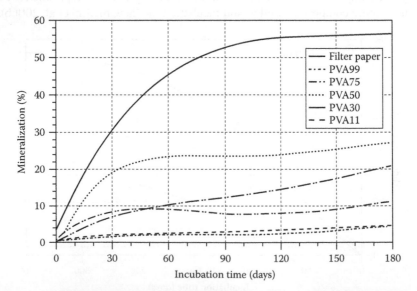

FIGURE 5.11 Time profiles of mineralization of reacetylated PVA samples in a mature compost respirometric test.

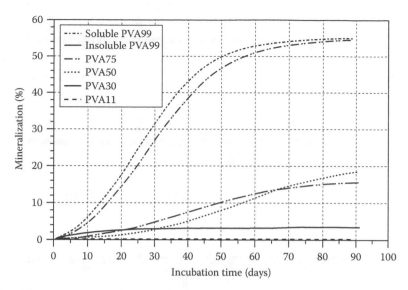

FIGURE 5.12 Time profiles of mineralization of reacetylated PVA samples in an aqueous medium in the presence of acclimated PVA-degrading microorganisms.

For comparison, reacetylated PVA samples were also submitted to a biodegradation test in aqueous medium in the presence of selected microorganisms. As expected, they dissolved in the aqueous medium at rate depending upon the degree of hydrolysis. Accordingly, PVA75 was found to be readily soluble in cool water. PVA50 was shown to disintegrate and partially solubilize, whereas the others resulted at least swellable (PVA30) or almost completely insoluble (PVA11). Indeed, for comparison, PVA98 was also tested either as insoluble film or as water solution as attainable after heating at 90°C.

Under these conditions, soluble PVA75 and PVA98 samples were promptly mineralized in a similar extent by the selected microbial inoculum (Figure 5.12). Lower but appreciable biodegradation degrees were recorded in the cultures fed with PVA30 and PVA50 even though after a prolonged lag phase, whereas the insoluble reacetylated PVA sample having the highest content of acetyl residues (PVA11) and insoluble PVA98 sample resulted almost completely recalcitrant to the microbial attack (Figure 5.12).

These data suggest that a certain grade of hydrophobicity achieved by a dominant content of vinyl acetate units in PVA samples may be taken as practical parameter for indication of the propensity of a PVA to be biodegraded in solid media. A similar influence was recognized for the degree of polymerization resulting, as expected, in a higher mineralization degree for the sample with lower molecular weight.

An opposite trend was instead observed in the biodegradation tests carried out in aqueous medium in the presence of PVA-acclimated microorganisms. In these conditions, the driving force in the biodegradation of PVA appears to be its solubility that makes the solvated chains vulnerable to the endocleavage of the polymer chains by specific enzymatic systems.

5.7.2 Oxo-Biodegradation of Poly(Ethylene)

Free radical oxidation, as induced by thermal and/or photolytic pre-abiotic treatment constitutes the first step for promoting the eventual biodegradation of both LDPE and LDPE-containing prooxidant additives (Scheme 5.3).

This can be accomplished by monitoring the initial variation of sample weight, molecular weight, and other structural parameters (tensile strength, degree of crystallinity, and spectroscopic characteristics). Biodegradation is then observed when degraded oxidized polymer fragments are exposed to biotic environments (Albertsson et al. 1998, Volke-Sepúlveda et al. 1999, Contat-Rodrigo and Ribes Greus 2002, Volke-Sepúlveda et al. 2002).

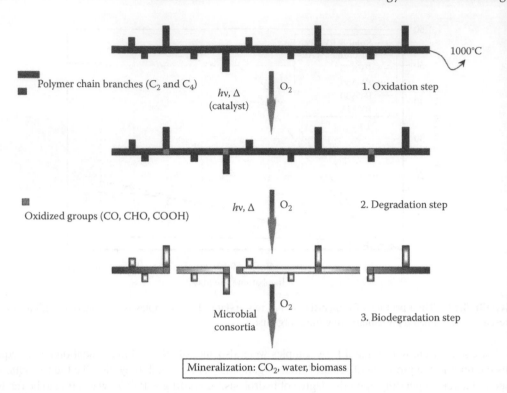

SCHEME 5.3 Schematic representation of free radical oxidative steps of polyethylene degradation.

Additionally, the required degree of macromolecular breakdown for the microbial assimilation of polyethylene to occur is substantiated by the observation of the propensity toe biodegradation of lower molecular weight hydrocarbon molecules. It has been demonstrated that linear hydrocarbon having molecular weight below 500 (Potts 1972) or n-alkanes up to tetratetracontane ($C_{44}H_{90}$, Mw = 618) (Haines and Alexander 1975) can be utilized as carbon source by microorganisms. The degradation of untreated high-density poly(ethylene) with molecular weights up to 28,000 by a *Penicillium simplicissimum* isolate it has been also reported (Yamada-Onodera et al. 2001) even though the extent of fungal attack of poly(ethylene matrix) has been only monitored by physicochemical tools (FT-IR and HT-GPC), as well as by growth-proliferation assays in agar plates containing the polyolefin sample. No respirometric data have been indeed reported.

Therefore, it has been generally accepted that the initial abiotic degradation step represents the major prerequisite in the induction of potential microbial assimilation of poly(ethylene). Many studies have been undertaken as aimed at getting an insight into the mechanisms of radical oxidation of PE (Scott 1994, Karlsson et al. 1997, Khabbaz et al. 1999, Gugumus 2001).

On the other hand, relatively little information on the influence of physical parameters, such as humidity, oxygen pressure, as well as the whole biological environmental conditions on the propensity of thermal oxidation of degradable PE films, are available (Weiland and David 1994).

Accordingly, the oxidation propensity of PE samples from EPI Inc. (Canada) containing TDPA™ prooxidant additives, as induced by heat, has been assayed in oven at different temperatures (55°C and 70°C) mimicking the thermophilic conditions of the composting process. The influence of the vapor pressure of water has been also investigated by comparing the thermal aging under dry condition and in atmosphere conditioned at approximately 75% relative humidity.

As the carbonyl group is representative of most of the oxidation products (carboxylic acids, ketones, aldehydes, and lactones) (Albertsson et al. 1998) of the oxidative degradation of polyethylene, the concentration of carbonyl groups, as determined by the carbonyl index (COi) can be used to monitor the progress of oxygen uptake and degradation (Gugumus 1996). COi determinations have been therefore utilized in order to compare the thermal oxidation behavior of test samples under dry and 75% relative humidity conditions at 55°C and 70°C.

First of all, the oxidation of the poly(ethylene) matrix in the thermally aged samples has been clearly ascertained by the FT-IR spectroscopy. The increasing absorption and broadening within time of the band in the carbonyl region has been recorded in all the samples aged at 70°C (Figure 5.13). Overlapping bands corresponding to acids (1712 cm⁻¹), ketones (1723 cm⁻¹), aldehydes (1730 cm⁻¹), and lactones (1780 cm⁻¹) (Hinsken et al. 1991) have also observed, thus indicating the presence of different oxidized species (Figure 5.13). Among these carboxylic acids and ester groups have been show to be produced in the early and later stages of polymer matrix oxidation, respectively (Khabbaz and Albertsson 2000). The increase of specimen weight up to a 4%–5% of the original value (Figure 5.14), as well as

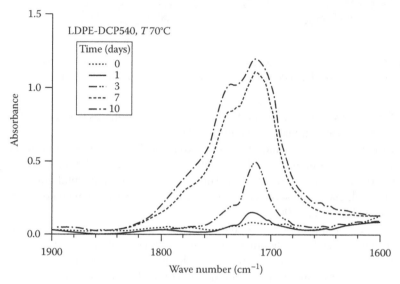

FIGURE 5.13 Time variation of carbonyl absorption band of LDPE sample containing prooxidant additives thermally treated in air in oven at 70°C.

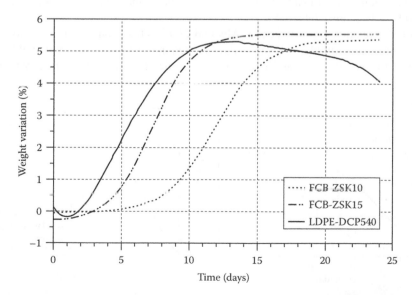

FIGURE 5.14 Weight variation profile of LDPE samples containing prooxidant additives thermally treated in air in oven at 70°C.

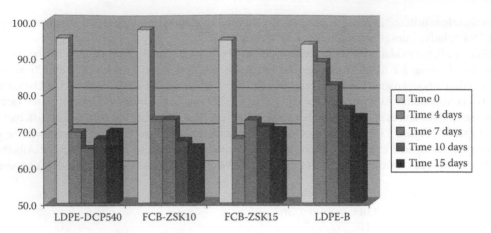

FIGURE 5.15 Contact angle determinations with distilled water as wetting agent of LDPE samples containing different EPI-TDPA prooxidant additives thermally treated in air in oven at 70°C.

the increase of wettability (Figure 5.15) of the thermally treated films, as determined by contact angle measurements, further demonstrated the polymer oxidation.

COi profiles recorded at 55°C and 70°C under dry and 75% RH condition, respectively, were showing that the predominant effect in the oxidation kinetic is played by the test temperature. The high humidity level, comparable to that occurring under real environmental conditions (e.g., composting) was only influencing, in some cases, the rate but not the overall extent of oxidation (Figures 5.16 and 5.17) (Chiellini et al. 2006a).

Thermally treated samples have been also characterized by HT-GPC, the recorded molecular weight and molecular weight distribution (ID) have been compared with the COi values detectable at the same time of thermal aging (Tables 5.9 through 5.11) (Chiellini et al. 2006a). A drastic decrease of Mw below 5 kDa and significant reduction of the ID have been thus recorded after a few days of thermal degradation under both dry and 75% RH conditions. In HT-GPC chromatograms relevant to LDPE samples retrieved at longest thermal degradation period, thus having the highest level of oxidation, elution peaks having a bimodal shape and very low (1.7 kDa) Mw fractions (Figure 5.18) have been also detected.

In addition, the relationship between the Mw and COi has been found to fit a mono-exponential trend (Figure 5.19). Accordingly, COi values may be used in order to predict the rate of Mw decrease as a function of the oxidation extent. Moreover, the recorded trend is in agreement with a statistical

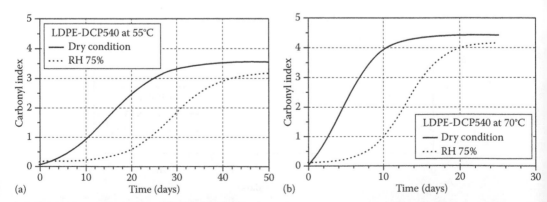

FIGURE 5.16 Carbonyl index (COi) variation of LDPE-DCP540 film sample aged in oven at 55°C (a) and 70°C (b), under both dry and 75% RH atmosphere.

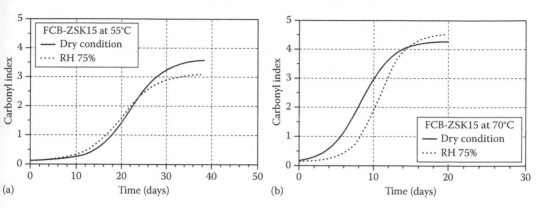

FIGURE 5.17 COi variation of FCB-ZSK15 film sample aged in oven at 55°C (a) and 70°C (b), under both dry and 75% RH atmosphere.

TABLE 5.9

Molecular Weight Analysis by HT-GPC of LDPE-DCP540 Sample Thermally Treated at 70°C

	Test Condition						
	Dry (Open Air)				RH ~ 75%		
Time (Days)	COi[a]	Mw (kD)	ID[b]	Time (Days)	COi[a]	Mw (kD)	ID[b]
0	0.61	39.4	4.24	2	1.60	10.7	2.92
1	1.14	19.5	2.96	3	2.23	10.4	2.88
2	2.32	9.7	2.59	9	5.37	4.8	1.37
9	5.44	4.5	1.27	N/A	N/A	N/A	N/A

[a] Carbonyl index as $D_{B\ 1720}/D_{B\ 1435}$.
[b] Dispersity index.

TABLE 5.10

Molecular Weight Analysis by HT-GPC of FCB-ZSK10 Sample Thermally Treated at 70°C

	Test Condition						
	Dry (Open Air)				RH ~ 75%		
Time (Days)	COi[a]	Mw (kD)	ID[b]	Time (Days)	COi[a]	Mw (kD)	ID[b]
1	0.63	45.7	3.94	2	0.57	28.6	3.63
2	1.40	16.4	2.75	6	2.54	9.9	2.59
3	2.33	10.1	2.47	11	6.20	4.9	1.33
9	5.35	4.4	1.25	N/A	N/A	N/A	N/A

[a] Carbonyl index as $D_{B\ 1720}/D_{B\ 1435}$.
[b] Dispersity index.

TABLE 5.11

Molecular Weight Analysis by HT-GPC of FCB-ZSK15 Sample Thermally Treated at 70°C

Test Condition						
Dry (Open Air)				RH ~ 75%		
Time (Days)	COi[a]	Mw (kD)	ID[b]	Time (Days)	COi[a]	Mw (kD)	ID[b]
1	0.67	34.4	3.66	3	1.49	18.1	3.53
3	1.59	14.9	2.91	5	2.89	8.1	2.68
5	2.83	7.6	2.44	11	6.20	4.2	1.44
6	4.53	5.1	1.32	N/A	N/A	N/A	N/A

[a] Carbonyl index as $D_{B\ 1720}/D_{B\ 1435}$.

[b] Dispersity index.

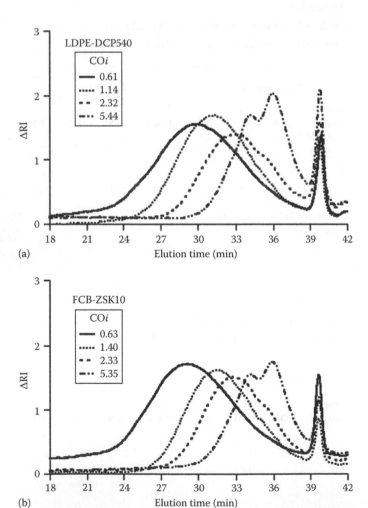

(a)

(b)

FIGURE 5.18 HT-GPC chromatograms of LDPE samples containing prooxidant additives. (a) DCP540 and (b) FCB-ZSK10.

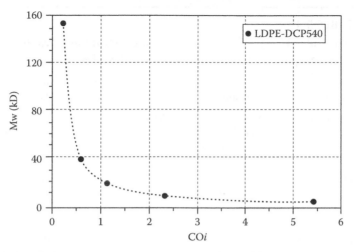

FIGURE 5.19 Molecular weight vs. COi relationship in LDPE-DCP540 film sample thermally treated in air in oven at 70°C.

chain scission mechanism, repeatedly suggested in the photophysical and thermal degradation of polyolefin (Scott 1994, Karlsson et al. 1997, Gugumus 2001).

The amount of oxidized degradation intermediates extractable by acetone, which can be considered more prone to be diffused in the environment, has been shown to positively correlated to the level of oxidation, thus reaching fairly high quantity corresponding to 25%–30% specimen weight (Table 5.12).

The recorded data have to be considered as partially representative of the overall amount of low Mw fractions produced during the thermal oxidation of the test material. Indeed, molecular weights

TABLE 5.12
Acetone-Extractable Fractions from Original and Thermally Treated LDPE Samples

Sample	Aging Time (Days) 70°C	55°C	COi[a]	Extract (%)	Residue (%)
FCB-ZSK10	0[b]	0[b]	0.534	7.7	91.9
	4	—	0.453	6.5	92.6
	—	42	3.583	17.9	82.2
	24		6.816	27.1	72.7
FCB-ZSK15	0[b]	0[b]	0.212	5.9	93.9
	5	—	2.864	9.2	88.4
	—	42	5.193	23.8	75.6
	24	—	7.256	22.6	74.4
LDPE-DCP540	0[c]	0[c]	0.627	5.5	94.2
	4	—	2.243	11.3	88.5
	—	42	4.818	21.1	78.2
	24	—	5.441	27.7	71.9

[a] Carbonyl index as $D_{B\ 1720}/D_{B\ 1435}$.
[b] Six months storage at room temperature.
[c] Seven months storage at room temperature.

TABLE 5.13

Molecular Weight Analysis of Acetone-Extractable Fractions from Original and Thermally Aged LDPE Samples

Sample	Treatment	COi[a]	Extract (% Weight)	Mw (kD)	ID
FCB-ZSK10	None[b]	0.534	7.7	1.52	1.49
	4 days at 70°C	0.453	6.5	1.47	1.46
	42 days at 55°C	3.583	17.9	1.30	1.39
	24 days at 70°C	6.816	27.1	0.92	1.32
FCB-ZSK15	None[b]	0.212	5.9	1.58	1.46
	5 days at 70°C	2.864	9.2	1.67	1.52
	42 days at 55°C	5.193	23.8	1.27	1.43
	24 days at 70°C	7.256	22.6	1.03	1.36
LDPE-DCP540	None[c]	0.627	5.5	1.08	1.27
	4 days at 70°C	2.243	11.3	1.49	1.41
	42 days at 55°C	4.818	21.1	1.08	1.37
	24 days at 70°C	5.441	27.7	0.89	1.33

[a] Carbonyl index as $D_{B\ 1720}/D_{B\ 1435}$.

[b] Six months storage at room temperature.

[c] Seven months storage at room temperature.

as low as 0.8–1.6 kDa for the acetone extracts of thermally treated samples have been recorded by SEC (Table 5.13).

It is also worth noting that at higher level of oxidation, increasing amounts of the extractable fractions are recorded, and in the meantime these fractions are characterized by very low Mw (Table 5.13) (Chiellini et al. 2006a). Therefore, the reported data suggest that cross-linking reactions do not affect notably the oxo-degradation behavior of the analyzed samples, which seems to proceed with cleavage of the macromolecules up to low molecular weight fractions capable to be assimilated by microorganisms (Potts 1972, Kawai et al. 1999, Volke-Sepúlveda et al. 2002). Indeed, earlier studies have demonstrated the microbial utilization as carbon source of the oxidation products formed during the thermal and photooxidation of polyethylene (Arnaud et al. 1994, Albertsson et al. 1995). Accordingly, in a respirometric experiment, aimed at evaluating the biodegradation behavior of low molar mass aliphatic hydrocarbons in soil, including linear docosane, (C22), branched 2,6,10,15,19,23-hexamethyltetracosane, (squalane—C30), and α,ω docosandioic acid, (C22), a fairly high rate and extent of mineralization (70%) of the acetone-extractable fraction from a thermally oxidized LDPE-TDPA sample was recorded (Figure 5.20) (Chiellini et al. 2003a).

This suggests that, similarly to low molar mass hydrocarbons, the oxidized fragments from LDPE can be rapidly biodegraded in natural soil.

Nevertheless, also the ultimate biodegradation propensity of the whole thermally treated LDPE-TDPA samples is a fundamental issue that has to be assessed. Thermally treated LDPE samples containing TDPA™ prooxidant additives from EPI Environmental Products Inc. Canada have been therefore assayed in respirometric tests aimed at simulating soil and mature compost incubation media.

There are several studies dealing with the evaluation of biodegradation of LDPE samples containing prooxidant and natural fillers that show only limited and slow conversion to carbon dioxide. The mineralization rate from long-term biodegradation experiments of both UV-irradiated samples (Albertsson and Karlsson 1988), non-pretreated, and additive-free LDPE samples, in natural soils indicates more than 100 years for the ultimate mineralization of polyethylene (Ohtake et al. 1998).

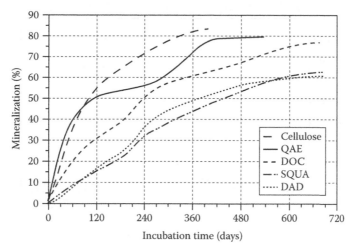

FIGURE 5.20   Mineralization profile of carbon substrates in soil burial respirometric tests. Cellulose = pure Whatman cellulose; DOC = docosane; SQUA = squalane; DAD = α,ω docosandioic acid; QAE = acetone-extractable fraction from a thermally oxidized LDPE-TDPA sample.

In a soil burial test, carried out in the presence of forest soil as incubation media, the biodegradation profile of a thermally pretreated LDPE-TDPA sample was monitored during 830 days of incubation (Chiellini et al. 2003a). The recorded mineralization kinetic (Figure 5.21) was characterized by the presence of a first exponential step that approached a 4% degree of biodegradation within 30 days of incubation. Afterward, a prolonged (120 days) stasis in the microbial respiration was recorded. At approximately 5 months of incubation, it was noted that a further and marked exponential increase in the biodegradation profile took place, thus reaching the highest extent of biodegradation 63.0% after 85 weeks of incubation (Figure 5.21). This two-step mineralization profile of thermally fragmented LDPE-TDPA samples has been repeatedly observed either in soil and mature compost biodegradation tests (Figure 5.22). In contrast with previous studies (Albertsson and Karlsson 1988, Ohtake et al. 1998), very large degree of mineralization

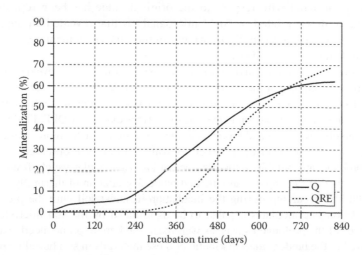

FIGURE 5.21   Mineralization profile LDPE-TDPA samples in soil burial respirometric tests. Q = LDPE sample containing a prooxidant additive, aged in an air oven at 70°C for 14 days prior to testing; Q-RE = residue of the aged Q-LDPE sample after extraction of the fraction (ca 25 wt.%) soluble in refluxing acetone.

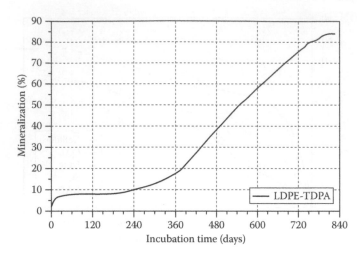

FIGURE 5.22 Mineralization profile LDPE-TDPA thermally oxidized sample in mature compost respirometric tests.

(70%–80%) has been then recorded, even though in a relatively long time frame (more than 800 days) (Chiellini et al. 2006b).

The first step in the biodegradation kinetic of thermally degraded LDPE in soil could be attributed to the preferential microbial assimilation of low molecular weight oxidized fragments present on the surface of the LDPE specimens, as previously suggested (Albertsson et al. 1987, 1995, Volke-Sepúlveda et al. 2002, Bonhomme et al. 2003). Albertsson et al. (1998) reported that abiotically aged pure LDPE, LDPE/starch, and LDPE-additives promoting degradation were characterized by the presence of several degradation products such as mono- and dicarboxylic acids and ketoacids. These almost completely disappeared after the incubation of the polymer samples in the presence of *Arthrobacter paraffineus* as consequence of the assimilation of the degradation products by the bacterial strain.

This hypothesis appears to be confirmed by the FT-IR characterization of soil-buried specimens retrieved after a few months of incubation. Indeed, a significant reduction of the intensity of carbonyl absorption band with respect to the original value has been repeatedly recorded. This result has been also validated by SEM microanalysis that revealed an appreciable reduction of the oxygen from 20.6% to 16.9% by weight in the surface elemental composition of soil-incubated sample (Chiellini et al. 2003a).

By contrast, a substantial increase in the carbon–carbon unsaturation, as revealed by the increase of the double bond index from 0.55 at the beginning to 0.88 after 62 weeks of incubation, was recorded.

This can be attributed to enzymatic dehydrogenation (Albertsson et al. 1998, Ohtake et al. 1998). Hence, it is likely that the macromolecular cleavage of thermal-oxidized LDPE-TDPA can be achieved concomitantly by abiotic oxidation and biotic, enzymatic scission. Furthermore, a dramatic change in the fingerprint region of the IR spectrum between 1300 and 950 cm^{-1} was observed with increasing the incubation time probably attributable to the presence of lower molecular weight fragments.

Therefore, it seems that the ongoing abiotic and biotic degradation of thermally degraded LDPE-TDPA polymer bulk is occurring during the incubation in forest soil with the production of large amount of degradation intermediates capable to be assimilated by the soil microflora. This might explain the presence of the prolonged stasis, as well the second, more pronounced, exponential phase repeatedly observed in the biodegradation kinetic of these materials in soil burial respirometric tests.

A similar approach has been utilized in order to evaluate the propensity to biodegradation of photo-degradable LLDPE samples for mulching purposes. Two films with prooxidant additives (LLDPE-TD1 and LLDPE-TD2) and a control (LLDPE-TD0) were firstly exposed to sunlight for

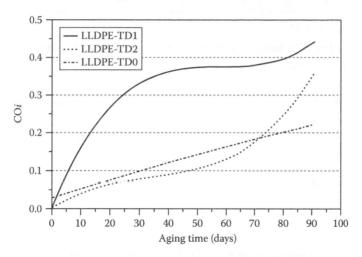

FIGURE 5.23 COi profiles of LLDPE-TD films exposed to sunlight for 3 months.

3 months during the summer season and analyzed for oxidation by measuring the changes in their COi. An appreciable increase in the carbonyl absorption peaks was apparent in films with additives compared to the control without any additive, indicating a much higher level of oxidation in films with prooxidants. Several studies already indicated that environmental degradation of LLDPE can be improved by the use of prooxidants (Chiellini et al. 2003a, 2006a, Jakubowicz 2003).

As expected, control sample without prooxidant additives also showed a steady increase in the COi but at a significantly slower rate compared to the level of oxidation achieved in films with additives. The observed increase of COi in LLDPE films under sunlight (Figure 5.23) is attributed to the initiation of chain scission as a result of photooxidation producing fragments in the form of shorter, more readily crystallizable molecules (Volke-Sepúlveda et al. 2002, Chiellini et al. 2003a, 2006a).

Further characterization of these samples by DSC showed a slight increase in the crystallinity as well as in the melting temperature (T_m) in both films with additives after sunlight exposure in comparison to unexposed samples (Table 5.14). The increase in crystallinity was observed after both the first and the second heating, but the increase was much more pronounced after the first heating.

The mere exposure to sunlight led to a considerably lowered value in the onset degradation temperatures (T_{ON}) in films with additives when compared to unexposed controls during TGA analysis (Table 5.15). Particularly, the decrease in T_{ON} was more prominent (about 119°C) in films with TD2

TABLE 5.14

DSC Characterization of Crystallinity in Films Exposed to Sunlight

Test Sample (Type)	Aging Time (Days)	First Heating		Second Heating	
		$T_m{}^a$ (°C)	Crystallinity (%)	$T_m{}^a$ (°C)	Crystallinity (%)
LLDPE-TD1	0	115.0, 121.8	42.7	110.9, 122.3	44.7
	93	116.0, 121.8	48.5	121.1	46.3
LLDPE-TD2	0	121.7	41.9	110.0, 121.5	44.2
	93	117.7, 122.0	52.4	121.5	47.5
LLDPE-TD0	0	122.0	43.4	111.6, 121.2	42.9
	93	—	—	—	—

a Melting temperature.

TABLE 5.15

Thermal Properties of Sunlight-Exposed Films Obtained by TGA Analysis

Test Sample (Type)	Aging Time (Days)	T_{ON} (°C)	Residue (Weight %)	CO_i
LLDPE-TD1	0	400.6	0.24	0.11
	93	353.5	0.88	0.46
LLDPE-TD2	0	422.9	1.03	0.04
	93	304.2	1.97	0.29
LLDPE-TD0	0	409.3	0.79	0.11
	93	381.7	1.20	0.15

additive. Additionally, a concomitant increase in the percentage weight of the residues as well as in CO_i upon thermal degradation of aged (93 days in sunlight) films was also apparent. As repeatedly ascertained, sunlight-induced aging resulted in the formation of low molecular weight products which readily degraded further in subsequent thermal degradation during the TGA analysis. Because oxidation is primarily confined to the amorphous portion of the polymer bulk, the remainder of the polymer is more susceptible to molecular reorganization which may explain the increase in the crystallinity of the prooxidant-containing photooxidized films. Any apparent change in the crystallinity or in the amount of residue was noticeable in LLDPE not containing prooxidant additive.

Photodegradable LLDPE films (LLDPE-TD1 and LLDPE-TD2) preexposed to sunlight for 93 days were submitted to biodegradation in the presence of single fungal strains previously isolated from thermally oxidized LDPE fragments confined in forest soil (Corti et al. 2010).

Controls containing unexposed films were also incubated with fungal strains along with test films under identical condition. Films exhibited profuse fungal growth; however, they were distinct in their growth behavior. While one strain that produced mycelia preferentially colonized the film surface directly, the other strain that produced spores only slightly colonized the surface but grew mostly around the periphery of the film (Figure 5.24). The film colonized by the first strain was further analyzed by SEM (Figure 5.25). A vibrant growth of mycelia on the surface of the film aged

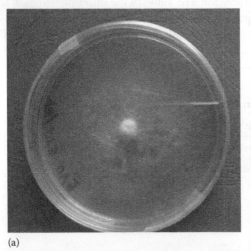 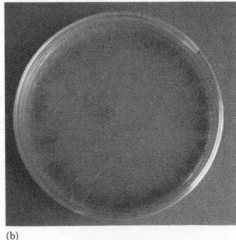

(a)   (b)

FIGURE 5.24   F2 (a) and F3 (b) soil fungal strains growth on the LLDPE-TD1 films preexposed to sunlight. The incubation period comprised of 20 days.

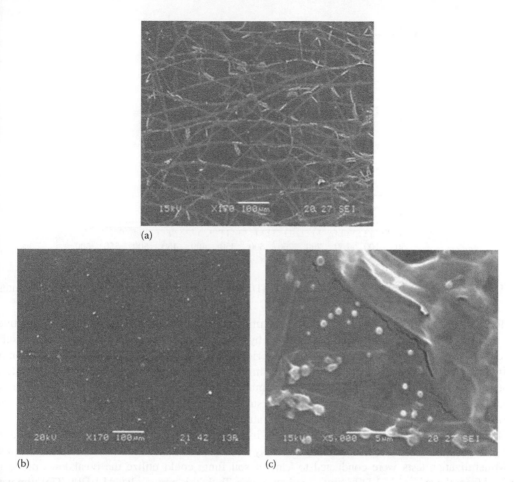

FIGURE 5.25 SEM micrographs of the fungal strain F2 mycelium growing on the surface of outdoor-exposed LLDPE-TD2 specimens (a). F3 strain spore-forming fungus that preferred not to colonize the surface (b) but grew around the periphery of the film; (c) is a higher magnification of (a), showing fungal hyphae colonizing the film surface.

in sunlight was apparent (6A) indicating a robust fungal metabolic activity. The growth of mycelia on the film surface showing cracks in the film developing around the edge of mycelium was also visible at a higher magnification (6B).

A similar behavior has been also observed by Weiland et al. (1995) for thermally oxidized PE films that were further treated in compost for a period of time. Film surfaces were increasingly colonized by fungi as examined by SEM, and the increase in fungal presence on the surface was accompanied by a decrease in molecular weight of the film, providing a strong evidence of fungal bioassimilation of the amorphous regions of thermally oxidized films.

The FT-IR spectra of the test and control samples reveal the prominence of absorption bands attributed to the C=C within the polymer chain in films incubated with fungal strains. Additionally, new absorption bands in the fingerprint region between 1700 and 1100 cm^{-1} of the spectra and a considerably elevated –OH absorption peak in the v_{OH} region were also noticeable in the samples incubated with fungi for 65 days but not in the controls. The appearance of new bands can be then associated with the formation of metabolites generated by the fungal attack on the pre-oxidized fragments.

The TGA traces also revealed that fungal biodegradation of films led to a decrease in their thermal stability, while control films remained unchanged (Figure 5.26).

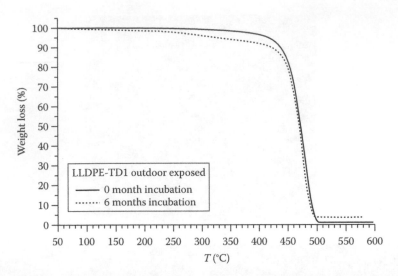

FIGURE 5.26 TGA traces of pre-oxidized LLDPE-TD1 films incubated with and without fungi for 6 months.

To examine condition that could bring a maximal deterioration in PE-LLD films with additive, films that were exposed to sunlight and followed by fungal incubation for 65 days were further subjected to thermal aging in a oven at 45°C for a period of 40 days and were analyzed for variation in their COi. Controls consisted of films that went through the same treatments, except the incubation treatment with fungi. This scenario closely resembled the PE-based agricultural mulch films used by growers to achieve higher crop yields that are subjected to similar treatments, that is, abiotic and biotic treatments followed by further low temperature aging.

Both test and control films showed a steady increase in the COi during the 40-day period of aging, but the test samples exhibited a significantly higher extent of oxidation reached at higher rate.

Mineralization tests were conducted to learn if soil fungi could utilize the breakdown products derived from the oxidized LLDPE film as carbon source. To do this, pre-oxidized LLDPE-TD1 film was incubated on a solid-agar media with soil fungi in the presence of glucose as secondary carbon source. The control consisted of pristine LLDPE films that had not been preexposed to sunlight (Figure 5.27).

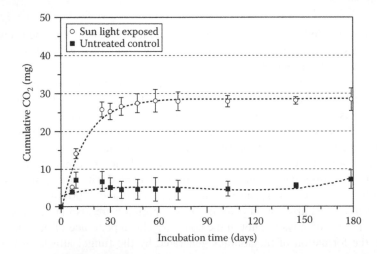

FIGURE 5.27 Average cumulative CO_2 production in pre-oxidized LLDPE-TD1 films incubated with soil fungi on agar containing glucose. Control consisted of similar films not submitted to any pretreatment (pristine).

The CO_2 production resulting from the microbial breakdown of the polymer and/or residue was monitored via respirometry for 180 days. Expectedly, both samples (test and control films) showed CO_2 production because a small amount of glucose was present in the media. But, the amount of CO_2 production in pre-oxidized samples was remarkably elevated compared to similar unoxidized control films, even though the same amount of glucose or co-metabolite was used in both samples (Figure 5.27). These results clearly demonstrated that the reason for a much higher respiration (CO_2 production) in oxidized films was bound to the ability of fungi to utilize the oxidation products present in the films as an additional carbon source. Weiland et al. (1995) used respirometry to study the consumption of O_2 in a highly oxidized PE sample at two different polymer concentrations (0.05% and 0.006%) incubated in the presence of microorganism from compost. Interestingly, they found that the polymer degradation was significantly higher at lower polymer concentration. Based on theoretical values, 54%–64% of the total polymer degraded at lower substrate concentration (0.006%) compared to 28–34 at higher substrate concentration (0.05%), suggesting that microbial bioassimilation of a photooxidized polymer is concentration dependent and is much higher at lower polymer concentration.

The pre-oxidized LLDPE films submitted to the fungal metabolism experienced an enhancement in the amount breakdown products as reflected by the increase in the FT-IR peak intensities typical of several chemical functional groups. Interestingly, these increases were only visible in pre-oxidized films incubated with fungi but not in control films. This fact once again is pointing toward the synergistic impact of biotic and abiotic factors in achieving efficient LLDPE degradation. Figure 5.28 depicts a magnified region of the spectrum between 1800 and 900 cm^{-1}, where notable increase in several peaks attributed to various functional groups such as –C=O (1740 cm^{-1}), –C=C– (1640 cm^{-1}), –OH bending vibration (1080 cm^{-1}), and –OH stretching vibration (3400 cm^{-1}) is prominently displayed.

Additionally, an increase in the carbonyl peak intensity and relevant COi as detectable in the FT-IR spectrum of Figure 5.28 in pre-oxidized films was observed (Table 5.16). Furthermore, an increase in the crystallinity of the LLDPE samples exposed to microbial action observed during the first DSC heating analysis (Table 5.17) gave further substantiation to the assimilation of low molecular weight-oxidized fractions of the amorphous polymer regions.

The data indicated that LLDPE degradation was accompanied by changes in chemical, physical, and mechanical properties, as observed by changes in the thermal degradation, degree of crystallinity, COi, and functionality to the polymer degradation products. In particular, there is considerable evidence that scores of smaller molecular weight products are generated during the abiotic and biotic degradation of PE polymer-loaded prooxidants that support microbial growth, with ultimate

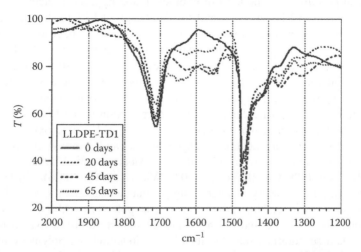

FIGURE 5.28 A magnified region of the spectrum between 900 and 1800 cm^{-1} of outdoor-exposed LLDPE-TD1 films incubated with fungal strain.

TABLE 5.16

TGA Analyses of Control and Pre-Oxidized LLDPE-TD1 Films Incubated with Fungi

LLDPE-TD1	Incubation Time (Days)	T_{ON} (°C)	Residue at 600°C (%)	COi
Pristine	0	400.6	0.24	0.11
	178	381.5	1.49	0.21
Sunlight exposed	0	353.5	0.88	0.46
	178	235.7	3.33	0.70

TABLE 5.17

DSC Analyses of Control and Pre-Oxidized LLDPE-TD1 Films Incubated with Fungi

		First Heating		Second Heating	
LLDPE-TD1	Incubation Time (Days)	T_m (°C)	Crystallinity (%)	T_m (°C)	Crystallinity (%)
Pristine	0	115.0, 121.8	42.7	110.9, 122.3	44.7
	178	113.8, 120.6	42.1	111.2, 122.6	43.4
Sunlight exposed	0	116.0, 121.8	48.5	121.1	46.3
	178	117.5, 121.4	55.4	111.5, 121.4	51.2

polymer biodegradation (Chiellini et al. 2006a). Albertsson et al. (1995) utilized GC-MS to identify many of the degradation products generated during the abiotic and biotic attacks of an LDPE film containing starch and prooxidant. Interestingly, many of lower molecular weight products that were present after the abiotic oxidation conspicuously disappeared or significantly reduced in amount when these films were subjected to biotic treatment. In particular, carboxylic acid moieties that were the major products of an abiotic oxidation completely disappeared during the biotic degradation.

The information pertaining to the level of photooxidation required to achieve an effective and sustained biodegradation in LLDPE is crucial for the design of LLDPE-based products and prediction of their environmental fate. Nowadays, information with respect to synergistic effects of microbial/enzymatic attack and physical-chemical parameters in promoting the degradation of partially oxidized LLDPE can be provided. The data suggest that the degradation of oxo-biodegradable LLDPE samples is enhanced by the synergistic action of both abiotic and biotic factors after their initial oxidation by exposure to direct sunlight. Full carbon backbone polymers, including polyethylene, are prone to oxidation, particularly to sunlight irradiation in the UV region. The oxidation process results in the formation of intermediate groups, such as hydroperoxides and peroxides, as well as functional groups such as alcohol, aldehyde, ketone, and carboxyl groups from the oxidative cleavage of the polymer chains (Arnaud et al. 1994, Albertsson et al. 1995, Weiland et al. 1995, Karlsson and Albertsson 1998, Chiellini et al. 2003c, Wiles and Scott 2006). The level of oxidation, however, is determined by a variety of interacting parameters such as the diffusion of oxygen in the polymer bulk, presence of tertiary carbon atoms in the polymer backbone, intensity of sunlight, length of exposure, and the type and extent of pro-degradant additives loaded in the film, and temperature. The data substantiate the validity of oxo-biodegradable polymers and confirm that LLPDE films with prooxidant additives are environmentally degradable by means of a combination of abiotic and biotic factors. Thus, several conclusions can be drawn from the fungal biodegradation studies on LLDPE films with prooxidant additives: (a) prolonged sunlight exposure causes considerable oxidation in these films leading to their deterioration and the production of degradation products, (b) soil-borne fungi are fully capable of utilizing degradation products as a carbon source, (c) presence of an additional carbon source (under co-metabolic conditions) could stimulate fungal metabolic activities to further promote the degradation of the LLDPE polymer matrix, possibly mediated by hydrolytic enzymes produced by fungi, and (d) more effective degradation of films with prooxidant additives could be achieved by a combined impact of abiotic and biotic factors.

Currently, LLDPE is used as agricultural mulch films, but it is extremely recalcitrant and its abiotic oxidation and physical integrity damages are negligible unless biodegradable fillers and/or prooxidants are added. In case of the presence of only biodegradable fillers, once these are metabolized, large volume of mostly used PE films is left behind for farmers to deal with causing fairly huge economic efforts bound to their disposal. The LLDPE with totally degradable (TD) prooxidant additives offers a much better alternative to traditional PE-based products. With the exception of sunlight exposure, most of the data were obtained in laboratory setting and further testing of these materials in a real-world situation is suggested, as needed for a general acceptance by customers and public opinion.

5.7.3 Microbial Cell Biomass Production during the Biodegradation of Full Carbon Backbone Polymers

The increasing development of biodegradable plastic items introduced since late 1980s has promoted in government authorities and stakeholders initiatives aimed at issuing by technical standardization bodies norms and laboratory test methods to assess the ultimate environmental behavior of plastics.

All the aforementioned procedures are stating that the biodegradation of a test material, under aerobic conditions, has to be determined as the extent of carbon substrate evolved to carbon dioxide as a consequence of microbial attack. That amount of carbon dioxide evolved is quantitatively referred to the corresponding value reached by cellulose taken as positive control, under the same operative conditions and time frame.

It has to be taken into consideration, however, that under aerobic conditions, heterotrophic soil microbial community metabolizes carbon substrates either as carbon dioxide and for the production of new biomass, while some of it is converted, through chemo-enzymatic reactions, to humic substances.

It is generally accepted that in relatively short term, 50% carbon content of most organic substrates is converted to CO_2, the remaining part being assimilated as biomass or humified.

The relationship between the structural features of low molar mass carbon substrates and the growth efficiency of microorganisms incubated on them has been thoroughly investigated, also by thermodynamic studies. In particular, relatively simple, closed systems, such as those represented by liquid cultures of certain microorganisms fed with specific carbon sources, have been assessed.

In aqueous media, heterotrophic bacteria utilize the organic substrate either as carbon and energy source, that is, the input carbon is driven into the biosynthetic pathway by anabolism and can as well be utilized for energetic demand through catabolism.

Metabolic efficiency is therefore strictly correlated to the cellular metabolic network that includes interrelated catabolic and anabolic pathways, especially under substrate-limited conditions (Forrest 1969, Lehninger 1975, Bitton 1994), being the carbon converted to biomass or the growth yield proportional to the anabolic activity. For instance, it has been repeatedly established, that anabolic processes, and consequently growth efficiency, are positively correlated with the standard free energy of oxidation ($\Delta G°$) of a particular carbon substrate (Schroeder and Bush 1968, Burkhead and McKinney 1969, Sykes 1975, Heijnen and van Dijken 1992, Schill et al. 1999). This correlation has been recently confirmed by studying the ratio between organic carbon channeled into biomass and that converted to CO_2 under substrate-limited cultures when fed with different types of organic substrates (Turner et al. 2001).

In the presence of carbon substrate characterized by a relatively low free energy content, a large portion of carbon can be converted to CO_2, and the energy liberated from catabolic reactions is not substantially utilized for cell growth and proliferation, but it is rather used to sustain the living activities of bacterial cells. Therefore, in the presence of low free energy (e.g., oxidized substrates) carbon sources, large amounts of carbon appear to be utilized for the maintenance functions, thus providing great production of carbon dioxide (e.g., high S_{CO_2}/S_B ratio), rather than consumed for growth purpose (e.g., low Ys and hence limited cell growth) (Table 5.18).

The role of free energy content of carbon substrate repeatedly evidenced in the metabolic efficiency of heterotrophic microorganisms in aqueous media and under substrate-limited conditions could be also reasonably holding firm in the soil microbial transformation of organic substances.

The substrate carbon channeling between the anabolic and catabolic pathways was evaluated by determining the gross cell growth yield (GCY) in the aqueous medium, as well as the net soil microbial biomass (SMB-C) as a function of the substrate addition in soil burial respirometric tests. The collected data had been analyzed against the amounts of carbon converted to CO_2 (mineralization).

The GCY was determined by gravimetric analysis of dry cell biomass divided by the amount of organic carbon in the substrate submitted to biodegradation test. Accordingly, the liquid cultures that once approached the plateau phase were submitted to centrifugation, and the resulting pellets washed with distilled water and dried at 105°C up to constant weight.

TABLE 5.18

Growth Yield (Y_s) and S_{CO_2}/S_B Ratio in Substrate-Limited Cultures of Activated Sludge

Substrate	Formal C-Oxidation Number	$\Delta H_C^{\circ a}$ (kJ g^{-1})	Y_s^b	S_{CO_2}/S_B^c
Carbohydrates				
Glucose	−0.16	−15.6	0.80	0.25
Fructose	−0.16	−15.6	0.71	0.41
Sucrose	−0.16	−16.5	0.72	0.40
Lactose	−0.16	−16.5	0.71	0.41
Xylose	−0.16	−15.6	0.63	0.59
Amino acids				
Glycine	+1	−12.9	0.38	1.67
Alanine	0	−18.2	0.53	0.88
Glutamic acid	+0.4	−13.5^d	0.48	1.08
Phenylalanine	−0.25	−28.1	0.55	0.82
Fatty acids				
Acetic acid	0	−14.6	0.53	0.90
Propionic acid	−0.6	−20.6	0.63	0.58
Butyric acid	−1	−24.8	0.84	0.19
Miscellaneous				
Butanol	−1.75	−36.1	0.78	0.29
Benzoic acid	+0.3	−26.4	0.46	1.19

[a] Values taken from Lide, D.R., *Handbook of Chemistry and Physics*, 85th edn., CRC Press, Boca Raton, FL, 2004.

[b] Determined as weight of carbon incorporated in cell biomass/overall weight carbon in substrate (g C_{MB}/mg C_S).

[c] Ratio of carbon evolved as CO_2 (S_{CO_2}) to that converted to biomass (S_B) (mg C/mg C).

[d] ΔG^0 of oxidation.

In soil burial respirometric tests, the organic C fixed in SMB-C was estimated at the end of the tests according to the chloro-fumigation, followed by K_2SO_4 extraction procedure reported by Turner et al. (2001).

The net CO_2 emission (e.g., mineralization) and the GCY associated to the metabolization of a water-soluble polymeric material like PVA grade 98 and various low molecular weight compounds were evaluated in respirometric biodegradation experiments carried out in a mineral liquid medium (0.4–0.6 g substrate/L), under substrate-limited condition, in the presence of a selected microbial inoculum (Chiellini et al. 1999a).

In particular, the metabolization pattern of aliphatic alcohols, such as 1-decanol and 2-octanol, was compared with the carbon utilization profile of compounds with lower free energy content that is with carbon in an average higher oxidation state, such as 3 hydroxybutyric acid and 2,4-pentanediol. The percentage of theoretical cumulative net CO_2 emissions of tested substances is reported in Figure 5.29.

3-Hydroxybutyric acid, with 0.5 formal average oxidation number of carbon, was shown to be easily metabolized, without a significant lag phase, throughout an extensive catabolic conversion to CO_2. On the contrary, the other substrates required a longer time (6–10 days) before appreciation of the onset of the mineralization process.

The overall extension of mineralization was also significantly different for the used substrates, being the highest value recorded in the culture fed with 3-hydroxybutyric acid (57%), and the lowest (22%) in those supplemented with 2-octanol (Figure 5.29, Table 5.19). Interestingly, complementary GCY values were recorded as in the case of 2-octanol whose assimilation accounted to 0.61 GCY per gram of added organic carbon in spite of fairly low mineralization extent (22.0%) (Table 5.19).

It is also worth noting that a positive correlation between the recorded GCY value and the enthalpy of combustion of the selected organic substrates can be detected (Table 5.19), thus confirming the existence of a correlation between the metabolic efficiency and the free energy content of the carbon substrate (Schroeder and Bush 1968, Burkhead and McKinney 1969, Sykes 1975, Heijnen and van Dijken 1992, Schill et al. 1999).

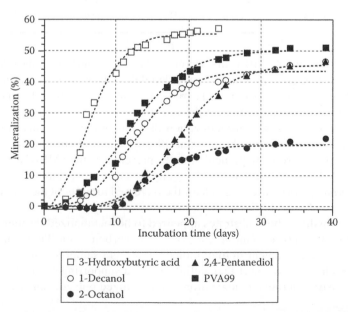

FIGURE 5.29 Mineralization profiles of different carbon substrates in the aqueous respirometric test.

TABLE 5.19
Growth Yield, Maximum Mineralization Extent and Heat of Combustion of Carbon Substrate Used in Aqueous Respirometric Test

Carbon Substrate	$\Delta H_C^{\circ a}$ (kJ g^{-1})	Th. CO$_2$ (%)	Growth Yield (mg MB/g C)
3-Hydroxybutyric acid	−21.5	57.1	0.39
1-Decanol	−41.5	46.6	0.56
2-Octanol	−40.6	22.0	0.61
2,4-Pentanediol	−31.0	46.5	0.41
PVA99	−19.0	51.0	0.34

MB, dry cell biomass.

[a] Values taken from Lide, D.R., *Handbook of Chemistry and Physics*, 85th edn., CRC Press, Boca Raton, FL, 2004.

TABLE 5.20
SMB-C Assessment in Soil Burial Respirometric Tests

Soil Culture	SMB-C (µg g^{-1} Dry Soil)	SMB-C Total (mg)	Sample-C Input (mg)	Sample-C Mineralization (%)	Sample-C Assimilation (%)	ΔH_C° (kJ g^{-1})
Blank	131.1 ± 11.0	3.3 ± 0.3	—	—	—	—
Cellulose	83.2 ± 3.0	2.1 ± 0.1	150.7	85.9	−0.7	−14.7
DOC	780.0 ± 14.0	19.5 ± 0.4	307.2	77.5	5.3	−47.2
SQUA	594.4 ± 13.0	14.9 ± 0.3	294.3	63.1	3.9	−47.3
DAD	494.2 ± 13.0	12.4 ± 0.3	277.3	61.1	3.3	−37.0
Q-LDPE	1034.1 ± 46.6	25.8 ± 1.2	278.1	63.8	8.5	−46.5
Q-RE	1105.5 ± 54.7	27.6 ± 1.4	312.7	69.6	7.8	−46.5

In Table 5.20, the amount of SMB-C in different soil cultures, along with the relevant overall fate of substrate carbon input, are reported. The recorded data indicate that the observed considerable differences in both mineralization extent and carbon substrate conversion to biomass are also in this case correlated to the free energy content of the substrate.

As previously reported, the soil metabolization of a glucosidic-like material such as cellulose does not allow the fixation of appreciable amounts of carbon as microbial biomass, being the major part converted to carbon dioxide. In particular, in the present test, the SMB-C associated to the soil cultures supplemented with cellulose resulted slightly lower than the blank, thus substantiating the promotion of the *priming effect* involving at least the increased turnover of the original soil microbial biomass (Dalenberg and Jager 1989).

On the contrary, hydrocarbon substrates appeared to be metabolized throughout a substantial conversion to SMB-C, thus accounting for 10%–15% carbon assimilation in the presence of relatively high free energy content in high molecular weight samples, whereas the carbon conversion to CO$_2$ was noticeably lower than cellulose. As a supporting element to this behavior, a picture of the extensive microbial colonization of an oxidized polyethylene fragment is reported in Figure 5.30.

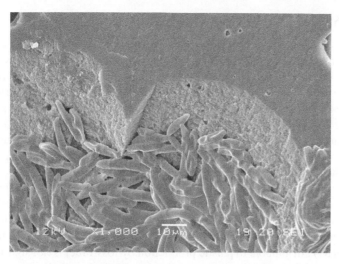

FIGURE 5.30 SEM micrograph of thermally oxidized LDPE fragment from soil burial test (fungal colonization).

There are some evidences that biodegradation in soil of carbon substrates with low free energy content (i.e., high level of oxidation) as measured only by net CO_2 evolution could be to some extent overestimated.

Actually, these compounds have been found to provoke the soil microbial consortia shift toward fast-growing species (e.g., zymogeneous), with low metabolic efficiency (i.e., high rate of carbon substrate conversion to CO_2). In these microorganisms, carbon substrate utilization patterns are correlated with relatively simple enzymatic tools, which account for the almost exclusive metabolization of easily assimilable compounds, especially carbohydrates and amino acids, whereas more complex carbon substrates cannot be utilized. As a consequence, once the substrate (e.g., glucose) is consumed, zymogeneous microbial cells die, thus determining a fast microbial biomass turnover, accompanied by the release of additional CO_2 deriving from the cellular component. As a result, in relatively short time, the addition of glucosidic material to soil is characterized by a great extent of respiration accompanied by a very little conversion to biomass or humic substance. Furthermore, the induction of the extra-mineralization of soil organic matter (e.g., *priming effect*) by glucosidic materials may increase as well the level of soil respiration with depletion of carbon in the soil.

These effects, that may act synergistically, could be crucial in biodegradation tests wherever the reference standard materials like cellulose or starch are used.

Indeed, this aspect can account for a potential underestimation of the biodegradation level of test compounds, especially when they consist of carbon atoms in fairly low formal oxidation state and hence in relatively high free energy content.

According to these considerations, the use of different reference materials representative of the different classes of carbon substrates in standardized soil biodegradation tests is recommended. Nevertheless, a front of the various interacting factors affecting the microbial transformation of carbon substrates in soil, the carbon balance including both biomass and carbon dioxide evolution should be also taken into account in standardized test procedures.

The substrate carbon channeling between the anabolic and catabolic pathways was evaluated by determining GCY in the aqueous medium, as well as the net SMB-C as a function of the substrate addition in soil burial respirometric tests. The collected data had been analyzed against the amounts of carbon converted to CO_2 (mineralization).

5.8 CONCLUSIONS

Plastic waste accumulation and problems connected with waste collection and eventual recycling are promoting the interest for plastic materials suitable to be bio-recycled not only in compost plants but also in soil and water media. The worldwide available biomass could largely supply low-cost polymeric matrices and fillers for bio-based production without negative interference with food chain. Thus, the use of by-products and waste collected in agriculture, forestry, and diary industry is gaining growing attention in the bio-based plastic production.

Gelatin from tannery, and lignocellulosic fibers derived from fruit juice production or fermentation to ethanol represent a valuable, low-cost, and largely available source of natural materials for bio-based composite materials.

In consideration of the modest percentage of fossil fuel devoted to plastic production (4% on worldwide oil usage), the utilization of degradable petro-derived polymeric matrices consisting of full carbon backbone chain-like polyolefins (PE, PP, PS, and rubber) and poly(vinylalcohol) (PVA) appears very convenient for the production of ecocompatible plastic items. Thus, the effect on green house emission (GHG) attributable to plastic production by fossil fuel is not significant if compared with the impact of fossil fuel used for energy production (93% of total oil usage).

Hydro-biomulching formulations realized by water suspensions of either gelatin or PVA and natural fillers resulted applicable on field trials by conventional machinery and outlined a mulching and soil structuring positive effect both on laboratory scale tests as well as in open field trials.

In hybrid bio-based composites production, PVA has been successfully blended with waste gelatin from pharmaceutical industry, SCB, orange and apple waste, by-products of juice extractions, as well as corn starch and CF by-products of ethanol industry. Compression-molded, injection–molded, and foamed items were prepared and presented valuable thermal and mechanical properties accomplished with a good resistance to aging. For example, injection-molded samples conserved consistence and mechanical properties with over 1 year of storing at room conditions. Improvements on processing and water resistance of starch-based foamed trays were achieved by the introduction of CFs in the formulations.

Biodegradation in soil of the prepared items was promoted by the natural components presence in the composites. This behavior will open new sceneries in the design and formulation of novel polymer blends and composites susceptible to biodegradation at the end of their service life occurring either in controlled infrastructures or in environmental sites wherever not intentionally disposed.

Studies on the biodegradation of PVA samples outlined that a certain grade of hydrophobicity achieved by a discrete content of vinyl acetate units may be taken as a practical parameter for indication of the propensity of PVA samples to be biodegraded in solid media. A similar influence was recognized for the degree of polymerization, resulting in a higher mineralization degree for the sample with lower molecular weight. On the contrary, in the biodegradation tests carried out in aqueous medium, in the presence of PVA-acclimated microorganisms, the driving force in PVA biodegradation appears to be its solubility that accounts for the endocleavage of the polymer chains by specific enzymatic systems.

Thermally promoted oxidation of reengineered polyethylene samples as introduced by EPI Inc. (Canada) containing TDPA™ prooxidant additives and successively by Symphony as d2w additives can be assayed in oven at different temperatures (55°C and 70°C) mimicking the thermophilic phase of the composting process.

Oxidation of the polyethylene matrix in the thermally aged samples has been clearly ascertained by the FT-IR spectroscopy. The relationship between the molecular weight and carbonyl index (COi) has been found to fit a mono-exponential trend; thus, COi values can be conveniently used in order to predict the rate of Mw decrease as a function of the oxidation extent. Moreover, the recorded trend is in agreement with a statistical chain scission mechanism, repeatedly suggested in the photo and thermal degradation of polyolefin.

Therefore, it seems that the ongoing abiotic degradation followed by the biotic one of thermally degraded of different grades of PE-TDPA polymer bulk is occurring during the incubation in forest soil with the production of large amount of degradation intermediates capable to be assimilated by the soil microflora.

It is also finally inferred that in biodegradation-level assessment of different polymeric as well as low molar mass components, it is important to perform a mass balance taking into account the amount of carbon that is converted to biomass.

This approach may allow then to pose all the materials submitted to biodegradation tests on the same level for a fair comparison consisting of carbon atoms in a different state of oxidation with hence the appropriate choice of the reference standard material employed for the validation of the test. The commonly used cellulose as a reference standard does not properly apply in biodegradation trials to be performed on macromolecular hydrocarbon matrices consisting of carbon atoms at lower level of the formal oxidation number.

Therefore, in standardized soil biodegradation tests, the use of different reference materials representatives of the different classes of carbon substrates should be recommended with the assessment of the carbon balance including both biomass formation and carbon dioxide evolution.

In almost 20 years of intensive activity spent by the industrial and academic world in the design and promotion of bio-based plastic items, susceptible to be biodegraded under composting conditions, the production of such kind of plastics as obtained by processing of first- or second-generation bio-based polymeric materials is to be generous not higher than 0.2% of the overall worldwide plastic consumption that nowadays is close to 300 million tons. This fact should sound like a warning signal or better an alarm bell for moving toward reengineering of the consolidated hydro-biodegradable and the more recently introduced oxo-biodegradable polymeric materials obtained from fossil fuel feedstock.

ACKNOWLEDGMENT

This chapter is an update of a former contribution stemming from the continuing action undertaken from 1996 to 2008 by ICS-UNIDO on environmentally degradable polymeric materials and plastics as a sustainable option in the plastic waste managements. The Environmental Products Inc., EPI Company Canada, and the Ecopol srl Italy are acknowledged for the support provided in the evaluation of TDPA-PE samples and hydrolene-PVA samples, respectively. The contribution provided by the Inter University Consortium of Material Science & Technology (INSTM) through the Reference Centre of Biomaterials & Ecocompatible Materials located in the BioLab at San Piero a Grado section of the Department of Chemistry and Industrial Chemistry of the University of Pisa, Italy is finally acknowledged.

REFERENCES

Albertsson, A.-C., Andersson, S. O., and Karlsson, S. 1987. The mechanism of biodegradation of polyethylene. *Polym. Degrad. Stab.* 18: 73–87.

Albertsson, A.-C., Barenstedt, C., Karlsson, S., and Lindberg, T. 1995. Degradation product pattern and morphology changes as means to differentiate abiotically and biotically aged degradable polyethylene. *Polymer* 36: 3075–3083.

Albertsson, A.-C., Erlandsson, B., Hakkarainen, M., and Karlsson, S. 1998. Molecular weight changes and polymeric matrix changes correlated with the formation of degradation products in biodegraded polyethylene. *J. Environ. Polym. Degrad.* 6: 187–195.

Albertsson, A.-C. and Karlsson, S. 1988. The three stages in degradation of polymers—Polyethylene as a model substance. *J. Appl. Polym. Sci.* 35: 1289–1302.

Allen, N. S. 1980. Influence of sensitisers on the photostabilising action of a hindered piperidine compound in polypropylene. *Makromol. Chem.* 181: 2413–2420.

Al-Malaika, S., Coker, M., and Scott, G. 1988. Mechanism of antioxidant action: Nature of transformation products of dithiophosphates—Part 1. Their role as antioxidant in polyolefins. *Polym. Degrad. Stab.* 22: 147–159.

Al-Malaika, S. and Scott, G. 1983. Mechanisms of antioxidant action: The effect of the thermal processing on the antioxidant activity of sulphur compounds. *Polym. Degrad. Stab.* 5: 415–424.

Arnaud, R., Dabin, P., Lemaire, J., Al-Malaika, S., Chohan, S., Coker, M., Scott, G., Fauve, A., and Maaroufi, A. 1994. Photooxidation and biodegradation of commercial photodegradable polyethylenes. *Polym. Degrad. Stab.* 46: 211–224.

Bitton, G. 1994. *Wastewater Microbiology*. New York: John Wiley & Sons.

Bloembergen, S., David, J., Geyer, D., Gustafson, A., Snook, J., and Narayan, R. 1994. Biodegradation and composting studies of polymeric materials. In *Biodegradable Plastics and Polymers*, eds. Doi, Y. and Fukuda, K., pp. 601–609. Amsterdam, the Netherlands: Elsevier.

Bohlmann, G. M. 2004. Process economics of biodegradable polymers from plants. Paper presented at the *227th ACS National Meeting*, Anaheim, CA.

Bonhomme, S., Cuer, A., Delort, A.-M., Lemaire, J., Sancelme, M., and Scott, G. 2003. Environmental biodegradation of polyethylene. *Polym. Degrad. Stab.* 81: 441–452.

Bozell, J. J. 1991. Chemicals and materials from renewable resources. In *Chemicals and Materials from Renewable Resources*, ed. Bozell, J. J., ACS Symp. Ser. 784, pp. 1–9. Washington, DC: ACS Publications.

Briassoulis, D. 2005. The effects of tensile stress and the agrochemical Vapam on the ageing of low density polyethylene (LDPE) agricultural films. Part I. Mechanical behaviour. *Polym. Degrad. Stabil.* 88: 489–503.

Burkhead, C. E. and McKinney, R. E. 1969. Energy concepts of aerobic microbial metabolism. *J. Sanitary Eng.* 95: 253–268.

Burman, L. and Albertsson, A.-C. 2005. Chromatographic fingerprint—A tool for classification and for predicting the degradation state of degradable polyethylene. *Polym. Degrad. Stab.* 89: 50–63.

Chen, L., Imam, S. H., Gordon, S. H., and Greene, R. V. 1997. Starch-polyvinyl alcohol crosslinked film—Performance and biodegradation. *J. Environ. Polym. Degr.* 5: 111–117.

Chiellini, E., Chiellini, F., Cinelli, P., and Ilieva, V. I. 2003a. Biobased polymeric materials for agriculture applications. In *Biodegradable Polymers and Plastics*, eds. Chiellini, E. and Solaro, R., pp. 185–210. London, U.K.: Kluwer Academic/Plenum Press.

Chiellini, E., Cinelli, P., Chiellini, F., and Imam, S. H. 2004a. Environmentally degradable biobased polymeric blends and composites. *Macromol. Biosci.* 4: 218–231.

Chiellini, E., Cinelli, P., Corti, A., and Kenawy, E. R. 2002. Composite films based on waste gelatin, thermal-mechanical properties and biodegradation testing. *Polym. Degrad. Stab.* 73: 549–555.

Chiellini, E., Cinelli, P., Corti, A., Kenawy, E. R., Fernandes Grillo, E., and Solaro, R. 2000a. Environmentally sound blends and composites based on water-soluble polymer matrices. *Macromol. Symp.* 152: 83–94.

Chiellini, E., Cinelli, P., Grillo Fernandes, E., Kenawy, El-R. K., and Lazzeri, A. 2001a. Gelatin based blends and composites. Morphological and thermal mechanical characterization. *Biomacromolecules* 2: 806–811.

Chiellini, E., Cinelli, P., Ilieva, V. I., Ceccanti, A., Alexy, P., and Bakos, D. 2003b. Biodegradable hybrid polymer films based on poly(vinyl alcohol) and collagen hydrolyzate. *Macromol. Symp.* 197: 125–132.

Chiellini, E., Cinelli, P., Imam, S. H., and Mao, L. 2001b. Composite films based on biorelated agroindustrial waste and poly(vinyl alcohol). Preparation and mechanical properties characterization. *Biomacromolecules* 2: 1029–1037.

Chiellini, E., Cinelli, P., Solaro, R., and Laus, M. 2004b. Thermo-mechanical behaviour of poly(vinylalcohol) and sugarcane bagasse blends. *J. Appl. Polym. Sci.* 92: 426–432.

Chiellini, E., Corti, A., D'Antone, S., and Baciu, R. 2006a. Oxo-biodegradable carbon backbone polymers. Oxidative degradation of polyethylene under accelerated test conditions. *Polym. Degrad. Stab.* 91: 2739–2747.

Chiellini, E., Corti, A., Del Sarto, G., and Antone, S. D. 2006b. Oxo-biodegradable polymers. Effect of hydrolysis degree on biodegradation behaviour of poly(vinyl alcohol). *Polym. Degrad. Stab.* 91: 3397–3406.

Chiellini, E., Corti, A., Politi, B., and Solaro, R. 2000b. Adsorption/desorption of poly(vinyl alcohol) on solid substrates and relevant biodegradation. *J. Polym. Environ.* 8: 67–79.

Chiellini, E., Corti, A., and Solaro, R. 1999a. Biodegradation of poly(vinylalcohol) based blown films under different environmental conditions. *Polym. Degrad. Stab.* 64: 305–312.

Chiellini, E., Corti, A., Solaro, R., and Ceccanti, A. 1999b. Environmentally degradable plastics: Poly(vinyl alcohol)—A case study. In *Proceedings of the ICS-UNIDO International Workshop on Environmentally Degradable Polymers: Plastica Materials and the Environment*, eds. Said, Z. F. and Chiellini, E., pp. 103–118, Doha, Qatar.

Chiellini, E., Corti, A., and Swift, G. 2003c. Biodegradation of thermally-oxidized, fragmented low-density polyethylenes. *Polym. Degrad. Stab.* 81: 341–351.

Choi, J. and Lee, S. Y. 1997. Process analysis and economic evaluation for poly(3-hydroxybutyrate) production by fermentation. *Bioprocess Eng.* 17: 335–342.

Cinelli, P. 1996. Formulation and characterization of environmentally compatible polymeric materials for agriculture applications. PhD dissertation, Pisa University, Pisa, Italy.

Cinelli, P., Chiellini, E., Lawton, J. W., and Imam, S. H. 2006. Foamed articles based on potato starch, corn fibers and poly(vinyl alcohol). *Polym. Degrad. Stab.* 91: 1147–1155.

Cinelli, P., Lawton, J. W., Gordon, S. H., Imam, S. H., and Chiellini, E. 2003. Injection molded hybrid composites based on corn fibers and poly(vinylalcohol). *Macromol. Symp.* 197: 115–124.

Coffin, R., Fishman, M. L., and Ly, T. V. 1996. Thermomechanical properties of blends of pectin and poly(vinyl alcohol). *J. Appl. Polym. Sci.* 61: 71–79.

Contat-Rodrigo, L. and Ribes Greus, A. 2002. Biodegradation studies of LDPE filled with biodegradable additives: Morphological changes. I. *J. Appl. Polym. Sci.* 83: 1683–1691.

Corti, A., Sudhakar, M., Vitali, M., Imam, S. H., and Chiellini, E. 2010. Oxidation and biodegradation of polyethylene films containing pro-oxidant additives: Synergistic effects of sunlight exposure, thermal aging and fungal biodegradation. *Polym. Degrad. Stab.* 95: 1106–1114.

Dalenberg, J. W. and Jager, G. 1989. Priming effect of some organic additions to ^{14}C-labelled soil. *Soil Biol. Biochem.* 21: 443–448.

Doane, W. M. 1994. Biodegradable plastics. *J. Polym. Mater.* 11: 229–237.

EC-Funded Project PACKTECH GIRT-CT-2002-050681. 2002. Assimilation and Standardisation of Environmentally Friendly Packaging Technologies within the Food Industry.

EC-Funded Project Polyver-COOP/CT2006-032967/. 2006. Production of poly(hydroxyl alkanoate) from olive oil mill wastewaters.

EC-Funded Project Wheypol-G5RD/CT2001/00591. 2001. Diary industry waste as source for sustainable polymeric materials.

El Bassam, N. 2001. Renewable energy for rural communities. *Renew. Energy* 24: 401–408.

Ellerman, A. D. and Buchner, B. K. 2007. The European union emissions trading scheme: Origins, allocations and early results. *Rev. Environ. Econ. Policy* 1: 66–87.

Erlandsson, B., Karlsson, S., and Albertsson, A-C. 1997. The mode of action of corn starch and a pro-oxidant system in LDPE: Influence of thermooxidation and UV-irradiation on the molecular weight changes. *Polym. Degrad. Stab.* 55: 237–245.

Espi, E., Salmeron, A., Fontecha, A., Garcia, Y., and Real, A. I. 2006. New ultrathermic films for greenhouse covers. *J. Plast. Film Sheet.* 22: 85–102.

Forrest, W. W. 1969. Energetic aspects of microbial growth. In *Microbial Growth*, eds. Meadow, P. M. and Pirt, S. J., pp. 65–86. London, U.K.: Cambridge University Press.

Grillo Fernandes, E., Cinelli, P., and Chiellini, E. 2004. Thermal behavior of composites based on poly(vinyl alcohol) and sugar cane bagasse. *Macromol. Symp.* 218: 231–240.

Gugumus, F. 1996. Thermooxidative degradation of polyolefins in the solid state: Part 1. Experimental kinetics of functional group formation. *Polym. Degrad. Stab.* 52: 131–144.

Gugumus, F. 2001. Re-examination of the thermal oxidation reactions of polymers 1. New views of an old reaction. *Polym. Degrad. Stab.* 74: 327–339.

Gugumus, F. 2002. Re-examination of the thermal oxidation reactions of polymers 3. Various reactions in polyethylene and polypropylene. *Polym. Degrad. Stab.* 77: 147–155.

Guillet, J. E. 1973a. *Polymers and Ecological Problems*. New York: Plenum Press.

Guillet, J. E. 1973b. Photodegradable Composition. US Patent 3, 753, 952.

Guillet, J. E. 1973c. Polymers with controlled life times. In *Polymers and Ecological Problems*. ed. Guillet, J. E., pp. 1–7. New York: Plenum.

Guillet, J. E. 1980. Studies of the mechanism of polyolefin photodegradation. *Pure Appl. Chem.* 52: 285–294.

Haines, J. R. and Alexander, M. 1975. Microbial degradation of high-molecular weight alkanes. *Appl. Microbiol.* 28: 1084–1085.

Harlan, G. and Kmiec, C. 1995. Ethylene-carbon monoxide copolymers. In *Degradable Polymers: Principles and Applications*, eds. Scott, G. and Gilead, D., pp. 153–168. London, U.K.: Chapman & Hall.

Hatanaka, T., Kawahara, T., Asahi, N., and Tsuji, M. 1995. Effects of the structure of poly(vinyl alcohol) on the dehydrogenation reaction by poly(vinyl alcohol) dehydrogenase from *Pseudomonas* sp. 113P3. *Biosci. Biotechnol. Biochem.* 59: 1229–1231.

Heijnen, J. J. and van Dijken, J. P. 1992. In search of a thermodynamic description of biomass yields for the chemotrophic growth of microorganisms. *Biotechnol. Bioeng.* 39: 833–858.

Hellmann, B., Zelles, L., Palojarvi, A., and Quingyun, B. 1997. Emission of climate-relevant trace gases and succession of microbial communities during open-windrow composting. *Appl. Environ. Microbiol.* 63: 1011–1018.

Heskins, M. and Guillet, J. E. 1968. Mechanism of ultraviolet stabilization of polymers. *Macromolecules* 1: 97–98.

Heyde, M. 1998. Ecological consideration on the use production of biosynthetic and synthetic biodegradable polymers. *Polym. Degrad. Stab.* 59: 3–6.

Hinsken, H., Moss, S., Pauquet, J.-R., and Zweifel, H. 1991. Degradation of polyolefins during melt processing. *Polym. Degrad. Stab.* 34: 279–293.

Hocking, P. J. and Marchessault, R. H. 1994. Biopolyesters. In *Chemistry and Technology of Biodegradable Polymers*, ed. Griffin, G. J. L., pp. 48–96. New York: Chapman & Hall.

http://en.wikipedia.org/wiki/plasticulture

IFEU/BIFA: LCA on loose-fill-packaging from starch and polystyrene. Final report 2003 S.76.

Imam, S. H., Cinelli, P., Gordon, S. H., and Chiellini, E. 2005. Characterization of biodegradable composite films prepared from blends of poly(vinyl alcohol), cornstarch, and lignocellulosic fiber. *J. Polym. Environ.* 13: 47–55.

Inés Mejía, A. G., Lucy López, B. O., and Mulet, A. P. 1999. Biodegradation of poly(vinyl alcohol) with enzymatic extracts of *Phanerochaete chrysosporium. Macromol. Symp.* 148: 131–147.

Iring, M., Földes, E., Barabás, K., Kelen, T., Tüdós, F., and O'dor, L. 1986. Thermal oxidation of linear low density polyethylene. *Polym. Degrad. Stab.* 14: 319–332.

Italian Republic D. Lgs. 2006. 152/06 "Norme in materia ambientale."

Jakubowicz, I. 2003. Evaluation of degradability of biodegradable polyethylene (PE). *Polym. Degrad. Stab.* 80: 39–43.

Karlsson, S. and Albertsson, A.-C. 1998. Biodegradable polymers and environmental interaction. *Polym. Eng. Sci.* 38: 1251–1253.

Karlsson, S., Hakkarainen, M., and Albertsson, A.-C. 1997. Dicarboxylic acids and ketoacids formed in degradable polyethylenes by zip depolymerization through a cyclic transition state. *Macromolecules* 30: 7721–7728.

Kawagoshi, Y. and Fujita, M. 1998. Purification and properties of the polyvinyl alcohol-degrading enzyme 2,4-pentanedione hydrolase obtained from *Pseudomonas vesicularis* var. *povalolyticus* PH. *World J. Microbiol. Biotechnol.* 14: 95–100.

Kawai, F. and Hu, X. 2009. Biochemistry of microbial polyvinyl alcohol degradation. *Appl. Microbiol. Biotechnol.* 84: 227–237.

Kawai, F., Shibata, M., Yokoyama, S., Maeda, S., Tada, K., and Hayashi, S. 1999. Biodegradability of Scott–Gilead photodegradable polyethylene and polyethylene wax by microorganisms. *Macromol. Symp.* 144: 73–84.

Kay, B. L., Evans, R. A., and Young, J. A. 1977. Effect of hydroseeding on bermuda grass. *Agronomy J.* 69: 555–557.

Kenawy, E. R., Cinelli, P., Corti, A., Miertus, S., and Chiellini, E. 1999. Biodegradable composite films based on waste gelatin. *Macromol. Symp.* 144: 351–364.

Khabbaz, F. and Albertsson, A.-C. 2000. Great advantages in using a natural rubber instead of asynthetic SBR in a pro-oxidant system for degradable LDPE. *Biomacromolecules* 1: 665–673.

Khabbaz, F., Albertsson, A.-C., and Karlsson, S. 1999. Chemical and morphological changes of environmentally degradable polyethylene films exposed to thermo-oxidation. *Polym. Degrad. Stab.* 63: 127–138.

Krupp, L. R. and Jewell, W. J. 1992. Biodegradability of modified plastic films in controlled biological environments. *Environ. Sci. Technol.* 26: 193–198.

Kunstoff Plastics Business Data and Chart. 2004. http://www.vke.de

Lahalih, S. M., Akashah, S. A., and Al-Hajjar, F. H. 1987. Development of degradable slow release multinutritional agricultural mulch film. *Ind. Eng. Chem. Res.* 26: 2366–2372.

Lamont, W. J. 1993. Plastic mulch for the production of vegetable crops. *Hort Technol.* 3:35–39.

Larkin, D. M., Crawford, R. J., Christie, G. B. Y., and Linergan, G. T. 1999. Enhanced degradation of polyvinyl alcohol by *Pycnoporus cinnabarinus* after pretreatment with Fenton's reagent. *Appl. Environ. Microbiol.* 65: 1798–1800.

Lawton, W. and Fanta, G. F. 1994. Glycerol-plasticized films prepared from starch—Poly(vinyl alcohol) mixtures: Effect of poly(ethylene-co-acrylic acid). *Carbohyd. Polym.* 23: 275–280.

Lawton, J. W., Shogren, R. L., and Tiefenbacher, K. F. 1999. Effect of batter solids and starch type on the structure of baked starch foams. *Cereal Chem.* 76: 682–687.

Lee, S. Y. 1996. Bacterial polyhydroxyalkanoates. *Biotechnol. Bioeng.* 49: 1–14.

Lehninger, A. L. 1975. *Biochemistry*. New York: Worth.

Lenz, R. W., Fuller, R. C., Scholz, C., and Touraud, F. 1994. Bacteria synthesis of poly-β-hydroxyalkanoates with functionalized side chains. In *Biodegradable Plastics and Polymers*, eds. Doi, Y. and Fukuda, K., pp. 109–119. Amsterdam, the Netherlands: Elsevier Science.

Lide, D. R. 2004. *Handbook of Chemistry and Physics*, 85th edn. Boca Raton, FL: CRC Press.

Lin, Y. 1997. Study of photooxidative degradation of LDPE film containing cerium carboxylate photosensitizer. *J. Appl. Polym. Sci.* 63: 811–818.

Madison, L. L. and Huisman, G. W. 1999. Metabolic engineering of poly(3-hydroxyalkanoates): From DNA to plastic. *Microbiol. Mol. Biol. Rev.* 63: 21–53.

Mao, L., Imam, S. H., Gordon, S., Cinelli, P., and Chiellini, E. 2000. Extruded cornstarch–glycerol–polyvinyl alcohol blends: Mechanical properties, morphology, and biodegradability. *J. Polym. Environ.* 8: 205–211.

Matsumura, S., Shimura, Y., Terayama, K., and Kiyohara, T. 1994. Effects of molecular weight and stereoregularity on biodegradation of poly(vinyl alcohol) *by Alcaligenes faecalis. Biotechnol. Lett.* 16: 1205–1210.

Matsumura, S. and Toshima, K. 1992. Biodegradation of poly(vinyl alcohol) and vinyl alcohol block as biodegradable segments. In *Hydrogels and Biodegradable Polymers for Bioapplications*, eds. Ottenbrite, R. M., Huang, S. J., and Park, K., pp. 767–772. New York: John Wiley & Sons.

Mayer, J. M. and Kaplan, D. L. 1994. Biodegradable materials: Balancing degradability and performance. *Trends Polym Sci.* 2: 227–235.

Mergaert, J., Anderson, C., Wouters, A., Swings, J., and Kersters, K. 1992. Biodegradation of polyhydroxyalkanoates. *FEMS Microbiol. Rev.* 103: 317–321.

Metabolix. Website 2001. Meatabolix purchase Biopol assets from Monsanto. http://www.metabolix.com/publications/pressreleases/PRbiopol.htlm

Mori, T., Sakimoto, M., Kagi, T., and Sakai, T. 1996. Isolation and characterization of a strain of *Bacillus megaterium* that degrades poly(vinyl alcohol). *Biosci. Biotech. Biochem.* 60: 330–332.

Mori, T., Sakimoto, M., Kagi, T., and Sakai, T. 1998. Secondary alcohol dehydrogenase from a vinyl alcohol oligomer-degrading *Geotrichum fermentans*; stabilization with Triton X-100 and activity toward polymers with polymerization degrees less than 20. *World J. Microbiol. Biotechnol.* 14: 349–356.

Narayan, R. 1992. Biomass (renewable) resources for production of materials, chemicals, and fuels/a. paradigm shift. In *Emerging Technologies for Materials and Chemicals from Biomass*, eds. Rowell, R. M., Schultz, T. P., and Narayan, R., ACS Symp. Ser. 476, pp. 1–10. Washington, DC: ACS.

Nord, F. F. 1936. Dehydrogenation activity of *Fusarium lini* B. *Naturwiss.* 24: 763.

Oades, J. M. 1976. Prevention of crust formation in soils by poly(vinyl alcohol). *Aust. J. Soil. Res.* 12: 139–148.

Ohtake, Y., Kobayashi, T., Asabe, H., and Murakami, N. 1998. Studies on biodegradation of LDPE-observation of LDPE films scattered in agricultural fields or in garden soil. *Polym. Degrad. Stab.* 60: 79–84.

Orts, W. J., Sojka, R. E., and Glenn, G. M. 2000. Biopolymer additives to reduce erosion-induced soil losses during irrigation. *Ind. Crops Prod.* 11: 19–29.

Osawa, Z. 1988. Role of metals and metal deactivators in polymer degradation. *Polym. Degrad. Stab.* 20: 203–236.

Osawa, P., Kurisu, N., Nagashima, K., and Nankano, K. 1979. The effect of transition metal stearate on the photodegradation of polyethylene. *J. Appl. Polym. Sci.* 23: 3583–3590.

Painuli, D. K. and Pagliai, M. 1990a. Effect of polyvinyl alcohol, dextran and humic acid on some physical properties of a clay and loam soil. I. Cracking and aggregate stability. *Agrochimica* 34: 117–130.

Painuli, D. K. and Pagliai, M. 1990b. Effect of polyvinyl alcohol, dextran and humic acid on some physical properties of a clay and loam soil. II. Hydraulic conductivity and porosity. *Agrochimica* 34: 131–146.

Patel, M. 1999. Closing carbon cycles. PhD dissertation, Utrecht University, Utrecht, the Netherlands.

Porter, J. J. and Snider, E. H. 1976. Long term biodegradability of textile chemicals. *J. Water Pollut. Control Fed.* 48: 2198–2210.

Potts, J. E., Clendinning, R. A., and Ackart, W. B. 1972. An investigation of the biodegradability of packaging plastics. EPA-R2-72-046. Washington, DC: US Environmental Protection Agency.

Reddy, C. S. K., Ghai, R., and Kalia, V. C. 2003. Polyhydroxyalkanoates: An overview *Bioresour. Technol.* 87: 137–146.

Reijnders, L. A. 2000. A normative strategy for sustainable resource choice and recycling. *Resour. Conserv. Recycl.* 28. 121–133.

Roper, H. 2002. Renewable raw materials in Europe—Industrial utilisation of starch and sugar. *NutraCos.* 1: 37–44.

Rowell, R. M., Sanadi, A. R., Caulfield, D. F., and Jacobson, R. E. 1997. Utilisation of natural fibres in plastic composites: Problems and opportunities. In *Lignocellulosic-Plastic Composites*, eds. Leao, A., Carvalho, F. X., and Frollini, E. pp. 23–52. Sao Paulo, Brazil: USP and UNESP.

Roy, P. K., Sureka, P., Rajagopal, C., Chatterejee, S. N., and Choudhary, V. 2006a. Accelerated aging of LDPE films containing cobalt complexes as prooxidants. *Polym. Degrad. Stab.* 91: 1791–1799.

Roy, P., Sureka, P., Rajagopal, C., and Choudhary, V. 2006b. Effect of cobalt carboxylates on the photo-oxidative degradation of low-density polyethylene. Part I. *Polym. Degrad. Stab.* 91: 1980–1988.

Sakai, K., Fukuba, M., and Hasuy, Y. 1998. Purification and characterization of an esterase involved in poly(vinyl alcohol) degradation by *Pseudomonas vesicularis* PD. *Biosci. Biotech. Biochem.* 62: 2000–2007.

Sakai, K., Hamada, N., and Watanabe, Y. 1986. Studies on the poly(vinyl alcohol)-degrading enzyme. Part VI. Degradation mechanism of poly(vinyl alcohol) by successive reactions of secondary alcohol oxidase and β-diketone hydrolase from *Pseudomonas* sp. *Agric. Biol. Chem.* 50: 989–996.

Sawada, H. 1994. Field testing of biodegradable plastics. In *Biodegradable Plastics and Polymers*, eds. Doi, Y. and Fukuda, K., pp. 298–310. Amsterdam, the Netherlands: Elsevier.

Schill, N. A., Liu, J. S., and von Stockar, U. 1999. Thermodynamic analysis of growth of *Methanobacterium thermoautotrophicum*. *Biotechnol. Bioeng.* 64: 74–81.

Schroeder, E. D. and Bush, A. W. 1968. Validity of energy change as a growth parameter. *J. Sanitary Eng.* 94: 193–200.

Scott, G.1965. *Atmospheric Oxidation and Antioxidants*. Amsterdam, the Netherlands: Elsevier.

Scott, G. 1981. Mechanism of antioxidant action. In *Developments in Polymer Stabilisation*, ed. Scott, G., pp. 1–21. London, U.K.: Applied Science Publishers.

Scott, G. 1994. Environmental biodegradation of hydrocarbon polymers: Initiation and control. In *Biodegradable Plastic and Polymers*, eds. Doi, Y. and Fukuda, K., pp. 79–92. Amsterdam, the Netherlands: Elsevier.

Scott, G. 2000. Green polymers. *Polym. Degrad. Stab.* 68: 1–7.

Scott, G. 1995. Photo-biodegradable plastics. In *Degradable Polymers. Principle and Applications*, eds. Scott, G., Gilead, D. pp. 169–185. London, U.K.: Chapman & Hall.

Scott, G. and Gilead, D. 1982. Time-controlled stabilisation of polymers. In *Developments in Polymer Stabilisation*, eds. Scott, G., pp. 71–106. London, U.K.: Elsevier Applied Science Publications.

Scott, G. and Wiles, D. M. 2001. Programmed-life plastics from polyolefins: A new look at sustainability. *Biomacromolecules* 2: 615–622.

Shimao, M., Saimoto, H., Kato, N., and Sakazawa, C. 1983. Properties and roles of bacterial symbionts of polyvinyl alcohol-utilizing mixed cultures. *Appl. Environ. Microbiol.* 46: 605–610.

Shogren, R. L., Lawton, J. W., Tiefenbacher, K. F., and Chen, L. 1998. Starch-poly(vinyl alcohol) foamed articles prepared by a baking process. *J. Appl. Polym. Sci.* 68: 2129–2140.

Snell, K. D. and Peoples, O. P. 2002. Polyhydroxyalkanoate polymers and their production in transgenic plants. *Metab. Eng.* 4: 29–40.

Solaro, R., Corti, A., and Chiellini, E. 2000. Biodegradation of poly(vinyl alcohol) with different molecular weights and degree of hydrolysis. *Polym. Adv. Technol.* 11: 873–878.

Sperling, L. H. and C. E. Carraher. 1988. Polymers from renewable resources. In *Encyclopedia of Polymer Science and Engineering*, eds. Mark, H. F., Bikales, N. M. Overberger, C. G., and Menges, G., Vol. 12, pp. 658–690. New York: Wiley.

Stefanson, R. C. 1974. Soil stabilization by poly(vinyl alcohol) and its effect on the growth of wheat. *Aust. J. Soil Res.* 12: 59–62.

Steinbüchel, A. and Füchtenbusch, B. 1998. Bacteria and other biological systems for polyester production. *TIBTECH* 16: 419–426.

Steinbüchel, A., Hustede, E., Liebergesell, M., Piepr, U., Timm, A., and Valentin, H. 1992. Molecular basis for bio-synthesis and accumulation of polyhydroxyalkanoic acid in bacteria. *FEMS Microbiol. Rev.* 103: 217–230.

Suzuki, T. 1979. Degradation of poly(vinyl alcohol) by microorganisms. *J. Appl. Polym. Sci., Appl. Polym. Symp.* 35: 431–437.

Suzuki, T., Ichihara, Y., Yamada, M., and Tonomura, K. 1973. Some characteristics of *Pseudomonas* O-3 which utilizes polyvinyl alcohol. *Agric. Biol. Chem.* 37: 747–756.

Sykes, R. M. 1975. Theoretical heterotrophic yields. *J. Water Pollut. Control Fed.* 47: 591–600.

Takasu, A., Aoi, K., Tsuchiya, M., and Okada, M. 1999. New chitin-based polymer hybrids, 4: Soil burial degradation behavior of poly(vinyl alcohol)/chitin derivative miscible blends. *J. Appl. Polym. Sci.* 73: 1171–1179.

Tavares, F. and Marreiros, A. 2009. Biomass: new feedstock for the plastic industry. Biomass Magazine (November) Web site: www.biomassmagazine.com. 2009.

Tokiwa, Y. and Iwamoto, A. 1992. Establishment of biodegradability evaluation for plastics of PVA series. Paper presented at the *Annual Meeting of the Fermentation and Bioengineering Society*, Tokyo, Japan.

Turner, B. L., Bristow, A. W., and Hygarth, P. M. 2001. Rapid estimation of microbial biomass in grassland soils by ultra-violet absorbance. *Soil Biol. Biochem.* 33: 913–919.

Van Os, G. 2001. The European plastic industry—A sunset industry, *EPF* Special Issue July 2001: 19.

Van Wyk, J. P. H. 2001. Biotechnology and the utilization of biowaste as a resource for bioproduct development. *Trends Biotechnol.* 19: 172–177.

Various web sites. 1995. http://www.ibaw.org, http:/www.mli.kvl.dk/foodchem/special/biopack, http://www.ienica.net/bioplastics/bioplasticsindex.htm, http://www.bio-pro.de/en/region/ulm/magazin/01391/Diepenbrock/Pelzer/Radtke: Energiebilanz im Ackerbaubetrieb.

Various web sites. 2006. http://plasticsresource.com/s_plasticsresource/docs/900/840.pdf, http://www. plasticseurope.org

Volke-Sepúlveda, T., Favela-Torres, E., Manzur-Guzman, A., Limon-Gonzalez, M., and Trejo-Quintero, G. 1999. Microbial degradation of thermo-oxidized low-density polyethylene. *J. Appl. Polym. Sci.* 73: 1435–1440.

Volke-Sepúlveda, T., Saucedo-Castañeda, G., Gutiérrez-Rojas, M., Manzur, A., and Favela-Torres, E. 2002. Thermally treated low density polyethylene biodegradation by *Penicillium pinophilum* and *Aspergillus niger. J. Appl. Polym. Sci.* 83: 305–314.

Warwel, S., Brüse, F., Dermes, S., Kunz, M., and Klaas, M. R. 2001. Polymers and surfactants on the basis of renewable resources. *Chemosphere* 43: 39–48.

Watanabe, Y., Hamada, N., Morita, M., and Tsujisaka, Y. 1976. Purification and properties of a polyvinyl alcohol-degrading enzyme produced by a strain of *Pseudomonas. Arch. Biochem. Biophys.* 174: 575–581.

Weber, C. J. 2000. Biobased packaging materials for the food industry. Status and Perspectives. Report of the Food Biopack project PL984046. http://www.biodeg.net/fichiers/Book%20on%20biopolymers%20(Eng).pdf

Weiland, M., Daro, A., and David, C. 1995. Biodegradation of thermally oxidized polyethylene. *Polym. Degrad. Stab.* 48: 275–289.

Weiland, M. and David, C. 1994. Thermal oxidation of polyethylene in compost environment. *Polym. Degrad. Stab.* 45: 371–377.

Wiles, D. M. and Scott, G. 2006. Polyolefins with controlled environmental degradability. *Polym. Degrad. Stab.* 91: 1581–1592.

Winslow, F. H. 1977. Photooxidation of high polymers. *Pure Appl. Chem.* 49: 495–502.

Yamada-Onodera, K., Makumoto, H., Katsuyaya, Y., Saiganji, A., and Tani, Y. 2001. Degradation of polyethylene by a fungus, *Penicillium simplicissimum* YK. *Polym. Degrad. Stab.* 72: 323–327.

Yamane, T. 1992. Cultivation engineering of microbial bioplastics production. *FEMS Microbiol. Rev.* 103: 257–264.

Young, R. A. 1997. Utilization of natural fibres: Characterization, modification, and applications. In *Lignocellulosic-Plastic Composites*, eds. Leao, A. Carvalho, F. X., and Frollini, E., pp. 1–22. Sao Paulo, Brazil: USP and UNESP.

Young, J. A., Kay, B. L., and Evans, R. A. 1977. Accelerating the germination of common bermudagrass seed for hydro-seeding. *Agronomy J.* 69: 115–119.

6 Fish Gelatin
Material Properties and Applications

Bor-Sen Chiou, Roberto J. Avena-Bustillos,
Peter J. Bechtel, Syed H. Imam, Greg M. Glenn,
Tara H. McHugh, and William J. Orts

CONTENTS

6.1 INTRODUCTION

Gelatin is produced from partial hydrolysis of collagen, usually from skins and bones of pigs and cows (Ward and Courts 1977). Collagen is usually found in the skin and connective tissues of animals and exists as a triple helical structure comprising two α1-chains and one α2-chain. Figure 6.1 shows a schematic diagram of collagen and gelatin. These triple helical structures can form into larger aggregates called fibrils. Both types of α-chains have molecular weights of approximately 100 kDa with the α1-chains having slightly higher molecular weights than α2-chains. Also, some fish species have an additional α3-chain (Piez 1964). During gelatin production, the collagen becomes partially hydrolyzed to form α-, β-, and γ-chains. The β- and γ-chains consist of two and three α-chains covalently attached to each other, respectively. In addition, low-molecular-weight chains, with molecular weights lower than α-chains, as well as high-molecular-weight chains, with molecular weights higher than γ-chains, might also be produced during gelatin production. At temperatures above gelation temperature of gelatin, these chains exist as random coils in solution. When the temperature is cooled below gelation temperature, the gelatin chains try to reform the triple helical structures originally present in collagen. However, due to the irreversible hydrolysis of collagen during gelatin production, the gelatin chains can only partially reform the original helical structures.

In 2005, the worldwide production of gelatin was 305,000 metric tons per year (Schrieber and Gareis 2007). Western Europe produced 42% of the world's gelatin, followed by 22% from North America, 18% from Asia, and 16% from South America. Approximately 42% of gelatin was derived from porcine skin, 29% from bovine skin, 28% from bones, and less than 1% from other sources (www.geafiltration.com 2010). Gelatin is typically produced from treatment of skins or bones with either acid or alkali washings. Porcine skin is usually treated with acid to produce Type A gelatin,

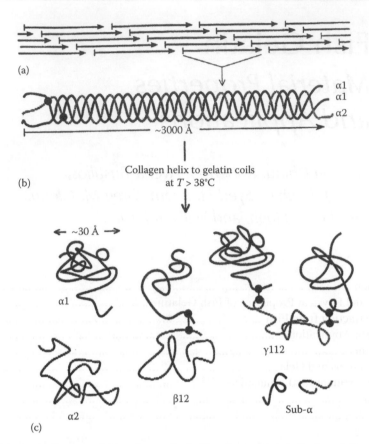

FIGURE 6.1 Schematic diagram of (a) collagen fibril, (b) collagen triple-helical, and (c) gelatin structures. (Reprinted from Rose, P.I., Inedible gelatin and glue, in A.M. Pearson and D.R. Dutson, eds., *Inedible Meat By-Products*, Elsevier Applied Science, London, U.K., 1992, pp. 217–263. With permission.)

whereas the more complex collagens found in bovine skins and bones are treated with alkali to produce Type B gelatin. Acid treatments generally require hours to complete, whereas alkali treatments require more than 1 week.

Gelatin is primarily used in the food, pharmaceutical, and photographic industries, with the food industry using a majority of the gelatin. In the food industry, it is used as a gelling agent and can be found in such food as yogurts, jams, and marshmallows. It is also a primary ingredient in soft and hard capsules for pharmaceutical applications since gelatin extracted from mammalian sources has a melting temperature close to the human body temperature. In the photographic industry, it is used as an emulsifier to stabilize silver halide in an emulsion. Over the past decade, there has been a decline in demand for photographic gelatin and an increase in demand for soft and hard capsules. In addition, gelatin is used in other applications, such as rheology modifiers for cosmetic creams, as adhesives, and as an ingredient in detergents (Ward and Courts 1977, Schrieber and Gareis 2007).

There has recently been an interest in using gelatin derived from fish rather than porcine or bovine sources. One main reason is due to religious considerations, since people of the Islamic and Jewish faith do not consume porcine products. Also, there have been concerns with bovine spongiform encephalopathy (BSE) appearing in gelatin from bovine sources. However, recent studies have indicated that no BSE remained after infected samples had undergone the gelatin extraction process (Grobben et al. 2004). In addition, the fishing industry in various parts of the world generates lots of by-products, which have the potential to be major sources of fish gelatin. These by-products include everything else besides filets, and can contain fish skins and bones from which gelatin can be extracted. For instance, the Alaskan fishing industry catches approximately 1 million ton of Alaska pollock annually with 66% eventually becoming by-products. These by-products consist of 32% viscera, 26% heads, 33% frames, and 9% fish skins (Crapo and Bechtel 2003).

This chapter focuses mainly on fish gelatin properties and potential applications. Fish gelatin properties are compared to those of mammalian gelatin. Also, the many different uses of fish gelatin are highlighted as well. This review emphasizes current literature on fish gelatin, but includes studies related to mammalian gelatin if those studies are pertinent to the discussions.

6.2 CHEMICAL AND PHYSICAL PROPERTIES OF FISH GELATIN

Fish gelatin has not been as widely used as mammalian gelatin primarily because of its lower gelation temperature. Mammalian gelatins have gelation temperatures of approximately 25°C. In contrast, warm-water fish gelatins have gelation temperatures of approximately 19°C and cold-water fish gelatins have gelation temperatures below 10°C (Karim and Bhat 2009). Consequently, mammalian gelatin behaves as a gel at room temperature, whereas fish gelatin remains as a liquid. The ability of mammalian gelatin to form a gel at room temperature offers advantages over fish gelatin. For instance, mammalian gelatin can be used as a gelling agent in many more applications. Also, the advantages extend to applications requiring solid forms of gelatin, such as gelatin films. Mammalian gelatin films dried at room temperature contained helical structures since the films were dried below their gelation temperature. These helical structures became locked in place as water evaporated during the drying process. In contrast, fish gelatin films dried at room temperature remained amorphous and did not contain helical structures, since the films were dried above their gelation temperature. This resulted in mammalian gelatin films having superior tensile strength and higher elasticity compared to fish gelatin films (Chiou et al. 2009). Consequently, soft and hard capsules for drug encapsulation are made from mammalian rather than fish gelatin.

Fish gelatins have lower gelation temperatures than mammalian gelatins, because they have lower concentrations of the amino acids proline and hydroxyproline. Proline and hydroxyproline have been shown to help stabilize the triple helical structures. These triple helical structures act as junctions in a network that eventually forms the gel. Mammalian gelatins have the highest concentration of proline and hydroxyproline, followed by warm-water fish gelatins and then cold-water fish gelatins. The proline and hydroxyproline concentrations have been shown to correlate well with the gelation temperatures of different gelatins (Gilsenan and Ross-Murphy 2000a,b, Joly-Duhamel et al. 2002b).

Even though fish gelatins have lower gelation temperatures than mammalian gelatins, their mechanical properties are not necessarily inferior when helical content is taken into account. The elastic modulus, G', had been shown to depend only on helical content, irrespective of gelatin source, molecular-weight distribution, and thermal history (Joly-Duhamel et al. 2002a, Eysturskard et al. 2009a). The modulus values of both mammalian and fish gelatin samples had been found to fit a master curve of G' as a function of helical content (Joly-Duhamel et al. 2002a). This master curve is shown in Figure 6.2. The comparable mechanical properties of fish and mammalian gelatin with similar helical content also seemed to apply to solid gelatin as well. Pollock and salmon gelatin films dried (at 4°C) below their gelation temperatures had comparable enthalpies of fusion and tensile properties to mammalian gelatin films dried (at 23°C) below their gelation temperature (Chiou et al. 2009).

The mechanism for helix formation in fish gelatin has not been fully determined, but several research groups have tried to correlate modulus with concentration of different molecular-weight species. Eysturskgard et al. (2010) showed that G' of cold-water saithe gelatin could be correlated to fractions of α-, β-, γ-, low-molecular-weight (MW < 70 kg/mol), and high-molecular-weight (MW > 230 kg/mol) chains. The authors found that G' positively correlated with β- and high-molecular-weight chains, but negatively correlated with α- and low-molecular-weight chains. The β-chains had the largest effect on G' values. Muyonga et al. (2004) also found similar results with Nile perch gelatin. They found Bloom and hardness values had positive correlations with chains that had higher molecular weights than β-chains (>β-chains), but negative correlations with α-chains

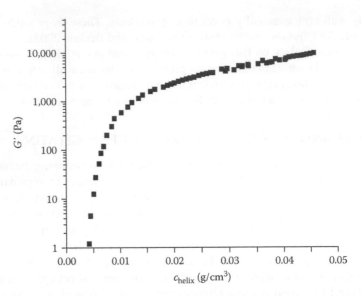

FIGURE 6.2 Master curve of G' as a function of helix concentration for gelatin. (Reprinted from Joly-Duhamel, C. et al., *Langmuir*, 18, 7158, 2002. With permission.)

and chains that had lower molecular weights than α-chains (<α-chains). The greater β-chains had the largest effect on Bloom and hardness values. The low-molecular-weight chains had a negative effect on G', Bloom, and hardness values because the chains were too short to participate in creating elastic segments in the gel network (Eysturskard et al. 2009b).

6.3 GELATIN EXTRACTION FROM FISH

There have been an increasing number of studies involving extraction of gelatin from different fish species. These studies include extracting gelatin from skins of both warm-water and cold-water fish species. Table 6.1 shows a list of some of these studies. In addition to skins, gelatin had been extracted from bones of king weakfish (*Macrodon ancylodon*) (da Trindade Alfaro et al. 2009), greater lizardfish (*Saurida tumbil*) (Taheri et al. 2009), and Nile perch (*Lates niloticus*) (Muyonga et al. 2004), as well as scales of grass carp (*Catenopharyngodon idella*) (Zhang et al. 2010) and silver carp (*Hypophthalmichthys molitrix*) (Wang and Regenstein 2009). Generally, gelatin yields from bones were lower than those from skins (Muyonga et al. 2004, Taheri et al. 2009). Also, bone gelatin had lower gel strength than skin gelatin due in part to the bone gelatin having lower concentrations of greater β-chains in Nile perch (Muyonga et al. 2004) or having greater amounts of low-molecular-weight chains in greater lizardfish (Taheri et al. 2009).

Various methods have been used to extract gelatin from different fish species. One reason for the different methods involves the unique skin structure present in each species. Collagen chains in some species might be more cross-linked than others and require harsher treatments to extract gelatin. Also, varying the extraction conditions such as acid concentration, alkali concentration, extraction time, and extraction temperature have been shown to affect final gelatin properties. For instance, Montero and Gomez-Guillen (2000) and Gomez-Guillen and Montero (2001) used different acids and extraction times to extract gelatin from megrim skin. They found that acetic acid generally produced gelatins with the best properties. Muyonga et al. (2004) used different extraction temperatures to extract gelatin from Nile perch skins and bones. Gimenez et al. (2005b,c) used various salts and examined lactic acid as an alternative to acetic acid for extracting gelatin from sole skin. Also, Gomez-Guillen et al. (2005) used high-pressure treatment during extraction of gelatin from sole skin to shorten the extraction process. Zhou and Regenstein (2005) examined the effects of different acids

TABLE 6.1
Fish Gelatin from Skins of Different Species

Fish Species	References
Alaska pink salmon (*Oncorhynchus gorbuscha*)	Avena-Bustillos et al. (2006) and Chiou et al. (2006, 2008, 2009)
Alaska pollock (*Theragra chalcogramma*)	Zhou and Regenstein (2004, 2005), Avena-Bustillos et al. (2006), Chiou et al. (2006, 2008, 2009) and Zhou and Regenstein (2007), Bower et al. (2010)
Atlantic halibut (*Hippoglossus hippoglossus*)	Carvalho et al. (2008)
Atlantic salmon (*Salmo salar*)	Arnesen and Gildberg (2007)
Bigeye snapper (*Priacanthus hamrur*)	Binsi et al. (2009)
Bigeye snapper (*Priacanthus macracanthus*)	Jongjareonrak et al. (2006a–c) and Benjakul et al. (2009)
Bigeye snapper (*Priacanthus tayenus*)	Nalinanon et al. (2008), Benjakul et al. (2009), and Rattaya et al. (2009)
Black tilapia (*Oreochromis mossambicus*)	Jamilah and Harvinder (2002)
Blue shark (*Prionace glauca*)	Yoshimura et al. (2000) and Limpisophon et al. (2009)
Brownstripe red snapper (*Lutjanus vitta*)	Jongjareonrak et al. (2006a–c)
Channel catfish (*Ictalurus punctatus*)	Zhang and Regenstein (2007), Yang et al. (2007a,b), Liu et al. (2008), Yang and Wang (2009), and Bao et al. (2009)
Cobia (*Rachycentron canadum*)	Yang et al. (2008)
Cod (*Gadus morhua*)	Fernandez-Diaz et al. (2001), Gomez-Guillen et al. (2002), Montero et al. (2002), Kolodziejska et al. (2004, 2006, 2008), Kolodziejska and Piotrowska (2007), Piotrowska et al. (2008), and Szutka et al. (2008, 2009)
Croaker (*Johnius dussumeiri*)	Cheow et al. (2007)
Dover sole (*Solea vulgaris*)	Gomez-Guillen et al. (2002, 2005) and Gimenez et al. (2005a–c)
Flounder (*Platichthys flesus*)	Fernandez-Diaz et al. (2003)
Giant catfish (*Pangasianodon gigas*)	Jongjareonrak et al. (2010)
Grass carp (*Catenopharyngodon idella*)	Kasankala et al. (2007)
Greater lizardfish (*Saurida tumbil*)	Taheri et al. (2009)
Hake (*Merluccius merluccius*)	Fernandez-Diaz et al. (2001) and Gomez-Guillen et al. (2002)
Herring (*Clupea harengus*)	Kolodziejska et al. (2008)
Herring (*Tenualosa ilisha*)	Norziah et al. (2009)
Hoki (*Johnius belengerii*)	Mendis et al. (2005)
Hoki (*Macruronus novaezelandiae*)	Mohtar et al. (2010)
Megrim (*Lepidorhombus boscii*)	Montero et al. (2000, 2002) and Gomez-Guillen et al. (2001, 2002)
Nile perch (*Lates niloticus*)	Muyonga et al. (2004)
Red tilapia (*Oreochromis nilotica*)	Jamilah ct al. (2002)
Saithe (*Pollachius virens*)	Eysturskard et al. (2009a, 2010)
Sea horse mackerel (*Trachurus trachurus*)	Badii et al. (2006)
Shortfin scad (*Decapterus macrosoma*)	Cheow et al. (2007)
Silver carp (*Hypophthalmichthys molitrix*)	Boran et al. (2009)
Rainbow trout (*Onchorhynchus mykiss*)	Tabarestani et al. (2010)
Yellowfin tuna (*Thunnus albacares*)	Cho et al. (2005)

and alkali pretreatments on gelatin extracted from pollock skin. Zhang et al. (2007) examined six different pretreatment conditions with various acid and alkali washes for extracting gelatin from catfish skin. Kolodziejska et al. (2008) varied the extraction time and temperature for extracting gelatin from cod, smoked salmon, and salted herrings. Nalinanon et al. (2008) used different enzyme concentrations to improve the extraction yield and strength of gelatin from snapper skins. Eysturskard et al. (2009a) examined the effects of acid concentration, extraction time, and extraction temperature on gelatin properties extracted from saithe skin. In addition, Wang and Regenstein (2009) used different acids to remove calcium salts before gelatin extraction from carp scales.

Most studies on the effects of extraction conditions on gelatin properties have not involved design of experiments. However, there has recently been an increase in studies using design of experiments and response surface methodology to optimize extraction conditions. The independent variables in these studies have included combinations of OH^- concentration (Zhou and Regenstein 2004, Cho et al. 2005, Yang et al. 2007a, Boran and Regenstein 2009, da Trindade Alfaro et al. 2009, Tabarestani et al. 2010), H^+ concentration (Zhou and Regenstein 2004, Kasankala et al. 2007, Yang et al. 2007a, Boran and Regenstein 2009, Tabarestani et al. 2010), pretreatment temperature (Zhou and Regenstein 2004, Boran and Regenstein 2009), pretreatment time (Cho et al. 2005, Kasankala et al. 2007, Yang et al. 2007a, Boran et al. 2009, Mohtar et al. 2010, Tabarestani et al. 2010), maceration time (da Trindade Alfaro et al. 2009), extraction temperature (Zhou et al. 2004, Cho et al. 2005, Kasankala et al. 2007, Yang et al. 2007a, Boran and Regenstein 2009, da Trindade Alfaro et al. 2009, Mohtar et al. 2010), extraction time (Cho et al. 2005, Kasankala et al. 2007, Boran and Regenstein 2009, Mohtar et al. 2010), and water/skin ratio (Boran and Regenstein 2009). The dependent variables have included gelatin yield (Zhou and Regenstein 2004, Kasankala et al. 2007, da Trindade Alfaro et al. 2009, Boran and Regenstein 2009, Mohtar et al. 2010, Tabarestani et al. 2010), protein yield (Yang et al. 2007a), gelatin content (Cho et al. 2005), gel strength (Zhou and Regenstein 2004, Cho et al. 2005, Kasankala et al. 2007, Yang et al. 2007a, Boran and Regenstein 2009, da Trindade Alfaro et al. 2009, Tabarestani et al. 2010), viscosity (Zhou and Regenstein 2004, Yang et al. 2007a, Boran and Regenstein 2009, Tabarestani et al. 2010), molecular-weight distribution (Tabarestani et al. 2010), and melting point (Tabarestani et al. 2010).

Some research groups have examined how stabilizing fish skins by freezing or drying have affected the extracted gelatin properties. Fresh skins are not always available, so stabilizing them ensures consistent gelatin properties. Fernandez-Diaz et al. (2003) found that gelatin extracted from frozen flounder skins (at $-12°C$ and $-20°C$) had lower amounts of β- and γ-chains than gelatin from fresh skins. Also, the $-12°C$ sample had notable amounts of low-molecular-weight chains. Both gelatins extracted from frozen skins had lower gel strengths than the fresh skin sample. In contrast, Liu et al. (2008) found that gelatin extracted from frozen catfish skin ($-18°C$) had comparable gel strength and higher melting and gelation temperatures than gelatin extracted from fresh skin. Also, Gimenez et al. (2005a) dried sole skin using ethanol, ethanol/glycerol mixture, and marine salt. The authors periodically extracted gelatin from the dried skins over a 160 day period. They found that gelatin from the dried skins had slightly lower G' values, lower gelation, and melting temperatures, but comparable gel strengths compared to gelatin from frozen skins. Bower et al. (2010) used various desiccants to dry pollock skin and compared gelatin properties from the desiccant-dried skins to air-dried skins. They found that all gelatin samples had comparable gel strengths and molecular-weight distributions. In addition, Liu et al. (2008) dried catfish skins using air and found that gelatin extracted from dried skins had greater gel strength as well as comparable gelation and melting temperatures to gelatin extracted from fresh skins.

6.4 FISH GELATIN APPLICATIONS

6.4.1 FILMS

Gelatin in a solid film form has been widely used as capsules in the pharmaceutical industry. Gelatin films have also been examined as possible coatings for foods and as packaging materials to replace petroleum-based films (Bae et al. 2009b). However, gelatin has several drawbacks that limit their potential applications. First, gelatin is soluble in water, so any contact with water results in gelatin films loosing their integrity. Also, gelatin films have inferior water vapor barrier properties compared to petroleum-based films (Kolodziejska and Piotrowska 2007, Sztuka and Kolodziejska 2009). In addition, gelatin films are more brittle and less elastic than synthetic films.

One method that has been used to improve physical properties of gelatin films has been to cross-link them. Cross-linkers that have been used in fish gelatin films include formaldehyde (Fraga and Williams 1985), glutaraldehyde (Liu et al. 2007, Chiou et al. 2008), transglutaminase

(Kolodziejska et al. 2006, Yi et al. 2006, Kolodziejska and Piotrowska 2007, Piotrowska et al. 2008, Szutka and Kołodziejska 2008, 2009), and 1-ethyl-3-(3-dimethylaminopropyl) carbodiimide (EDC) (Kolodziejska et al. 2006, Kolodziejska and Piotrowska 2007, Piotrowska et al. 2008, Szutka and Kolodziejska 2008, 2009). Formaldehyde, glutaraldehyde, and EDC are toxic, whereas transglutaminase is a food grade additive. Each type of cross-linker has a different mechanism for cross-linking gelatin. Glutaraldehyde and formaldehyde react with amino groups to form imine linkages. EDC connects carboxyl groups in one amino acid to amino groups in another amino acid. Transglutaminase catalyzes the formation of covalent bonds between γ-carboxyamide groups in glutamine with ε-amino groups in lysine. Various research groups have determined that cross-linking fish gelatin films reduced their solubility in water. Kolodziejska et al. (2006) and Kolodziejska and Piotrowska (2007) cross-linked cod gelatin-chitosan films with transglutaminase and EDC. The authors found that both cross-linkers reduced solubility of the films in water, with EDC being more effective than transglutaminase. Piotrowska et al. (2008) found similar behavior for cod gelatin films containing transglutaminase and EDC cross-linkers. Also, Liu et al. (2007) determined that fish gelatin-pectin films cross-linked with glutaraldehyde had lower solubility compared to non-cross-linked films. In addition, Szutka and Kołodziejska (2008) found that adding transglutaminase and EDC cross-linkers increased some cod gelatin films' resistance to enzyme degradation. There have been mixed results for using cross-linkers to improve water vapor barrier properties of fish gelatin films. Liu et al. (2007) found that adding glutaraldehyde increased water vapor transmission rates for fish gelatin films containing plasticizer, but reduced transmission rates for fish gelatin-pectin films. Szutka and Kolodziejska (2009) found that cod gelatin films cross-linked with EDC had lower water vapor permeability values than neat films. Also, Chiou et al. (2008) showed that salmon gelatin films cross-linked with glutaraldehyde had slightly lower water vapor permeability values compared to neat films. However, others have shown that adding cross-linkers had no effect on water vapor permeability values for cod gelatin films (Piotrowska et al. 2008), pollock gelatin films (Chiou et al. 2008), cod gelatin-chitosan films (Kolodziejska and Piotrowska 2007), and fish gelatin films containing plasticizers (Yi et al. 2006). There have also been mixed results for using cross-linking to improve oxygen barrier properties. Chiou et al. (2008) found no or slight increases in oxygen permeability values of films after addition of cross-linkers. In contrast, Yi et al. (2006) found that cross-linked films had twice the oxygen permeability values compared to neat films. The addition of cross-linkers has also produced mixed results for improving tensile properties of fish gelatin films. Kolodziejska and Piotrowska (2007) determined that incorporating cross-linkers into fish gelatin-chitosan films resulted in a decrease in tensile strength, but an increase in elongation. For fish gelatin films without plasticizers, adding cross-linkers resulted in little change in modulus (Chiou et al. 2008), tensile strength (Chiou et al. 2008, Piotrowska et al. 2008), and elongation (Chiou et al. 2008, Piotrowska et al. 2008) values. For fish gelatin films containing plasticizers, adding cross-linkers led to an increase in modulus (Liu et al. 2007) and tensile strength (Yi et al. 2006, Liu et al. 2007), but a decrease in elongation values (Yi et al. 2006, Liu et al. 2007).

Various research groups have also blended fish gelatin with polysaccharides to improve their material properties. Some of these polysaccharides include chitosan (Lopez-Caballero et al. 2005, Kolodziejska et al. 2006, Kolodziejska and Piotrowska 2007, Szutka et al. 2008), pectin (Liu et al. 2007), gellan (Pranoto et al. 2007), and κ-carrageenan (Pranoto et al. 2007). Lopez-Caballero et al. (2005) incorporated chitosan into megrim gelatin films to improve antimicrobial properties. The authors used chitosan-gelatin films to coat fish patties and found that the coatings generally inhibited bacterial growth during 8–12 days of storage. Liu et al. (2007) found that adding fish gelatin improved the modulus and tensile strength of pectin films. Also, Pranoto et al. (2007) found that adding gellan and κ-carrageenan to tilapia gelatin films resulted in improved tensile properties and increased melting temperatures. In addition, adding gellan resulted in lower water vapor permeability values, whereas adding κ-carrageenan did not affect water vapor barrier properties.

Another method to improve barrier properties of gelatin films involved adding hydrophobic materials. Bae et al. (2009a–c) incorporated nanoclays into fish gelatin films. Nanoclays have been found to improve barrier properties of composites since the exfoliated nanoclays force the gas molecules to travel a more tortuous path through the material. Bae et al. (2009c) found that an increase in nanoclay concentration resulted in decreases in water vapor and oxygen permeability values of fish gelatin films. Bae et al. (2009a) also used transglutaminase to cross-link the gelatin-nanoclay films, but cross-linking had no effect on barrier properties. Also, Bae et al. (2009b) sandwiched a fish gelatin-nanoclay film between polyethylene terephthalate and polyethylene films. The authors found that these laminate films had low oxygen permeability, except at high humidity. They postulated that in a high-humidity environment, gelatin absorbs water and swells the film, resulting in reduced resistance to oxygen transport. Sztuka and Kolodziejska (2009) added various hydrophobic materials, such as amaranth oil, rapeseed oil, lanolin, beeswax, and ozococerite, to cod gelatin films to improve their water vapor barrier properties. They found that beeswax was generally more effective than the other materials. Jongjareonrak et al. (2006b) also incorporated several fatty acids and their esters into snapper gelatin films and found that films containing these hydrophobic materials had lower water vapor permeability values.

The water vapor barrier properties of gelatin films can be affected by the amount of helical structures present in the film. Chiou et al. (2009) examined the water vapor permeability of pollock and salmon gelatin films dried at temperatures below and above their gelation temperatures. Films dried below gelation temperature (cold cast) contained helical structures because the gelatin chains formed these structures during the drying process. These helical structures became locked in place as the film became fully dried over time. In contrast, films dried above gelation temperature (hot-cast) remained amorphous since the gelatin chains did not form helical structures. The authors found that cold-cast films had water vapor permeability values two to three times those of hot-cast films. This might be due in part to greater water sorption by helical structures compared to amorphous chains. This is shown in Figure 6.3, which presents isotherm sorption curves for cold- and hot-cast fish and porcine films. The cold-cast films had greater water sorption values than hot-cast films over much of the relative humidity range.

Several research groups have incorporated antioxidant (Bao et al. 2009, Gomez-Estaca et al. 2009a) and antimicrobial (Bower et al. 2006, Gomez-Estaca et al. 2009b) additives into fish gelatin films to improve their functionality. Gomez-Estaca et al. (2009a) found that catfish gelatin films

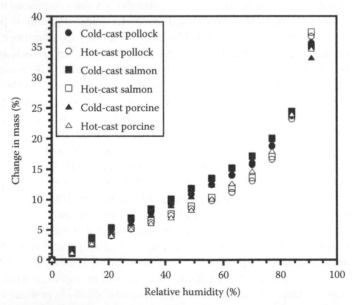

FIGURE 6.3 Sorption isotherms for cold- and hot-cast pollock, salmon, and porcine (Gelita 250A) films. Cold-cast fish and porcine films were dried at 4°C and 23°C, respectively, whereas hot-cast fish and porcine films were dried at 23°C and 60°C, respectively.

containing borage extracts had better antioxidant activity than those containing common food antioxidants, such as butylated hydroxytoluene (BHT) and α-tocopherol. Bao et al. (2009) incorporated tea polyphenol into chitosan nanoparticles and added them to catfish gelatin films. The authors found that these nanoparticles increased antioxidant activity in the films and inhibited oxidation of fish oil. In addition, Bower et al. (2006) and Gomez-Estaca et al. (2009b) incorporated antimicrobial lysozyme and clove essential oil, respectively, to fish gelatin films. Both studies showed improved antimicrobial activity against certain bacteria.

6.4.2 SOLUTIONS AND GELS

Salts and sugars have been added to fish gelatin solutions to increase their strength and melting temperatures (Choi and Regenstein 2000, Sarabia et al. 2000, Fernandez-Diaz et al. 2001). Sarabia et al. (2000) added various sulfates, phosphates, and chlorides to megrim gelatin solutions. The authors found that sulfates and phosphates increased melting temperatures, whereas chlorides decreased them. Also, $MgSO_4$ was the only additive that generally increased G' of the samples. Choi and Regenstein (2000) also added NaCl to fish gelatin solutions and found similar results. They reported that an increase in NaCl concentration resulted in a decrease in gel strength and melting temperature. The authors attributed this behavior to NaCl disrupting hydrophobic and hydrogen bonds, which led to reduced stabilization of gelatin chains. Choi and Regenstein (2000) also added sucrose and found that an increase in sucrose concentration led to an increase in gel strength and melt temperatures. They attributed this to sucrose-stabilizing hydrogen bonds. Fernandez-Diaz et al. (2001) added $MgSO_4$ and glycerol to cod and hake gelatin solutions and found that both additives increased gel strength of the samples. In addition, glycerol increased G' of both cod and hake samples.

Different cross-linkers, such as transglutaminase (Fernandez-Diaz et al. 2001, Kolodziejska et al. 2004, Jongjareonrak et al. 2006c, Norziah et al. 2009), glutaraldehyde (Chiou et al. 2006), and genipin (Chiou et al. 2006), have been added to fish gelatin solutions to convert them from liquids to gels at room temperature. Fernandez-Diaz et al. (2001) found that adding transglutaminase improved the gel strength and G' of cod and hake gelatin samples. Kolodziejska et al. (2004) also found that an increase in transglutaminase concentration reduced the penetration depth of plungers in cod gelatin samples. Jongjareonrak et al. (2006c) determined an optimal transglutaminase concentration required to produce snapper gelatin samples with the highest Bloom strength. Also, Norziah et al. (2009) determined that an increase in transglutaminase concentration led to an increase in G' values as well as melting and gelation temperatures. Chiou et al. (2006) compared the cross-linking behavior of pollock and salmon gelatin samples containing glutaraldehyde and genipin. Genipin is a natural extract that is considered less toxic than glutaraldehyde and reacts with amino acids that contain amino side groups, such as lysine and arginine. The reaction involves two different mechanisms. One mechanism is a nucleophilic attack by an amino group that eventually leads to the formation of a heterocyclic amine. The second mechanism involves a nucleophilic substitution reaction that results in the replacement of the ester group on genipin with a secondary amide linkage. The second reaction is thought to be slower than the first one. The authors found that samples containing glutaraldehyde cross-linked faster than those containing genipin.

Various polysaccharides and proteins have been blended with fish gelatin solutions to improve their material properties. Haug et al. (2004) examined blends of κ-carrageenan with fish gelatin. The authors found that adding fish gelatin and KCl to κ-carrageenan resulted in samples with higher modulus values compared to that of neat κ-carrageenan. However, adding fish gelatin had little effect on κ-carrageenan's melting and gelation temperatures. Badii and Howell (2006) blended fish gelatin with egg albumen and found that gelatin interacted with the albumen to increase G' values. The albumen/gelatin sample had smaller aggregates than the neat albumen sample when examined by phase contrast microscopy. This indicated that fish gelatin might have interacted with albumen to produce more uniform and stronger gels. Cheng et al. (2008) blended fish gelatin with pectin to produce a low-fat spread. The authors found that samples containing higher pectin to gelatin ratios had greater G' values. Chen et al. (2009) blended hydroxypropylmethylcellulose (HPMC) with tilapia gelatin and found that adding HPMC increased gel strength by two to five times over that of neat gelatin, even though HPMC does not gel until 50°C–90°C. In addition, samples containing HPMC had slightly higher melting temperatures and enthalpies compared to neat gelatin, indicating HPMC improved thermal stability.

High-pressure treatments have shown some potential to improve gel strength of fish gelatin samples. Montero et al. (2002) used pressures of 200–400 MPa to treat cod and megrim samples. The authors found that an increase in pressure resulted in an increase in gel strength for all treated samples. They speculated that high pressures can stabilize the gels by inducing more hydrogen bonds. All pressure-treated cod samples had higher gel strengths than the untreated sample. In contrast, most of the treated megrim samples had lower gel strength than the untreated one.

Gelatin desserts have also been made from fish gelatin. Zhou and Regenstein (2007) used a cold-water (pollock) and a warm-water fish (tilapia) gelatin to make the desserts. They then compared textural properties of fish gelatin desserts to those made from porcine gelatin. The authors examined the behavior of all samples under 75% compression and reported that the pollock desserts were the only ones that did not break into smaller pieces, indicating the pollock samples had higher cohesiveness.

6.4.3 ANTIOXIDANTS AND EMULSIFIERS

Gelatin hydrolysates have been characterized for radical scavenging and antioxidant properties. Mendis et al. (2005) used several enzymes to hydrolyze hoki gelatin. Hydrolysates from trypsin showed the best radical scavenging and antioxidant activities. In fact, the trypsin sample exhibited better antioxidant activity than the natural antioxidant α-tocopherol. The authors then fractionated the trypsin samples further and found fractions that showed comparable antioxidant activity to the synthetic antioxidant butylated hydroxytoluene. Yang et al. (2008) also used various enzymes to hydrolyze cobia gelatin to produce hydrolysates. The authors found that hydrolysates with lower molecular weights had better radical scavenging and antioxidant activities than those with higher molecular weights. In fact, samples containing hydrolysates with molecular weights less than 700 Da had the highest activities.

Fish gelatin has been examined as emulsifiers in oil-in-water emulsions. Dickinson and Lopez (2001) compared the emulsifying properties of low-molecular-weight (~40 kDa) fish gelatin, sodium caseinate, and whey protein in n-tetradecane or triglyceride oil-in-water emulsions. For fish gelatin in both emulsions, the oil droplet size increased in value and some creaming and flocculation occurred in the samples. In contrast, sodium caseinate performed the best in both emulsions, with little change in droplet size, little creaming, and no flocculation. However, the authors found that fish gelatin required a lower optimal concentration than sodium caseinate and whey protein to produce the best emulsions. Surh et al. (2006) compared the emulsifying properties of low-molecular-weight fish gelatin (LMW-FG, ~55 kDa) to those of high-molecular-weight fish gelatin (HMW-FG, 120 kDa). The authors used a corn oil-in-water emulsion. They found that low-concentration LMW-FG and HMW-FG samples had multimodal droplet size distributions, whereas high gelatin concentration samples had mono-modal distributions. Optical microscopy results indicated that large droplets appeared in all samples, although fewer large droplets remained at higher gelatin concentrations. The HMW-FG samples provided slightly better stability compared to the LMW-FG samples. Also, droplet diameters showed no dependence on ionic concentration, indicating that polymeric steric repulsion rather than electrostatic repulsion played the major role in preventing aggregation of droplets. Fish gelatin had also been used as part of electrostatically deposited multilayer membranes to improve stability of oil-in-water emulsions. Gu et al. (2005) compared emulsifying properties of samples containing primary (β-lactoglobulin), secondary (β-lactoglobulin-ι-carrageenan), and tertiary (β-lactoglobulin-ι-carrageenan-fish gelatin) layers. The authors found that the multilayer emulsions generally had greater stability to changes in ionic strength and temperature than single-layer emulsions under certain conditions. Also, certain layers could be detached from each other, indicating possible use for control release applications. In addition, Surh et al. (2005) examined sodium dodecyl sulfate (sds) and sds-fish gelatin emulsifiers for stabilizing corn oil-in-water emulsions. The authors found that temperature had little effect on stability of droplets in all samples. However, sds-fish gelatin samples became more stable than sds samples under different ionic strength and pH conditions.

6.5 CONCLUSIONS

Fish gelatin has become an attractive alternative to mammalian gelatin due to religious considerations and the large potential sources available from by-products generated by the fishing industry. However, the low gelation temperatures of fish gelatin pose some challenges for possible applications. Nevertheless, fish gelatin has been examined for various applications, including protective coatings for foods, biodegradable packaging to replace petroleum-based ones, edible antimicrobial films, desserts and low-fat spreads, antioxidant additives, and emulsifiers. Various methods have also been used to improve material properties of fish gelatin films and gels. These include adding cross-linkers, adding salts, blending with polysaccharides, and using high-pressure treatments. In the future, fish gelatin use will most likely depend on development of new techniques to improve their properties or just by taking advantage of their unique qualities.

REFERENCES

Arnesen, J.A. and Gildberg, A. 2007. Extraction and characterisation of gelatine from Atlantic salmon (*Salmo salar*) skin. *Bioresource Technology* 98:53–57.

Avena-Bustillos, R.J., Olsen, C.W., Olson, D.A., Chiou, B., Yee, E., Bechtel, P.J., and McHugh, T.H. 2006. Water vapor permeability of mammalian and fish gelatin films. *Journal of Food Science* 71(4):E202–E207.

Badii, F. and Howell, N.K. 2006. Fish gelatin: Structure, gelling properties and interaction with egg albumen proteins. *Food Hydrocolloids* 20:630–640.

Bae, H.J., Darby, D.O., Kimmel, R.M., Park, H.J., and Whiteside, W.S. 2009a. Effects of transglutaminase-induced cross-linking on properties of fish gelatin–nanoclay composite film. *Food Chemistry* 114:180–189.

Bae, H.J., Park, H.J., Darby, D.O., Kimmell, R.M., and Whiteside, W.S. 2009b. Development and characterization of PET/fish gelatin–nanoclay composite/LDPE laminate. *Packaging Technology and Science* 22:371–383.

Bae, H.J., Park, H.J., Hong, S.I., Byun, Y.J., Darby, D.O., Kimmel, R.M., and Whiteside, W.S. 2009c. Effect of clay content, homogenization RPM, pH, and ultrasonication on mechanical and barrier properties of fish gelatin/montmorillonite nanocomposite films. *LWT-Food Science and Technology* 42:1179–1186.

Bao, S., Xu, S., and Wang, Z. 2009. Antioxidant activity and properties of gelatin films incorporated with tea polyphenol-loaded chitosan nanoparticles. *Journal of Science of Food and Agriculture* 89:2692–2700.

Benjakul, S., Oungbho, K., Visessanguan, W., Thiansilakul, Y., and Roytrakul, S. 2009. Characteristics of gelatin from the skins of bigeye snapper, *Priacanthus tayenus* and *Priacanthus macracanthus*. *Food Chemistry* 116:445–451.

Binsi, P.K., Shamasundara, B.A., Dileep, A.O., Badii, F., and Howell, N.K. 2009. Rheological and functional properties of gelatin from the skin of bigeye snapper (*Priacanthus hamrur*) fish: Influence of gelatin on the gel-forming ability of fish mince. *Food Hydrocolloids* 23:132–145.

Boran, G. and Regenstein, J.M. 2009. Optimization of gelatin extraction from silver carp skin. *Journal of Food Science* 74(8):E432–E441.

Bower, C.K., Avena-Bustillos, R.J., Hietala, K.A., Bilbao-Sainz, C., Olsen, C.W., and McHugh, T.H. 2010. Dehydration of pollock skin prior to gelatin production. *Journal of Food Science* 75(4):C317–C321.

Bower, C.K., Avena-Bustillos, R.J., Olsen, C.W., McHugh, T.H., and Bechtel, P.J. 2006. Characterization of fish-skin gelatin gels and films containing the antimicrobial enzyme lysozyme. *Journal of Food Science* 71(5):M141–M145.

Carvalho, R.A., Sobral, P.J.A., Thomazine, M., Habitante, A.M.Q.B., Gimenez, B., Gomez-Guillen, M.C., and Montero, P. 2008. Development of edible films based on differently processed Atlantic halibut (*Hippoglossus hippoglossus*) skin gelatin. *Food Hydrocolloids* 22:1117–1123.

Chen, H., Lin, C., and Kang, H. 2009. Maturation effects in fish gelatin and HPMC composite gels. *Food Hydrocolloids* 23:1756–1761.

Cheng, L.H., Lim, B.L., Chow, K.H., Chong, S.M., and Chang, Y.C. 2008. Using fish gelatin and pectin to make a low-fat spread. *Food Hydrocolloids* 22:1637–1640.

Cheow, C.S., Norizah, M.S., Kyaw, Z.Y., and Howell, N.K. 2007. Preparation and characterisation of gelatins from the skins of sin croaker (*Johnius dussumieri*) and shortfin scad (*Decapterus macrosoma*). *Food Chemistry* 101:386–391.

Chiou, B., Avena-Bustillos, R.J., Bechtel, P.J., Imam, S.H., Glenn, G.M., and Orts, W.J. 2009. Effect of drying temperature on barrier and mechanical properties of cold-water fish gelatin films. *Journal of Food Engineering* 95(2):327–331.

Chiou, B., Avena-Bustillos, R.J., Bechtel, P.J., Jafri, H., Narayan, R., Imam, S.H., Glenn, G.M., and Orts, W.J. 2008. Cold water fish gelatin films: Effects of cross-linking on thermal, mechanical, barrier, and biodegradation properties. *European Polymer Journal* 44:3748–3753.

Chiou, B., Avena-Bustillos, R.J., Shey, J., Yee, E., Bechtel, P.J., Imam, S.H., Glenn, G.M., and Orts, W.J. 2006. Rheological and mechanical properties of cross-linked fish gelatins. *Polymer* 47:6379–6386.

Cho, S.M., Gu, Y.S., and Kim, S.B. 2005. Extracting optimization and physical properties of yellowfin tuna (*Thunnus albacares*) skin gelatin compared to mammalian gelatins. *Food Hydrocolloids* 19:221–229.

Choi, S.-S. and Regenstein, J.M. 2000. Physicochemical and sensory characteristics of fish gelatin. *Journal of Food Science* 65(2):194–199.

Crapo, C. and Bechtel, P. 2003. Utilization of Alaska's seafood processing byproducts. In *Advances in Seafood Byproducts: 2002 Conference Proceedings*, ed. P.J. Bechtel, pp. 105–119. University of Alaska, Fairbanks, AK: Alaska Sea Grant College Program.

da Trindade Alfaro, A., Simões da Costa, C., Fonseca, G.G., and Prentice, C. 2009. Effect of extraction parameters on the properties of gelatin from king weakfish (*Macrodon ancylodon*) bones. *Food Science and Technology International* 15(6):553–562.

Dickinson, E. and Lopez, G. 2001. Comparison of the emulsifying properties of fish gelatin and commercial milk proteins. *Journal of Food Science* 66(1):118–123.

Eysturskard, J., Haug, I.J., Elharfaoui, N., Djabourov, M., and Draget, K.I. 2009a. Structural and mechanical properties of fish gelatin as a function of extraction conditions. *Food Hydrocolloids* 23:1702–1711.

Eysturskard, J., Haug, I.J., Ulset, A., and Draget, K.I. 2009b. Mechanical properties of mammalian and fish gelatins based on their weight average molecular weight and molecular weight distribution. *Food Hydrocolloids* 23:2315–2321.

Eysturskard, J., Haug, I.J., Ulset, A., Joensen, H., and Draget, K.I. 2010. Mechanical properties of mammalian and fish gelatins as a function of the contents of α-chain, β-chain, and low and high molecular weight fractions. *Food Biophysics* 5:9–16.

Fernandez-Diaz, M.D., Montero, P., and Gomez-Guillen, M.C. 2001. Gel properties of collagens from skins of cod (*Gadus morhua*) and hake (*Merluccius merluccius*) and their modification by the coenhancers magnesium sulfate, glycerol and transglutaminase. *Food Chemistry* 74:161–167.

Fernandez-Diaz, M.D., Montero, P., and Gomez-Guillen, M.C. 2003. Effect of freezing fish skins on molecular and rheological properties of extracted gelatin. *Food Hydrocolloids* 17:281–286.

Fraga, A.N. and Williams, R.J.J. 1985. Thermal properties of gelatin films. *Polymer* 26(1):113–118.

Gilsenan, P.M. and Ross-Murphy, S.B. 2000a. Rheological characterization from mammalian and marine sources. *Food Hydrocolloids* 14:191–195.

Gilsenan, P.M. and Ross-Murphy, S.B. 2000b. Viscoelasticity of thermoreversible gelatin gels from mammalian and piscine collagens. *Journal of Rheology* 44(4):871–883.

Gimenez, B., Gomez-Guillen, M.C., and Montero, P. 2005a. Storage of dried fish skins on quality characteristics of extracted gelatin. *Food Hydrocolloids* 19:958–963.

Gimenez, B., Gomez-Guillen, M.C., and Montero, P. 2005b. The role of salt washing of fish skins in chemical and rheological properties of gelatin extracted. *Food Hydrocolloids* 19:951–957.

Gimenez, B., Turnay, J., Lizarbe, M.A., Montero, P., and Gomez-Guillen, M.C. 2005c. Use of lactic acid for extraction of fish skin gelatin. *Food Hydrocolloids* 19:941–950.

Gomez-Estaca, J., Gimenez, B., Montero, P., and Gomez-Guillen, M.C. 2009a. Incorporation of antioxidant borage extract into edible films based on sole skin gelatin or a commercial fish gelatin. *Journal of Food Engineering* 92:78–85.

Gomez-Estaca, J., Lopez de Lacey, A., Gomez-Guillen, M.C., Lopez-Caballero, M.E., and Montero, P. 2009b. Antimicrobial activity of composite edible films based on fish gelatin and chitosan incorporated with clove essential oil. *Journal of Aquatic Food Product Technology* 18:46–52.

Gomez-Guillen, M.C., Gimenez, B., and Montero, P. 2005. Extraction of gelatin from fish skins by high pressure treatment. *Food Hydrocolloids* 19:923–928.

Gomez-Guillen, M.C. and Montero, P. 2001. Extraction of gelatin from megrim (*Lepidorhombus boscii*) skins with several organic acids. *Journal of Food Science* 66(2):213–216.

Gomez-Guillen, M.C., Turnay, J., Fernandez-Diaz, M.D., Ulmo, N., Lizarbe, M.A., and Montero, P. 2002. Structural and physical properties of gelatin extracted from different marine species: A comparative study. *Food Hydrocolloids* 16:25–34.

Grobben, A.H., Steele, P.J., Somerville, R.A., and Taylor, D.M. 2004. Inactivation of the bovine-spongiform-encephalopathy (BSE) agent by the acid and alkaline processes used in the manufacture of bone gelatin. *Biotechnology and Applied Biochemistry* 39(3):329–338.

Gu, Y.S., Decker, E.A., and McClements, D.J. 2005. Production and characterization of oil-in water emulsions containing droplets stabilized by multilayer membranes consisting of β-lactoglobulin, ι-carrageenan and gelatin. *Langmuir* 21:5752–5760.

Haug, I.J., Draget, K.I., and Smidsrod, O. 2004. Physical behaviour of fish gelatin-κ-carrageenan mixtures. *Carbohydrate Polymers* 56:11–19.

Jamilah, B. and Harvinder, K.G. 2002. Properties of gelatins from skins of fish-black tilapia (*Oreochromis mossambicus*) and red tilapia (*Oreochromis nilotica*). *Food Chemistry* 77:81–84.

Joly-Duhamel, C., Hellio, D., Ajdari, A., and Djabourov, M. 2002a. All gelatin networks: 2. The master curve for elasticity. *Langmuir* 18:7158–7166.

Joly-Duhamel, C., Hellio, D., and Djabourov, M. 2002b. All gelatin networks: 1. Biodiversity and physical chemistry. *Langmuir* 18:7209–7217.

Jongjareonrak, A., Benjakul, S., Visessanguan, W., and Tanaka, M. 2006a. Effects of plasticizers on the properties of edible films from skin gelatin of bigeye snapper and brownstripe red snapper. *European Food Research and Technology* 222:229–235.

Jongjareonrak, A., Benjakul, S., Visessanguan, W., and Tanaka, M. 2006b. Fatty acids and their sucrose esters affect the properties of fish skin gelatin-based film. *European Food Research and Technology* 222:650–657.

Jongjareonrak, A., Benjakul, S., Visessanguan, W., and Tanaka, M. 2006c. Skin gelatin from bigeye snapper and brownstripe red snapper: Chemical compositions and effect of microbial transglutaminase on gel properties. *Food Hydrocolloids* 20:1216–1222.

Jongjareonrak, A., Rawdkuen, S., Chaijan, M., Benjakul, S., Osako, K., and Tanaka, M. 2010. Chemical compositions and characterisation of skin gelatin from farmed giant catfish (*Pangasianodon gigas*). *LWT-Food Science and Technology* 43:161–165.

Karim, A.A. and Bhat, R. 2009. Fish gelatin: Properties, challenges, and prospects as an alternative to mammalian gelatins. *Food Hydrocolloids* 23:563–576.

Kasankala, L.M., Xue, Y., Weilong, Y., Hong, S.D., and He, Q. 2007. Optimization of gelatin extraction from grass carp (*Catenopharyngodon idella*) fish skin by response surface methodology. *Bioresource Technology* 98:3338–3343.

Kołodziejska, I., Kaczorowski, K., Piotrowska, B., and Sadowska, M. 2004. Modification of the properties of gelatin from skins of Baltic cod (*Gadus morhua*) with transglutaminase. *Food Chemistry* 86:203–209.

Kołodziejska, I. and Piotrowska, B. 2007. The water vapour permeability, mechanical properties and solubility of fish gelatin–chitosan films modified with transglutaminase or 1-ethyl-3-(3-dimethylaminopropyl) carbodiimide (EDC) and plasticized with glycerol. *Food Chemistry* 103:295–300.

Kołodziejska, I., Piotrowska, B., Bulge, M., and Tylingo, R. 2006. Effect of transglutaminase and 1-ethyl-3-(3-dimethylaminopropyl) carbodiimide on the solubility of fish gelatin–chitosan films. *Carbohydrate Polymers* 65:404–409.

Kołodziejska, I., Skierka, E., Sadowska, M., Kołodziejski, W., and Niecikowska, C. 2008. Effect of extracting time and temperature on yield of gelatin from different fish offal. *Food Chemistry* 107:700–706.

Limpisophon, K., Tanaka, M., Weng, W., Abe, S., and Osako, K. 2009. Characterization of gelatin films prepared from under-utilized blue shark (*Prionace glauca*) skin. *Food Hydrocolloids* 23:1993–2000.

Liu, H., Li, D., and Guo, S. 2008. Rheological properties of channel catfish (*Ictalurus punctaus*) gelatin from fish skins preserved by different methods. *LWT-Food Science and Technology* 41:1425–1430.

Liu, L., Liu, C., Fishman, M.L., and Hicks, K.B. 2007. Composite films from pectin and fish skin gelatin or soybean flour protein. *Journal of Agricultural and Food Chemistry* 55:2349–2355.

Lopez-Caballero, M.E., Gomez-Guillen, M.C., Perez-Mateos, M., and Montero, P. 2005. A chitosan–gelatin blend as a coating for fish patties. *Food Hydrocolloids* 19:303–311.

Mendis, E., Rajapakse, N., and Kim, S. 2005. Antioxidant properties of a radical-scavenging peptide purified from enzymatically prepared fish skin gelatin hydrolysate. *Journal of Agricultural and Food Chemistry* 53:581–587.

Mohtar, N.F., Perara, C., and Quek, S. 2010. Optimisation of gelatine extraction from hoki (*Macruronus novaezelandiae*) skins and measurement of gel strength and SDS–PAGE. *Food Chemistry* 122:307–313.

Montero, P., Fernandez-Diaz, M.D., and Gomez-Guillen, M.C. 2002. Characterization of gelatin gels induced by high pressure. *Food Hydrocolloids* 16:197–205.

Montero, P. and Gomez-Guillen, M.C. 2000. Extracting conditions for megrim (*Lepidorhombus boscii*) skin collagen affect functional properties of the resulting gelatin. *Journal of Food Science* 65(3):434–438.

Muyonga, J.H., Cole, C.G.B., and Duodu, K.G. 2004. Extraction and physico-chemical characterization of Nile perch (*Lates niloticus*) skin and bone gelatin. *Food Hydrocolloids* 18:581–592.

Nalinanon, S., Benjakul, S., Visessanguan, W., and Kishimura, H. 2008. Improvement of gelatin extraction from bigeye snapper skin using pepsin-aided process in combination with protease inhibitor. *Food Hydrocolloids* 22:615–622.

Norziah, M.H., Al-Hassan, A., Khairulnizam, A.B., Mordi, M.N., and Norita, M. 2009. Characterization of fish gelatin from surimi processing wastes: Thermal analysis and effect of transglutaminase on gel properties. *Food Hydrocolloids* 23:1610–1616.

Piez, K.A. 1964. Nonidentity of the three α chains in codfish skin collagen. *Journal of Biological Chemistry* 239:PC4315–PC4316.

Piotrowska, B., Sztuka, K., Kolodziejska, I., and Dobrosielska, E. 2008. Influence of transglutaminase or 1-ethyl-3-(3-dimethylaminopropyl) carbodiimide (EDC) on the properties of fish-skin gelatin films. *Food Hydrocolloids* 22:1362–1371.

Pranoto, Y., Lee, C.M., and Park, H.J. 2007. Characterizations of fish gelatin films added with gellan and k-carrageenan. *LWT-Food Science and Technology* 40:766–774.

Rattaya, S., Benjakul, S., and Prodpran, T. 2009. Properties of fish skin gelatin film incorporated with seaweed extract. *Journal of Food Engineering* 95:151–157.

Rose, P.I. 1992. Inedible gelatin and glue. In *Inedible Meat By-Products*, A.M. Pearson and D.R. Dutson, eds., London, U.K.: Elsevier Applied Science, pp. 217–263.

Sarabia, A.I., Gomez-Guillen, M.C., and Montero, P. 2000. The effect of added salts on the viscoelastic properties of fish skin gelatin. *Food Chemistry* 70:71–76.

Schrieber, R. and H. Gareis. 2007. *Gelatin Handbook. Theory and Industrial Practice*. Weinheim, Germany: Wiley-VCH Verlag GmbH & Co.

Surh, J., Decker, E.A., and McClements, D.J. 2006. Properties and stability of oil-in-water emulsions stabilized by fish gelatin. *Food Hydrocolloids* 20:596–606.

Surh, J., Gu, Y.S., Decker, E.A., and McClements, D.J. 2005. Influence of environmental stresses on stability of o/w emulsions containing cationic droplets stabilized by SDS-fish gelatin membranes. *Journal of Agricultural and Food Chemistry* 53:4236–4244.

Sztuka, K. and Kołodziejska, I. 2008. Effect of transglutaminase and EDC on biodegradation of fish gelatin and gelatin-chitosan films. *European Food Research and Technology* 226:1127–1133.

Sztuka, K. and Kołodziejska, I. 2009. The influence of hydrophobic substances on water vapor permeability of fish gelatin films modified with transglutaminase or 1-ethyl-3-(3-dimethylaminopropyl) carbodiimide (EDC). *Food Hydrocolloids* 23:1062–1064.

Tabarestani, H.S., Maghsoudlou, Y., Motamedzadegan, A., and Mahoonak, A.R.S. 2010. Optimization of physico-chemical properties of gelatin extracted from fish skin of rainbow trout (*Onchorhynchus mykiss*). *Bioresource Technology* 101:6207–6214.

Taheri, A., Kenari, A.M.A., Gildberg, A., and Behnam, S. 2009. Extraction and physicochemical characterization of greater lizardfish (*Saurida tumbil*) skin and bone gelatin. *Journal of Food Science* 74(3):E160–E165.

Wang, Y. and Regenstein, J.M. 2009. Effect of EDTA, HCl, and citric acid on Ca salt removal from Asian (silver) carp scales prior to gelatin extraction. *Journal of Food Science* 74(6):C426–C431.

Ward, A.G. and Courts, A.G. 1977. *The Science and Technology of Gelatin*. London, U.K.: Academic Press.

www.geafiltration.com/library/gelatin_processing_aid.asp, accessed on 9/1/2010.

Yang, J., Ho, H., Chu, Y., and Chow, C. 2008. Characteristic and antioxidant activity of retorted gelatin hydrolysates from cobia (*Rachycentron canadum*) skin. *Food Chemistry* 110:128–136.

Yang, H. and Wang, Y. 2009. Effects of concentration on nanostructural images and physical properties of gelatin from channel catfish skins. *Food Hydrocolloids* 23:577–584.

Yang, H., Wang, Y., Jiang, M., Oh, J., Herring, J., and Zhou, P. 2007a. 2-Step optimization of the extraction and subsequent physical properties of channel catfish (*Ictalurus punctatus*) skin gelatin. *Journal of Food Science* 72(4):C188–C195.

Yang, H., Wang, Y., Regenstein, J.M., and Rouse, D.B. 2007b. Nanostructural characterization of catfish skin gelatin using atomic force microscopy. *Journal of Food Science* 72(8):C430–C440.

Yi, J.B., Kim, Y.T., Bae, H.J., Whiteside, W.S., and Park, H.J. 2006. Influence of transglutaminase-induced cross-linking on properties of fish gelatin films. *Journal of Food Science* 71(9):E376–E383.

Yoshimura, K., Terashima, M., Hozan, D., Ebato, T., Nomura, Y., Ishii, Y., and Shirai, K. 2000. Physical properties of shark gelatin compared with pig gelatin. *Journal of Agricultural and Food Chemistry* 48:2023–2027.

Zhang, S., Wang, Y., Herring, J.L., and Oh, J.-H. 2007. Characterization of edible film fabricated with channel catfish (*Ictalurus punctatus*) gelatin extract using selected pretreatment methods. *Journal of Food Science* 72(9):C498–C503.

Zhang, F., Xu, S., and Wang, Z. 2011. Pre-treatment optimization and properties of gelatin from freshwater fish scales. *Food and Bioproducts Processing* 89(3):185–193.

Zhou, P. and Regenstein, J.M. 2004. Optimization of extraction conditions for pollock skin gelatin. *Journal of Food Science* 69(5):C393–C398.

Zhou, P. and Regenstein, J.M. 2005. Effects of alkaline and acid pretreatments on Alaska pollock skin gelatin extraction. *Journal of Food Science* 70(6):C392–C396.

Zhou, P. and Regenstein, J.M. 2007. Comparison of water gel desserts from fish skin and pork gelatins using instrumental measurements. *Journal of Food Science* 72(4):C196–C201.

7 Polymeric Materials from Renewable Resources

Blends of Poly(3-Hydroxybutyrate) and Cellulose Acetate Derived from Rice Straw and Bagasse

Mohamed El-Newehy, Arianna Barghini, Stefania Cometa, Stanislav Miertus, and Emo Chiellini

CONTENTS

LIST OF ABBREVIATIONS

APSH	Alkaline peroxide–soluble hemicellulose
APSL	Alkaline peroxide–soluble lignin
CA	Cellulose acetate
Cell	Cellulose
C-RS	Chopped rice straw
C-SCB	Chopped sugarcane bagasse
DC-RS	Dewaxed chopped rice straw
DC-SCB	Dewaxed chopped sugarcane bagasse
DM-RS	Dewaxed milled rice straw
DM-SCB	Dewaxed milled sugarcane bagasse
D-RS	Dewaxed rice straw
DS	Degree of substitution
D-SCB	Dewaxed sugarcane bagasse
DTGA	Derivative thermogravimetric analysis
FTIR	Transmission Fourier transform infrared spectroscopy
HCell	Hemicellulose
Lig	Lignin
Moist	Moisture
M-RS	Milled rice straw
M-SCB	Milled sugarcane bagasse
NDF	Not dissolved fibers
NMR	Nuclear magnetic resonance
PHB	Poly(hydroxyl butyrate)
RS	Rice straw
SCB	Sugarcane bagasse
SEM	Scanning electron microscopy
Si	Silica
T_g	Glass transition temperature
TGA	Thermogravimetric analysis
T_{on}	Onset temperature
T_p	Peak decomposition temperature
WAXD	Wide-angle x-ray diffraction

7.1 INTRODUCTION

Agro-industrial residues represent the most abundant source of renewable resources on Earth.

From one point of view, accumulation of this biomass in large quantities every year results in the deterioration of the environment; therefore, its reuse could help in solving pollution problems, which their disposal may otherwise cause (Sun et al. 2004a).

On the other hand, the search for new materials, having high performance at affordable costs, provides the interesting opportunity to turn a valueless waste into a valuable substrate for new value-added materials, with unique characteristics that could include the eco-friendliness, the large availability in nature, and the resulting reduced dependence from petroleum-based materials.

In recent years, there has been an increasing trend toward more efficient utilization of agro-industrial residues, such as sugarcane bagasse (SCB) (Sun et al. 2004a) and rice straw (RS) (Abdel-Mohdy et al. 2009). These natural fibers are gaining progressive account as renewable, environmentally acceptable, and biodegradable starting material for industrial applications, technical textiles, composites, pulp and paper, and civil engineering and building activities (Habibi et al. 2008), as well as a source of chemicals or building blocks (Sun et al. 2004b).

Natural fibers offer a number of well-known advantages that include availability of renewable natural resources at low cost and biodegradability.

In particular, the use of waste products such as SCB arises from the urgent need of demand of alternative fuels that was in turn traduced in an increased production of ethanol from sugarcane. The bagasse derived as a residue from this activity is usually burned to produce the electricity required by the sugar mills and, in some cases, excess energy is even sold in the electricity market. Moreover, SCB was used for pulp and paper production, and products based on fermentation (Pandey et al. 2000) However, a great amount of bagasse is still wasted.

Egypt is a country with a large agricultural production. It is estimated that 23 million ton/year of agricultural residues result from cotton and rice. Besides that, there are wastes from sugarcane and banana plant, which pose major disposal problems (Habibi et al. 2008).

The total production of sugar in Egypt in 2007–2008 is 1582 million tonnes and the consumption is 2485 million tonnes. Sugar beet is grown in 247 million ha and produces 482 mt of white sugar with an average of 50.7 ton/ha. However, the sugarcane is grown in 136 million hectares and produces 1,075,184 mt of sugar (Foly et al. 2008).

SCB is the fibrous residue that remains after the crushing and extraction of juice from the sugarcane. SCB is a lignocellulosic material that is currently a low value product burned for its energy value (Doherty et al. 2007). Chemically, SCB contains about 40%–50% of the dry residue of the glucose polymer cellulose, much of which has a crystalline structure. Another 25%–35% are hemicelluloses, an amorphous polymer usually composed of xylose, arabinose, galactose, glucose, and mannose. The remainder is mostly lignin plus lesser amounts of minerals, waxes, and other components (Sun et al. 2004a). Because of its low ash content (about 2.4%), bagasse can be used in bioconversion processes by microbial cultures and can be considered as a rich solar energy reservoir due to its high yields and annual regeneration capacity (Pandey et al. 2000).

Different studies have been carried out by Indian researchers about the use of (1) bagasse after different pretreatments for cellulase enzyme production (Adsula et al. 2004, Singh et al. 2009); (2) delignified SCB as a source for the production of different sugars (glucose, xylose, and arabinose) using enzymes that were produced by treating delignified bagasse polysaccharides with mutant microorganisms (Adsula et al. 2005); and (3) hemicellulose from SCB as a source of xylose for the microbial production of xylitol, which could have important applications in pharmaceuticals and food industry (Prakash et al. 2011).

RS is an important lignocellulosic biomass with nearly 800 million dry tons produced annually worldwide. It therefore has great potential as a lignocellulosic feedstock for making renewable fuels and chemicals. However, it appears to be more recalcitrant than other agricultural residues (Balan et al. 2008). Chemically, it contains 37% of cellulose, 45% of hemicellulose, 5% of lignin, and 13% of silicon ash. The disposal of RS by open-field burning frequently causes serious air pollution, hence new economical technologies for RS disposal and utilization must be developed (Gong et al. 2007).

However, there are some inherent drawbacks in the use of the natural fibers: poor moisture resistance (Bismarck et al. 2002), low resistance to microbial attack, tendency to form aggregates during processing, and poor surface adhesion for association with a polymer matrix (Mwaikambo et al. 1999). These disadvantages could be solved by chemical modification of fiber surfaces.

Dewaxing or defatting, delignification, bleaching, acetylation, mercerization, salinization, and chemical grafting represent examples of chemical modifications of agro-industrial residues.

These lipophilic extractives are difficult to remove, even if during the alkaline process such as kraft pulping, the triglycerides are completely saponified and fatty acids dissolved. In contrast to glycerol esters, sterols and some steryl esters and waxes do not form soluble soap under the alkaline conditions and therefore have a tendency to deposit on equipment and cause pitch problems (Chen et al. 1994, Gutierrez et al. 1999). Moreover, these extractives have also a decisive impact on the environment as contributors to the effluent toxicity (Gutierrez et al. 1999, Wallis and Wearne 1997).

On the other hand, during chip storage, extractives undergo voltalization, enzymatic hydrolysis, and air oxidation. Unfortunately, these reactions are slow, particularly in the colder weather conditions (Chen et al. 1994, Gutierrez et al. 1999). Recently, new biological methods using fungal pretreatment have been developed to reduce extractives levels with a minimal period of storage (Patel et al. 2007, Sukumaran et al. 2009), but the fungi are poorly adapted to low temperatures.

The extraction and characterization of these lipophilic extractives (Gutierrez et al. 1999) comprises a Soxhlet extraction with toluene/ethanol (2:1, v/v), chloroform, petroleum ether, dichloromethane, or hexane, silylation, and gas chromatography.

Among the several options for recycling SCB or RS to obtain value-added products, cellulose and its derivatives represent an important feedstock for a number of industries and have created a great deal of research interest.

Cellulose is the major constituent of all plant materials, forms about half to one-third of plant tissues, and is constantly replenished by photosynthesis, with estimates of annual world biosynthesis of about 1.56×10^{12} ton (Klemm et al. 2005). It is indeed an abundant resource, which has found wide applications. Besides its domestic use, cellulose has also found wide industrial applications. For example, cellulose is widely used to prepare precursors and gel casting for different materials, and to fabricate hollow fiber membranes (Shao et al. 2002, Shieh and Chung 2000, Stading et al. 2001).

Cellulose and all the carbohydrates obtainable from renewable natural sources represent an important source of building blocks for the synthesis of biodegradable polymers and for the design of optically active polymers, containing stereocenters in the repeating unit, with a rich variety of chemical structures (Galbis and Garcia-Martin 2008).

Particular interest was devoted to the chemical modification of cellulose, such as its hydrolysis for producing ethanol (Sanchez and Cardona 2008), the production of cellulose derivatives such as carboxymethylcellulose (Ruzene et al. 2007), methylcellulose (Viera et al. 2007), or cellulose acetate (CA) (Cerqueira et al. 2008, Gamez et al. 2006, Liu et al. 2007, Meireles et al. 2007, Sarrouh et al. 2007).

CA is one of the oldest man-made macromolecular compounds of renewable origin (Barkalow et al. 1989) used extensively in the textile industry, or to produce filters, photographic films, and transparent and pigmented sheeting. Moreover, it was used in plastic compositions such as those used for compression, extrusion, injection molding, and, to a lesser extent, surface coating (Reverley, 1985, Samois et al. 1997). CA has an inherent advantage in that the starting material, cellulose, is a renewable natural resource (Barkalow et al. 1989).

CA is produced by cellulose acetylation, in which cellulose reacts with acetic anhydride that is used as acetylating agent, in the presence of acetic acid as solvent, and sulfuric acid or perchloric acid used as catalyst (Cerqueira et al. 2007, Edgar et al. 2001, Steinmeier 2004). The degree of substitution (DS) of CA can be defined as the number of acetyl groups per anhydroglucose unit, and can range from 0 for cellulose to 3 for the cellulose triacetate (CTA). Higher-DS polymers are acetone soluble and the CA polymers with a DS below 1.7 do not dissolve in acetone. Moreover, there is a strong link between the DS and biodegradability; the lower the DS the more biodegradable CA becomes which will make the lower-DS polymers very important in CA copolymers and/or blends. Also, as the DS decreased, an increase in the crystallinity was observed due to the fact that as the acetyl content of low DS decreased, a more "celluloselike," semicrystalline structure was adopted (Samois et al. 1997).

On the other hand, poly(hydroxyalkanoate)s (PHAs) are a family of natural polyesters produced by a wide variety of microorganisms, which use them as intracellular carbon and energy storage compounds (Scholz and Gross 2000). PHAs are generated and stored in bacteria in the form of fine powder particles. They may be obtained from recycling materials, renewable agricultural sources, or through the action of microorganisms to get biodegradable materials.

Poly(3-hydroxy butyrate) (PHB) belongs to the PHAs family, and it is one of the typical natural biopolyesters produced by many microorganisms as intracellular carbon and energy storage compounds (Chen et al. 2000, Greco and Martuscelli 1989, Xu et al. 2006). It was produced from renewable resources via a classical biotechnological process. It has attracted industrial attention as an environmentally friendly polymer for agricultural, marine, and medical applications due to its biocompatibility and biodegradability (Inoue and Yoshie 1992, Luo et al. 2007). Moreover, it possesses high melting temperature, good resistance to organic solvents, and excellent mechanical strength and modulus (Chodak 2008).

This is attracting attention due to both economic and ecological considerations, since PHB is obtained through biosynthesis and its biodegradability enhances environmental safety.

However, PHB has several inherent deficiencies for use as a practical polymer material, such as its brittleness due to high crystallinity and thermal instability near its melting point of 180°C. To overcome the drawbacks of PHB and obtain some useful new material properties, physical blending and chemical modification have been adopted (Xu et al. 2006, 2002, Yang et al. 2005).

There are many attempts to blend PHB with other flexible polymers or low molecular weight plasticizers to turn PHB into materials with improved properties in impact strength, film formation, processing, mechanical strength, amphiphilicity, biodegradability, and biocompatibility (Chen and Wu 2005, Rodringues et al. 2005, Xu et al. 2006).

Thus, several polymers, such as poly(ethylene glycol) (PEG) (Rodrigues et al. 2005), poly(ethylene oxide) (PEO) (Avella et al. 1991), poly(vinyl acetate) (PVAc) (Chiu 2006), poly(4-vinylphenol) (PVPh) (Xing et al. 1997), poly(vinyl alcohol) (PVA) (Huang et al. 2005), and polylactide (PLLA) (Zhang et al. 2006), have been selected to be blended for lowering the melting temperature and enhancing the processability.

In view of actual applications, blending between PHB and cellulose esters (CEs) has been proposed (Ceccoruli et al. 1993, El-Shafee et al. 2001, Park et al. 2005, Pizzoli et al. 1994, Scandola et al. 1992, Suttiwijitpukdee et al. 2010, Yu et al. 2006).

We report the recent advances in the dewaxing process of RS and SCB as agro-waste materials using toluene/ethanol (2:1, v/v) in the Kumagawa extractor apparatus in order to improve the thermal stability of these renewable materials. The effect of the particle sizes of the agro-waste materials on the dewaxing process has also been assessed.

Then, cellulose was isolated from SCB and was modified by acetylation to CA, and flat membranes of CA and cellulose were produced. Cellulose and CA were selected to blend with PHB with the aim to improve the application properties of PHB. PHB–CA blending was found to be a more convenient and well-developed technology with lower cost for improving PHB properties.

7.2 EXPERIMENTAL

7.2.1 MATERIALS

The raw material, SCB (Figure 7.1), was collected from a local sugar factory (Tanta, Egypt).

The SCB was dried by exposure to the sunlight and then cut into small pieces (1–3 cm), was dried in an oven under vacuum at 55°C for 48 h, and then pulverized with a blade grinder (Tomaz Duarte et al. 2006). The ground bagasse was sieved and the fractions passing through a mesh sieve of 0.425 mm (not passing through a mesh sieve of 0.3 mm) were collected.

FIGURE 7.1   Chopped sugarcane bagasse (C-SCB).

FIGURE 7.2   Chopped rice straw (C-RS).

The raw material RS (Figure 7.2) was collected from Egypt, from the farmers after harvesting the rice.

The RS was dried by exposure to the sunlight and then cut into small pieces (1–2 cm), was dried in an oven under vacuum at 55°C for 48 h, and then pulverized with a blade grinder. The ground RS was sieved and the fraction passing through a mesh sieve of 0.425 mm (not passing through a mesh sieve of 0.3 mm) was collected.

PHB was supplied by PHB Industrial S.A., Usina da Pedra, Serrana, SP, Brasile, with molecular weight around 200 kDa, and was dried in oven under vacuum at 50°C–55°C for 72 h prior use.

Toluene was purchased from BDH. Ethanol 96% was purchased from Carlo Erba reagents.

Dichloromethane, chloroform, glacial acetic acid (99%–100%), and hydrogen peroxide (H_2O_2) (30%) were purchased from J.T. Baker. Acetic anhydride was purchased from Aldrich. All solvents and reagents were used as received.

7.2.2   Dewaxing Process

7.2.2.1   Sugarcane Bagasse and Rice Straw

SCB and RS were exposed to open air and sunlight and then cut into small pieces. The fibers were dried in an oven under vacuum at 55°C for 48 h and finally grinded to fine powder with a blade grinder (Chiellini et al. 2001).

The powder was sieved and the fractions between the following mesh sieves were collected: 0.60–0.425, 0.425–0.3, 0.212–0.15, 0.15–0.106, and 0.106–0.053 mm.

Wax was removed from milled and chopped lignocellulosic materials by using a Kumagawa extractor (Figure 7.3).

The solvent was a mixture of toluene/ethanol (2:1). When the extraction was completed, dewaxed materials (D-SCB or D-RS) were dried under vacuum at room temperature overnight. A sample of extracted solvent in the dean stark stopcock was taken every 3 h and evaporated using a rotavapor to check the progress of extraction process and to dissolve wax.

7.2.2.2   Isolation of Cellulose from Sugarcane Bagasse

Isolation of cellulose from SCB takes place in three main steps.

The first step includes the dewaxing process (Figure 7.4) by using Kumagawa apparatus, in which SCB (15 g) was extracted for wax using a mixture of toluene/ethanol (2:1) for 9 h (Raveendran et al. 1996, Sun et al. 2004a,b).

After the extraction was completed, dewaxed sugarcane bagasse (D-SCB) was dried under vacuum at room temperature for 48 h. The second step includes the removal of water-soluble fraction,

FIGURE 7.3   Kumagawa extractor.

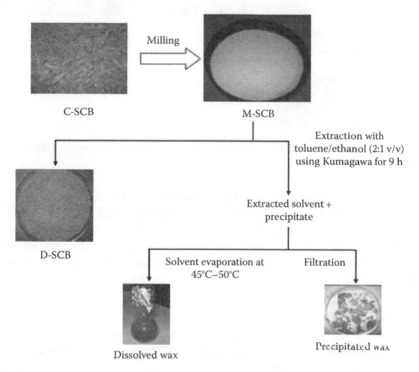

FIGURE 7.4   Dewaxing process for sugar cane bagasse (SCB).

water-soluble hemicellulose (WSH), and water-soluble lignin (WSL), according to the procedure described by Sun et al. (Raveendran et al. 1996, Sun et al. 2004a,b); the D-SCB (10 g) was soaked in 300 mL distilled water at 55°C for 2 h with stirring (Figure 7.5).

The water-soluble free residue (WSFR) was recovered by vacuum filtration, washed with distilled water, and was dried under vacuum at 50°C for 48 h.

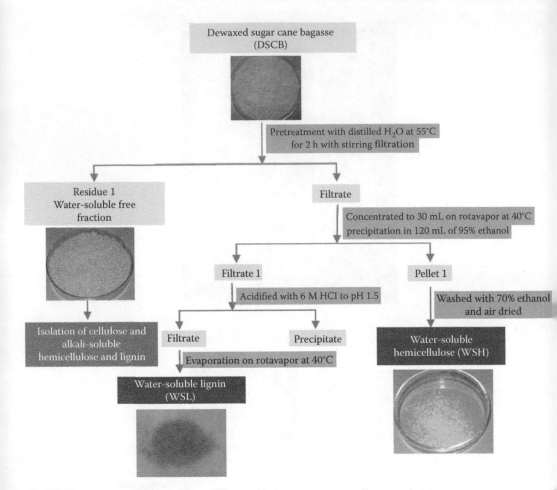

FIGURE 7.5 Isolation of water-soluble fractions (hemicelluose [WSH] and lignin [WSl]) from dewaxed sugarcane bagasse (D-SCB).

The last step includes the removal of alkaline peroxide–soluble fractions, alkaline peroxide–soluble hemicellulose (APSH), and alkaline peroxide–soluble lignin (APSL), by soaking the water-soluble free D-SCB (9 g) in 300 mL of 2% (w/v) H_2O_2 adjusted to pH 11.5 with 4 M NaOH at 50°C for 16 h with stirring (Figure 7.6). Cellulose was recovered by vacuum filtration, washed with distilled water till neutral (pH 6–7), and was dried under vacuum at 50°C for 48 h. All experiments were performed in duplicate.

7.2.2.3 Acetylation of Cellulose

Acetylation of cellulose was carried out according to the following procedure (UNI EN ISO 1183-1): cellulose (3.0 g) was soaked in 75 mL of glacial acetic acid and was stirred at room temperature for 30 min in a 250 mL round-bottom flask.

Then, 0.24 mL of concentrated H_2SO_4 in 27 mL of glacial acetic acid was added under stirring and stirring was continued for further 30 min at room temperature after the H_2SO_4 addition was completed. Cellulose pulp solution was separated by tilting the flask and pouring it into a beaker.

An amount of 96 mL of acetic anhydride was added to the solution that was poured back into the flask containing the cellulose. The reaction mixture was stirred for 90 min and the flask was covered with aluminum foil and left to stand at room temperature for 17 h. The reaction mixture was then stirred at 30°C for 6 h and then stood at room temperature for further 16 h.

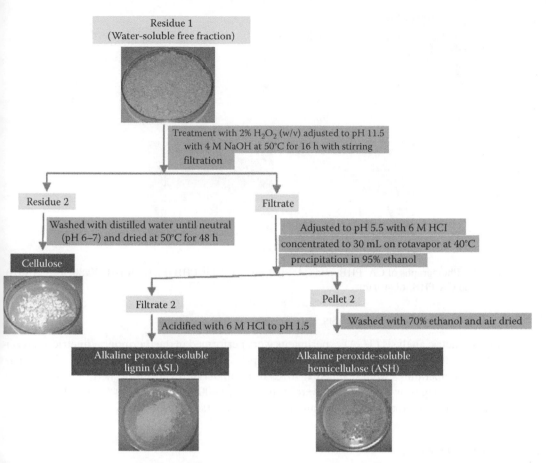

FIGURE 7.6 Isolation of alkaline-peroxide-soluble fractions (hemicellulose (ASH) and ligin (ASL)) from water-soluble free dewaxed sugar cane bagasse (WSFR).

The undissolved material was removed by filtration, and water was added to the filtrate to stop the reaction and precipitate CA.

The produced CA was filtered and washed with distilled water till neutral pH and then dried at room temperature for 48 h in a desiccator under vacuum.

CA was dissolved in dichloromethane (5 mL) with stirring for 1 week at room temperature (20°C), and PHB was dissolved in chloroform (5 mL) by heating for 1 h. The solution of CA/dichloromethane was added to PHB/chloroform, and the mixture was stirred for further 2 h to give a total concentration of 2.5% in a mixture of dichloromethane/chloroform (1:1). The solution was cast on a glass Petri dish at 20°C, the solvent was evaporated at 20°C. After the evaporation of the solvent, the membrane was dried under vacuum at room temperature till constant weight. Blends were produced utilizing 0, 20, 40, 50, 60, and 100 w/w of CA/PHB, which hereafter will be referred to as PHB, CA/PHB (20/80), CA/PHB (40/60), CA/PHB (50/50), CA/PHB (60/40), and CA, respectively. In Figure 7.7, the photographs of CA/PHB film are reported.

7.2.3 SCANNING ELECTRON MICROSCOPY

The cross-section morphologies of films were recorded using a JEOL (JSM-5600LV) SEM at the required magnification and with accelerating voltage of 14 kV. The film samples frozen in liquid nitrogen were fractured and sputtered with gold before SEM observation.

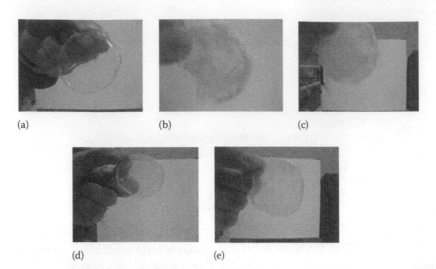

FIGURE 7.7   Photographs of CA, PHB, and CA/PHB films: CA (a), PHB (b), CA/PHB (20/80) (c), CA/PHB (50/50) (d), and CA/PHB (60/40) (e).

7.2.4 THERMOGRAVIMETRIC ANALYSIS

Thermogravimetric analysis (TGA) experiments were performed in the thermogravimetric analyzer Series Q500 of the TA Instruments. Generally, sample size was between 10 and 20 mg. Two different ranges of temperatures were scanned at $10°C \cdot min^{-1}$ under nitrogen atmosphere at $60 mL \cdot min^{-1}$ flow rate. These ranges of temperatures were from 30°C to 800°C and from 30°C to 600°C depending on the type of filler (organic or inorganic) in the formulations, respectively.

7.2.5 FIBER ANALYZER

The instrument used for the determination of the chemical composition of used lignocellulosic materials is a FIBER ANALYZER MOD. ANCOM located in the Department of Agronomy and Agrosystem Management (DAGA), University of Pisa, Pisa, Italy.

The system is a pressure pot with a stirrer, where the pouches similar to tea bags and made of synthetic material are immersed in a boiling solution and stirred for 1 h.

After that, the solvent is removed and the sample is washed with distilled water and acetone and is dried in a ventilated oven. For the not dissolved fibers (NDFs) determination, α-amylase is used for eliminating starches. The sample incineration is made by a muffola oven and the material is brought at constant weight with an analytic balance.

7.2.6 TRANSMISSION FOURIER TRANSFORM INFRARED SPECTROSCOPY (FTIR)

Transmission infrared spectra of all samples were recorded with a Jasco FTIR spectrometer mod. FT/IR-410 in the mid-IR region ($4000–400 cm^{-1}$) at $4 cm^{-1}$ resolution using 32 scans. Samples were casted from $CHCl_3$ solutions on a KBr crystal plate.

7.2.7 WIDE-ANGLE X-RAY DIFFRACTION (WAXD)

Wide-angle x-ray diffraction (WAXD) patterns were performed at room temperature with a Kristalloflex 810 diffractometer (SIEMENS) using a Cu Kα ($\lambda = 1.5406 Å$) as x-ray source. Scans were run in the high-angle region from 5° to 40° at scan rate of 0.016°/min and a dwell time of 1 s.

7.2.8 NUCLEAR MAGNETIC RESONANCE (NMR)

^{1}H-NMR and ^{13}C-NMR spectra of CA samples were recorded on Varian Mercury 200 MHz NMR, using DMSO-d_6.

7.3 RESULTS AND DISCUSSION

7.3.1 CHEMICAL COMPOSITION

SCB as received was chopped (C-SCB) contained about 31 wt-% of cellulose, 27 wt-% of hemicellulose, and 8 wt-% of lignin. In general, these values slightly decreased for the fraction of milled SCB (M-SCB) and collected from the sieve that pass 0.425 mm. Exception was verified for the amount of silica.

The behavior of the samples C-SCB and M-SCB was somewhat different after dewaxing. Both samples presented higher composition values of the components analyzed in relation to undewaxed samples.

However, after milling, only cellulose and hemicellulose decreased and the other components increased. Besides the samples that were milled and after dewaxing (DM-SCB) resulted to have about 13% of lignin that represented an increasing of 66% compared to C-SCB.

Chemical compositions of C-SCB, M-SCB, DC-SCB, and DM-SCB are reported in Table 7.1.

Based on the weight of the starting C-SCB, 77.4% of DC-SCB, 12.0% of precipitated wax, and 8.9% of dissolved wax were obtained with a total 20.9% of wax. Based on the weight of the starting M-SCB, 68.3% of DM-SCB, 5.8% of precipitated wax, and 26.3% of dissolved wax were obtained with a total of 32.1% of wax.

It was clear that the dewaxing of M-SCB took place in a shorter time (9 h) than C-SCB (66 h). Moreover, based on the weight of the starting material, the obtained DC-SCB and DM-SCB represented 77.4% and 68.3%, respectively. This may be due to the high surface area of the M-SCB compared to the C-SCB which makes the extraction process faster and more effective.

The behavior of RS as a function of milling and dewaxing was quite different to that of SCB (Table 7.2).

RS as received was chopped (C-RS) and contained 30 wt-% of cellulose, 13 wt-% of hemicellulose, and 13 wt-% of lignin. After milling (M-RS), only lignin decreased of about 45%, compared to C-RS. On the other hand, the values of cellulose and hemicellulose amounts were higher than those of C-RS about 6% and 17%, respectively. This behavior was contrary to that observed in M-SCB.

TABLE 7.1

Chemical Composition (wt-%) of Egyptian Sugarcane Bagasse

Sample	Wax[a]	Moist.[b]	Cel.[c]	Hcel.[d]	Lig.[e]	Si[f]	Ash
C-SCB	20.9	9.1	31.1	27.5	8.0	0.8	2.6
DC-SCB	—	11.6	42.1	33.2	9.5	1.1	2.5
M-SCB	32.1	5.5	27.6	24.0	7.3	1.3	2.2
DM-SCB	—	10.0	39.8	32.2	13.3	1.9	2.8

[a] Wax: Toluene/ethanol extractive.
[b] Moist.: Moisture.
[c] Cel.: Cellulose.
[d] Hcel.: Hemicellulose.
[e] Lig.: Lignin.
[f] Si: Silica. Hemicellulose, cellulose, lignin, and silica have been, respectively, determined, following the Van Soest procedure while moisture has been determined gravimetrically.

TABLE 7.2

Chemical Composition (wt-%) of Egyptian Rice Straw

Sample	Wax[a]	Moist.[b]	Cel.[c]	Hcel.[d]	Lig.[e]	Si[f]	Ash
C-RS	9.4	9.8	29.7	12.6	12.7	10.7	15.1
DC-RS	—	10.5	39.6	15.4	7.0	12.3	15.2
M-RS	8.1	11.2	31.6	14.7	7.0	10.5	15.0
DM-RS	—	12.8	31.2	15.9	12.4	12.4	15.3

[a] Wax: Toluene/ethanol extractive.
[b] Moist.: Moisture.
[c] Cel.: Cellulose.
[d] Hcel.: Hemicellulose.
[e] Lig.: Lignin.
[f] Si: Silica. Hemicellulose, cellulose, lignin, and silica have been, respectively, determined, following the Van Soest procedure while moisture has been determined gravimetrically.

As received RS after dewaxing (DC-RS) showed to contain larger amount of the analyzed components than in C-RS, except for lignin. These changes corresponded to 33%, 22%, and 15% increasing in cellulose, hemicellulose, and silica, respectively, and 45% decreasing in lignin.

However, for the composition of RS that was firstly milled (M-RS) and then dewaxed (DM-RS), only lignin amount showed significant changes with the performed treatment. These changes corresponded to an increasing of 77%, in comparison to M-RS.

For the C-RS, 90.3% of DC-RS, 5.5% of precipitated wax, and 3.9% of dissolved wax were obtained with a total 9.4% of wax. Based on the weight of the starting M-RS, 91.8% of DM-RS, 3.3% of precipitated wax, and 4.8% of dissolved wax were obtained with a total 8.1% of wax.

Dewaxing of M-RS took place in shorter times (12 h) than C-RS (96 h). Based on the weight of the starting materials, the dewaxing process resulted to be effective in an almost similar manner for M-RS and C-RS, allowing us to conclude that no evident effect of the particle sizes on the efficiency of the dewaxing process was observed. Surely, the use of M-RS for dewaxing process brought the advantage of saving time.

Both SCB and RS showed that ash amounts, which were around 3% and 15% respectively, were basically invariable with the milling and dewaxing treatments. On the other hand, these biomass treatments had a significant effect on their apparent composition as previously showed.

7.3.2 FTIR Spectroscopy

Figure 7.8 shows the FTIR spectra C-SCB, DC-SCB, M-SCB, and DM-SCB.

The results resulted in agreement with literature data (Bilba and Ouensanga 1996, Hoareau et al. 2004, Sun et al. 2004b, Yang et al. 2007).

Solubility of cellulose-based materials is a crucial parameter since, in the case of cellulose, which forms highly ordered fibrils in which the polar macromolecules attach strongly through hydrogen bonds, the tightly packed structures prevent the intrusion of solvents. This behavior makes the cellulose-based residues particularly difficult to analyze with techniques where solubilization is necessary. In this respect, we proceed to analyze the samples in the solid form. The preparation of the samples for the FTIR examination resulted particularly difficult in the case of the chopped SBF or RS, due to the coarse and poorly homogeneous particles. However, qualitatively speaking, some conclusions have been anyway achieved.

Table 7.3 shows the main IR bands for SCB before and after the milling and the dewaxing processes.

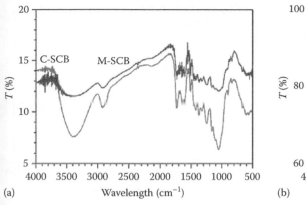

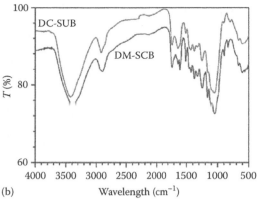

(a) Wavelength (cm⁻¹)   (b) Wavelength (cm⁻¹)

FIGURE 7.8   FTIR spectra of C-SCB, M-SCB (a) and DC-SCB, DM-SCB (b).

TABLE 7.3

Main IR Bands for SCB before and after Milling and Dewaxing Process

Sample	γ (cm⁻¹)
C-SCB	3295 (γ_{OH}), 2937 (γ_{CH}), 1734 ($\gamma_{C=O}$), 1593 and 1452 ($\gamma_{C6H6\ ring}$), 1223 ($\gamma_{C-O-C\ pyranose\ skeletal}$), 1380 ($\delta_{CH2}$), 1161, 959 ($\gamma_{OH}, \gamma_{C-C}, \gamma_{C-O}$)
DC-SCB	3423 (γ_{OH}), 2920 (γ_{CH}), 1730 ($\gamma_{C=O}$), 1635 and 1429 ($\gamma_{C6H6\ ring}$), 1253 ($\gamma_{C-O-C\ pyranose\ skeletal}$), 1374 ($\delta_{CH2}$), 1055 ($\gamma_{OH}, \gamma_{C-C}, \gamma_{C-O}$)
M-SCB	3406 (γ_{OH}), 2914 (γ_{CH}), 1727 ($\gamma_{C=O}$), 1605, 1424 ($\gamma_{C6H6\ ring}$), 1243 ($\gamma_{C-O-C\ pyranose\ skeletal}$), 1374 ($\delta_{CH2}$), 1051 ($\gamma_{OH}, \gamma_{C-C}, \gamma_{C-O}$)
DM-SCB	3424 (γ_{OH}), 2897 (γ_{CH}), 1734 ($\gamma_{C=O}$), 1604, 1513 ($\gamma_{C6H6\ ring}$), 1246 ($\gamma_{C-O-C\ pyranose\ skeletal}$), 1374 ($\delta_{CH2}$), 1057 ($\gamma_{OH}, \gamma_{C-C}, \gamma_{C-O}$)

It can be observed that the three main components of SCB (cellulose, hemicellulose, and lignin) most likely consist of alkene, ester, aromatic, ketone, and alcohol functional groups.

By comparing FTIR spectra of SCB, as received and milled (i.e., C-SCB and M-SCB, in Figure 7.8a), and the same before and after the dewaxing process (i.e., DC-SCB and DM-SCB, in Figure 7.8b), we can observe for all the samples the presence of a broad peak at around 3400 cm⁻¹ that represents OH groups either from cellulose or lignin.

The peak at around 2920–2930 cm⁻¹ represents the C-H asymmetric stretching in aliphatic methyl. An intense peak appears at around 1730 cm⁻¹ due to ester carbonyl stretching in unconjugated ketone, ester, or carboxylic groups in carbohydrates (hemicelluloses) and not from lignin (Hoareau et al. 2004).

Figure 7.9 shows the FTIR spectra C-RS and M-RS (in Figure 7.9a) and DC-RS and DM-RS (in Figure 7.9b).

Table 7.4 reports the main IR bands for the C-RS, DC-RS, M-RS, and DM-RS samples.

As far as cellulose and CA obtained by SCB are concerned, the FTIR spectra of the cellulose and CA are shown in Figure 7.10.

In particular, in the FTIR spectrum of unacetylated cellulose, a broad band at around 3477 cm⁻¹, which is due to the stretching of the O–H group, was evident.

This band decreased in CA spectrum, as a result of partial esterification of the hydroxyl groups present on the repeating cellobiose units. Moreover, in the CA FTIR spectrum, three bands absent in the cellulose sample were detected: the first at 1752 cm⁻¹ assigned to the C=O stretching vibration mode of the acetyl group, the second at 1378 cm⁻¹ assigned to the CH₃ asymmetric bending

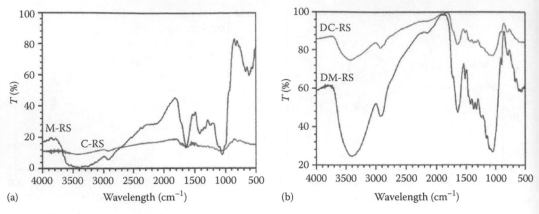

FIGURE 7.9   FYIR spectra of C-RS (a) and DC-RS, DM-RS (b).

TABLE 7.4

Main IR Bands for the RS before and after Milling and Dewaxing Process

Sample	γ (cm^{-1})
C-RS	3425 (γ_{OH}), 2921 (γ_{CH}), 1727 ($\gamma_{C=O}$), 1646 and 1510 (γ_{OH}, γ_{c-c}, γ_{C-O}), 1245 ($\gamma_{C-O-C\ pyranose\ skeletal}$), 1430 ($\gamma_{C6H6\ ring}$), 1373 ($\delta_{CH2}$)
DC-RS	3400 (γ_{OH}), 2922 (γ_{CH}), 1645 ($\gamma_{C=O}$), 1463 ($\gamma_{C6H6\ ring}$), 1028 (γ_{OH}, γ_{c-c}, γ_{C-O})
M-RS	3386 (γ_{OH}), 2924 (γ_{CH}), 1638, 1419 ($\gamma_{C6H6\ ring}$), 1372 (δ_{CH2}), 1057 (γ_{OH}, γ_{C-C}, γ_{C-O}), 1205 ($\gamma_{C-O-C\ pyranose\ skeletal}$)
DM-RS	3407 (γ_{OH}), 2925 (γ_{CH}), 1638 ($\gamma_{C6H6\ ring}$), 1510 ($\gamma_{C=O}$), 1374 (δ_{CH2}), 1058 (γ_{OH}, γ_{C-C}, γ_{C-O}), 1321 ($\gamma_{C-O-C\ pyranose\ skeletal}$)

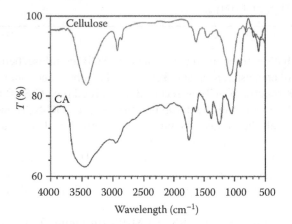

FIGURE 7.10   FTIR spectra of cellulose and CA.

vibration mode of the acetyl group, and the third at 1252 cm^{-1} assigned to the C–O stretching vibration mode of the acetyl group (Cerqueira et al., 2007, Ifuku et al., 2007).

7.3.2.1   Determination of the Degree of Substitution

The FTIR technique made the determination of the *DS* easier and faster than the chemical titration method. The other problem with the titration method was the fact that below a certain *DS*, the method broke down and gave incorrect results. Two peaks were of interest: the carbonyl group at 1751 cm^{-1} and the hydroxyl group at 3464 cm^{-1}. With the FTIR technique, the *DS* of any CA polymer could be determined with the aid of the calibration curve. The DS can be determined as the ratio of the hydroxyl peak height to the carbonyl peak area in absorbance units (Cerqueira et al. 2007, Hurtubise, 1962, Samois et al. 1997). Based on the ratio of hydroxyl peak height (0.094) to the carbonyl peak area (36.413) with respect to the calibration curve done by Samios (Samois et al. 1997), it was found that the *DS* was equal to 2.5.

7.3.3 MORPHOLOGY

Comparative SEM photomicrographs of C-SCB, DC-SCB, M-SCB, and DM-SCB, and C-RS, DC-RS, M-RS, and DM-RS are reported in Figures 7.11 and 7.12, respectively.

Pictures (a) and (b) in Figures 7.11 and 7.12 showed the characteristic cylinder and/or vascular bundles of the grass plants like SCB and RS.

The morphology of C-SCB and DC-SCB samples (Figure 7.11a and b) clearly showed that modifications on the fiber structure resulting from the dewaxing occurred. On the other hand, a minor evidence of the fiber modification due to dewaxing on the RS was found. This morphological evidence was also in confirmation of the different extraction yields observed for the different materials used.

The milling process seemed to be more effective for SCB (Figure 7.11c) than for RS (Figure 7.12c), as guessed from the attainment of fibers characterized by a reduced size for M-SCB with respect to M-RS. After dewaxing (Figure 7.11d), these particles resulted to have a further decrease in size with respect to undewaxed ones (Figure 7.8c). However, for milled RS samples, dewaxing was not traduced in a significant variation of the morphological features (Figure 7.12c and d).

By SEM analysis, we can confirm the conclusions exposed about the different dewaxing process efficiency found for SCB and RS materials.

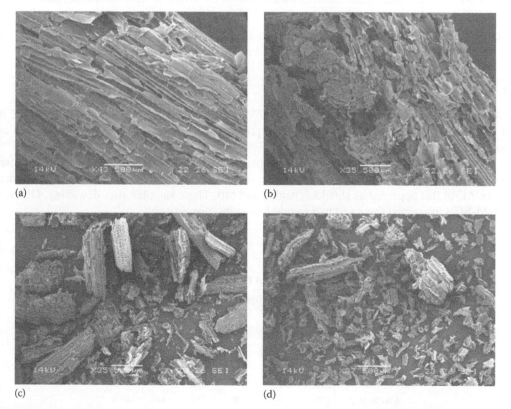

(a)　　　　　　　　　　　　　　　　　　　(b)

(c)　　　　　　　　　　　　　　　　　　　(d)

FIGURE 7.11 SEM photomicrographs for C-SCB-43X (a), DC-SCB-35X (b), M-SCB-35X (c), and DM-SCB-37X (d).

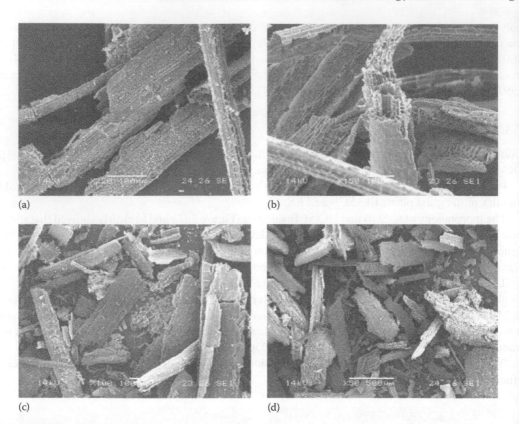

(a)   (b)

(c)   (d)

FIGURE 7.12   SEM photomicrographs for C-RS-220X (a), DC-RS-150X (b), M-RS-100X (c), and DM-RS-50X (d).

7.3.4   TGA ANALYSIS

TGA pyrolysis traces and their derivatives (DTGA) of SCB as received (C-SCB), milled (M-SCB), and dewaxed (DC-SCB and DM-SCB) were represented in Figure 7.10 and their TGA data are reported in Table 7.5.

The weight loss traces observed in the samples C-SCB and M-SCB (Figure 7.13a) are characterized at least by six steps, of which five of them are overlapped as shown in DTGA traces (Figure 7.13b) that appeared as shoulder, peaks, and tail. These samples after dewaxing, DC-SCB and DM-SCB, presented three and two weight loss steps, respectively.

TABLE 7.5
Thermogravimetric Data of SCB-Based Materials as Recorded under Nitrogen Atmosphere

Sample	T_{on} (°C)	T_{p1} (°C)	T_{p2} (°C)	T_{p3} (°C)	T_{p4} (°C)	R_{800} (%)
C-SCB	221	220	282	327	399	18.1
DC-SCB	277	—	317	351	416	4.4
M-SCB	206	222	281	335	400	15.3
DM-SCB	275	—	306	352	447	19.6

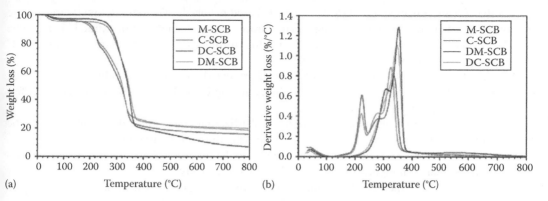

FIGURE 7.13 TGA (a) and DTGA (b) traces of C-SCB, DC-SCB, M-SCB, and DM-SCB as recorded under nitrogen atmosphere.

The first step (25°C–120°C) of all samples corresponded to the moisture volatilization and was 3.1%, 2.1%, 4%, and 4.7% in the C-SCB, DC-SCB, M-SCB, and DM-SCB samples, respectively.

The second and third overlapped steps of weight loss in the samples C-SCB and M-SCB occurred between ca 120°C and 240°C. These steps can be assigned to the low molecular weight extractives that in the case of SCB can be starch, sugars (principally saccarose), phenolics, and tannins (Azuma et al. 1992).

As the second step seemed to be a tail of the third one, the weight loss step was calculated as a single step and it corresponded to 18.5% and 20.1% for C-SCB and M-SCB, respectively.

This slight increase on extractive amounts in the sample M-SCB suggested that probably milling-sieving process concentrates something more of the fragile component in the fraction that pass 0.425 mm sieve.

Fourth weight loss step was recorded between 240°C and 300°C, and the values for C-SCB and M-SCB were 14.2% and 18%, respectively. The principal chemical component of lignocellulosic biomass that decomposes in this range of temperature was the hemicelluloses (Day et al. 1998).

The higher amount of hemicellulose in C-SCB with respect to M-SCB, obtained by chemical assessment (Table 7.1), was verified for the first one that apparently disagreed with the TGA results. Probably, the milling-sieving process changed the proportion between components.

The last weight loss step was a peak followed by a tail up to the end of the experiment. The temperature range of the step corresponding to the DTGA peak was comprised in 300°C–380°C range, and the principal decomposing component was represented by the cellulose. A small fraction of lignin that continues to degrade at higher temperatures was also present. The values of weight loss for this step were 37% and 35% for C-SCB and M-SCB, respectively.

Finally, in the 400°C–800°C temperature range, the weight loss was 6.6% and 5.2%, leaving a recovered residue of 18.1% and 15.3% for C-SCB and M-SCB, respectively. This last step corresponded principally to the lignin decomposition and the residue to the inorganic materials and carbon.

C-SCB and M-SCB after extraction with the solvent mixture toluene/ethanol (samples DC-SCB and DM-SCB, respectively) displayed no weight loss steps in the temperature range of 120°C–240°C. Therefore, DC-SCB and DM-SCB TGA traces suggested at least four weight loss steps, where the first one corresponded to the moisture volatilization, as previously reported.

The first and second degradation weight loss steps, in the 200°C–380°C temperature range, were overlapped at around 320°C.

The values of these steps for DC-SCB were 30% and 44% and for DM-SCB were 27% and 44%, respectively. The last step that resulted principally from lignin pyrolysis corresponded to the weight loss in the 380°C–800°C temperature range and that was 8% and 3% for DC-SCB and DM-SCB, respectively. The residues recovered at 800°C were ca. 4% and 20%, respectively.

In Table 7.5 are collected the TGA data of SCB-based materials. The onset temperatures for the pyrolysis (T_{on}) depended on both milling-sieving process and toluene/ethanol extraction. T_{on} value of C-SCB was 221°C, that decreased by ca. 15°C after milling (M-SCB) and increased by ca. 56°C for the sample without extractives (DC-SCB).

On the other hand, both DC-SCB and DM-SCB presented equivalent T_{on} values around 276°C. Consequently, the difference in the thermal stability among C-SCB and M-SCB was due to different extractive compositions.

The maximum rate of decomposition product volatilization (T_p) depended principally from the toluene/ethanol extraction, which values were higher than those of pristine materials.

TGA and DTGA experiments were conducted also under air atmosphere: the traces and the relative data are presented in Figure 7.11 and in Table 7.6, respectively.

The samples C-SCB and M-SCB (Figure 7.14) showed at least five steps, while the same samples after dewaxing process (DC-SCB and DM-SCB) presented four weight loss steps.

The first step (25°C–120°C) of all the samples corresponded to the moisture and volatiles elimination and it was 2.3%, 1.7%, 1.6%, and 2.5% in the C-SCB, DC-SCB, M-SCB, and DM-SCB, respectively.

The second, third, and fourth steps of weight loss in the samples C-SCB and M-SCB occurred between 120°C and 500°C.

The third step for M-SCB and the fourth one for C-SCB presented shoulders so as weight loss we considered the total effect finding 30.3% (C-SCB) and 18.1% (M-SCB).

For both samples, the second step felt down in the 180°C–250°C temperature range and the corresponding weight losses were 12.5% and 19.9% for C-SCB and M-SCB, respectively. The same trend was observed also for the fourth step that occurred in the 350°C–500°C range for C-SCB and M-SCB. These steps corresponded to the weight losses of 30.3% and 33.0% for C-SCB and M-SCB, respectively.

TABLE 7.6
Thermogravimetric Data for RS-Based Materials
as Recorded under Air Atmosphere

Sample	T_{on} (°C)	T_{p1} (°C)	T_{p2} (°C)	T_{p3} (°C)	T_{p4} (°C)	T_{p5} (°C)	R_{800} (%)
C-SCB	229	222	310	440	600	—	2.3
DC-RS	265	—	295	322	437	469	2.9
M-RS	203	222	292	321	445	—	4.0
DM-RS	259	—	224	293	325	439	3.4

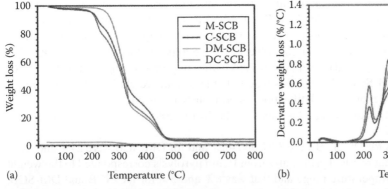 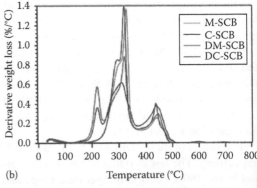

FIGURE 7.14 TGA (a) and DTGA (b) traces of C-SCB, DC-SCB, M-SCB, and DM-SCB as recorded under air atmosphere.

A difference in the intensity of these peaks was observed, due to the milling process, which caused an increase of the peak with respect to that of the pristine material.

Finally, in the 500°C–800°C temperature range, the weight losses were 0.9% and 22.4%, leaving a residue of 2.3% and 4.0% for C-SCB and M-SCB, respectively.

C-SCB and M-SCB after extraction with the solvent mixture toluene/ethanol (samples DC-SCB and DM-SCB, respectively) displayed no weight loss steps in the 120°C–240°C temperature range.

Therefore, DC-SCB and DM-SCB TGA traces were characterized by four weight loss steps, where the first one corresponded to the moisture and volatiles elimination, as previously reported.

The shoulder in the first degradation weight loss step was more evident for DC-SCB with respect to DM-SCB, and the values were 37.5% and 25.1%, respectively.

The weight losses for the real steps were 32.8% and 45.1%. The same effect was also evident in the second degradation weight loss step, mainly due to the decomposition of cellulose and lignin. The values were 23.8% for DC-SCB and 21.0% for DM-SCB. The residues recovered at 800°C were ca. 3.0% and 3.4%, respectively.

The onset temperature for C-SCB, reported in Table 7.6, was 229°C. This value decreased of 16°C after milling (M-SCB) and increased of 36°C for the sample not submitted to mixed solvent extraction (DC-SCB).

DC-SCB and DM-SCB presented similar onset temperatures: the difference of 6°C suggested that the thermal stability depended on the extractive composition as recorded under nitrogen atmosphere.

Figure 7.15 reports the TGA (a) and DTGA (b) traces for RS before and after milling and dewaxing processes.

The first step (25°C–120°C) of all samples corresponding to the moisture volatilization was 3.3%, 3.0%, 1.9%, and 4.4% for C-RS, DC-RS, M-RS, and DM-RS, respectively.

The second and third weight loss steps for the samples C-RS and M-RS occurred in the 120°C–240°C temperature range, and the corresponding values were 4.7% and 18.5% for C-RS and 6.8% and 46.7% for M-RS. The residues at 800°C were 31.8% and 32.2% for C-RS and M-RS, respectively.

The extraction process shifted the shoulder of the chopped material (DC-RS) from around 200°C to 300°C and a third peak around at 480°C appeared in the relative trace. The corresponding weight losses were 22.1%, 34.2%, 4.2% and the residue recovered at 800°C was 27.0%.

In the DM-RS sample, the shoulder was around at 190°C and was associated with a weight loss of 17.6%. Only one degradation weight loss step was present in the 200°C–400°C temperature range, corresponding to 38.2% weight loss and a second effect smaller than the first one in the 420°C–490°C temperature range, with a weight loss of 4.6%, was also detected. The residue recovered at 800°C was 29.5%.

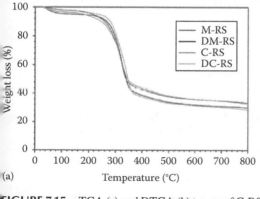
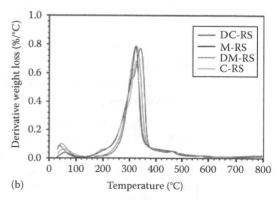

FIGURE 7.15 TGA (a) and DTGA (b) traces of C-RS, DC-RS, M-RS and DM-RS as recorded under nitrogen atmosphere.

TABLE 7.7

Thermogravimetric Data for RS-Based Materials as Recorded under Nitrogen Atmosphere

Sample	T_{on} (°C)	T_{p1} (°C)	T_{p2} (°C)	T_{p3} (°C)	T_{p4} (°C)	T_{p5} (°C)	R_{800} (%)
C-RS	258	194	306	325	465	—	31.8
DC-RS	277	—	182	313	341	462	27.0
M-RS	259	—	325	466	568	—	32.2
DM-RS	272	—	124	324	468	—	29.5

Table 7.7 shows the onset temperatures for the RS-based materials.

RS before and after milling processes (C-RS and M-RS) presented the same T_{on}, while this value increased by 18°C for the sample after extraction (DC-RS). We can conclude that, also for RS, the extractive composition influenced the thermal stability: the presence of wax, sugars, and phenols made the material less stable than the undewaxed pristine sample.

The TGA (a) and DTGA (b) traces for the RS-based materials as recorded under air atmosphere are reported in Figure 7.16, and the relative data are presented in Table 7.8.

Under air atmosphere, the weight loss steps observed in the samples C-RS and M-RS were at least four, where two of them were overlapped as shown by the DTGA traces (Figure 7.16b).

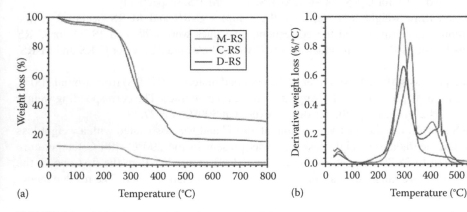

(a)   Temperature (°C)   (b)   Temperature (°C)

FIGURE 7.16 TGA (a) and DTGA (b) traces for RS-based materials as recorded under air atmosphere.

TABLE 7.8

Thermogravimetric Data for RS-Based Materials as Recorded under Air Atmosphere

Sample	T_{on} (°C)	T_{p1} (°C)	T_{p2} (°C)	T_{p3} (°C)	T_{p4} (°C)	T_{p5} (°C)	T_{p6} (°C)	T_{p7} (°C)	T_{p8} (°C)	R_{800} (%)
C-RS	248	195	298	413	437	450	488	583	737	15.9
M-RS	264	193	323	468	—	—	—	—	—	28.9
DM-RS	258	293	405	609	—	—	—	—	—	13.7

The first step, comprised within 25°C–120°C temperature range, corresponding to the moisture and volatiles elimination, was characterized by relative weight losses 3.3%, 4.7%, and 3.9% in the C-RS, M-RS, and DM-RS, respectively.

The second step appeared as shoulder and it occurred in the 120°C–240°C temperature range with weight losses of 22.6%, 25.4%, and 16.7% for C-RS, M-RS, and DM-RS, respectively. The third weight loss step occurred in the 240°C–365°C temperature range and it principally represented the cellulose degradation.

The fairly broad range of decomposition was associated to the presence of some sugars that started to degrade at lower temperature with respect to the pure cellulose (Persenaire et al. 2001).

The recorded weight losses were 29.0% for C-RS, 25.0% for M-RS, and 41.3% for DM-RS.

The fourth step for C-RS occurred in the 365°C–500°C temperature range, with a weight loss of 24.6%; the M-RS presented a weight loss step of 8.8%.

The residues recorded at 800°C were 15.9% and 28.9% for C-RS and M-RS, respectively. The DM-RS sample, under air atmosphere, was degraded in four steps. The first step with a weight loss of about 3.9% in the 25°C–100°C temperature range associated with the moisture evolution. The shoulder characterized by a weight loss of 16.7% in the 100°C–240°C temperature range represented the extractives start decomposition of RS. The third step with a weight loss of 41.3% represented mainly the decomposition of cellulose. The fourth step, associated to a weight loss of 22.5%, principally represented the decomposition of hemicellulose and lignin. The residue recovered at 800°C was 13.7%.

Figure 7.17 shows the TGA (a) and DTGA (b) of cellulose and CA samples. Cellulose degraded in three steps. The first step, with a weight loss of about 4.3% in the 25°C–150°C temperature range, which represents the evaporation of the residual absorbed water. The second step, with a weight loss of about 71.7% in the 250°C–500°C temperature range, represented the main thermal degradation of cellulose. The third step, with a weight loss of about 3.3% in the 500°C–800°C temperature range, represented the total cellulose texture breakdown and its carbonization to ash (Hanna et al. 1999).

CA degraded in three steps. The first step, with a weight loss of about 5.0% in the 25°C–147°C temperature range, represented the evaporation of the residual absorbed water. The two peaks between 147°C and 403°C could be due to a cellulose degradation process such as depolymerization, dehydration, and the decomposition of glucosyl units followed by the formation of a charred residue (Roman and Winter 2004). Finally, the third step, with a weight loss of about 7.1% in the 403°C–800°C temperature range, represented the total CA texture breakdown and its carbonization to ash (Hanna et al. 1999). The aspect of the DTGA curve relevant to CA indicated a more heterogeneous structure than the untreated cellulose. The coexistence of microfibers of unmodified cellulose and preacetylated materials is possible. In this condition, the material shows

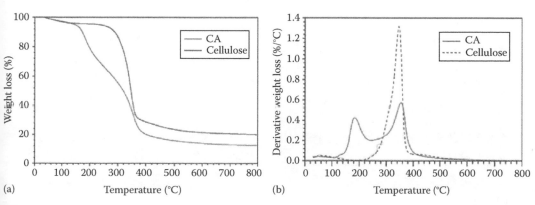

FIGURE 7.17 TGA (a) and DTGA (b) traces of cellulose and CA samples.

low thermal stability and the decomposition process is more complex due to the copresence of acetylated material having a different DS and unmodified cellulose. CA became less thermally stable than the original untreated cellulose: T_{on} for CA and cellulose were found to occur at 154.8°C and 303.6°C, respectively.

7.3.5 Cellulose Acetate/Poly(3-Hydroxy Butyrate) Blends Characterization

The prepared CA/PHB blends (obtained by using CA with a *DS* equal to 2.5) were characterized by SEM for their morphology features, by TGA for their thermal characteristics, whereas the assembly of the two components was assessed by WAXD. Moreover, the chemical structure of the modified cellulose used in the blends was also assessed by ¹H-NMR and ¹³C-NMR spectroscopies.

7.3.5.1 Morphology

SEM photomicrographs for CA/PHB blends are shown in Figure 7.18. The selected percentages for CA were 20%, 40%, 50%, and 60%. The CA/PHB blends 20/80 (Figure 7.18a) and 40/60 (Figure 7.18b) showed two phases where the particles of CA were dispersed in the PHB matrix. It can be observed that some spherical CA particles were pulled out probably due to the low adhesion between PHB and CA.

The adhesion between CA and PHB appeared to be improved when the CA contents were increased, as shown in CA/PHB 50/50 (Figure 7.18c) and CA/PHB 60/40 (Figure 7.18d) blends. Micrographs (c) and (d) of Figure 7.18 showed in fact a more homogeneous distribution of the two compounds.

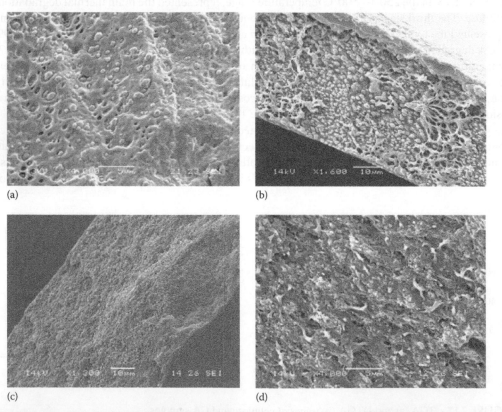

(a)   (b)

(c)   (d)

FIGURE 7.18 SEM photomicrographs for CA/PHB (20/80)-4000X (a), CA/PHB (40/60)-2200X (b), CA/PHB (50/50)-1300X (c), and CA/PHB (60/40)-4000X (d). Subset: high magnification 1000X.

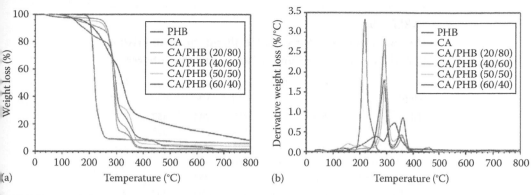

FIGURE 7.19 TGA (a) and DTGA (b) traces for CA/PHB blends as recorded under nitrogen atmosphere.

7.3.5.2 Thermal Properties

Thermal characterization of neat PHB and CA components and their blends was performed to obtain information about degradation, processing temperatures, and the components compatibility (Cerqueira et al. 2006, 2008, Shao et al. 2002).

In this respect, the thermal properties of PHB, CA, and CA/PHB blends were assessed by TGA.

Figure 7.19 shows the weight loss (a) and the derivative weight loss (b) traces of CA/PHB blends and of the neat polymers. After initial moisture volatilization in the 25°C–120°C temperature range, the blends started to degrade in the 200°C–280°C temperature range.

The weight losses associated to the humidity elimination were 2.4%, 4.4%, 4.5%, and 7.7% for CA/PHB (20/80), CA/PHB (40/60), CA/PHB (50/50), and CA/PHB (60/40), respectively. At least four degradation weight loss steps were apparent in the DTGA traces shown in Figure 7.16b. The first step corresponded to weight losses of 11.7% for CA/PHB (20/80), 18% for CA/PHB (40/60), 26.3% for CA/PHB (50/50), and 10.7% for CA/PHB (40/60). The second step corresponded to weight losses of 83.1% for CA/PHB (20/80), 74.3% for CA/PHB (40/60), 63.4% for CA/PHB (50/50), and 54% for CA/PHB (60/40). In the third step, the decomposition peak falling in the 400°C–500°C temperature range was associated with weight losses of 0.6% for CA/PHB (20/80), 1.2% for CA/PHB (40/60), 1.7% for CA/PHB (50/50), and 17.7% for CA/PHB (60/40). The residue recorded at 800°C was around 1%–3% for all the selected blends. PHB started to degrade at 202°C and the residue recorded at 800°C was 0.73%, while CA started to degrade at 160°C and the residue recovered at 800°C was 8.3%. As well documented in literature, four overlapped steps were involved in the pyrolysis of CA (Shao et al. 2002). The first one, correlated to the moisture volatilization, showed a relative weight loss of 7.6% and occurred in the 25°C–130°C temperature range. The second step, falling in the 130°C–210°C temperature range, showed a weight loss of 25.0%, while the third step, correlated to the total decomposition of the cellulose structure, took place in the 210°C–290°C temperature range with an associated weight loss of 38.7%. The fourth step, attributed to the remnant carbon, was comprised in the 290°C–410°C temperature range, with a relative weight loss of 9.5%.

The blending of PHB with the CA induced an increase in thermal stability with respect to the starting polymeric materials. Indeed, all the blends exhibited a T_{on} higher than the pristine PHB, which was around 231°C.

The CA/PHB (50/50) blend had a similar thermal behavior to that of pristine PHB. In fact, its onset temperature was around at 256°C, the closest value to the PHB onset temperature.

In Table 7.9 are collected the data relevant to the thermal properties of the CA/PHB blends and of the pristine components.

7.3.5.3 Wide-Angle X-Ray Diffraction

WAXD patterns were recorded on CA/PHB blends, where the CA percentage varied from 20 to 60. Figure 7.20 shows the WAXD diffraction patterns of pristine polymers and their binary blends.

TABLE 7.9

Thermal Properties for CA/PHB Blends and the Pristine Materials as Recorded under Nitrogen Atmosphere[a]

Sample	T_{on}[a] (°C)	T_{p1}[b] (°C)	T_{p2} (°C)	T_{p3} (°C)	T_{p4} (°C)	R_{800}[c] (%)
PHB	202	—	217	248[a]	457	0.53
CA	231	133	174	257	329	8.3
CA/PHB (20/80)	264	108	291	358	—	1.9
CA/PHB (60/60)	271	135	290	355	—	1.6
CA/PHB (50/50)	256	137	289	362	—	1.7
CA/PHB (60/40)	279	154	200	295	352	2.9

[a] T_{on} is the starting degradation temperature.
[b] T_p is the main degradation peak.
[c] R_{800} is the residue at 800°C.

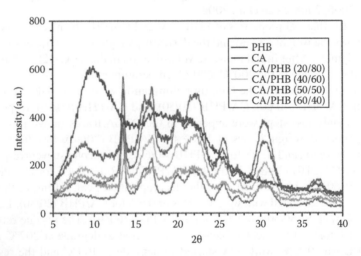

FIGURE 7.20 WAXD patterns of prittine PHB, cellulose acetate (CA), and their blends.

The pattern of diffraction peaks for PHB appeared to be typical of a semicrystalline polymeric material. Interestingly, the positions and intensities of these diffraction peaks remained unchanged when PHB was blended with CA, indicating that the addition of CA did not affect the self-assembly of PHB segments in the blends.

7.3.5.4 ¹H-NMR and ¹³C-NMR Spectroscopy

The ¹H-NMR and ¹³C-NMR spectroscopies were used to asses the chemical groups of CA samples.

In the ¹H-NMR, three methyl protons of acetyl groups were clearly evident as separated peaks in the range of 1.9–2.1 ppm, in agreement with literature data (Biswas et al. 2006). In particular, the assignment of the peaks was done at δ 1.91 ppm due to acetyl on C-3, δ 1.94 ppm due to acetyl on C-2, and δ 2.07 ppm due to acetyl on C-6. The other peaks in the range 3.5–5.06 ppm were assigned to cellulose protons.

In the ¹³C-NMR spectrum of CA, peaks were found at δ 170.42 ppm (ascribed to C=O), at δ 99.61 ppm (due to C-1), at δ 72.23 ppm (due to C-2, C-3, and C-5), at δ 63.12 ppm (due to C-4 and C-6), and finally at δ 20.63 ppm (ascribed to CH₃), in good agreement with reports by other authors (Sassi et al. 2000).

7.4   CONCLUSIONS

Biomass, including agro-industrial-based waste materials, is composed of a complex mixture of natural polymeric materials (such as cellulose, hemicelluloses, polysaccharides, lignin, polyphenolic materials, and other minor components such as minerals). Therefore, it represents a huge natural source of very valuable products, having different interesting properties that make it as a cheap raw material convertible either to low molar mass compounds or as a convenient source of medium and high molecular weight compounds.

On the other hand, a major limitation to the use of agro-industrial-based wastes as a chemical feedstock is the structural, physical, and chemical variety of their components and difficulties with their processing or chemical modification. In this respect, the dewaxing process of lignocellulosic materials could represent a key process to obtain valuable products that can find specific applications as polymer matrices to be chemically modified and/or used as fillers or continuous matrices in blends and composites.

In this respect, the dewaxing process of SCB and RS agro-industrial wastes from Egyptian farms was deeply studied within the framework of the action undertaken by ISC-UNIDO on the sustainable development based on the use of natural renewable resources as raw feedstock alternative to fossil fuel.

Each component of both types of lignocellulosic materials was isolated by physicochemical methodologies and chemically and thermally characterized.

Moreover, cellulose was isolated from SCB and was acetylated using acetic anhydride as acetylating agent, in the presence of acetic acid as solvent, and sulfuric acid used as catalyst.

The FTIR technique confirmed the structure of all SCB and RS samples and the structure of each isolated fraction, and permitted the determination of the substitution degree for CA.

The investigations of the chemical composition and structural features of SCB and RS samples, performed by Fiber Analyzer and SEM instruments, allowed for the determination of the percentage of each lignocellulosic component and for confirmation of the agglomerated distribution of the cellulose fiber bundles and their acetylated products.

The obtained results allowed us to conclude that the best results in the dewaxing process were obtained when the SCB samples were previously milled, due to the strong increase of the surface area after the milling process. The same process for RS did not lead to a substantial increase in the dewaxing efficacy process.

The most interesting result regarded the higher thermal stability as recorded under nitrogen atmosphere of the dewaxed materials when compared to the pristine raw materials, as attained by the TGA analysis.

The thermal behavior was connected to the waxes amount. T_{on} showed an increase from 205°C up to 275°C for the M-SCB and from 221°C up to 277°C for C-SCB in going from the pristine sample to the dewaxed one. This reflects also a drastic decreasing of the residue at 800°C (from 18% up to 4.4%) for C-SCB before and after dewaxing. The same trend for T_{on} and for the residue was also confirmed for the SCB analyses conducted under air atmosphere. Moreover, TGA analyses for RS samples conducted under nitrogen and air atmosphere gave the same information.

Also, the hemicellulose and lignin fractions soluble in alkaline solutions showed a higher T_{on} when compared to that recovered by TGA analysis performed under nitrogen atmosphere. The materials started to degrade when some oxidative processes were just verified during their weight loss steps.

Finally, CA at a degree of acetylation 2.5, obtained by SCB, was successfully blended with PHB, in different compositions.

CA/PHB blends, obtained by film casting, were analyzed for their morphological, structural and thermal characteristics.

SEM analysis confirmed a higher adhesion between CA and PHB when the CA contents were increased from 20% to 60%, while the TGA analysis pointed out that the blending of PHB with CA improved the thermal stability of the polymer, as confirmed by the higher value of T_{on} of the blends with respect to the pristine PHB. The 50/50 blend composition showed a T_{on} value similar to that of PHB sample, probably due to a miscibility of the two polymeric phases in the blend. Finally, WAXD analysis showed that the addition of CA did not affect the semicrystalline structure of PHB.

ACKNOWLEDGMENTS

The financial support was given by the ICS-UNIDO through the program of "Actions on environmentally degradable plastics (EDPS) and sustainable polymeric materials within ICS-UNIDO program 1996–2008" and the financial support of a fellowship to Mohamed El-Newehy.

REFERENCES

Abdel-Mohdy, F.A., Abdel-Halim, E.S., Abu-Ayana, Y.M., and El-Sawy, S.M. 2009. Rice straw as a new resource for some beneficial uses. *Carbohydrate Polymers* 75: 44–51.

Adsula, M.G., Ghule, J.E., Shaikh, H., Singh, R., Bastawde, K.B., Gokhale, A.J., and Varma, A.J. 2005. Enzymatic hydrolysis of delignified bagasse polysaccharides. *Carbohydrate Polymers* 62: 6–10.

Adsula, M.G., Ghule, J.E., Singh, R., Shaikh, H., Bastawde, K.B., Gokhale, D.V., and Varma, A.J. 2004. Polysaccharides from bagasse: Applications in cellulose and xylanase production. *Carbohydrate Polymers* 57: 67–72.

Avella, M., Martuscelli, E., and Greco, P. 1991. Crystallization behaviour of poly(ethylene oxide) from poly(3-hydroxybutyrate)/poly(ethylene oxide) blends: Phase structuring, morphology and thermal behaviour. *Polymer* 32: 1647–1653.

Azuma, Y., Yoshie, N., Sakuray, M., Inocue, Y., and Chujo, R. 1992. Thermal behaviour and miscibility of poly(hydroxybutyrate)/poly(vinyl alcohol) blends. *Polymer* 33: 4763–4767.

Balan, V., Sousa, L.C., Chundawat, S.P.S., Vismeh, R., Jones, A.D., and Dale, B.E. 2008. Mushroom spent straw: A potential substrate for an ethanol-based biorefinery. *Journal of Industrial Microbiology and Biotechnology* 35: 293–301.

Barkalow, D.G., Rowell, R.M., and Young, R.A. 1989. A new approach for the production of cellulose acetate: Acetylation of mechanical pulp with subsequent isolation of cellulose acetate by differential solubility. *Journal of Applied Polymer Science* 37: 1009–1018.

Bilba, K. and Ouensanga, A. 1996. Fourier transform infrared spectroscopic study of thermal degradation of sugar cane bagasse. *Journal of Analytical and Applied Pyrolysis* 38: 61–73.

Bismarck, A., Aranberri-Askargorta, I., and Springer, J. 2002. Surface characterization of flax, hemp and cellulose fibers; Surface properties and water uptake behavior. *Polymer Composites* 23: 872–889.

Biswas, A., Saha, B.C., Lawton, J.W., Shogren, R.L., and Willett, J.L. 2006. Process for obtaining cellulose acetate from agricultural by-products. *Carbohydrate Polymers* 64: 134–137.

Ceccoruli, G., Zizzoli, M., and Scandola, M. 1993. Effect of a low-molecular weight plasticizer on the thermal and viscoelastic properties of miscible blends of bacterial poly(3-hydroxybutyrate) with cellulose acetate butyrate. *Macromolecules* 26: 6722–6726.

Cerqueira, D.A., Filho, G.R., and Assuncao, R.M.N. 2006. A new value for the heat of fusion of a perfect crystal of cellulose acetate. *Polymer Bulletin* 56: 475–484

Cerqueira, D.A., Filho, G.R., and Meireles, C.D.S. 2007. Optimization of sugar cane bagasse cellulose acetylation. *Carbohydrate Polymer* 69: 597–582.

Cerqueira, D.A., Filho, G.R., Nascimento de Assuncao, R.M., Meireles, C.d.S., Toledo, L.C., Zeni, M., Mello, K., and Duarte, J. 2008. Characterization of cellulose triacetate membranes produced from sugar cane bagasse, using PEG 600 as additive. *Polymer Bulletin* 60: 397–404.

Chen, T., Breuil, C., Carriere, S., and Hatton, J.V. 1994. Solid phase extraction can rapidly separate lipid classes from acetone extracts of wood and pulp. *Tappi Journals* 77: 235–240.

Chen, G.Q. and Wu, Q. 2005. Polyhydrohyalkanoates as tissue engineering materials. *Biomaterials* 26: 6565–6578.

Chen, G.Q., Wu, Q., Zhao, K., Yu, H.P., and Chan, A. 2000. Chiral biopolyesters-polyhydroxyalkanoates synthesized by microorganisms. *Chinese Journal of Polymer Science* 18: 389–396.

Chiellini, E., Cinelli, P., Fernandes, E.G., Kenawy, E-R., and Lazzeri, A. 2001. Gelatin-based blends and composites. Morphology and thermal mechanical characterization. *Biomacromolecules* 2: 806–811.

Chiu, H.J. 2006. Miscibility and crystallization behaviour of poly(3-hydroxybutyrate-co-3-hydroxyvalerate)/poly(vinyl acetate) blends. *Journal of Applied Polymer Science* 100: 980–988.

Chodak, I. 2008. Polyhydroxyalkanoates. Origin, properties and applications. In *Monomers, Polymers and Composites from Renewable Resources,* eds. M.N. Belgacem and A. Gandini, pp. 451–477. Elsevier, Oxford, U.K.

Day, M., Cooney, J., Shaw, K., and Watts, J. 1998. Thermal analysis of some environmentally degradable polymers. *Journal of Thermal Analysis and Calorimetry* 52: 261–274.

Doherty, W., Halley, P., Edye, L., Rogers, D., Cardona, F., Park, Y., and Woo, T. 2007. Studies on polymers and composites from lignin and fiber derived from sugar cane. *Polymer for Advance Technology* 18: 673–678.

Edgar, K.J., Buchanan, C.M., Debenham, J.S., Rundquist, P.A., Seiler, B.D., and Shelton, M.C. 2001. Advances in cellulose ester performance and application. *Progress in Polymer Science* 26: 1605–1688.

El-Shafee, E., Saad, G.R., and Fahmy, S.M. 2001. Miscibility, crystallization and phase structure of poly(3-hydroxybutyrate)/cellulose acetate butyrate blends. *European Polymer Journal* 37: 2091–2104.

Foly, S., Nasr, H., and Nasr, M.I. 2008. Sugar industry in Egypt. *Sugar Tech* 10: 204–209.

Galbis, J.A. and Garcia-Martin, M.G. 2008. Cellulose chemistry: Novel products and synthesis paths. In *Monomers, Polymers and Composites from Renewable Resources,* eds. M.N. Belgacem and A. Gandini, pp. 89–114. Elsevier, Oxford, U.K.

Gamez, S., Gonzalez-Cabriales, J.J., Ramirez, J.A., Garrote, G., and Vazquez, M. 2006. Study of the hydrolysis of sugar cane bagasse using phosphoric acid. *Journal of Food Engineering* 74: 78–88.

Gong, R., Jin, Y., Chen, J., Hu, Y., and Sun, J. 2007. Removal of basic dyes from aqueous solution by sorption on phosphoric acid modified rice straw. *Dyes and Pigments* 73: 332–337.

Greco, P. and Martuscelli, E. 1989. Crystallization and thermal behaviour of poly(D(-)-3-hydroxybutyrate)-based blends. *Polymer* 30: 1475–1483.

Gutierrez, A., Del Ril, J.C., Gonzalez-Vila, F.J., and Martin, F. 1999. Spectroscopic characterization of extractives isolated with MTBE from straws. *Holzforschung* 53: 481–486.

Habibi, Y., El-Zawawy, W.K., Ibrahim, M.M., and Dufresne, A. 2008. Processing and characterization of reinforced polyethylene composites made with lignocellulosic fibers from Egyptian agro-industrial residues. *Composites Science and Technology,* 68: 1877–1885.

Hanna, A.A., Basta, A.H., El-Saied, H., and Abadir, I.F. 1999. Thermal properties of cellulose acetate and its complexes with some transition metals. *Polymer Degradation and Stability* 63: 293–296.

Hoareau, W., Trindade, W.G., Siegmund, B., Castellan, A., and Frollini, E. 2004. Sugar cane bagasse and curaua lignins oxidized by chlorine dioxide and reacted with furfuryl alcohol: Characterization and stability. *Polymer Degradation and Stability* 86: 567–576.

Huang, H., Yun, H., Zhang, J., Sato, H., Zhang, H., Noda, I., and Ozaki, Y. 2005. Miscibility and hydrogen bonding interactions in biodegradable polymer blends of poly (3-hydroxybutyrate) and a partially hydrolized poly(vinyl alcohol). *Journal of Physical Chemistry B* 109: 19175–19183.

Hurtubise, F.G. 1962. The analytical and structural aspects of the infrared spectroscopy of cellulose acetate. *Tappi* 45(6): 460–465.

Ifuku, S., Nogi, M., Abe, K., Handa, K., Nakatsubo, F., and Yano, H. 2007. Surface modification of bacterial cellulose nanofibers for property enhancement of optically transparent composites: Dependence on acetyl group DS. *Biomacromolecules* 8: 1973–1978.

Inoue, Y. and Yoshie, N. 1992. Structure and physical properties of bacterially synthesized polyesters. *Progress in Polymer Science* 17: 571–610.

Klemm, D., Heublein, B., Fink, H.-P., and Bohn, A. 2005. Cellulose: Fascinating biopolymer and sustainable raw material. *Angewandte Chemie—International Edition* 44: 3358–3393.

Liu, C.F., Sun, R.C., Zhang, A.P., and Ren, J.L. 2007. Preparation of sugar cane bagasse cellulosic phthalate using an anionic liquid as reaction medium. *Carbohydrate Polymer* 68: 17–25.

Luo, R., Xu, K., and Chen, G.Q. 2007. Study of miscibility, crystallization, mechanical properties, and thermal stability of blends of poly(3-hydroxy butyrate) and poly(3-hydroxy butyrate-co-4-hydroxy butyrate). *Journal of Applied Polymer Science* 105: 3402–3408.

Meireles, C.d.S., Filho, G.R., and Nascimento de Assuncao, R.M. 2007. Blend compatibility of waste materials. Cellulose acetate (from sugar cane bagasse) and polystyrene (from plastic cups)- diffusion of water, FTIR, DSC, TGA and SEM study. *Journal of Applied Polymer Science* 104: 909–914.

Mwaikambo, L.Y. and Ansell, M.P. 1999. The effect of chemical treatment on the properties of hemp, sisal, jute and kapok for composite reinforcement. *Macromolecular Materials and Engineering* 272: 108–116.

Pandey, A., Soccol, C.R., Nigam, P., and Soccol, V.T. 2000. Biotechnological potential of agro-industrial residues. I: Sugar cane bagasse. *Bioresource Technology* 74: 69–80.

Park, J.W., Tanaka, T., Doi, Y., and Iwata, T. 2005. Uniaxial drawing of poly[(R)-3-hydroxybutyrate]/cellulose acetate butyrate blends and their orientation behaviour. *Macromolecular Bioscience* 5: 840–852.

Patel, J.S., Onkarappa, R., and Shobha, K.S. 2007. Comparative study of ethanol production from microbial pretreated agricultural residues. *Journal of Applied Sciences and Environmental Management* 11(4): 131–135.

Persenaire, O., Alexandre, M., Degee, P., and Dubois, P. 2001. Mechanism and kinetics of thermal degradation of poly (ε-caprolactone). *Biomacromolecules* 2: 288–294.

Pizzoli, M., Scandola, M., and Ceccorulli, G. 1994. Crystallization kinetics and morphology of poly(3-hydroxybutyrate)/cellulose ester blends. *Macromolecules* 27: 4755–4761.

Prakash, G., Varma, A.J., Prabhune, A., Shouche, Y., and Rao, M. 2011. Microbial production of xylitol from D-xylose and sugarcane bagasse hemicellulose using newly isolated thermotolerant yeast *Debaryomyces hansenii*. *Bioresource Technology* 102(3):3304–3308.

Raveendran, K., Ganesh, A., and Khilar, K.C. 1996. Pyrolysis characteristics of biomass and biomass components. *Fuel* 75(8): 987–998.

Reverley, A. 1985. Review of cellulose derivatives and their industrial applications. In *Cellulose and Its Derivatives: Chemistry, Biochemistry and Applications*, eds. J.F. Kennedy, G.O. Philips, D.J. Wedlock, and P.A. Williams, p. 211. Ellis Horwood, New York.

Rodrigues, J.A.F.R., Parra, D.F., and Lugao, A.B.J. 2005. Crystallization and evaluation by DSC on films of PHB/PEG blends. *Thermal Analysis and Calorimetry* 79: 379–381.

Roman, M. and Winter, W.T. 2004. Effect of sulphate groups from sulfuric acid hydrolysis on the thermal degradation behaviour of bacterial cellulose. *Biomacromolecules* 5: 1671–1677.

Ruzene, D.S., Goncalves, A.R., Teixeira, J.A., and De Amorim, M.T.P. 2007. Carboxymethyl cellulose obtained by ethanol/water organosolv process under acid conditions. *Applied Biochemistry and Biotechnology* 137: 573–582.

Samois, E., Dart, R.K., and Dawkins, J.V. 1997. Preparation, characterization and biodegradation studies on cellulose acetates with varying degrees of substitution. *Polymer* 38(12): 3045–3054.

Sanchez, O. J. and Cardona, C. A. 2008. Trends in biotechnological production of fuel ethanol from different feedstocks. *Bioresource Technology* 99: 5270–5295.

Sarrouh, B.F., Silva, S.S., Santos, D.T., and Converti, A. 2007. Technical/economical evaluation of sugar cane bagasse hydrolysis for bioethanol production. *Chemical Engineering and Technology* 30: 270–275.

Sassi, J.F., Tekely, P., and Chanzy, H. 2000. Relative susceptibility of the Iα and Iβ phases of cellulose towards acetylation. *Cellulose* 7: 119–132.

Scandola, M., Ceccorulli, G., and Zizzoli, M. 1992. Miscibility of bacterial poly(3-hydroxybutyrate) with cellulose esters. *Macromolecules* 25: 6441–6446.

Scholz, C. and Gross, R.A. 2000. Poly(hydroxyalkanoates) as potential biomedical materials: An overview. In *Polymers From Renewable Resources*, pp. 328–334. American Chemical Society, Washington, DC.

Shao, Z., Li, G., Xiong, G., and Yang, W. 2002. Modified cellulose absorption method for the synthesis of conducting peovskite powders for membrane application. *Powder Technology* 122: 26–33.

Shieh, J.J. and Chung, T.S. 2000. Cellulose nitrate-based multilayer composite membrane for gas separation. *Journal of Membrane Science* 166: 259–269.

Singh, R., Varma, A.J., Laxman, S., and Rao, M. 2009. Hydrolysis of cellulose derived from steam exploded bagasse by Penicillium cellulases: Comparison with commercial cellulose. *Bioresource Technology* 100: 6679–6681.

Stading, M., Rindlav-Westling, A., and Gatenholm, P. 2001. Humidity induced structural transitions in amylose and amylopectin films. *Carbohydrate Polymers* 45: 209–217.

Steinmeier, H. 2004. Acetate manufacturing, process, and technology. *Macromolecular Symposia* 208: 49–60.

Sukumaran, R.K., Singhania, R.R., Mathew, G.M., and Ashok Pandey, A. 2009. Cellulase production using biomass feed stock and its application in lignocellulose saccharification for bio-ethanol production. *Renewable Energy* 34: 421–424.

Sun, J.X., Sun, X.F., Sun, R.C., and Su, Y.Q. 2004a. Fractional extraction and structural characterization of sugar cane bagasse hemicelluloses. *Carbohydrate Polymers* 56: 195–204.

Sun, J.X., Sun, X.F., Zhao, H., and Sun, R.C. 2004b. Isolation and characterization of cellulose from sugar cane bagasse. *Polymer Degradation and Stability* 84: 331–339.

Suttiwijitpukdee, N., Sato, H., Zhang, J., Hashimoto, T., and Ozaki, Y. 2010. Intermolecular interactions and crystallization behaviors of biodegradable polymer blends between poly (3-hydroxybutyrate) and cellulose acetate butyrate studied by DSC, FT-IR, and WAXD. *Polymer*, accepted manuscript, doi: 10.1016/j.polymer.2010.11.021.

Tomaz Duarte, M.A., Hugen, R.G., Martins Sant'Anna, E., Pezzin Testa, A.P., and Pezzin, S.H. 2006. Thermal and mechanical behaviour of injection molded poly (3-hydroxybutyrate)/poly (ε-caprolactone) blends. *Materials Research* 9: 25–27.

UNI EN ISO 1183-1. 2005. Methods for determining the density of non-cellular plastics. Part I: Immersion method, Liquid picnometer method and titration method. Engineering Library, University of Pisa, Pisa, Italy.

Viera, R.G.P., Filho, G.R., Assunção, R.M.N., Meireles, C.S., Vieira, J.G., and De Oliveira, G.S. 2007. Synthesis and characterization of methylcellulose from sugar cane bagasse cellulose. *Carbohydrate Polymers* 67: 182–189.

Wallis, A.F.A. and Wearne, R.H. 1997. Characterization of resin in radiata pine woods, bisulfite pulps and mill pitch samples. *Appita* 50: 409–414.

Xing, P., Dong, L., An, Y., Feng, Z., Avella, M., and Martuscelli, E. 1997. Miscibility and crystallization of poly(β-hydroxybutyrate) and poly(p-vinylphenol) blends. *Macromolecules* 30: 2726–2733.

Xu, J., Guo, B.H., Yang, R., Wu, Q., Chen, G.Q., and Zhang, Z.M. 2002. In situ FTIR study on melting and crystallization of polyhydroxyalkanoates. *Polymer* 43: 6893–6899.

Xu, S., Luo, R., Wu, L., Xu, K., and Chen, G.Q. 2006. Blending and characterizations of microbial poly(3-hydroxybutyrate) with dendrimers. *Journal of Applied Polymer Science* 102(4): 3782–3790.

Yang, H., Li, Z.S., Lu, Z.Y., and Sun, C.C. 2005. Computer simulation studies of the miscibility of poly (3-hydroxybutyrate)-based blends. *European Polymer Journal* 41: 2956–2962.

Yang, H., Yan, R., Chen, H., Lee, D.H., and Zheng, C. 2007. Characteristics of hemicellulose, cellulose, and lignin pyrolysis. *Fuel* 86: 1781–1788.

Yu, L., Dean, K., and Li, L. 2006. Polymer blends and composites from renewable resources. *Progress in Polymer Science* 31: 576–602.

Zhang, J., Sato, H., Furukawa, T., Tsuji, H., Noda, I., and Ozaki, Y. 2006. Polymer blends and composites from renewable resources. *Journal of Physical Chemistry B* 110: 24463–24471.

8 Using Life Cycle Assessment to Evaluate the Environmental Performance of Bio-Based Materials

Martin Weiss, Juliane Haufe, Barbara Hermann,
Miguel Brandão, Martin K. Patel, and Michael Carus

CONTENTS

8.1 INTRODUCTION

Bio-based materials comprise a large and heterogeneous group of materials produced partially or fully from biomass. Bio-based materials include conventional wood, paper, and textile products as well as novel chemicals, resins, lubricants, and composites. The manufacturing processes of bio-based materials are as versatile as the group of materials ranging from extraction and simple mechanical processing of naturally occurring polymers to fermentation and advanced enzymatic or catalytic conversions (Figure 8.1).

Twenty percent of all nonfood and nonfeed biomass is currently used for the production of bio-based materials, while 80% is used for energy and fuels [1]. Energy from biomass contributed in 2006 around 10% (50 EJ, exajoules) to the global total primary energy supply [2], while bio-based materials accounted for approximately 14% (900 Mt) of the global production of bulk materials [1–4].*

---

* The percentage accounts for the production of cement, iron and steel, bricks, glass, polymers, aluminum, textiles, wood, and paper.

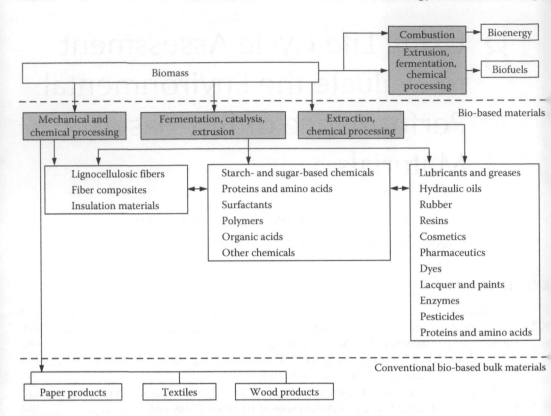

FIGURE 8.1   Overview of bio-based materials and their relevant production processes.

The largest share of both energy and material use of biomass is still attributed to conventional purposes, e.g., wood burning for cooking and heating in developing countries as well as the manufacturing of wood, paper, and textile products. In the past decade, however, the interest in using biomass as a feedstock for the production of a larger variety of materials has been growing. Biomass accounts already today for 8%–10% of the total feedstock used in the chemical industry [5]. The global production of bio-based polymers such as alkyd resins or polylactic acid (PLA) accounted in 2007 for 7% (20 Mt) of the production of conventional polymers (270 Mt) [6,7]. In parallel, the production of bio-based plastics, which represent the polymer fraction of high molecular mass, is growing. However, the global production capacity of 0.36 Mt accounted in 2007 for only 0.3% of the worldwide plastics production [7]. Likewise, bio-based lubricants represented only 1% (100 kton, kiloton) of the European lubricants market in 2006 [8].

Despite the currently low market shares, novel bio-based materials are expected to make an important contribution toward a more sustainable industrial production from the perspective of both resource use and climate change [7,9,10]. Substantial technological progress in recent years has opened a market for a wide range of new applications such as bio-based insulation and packaging as well as novel bio-based plastics for consumer goods (e.g., electronics and carpets). Technological breakthroughs are expected in the coming years for integrated biorefineries, which may provide a range of materials and energy products, thereby making the best possible use of the biomass feedstock.

The growth of bio-based materials, however, also presents scientists, policy-makers, and industry with a problem. Bio-based materials promise to reduce nonrenewable resource use and, GHG (greenhouse gas) emissions while creating substantial opportunities for process and technology innovation, which likely benefit other manufacturing industries [10]. Bio-based materials, however, expand agricultural land use and may enhance associated adverse environmental impacts, such as soil erosion, eutrophication, acidification, as well as the loss of natural habitats and biodiversity.

If bio-based materials production continues to increase, a thorough analysis of the associated environmental impacts is indispensable. To this end, many assessment tools are available, including life cycle assessment (LCA), strategic environmental assessment (SEA), environmental risk assessment (ERA), material flow analysis (MFA), or ecological footprinting [11]. In particular, LCA has been frequently applied in the past decade to quantify the environmental impacts of bio-based materials. Initially, LCA studies focused on first-generation bio-based materials such as fiber-composites and starch-based polymers (e.g., [12,13]). The substantial progress in research and development over the past decade also reflected by an increasing number of recent LCA studies on second-generation bio-based materials produced via advanced fermentation of lignocellulose, catalytic processes, and in integrated biorefineries (e.g., [9,14]).

This chapter seeks to report on the environmental impacts of bio-based materials. It provides an up-to-date review of LCA studies and aims at a comprehensive scope with regard to the stages of product life cycles and the variety of environmental impacts. However, the chapter does not claim absolute completeness with respect to all published LCA studies. Instead, it seeks to identify general pattern of environmental impacts and to discuss their implications for a sustainable production, use, and disposal of bio-based materials. The chapter excludes social and economic aspects of bio-based materials. These aspects obviously have to be considered before deriving conclusions regarding the suitability of bio-based materials for meeting the goals of sustainable production and consumption.

The following section provides a short overview of the LCA methodology and the most important peculiarities when applying it to bio-based materials. Section 8.3 quantifies the environmental performance of bio-based materials based on the results presented in the reviewed LCA studies. Section 8.4, discusses critical aspects in the application of LCA to bio-based materials. Section 8.5 summarizes the implications of our findings for scientists, policy makers, and industry.

8.2   LIFE CYCLE ASSESSMENT OF BIO-BASED MATERIALS

The environmental impacts of bio-based materials can be quantified and compared with conventional fossil-based product alternatives by applying an internationally standardized LCA methodology. LCA is not restricted to bio-based materials; instead, it has been used and modified for analyzing the environmental impacts of a large variety of products and services [15]. Yet, its application to bio-based materials and, in more general, to bioenergy and biofuels has received special attention due to the expected savings in nonrenewable energy use and GHG emissions. Any LCA follows a systematic approach that consists of four consecutive steps ([16–18]; Figure 8.2):

1. The *goal and scope definition* includes (i) the definition and description of the product or service under consideration, (ii) the definition of the functional unit to be analyzed, (iii) the precise definition of the product system and its boundaries, as well as (iv) making choices regarding allocation procedures, impact categories, and the methodology of the impact assessment.
2. The life cycle *inventory analysis* involves the actual data collection and calculation procedures to quantify all environmentally relevant inputs and outputs of the product system. The most prominent of these are typically resource use, in particular energy use and land use, as well as gaseous, liquid, and particulate emissions, and liquid and solid waste.
3. The life cycle *impact assessment* evaluates all inputs and outputs identified in the previous step with respect to their specific environmental impacts. To do so, the impact assessment aggregates data belonging to the same environmental impact category into a single environmental impact value by using so-called characterization or equivalence factors. As an optional step, the results of each impact category can be normalized with respect to a reference value (e.g., total greenhouse gas emissions of a country or total greenhouse gas emissions per capita and year).

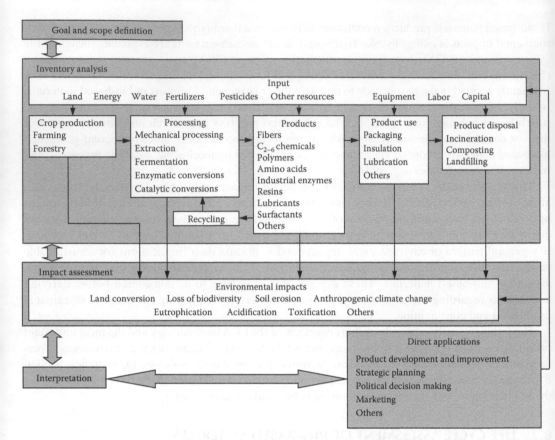

FIGURE 8.2 LCA in the context of bio-based materials. (Adapted from ISO, Environmental management—Life cycle assessment—Principles and framework, ISO 14040, ISO—International Organization for Standardization, Geneva, Switzerland, 2006a; Bringezu, S. et al., Towards sustainable production and consumption of resources: Assessing biofuels, International Panel for Sustainable Resource Management, UNEP—United Nations Environment Programme, Nairobi, Kenya, 2009, *Source:* http://www.unep.fr/scp/rpanel/pdf/Assessing_Biofuels_Full_Report.pdf).

4. The last step of an LCA is termed *interpretation*. Here, final conclusions are drawn based on the findings of the life cycle inventory analysis and the life cycle impact assessment. This step includes a check for completeness and consistency of inventory data and a sensitivity analysis as well as the reporting of results. Consequently, recommendations for producers, consumers, policy-makers, or environmental stakeholders can be formulated according to the original goal of the LCA.

When defining the goal and scope in the LCA of bio-based materials, several choices have to be made, which are particularly critical for the final results. These choices include the selection of

1. *System boundary*—LCA ideally considers the entire life cycle of products or services. Adopting this comprehensive approach is referred to as a "cradle-to-grave" analysis. However, specific research questions might make it desirable to analyze individual life cycle stages separately. Analyzing only the environmental impacts of the production stage is generally referred to as a "cradle-to-factory-gate" analysis and has the advantage of lowering data requirements by excluding use-phase and end-of-life waste treatment options.

2. *End-of-life waste treatment*—A "cradle-to-grave" LCA is comprehensive but can become complex because of the necessity to select and analyze appropriate end-of-life waste treatment options. In the case of bio-based materials, these options might include product reuse, recycling, incineration with or without energy recovery, composting, digestion, or landfilling. Depending on the choice of either of these options, the environmental impacts of bio-based materials can vary substantially [13,20,83].

3. *Functional unit*—LCAs on bio-based materials typically chose for either (i) a product-based approach that ascribes all environmental impacts consistently to a physical product quantity, e.g., 1 kg of PLA, 1 m³ of loose-fill packaging material or (ii) a land use–based approach that refers all environmental impacts consistently to a defined area of agricultural land, usually 1 hectare (ha), which is used for producing the biomass feedstock. Both approaches have their justifications and limitations (see discussion below).

4. *Treatment of co-products*—The production of bio-based materials often yields useful by-products, to which part of the environmental impacts should be allocated. According to ISO [17], allocation should be avoided whenever possible. Instead, processes should be divided into subprocesses, or the product system should be expanded. The choice regarding either of these options is controversial and can have a substantial impact on the results of LCAs [21,22].

5. *Inclusion of secondary effects*—Every LCA needs to define a cut-off criterion according to which environmental impacts are included in or excluded from the analysis. In the case of bio-based materials, the inclusion or exclusion of direct and indirect land-use change is particularly critical for the product-related GHG emissions (see discussion below).

LCA studies of bio-based materials generally differ from each other in several, if not all, choices and assumptions. In our review, we do not correct for potential differences. Instead, we present average values for the various environmental impact categories separately for individual product groups. We present uncertainty intervals that indicate the standard deviation of the data sample and thereby also reflect the effect of diverse choices and assumptions on the outcome of LCA studies. This chapter reviews around fifty LCA studies on bio-based materials. The reviewed LCAs focus almost exclusively on bio-based materials of European origin being manufactured by both small pilot installations and large-scale industrial plants. The selected studies differ considerably with respect to the amount of published background data and the degree of detail regarding explanations about methodological choices and results. The reviewed LCA studies typically compare the environmental impacts of bio-based materials with the impacts of conventional fossil-based materials. We follow this approach and express the product-specific environmental impacts uniformly as

$$D_{ij} = EI_{\text{bio-based},ij} - EI_{\text{fossil-based},ij}$$

where

D_{ij} stands for the difference in the environmental impact j between the bio-based material i and fossil-based material i

$EI_{\text{bio-based},ij}$ represents the environmental impact j of the bio-based material i

$EI_{\text{fossil-based},ij}$ stands for the environmental impact j of the fossil-based product alternative i

This approach results in negative values if bio-based materials exert lower impacts on the environment and in positive values if bio-based materials exert higher impacts on the environment relative to their fossil-based product alternatives. We exclude from our review LCA studies of traditional biomass use for the production of cosmetics, textiles, as well as wood and paper products.

8.3 ENVIRONMENTAL IMPACTS OF BIO-BASED MATERIALS

A comprehensive analysis of all environmental impacts associated with bio-based materials is a time- and resource-intensive task, requiring an extensive compilation of life cycle inventory data. Most LCA studies therefore focus only on a few impact categories, with the most prevalent ones being nonrenewable energy use (i.e., the total use of fossil and nuclear energy) and GHG emissions.

8.3.1 NONRENEWABLE ENERGY USE AND GHG EMISSIONS

Past reviews of LCA studies have shown that bio-based materials generally use less nonrenewable energy and cause lower GHG emissions than their fossil-based product alternatives [1,10,23–27]. This chapter confirms these findings (Figures 8.3 and 8.4). The savings in nonrenewable primary energy use and GHG emissions range from 36 ± 17 GJ (gigajoules) and 0.9 ± 1.3 t CO_2-eq. (tons of carbon dioxide equivalents) per ton of bio-based composites to 130 ± 81 GJ and 5.6 ± 4.4 t CO_2-eq. per ton of bio-based lacquer and paints.* Figures 8.3 and 8.4 indicate a large variability of savings both between and within individual groups of bio-based materials. This finding suggests that

1. Bio-based materials contribute in general but not always to savings of nonrenewable primary energy and GHG emissions—they are thus suitable to decrease both the dependency on fossil resources and the carbon footprint of the manufacturing industry.
2. The ranges of results within individual groups of bio-based materials (e.g., lacquer and paints, composites, or plastics) are often as large as the ranges between different groups of bio-based materials (e.g., compare plastics, surfactants, and textiles)—thus, it is difficult to identify individual groups, which are environmentally superior with respect to these two impact categories.

These results neglect direct and indirect land-use change for biomass production. In Europe, direct land-use change, that is, the shift from one type of land use (e.g., forestry) to another (e.g., farming), is negligible, while indirect land use change is an issue of concern. Expanding the domestic feedstock base for the production of bio-based materials might, in particular, increase the imports of agricultural products for food or nonfood use and may exert additional land requirements in ecologically sensitive regions elsewhere, for example, in tropical rainforests (see discussion below and in the following section). Such a scenario includes a variety of additional environmental impacts, among them increasing GHG emissions.

Special attention has been paid in recent years to the production of bio-based chemicals. Patel et al. [9] conducted an extensive analysis on the nonrenewable primary energy use and GHG emissions of a wide range of bio-based chemicals produced by white biotechnology (Figures 8.5 and 8.6). Their results indicate that both nonrenewable energy use (−99 ± 25 − 7 ± 37 GJ/ton) and GHG emissions (−7 ± 1 − 0 ± 2 GJ/ton) are generally lower for bio-based chemicals than for their conventional fossil-based counterparts. Likewise, Ren and Patel [68] find that CO_2 emissions from the production of bio-based ethylene, propylene, and aromatics (−4–2 ton CO_2/ton basic chemicals) are generally lower than the emissions of basic chemicals produced from methane feedstock (4–5 ton CO_2/ton) or coal feedstock (8–11 ton CO_2/ton). Variability in the nonrenewable primary energy use and GHG emissions bio-based chemicals results from differing assumptions regarding (i) the type of biomass feedstock (i.e., corn starch, lignocellulosic biomass, or sucrose from sugarcane) and (i) the applied production technology (e.g., current versus future technologies; aerobic versus anaerobic processes, or continuous versus batch processes).

Using corn starch as a feedstock for the production of bio-based chemicals saves on average less nonrenewable primary energy than using lignocellulosic biomass or sugarcane; the latter may offer savings of up to 75%–85% when compared to conventional fossil-based chemicals, if future

---

* We consider here the global warming potential over a time horizon of 100 years.

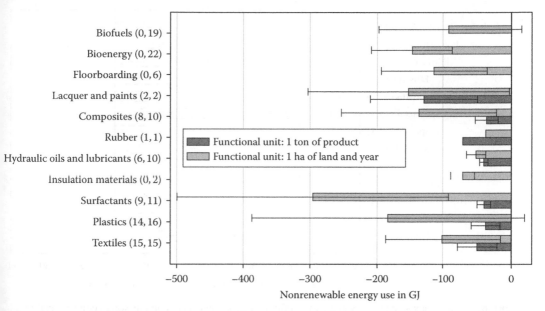

FIGURE 8.3 Nonrenewable primary energy use of bio-based materials as well as bioenergy and biofuels in comparison to conventional fossil-based product alternatives on a cradle-to-factory-gate basis; numbers in parentheses indicate the sample sizes for the functional unit of 1 ton of product and 1 ha and year; uncertainty intervals represent the standard deviation of data. (Primary data sources: [9,12,13,21,23,28–67].)

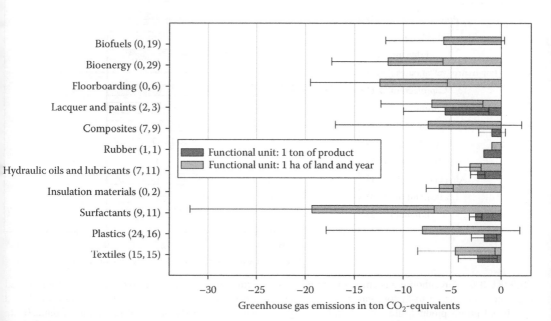

FIGURE 8.4 Greenhouse gas emissions of bio-based materials as well as bioenergy and biofuels in comparison to conventional fossil-based product alternatives on a cradle-to-factory-gate basis; numbers in parentheses indicate the sample sizes for the functional unit of 1 ton of product and 1 ha and year; uncertainty intervals represent the standard deviation of data. (Primary data sources: [9,12,13,28–67].)

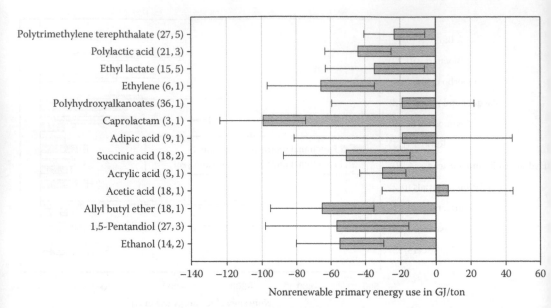

FIGURE 8.5 Nonrenewable primary energy use of bio-based chemicals in comparison to conventional fossil-based chemicals on a cradle-to-factory-gate basis; numbers in parentheses indicate the sample sizes for the bio-based and fossil-based product alternatives; uncertainty intervals represent the standard deviation of data. (Data from Patel, M.K. et al., Medium- and long-term opportunities and risks of the biotechnological production of bulk chemicals from renewable resources—The potential of white biotechnology, The BREW Project, Utrecht University, Utrecht, the Netherlands, 2006, source: http://www.chem.uu.nl/brew/BREWProjectProfile_FINAL.pdf, accessed: September 15, 2010.)

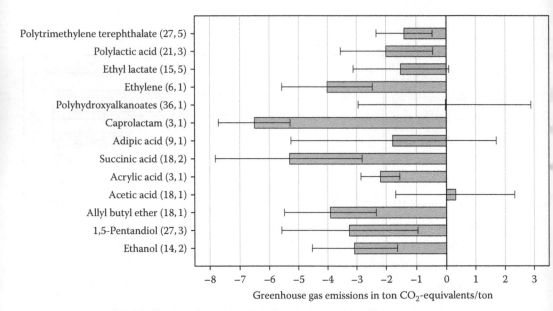

FIGURE 8.6 Greenhouse gas emissions of bio-based chemicals in comparison to conventional fossil-based chemicals on a cradle-to-factory-gate basis; numbers in parentheses indicate the sample sizes for the bio-based and fossil-based product alternatives; uncertainty intervals represent the standard deviation of data. (Data from Patel, M.K. et al., Medium- and long-term opportunities and risks of the biotechnological production of bulk chemicals from renewable resources—The potential of white biotechnology, The BREW Project, Utrecht University, Utrecht, the Netherlands, 2006, source: http://www.chem.uu.nl/brew/BREWProjectProfile_FINAL.pdf, accessed: September 15, 2010.)

technologies are employed. Patel et al. [9] conclude that bio-based chemicals provide substantial opportunities for the chemical industry to reduce its nonrenewable primary energy use and GHG emissions. However, the identified savings exclude the potential effects of indirect land-use change; additional research is needed to quantify these, in particular, in regards to the GHG emissions of bio-based materials.

The findings in Figures 8.3 and 8.4 indicate that bio-based materials alongside with bioenergy and biofuels save nonrenewable primary energy use and GHG emissions. The large variability of data, however, does not permit to draw definite conclusions on whether bio-based materials are preferable over biofuels and bioenergy with respect to these two impact categories. Conclusions must therefore be drawn based on the analysis of individual cases.

8.3.2 EUTROPHICATION AND ACIDIFICATION POTENTIAL

Nonrenewable primary energy use and GHG emissions are by far the most frequently analyzed environmental impacts of bio-based materials. However, relatively low environmental impacts in these two impact categories are often associated with high impacts in other categories. Figure 8.7 presents eutrophication and acidification potentials of bio-based materials on the basis of 1 ha of agricultural land. Despite large uncertainty intervals, the results indicate that bio-based materials (as well as bioenergy and biofuels) induce higher eutrophication potentials than their conventional fossil-based counterparts. The relatively poor performance of bio-based materials in this impact category results from biomass production with conventional agricultural practices, in particular from ammonia emissions as well as nitrate and phosphate leaching from the applied nitrogen fertilizers (e.g., [1,13,14]).

Alternative farming practices that improve the management of fertilizer application or shift to extensive farming practices can substantially reduce the eutrophication potential of biomass production (see also discussions in Section 8.4). The use of biomass to manufacture plastics and fiber composites generally decreases acidification. In contrast, acidification typically increases when biomass

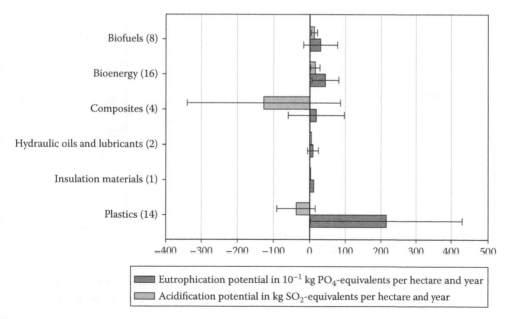

FIGURE 8.7 Eutrophication and acidification potential of bio-based materials as well as bioenergy and biofuels in comparison to conventional fossil-based product alternatives on a cradle-to-grave basis; eutrophication potential represents the sum of terrestrial and aquatic eutrophication; numbers in parentheses indicate the sample sizes for the bio-based and conventional product alternatives; uncertainty intervals represent the standard deviation of data. (Primary data sources: [12,13,21,28,33,34,37,40,45,69,70].)

is used to replace fossil energy and fuels. Acidification is mainly caused by ammonia emissions from fertilizers as well as sulfur and nitrogen oxide emissions from incineration processes. The first source is relevant for biomass production in general; the second one, however, refers mainly to bioenergy and biofuels. Even if bio-based materials are combusted at the end of their life cycle, emissions from incineration constitute only a minor portion of the total acidifying emissions along the entire life cycle of bio-based materials because process chains are typically longer than those of bioenergy and biofuels. Thus, bio-based materials typically cause a lower acidification potential relative to their conventional fossil-based counterparts than bioenergy and biofuels do.

8.3.3 Tropospheric Ozone Creation and Stratospheric Ozone Depletion Potential

A relatively limited number of LCA studies indicate that bio-based materials might potentially (i) decrease photochemical ozone creation in the troposphere but (ii) increase ozone depletion in the stratosphere relative to their fossil-based counterparts (Figure 8.8). The relatively high potentials of bio-based materials for stratospheric ozone depletion largely result from N_2O (nitrous oxide) emissions due to fertilizer application in agriculture but not from emissions of other ozone-depleting substances such as halogenated hydrocarbons (e.g., [13,40]). After the successful restriction of the production of ozone-depleting substances under the Montreal Protocol [71], N_2O emissions constitute the single most important driver of stratospheric ozone depletion at a global scale. Limiting future N_2O emissions would enhance the recovery of the ozone layer and additionally reduce the anthropogenic forcing of the climate system [72]. Tropospheric ozone creation and stratospheric ozone depletion have only been analyzed for a limited number of bio-based materials. Although industrial farming practices, typically characterized by the excessive application of fertilizers, is common to biomass production in general, we suggest to study a wider range of bio-based materials before drawing definite conclusions.

This Chapter has shown so far that bio-based materials save nonrenewable energy resources and exert lower environmental impacts than their conventional fossil-based counterparts with respect

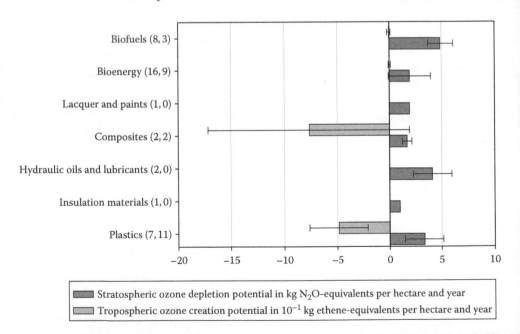

FIGURE 8.8 Tropospheric ozone creation potential and stratospheric ozone depletion potential of bio-based materials as well as bioenergy and biofuels in comparison to conventional product alternatives on a cradle-to-grave basis; numbers in parentheses indicate the sample sizes for the bio-based and conventional product alternatives; uncertainty intervals represent the standard deviation of data. (Primary data sources: [14,22,29, 34,35,38,41,46,87,88].)

to GHG emissions (if indirect land-use change is disregarded) and photochemical ozone creation in the troposphere. Our results are inconclusive with regard to acidification. Bio-based materials generally cause higher environmental impacts with respect to eutrophication and stratospheric ozone depletion. With the exception of acidification, similar patterns of environmental impacts have also been identified for bioenergy and biofuels [34,73,74].

The results of individual LCA studies come with errors and uncertainties related to, for example, the functionality and life time of products, coproduct allocation, as well as the quality of LCA inventory data. Dinkel et al. [21] quantify error ranges of inventory data with 40% and deviations resulting from differences in the applied allocation method with up to 90% of the final result. These uncertainties might introduce a systematic bias into the results of individual LCA studies but present a random error in review analysis that use a sufficiently large data base.

Bio-based materials exert various impacts on the environment; only a few have been addressed here. Both land and water use for biomass production have recently received special attention. Impacts such as environmental and human toxicity, carcinogenic potential, soil erosion potential, and the depletion of biodiversity are excluded from this review. Data availability for these categories is generally limited to the extent that it does not allow drawing general conclusions regarding the performance of bio-based materials [1]. Yet, these categories are particularly relevant at local and regional scales and call for a more comprehensive evaluation of the environmental impacts of bio-based materials. The next section will analyze in greater detail the sensitivity of environmental impacts with respect to several critical aspects in the LCA of bio-based materials.

8.4 CRITICAL ASPECTS IN THE LIFE CYCLE ASSESSMENT OF BIO-BASED MATERIALS

The previous section has shown that the environmental impacts of bio-based materials scatter over a wide range. The diversity of life cycle assessment (LCA) results stems from the diversity of products as well as assumptions and methodological choices. Usually, a combination of the factors summarized in Figure 8.9 potentially leads to substantial differences in the outcome of LCA studies, even if similar product scenarios are analyzed.

The debate about the LCA of bio-based materials focuses, in particular, on: (i) the choice of an appropriate functional unit, (ii) the selection of crops, (iii) secondary effects such as impacts of indirect land-use change on climate and ecosystems, (iv) the treatment of agricultural residues, (v) assumptions regarding agricultural practices, and (vi) the choice of end-of-life waste treatment scenarios [82]. This section discusses in more detail the implications of each of these aspects on the outcome of LCAs.

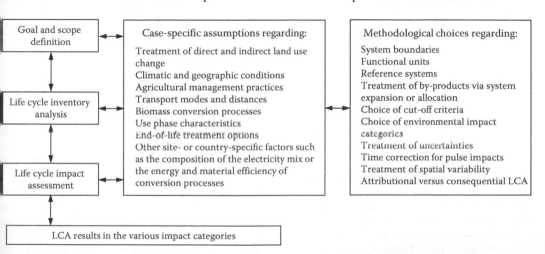

FIGURE 8.9 Overview of factors affecting the outcome of LCA studies on bio-based materials.

8.4.1 Choice of Functional Unit

The production of biomass feedstocks requires land. To account for land requirements, the environmental impacts of bio-based materials can be expressed per hectare of agricultural land instead of per physical product quantity (e.g., ton or cubic meter of final product). The former approach quantifies the hectare-specific environmental impacts by dividing the environmental impacts of a bio-based material by the area that is required to produce the analyzed bio-based product. The results of this approach provide an indication of the *land-use efficiency* of bio-based materials. The choice of either agricultural area or physical product quantity does not change the direction of the results, but it has a considerable effect on the magnitude of environmental impacts. It is therefore important to clearly define the goal of an LCA before conducting the actual analysis. If the environmental impacts of products are to be assessed, the functional unit should represent a defined product quantity such as one ton of PLA or a cubic meter of loose-fill packaging material. Alternatively, if the focus is on land use, the functional unit should be one hectare of agricultural land. To adequately address the problem of choosing a suitable functional unit, Bringezu et al. [19] recommend that LCAs should take two perspectives, namely, a product-based and spatial perspective and discuss the outcomes of both perspectives in an integrated manner.

8.4.2 Selection of Crops

The environmental impacts of bio-based materials depend on the crop that is chosen for feedstock supply (Figure 8.10). Differences in environmental impacts result here from crop-specific farming practices, climatic conditions, agricultural yields, transport distances, as well as plant characteristics such as starch or fiber content.

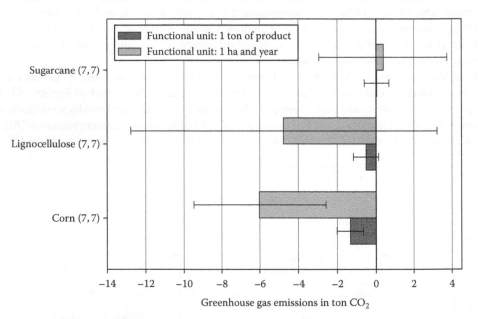

FIGURE 8.10 The effect of alternative functional units on the GHG emissions of bio-based PLA in comparison to conventional fossil-based polymers on a cradle-to-factory-gate basis. (numbers in parenthesis indicate the sample sizes for PLA and the conventional polymers; uncertainty intervals represent the standard deviation of data). (Data from Patel, M.K. et al., Medium- and long-term opportunities and risks of the biotechnological production of bulk chemicals from renewable resources—The potential of white biotechnology, The BREW Project, Utrecht University, Utrecht, the Netherlands, 2006, source: http://www.projects.science.uu.nl/brew/BREW_Final_Report_September_2006.pdf, accessed: September 15, 2011.)

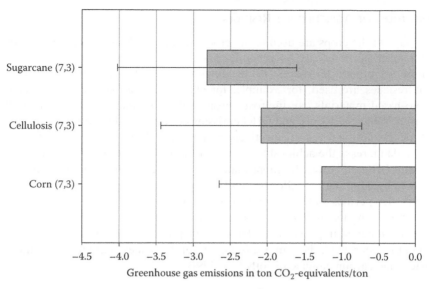

FIGURE 8.11 GHG emissions of PLA production relative to conventional polymers; numbers in parentheses indicate the sample sizes for PLA and the conventional polymers; uncertainty intervals represent the standard deviation of data. (Data from Patel, M.K. et al., Medium- and long-term opportunities and risks of the biotechnological production of bulk chemicals from renewable resources—The potential of white biotechnology, The BREW Project, Utrecht University, Utrecht, the Netherlands, 2006, source: http://www.projects.science.uu.nl/brew/BREW_Final_Report_September_2006.pdf, accessed: September 15, 2011.)

The product-specific environmental impacts typically decrease if crop-specific feedstock yields are high, environmental impacts of agricultural practices are low, and in cases where agricultural crops yield substantial amounts of useful coproducts (e.g., bagasse in the case of sugarcane). Figure 8.11 indicates that sugarcane might be the crop of choice for producing PLA considering solely GHG emissions and disregarding secondary effects such as indirect land-use change. Choosing suitable crops should account for these effects and consider also the existing potentials for growing biomass crops on abandoned or set-aside farmland [14,75].

8.4.3 Secondary Effects: The Case of Indirect Land-Use Change

The currently growing in the demand for biomass is mainly driven by the expansion of biofuels and bioenergy. Still, an additional biomass demand for the production of bio-based materials may accelerate the direct and indirect land-use change at global scale. In particular indirect land-use change, that is the unintended expansion of crop land elsewhere due to the use of existing agricultural land for the production of nonfood biomass, may add substantially to the environmental impacts of bio-based materials. Indirect land-use change may occur predominantly in the tropics, raising concerns about the expansion of farmland at the expense of forests, peat lands, and other natural habitats. The effects of direct and indirect land-use change are not considered by the reviewed LCA studies on bio-based materials but have been extensively analyzed in the context of bioenergy and biofuels. Wicke et al. [75] suggest that land-use change is the most decisive factor in the GHG emissions of palm oil energy chains. Hoefnagels et al. [22] showed that GHG emissions associated with the production of biodiesel from palm oil quadruple if plantations are located on former peat lands and rainforests instead of on degraded land or logged-over forests. Similar effects have been identified for the indirect land-use change induced by the biofuels production in the United States [76] and Brazil [77]. Potential solutions to this problem include (i) biomass production on abandoned land, (ii) productivity and yield increases for both food and nonfood crop production, and (iii) the introduction of comprehensive land-use management guidelines. Further research is highly needed, in particular to determine the economic and social feasibility of such solutions.

8.4.4 TREATMENT OF AGRICULTURAL RESIDUES

Often, only parts of the crops are used in the production of bio-based materials, leaving residues available for other purposes. LCAs generally assume that residues either remain on the field as a substitute for mineral fertilizers [13] or are simply excluded from the product system [37]. However, if residues are used, for example, for energy generation, the environmental performance of bio-based materials can improve substantially. Dornburg et al. [24] identified reductions in nonrenewable energy use and GHG emissions of bio-based polymers of up to 190 GJ and 15 ton CO_2 per hectare and year, respectively, if agricultural residues are used for energy. Such a scenario would increase the achievable savings of bio-based materials regarding nonrenewable energy use and GHG emissions by more than a factor of 2. Additional environmental benefits can be realized by adopting so-called agricultural intercrops, which cover agricultural land in-between vegetation periods and serve for energy production afterward [78]. Integrated biorefineries present a promising technological solution to make use of the whole crop and subsequently reduce the environmental impacts associated with the production of bio-based materials, energy, and fuels. On the other hand, the removal of agricultural residues could have substantial impacts on the nutrient and carbon balance of soils (e.g., [79]).

8.4.5 AGRICULTURAL PRACTICES

The relatively high environmental impacts of bio-based materials in the categories of eutrophication and stratospheric ozone depletion arise from biomass production with conventional agricultural practices. Würdinger et al. [13] demonstrated that extensive farming, characterized by the absence of mineral fertilizers and pesticides, can reduce the environmental impacts of bio-based loose-fill packaging materials (Figure 8.12). Extensive farming scores better than conventional farming in many of the analyzed impact categories. Striking differences between the two farming practices exist with respect to ozone depletion potential, which is estimated by Würdinger et al. [13] solely based on nitrous oxide (N_2O) emissions. Loose-fill packaging materials from

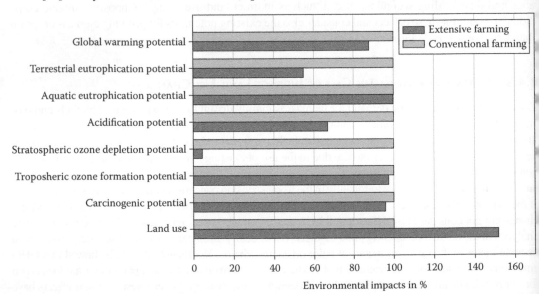

FIGURE 8.12   The relative environmental impacts of loose-fill packaging materials produced from wheat-starch. (Data from Würdinger, E. et al., Kunststoffe aus nachwachsenden Rohstoffen: Vergleichende Ökobilanz für Loose-fill-Packmittel aus Stärke bzw, Polystyrol, Projektgemeinschaft BIfA/IFEU/Flo-Pak, Final report 2002 (DBU-Az. 04763), Institut für Energie und Umweltforschung Heidelberg, Heidelberg, Germany, *Source:* http://www.ifeu.de, accessed: 29 July 2003.)

extensively grown wheat reduce N_2O emissions by 95% compared loose-fills from conventionally grown wheat (Figure 8.12).

The findings of Würdinger et al. [13] highlight the importance of reducing fertilizer application for cutting the N_2O emissions of industrial biomass production. N_2O is furthermore a potent greenhouse gas that might not always be accounted for appropriately in the assessment of GHG emissions. In the case of excessive fertilizer application, N_2O might substantially add to the GHG emissions of bio-based materials. Results for biofuels indicate that N_2O emissions might contribute 10%–80% to the total production-related GHG emissions, depending on crop type, climate conditions, and land-use scenario [80]. However, Smeets et al. [80] stress that optimized crop management, can substantially reduce N_2O emissions. These findings show that biomass production for materials, energy, and fuels can become more environmentally friendly if agricultural management practices change. However, the required changes are generally constrained by prevailing economic and social factors; it remains doubtful whether a substantial reduction in the environmental impacts of biomass production are achievable at a large scale. Figure 8.12 indicates in particular that extensive farming practices come at the costs of decreasing yields and larger land requirements for biomass production; for example, yields of wheat decline by roughly 30% in the case analyzed by Würdinger et al. [13]. This way, the expansion of extensive farming might further increase direct and indirect land-use change and the associated environmental impacts.

8.4.6 END-OF-LIFE WASTE TREATMENT

The choice of an appropriate end-of-life waste treatment scenario affects the environmental impacts of bio-based materials. In particular, when assessing the global warming potential of bio-based materials, there are two ways of dealing with biogenic carbon:

1. Considering biogenic carbon as neutral because the carbon withdrawn from the atmosphere during photosynthesis is returned to the atmosphere within a limited period of time. However, if allocation of environmental impacts becomes necessary, this approach implies allocation by carbon-content in practice [81].
2. Assuming biogenic carbon is sequestered by bio-based materials that function as a temporary carbon sink.

The latter approach implies that both the uptake of carbon from the atmosphere as well as its release during waste treatment must be accounted for as positive and negative carbon emissions. The period of time in which the carbon is stored outside of the atmosphere may also be regarded as a climate benefit due to the temporarily foregone radiative forcing. The method of carbon accounting can result in substantial differences of product-specific carbon emissions depending on whether only the production stage or the entire product life cycle has been analyzed and, in the case of the latter, which waste treatment option is chosen.

Analyzing bio-based materials on a cradle-to-grave basis allows for a variety of waste treatment scenarios, which are associated with a wide range of GHG emissions (Figure 8.13). For PLA, product-specific GHG emissions vary, for example, by approximately 20% depending on whether or not energy is recovered during waste incineration (Figure 8.13).

The LCA study of Würdinger et al. [13] on loose-fills suggests the differences in GHG emissions between various waste treatment scenarios are comparable to the differences between bio-based and conventional loose-fill packaging materials, based on a "cradle-to-factory-gate" comparison. Optimizing the waste management of bio-based materials can hence effectively reduce the product-specific environmental impacts. This finding calls for a detailed assessment of all major waste management options including landfilling, composting, waste-to-energy facilities, municipal waste incineration, digestion, and recycling [84,85]. In particular, carbon cascading by using

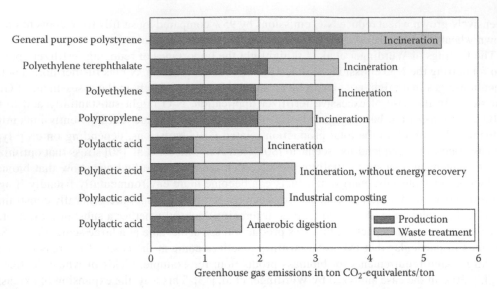

FIGURE 8.13   GHG emissions of various end-of-life waste treatment scenarios for bio-based polymers relative to conventional polymers. (Data from Hermann, B.G. et al., *Polym. Degrad. Stabil.*, 96, 1159, 2010b.)

biomass first for material purposes and second for recovering energy at the end of the product life cycle can maximize the CO_2 mitigation potential of bio-based materials [19,24]. Oertel [10] suggests that carbon cascading via the energetic use of bio-based materials at the end of their life cycle is likely to be preferable over composting from a GHG perspective. However, Hermann et al. [83] have shown that anaerobic digestion of postconsumer waste can be more attractive than incineration, if the resulting compost is used to replenish carbon stocks in agricultural soils.

8.5   CONCLUSIONS AND OUTLOOK

This chapter has shown that the environmental impacts of bio-based materials span a large value range; the assessment of these impacts are typically associated with uncertainties and limitations. The diversity of results suggests that the environmental performance of bio-based materials should be evaluated on a case-by-case basis, considering and weighing global concerns (anthropogenic climate change and stratospheric ozone depletion) against local and regional concerns (eutrophication and land use) [86]. The magnitude of the environmental impacts of bio-based materials is highly sensitive to assumptions and methodological choices in the LCA. The following conclusions can be drawn [82]:

1. Bio-based materials save nonrenewable energy resources; expanding their production allows the manufacturing industry to substitute parts of the nonrenewable energy and feedstock use by renewable resources and thereby to attain a more secure energy and feedstock supply.
2. Bio-based materials are generally associated with lower environmental impacts than their fossil-based counterparts with respect to GHG emissions and tropospheric ozone creation.
3. The LCA results are inconclusive with regard to environmental acidification but indicate that bio-based materials cause higher environmental eutrophication and stratospheric ozone depletion than their fossil-based counterparts.

4. The potential of bio-based materials to save GHG emissions refers in general to European feedstock use and European technology. Expanding bio-based materials production may, however, induce additional land requirements and cause GHG emissions from direct and indirect land-use change elsewhere, e.g., in the tropics. If biomass production results directly or indirectly in deforestation or drainage of peat lands, the GHG emissions associated with bio-based materials will increase substantially. Further research on quantifying these effects is highly recommended.

5. Biomass production at current agricultural practices accounts to a large extent for the high environmental impacts of bio-based materials in the category of eutrophication and stratospheric ozone depletion. These impacts can be largely reduced by improving the management of fertilizers and employing extensive agricultural practices. The latter measure, however, comes at the cost of declining agricultural yields and thus higher land requirements for biomass production. Nonetheless, the resource use in agriculture should be subject to a more detailed and critical analysis in the discussions about bio-based materials [1].

6. Bio-based materials exert a large variety of impacts on the environment. The current emphasis on nonrenewable energy use and GHG emissions might be justified but excludes land use impacts that are mainly relevant at local and regional scale, such as the potential loss of biodiversity, soil erosion, deforestation, as well as environmental and human toxicity. Difficulties in quantifying these impacts may prevail; LCA studies should therefore include a qualitative discussion to allow a more complete evaluation of the environmental performance of bio-based materials.

7. Each step in the life cycle of bio-based materials offers substantial potentials for reducing environmental impacts [9,13,24]. Effective measures include: (i) reducing the environmental impacts of agriculture, (ii) selecting the most suitable crops for bio-based materials production, (iii) planting of intercrops, (iv) utilizing residues and by-products for energy, fuel or material production, (v) carbon cascading and (vi) optimizing the end-of-life waste treatment. At the same time, additional environmental impacts may arise, for example, from the use of genetically modified crops and microorganisms in fermentation processes. Risks related to both have to be taken into account when evaluating bio-based materials produced by biotechnology [9].

8. The increasing production of bio-based materials might lead to competition for agricultural land between the production of food and feed and the production of bio-based materials, bioenergy, and biofuels. While the latter can be also supplied by carbon-free renewable resources, materials often rely on carbon as feedstock. This may justify a focus on biomass use for materials in view of the limited land availability. In addition, two compatible strategies might offer promising solutions: (i) *carbon cascading* by using biomass first for material purposes and at the end of the product life cycle for energy [1] and (ii) *biorefineries* by allowing for more complete use of the biomass during the integrated production of bio-based materials, energy, and fuels [14].

9. Many bio-based materials are currently produced by small-scale pilot plants. Progress in biotechnology, plant up-scaling, and further process integration promises to reduce the environmental impacts of bio-based materials [20]. To exploit these potentials, technological, economic, as well as logistical shortcomings need to be overcome.

10. Sustainable biomass production is a key for the sustainability of bio-based materials. Options for a more efficient and sustainable biomass production differ among regions. In developing countries, increasing yields are paramount to sustainable biomass production. To realize existing potentials requires innovative investment solutions for both food and nonfood farming [87]. Increments in crop and land productivity due to improvements in agricultural technology and institutional settings potentially generate spill-over effects for food production.

11. Whether bio-based materials can make a noteworthy contribution to an industry-wide reduction of nonrenewable energy use will depend on their market diffusion. In spite of current growth rates, the large-scale production of conventional fossil-based materials implies that bio-based materials will require at least a decade to reach substantial market shares, even at high-fossil-fuel prices [7]. While bio-based materials might potentially contribute to substantial savings of nonrenewable energy (and, hence, of GHG emissions) in the manufacturing industry [88], their economy-wide impacts are likely to remain limited because at global scale, only 6% of fossil fuels are used for material purposes, while the remainder is consumed for fuel and energy [89].

DISCLAIMER

The views expressed in this chapter are purely those of the authors and may not, under any circumstances, be regarded as an official position of the European Commission.

REFERENCES

1. Deimling, S., M. Goymann, M. Baitz, and T. Rehl. 2007. Auswertung von Studien zur ökologischen Betrachtung von nachwachsenden Rohstoffen bei einer stofflichen Nutzung. Commissioned by Fachagenturfür Nachwachsende Rohstoffe e.V., FKZ 114-50.10.0236/06-E. Gülzow, Germany.
2. IEA. 2009. IEA bioenergy annual report 2009. IEA—International Energy Agency. Paris, France. Source: http://www.ieabioenergy.com/DocSet.aspx?id=6506&ret=lib. Accessed: April 7, 2010.
3. UN. 2008. Industry commodity production data 1950–2005. CD-ROM. UN—United Nations, Statistics Division, New York.
4. IAI. 2010. Source: Historic IAI statistics. IAI—International Aluminium Institute. London, U.K. Source: https://stats.world-aluminium.org/iai/stats_new/index.asp. Accessed: August 3, 2010.
5. Rothermel, J. 2008. Raw material change in the chemical industry—The general picture. *HLG Chemicals—Working Group Feedstock*, Energy & Logistics. Presentation given in Brussels, Belgium for the German Chemical Industry Association (VCI).
6. Simon, C.J. and F. Schnieders, Eds. 2007. *Business Data and Charts 2006*. Prepared by Plastics Europe Market Research Group (PEMRG), PE—Plastics Europe, Brussels, Belgium.
7. Shen, L., J. Haufe, and M.K. Patel. 2009. Product overview and market projection of emerging bio-based plastics. PRO-BIP 2009, Report. Utrecht University, Utrecht, the Netherlands.
8. Bremmer, B.J. and L. Plonsker. 2008. Bio-based lubricants—A market opportunity study update. Unite Soybean Board. Chesterfield, MO. Source: http://www.soynewuses.org/downloads/reports/ BioBasedLubricantsMarketStudy.pdf. Accessed: April 13, 2010.
9. Patel, M.K., M. Crank, V. Dornburg et al. 2006. Medium- and long-term opportunities and risks of the biotechnological production of bulk chemicals from renewable resources—The potential of white biotechnology. The BREW Project. Utrecht University, Utrecht, the Netherlands. Source: http://www.chem. uu.nl/brew/BREWProjectProfile_FINAL.pdf. Accessed: September 15, 2010.
10. Oertel, D. 2007. Industrielle stoffliche Nutzung nachwachsender Rohstoffe. Sachstandsbericht zum Monitoring Nachwachsende Rohstoffe. Arbeitsbericht 114. TAB—Büro für Technikfolgen-Abschätzung beim Deutschen Bundestag. Berlin, Germany. Source: http://www.tab.fzk.de/de/projekt/zusammenfassung/ ab114.pdf. Accessed: April 13, 2010.
11. Ness, B., E. Urbel-Piirsalu, S. Anderberg, and L. Olsson. 2007. Categorising tools for sustainability assessment. *Ecol Econ* 60: 498–508.
12. Wötzel, K., R. Wirth, and M. Flake. 1999. Life cycle studies on hemp fibre reinforced components and ABS for automotive sections. *Angew Makromol Chem* 272: 121–127.
13. Würdinger, E., U. Roth, A. Wegener et al. 2002. Kunststoffe aus nachwachsenden Rohstoffen: Vergleichende Ökobilanz für Loose-fill-Packmittel aus Stärke bzw. Polystyrol. Projektgemeinschaft BIfA/IFEU/Flo-Pak. Final report 2002 (DBU-Az. 04763). Institut für Energie und Umweltforschung Heidelberg, Heidelberg, Germany. Source: http://www.ifeu.de. Accessed: July 29, 2003.
14. Cherubini, F. and G. Jungmeier. 2010. LCA of a biorefinery concept producing bioethanol, bioenergy, and chemicals from switchgras. *Int J LCA* 15: 53–66.

15. Finnveden, G., M.Z. Hauschild, and T. Ekvall. 2009. Recent developments in life cycle assessment. *J Environ Manage* 91: 1–21.
16. ISO. 2006. Environmental management—Life cycle assessment—Principles and framework. ISO 14040. ISO—International Organization for Standardization, Geneva, Switzerland.
17. ISO. 2006. Environmental management—Life cycle assessment—Requirements and guidelines. ISO/FDIS 14044. ISO—International Organization for Standardization, Geneva, Switzerland.
18. EC. 2010. *ILCD Handbook—International Reference Life Cycle Data System. General Guide for Life Cycle Assessment—Detailed Guidance*, 1st edn. EC—European Commission Directorate General—Joint Research Centre, Ispra, Italy. Source: http://lct.jrc.ec.europa.eu/pdf-directory/ILCD-Handbook-General-guide-for-LCA-DETAIL-online-12March2010.pdf. Accessed: August 4, 2010.
19. Bringezu, S., H. Schütz, M. O'Brien, L. Kauppi, R.W. Howarth, and J. McNeely. 2009. Towards sustainable production and consumption of resources: Assessing biofuels. International Panel for Sustainable Resource Management. UNEP—United Nations Environment Programme, Nairobi, Kenya. Source: http://www.unep.fr/scp/rpanel/pdf/Assessing_Biofuels_Full_Report.pdf. Accessed: April 4, 2010.
20. Hermann, B.G., K. Blok, and M.K. Patel. 2010. Twisting bio-based materials around your little finger: Environmental impacts of bio-based wrappings. *Int J LCA* 15: 346–458.
21. Dinkel, F., C. Pohl, M. Ros, and B. Waldeck. 1996. Ökobilanz stärkehaltiger Kunststoffe. Carbotech AG for: Bundesamt für Umwelt, Wald und Landschaft (BUWAL), Bundesamt für Landwirtschaft (BLW) and Fluntera AG. Schriftenreihe Umwelt NR. 271/I-II. Bern, Switzerland.
22. Hoefnagels, R., E. Smeets, and A. Faaij. 2010. Greenhouse gas footprints of different production systems. *Renew Sust Energ Rev* 14: 1661–1694.
23. Patel, M.K., C. Bastioli, L. Marini, and E. Würdinger. 2003. Life-cycle assessment of bio-based polymers and natural fiber composites. In Steinbüchel, A. (Ed.) *Biopolymers*, vol. 10, 1st edn. Wiley-VCH, Weinheim, Germany.
24. Dornburg, V., I. Lewandowski, and M.K. Patel. 2004. Comparing the land requirements, energy savings, and greenhouse gas emissions reduction of bio-based polymers and bioenergy. *J Ind Ecol* 7: 93–116.
25. Kaenzig, J., G. Houillon, M. Rocher et al. 2004. Comparison of the environmental impacts of bio-based products. *Proceedings of the 2nd World Conference and Technology Exhibition on Biomass for Energy, Industry and Climate Protection*, May 10–14, Rome, Italy.
26. Weiss, M. and M.K. Patel. 2007. On the environmental performance of bio-based energy, fuels, and materials: A comparative analysis of life cycle assessment studies. In Graziani, M. and Fornasiero, P. (Eds.) *Renewable Resources and Renewable Energy: A Global Challenge*, 1st edn. Taylor & Francis, CRC Press, London, U.K., 368pp.
27. Carus, M., S. Piotrowski, A. Raschka et al. 2010. Entwicklung von Förderinstrumenten für die stoffliche Nutzung von Nachwachsenden Rohstoffen in Deutschland—Volumen, Struktur, Substitutionspotenziale, Konkurrenzsituation und Besonderheiten der stofflichen Nutzung sowie Entwicklung von Förderinstrumenten. Nova-Institute, Hürth, Germany, Commissioned by the Agency for Renewable Resources (Fachagentur Nachwachsende Rohstoffe e.V). Report-FKZ: 22003908.
28. Diener, J. and U. Siehler. 1999. Ökologischer Vergleich von NMT- und GMT-Bauteilen. *Angew Makromol Chem* 272: 1–4.
29. Wightman, P., R. Eavis, S. Batchelor et al. 1999. Comparison of rapeseed and mineral oils using life-cycle assessment and cost-benefit analysis. *OCL* 6: 384–388.
30. Diehlmann, A., G. Kreisel, D. Bartmann et al. 2000. Strahlenpolymerisierbare lösemittelfreie Schutz- und Dekorationsbeschichtungen für Holz und Holzwerkstoffe auf Basis nachwachsender heimischer Rohstoffe. DREISOL GmbH & Co KG, Oldendorf-Holzhausen in cooperation with: Lott-Lacke Produktions- und Handels-GmbH, Bielefeld, University of Applied Sciences Osnabrück, Carl-von-Ossietzky-University Oldenburg, Oldenburg, Germany.
31. Estermann, R., B. Schwarzwälder, and B. Gysin. 2000. Life cycle assessment of Mater-Bi and EPS loose fills. Study prepared by COMPOSTO for Novamont, Novara, Italy.
32. Flake, M., T. Fleissner, and A. Hansen. 2000. Ökologische Bewertung des Einsatzes nachwachsender Rohstoffe für Verkleidungskomponenten im Automobilbau. University Braunschweig, Braunschweig, Germany.
33. Reinhardt, G.A. and G. Zemanek. 2000. Ökobilanz Bioenergieträger: Basisdaten, Ergebnisse, Bewertungen. Erich Schmidt Verlag, Berlin, Germany.
34. Reinhardt, G.A., J. Calzoni, N. Caspersen et al. 2000. Bioenergy for Europe: Which one fits best?—A comparative analysis for the community. Final Report 2000. Heidelberg, Germany. Source: http://www.ifeu.de. Accessed: July 25, 2003.

35. Reinhardt, G.A., R. Herbener, and S.O. Gärtner. 2001. Life-cycle analysis of lubricants from rape seed oil in comparison to conventional lubricants. Institut für Energie- und Umweltforschung (IFEU), Heidelberg and Berlin, Deutschland. http://www.brdisolutions.com/pdfs/bcota/abstracts/13/z312.pdf. Accessed: May 2, 2010.
36. Reinhardt, G.A., A. Detzel, S.O. Gärtner, N. Rettenmaier, and M. Krüger. 2007. Nachwachsende Rohstoffe für die chemische Industrie: Optionen und Potenziale für die Zukunft. Institut für Energie- und Umweltforschung GmbH (IFEU), Heidelberg, Germany.
37. Corbière-Nicollier, T., B. Gfeller-Laban, L. Lundquist, Y. Leterrier, J.-A.E. Månson, and O. Jolliet. 2001. Life cycle assessment of bio fibres replacing glass fibres as reinforcement in plastics. *Resour Conserv Recy* 33: 267–287.
38. Environment Australia. 2001. A national approach to waste tyres. Prepared by Atech Group for Environment Australia, Canberra, Australia. Source: http://www.environment.gov.au/settlements/publications/waste/tyres/national-approach/pubs/national-approach.pdf. Accessed: May 2, 2010.
39. Pless, P.S. 2001. Technical and environmental assessment of thermal insulation materials from bast fiber crops. University of California at Los Angeles, Los Angeles, USA.
40. Müller-Sämann, K.M., G. Reinhardt, R. Vetter, and S.O. Gärtner. 2002. Nachwachsende Rohstoffe in Baden-Württemberg: Identifizierung vorteilhafter Produktlinien zur stofflichen Nutzung unter Berücksichtigung umweltgerechter Anbauverfahren. Report FZKA-BWPLUS, Heidelberg, Germany. Source: http://www.fachdokumente.lubw.baden-wuerttemberg.de/servlet/is/40139/BWA20002SBer.pdf?command=downloadContent&filename=BWA20002SBer.pdf&FIS=203. Accessed: May 2, 2010.
41. Sharai-Rad, M. and J. Welling. 2002. Environmental and energy balances of wood products and substitutes. Food and Agriculture Organization of the United Nations (FAO), Rome, Italy.
42. Akiyama, M., T. Tsuge, and Y. Doi. 2003. Environmental life cycle comparison of polyhydroxyalkanoates produced from renewable carbon resources by bacterial fermentation. *Polym Degrad Stabil* 80: 183–194.
43. Havinga, J. 2003. How green is natural rubber? In *Natuurrubber* (*Natural Rubber*–Newsletter of the Rubber Foundation 30, 2nd quarter 2003). Information Center for Natural Rubber, Eindhoven, the Netherlands. Source: http://www.rubber-stichting.info/. Accessed: May 2, 2010.
44. Pervaiz, M. and M.M. Sain. 2003. Carbon storage potential in natural fiber composites. *Resour Conserv Recy* 39: 325–340.
45. Reinhardt, G.A. and S.O. Gärtner. 2003. Biodiesel or pure rape-seed oil for transportation: Which one is best for the environment? Study of IFEU—Institut für Energie- und Umweltforschung Heidelberg GmbH. In Bartz, W.J. (Ed.) *Proceedings of the 4th International Colloquium "Fuels 2003,"* Ostfildern (D), January 15–16, 2003, pp. 111–114. Technische Akademie Esslingen, Esslingen, Germany.
46. McManus, M.C., G.P. Hammond, and C.R. Burrows. 2004. Life-cycle assessment of mineral and rapeseed oil in mobile hydraulic systems. *J Ind Ecol* 7: 163–177.
47. Quirin, M., S.O. Gärtner, M. Pehnt, and G. Reinhard. 2004. CO_2-neutrale Wege zukünftiger Mobilität durch Biokraftstoffe: Eine Bestandsaufnahme. Institut für Energie- und Umweltforschung GmbH (IFEU). Commissioned by Forschungsvereinigung Verbrennungskraftmaschinen (FVV), Union zur Förderung von Öl und Proteinpflanzen (UFOP), and Forschungsvereinigung Automobiltechnik (FAT). Heidelberg, Germany. Source: http://www.ufop.de/downloads/Co2_neutrale_Wege.pdf. Accessed: May 2, 2010.
48. Kim, S. and B.E. Dale. 2005. Life cycle assessment study of biopolymers (polyhydroxy-alkanoates) derived from no-tilled corn. *Int J LCA* 10: 200–210.
49. Nebel, B., B. Zimmer, and G. Wegener. 2006. Life cycle assessment of wood floor coverings: A representative study for the German flooring industry. *Int J LCA* 11: 172–182.
50. Turunen, L. and H.M.G. van der Werf. 2006. Life cycle assessment of hemp textile yarn—Comparison of three hemp fiber processing scenarios and a flax scenario. Institut National de la Recherche Acronomique (INRA). Paris, France.
51. Althaus, H.J., F. Dinkel, C. Stettler, and F. Werner. 2007. Life-cycle inventories of renewable materials. Final report ecoinvent data v2.0 No. 21. EMPA, Swiss Centre for Life-Cycle Inventories, Dübendorf, Schweiz.
52. Harding, K.G., J.S. Dennis, H. von Blottnitz, and S.T.L. Harrison. 2007. Environmental analysis of plastic production processes: Comparing petroleum-based polypropylene and polyethylene with biologically-based poly-β-hydroxybutyric acid using life cycle analysis. *J Biotechnol* 130: 57–66.
53. Miller, S.A., A.E. Landis, T.L. Theis, and R.A. Reich. 2007. A comparative life cycle assessment of petroleum and soybean-based lubricants. *Environ Sci Technol* 41: 4143–4149.

54. Reis Neto, O.P. 2007. Braskem, S.A. (Ed.) *Bio-Based Polyolefin Initiative.* Sao Paulo, Brazil.
55. Vink, E.T.H., D.A. Glassner, J.J. Kolstad, R.J. Wooley, and R.P. O'Connor. 2007. The eco-profiles for current and near-future NatureWorks® polylactide (PLA) production. *Ind Biotechnolo* 3: 58–81.
56. Zah, R., R. Hischier, A.L. Leão, and I. Braun. 2007. Curauá fibers in the automobile industry—A sustainability assessment. *J Clean Prod* 15: 1032–1040.
57. Albrecht, S., S. Rüter, J. Welling et al. 2008. ÖkoPot—Ökologische Potenziale durch Holznutzung gezielt fördern. Final report BMBF-Project, FKZ 0330545, Stuttgart, Deutschland. Source: http://www.holzundklima.de/doen/albrecht_etal_2008_oekopot_bericht.pdf. Accessed: May 2, 2010.
58. Egger. 2008. Umweltproduktdeklaration von Laminatböden nach ISO 14025: Egger Laminatboden. Egger Floor Products GmbH. Wismar, Germany. Source: http://www.egger.com/pdf/ZF_EPD_Laminatfussboden_DE.pdf. Accessed: May 2, 2010.
59. KTBL—Kuratorium für Technik und Bauwesen in der Landwirtschaft E.V. (Eds.). 2008. *Ökologische und ökonomische Bewertung nachwachsender Energieträger. KTBL-Vortragstagung vom 8. bis 9. September 2008 in Aschaffenburg.* Darmstadt, Germany.
60. Rettenmaier, N., G.A. Reinhardt, S.O. Gärtner, and J. Münch. 2008. Bioenergie aus Getreide und Zuckerrübe: Energie- und Treibhausgasbilanzen. Institut für Energie- und Umweltforschung GmbH (IFEU). Heidelberg and Berlin, Germany. Source: http://www.ifeu.org/landwirtschaft/pdf/IFEU%20-%20Bioenergie%20LAB%20-%20Kurzversion.pdf. Accessed: May 2, 2010.
61. Vidal, R., P. Martínez, and D. Garraín. 2008. Life cycle assessment of composite materials made of recycled thermoplastics combined with rice husks and cotton linters. *Int J LCA* 14: 1–10.
62. Macedo, I.C., J.E.A. Seabra, and J.E.A.R. Silva. 2008. Green house gas emissions in the production and use of ethanol from sugarcane in Brazil: The 2005/2006 averages and a prediction for 2020. *Biomass Bioenerg* 32: 582–595.
63. Shen, L. and M.K. Patel. 2008. Life cycle assessment of man-made cellulose fibres. Utrecht University. Utrecht, the Netherlands.
64. Shen, L. and M.K. Patel. 2008. Life cycle assessment of polysaccharide materials: A review. *J Polym Environ* 16: 154–167.
65. Tufvesson, L. and P. Börjesson. 2008. Wax production from renewable feedstock using biocatalysts instead of fossil feedstock and conventional methods. *Int J LCA* 13: 328–338.
66. WBGU. 2008. Welt im Wandel: Zukunftsfähige Bioenergie und nachhaltige Landnutzung. Wissenschaftlicher Beirat der Bundesregierung für Globale Umweltveränderungen (WBGU), Berlin, Germany.
67. Schmitz, N., J. Henke, and G. Klepper. 2009. Biokraftstoffe—Eine vergleichende Analyse. Agency for Renewable Resources (Fachagentur Nachwachsende Rohstoffe e.V.), Gülzow, Germany. Source: http://www.ee-direkt.de/pdf/Biokraftstoffvergleich_FNR_2006.pdf. Accessed: May 2, 2005.
68. Ren T. and M.K. Patel. 2009. Basic petrochemicals from natural gas, coal and biomass: Energy use and CO_2 emissions. *Resour Conserv Recy* 53: 513–528.
69. Dinkel, F. and B. Waldeck. 1999. Ökologische Beurteilung verschiedener Geschirrtypen mit Empfehlungen. Arbeitspapier 4/99. Carbotech AG. Basel, Switzerland. Source: www.kompost.ch. Accessed: Octobre 27, 2003.
70. Gärtner, S.O., K. Müller-Sämann, G.A. Reinhardt, and R. Vetter. 2002. Corn to plastics: A comprehensive environmental assessment. In Pala, W. et al. (Eds.) *Proceedings of the 12th European Conference on Biomass for Energy, Industry and Climate Protection*, June 17–22, 2002, Vol. II, pp. 1324–1326, Amsterdam, the Netherlands.
71. UNEP. 2000. The Montreal protocol on substances that deplete the ozone layer. UNEP—United Nations Environmental Programme, Nairobi, Kenya. Source: http://www.unep.org/ozone/pdfs/montreal-protocol2000.pdf. Accessed: September 3, 2010.
72. Ravishankara, A.R., J.S. Daniel, and R.W. Portmann. 2009. Nitrous oxide (N_2O): The dominant ozone-depleting substance emitted in the 21st century. *Science* 326: 123–125.
73. Larson, E.D. 2006. A review of life-cycle analysis studies on liquid biofuel systems for the transport sector. *Energy for Sustainable Development* 10: 109–126.
74. von Blottnitz, H. and M.A. Curran. 2007. A review of assessments conducted on bio-ethanol as a transportation fuel from a net energy, greenhouse gas, and environmental life cycle perspective. *J Clean Prod* 15: 607.
75. Wicke, B., V. Dornburg, M. Junginger, and A. Faaij. 2008. Different palm oil production systems for energy purposes and their greenhouse gas implications. *Biomass Bioenerg* 32: 1322–1337.
76. Searchinger, T., R. Heimlich, R.A. Houghton et al. 2010. Use of U.S. cropland for biofuels increases greenhouse gases through emissions from land-use change. *Science* 319: 1238–1240.
77. Lapola, D.M., R. Schaldach, J. Alcamo et al. 2010. Indirect land-use changes can overcome carbon savings from biofuels in Brazil. *PNAS* 107: 3388–3393.

78. Karpenstein-Machan, M. 2001. Sustainable cultivation concepts for domestic energy production from biomass. *Cr Rev Plant Sci* 20: 1–14.
79. Blanco-Canqui, H. 2010. Energy crops and their implications on soil and environment. *Agron J* 102: 403–419.
80. Smeets, E.M.W., L.F. Bouwman, E. Stehfest, D.P. van Vuuren, and A. Posthuma. 2009. Contribution of N_2O to the greenhouse gas balance of first-generation biofuels. *Glob Change Biol* 15: 1–23.
81. Guinée, J.B., R. Heijungs, and E. van der Voet. 2009. A greenhouse gas indicator for bioenergy: Some theoretical issues with practical implications. *Int J LCA* 14: 328–339.
82. Weiss, M., J. Haufe, M. Carus, M. Brandão, S. Bringezu, and M.K. Patel. 2011. The environmental impacts of bio-based materials. *Journal of Industrial Ecology*. Accepted for publication.
83. Hermann, B.G., L. Debeer, B. de Wilde, K. Blok, and M.K. Patel. 2010b. To compost or not to compost: Carbon and energy footprints LCA of biodegradable materials' waste treatment. *Polym. Degrad. Stabil.* 96:1159–1171.
84. Edelmann, W. and K. Schleiss. 2001. Ökologischer, energetischer und ökonomischer Vergleich von Vergärung, Kompostierung und Verbrennung fester biogener Abfallstoffe. Bundesamt für Energie (BFE) und Bundesamt für Umwelt, Wald und Landschaft (BUWAL), Baar, Switzerland.
85. Amlinger, F., S. Peyr, and C. Cuhls. 2008. Greenhouse gas emissions from composting and mechanical biological treatment. *Waste Manag Res* 26: 47–60.
86. Weiss, M., M.K. Patel, H. Heilmeier, and S. Bringezu. 2007. Applying distance-to-target weighing methodology to evaluate the environmental performance of bio-based energy, fuels, and materials. *Resour Conserv Recy* 50: 260–281.
87. Herrero, M., P.K. Thornton, A.M. Notenbaert et al. 2010. Smart investments in sustainable food production: Revisiting mixed crop-livestock systems. *Science* 327: 822–825.
88. Hermann, B.G., K. Blok, and M.K. Patel. 2007. Producing bio-based bulk chemicals using industrial biotechnology saves energy and combats climate change. *Environ Sci Technol* 41: 7915–7921.
89. IEA. 2009. *Energy Balances of Non-OECD Countries*, 2009 edn. IEA—International Energy Agency. Paris, France.

Part III

Technologies for Renewable Energy

9 Biomass Gasification for Second-Generation Fuel Production

Francesco Basile and Ferruccio Trifirò

CONTENTS

9.1 INTRODUCTION

Nowadays, fossil fuels such as coal, oil, and natural gas are used for generating 80% of the energy proportion in the world. The use of the fossil resources in the energy production processes gives rise to an amount of carbon dioxide emission amount equal to the carbon transformed in the processes. So there is keen interest in reducing the CO_2 emissions either modifying the efficiency of the processes using fossil fuels or using renewable resources, such as biofuels for which the carbon dioxide generated in the combustion is not considered to give any net contribution to the CO_2 emissions in the atmosphere, since the carbon of the biomasses is originated from the CO_2 transformed by photosynthesis [1,2]. The use of biofuels for power generation is limited even if, recently, several plants have been built based on conventional technology, that is, a boiler plant and a steam turbine cycle. A number of large plants have been built especially in the northern countries where the heat can be efficiently used for heating system acting for large part of the year. On the other hand, smaller combined heat and powers (CHP) plants have been built in recent years at any latitude based on the availability of biomasses [3]. The electrical efficiency of these plants is around 30% [4], and the ratio of electrical energy to thermal energy generated (called alpha value) is around 0.5 or below. Although there is potential development for these plants, the electrical efficiency and the alpha value cannot be expected to increase to any significant extent. A technique that offers opportunities for achieving higher electrical efficiencies is based on the gasification of solid fuels and combustion of gas: (1) in a CHP engine such as the Güssing plant or (2) in a gas turbine combined with a steam cycle such as the Värnamo Plant. These integrated gasification combined cycle (IGCC) plants were originally developed for fossil fuels, but the principle can also be applied to biofuels and requires pressurized gasification technology. Several studies [5] have shown that well-optimized generation plants rated at 30–60 MW$_e$, based on pressurized gasification of wood fuel and integrated into a combined cycle, can achieve net electrical efficiencies of 40%–50% and an overall efficiency of 85%–90% with competitive generation costs and low emission levels. An interesting alternative to the production of power from biomass gasification is the gas upgrading to obtain a mixture of carbon monoxide (CO) and H$_2$, having a high added value. The syngas production by gasification is an interesting way to transform biomasses and maintain the chemical energy in a single phase and, specifically, the gas phase, where it can be easily and efficiently transformed into fuel maintaining the most of the chemical energy (Figure 9.1) [6].

Different from the combustion where the energy is mainly present as sensitive heat and different from the pyrolysis where the chemical energy is distributed in three phases produced during the processes: a gas phase (H$_2$, CO, CH$_4$, CO$_2$, and C$_2$, in order of concentration), a condensable phase (pyro-oil), and a solid phase (char) that at the present state is not suitable for fuel production.

The interest in increasing the fuel share from renewable can be clear from few considerations: the use of fossil raw materials for the production of fuel is responsible of more than 20% of the CO$_2$ emitted in the world, and its share is increasing and is forecast to be close to 30% in 2020. While ready alternatives with increasing share are available on the power energy market world-wide and integrated network are planned to be installed among different countries, the alternatives in the fuel sector are not yet well defined, and the use of first-generation fuels has collected increasing criticism from an environmental and social perspective. The possible definition of a trademark as guaranty of its social and environmental friendly production is only a partial solution. Second-generation biofuels are still not in the commercialization phase; nevertheless, different processes are in a development stage and will be soon ready for market. Most of them are

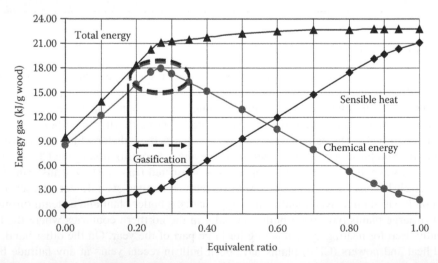

FIGURE 9.1 Distribution between sensitive energy (heat) and chemical energy in the gas stream derived from biomasses after thermochemical conversion.

Gas to fuel/chemicals processing routes

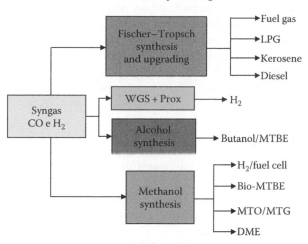

FIGURE 9.2 Product derived from synthesis gas to be used as fuel in the energy transportation sector.

based on gasification processes and have the synthesis gas as intermediate for the production of hydrogen, methanol dimethilether, Fisher–Tropsch diesel, etc.

The production of synthesis gas is well known from fossil fuels, such as natural gas (via steam reforming/partial oxidation) and coal (via gasification). However, these processes lead to net CO_2 emissions into the environment corresponding roughly to the carbon used in the processes. Each process developed with specific raw material has faced the challenges to obtain an efficient FT or MeOH production process. Biomasses are in that perspective very peculiar and requires significant change in the process development and in its the approach. Nevertheless they can generate the same intermediate than fossil fuel and therefore share the same distribution network and products pool. A scheme of the different possibilities derived from syngas utilization after biomass gasification is shown in Figure 9.2.

The fuel production form syngas can be used as bridge strategy through a hydrogen-based fuel system as a clean fuel that can be used in fuel cells and internal combustion engines [7]. The shift from synfuel to hydrogen production can be easily obtained in terms of plant modification while more challenging are the long-term changes in the distribution systems. Nowadays, the hydrogen is an important raw material for the chemical industry and refinery, with increasing interest due to the progressive reduction of quality of oil. The main process for hydrogen production is currently the catalytic steam reforming of methane, light hydrocarbons, and naphtha. Partial oxidation of heavy oil residues and coal gasification are also alternative processes to produce hydrogen [8]. Renewable lignocellulosic biomass can be used as an alternative feedstock for hydrogen production. Two possible technologies that have been explored in recent years are steam gasification [9–13] and catalytic steam reforming of pyrolysis oils [14,15]. The latter route begins with fast pyrolysis of biomass to produce bio-oil, which can be converted to hydrogen via catalytic steam reforming followed by, if necessary, a shift-conversion step. Moreover, the economy of scale can make possible a more efficient but more complex utilization of the pyrolysis slurry. Since the costs per kilogram of transporting low-density biomass (such as straw, wood chips, and bagasse) are much higher than those of transporting high-density liquor, the slurries can be transported from a wide number of small and local pyrolysis plants to a large central gasification facility to produce the most valuable products [16]. The demand for hydrogen as a fuel derived from renewable sources will be increasing due to major advances in the field of hydrogen-based fuel cell research [17]. While the technology for efficiently converting hydrogen to electricity at the required power levels is approaching commercialization, attempts have been made to directly power fuel cells with biogas (internal reforming), but these were mostly unsuccessful [18,19]. One major drawback in the reforming process is the formation of CO as a gaseous by-product that can poison the fuel cell by disproportionation and carbon deposition. Moreover, the variability of the biogas composition as well as the presence of trace quantities of sulfur is difficult to be managed and together with the high cost of the fuel cell has delayed its application. Therefore, maximization of the hydrogen yield from biogas through controlled steam reforming followed by CO shift reactions (high temperature HT and low temperature LT) appears to be a less elegant but more economical and feasible solution, particularly with respect to CO minimization.

9.2 BIOMASS FEEDSTOCK

Biomass is mainly composed of the following constituents [20]:

1. Cellulose, a linear polymer of anhydroglucose C6 units with a degree of polymerization of up to 10, 000, and the long strains form fibers that give biomass its mechanical strength but also a strongly anisotropic character
2. Hemi-cellulose, a linear polymer of C6 and C5 units with a degree of polymerization of less than 200
3. Lignin, which is a random three-dimensional structure of phenolic compounds
4. Minor constituent such as oil, protein, and terpenes usually obtained by extraction techniques

Traditionally, biomass (mainly in the form of wood) has been utilized by humans through direct combustion, and this process is still widely used in many parts of the world. Biomass is a dispersed, labor intensive, and land intensive source of energy. Therefore, as industrial activity has increased in countries, more concentrated and convenient sources of energy have been substituted for biomass. Nowadays, biomass represents only 3% of primary energy consumption in industrialized countries [21]. However, much of the rural population in developing countries, which represents about 50% of the world's population, is reliant on biomass for fuel. Biomass accounts for 35% of primary energy consumption in developing countries, raising the world total to 14% of primary energy consumption. The earth's natural biomass replacement represents an energy supply of around 3000 EJ $(3 \times 10^{21}$ J) a year, of which just under 2% is currently used. However, it is not possible to use all of the annual production of biomass in a sustainable manner. The main potential benefit in growing biomass especially for fuel is that, provided the right crops are chosen, it is possible to use poor quality land, which is unsuitable for growing food. Burning biomass produces some pollutants, including dust and the acid gases such as sulfur dioxide (SO_2) and nitrogen oxides (NO_x), the amount of SO_x is related to the content present in the raw material, and, therefore, the biomass combustion produces less SO_x than burning coal. Carbon dioxide, the greenhouse gas, is also released. However, as this originates from harvested or processed plants, which have absorbed it from the atmosphere in the first place, no additional amounts are involved. A wide variety of biomass resources can be used as feedstock (Figure 9.3, [2]), divided into three general categories [22]:

1. *Wastes*: Large quantities of agricultural plant residues are produced annually worldwide and are vastly under utilized. The most common agricultural residue is the rice husk, which makes up 25% of rice by mass. Other plant residues include sugar cane fiber (bagasse), coconut husks and shells, groundnut shell, and straw. Included in agricultural residue is waste, such as animal manure (e.g., from cattle, chicken, and pigs). Residues are also widely produced in the agroindustry sectors. Refuse derived fuel (RDF) is the combustible material in domestic or industrial refuse. It consists mainly of biomass-derived material but may also include some plastics. RDF may be used raw and unprocessed, partially processed or highly processed in the form of pellets. These burn more efficiently with lower emissions.

FIGURE 9.3 Biomass types from energy crops and forest products for thermochemical conversion.

2. *Standing forests*: Wood fuels are fuels derived from natural forests, natural woodlands, and forestry plantations (fuelwood and charcoal). These fuels include sawdust and other residues from forestry and wood processing activities. Fuelwood is the principal source for small-scale industrial energy in rural areas of developing countries. However, large reforestation programs will be required to meet future energy demands as the world population grows. In industrialized countries, the predominant wood fuels used in the industrial sector are from wood processing industries. The utilization of this residue for energy production at or near its source has the advantage of avoiding expensive transport costs. Domestic wood fuels are sources principally from land clearing and logging residues.

3. *Energy crops*: Energy crops are grown specially for the purpose of producing energy. These include short rotation plantations (or energy plantations), such as eucalyptus, willows, and poplars, algae, herbaceous crops, such as sorghum, sugarcane, and artichokes, and vegetable oil–bearing plants, such as soya beans, sunflowers, cotton, and rapeseed. Plant oils are important, as they have a high energy density. The oil extraction technology and the agricultural techniques are simple, and the crops are very hardy.

9.3 PRETREATMENT OF THE FEEDSTOCK

The characteristics of feedstocks as they are collected are often very different from the feed characteristics demanded by the conversion reactor, and several steps are usually required to match the feedstock characteristics with the process conditions. The key requirements of the feed pretreatment system are [23] as follows:

1. The reception and storage of incoming biomass until it is required by the conversion step. The logistics of ensuring a constant feed supply is very important, because most of the biomass is only available on a seasonal basis. In such cases, continuous operation of the conversion facility will require either extensive long-term storage of the feedstock or a feed reactor and pretreatment system that is flexible enough to accommodate multiple feedstocks.

2. The screening of the feedstock to keep particle sizes within appropriate limits and prevent contamination of the feedstock by metal or rocks.

3. The drying of the feedstock to a moisture content suitable for the conversion technology. Drying is generally the most important pretreatment operation, necessary for high cold gas

efficiency at gasification. Drying reduces the moisture content to 10%–15%. Drying can either be done with flue gas or with steam. Steam drying results in very low emissions, and it is safer with respect to risks for dust explosion. However, using flue gas is the cheapest way to dry the feedstock.

4. The comminution of the feedstock to an appropriate particle size. To avoid pressure drop, sawdust and other small particles must be pelletized. Smaller particle can be used in fluidized bed gasifiers.
5. The buffer storage of prepared feed immediately prior to the reactor.

The pyrolysis of biomass to produce the slurry to be pumped into a gasifier to produce hydrogen or syngas instead of burning can be classified as a pretreatment step. Pyrolysis of biomass can be described as the direct thermal decomposition of the organic matrix in the absence of oxygen to obtain an array of solid (char), liquid (oil), and gas products, depending on the pyrolysis conditions. The solid char can be used as a fuel in the form of briquettes or as a char–oil/water slurry or it can be upgraded to activated carbon and used in purification processes. The gases generated have a low-to-medium heating value, but may contain sufficient energy to supply the energy requirements of a pyrolysis plant. The pyrolysis liquid is a homogenous mixture of organic compounds and water in a single phase, and it is commonly burned in a diesel stationary engine, but extraction can be carried out to obtain chemicals and other valuable products (food additives and perfumes). Pyrolysis processes can be classified in slow, fast, and flash depending on the applied residence time and heating rate. Slow pyrolysis has traditionally been used for the production of charcoal. Fast and flash pyrolysis of biomass at moderate temperatures has generally been used to obtain high yield of liquid products (up to 75% wt on a dry biomass feed basis), and it is characterized by high heating rates and rapid quenching of the liquid products to terminate the secondary conversion of the products [24]. High heating rates of up to 10^4 K s^{-1}, at temperatures <650°C and with rapid quenching, favor the formation of liquid products and minimize char and gas formation; these process conditions are often referred to as "flash pyrolysis." High heating rates to temperatures >650°C tend to favor the formation of gaseous products at the expense of liquids. Slow heating rates coupled with low maximum temperatures maximize the yield of char. In slow pyrolysis, the reactions taking place are always in equilibrium, because the heating period is sufficiently slow to allow equilibration during the thermochemical process. In this case, the ultimate yield and product distribution are limited by the heating rate. In fast and flash pyrolysis, there are a negligible number of reactions during the heat-up period. Volatile residence time is a very important factor to affect yields of gaseous and liquid products in a biomass sample [25]. A low-volatile residence time, obtained by rapid quenching, increases the liquid fraction. Also, the particle size is known to influence pyrolysis yield. This effect may be related to heating rate, in that larger particles will heat up more slowly, so the average particle temperatures will be lower, and hence volatile yields may be expected to be less. If the particle size is sufficiently small, it will be heated uniformly. Finally, the use of a sweeping gas allows a faster removal of the compounds from the hot zone, minimizing the unwished secondary reactions of cracking and polycondensation, thus achieving higher yields in the oil.

9.4　GASIFICATION

Gasification is the conversion by partial oxidation at elevated temperature of a carbonaceous feedstock into a gaseous energy carrier consisting of permanent, noncondensable gases. Development of gasification technology dates back to the end of the eighteenth century when hot gases from coal and coke furnaces were used in boiler and lighting applications [26]. Gasification of coal is now well established, and biomass gasification benefited from that sector [27]. However, the two technologies are not directly comparable due to differences between the feedstocks (e.g., char reactivity, proximate composition, ash composition, moisture content, and density). Gasifiers have been designed in various configurations, but only the fluidized bed configurations are being considered

TABLE 9.1

Gasifier Product Gas Characteristics

| Gasifier Configuration | Gas Composition (%v/v) | | | | | HHV |
	H_2	CO	CO_2	CH_4	N_2	(MJ/N m³)
Fluid bed (air-blown)	9	14	20	7	50	5.4
Updraft (air-blown)	11	24	9	3	53	5.5
Downdraft (air-blown)	17	21	13	1	48	5.7
Downdraft (O_2-blow)	32	48	15	2	3	10.4
Multi-solid fluid bed	15	47	15	23	0	16.1
Twin fluid bed	31	48	0	24	0	17.4

Source: Bridgewater, A.V. et al., *Renew. Sust. Energy Rev.*, 6, 181, 2002.

in applications that generate over 1 MW$_e$ [4,28]. Fluid bed gasifiers are available from a number of manufacturers in thermal capacities ranging from 2.5 to 150 MWth for operations at atmospheric or elevated pressures, using air or oxygen as gasifying agent. Ideally, the process produces only a non-condensable gas and an ash residue. However, incomplete gasification of char and the pyrolysis tars produce a gas containing several contaminants such as particulate, tars, alkali metals, fuel-bound nitrogen compounds, and an ash residue containing some char. The composition of the gas and the level of contamination vary with the feedstock, reactor type, and operating parameters (Table 9.1).

Since the mid-1980s, interest has grown on the subject of catalysis for biomass gasification. The advances in this area have been driven by the need to produce a tar-free product gas from the gasification of biomass, since the removal of tars and the reduction of the methane content increase the economic viability of the biomass gasification process.

9.4.1 GASIFICATION REACTORS AND PROCESSES

Three different strategies can be used to carry on the gasification processes [28–30]:

1. The entrained flow reactor on a slurry formed by the solid and vapor produced by pyrolysis (the Karlsruhe concept). While the gas can be used as source of heat in the pyrolysis (Figure 9.4).
2. A three-stage gasification with chemical quenching in which the gas produced by low-temperature gasification (400°C–500°C) is oxidized at 1300°C–1400°C and is quenched with bio-coke produced during the first stage of gasification in a third reactor working as endothermic gasification (Choren) producing a gas with high LHV and a vitrified slag.
3. The fluidized bed gasifier direct on biomasses.

The gasification process 1 and 2 coupled with a chemical pretreatment (pyrolysis or low-temperature gasification) are based on the idea of increasing the LHV of the fraction entering the main gasifier; the high LHV allows high-temperature processes, and a low tar and light hydrocarbon at the exit of the gasifier are produced. Similar approach is also at the bases of the gasification processes with a torrefaction pretreatment that have been proposed in several projects [31,32]. The main problems with these technologies are the number of reactor required for the biomass transformation. In the Karlsruhe concept, the pyrolysis and gasification are carried out in two different places; the pyrolysis step is delocalized in small plant of 50–100 MWth and then concentrated by transporting the slurry in a centralized entrained flow gasification of at least 2000–5000 MWth at 50–60 bar and 1200°C–1400°C with a very high conversion (99%) and low methane slip 0.1%. The pyrolysis is aimed to densify the biomasses and increase the useful radio for the biomass crop avoiding long

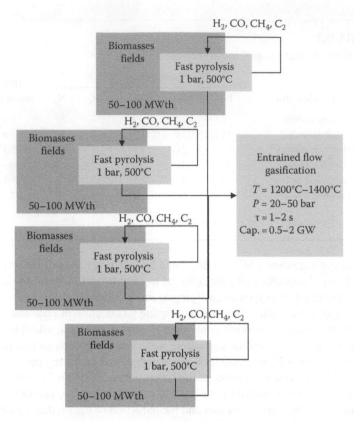

FIGURE 9.4   Scheme of the Karlsruhe concept of biomass delocalized pyrolysis and centralized gasification.

distance transportation, which is not convenient with low-density biomasses (such as the herbaceous one) above 30–50 km and close to 200 km thanks to the high density of the slurry (Figure 9.5) [33].

The increase of the area is also important to reduce the social and environmental impact in a region by increasing the degree of freedom for the farmers in production and selling different products, which also means decreasing the percentage of biomass collection in a defined area leaving biomass available for soil fertilization. The main problem of this approach is the production of a stable and transportable slurry produced by the pyrolysis and the costs of building several delocalized plants (investment and personnel), which maybe compensated by the larger gasifier plant running at elevated pressure (50–60 bar), which may use other hydrocarbon sources as co-gasification feedstock.

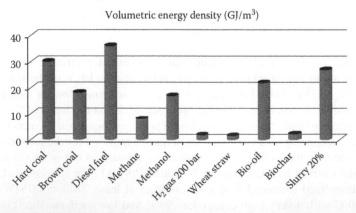

FIGURE 9.5   Volumetric energy density for fossil and renewable fuels.

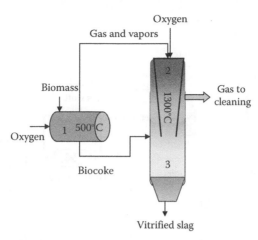

FIGURE 9.6 Choren scheme of gasification: (1) low-temperature gasification; (2) high-temperature exothermic partial oxidation; (3) endothermic gasification.

In the case of the Choren, the development of the Carbon-V process is based on a low-temperature and high-temperature gasification followed by an endothermic entrained flow gasification carried out in the same site (Figure 9.6) [34]. In this case, the first stage will be carried out for a size of 45 MWth, which will be proven as full scale in the next plant in Freiberg (beta plant) while the alfa plant is a 1 MWth plant. The beta plant reveals the needs to prove at almost full scale the low-temperature gasification and indicate the importance to define the operability of this part of the plant, characterized by mechanical stirring and not easy to scale up for the presence of solid phase in physical and chemical transformation. The beta plant is claimed to produce an 18 ML of biomass-to-liquid (BTL) equal to an annual consumption of 15,000 cars. The second part of the gasification is a 200–400 MW reactor and therefore will require from 5 to 10 low-temperature stages to be used in a continuous processes. The process runs at moderate pressure 5 bar and is claimed to produce a gas stream with a methane content of 0.5%, tar under detection limit, and a vitrified slag.

Therefore, with respect to the previous processes, the process loses the advantage of large area for biomasses provisioning and the advantage of large gasification plant even if is gaining in terms of reliability for the ready and in situ use of the low-temperature-gasification products.

The fluidized bed biomass gasification will be described in detail. The fluidized reactor is the winning technology in the combustion and gasification for heat and power production. The differences in fuel production are as follows: (1) the need, in the gasification for fuel production, to reduce the methane, light hydrocarbon, and the tar content; and (2) the need to use oxygen instead of air, and, as a consequence, the need to have large plant size to balance the investment cost of the oxygen purification plant.

An analysis of the last type of gasification processes requires taking into account the main problems related to the LHV of the biomasses with respect to other resources such as fossil fuels (coke and heavy vacuum distillation residues) due to the presence of oxygen that leads to reaching a maximum of LHV in the gas phase at low lambda value, that is, at an autothermal temperature between 850°C and 950°C in oxygen/steam gasification [35,36]. The relative low temperature of the transformation leads to several consequences in the degree of hydrocarbon transformation into syngas. Even if the water gas shift reaction can be considered close to the equilibrium at the operative temperature, other reactions such as hydrocarbon cracking and reforming are far from the equilibrium and significant amount of methane (5%–7%), C2/C3, or BTX are still present in the gas phase. The amount of high molecular weight hydrocarbons and tars present in the gas phase, along with the C1-C3 and BTX, are rich in energy and required to be transformed in syngas to have an high efficiency in terms of chemical energy (as LHV products/LHV of reactants).

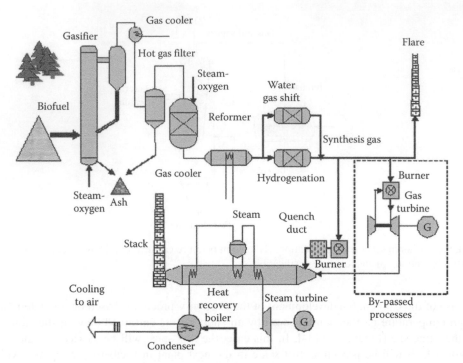

FIGURE 9.7   Värnamo plant rebuilt scheme from CHRISGAS Project (IP EU, VI FP).

In a fluidized biomass gasification such as the one represented by the Värnamo plant (Figure 9.7) [37], the overall tars and BTX products account for more than 13% of the gasification product heating value. The second constrain deals with the available biomasses as discussed previously and is less feasible for low-energy-dense biomasses as in the case of the Choren process. On the other hand, a fluidized bed biomass reactor with a 200–400 MW thermal seems not to be a problem especially using a circulating or bubbling fluidized reactor operating at elevated pressure as proven at lower scale in the Varnamo plant and hypothesized in the Chrisgas project (30 bar 200–400 MWth).

The possibility to reduce the size of the plant is therefore connected with the utilization of an oxygen stream and the purification stage to obtain it. A possible solution is the use of allothermal fluidized bed process such as that carried out in a double-fluidized bed reactor (Güssing plant) with the solid circulating between the two reactors (Figure 9.8). In this configuration, the first

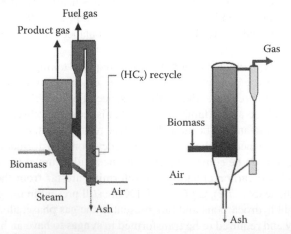

FIGURE 9.8   Circulating fluidized bed gasifier and dual bed gasifier (Güssing type).

fluidized reactor uses air to complete oxidizing the feed while the second reactor is used in an endothermic configuration using steam to gasify the biomasses. The inlet in the first reactor is constituted by the char and carbon present on the solid that can be burned at a temperature of 1200°C, the hot solid heated in the reactor, and circulating in the endothermic riser is responsible of the heat required by the endothermic gasification process. The exit temperature is below 900°C, and the unconverted hydrocarbon in the gasification step is higher than the one obtained in the gasification with air (methane concentration between 8% and 11% instead of 5% and 7%). The aim of Gussing plant is a CHP cogeneration system, and the energy of the light hydrocarbons is used to produce heat and power while tars are absorbed using the Olga process. In the case of syngas production for automotive fuel, the yield in syngas needs to be maximized to have high atom and chemical energy efficiency. The main advantage of the process is the absence of air purification, and, as a consequence, the size of the plant that is not limited from scale factor due to the oxygen production. The main problems are related to two main constrains: a lack of experience in high pressure system, which may limit the upper scale of the plant, and the yield in syngas, which requires to be improved.

The use of the double-fluidized bed gasifier is enlarging the possible use of catalyst inside the gasifier, since the solid is going in cyclic reductive and oxidative cycle, which can be seen as conversion and regeneration cycle in analogy with the dehydrogenation of C4 and ethylbenzene developed by Snamprogetti and based on the same double-fluidized reactor concept in which the endothermic dehydrogenation is carried out using the heat transported by the catalyst and produced in the oxidative reactor, burning the coke placed on the catalyst surface. The presence of a catalyst enhancing the reforming reaction also requires a change in the fluidized parameters, since the presence of the catalyst promoting cracking and reforming may change the carbon on the recirculating solid and therefore the reaction parameters of the dual fluidized bed reactor. After this consideration, it is possible to draw a preliminary conclusion from which it seems that the winning technology is still not established and demonstration activity is required to better identify advantage and drawbacks of each technology and the improvements of the pretreatment, gasification, and gas cleaning steps. Furthermore, we can also state that the winning technology is site and biomasses dependent, especially the biomass availability in quantity, distribution, and quality seems to be the driving force for a technology choice [38,39]. Recently, even wet biomasses have been efficiently treated by gasification technology [40].

A general trend to be noted is an increasing use of catalysts for gas cleaning and upgrading even using entrained flow gasifier at high temperature [41].

Furthermore, inside the same process technology, different strategy can be used to address the differences in the biomass quality.

9.5   GAS CLEAN UP

Tar formation is one of the major problems to deal with during biomass gasification [42,43]. Tar condenses at reduced temperature, thus blocking and fouling process equipment such as engines and turbines. Tar removal technologies can be divided into gas cleaning after the gasifier (secondary methods) and treatments inside the gasifier (primary methods). Although secondary methods are proven to be effective, treatments inside the gasifier are gaining much attention as these may reduce the downstream cleanup. In primary treatment, the gasifier is optimized to produce a fuel gas with minimum tar concentration. The different approaches of primary treatment are (a) proper selection of operating parameters, (b) use of bed additive/catalyst, and (c) gasifier modifications. The operating parameters such as temperature, gasifying agent, equivalence ratio, and residence time play an important role in formation and decomposition of tar (Figure 9.9).

The main strategies to increase the tar conversion are based on the uses of a solid as catalyst for the reforming and cracking reactions. Potential bed additives can be used as active catalyst inside the gasifier for tar cracking. The active sites can be the ashes and char produced in the gasification

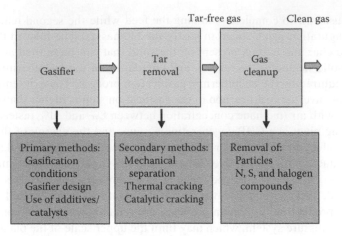

FIGURE 9.9 Tar removal approach in biomass gasification.

processes, the "inert" material used as gasifier agent such as dolomite and olivine. If the catalyst is used as gasifier agent, the criteria for its selection may be summarized as follows:

1. The catalysts must be effective in the removal of tars.
2. If the desired product is syngas, the catalysts must be capable of reforming methane, also providing a suitable syngas ratio for the intended process.
3. The catalysts should be resistant to deactivation as a result of carbon fouling and sintering.
4. The catalysts should be easily regenerated.
5. The catalysts should be resistant to abrasion and attrition.
6. The catalysts should be cheap.

This first group of catalyst (primary catalysts), directly added to the biomass prior to gasification, catalyzes the reactions listed in Table 9.2. The addition is performed either by wet impregnation of the biomass material or by dry mixing of the catalyst with it. These catalysts have a specific purpose of reducing the tar content and have little effect on the conversion of methane and C_{2-3} hydrocarbons in the product gas. They operate under the same conditions of the gasifier and usually consist of cheap disposable material. Dolomite, an ore with the general formula $MgCO_3 \cdot CaCO_3$, is a suitable

TABLE 9.2

Main Chemical Reactions of Biomass Gasification

Main Reactions Occurring in a Gasifier	ΔH^0_{298} (kJ/mol)
$C_nH_mO_y \leftrightharpoons yH_2O + C_nH_{m-2y}$	<0
$C_nH_m \leftrightharpoons C_nH_y + (m - y)/2H_2$	>0
$C_nH_m \to (m/4)CH_4 + (n - m/4)C$	<0
$C + \frac{1}{2}O_2 \to CO$	−111
$CO + \frac{1}{2}O_2 \to CO_2$	−254
$H_2 + \frac{1}{2}O_2 \leftrightharpoons H_2O$	−242
$C + H_2O \leftrightharpoons CO + H_2$	+131
$C + CO_2 \leftrightharpoons 2CO$	+172
$C + 2H_2 \leftrightharpoons CH_4$	−75
$CO + 3H_2 \leftrightharpoons CH_4 + H_2O$	−206
$CO + H_2O \leftrightharpoons CO_2 + H_2$	−41
$CO_2 + 4H_2 \leftrightharpoons CH_4 + 2H_2O$	−165

catalyst for the removal of hydrocarbons evolved in the gasification of biomass. Dolomites increase gas yields at the expense of liquid products. With suitable ratios of biomass feed to oxidant, almost 100% elimination of tars can be achieved. The dolomite catalyst deactivates due to carbon deposition and attrition, but it is cheap and easily replaced. The catalyst is most active if calcined and placed downstream of the gasifier in a fluidized bed at temperatures above 800°C. The reforming reaction of tars over dolomite occurs at a higher rate with carbon dioxide instead of steam. Dolomite activity can be directly related to the pore size and distribution. A higher activity is also observed when iron oxide is present in significant amounts. Moreover, dolomite is basic and does not react with alkali from the fuel. However, this material is not active for reforming the methane present in the product gas, and hence, they are not suitable catalysts if syngas is required. Therefore, the main function of dolomite is acting as a guard bed for the removal of heavy hydrocarbons prior to the reforming of the lighter hydrocarbons to produce a product gas of syngas quality. Furthermore, using dolomite or CaO-containing materials, attention has to be paid at the equilibrium between $CaCO_3$ and CaO; the latter increases with T and with decrease by increasing the CO_2 concentration in the feed (e.g., changing from air to pure oxygen change significantly the species present in the reactor).

An alternative of dolomite is olivine, a mineral containing magnesium oxide, iron oxide, and silica. Olivine is advantageous because of its higher attrition resistance than that of dolomite. Moreover, pretreated olivine has a good performance in tar reduction, and the activity is comparable to calcined dolomite [44]. Natural olivine presents good characteristics to be used as biomass gasification catalyst in a fluidized bed reactor but also as nickel support [45]. Iron presence helps in stabilizing nickel in reducing conditions. One part of nickel oxide seems to be included within olivine structure and maintains the reducible nickel oxide on the olivine surface. On the other hand, nickel integration in the olivine structure leads to an increase of free iron oxide, which favors reverse water gas shift reaction. Ni-based catalysts are very effective not only for tar reduction but also for decreasing the amount of nitrogenous compounds such as ammonia. Concluding, this catalytic system seems to meet all the activity requirements and have good attrition resistance for use in a fluidized bed for biomass steam gasification while stability is still not well addressed and the compatibility of the Ni in the recycling of ashes will not be feasible.

Alkali catalysts directly added to the biomass by wet impregnation or dry mixing reduce tar content significantly and also reduce the methane content of the product gas. However, the recovery of the catalyst is difficult and costly. Variable concentrations of alkali metals are included in the ash of several biomass types. Ash is an effective catalyst for the removal of tar when mixed with the biomass. Alkali catalysts directly added to the biomass in a fluidized bed gasifier are subject to particle agglomeration. Alkali metal catalysts are also active as secondary catalysts. Potassium carbonate supported on alumina is more resistant to carbon deposition although not as active as nickel, having a much lower hydrocarbon conversion. Very often, alkali oxide and chloride are present in the ashes as results of the gasification process.

The fields of research dealing with catalytic transformation have just began and, so far, have been devoted to the use of natural derived minerals such as dolomite, olivine, or MgO, which are active in tar cracking, but their activity can hardly be standardized due to the several impurities characterizing the natural material. Between 3% and 5% amount of iron have been found in the analysis of MgO used in Värnamo plant during demonstration activity in the IGCC program. Even if the tar cracking inside the reactor can be seen less critical in terms of catalytic activity, it became a serious challenge if the recycling of the ashes in the field as nutrient for biomass crop is required. Since the bed material is hard to be separated after the biomass gasification, the elements present in the tar cracking catalyst are limited to the harmless ones, and, therefore, Alkali and Fe, Cu-based materials seem to be the candidate as active catalysts, but their activity is limited to tar cracking.

However, primary measures cannot solve the purpose of tar reduction without affecting the useful gas composition and heating value. Combination of proper primary measures with downstream methods is observed to be very effective in all respect. Secondary methods are conventionally used as treatments to the hot product gas from the gasifier [2]. These methods can be classified into two distinct routes: "wet" low-temperature cleaning and "dry" high-temperature cleaning. Conventional "wet" low-temperature syngas cleaning is the preferred technology in the short term. This technology has some efficiency penalties and requires additional wastewater treatment but it is well established. Hot gas cleaning consists of several filters and separation units in which the high temperature of the syngas can partly be maintained, achieving efficiency benefits and lower operational costs. Hot gas cleaning is specifically advantageous when preceding a reformer or shift reactor, because these process steps have high inlet temperatures. Hot gas cleaning after atmospheric gasification does not improve efficiency, because the subsequent essential compression requires syngas cooling anyway. Hot gas cleaning is not a commercial process yet, since some unit operations are still in the experimental phase. However, within few years, hot gas cleaning has become a promising technology and soon will be commercially available. High-temperature filtration is required if a high-temperature reactor is placed downstream the gasifier to avoid to consume chemical energy to increase the temperature after the cooling required by low-temperature filtration. Metal candle filter is widely used after gasification for heat and power purpose. The metal filter works below 600°C and is not sufficient to be employed in the gasification coupled with secondary reforming that runs at a temperature above 800°C. Ceramic candle has been proposed in many plant, and significant tests have been carried out in the Chrisgas project by Delft University and developed by Pall Schumacheer with a 100 kW gasifier obtaining stable performances at 800°C with a maximum operation temperature at 850°C. A hot gas filter with catalytic cracking activity was also proposed [46]. It has to be noted that during the filtration above 800°C, most of the alkali, especially in the form of chloride, are present in gas phases and will be present in the downstream reforming process, while the rest of the ashes can be efficiently separated. Between 650°C and 750°C, a sticky slag can be formed due to the alkali condensation, and the filtration became very problematic.

9.6 GAS UPGRADING BY REFORMING

The exit gas from the gasifier needs to be improved to a rather clean syngas in order to produce fuels or other products. Nowadays, the predominant commercial technology for syngas generation is steam reforming (SRM), in which methane and steam are catalytically and endothermically converted to hydrogen and carbon monoxide [47]. An alternative approach to produce a syngas mixture is partial oxidation (PO), the exothermic, noncatalytic reaction of methane, and oxygen. SRM and PO produce syngas mixtures having appreciably different compositions. In particular, SRM produces a syngas having a much higher H_2/CO ratio. This, of course, represents a distinct advantage for SRM in hydrogen production applications; nevertheless, high-temperature SMR is a difficult application since it requires a source of external heat obtained by burning gaseous or liquid fuel, which decreases the overall efficiency or increases the CO_2 emission (if a fossil fuel is used). A further innovation is the catalytic partial oxidation (CPO) process, in which the oxidation reactions and the reforming ones occur on the catalytic bed. This allows working at low residence time (few ms), with the advantages of small reactor dimension and high productivity. This process is particularly interesting for small-medium size applications, and it is under development by many companies (Air Liquide, Shell, Amoco), and at the moment it is at a demonstrative step by Snamprogetti. Catalyst containing Ni, Ni/Rh, or just Rh dispersed on high-temperature (above 1000°C) resistant supports is suitable for this reaction. A different approach is autothermal reforming (ATR), which combines PO with SRM in one reactor. The process is "autothermal" because of the endothermic reforming reactions proceed with the assistance of the internal combustion (or oxidation) of a portion of the feed hydrocarbons, in contrast to the external combustion of fuel characteristic of conventional tubular reforming. ATR properly refers to a stand-alone, single-step process for feedstock conversion to syngas. However, the same basic idea can be applied to reactors fed by partially reformed gases from a primary reformer. Such reactors constitute a subcategory of ATR that is commonly called secondary reforming. Due to feed composition differences, in particular, the lower concentration of combustibles in secondary reformer feeds, ATR reactors, and secondary reformers has different thermal and soot formation characteristics that require different burner and reactor designs. Nonetheless, the distinction between ATR and secondary reforming is not consistently drawn by technology users and vendors, with the result that secondary reformers often are referred to as ATRs. The prereformed feedstock of the secondary ATR may be assimilated to the stream coming from a gasifier. In practice, oxygen-blown ATR has yet to see application in a large-scale methanol plant, although oxygen-blown secondary reformers have seen operation in a limited number of plants, such as the 2400 MTPD Conoco/Statoil methanol plant, of Haldor Topsøe design that started up in Norway in 1997. This plant, which also contains a prereformer upstream of the SMR, is said to be operating well. It appears that considerable confidence is being placed in advances in the engineering tools now available for designing autothermal reforming burners and reactors [48,49]. Both Lurgi and Haldor Topsøe now claim to have rigorous computer models to facilitate the scale-up and design of oxygen-blown ATRs. ICI claimed a similar capability with respect to their oxygen-blown secondary reformers. Accordingly, further dramatic cost reductions may require the application of still newer reforming technologies. One such development to watch is Exxon's oxygen-blown, fluidized bed ATR, which could offer increased potential for economies of scale. Finally, it was found [50] that high reforming conversion could be attained by using a fluidized bed reactor in ATR of methane under pressurized condition. Fluidization of the catalysts could reduce and activate the catalyst, which was oxidized by oxygen near the inlet of the catalyst bed with the produced syngas. Since catalyst with higher Ni content had higher reducibility, they exhibited high conversion in ATR using fluidized bed reactor. Moreover, it was found [51] that carbon deposition could be inhibited in a fluidized bed reactor through the gasification of carbon with oxygen. The ATR technology has considerable potential for further optimization, especially in combination with Gas Heated Reforming (GHR). By GHR, part of the heat in the ATR effluent is used for steam reforming and feed preheats in a heat exchange type reactor. There are two principally different layouts for incorporating GHR in combination with ATR [52], a parallel and a series arrangement. In the parallel arrangement, the two reformers are fed independently, giving freedom to optimize the S/C ratio individually. In the series arrangement, all gas passes through the GHR unit and then the ATR. The commercial plants commonly use supported nickel catalysts [48,49]. The catalyst contains 15–25 wt% nickel oxide on a mineral carrier (α-Al_2O_3, alumina-silicates, calcium-aluminate, and magnesia). Before start-up, nickel oxide must be reduced to metallic nickel. This is preferably done with hydrogen but also with natural gas or even with the feedgas itself at high temperature (above 600°C, depending of the reducing stream). Required properties of the catalyst carriers are relatively high specific surface area, low pressure drop, and high mechanical resistance at temperatures up to 1000°C. The main catalyst poison in SMR plants is sulfur. Concentrations as low as 50 ppm give rise to a deactivating layer on the catalyst surface [49]. To some extent, activity loss can be offset by raising the reaction temperature. This helps to reconvert the inactive nickel sulfide to active nickel sites. Moreover, in the specific case of biomass gasification, a number of alkaline salts and heavy metals and metal oxides particles may act as additional poisons. Another cause of activity loss is the carbon deposition that can be avoided if a high steam(S)/C ratio is employed. However, economic evaluations indicate that the optimum steam/C ratio tends to be low (2.5–3 v/v).

In real fluidized gasifier, two distinct cases can be present: in processes using oxygen as gasifying agent, the autothermal reforming as well as partial oxidation processes can be used [53] to convert the hydrocarbons present in the gasification stream (Table 9.3). In particular, the two possibilities foreseen in the Chrisgas project for the Varnamo plant were (1) the thermal partial oxidation at 1300°C with a calculated loss of LHV/kg of 23% with respect to the entrance gases and (2) the autothermal reforming at 1000°C (outlet temperature) with a LHV/kg loss with respect of the inlet gas of 15% but requiring 30% of oxygen. The calculation is based on a filter temperature of 800°C and a final syngas concentration at the equilibrium (70% on dry and nitrogen free based).

Although the processes mentioned earlier are well known for the production of H_2 and/or syngas [54], their feasibility applied to a gasification-generated gas depends on the activity and stability of the catalysts. The main problem of the reforming process after hot gas filtration is

TABLE 9.3

Calculated Exit Gas Composition after Pressurized CFB with Oxygen Feed and Comparison of the Gas Composition after Partial Oxidation and ATR Carried Out Downstream a Hot Gas Filtration at 800°C

Component	After Gasifier (vol%)	After ATR 1000°C	After POX 1300°C
Inlet O_2		7	10
Inlet T (°) C		800	800
C2-hydrocarbons	1.5	—	—
CH_4	8	—	—
CO	12	24	24.5
CO_2	28	20	19
H_2	12	23	16
H_2O	37.5	33	39.5
NH_3	0.3	0.2	0.2
H_2S	0.01	0.01	0.01
Tars	0.3	—	—
LHV MJ/kg	6.6	5.6 (85% inlet)	4.8 (73% inlet)
LHV MJ/N m³	7.3	5.4	4.8

connected with the deactivation and especially the sulfur poisoning while some long-term effect of the alkali, passing throughout the hot gas filter, cannot be excluded especially on the Ni sintering [55]. In previous work, the effects of these contaminants in the deactivation and catalytic performances have been studied under laboratory conditions by exposure of the catalyst to a simulated atmosphere [56]. A recent research illustrates the steam-reforming tests on an Ni-based commercial-like catalyst carried out after a 100 kWth CFB gasifier and a hot gas filter fed with three types of biomass ranging from clean wood to miscanthus [57].

The use of clean wood (mainly saw dust) leads to high tar conversion already at low temperature (550°C) and high methane conversion at 750°C. On the other hand, the use of woody residues (mainly bark) with a higher sulfur content ($H_2S = 100$ ppm in the inlet gas) required higher temperature to reach a similar conversion. Nevertheless, an enriched syngas production (CO + H_2 > 70% dry and N_2 free bases) was produced at moderate temperature (975°C–1000°C) demonstrating the possibility to run the process with a large advantage in terms of efficiency and fuel production with respect to the thermal partial oxidation option (Figure 9.10).

Finally, using an herbaceous biomass (miscanthus), an H_2S content well above 200 ppm is present in the inlet gas, after gasification. In these conditions, only approximately half of the methane

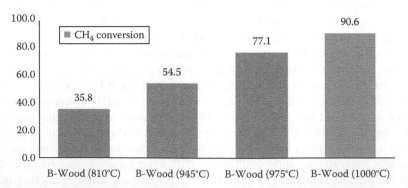

FIGURE 9.10 Methane conversion obtained by reforming on Ni catalyst of gasified wood residues (H_2S 50–150 ppm).

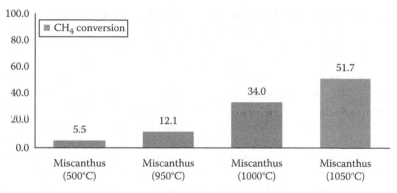

FIGURE 9.11 Methane conversion obtained by reforming on Ni catalyst of gasified miscanthus (H₂S 150–250 ppm).

was converted at 1050°C (with a CO + H₂ concentration = 63%). Therefore, using miscanthus, the advantage with respect to the thermal process is still present but needs to be evaluated in the long term, taking into consideration the deactivation risk due to the presence of alkali, which can induce Ni sintering (Figure 9.11) [58].

The catalyst has shown significant tar-cracking activity also at 550°C using woody biomasses while herbaceous biomasses require higher temperature. An Ni/Rh catalyst was prepared and used after calcinations and reduction in the same series of tests. Using wood B, the methane conversion was higher than the commercial Ni catalyst and reach 75% at 945°C T(out). The methane in the outlet of the reforming was below the detection limit using a T(out) of 1000°C. All the tests have been stabilized for 1 h and repeated after 2 h to detect any deactivation showing a rather stable results. The Ni/Rh-containing catalysts is therefore more resistant to sulfur deactivation and allows carrying out the reaction with woody and herbaceous biomasses at a temperature of 50°C or more below that of the catalysts containing Ni.

The second option is in the case of dual bed gasifier with not pure oxygen uses (Güssing plant); in this case, neither autothermal reforming nor partial oxidation is suitable, since both require oxygen that is not present in the plant site, and its delivery will increase the costs. Therefore, relative low-temperature reforming (800°C) has to be carried out for this purpose, and sulfur-resistant catalyst needs to be used. Rh and other noble metal catalysts seem to be good candidate for this purpose even if their cost needs to be taking into account in the economical evaluation.

The possibility to convert, after tar hydrocracking directly to methane the syngas obtained upstream will allow to produce bio synthetic natural gas (bioSNG) and hydrogen and save the cost of the reforming transformation while the water gas shift reaction will be still required (Greensyngas project VII FP, UE).

9.7 DOWNSTREAM OF THE REFORMER: WATER GAS SHIFT UNIT

According to the aim to convert the produced gas from biomass to synthesis gas for applications requiring different H₂/CO ratios, the reformed gas may be ducted to a conventional water gas shift (WGS) unit (7.1) to obtain the H₂ purity required for fuel cells or directly to a hydrogenation/hydrogenolysis unit (7.2) to convert the residual hydrocarbons, in particular, aromatic compounds and olefins, for applications requiring an H₂/CO ratio close to 2 (dimethyl ether, methanol, Fischer–Tropsch synthesis).

The WGS reaction is used to fix the optimal value of the H₂/CO ratio for the fuel production.

The WGS unit is also a critical step in hydrogen production for additional hydrogen generation and preliminary CO clean up and prior to the CO preferential oxidation (PROX) or methanation step [59]. WGS units are placed downstream of the reformer to further lower the CO content and improve the H_2 yield. Ideally, the WGS stage(s) should reduce the CO level to less than 5000 ppm. The WGS catalysts have to be active at low temperatures, 200°C–280°C, depending on the inlet concentrations in reformate. The reaction is moderately exothermic, with low CO levels resulting at low temperatures, however, with favorable kinetics at higher temperatures. Under adiabatic conditions, conversion in a single bed is thermodynamically limited (as the reaction proceeds, the heat of reaction increases the operating temperature), but improvements in conversion are achieved by using subsequent stages with cooling and perhaps CO_2 removal between the stages. Since the flow contains CO, CO_2, H_2O, and H_2, additional reactions can occur, depending on the H_2O/CO ratio and favored at high temperatures: methanation, CO disproportionation, or decomposition. In industrial applications, the classical catalyst formulations employed are Fe–Cr oxide for the first stage (high-temperature shift [HTS], typically in the range of 360°C–400°C), and Cu–ZnO–Al_2O_3 for subsequent stages (low-temperature shift [LTS], operating just above the dew point, the lowest possible inlet temperature, i.e., about 200°C), for good performance under steady-state conditions (CO exit concentration in the range of 0.1%–0.3%).

However, these catalysts are very sensitive to the contaminants from biomass gasification. The FeCr-catalyst can be used to shift the synthesis gas at the expected H_2S-levels (0–150 vppm). The presence of other poisons in the gas phase, like HCl and NH_3, must be considered since the experiments show a strong impact on the catalyst activity when exposed to HCl, NH_3, and gas from a biomass gasifier. Even if the NH_3 should be mainly transformed in N_2 and H_2 during the reforming step, residual content of NH_3 could affect the activity of the WGS catalysts. The FeCr catalyst can be activated in the produced synthesis gas at 350° C or above. The inlet reactor temperature should be at least 350°C [60].

WGS of gases containing appreciable amounts of sulfur and tar requires catalysts consisting mainly of Co and Mo oxides. Moreover, their activity is full only when Co and Mo are in the sulfided forms.

9.8 FISCHER–TROPSCH SYNTHESIS OF MIDDLE DISTILLATE

In 2007, two Fischer–Tropsch processes had a significant market share: the Shell Middle Distillate Synthesis (SMDS) process and the Sasol Slurry Phase Distillate (SPD) process. SMDS uses a tubular fixed-bed reactor [61].

SPD uses a slurry reactor, but a fixed-fluidized bed reactor has also been used [62]. Other processes have been designed by companies such as Syntroleum, but these are not yet applied commercially. The only FT fuel that was commercially available in 2007 in the EU is GTL product from the Bintulu plant, which is a constituent of the V-Power brand from Shell. As of 2007, companies like Shell, Sasol Chevron, ConocoPhillips, and Total are all working on GTL and CTL plants. Large-scale GTL activity is underway, especially in Qatar [62,63]. In addition to confirmed projects, large CTL plants in China (e.g., in Ningxia Hui) and India are being planned.

CHOREN industries in Germany and others experiment with the use of biomass as a feedstock for FT fuel production (BTL). Development of FT production could also benefit from improvements in technologies such as gasification and gas cleaning that are used in other commercial activities. However, due to the large projects and investments involved, the speed of implementation of new technologies is limited.

Synthesis gas entering Fischer–Tropsch synthesis must have gas contaminants below 200 ppb sulfur and 10 ppm ammonia. Since sulfur levels are above the contaminant limit at this point in the process, the syngas is polished for sulfur and trace contaminants with a zinc oxide. Further purification may be required to decrease the traces of poisoning and separate CO_2. A pressure swing absorption can be used to increase purity and separate the hydrogen to be used in the hydroprocessing step. In a FT synthesis, the syngas at 20–30 bar operating pressure reacts over a cobalt-based catalyst in a fixed-bed FT reactor at 200°C according to the reaction in Equation 9.1. FT product distribution followed the Anderson–Schulz–Flory alpha chain growth model [64].

Two types of catalysts can be used for the FT synthesis of two types of processes:

1. Co-based catalysts with several promoters and alpha value = 0.90–0.92 and high resistance to attrition
2. Fe-based catalysts with lower alpha value, lower mechanical stability, and lower cost than the Co catalyst

$$CO + 2.1H_2 \rightarrow -(CH_2)- + H_2O \qquad (9.1)$$

The products from FT of Co catalysts are mainly aliphatic straight-chain hydrocarbons (CxHy). Besides the CxHy, also branched hydrocarbons, unsaturated hydrocarbons, and primary alcohols are formed especially with Fe catalysts. The product distribution obtained from FT includes the light hydrocarbons methane (CH_4), ethane (C_2H_4) and ethane (C_2H_5), LPG (C_3–C_4, propane and butane), gasoline (C_5–C_{12}), diesel fuel (C_{13}–C_{22}), and light and waxes (C_{23}–C_{33}) (Figure 9.12) [64].

After the gas is cooled and liquid hydrocarbons and water separated, part of the unconverted syngas is recycled directly back into the FT reactor with another portion going to the acid gas removal system as another recycle stream. The balance of unconverted syngas is sent to the power generation area. Overall conversion of carbon monoxide is approximately 67% and accomplished via recycle streams. Fischer–Tropsch liquids which contain significant amounts of waxes are hydrocracked in a hydroprocessing unit. The product distribution by weight is 25% naphtha and 65% diesel with the balance being gaseous hydrocarbons that are used as fuel in the gas turbine. The liquid fuels can then be used as blendstock for the gasoline and diesel pool.

The use of iron-based catalysts requires higher temperature and gives rise to branched, oxygenated products characterized by alpha value below 0.9 therefore producing a distribution of products with increasing amount of low-molecular-weight hydrocarbons. The Fe catalyst can be used with a wider range of H_2/CO ration due to the intrinsic capacity of iron to promote the water gas shift reaction. The fuel gas produced as side stream can be recycled at the gasifier if the increase of the heating value of the feeding mixture is needed or can be used as by products as fuel or as chemical building block. Since a significant amount of methane is produced in the FT synthesis, specific consideration on the overall cost and efficiency needs to be done if the recycling of methane in the gasifier can avoid the high-temperature step of methane reforming downstream the gasifier maintaining just the step of tar hydrocracking requiring lower temperature.

Improvements in the FT synthesis are given by reactor modification going from a slurry reactor to a slurry bubble column and from a fixed multitubular reactor to advanced heat exchanger fixed bed.

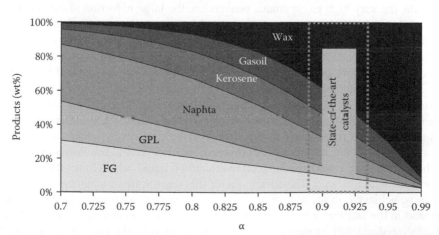

FIGURE 9.12   Fischer–Tropsch distribution products as function of the catalyst selectivity (α value).

9.9   SYNTHESIS OF METHANOL AND DIMETHYL ETHER

Methanol synthesis is the second largest process after ammonia that uses catalysts at high pressure

$$CO + 2H_2 \leftrightarrows CH_3OH \quad \Delta H^\circ_{298K} = -90.6 \text{ kJ mol}^{-1}$$

The mechanism is believed to be

$$CO + H_2O \leftrightarrows CO_2 + H_2 \quad \Delta H^\circ_{298K} = -41.2 \text{ kJ mol}^{-1}$$

$$CO + 2H_2 \leftrightarrows CH_3OH + H_2O \quad \Delta H^\circ_{298K} = -49 \text{ kJ mol}^{-1}$$

The catalysts composition and the main parameter involved in the process [65–67] are: CuO (60%–70%)-ZnO (20%–30%) with Al_2O_3 or Cr_2O_3 (5%–15%), temperature 220°C–300°C, pressure 5–10 MPa, H_2/CO ratio of 2.17, conversion of CO to methanol per pass 16%–40%, selectivity 99.8%.

Other catalysts that have been proposed are as follows:

1. Raney Cu catalyst is prepared by partial leaching of Cu/Zn/Al alloys with sodium hydroxide to produce a metallic sponge and have activity comparable to higher 50% than conventional CuO/ZnO.
2. $ThCu_4$, $ThCu_6$, and $CeCu_2$ are deactivated by CO_2.
3. Pd is supported on alumina, and MgO presents too low activity.

Today, there are four catalyst suppliers, and complete proprietary processes for methanol synthesis are available from six companies: ICI, Lurgi, Topsøe, Mitsubishi, M.W. Kellogg, and Uhde. The capacity of the reactors can be as high as 2,500–10,000 ton day⁻¹, and a good catalyst in a natural gas-based plant may have lifetime of about 4 years. In order to improve the yield in methanol, as the reaction is exothermic and favored at low temperature, it is necessary to remove the heat of reaction to keep the reaction temperature as low as possible in order to increase the conversion. In the future, to increase the yield on methanol, it is necessary to develop more active catalysts that operate at lower temperature, increasing the thermodynamically allowed conversion. There are several industrial processes; however, 55% of produced methanol uses the Lurgi Process, 25% the ICI, and 20% all the other processes. In the Lurgi two multitubular reactors [68], the heat of reaction is removed in the first reactor by boiling water around bed and in the second reactor by gas. The Lurgi reactor is nearly isothermal, and the heat of reaction is used to generate high-pressure steam, which is used to drive the compressor and as distillation steam. The advantages of this process are the optimum temperature profile, the very high gas synthesis conversion, the large reduction of catalyst volume, the lower gas recycle, and the high energy efficiency, which allow about 40% capital cost saving. In the ICI [69] adiabatic single bed reactor, the heat of reaction is removed by adding cold reagent at different heights in the bed, and the catalyst in the reactor is localized in shells, and between the shells the gas is cooled by fresh reaction mixture. The Haldor Topsoe [70] uses several adiabatic reactors arranged in series with intermediate cooler to remove the heat of reaction, and the synthesis gas flues radially through the catalyst bed, which results in reduced pressure drop compared to axial flow. The heat of reaction in the three reactors is eliminated by cooling outside the reactors. The Air product-Chem [71] system uses a three-phase fluidized bed, where an inert hydrocarbon liquid inside the reactor removes the heat. The main feature of this process (they have a demonstration plant in Texas) is the fact that the catalyst is suspended in inert hydrocarbon liquid, which limits the temperature rise and it adsorbs the heat liberated. The advantages of this reactor are in a higher single pass conversion that can be achieved, in the reduction of the syngas compression costs, in the increase of life of catalyst, and in the fact that it is possible to work with 50% CO entering feedstock. Commercial Cu/Zn/Al catalysts developed for the two-phase process are used for the three-phase process, the powdered catalyst particles typically measure 1–10 µm and are densely suspended in a thermostable oil, chemically resistant to components of the reaction mixture at process conditions, usually paraffin. The deactivation of the catalysts due to exposure to trace contaminants is still a point of concern.

In the Casale [72] isothermal reactor, the heat is removed by plates immersed in the catalyst bed and an axial–radial flow of reagents is used, and this arrangement can solve the problem of reducing the pressure drop of the converter. This design can be obtained easily with the use of plates as cooling surface area, and the flow of cooling gas inside the plates can have the same direction of the gas in the catalyst, that is, in a horizontal direction, co-current or counter-current. It is clear that an axial–radial design leads to a much slimmer vessel for the same catalyst volume, allowing to reach capacities above 7000 MTD in a single vessel converter. One Synergy [73] has developed a novel methanol synthesis loop configuration (patent pending), particularly suited to large capacity plants. This scheme utilizes two methanol converters (each generating medium pressure steam) in series and provides benefits over a conventional arrangement in terms of reduced catalyst volume. Compared to a conventional large capacity methanol synthesis loop with parallel converters, the series loop operates with approximately 30% lower recycle ratio. This is beneficial in terms of reducing piping size and maximizing circulator capacity. However, since the gas is circulated through the converters in series, the mass flux through each converter is actually increased compared to a parallel converter layout. This has the benefit of maximizing the reaction rate per unit volume, reducing catalyst volumes, and improving the heat transfer performance, providing better temperature control resulting in lower temperatures and therefore longer catalyst life. The series loop configuration is best suited to larger plants, where two or more converters operating in parallel would be required to meet the desired capacity. The full benefit of the series configuration can only be realized if product is separated after each converter. This ensures that equilibrium limitations do not come into play, and reaction rates are maximized in each converter, thereby minimizing the catalyst volume required. Gas flow differs between proprietary offerings. The ARC converter (JM Catalysts/Methanol Casale) is a quench-type converter, but with separate catalyst beds rather than a continuous bed to improve distribution. Single streams of up to 3000 ton/day are regarded as feasible by JM catalysts. One synergy also offers the Steam Raising converter (SRC) with catalyst contained on the shell side and steam in the tubes. The gas flows radially out through the catalyst bed from the inside to the outside. Radial flow reactors such as the SRC, Toyo MRF-Z, and DPT's proprietary design offer low pressure drops and can be scaled up to very large sizes by extending the height. Radial flow reactors can be combined with steam raising. For the established technology (the low-pressure methanol process), the selectivity is high, in excess of 99.8%. Conversion per pass depends on process conditions, but can be as high as 40%. The recycle of the nonreacted gas in the synthesis loop raises the overall conversion of carbon in the feedstock to 88%–94%.

9.9.1 Synthesis of Dimethyl Ether

The synthesis of dimethyl ether (DME) from synthesis gas involves three reactions:

1. $CO_2 + 3H_2 \leftrightarrows CH_3OH + H_2O$
2. $CO + H_2O \leftrightarrows CO_2 + H_2$
3. $2CH_3OH \leftrightarrows 2CH_3OCH_3 + H_2O$

The introduction of reaction (3), the DME synthesis, serves to minimize the equilibrium constraints inherent to the methanol synthesis by transforming the methanol into DME. Moreover, the water formed in reaction (3) is to some extent driving reaction (2) to produce more hydrogen, which in turn will drive reaction (1) to produce more methanol. Thus, the combination of these reactions results in a strong synergetic effect, which dramatically increases the synthesis gas conversion potential (Figure 9.13). The catalyst applied is a proprietary dual-function catalyst, catalyzing both steps (i.e., methanol and DME synthesis) in the sequential reaction. Therefore, significant advantages arise by permitting the methanol synthesis, the water–gas shift, and the DME synthesis reaction to take place simultaneously. A dual catalyst system is based on a combination of $Cu/ZnO/Al_2O_3$

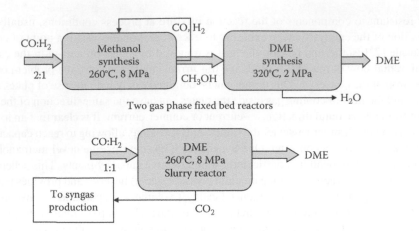

FIGURE 9.13 DME two-step and one-step synthesis from syngas.

catalyst and gamma-alumina catalyst. The use of axial–radial flow can solve the problem of reducing the pressure drop through the converter; this design can be obtained easily with the use of plates as cooling surface area; and the flow of cooling gas inside the plates can have the same direction of the gas in the catalyst, that is, in a horizontal direction, co-current or counter-current. It is clear that an axial–radial design leads to a much slimmer vessel for the same catalyst volume, allowing to reach capacities above 7000 MTD in a single vessel converter.

9.9.2 From Methanol to Gasoline

This process was the starting point of all further technologies of transformation of methanol to olefins (MTO) [67,74]. Mobil found a plant in New Zealand in 1970 (that was closed after some years, because it was not economic) to produce high-quality gasoline from methanol with a catalyst based on ZSM-5 zeolite. In the Mobil MTG process, methanol is dehydrated over the zeolite ZSM-5 catalyst, first to dimethyl ether and then to hydrocarbons that are predominantly in the gasoline boiling range, with some LPG and fuel gas. The complex reaction sequence can be represented as follows:

$$2CH_3OH \leftrightarrows (CH_3)_2O + H_2O$$

$$(CH_3)_2O \leftrightarrows (CH_2)_2 + H_2O$$

Light olefins → heavier olefins
Heavy olefins → aromatics, alkanes, and cycloalkanes

9.9.3 From Methanol to Olefins

It is possible to modify the MTG process in order to obtain only the olefins by operating at higher temperature, using zeolites with lesser acid strength and with narrow pore width. Since its discovery in 1977, the conversion of MTO and other hydrocarbons has been realized on microporous solid acids, especially the aluminosilicate HZSM-5 and more recently on the silico-aluminophosphate HSAPO-34. There are three types of MTO processes available [75]:

1. The UOP/HYDRO MTO process, which produces propylene and ethylene with minimal other by-products (a plant is in Norway).
2. The Lurgi's "methanol-to-propylene (MTP)" process, which produces propylene and gasoline (two plants will start in China).
3. The Dalian Institute of Chemical Physics (DICP), which starts from syngas to dimethylether and after transformation to olefins (SDTO), and the process is close to be commercialized [76,77].

The UOP/HYDRO MTO has been introduced by UOP and HYDRO of Norway (1997), in a large process demonstration plant, in which primarily the methanol is converted into ethylene and propylene in 80% yield with a methanol conversion of 99.8% and a stable product selectivity with time has been demonstrated. This process utilizes a fluidized bed reactor with a continuous fluidized bed regenerator, and the catalyst circulates and regenerates continuously and uses crude methanol as a feedstock. The advantages of MTO versus cracking are as follows:

1. Direct use of ethylene and propylene in chemical grade products with greater than 98% purity using a flow scheme that does not require expensive ethylene/ethane or propylene/propane splitters
2. Limited production of by products (H_2, CH_4 diolefins acetylenes) compared to a steam cracker, which results in a simplified product recovery section
3. Easy integration into existing naphtha cracker facilities due to low paraffin (ethane and propane) yields
4. Flexibility to change the propylene to ethylene product weight ratio from 0.77 to 1.33

The features of MTO are the exothermicity and the accumulation of Carbon or coke on the catalyst, which must be removed to maintain catalyst activity by combustion with air in a catalyst regenerator system. Other co-products include very small amounts of C1–C4 paraffins, hydrogen, CO, and CO_2 as well as parts per million levels of heavier oxygenates that are removed to ensure that the product olefins meet polymer-grade specifications. The reaction is catalyzed by a silico-alumino-phosphate synthetic molecular sieve with high degree of attrition resistance and stability required by the multiple regenerations and fluidized bed conditions over the long term. The catalyst is extremely selective toward the production of ethylene and propylene.

Lurgi methanol to propylene (MTP®) technology is based on the efficient combination of two main features: a very selective to propylene with low propane yield and a stable zeolite-based fixed-bed catalyst commercially manufactured. The high activity achieves thermodynamic equilibrium with conversion 99%. Fixed-bed reactor system is selected as the most suitable reaction system from a technological and economic point of view. High tendency to coking and need for regeneration methanol even at low grade is catalytically converted to hydrocarbons, predominantly propylene.

Methanol fed to the MTP® plant is first converted to DME and water in a DME prereactor. Using a highly active and selective catalyst, thermodynamic equilibrium is achieved resulting in the methanol/water/DME mixture at appropriate operating conditions. Propylene is the single main product, and there is no ethylene product. Gasoline, LPG, fuel gas, and water are by-products.

The highlights of MTP process are the propylene production only and low coking of catalyst results in low number of regeneration cycles. The regeneration of the catalyst realized in two reactors is operating in parallel, while the third one is in regeneration. It is necessary after about 500–600 h of cycle. Deactivation occurs when the catalysts' centers become blocked by coke formed by side reactions. By using diluted air, the regeneration is performed at mildest conditions avoiding thermal stress of the catalyst. Olefin recycle and steam made from water recycle are added to entering this mixture before it enters the first MTP® reactor of the multistage adiabatic reactor system. The methanol/DME conversion rate exceeds 99%, with propylene as the essential compound. Additional reaction proceeds in the downstream reactor stages. After product gas compression, traces of water, CO_2, and DME are removed, and the gas is further processed yielding chemical-grade propylene with a typical purity of more than 97%.

Catalyst regeneration is realized after a cycle of approximately 500–600 h of operation, and the catalyst has to be regenerated by burning the coke with a nitrogen/air mixture. The regeneration is carried out at temperatures similar to the reaction itself, and hence, the catalyst particles do not experience any unusual temperature stress during the in situ catalyst regeneration procedure.

The company also plans to construct a 5000 ton day^{-1} facility for Atlas Methanol Ltd in Trinidad, slated to come on stream about 2003. Another 5000 ton day^{-1} plant will also be built for Iran's National Petrochemical for commissioning around 2004. The two plants will feature Lurgi's novel Mega Methanol technology.

In the mid-1990s, DICP was awarded two patents in the United States concerned with the conversion of methanol/DME to light olefins. These patents are the basis for the syngas via dimethyl ether to olefin process (SDTO). Compared with the MTO process, SDTO directly converts synthesis gas to DME with high carbon monoxide conversion, thus exhibiting greater efficiency than the MTO process. The special features of the SDTO process that are the bifunctional metal (Cu, Zn, etc.,)-zeolite catalysts have been developed, which can convert syngas very selectively to DME with high carbon monoxide (CO) conversion (this reaction is far more favorable thermodynamically than methanol synthesis from syngas). SDTO 50 ton (methanol)/day unit for the conversion of methanol to lower olefins has a methanol conversion of close to 100% and a selectivity to lower olefins (ethylene, propylene, and butylenes) of higher than 90%. By utilizing a proprietary SAPO-34 catalyst system and a recycling fluidized bed reaction system for the production of lower olefins from methanol is the first unit in the world having a capacity of producing nearly 10,000 ton lower olefins (ethylene, propylene, and butene) for year. MTO and MTP plants can be located near or integrated with a methanol plant. If there is local demand for a portion of the light olefins or their derivatives or they can be located separately with a methanol plant located near the gas source and to ship the methanol the MTO plant located near the olefin markets or olefin derivative plants. If all of the light olefins or their derivatives are to be exported, in either case, the methanol plant is located with access to low-cost natural gas or coal or biomass.

9.10 SYNTHESIS OF BUTANOL

9.10.1 SYNTHESIS FROM BIOMASS

Three different methods for the preparation of butanol from biomass have been proposed:

1. The anaerobic fermentation of sugar substrates, available as such or derived from starch or cellulose in 1-butanol or isobutanol to butanol.
2. The synthesis by fermentation of intermediates such as ethanol, 1,3-butanediol, butyraldehyde, or butyric acid and their subsequent chemical transformation to one of the three biobutanol.
3. Chemical synthesis through gasification of lignocellulosic biomass (possibly via their prior pyrolysis) and the subsequent conversion of synthesis gas produced or to a mixture of alcohols from methanol to butanol or production of methanol and propylene, and then after oxosynthesis and hydrogenation to 1-butanol and isobutanol. In this work, the term butanol will refer to 1-butanol, 2-butanol, and 2-methyl-1-propanol (isobutanol) and mixtures thereof. We will first examine the actual processes from fossil fuels and later the possible future ones from biomass.

9.10.2 SYNTHESIS FROM FOSSIL FUEL

The synthesis routes (see Table 9.4) currently used are the hydroformylation of propylene for the production of aldehydes and their subsequent hydrogenation to 1-butanol and isobutanol and the hydration of butenes to produce 2-butanol. Currently, there are two industrial processes [78–81] for the production of butyraldehyde using catalysts Rh-based, that of the Dow LPO Oxo and RCHRP of Rurchemie and RhônePoulenc. There is still a system that produces directly butanol using more complex-based catalysts cobalt phosphines. LPO Oxo uses a catalyst based on Rh phosphine or phosphite, using as solvent the some aldehyde and their condensation products with a separation of aldehyde within the reactor by stripping with the synthesis gas, in situ regeneration of catalyst deactivated and a complete purification of the reagents to avoid deactivation of the catalyst.

TABLE 9.4

Scheme of the Chemical Butanol Synthesis

Hydroformylation and hydrogenation

$$CH_2=CHCH_3 + CO + 2H_2 \rightarrow CH_3CH_2CH_2CHO$$
$$CH_3CH_2CH_2CHO + H_2 \rightarrow CH_3CH_2CH_2CH_2OH$$

Reppe synthesis

$$CH_2=CHCH_3 + 3CO + 2H_2O \rightarrow CH_3CH_2CH_2CH_2OH + 2CO_2$$

Indirect hydration

$$CH_3CH=CHCH_3 + H_2SO_4(conc.) \rightarrow CH_3CH_2CH(CH_3)OSO_3$$
$$CH_3CH_2CH(CH_3)OSO_3 + H_2O \rightarrow CH_3CH_2CH(OH)CH_3 + H_2SO_4(dil.)$$

Direct hydration

$$CH_3CH=CHCH_3 + H_2O \rightarrow CH_3CH_2CH(OH)CH_3$$

Guerbet's synthesis

$$2CH_3CH_2OH \rightarrow CH_3CH_2CH_2CH_2OH + H_2O$$

Synthesis of higher alcohols

$$nCO + mH_2 \rightarrow CH_3OH + CH_3CH_2OH + CH_3CH_2CH_2OH + CH_3CH_2CH_2CH_2OH$$

Through MTP and hydroformylation

$$3CO + 6H_2 \rightarrow 3CH_3OH \rightarrow CH_2=CHCH_3 + 3H_2O$$

The use of phosphite ligands allowed to reach ratio N/ISO 30/1. The second process used is RCHRP with bis-phosphine amines as ligands soluble in water at 50 atm and 120°C, this catalysts is less active, but there are fewer losses Rh and the separation takes place downstream of the reactor for separation of the aqueous phase containing the catalyst and the organic one containing the product [82,83]. This process also has the advantage of being able to obtain a ratio normal/iso 30/1. The hydrogenation of butyraldehyde [84,85] takes place with heterogeneous catalysts based on Ni, as Cu-Zn-Ni or Ni-Raney, with conversion of 98.6%, and to a lesser extent with iridium complexes in acetic acid. Another process used only in the 1950s is a summary of Reppe [86,87] that operated between 5 and 20 atm at 100°C in the presence of a catalyst based on $Fe(CO)_5$ leading directly buta-nols a ratio normal/iso 9/1. This process, due to high costs due to recycling of the catalyst present in a concentration of 10% and high consumption of CO, has not been commercialized, despite the mild reaction conditions and low cost of the catalyst. The synthesis of 2-butanol is mainly indirect hydration of n-butenes, making them react first with concentrated sulfuric acid and subsequent hydration with the formation of 2-butanol and dilute sulfuric acid. It is also used directly with heterogeneous catalysts such as amines P_2O_5–SiO_2 or resins that operate at temperatures between 100°C and 250°C and pressures between 50 and 250 atm.

9.10.3 CHEMICAL SYNTHESIS

The chemical synthesis of bio-butanol from biomass is realized by biomass gasification [53,56], physical–chemical treatment of gas purification synthesis obtained and its transformation or to a mixture of alcohols and their separation for 1-butanol and/or isobutanol, or methanol and then to propylene and then to oxosynthesis to butyraldehyde and subsequent hydrogenation to 1-butanol and isobutanol. The catalysts for the production of higher alcohols from synthesis gas can be divided into four classes: (1) catalysts for methanol at high pressure to low amended by alkali metals (Zn–Cr oxides, CuZnCr, and CuZnAl) [91], (2) Fischer–Tropsch catalysts properly modified, (3) catalysts based on Cu, Co, ZnAl [92,93], and (4) catalysts based on Mo-K (oxide or sulfide) doped with metals Eighth group [18]. The catalysts of group 1 mainly provide methanol and isobutanol (e.g., 75% methanol, ethanol, propanol 1.8% 2.6% BuOH 13.7%), while the catalysts of the other classes follow distribution Schulz–Flory type of alcohol, giving essentially methanol and higher alcohols all the sliding scale. Butanol is obtained by separation from other types of alcohol by distillation. The second route starts at the synthesis gas and is more complex, but it is more selective in butanol. From the methanol synthesis, gas is obtained with the traditional catalysts Cu, Zn, and Al, which is converted to propylene with zeolite-type ZSM (5). Later, propylene can be obtained via oxo-synthesis to butanol as is presently realized in the petrochemical process.

The synthesis gas comes mainly from methane also from coal to a small extent, but may increase in the future the amount produced by biomass. Currently, the propylene used in hydroformylation is obtained by steam cracking of petroleum fractions, but in future will be produced both by natural gas and coal and even from biomass. In future, butenes used for the reaction of hydration as well as for propylene hydroformylation may be produced by dehydrogenation of light paraffins present as impurities in many natural gas and gas by-products of liquid fuels produced via Fischer–Tropsch from syngas (from coal, natural gas and biomass). So at least in the next hundred years, there will be no problems for raw materials for the synthesis of butanol. The reaction of hydroformylation in the future could be changed in direction simultaneously to achieve the hydrogenation aldehydes, coupled with the Rh complex, even one of Ru, always with triphenylphosphine as a binder, and these catalysts heterogenization on polymeric supports or inorganic. The use of biomass can become competitive with the use of fossil fuels using lignocellulosic from different sectors (agriculture, cleaning forests, municipal waste, or sludge from wastewater treatment). The way purely chemical from biomass using gasification (and possible after pyrolysis) has the advantage over fermentation processes you can use any type of raw material, including the lignin, cannot be used in biotechnological processes, for which also always start from the same type of biomass. Gasification requires production of more than 2500 ton day^{-1}, but studies are still needed to lower the cost of treatment synthesis gas and to improve the process of gasification or pyrolysis introducing catalysts that transform biomass into gas synthesis in a more selective and to lower costs. In addition, the direct route for the synthesis of higher alcohols from synthesis gas currently suffers from the fact that the selectivity in butanol is still low, while the path using the hydroformylation of propylene, even if selective needs two stages to produce it via methanol. The synthesis fermentation, in contrast, the advantage of producing in a single butanol stage: in fact, the most promising process is what makes hydrolysis of waste cellulose and fermentation in a single-stage and the separation of the product in situ in the reactor (the gas stripping seems ideal). We must however wait for the operation of the demonstration plants to really know what will be the winning technology in the future and when you develop the marketing of plant biomass.

9.11 POWER BY FUEL CELLS

Hydrogen, Syngas, or methanol or even methane can be used for power generation by fuel cell. Low-temperature fuel cell uses hydrogen or methanol. Both can be produced by biomass gasification. Hydrogen required to be purified after water gas shift by separation (membrane or pressure swing absorption) or by reaction (methanation or selective CO oxidation). The principle of fuel cell was first discovered in 1839 by Sir William R. Grove, who used hydrogen and oxygen as fuels catalyzed on platinum electrodes [18]. A fuel cell is defined as an electrochemical device in which the chemical energy stored in a fuel is converted directly into electricity. A fuel cell consists of an anode (negatively charged electrode) to which a fuel, commonly hydrogen, is supplied—and a cathode (positively charged electrode) to which an oxidant, commonly oxygen, is supplied. The oxygen needed by a fuel cell is generally supplied by feeding air. The two electrodes of a fuel cell are separated by an ion-conducting electrolyte. The input fuel is catalytically reacted (electrons removed from the fuel elements) in the fuel cell to create an electric current. The input fuel passes over the

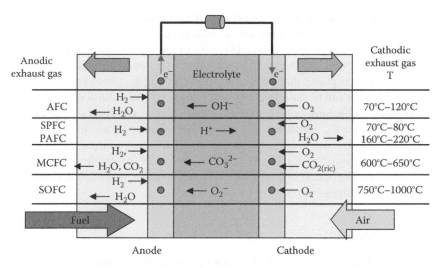

FIGURE 9.14 Fuel cell type and conditions.

anode where it catalytically splits into electrons and ions, and oxygen passes over the cathode. The electrons go through an external circuit to serve an electric load while the ions move through the electrolyte toward the oppositely charged electrode. At the electrode, ions combine to create by-products, primarily water and CO_2. Depending on the input fuel and electrolyte, different chemical reactions will occur. The main product of fuel cell operation is the DC electricity produced from the flow of electrons from the anode to the cathode. The amount of current available to the external circuit depends on the chemical activity and amount of the substances supplied as fuels and the loss of power inside the fuel cell stack. The current-producing process continues for as long as there is a supply of reactants because the electrodes and electrolyte of a fuel cell are designed to remain unchanged by the chemical reactions. The stack is the main component of the power section in a fuel-cell power plant. The by-products of fuel cell operation are heat, water in the form of steam or liquid water, and CO_2 in the case of hydrocarbon fuel.

There are five types of fuel cells, on the basis of the electrolyte employed. They are alkaline fuel cells (AFC), phosphoric acid fuel cells (PAFC), proton-exchange membrane fuel cells (PEMFC), molten carbonate fuel cells (MCFC), and solid oxide fuel cells (SOFC) (Figure 9.14). In all types, there are separate reactions at the anode and the cathode, and charged ions move through the electrolyte, while electrons move round an external circuit. Another common feature is that the electrodes must be porous, because the gasses must be in contact with the electrode and the electrolyte at the same time.

The advantages of the fuel cells are that they operate without combusting fuel and with few moving parts, but mainly because a fuel cell can be two to three times more efficient than an engine in converting fuel to electricity. A fuel cell resembles an electric battery, because both of them produce a direct current by using an electrochemical process. A battery contains only a limited amount of fuel material and oxidant, which are depleted with use. Unlike a battery, a fuel cell does not run down or require recharging; it operates as long as the fuel and the oxidizer are supplied continuously from outside the cell. The main advantages of fuel cells are reflected by the following desirable characteristics: (1) high energy conversion efficiency; (2) extremely low emissions of pollutants; (3) extremely low noise or acoustical pollution; (4) effective reduction of greenhouse gas (CO_2) formation at the source compared to low-efficiency devices; and (5) process simplicity for conversion of chemical energy to electrical energy. Depending on the specific types of fuel cells, other advantages may include fuel flexibility and existing infrastructure of hydrocarbon fuel supplies; cogeneration capability; modular design for mass production; relatively rapid load response. Therefore, fuel cells have great potential to penetrate into markets for stationary power plants (for industrial, commercial, and residential home applications) but also for mobile power plants for transportation by cars, buses, trucks, trains, and ships, as well as man-portable microgenerators. Due to these features, fuel cells have emerged in the last decade as one of the most promising new technologies for meeting the energy needs well into the twenty-first century for power generation and for transportation [91–93]. Unlike power plants that use combustion technologies, fuel cell plants that generate electricity and usable heat can be built in a wide range of sizes, from 200 kW suitable for powering commercial buildings to 100 MW suitable to utility power plants. The disadvantages to be overcome include the costs of fuel cells, which are still considerably higher than conventional power plants per kilowatt. Moreover, the fuel hydrogen is not readily available and thus on-site or on-board H_2 production via reforming is necessary. Finally, there are no readily available and affordable ways for on-board or on-site desulfurization of hydrocarbon fuels; therefore, the efficiency of fuel processing affects the over system efficiency.

9.12   PERSPECTIVE

Biomass gasification is a central enabling technology for the clean, sustainable use of biomass in serving both CHP and fuels generation as well as in the much smaller scale of operation needed in distributed generation and developing country village and small industry power applications. Concerning CHP, the successful demonstration of the IGCC in Värnamo [58] and the widespread demonstration worldwide have reduced the concerns about technical feasibilities while economic feasibility still depends on favorable legislation or carbon taxes.

However, for the hydrogen and/or syngas-based fuel production from biomass gasification, there are still space for large economical and technical improvements dealing with the use of catalytic system to decrease the temperature of the reaction while other improvements can be reached by integration with downstream fuel production system and upstream biomass production. In this perspective, highly efficiency in relative small plant (10–50 MW) would be a further step in enlarging the potential application in Europe of biomass gasification and second-generation biosynfuel production.

REFERENCES

1. Sutton, D., Kelleher, B., and Ross, J.H.R., *Fuel Processing Technology* 73 (2001) 155.
2. Tijmensen, M.J.A., Faaij, A.P.C., Hamelinck, C.N., and van Hardeveld, M.R.M., *Biomass and Bioenergy* 23 (2002) 129–152.
3. Devi, L., Ptasinski, K.J., and Janssen, F.J.J.G., *Biomass and Bioenergy* 24 (2003) 125.
4. Beenackers, A.A.C.M., *Renewable Energy* 16 (1999) 1180.
5. Bridgewater, A.V., *Fuel* 74 (1995) 631.
6. Reed, T.B., *Biomass Gasification: Principle and Technology*, Noyes Data Corporation, Park Ridge, NJ, 1981.
7. Larson, E.D. and Katofsky, R.E. In: Bridgwater, A.V. (Ed.), *Advances in Thermochemical Biomass Conversion*, Blackie, London, U.K., 1994, p. 495.
8. Garcia, L., French, R., Czernik, S., and Chornet, E., *Applied Catalysis A: General* 201 (2000) 225.
9. Aznar, M.P., Corella, J., Delgado, J., and Lahoz, J., *Industrial Engineering Chemistry Resource* 32 (1993) 1.
10. Aznar, M.P., Caballero, M.A., Gil, J., Olivares, A., and Corella, J., Biomass gasification: produced gas upgrading by in-bed use of dolomite. In: Overend, R.P. and Chornet, E. (Eds.), *Making a Business from Biomass*, Pergamon Press, New York, 1997, p. 859.
11. Rapagna, S., Jand, N., and Foscolo, P.U., *International Journal of Hydrogen Energy* 23 (1998) 551.
12. Turn, S., Kinoshita, C., Zhang, Z., Ishimura, D., and Zhou, J., *International Journal of Hydrogen Energy* 23 (1998) 641.
13. Garcia, L., Salvador, M.L., Arauzo, J., and Bilbao, R., *Energy and Fuels* 13 (1999) 851.
14. Wang, D., Czernik, S., Montané, D., Mann, M., and Chornet, E., *Industrial Engineering Chemistry Resource* 36 (1997) 1507.
15. Wang, D., Czernik, S., and Chornet, E., *Energy and Fuels* 12 (1998) 19.
16. Henrich, E. and Weirich, F., *Environmental Engineering Science* 21 (2004) 53.

17. Effendi, A., Hellgardt, K., Zhang, Z.G., and Yoshida, T., *Fuel* 84 (2005) 869.
18. Song, C., *Catalysis Today* 77 (2002) 17.
19. Kirby, K.W., Chu, A.C., and Fuller, K.C., *Sens Actuators* B95 (2003) 224.
20. Janse, A.M.C., Westerhout, R.W.J., and Prins, W., *Chemical Engineering and Processing* 39 (2000) 239.
21. Ramage, J. and Scurlock, J., Biomass. In: Boyle, G. (Ed.), *Renewable Energy-Power for a Sustainable Future*, Oxford University Press, Oxford, 1996.
22. Demirbaş, A., *Energy Conversion and Management* 42 (2001) 1357.
23. Bildgewater, A.V., Toft, A.J., and Brammer, J.G., *Renewable and Sustainable Energy Reviews* 6 (2002) 181.
24. Yaman, S., *Energy Conversion and Management* 45 (2004) 651.
25. Onay, O. and Kockar, O.M., *Renewable Energy* 28 (2003) 2417.
26. Buekens, A.G., Maniatis, K., and Bridgwater, A.V., Introduction. In: Buekens, A.G., Bridgwater, A.V., Ferrero, G.L., and Maniatis, K. (Eds.), *Commercial and Marketing Aspects of Gasifiers*, Commission of the European Communities, Brussels, Belgium, 1990, p. 8.
27. Maniatis, K., Progress in biomass gasification: An overview. In: Bridgwater, A.V. (Ed.), *Progress in Thermochemical Biomass Conversion*, Blackwell, Oxford, U.K., 2001.
28. Higman, C. and van der Burgt, M. Gasification processes, In: *Gasification* (2nd Edn.), Elsevier, Amsterdam, the Netherlands, 2008, pp. 91–191.
29. Kirkels, A.F. and Verbong, G.P.J., *Renewable and Sustainable Energy Reviews* 15 (2011) 471–481.
30. Zhang, W., *Fuel Processing Technology* 91 (2010) 866–876.
31. Couhert, C., Salvador, S., and Commandré, J.-M., *Fuel* 88 (2009) 2286–2290.
32. Svoboda, K., Pohořelý, M., Hartman, M., and Martinec, J., *Fuel Processing Technology* 90 (2009) 629–635.
33. Iliuta, I., Leclerc, A., and Larachi, F., *Bioresource Technology* 101 (2010) 3194–3208.
34. Seiler, J., Hohwiller, C., Imbach, J., and Luciani, J.F., *Energy* 35 (2010) 3587–3592.
35. Stahl, K. and Neergaard, M., *Biomass and Bioenergy* 15 (1998) 205–211.
36. Kitzler, H., Pfeifer, C., Hofbauer, H., *Fuel Processing Technology* 92 (2011) 908–914.
37. Albertazzi, S., Basile, F., Brandin, J., Einvall, J., Hulteberg, C., Fornasari, G., Rosetti, V., Sanati, M., Trifirò, F., and Vaccari, A., *Catalysis Today* 106 (2005) 297.
38. Ptasinski, K.J., Prins, M.J., and Pierik, A., *Energy* 32 (2007) 568.
39. Wang, L., Weller, C.L., Jones, D.D., and Hanna, M.A., *Biomass and Bioenergy* 32 (2008) 573.
40. van Rossum, G., Potic, B., Kersten, S.R.A., and van Swaaij, W.P.M., *Catalysis Today* 145 (2009) 10.
41. Leibold, H., Hornung, A., and Seifert, H., *Powder Technology* 180 (2008) 265.
42. Xu, C.C., Donald, J., Byambajav, E., and Ohtsuka, Y., *Fuel* 89 (2010) 1784.
43. Pfeifer, C. and Hofbauer, H., *Powder Technology* 180 (2008) 9.
44. Devi, L., Ptasinski, K.J., and Janssen, F.J.J.G., *Fuel Processing Technology* 86 (2005) 707.
45. Courson, C., Udron, L., Świerczyński, D., Petit, C., and Kiennemann, A., *Catalysis Today* 76 (2002) 75.
46. Nacken, M., Ma, L., Heidenreich, S., and Baron, G.V., *Applied Catalysis B: Environmental* 88 (2009) 292.
47. Wilhelm, D.J., Simbeck, D.R., Karp, A.D., and Dickenson, R.L., *Fuel Processing Technology* 71 (2001) 139.
48. *Ullmann's Encyclopedia of Industrial Chemistry*, (5th Edn.), John Wiley & Sons, Canada, A12, 1989, 238.
49. M. Turigg (Ed.), *Catalyst Handbook*, 2nd Ed., Wolfe, London, 1989.
50. Matsuo, Y., Yoshinaga, Y., Sekine, Y., Tomishige, K., and Fujimoto, K., *Catalysis Today* 63 (2000) 439.
51. Tomishige, K., Matsuo, Y., Sekine, Y., and Fujimoto, K., *Catalysis Communications* 2 (2001) 11.
52. Bakkerud, P.K., Gol, J.N., and Aasberg-Petersen, K., Dybkjall, lb., Preferred synthesis gas production routes for GTL. In: Bao, X. and Su, Y. (Eds.), *Studies in Surface Science and Catalysts*, Elsevier, 147 (2004) 13–18.
53. Albertazzi, S., Basile, F., Benito, P., Fornasari, G., Trifirò, F., and Vaccari, A., Technologies of syngas production from biomass generated gases. In: Kurucz, A. and Bencik, I. (Eds.), *Syngas: Production Methods, Post Treatment and Economics*, Nova Publishers, New York, 2009, pp. 409–416.
54. Holladay, J.D., Hu, J., King, D.L., and Wang, Y., An overview of hydrogen production technologies. *Catalysis Today* 139 (2009) 244–260.
55. Albertazzi, S., Basile, F., Fornasari, G., Trifirò, F., and Vaccari, A., Gasification of biomass to produce hydrogen. In: Graziani, M. and Fornasiero, P. (Eds.), *Gasification Renewable Resources and Renewable Energy*, Taylor & Francis, London, U.K., 2007, pp. 197–213.
56. Albertazzi, S., Basile, F., Brandin, J., Einvall, J., Hulteberg, C., Fornasari, G., Sanati, M., Trifirò, F., and Vaccari, A., *Biomass and Bioenergy* 32 (2008) 345–353.

57. Siedlecki, M., Nieuwstraten, R., Simeone, E., de Jong, W., and Verkooijen, A.H.M., *Energy Fuels* 23 (2009) 5643.
58. Basile, F., Albertazzi, S., Barbera, D., Benito, P., Einvall, J., Brandin, J., Fornasari, G., Trifirò, F., and Vaccari, A., *Biomass and Bioenergy* (2011), doi: 10.1016/j.biomboe.2011.06.047.
59. Ghenciu, A.F., *Current Opinion in Solid State and Materials Science* 6 (2002) 389.
60. Einvall, J., Parslanda, C., Benito, P., Basile, F., and Brandin, J., *Biomass and Bioenergy* (2011) doi: 10.1016 j.biomboe.2011.04.52.
61. Eilers, J., Posthuma, S.A., and Sie, S.T., *Catalysis Letter* 7 (1990) 253–269.
62. Dry, M.E., *Catalysis Today* 6 (1990) 183–206.
63. van Vliet, O.P.R., Faaij, A.P.C., and Wim Turkenburg, C., *Energy Conversion and Management* 50 (2009) 855–876.
64. Balat, M., Balat, M., Kırtay, E., and Balat, H., *Energy Conversion and Management* 50 (2009) 3158–3168.
65. Lange, J.P., *Catalysis Today* 64 (2001) 3–8.
66. Tijm, P.J.A., Waller, F.J., and Browna, D.M., *Applied Catalysis A: General* 221 (2001) 275–282.
67. Olah, G.A., Goeppert, A., and Suria Prakash, G.K., *Beyond Oil and Gas the Methanol Economy*, Wiley-VCH, Germany, 2009.
68. Lurgi Megamethanol®. Available at www.lurgi.com/website/fileadmin/user_upload/1_PDF/1_Broshures_Flyer/englisch/0312e_MegaMethanol.pdf (2010), Retrieved December 2010.
69. Development in Methanol Production Technology. Available at www.chemsystems.com/reports/search/docs/toc/96s14toc.pdf (1998). Retrieved December 2010.
70. Aasberg-Petersen, K., Nielsen, C.S., Dybkaier, Lb., and Perregard, J., Large Scale Methanol production from natural gas. Available at www.topsoe.com/business_areas/methanol/~/media/PDF%20files/Methanol/Topsoe_large_scale_methanol_prod_paper.ashx (2008). Retrieved December 2010.
71. Eydorn, E.C., Kornosky, R.M., Stein, V.E., Street, B.T., and Tijm, P.J.A. Liquid Phase Methanol (LPMEOH™) Project, Operational Experience. Available at www.gasification.org/uploads/downloads/Conferences/1998/gtc9807p.pdf 1998. Retrieved December 2010.
72. Filippi, E. and Zardi, F., Casale Technologies for Fertilizer and Methanol Grass-Roots Plants. Available at www.casale.ch/group/images/stories/Casale_Group/Downloads/Papers/Casale_Group/nitrogen-syngas_2009_casale_technologies_for_fert_and_methanol_grass-roots_plants.pdf 2008. Retrieved December 2010.
73. Sutton, M., Methanol Technology. Available at http://www.davyprotech.com/pdfs/Methanol%20Brochure.pdf 2010. Retrieved March 2010).
74. Ciscsery, S., *Pure and Applied Chemistry* 58 (1986) 841–856.
75. Methanol to olefins. Available at http://www.chemsystems.com/reports/search/docs/toc/00S9.pdf 2002 36–44. Retrieved December 2010.
76. Chena, Q., Bozzaga, A., and Glovera, G., *Catalysis Today* 106 (2005) 103–107.
77. Liu, Z., Sun, C., Wang, G., Wang, Q., and Cai, G., *Fuel Processing Technology* 52 (2000) 161–172.
78. Trifiro, F., *Chemistry and Industry* 92(4) (2010) 112.
79. Cascone, R., *Chemical Engineering Progress* 104 (8) (2008) 54.
80. Davies, E., *Chemistry World* 6 (4) (2009) 40.
81. Hahn, H.D., *Ulmann's Encyclopedia of Industrial Chemistry*, John Wiley & Sons, Hoboken, NJ, 2000.
82. Bahrmann, H. and Bach, H., *Ullman's Encyclopedia of Industrial Chemistry*, Wiley-VCH, Weinheim, Germany, 2000.
83. Bischoff, S. and Kant, M., *Catalysis Today* 66 (2001) 183.
84. Billig, E. and Othmer, K., *Encyclopedia of Chemical Technology*, John Wiley & Sons, Hoboken, NJ, 2000.
85. von Katepow, N. and Kindler, H., *Angewandte Chemie* 72 (22) (1960) 802.
86. Massoudi, R., *Journal of American Chemical Society* 109 (1987) 7428.
87. Majocchi, L., *Applied Catalysis A: General* 155 (1998) 393.
88. Xu, R., *Journal of Molecular Catalysis A: Chemical* 221 (2004) 51.
89. Courty, P., *Journal of Molecular Catalysis* 17 (1982) 241.
90. Mueller, H., Patent DD 256456, 1986.
91. Katikaneni, S., Gaffney, A.M., Ahmed, S., and Song, C., *Catalysis Today* 99 (2005) 255.
92. Nagel, F.P., Schildhauer, T.J., McCaughey, N., and Biollaz, S.M.A., *International Journal of Hydrogen Energy* 34 (2009) 6826–6844.
93. Seitarides, T., Athanasiou, C., and Zabaniotou, A., *Renewable and Sustainable Energy Reviews* 12 (2008) 1251–1276.

10 Future Perspectives for Hydrogen as Fuel in Transportation

Loredana De Rogatis and Paolo Fornasiero

CONTENTS

10.1 INTRODUCTION

The world's current energy system, based primarily on fossil fuels, does not correspond to the requirements of sustainability in many respects [1–2]. The expanding world population and the improving standards of living and demands for energy in developing countries is putting increasing pressure on diminishing fossil fuel resources and making them even more costly. Predictions based on extrapolation of the energy consumption show that the demand will soon exceed the supply. However, it is difficult to establish exactly how long the fossil fuels will last. New oil and gas fields are being still discovered, and the methods for retrieving oil from known fields are continuously improving although extraction energy costs would become higher than the actual energy yield due to the increased energy costs for research, deep drilling, as well as to lower quality and accessibility of the still available oil storages [3]. Vast reserves, like tar and gas hydrates, await novel technology to enable their economically and environmentally sound exploitation. Large coal reserves can also be exploited, for example, through gasification and Fischer–Tropsch synthesis.

In addition to these aspects, there is clearly a problem of worldwide energy dependence. The largest deposits of oil, gas, and coal are concentrated in particular regions of the globe often characterized by political instability in their international relationships. This highlights the necessity to move away from the exclusive dependence on fossil fuels for the major oil importers [4].

The damaging environmental effect of continuous fossil fuel usage is another key factor boosting the interest in new energy strategies. Although there is a considerable debate as to whether increased fossil fuel consumption is the primary cause of global climate change [5–8], there is a general agreement that a strong correlation exists between localized and regional air pollution and fossil fuel consumption.

Nowadays, it is generally accepted that the solution to the world's environmental and energy challenges is not one single energy source or technology, but all options need to be pursued and explored. Indeed, any single strategy can meet the world energy needs and fulfill acceptable exhaust emission at the same time. Each country thus should develop its own "local" strategy that must be economically and environmentally sustainable and be able to meet the demands for a broad range of its own services (household, commerce, industry, and transportation needs). The three "A" criteria of the World Energy Council is often adopted as guidance for assessing possible technological options and policy decision: *accessibility* to affordable energy, environmental *acceptability* of the energy sources, and reliable and secure *availability*.

The need for new and sustainable energy technologies is particularly urgent in the transport sector, where energy demand keeps growing. Indeed, over time as population has increased, cities have grown, and globalization and free trade have increased the movement of people and goods, our transportation infrastructure and systems have expanded dramatically.

Energy used in transportation sector includes the energy consumed in moving people and goods by road, rail, water, and pipeline. The transport is second only to the industrial sector in terms of total end-use energy consumption. Almost 20% of the world's total delivered energy is used for transportation, most of it in the form of liquid fuels. Transportation alone accounts for more than 50% of world consumption of liquid fuels. Although the 2008–2009 global economic recession has resulted in a slowdown in transportation sector activity, transport energy use is expected to continue to escalate in the coming decades in untenable way due to the continued growth in the world's population as well as the progressive industrialization of developing nations [9].

Transport is also one of the primary sources of air pollution (mainly CO, hydrocarbons, NO_x and particulate matter) and greenhouse gases (GHG) [10]. Indeed, it produces a significant share of the global fossil fuel combustion-related CO_2 emissions (~20%) and is the fastest growing sector in terms of GHG emissions especially in developing countries.

These unsustainable levels of growth in pollutant emissions must be cut in order to respect the more and more stringent environmental regulations. This requires action through technological innovation and travel demand-management strategies as well as the setting of ambitious targets and incentives for the introduction of low-emission vehicle technologies, clean fuels, and appropriate modal choices.

The development of a sustainable transport system is surely an enormously complex and dynamic subject. It does not merely mean the reduction of negative environmental impacts within the sector. It involves fundamental changes to our social, economic, and industrial systems. The major policy directions for dealing with environmental and energy challenges are (1) to reduce their use, (2) to clean up their emissions, and (3) to make them more fuel efficient.

The transition to new fuels and/or energy carriers is problematic in the transportation area because of the diffuse nature of the system and its complex public–private composition. Considering land vehicles only, there are more than 600 millions of motor vehicles worldwide, with an annual production rate in 2009 of 48 millions for passenger cars, 10 millions for light commercial vehicles, 3 millions for heavy trucks, and 313,000 for buses and coaches [11]. The geographically diffuse distribution of vehicles favors fuels that are easy to transport and store (i.e., fuels that are liquid at room temperature). Consider, for instance, that natural gas and electricity are generally less expensive and tend to be cleaner than liquid fuels, but they are much more difficult to transport, in the case of natural gas, and much more difficult to store, in the case of electricity. Alcohol fuels are easy to transport and store, but they tend to be more expensive than gasoline, petroleum, diesel, natural gas, and electricity. Most vehicular fuels continue to be gasoline and diesel.

Advances in engine and vehicle technology have been developed for reducing the toxicity of exhaust leaving the engine, but these alone are unable to meet zero emission goal [12]. In this respect, typical examples of existing end-of-pipe technologies for automotive pollution control are represented by the three-way catalytic (TWC) converter and diesel particulate filter.

The TWC converter represents one of the technological success stories of the past 40 years [13,14]. To date, it is the most advanced systems used in gasoline vehicle to remove exhaust pollutants such as carbon monoxide, unburnt or partially burnt hydrocarbons, and nitrogen oxides. Its name derives from the ability to simultaneously convert these three categories of pollutants. The physical arrangement of a modern automotive catalyst consists of a thin layer of the porous catalytic material (washcoat) coated on the channel walls of a ceramic or sometimes metal monolith. The key feature of this catalytic device is the presence of noble metals, as the catalytic activity occurs at the metal center. Specifically, Pt and Pd are chosen to promote oxidation reaction while Rh is added to promote nitrogen oxides dissociation. Pushed by ever more stringent legislation, the formulation of three-way catalysts have evolved significantly since they were first introduced as a result of a combination of several factors such as material advances, improved engine, and fuel characteristics. They convert more than 98% of the pollutants. The design of new three-way catalysts that are more active at low temperatures continues together with the investigation and development of new technological solutions [15].

In the case diesel engines, Particulate Matter (PM) remains a particular problem. Two technologies are generally used to control and remove diesel PM or soot from the exhaust gas: diesel oxidation catalysts for the liquid fraction of the PM [16] and particulate filters [17]. Wall-flow diesel particulate filters usually remove 85% or more of the soot and can at times attain soot removal efficiencies of close to 100%. A diesel-powered vehicle equipped with functioning filter emits no visible smoke from its exhaust pipe. Notably, smaller are the dimension (aerodynamic diameter) of the particles, more dangerous is the PM. Filter efficiency decreases with diminishing particle size. Therefore, further efforts must be made in order to improve the removing capacity of the small fraction of PM that is not filtered and oxidized yet.

The benefit of these technical measures to reduce vehicle emissions is outstripped by the increase in vehicle numbers, engine size, travel frequency, and trip length. Pollution prevention in many ways appears to be a more obvious solution, and it is indeed an approach that is actively pursued. In conjunction with end-of-pipe technologies, a whole range of measures aimed at pollution prevention are adopted.

One of the most controversially discussed long-term solutions to depleting oil reserves and environmental issues is the introduction of hydrogen as an energy vector in combination or not with fuel cell vehicles.

Hydrogen has not to be considered the ultimate solution but rather one of the potential energy vector and thus as part of sustainable energy future for coming years. Hydrogen used as energy carrier may not solve all energy problems related to the security of supply and environmental issues, but hydrogen energy in combination with fuel cell technology may fit well into a future energy system based on renewables. Moreover, hydrogen technology could penetrate into most of the energy cycle applied in our society and can provide sustainability, diversity, high efficiency, and flexibility. Future energy supply systems based on hydrogen could be designed for combined production of electricity and heat. Such systems can achieve very high overall efficiencies, and due to the modular nature of fuel cell technologies, they can be configured to be distributed, centralized, or anywhere in between.

The use of hydrogen as an energy carrier is not new. It has been known for approximately two centuries. One of the early internal combustion engines, developed by Isaac de Rivaz in 1805, was fueled by hydrogen. Ninety years later, Rudolf Diesel invented the diesel engine that was fueled by pulverized coal (carbon and hydrogen). NASA has used liquid hydrogen since the 1970s to propel the space shuttle and other rockets into orbit. NASA was the first to use fuel cells powered by hydrogen as a dependable source of both electric power and potable water on manned missions. In the United States, "town gas" was used to light city streets and was piped into homes to fuel lamps, cooking stoves, and gas heaters. Town gas was produced from coal and contained about 50% hydrogen along with methane, carbon dioxide, and carbon monoxide. Town gas is still used safely today in parts of Japan, China, and other Asian countries.

However, before a hydrogen-fueled future can become an integrated reality, many complex challenges must be still overcome. Unlike the conventional fuels, the inherent properties of hydrogen make it a difficult commodity to produce, store, and handle on a large scale. Nevertheless, several companies and researcher groups are inching ahead by proposing solutions to the problems in hydrogen technologies.

The development of "Hydrogen Economy" needs the design of an infrastructure comprising five key elements: production, delivery, storage, conversion, and applications, which are in different stages of technological advancement. While hydrogen production and conversion are already technologically feasible, its large-scale delivery and storage still face serious challenges. For example, due to possible embrittlement of steel, existing natural gas transmission systems may be unsuitable for the transportation of pure hydrogen gas. Therefore, other options are being considered. In comparison, prototype hydrogen-based fuel cell vehicles are already being tested and demonstrated in Europe, Japan, and the United States; and some major automotive manufacturers have announced that they will market fuel cell vehicles already by several years. Considerable efforts are being made to develop infrastructure solutions to support these vehicles, both through national R&D and demonstration programs and by the car industry. The development of hydrogen vehicles and, at least locally, of other emerging second-generation biofuels (e.g., biodiesel and bioethanol) brings an urgent need for common codes, standards, and regulations.

Apart from the technological challenges, hydrogen has to compete on price with the conventional fuels in order to have a wide user base. The cost of hydrogen is proportional to the volume demand. If the demand increases, then the price can increase. However, the technological advancements in hydrogen technologies and in the fuel cells can bring down the cost of these technologies.

10.2   HYDROGEN PRODUCTION AND PURIFICATION METHODS

Hydrogen is currently mainly used as raw material in the petroleum industry for refining and upgrading of crude oil and in the chemical industry for the manufacture of ammonia (e.g., for fertilizers), methanol, and a variety of organic chemicals [18]. Other important uses are found in the food industry for the hydrogenation of edible plant oils to fats (margarine) and in the plastics industry for making various polymers. Lesser applications occur in the metals, electronics, glass, electric power, and space industries (Figure 10.1).

Hydrogen world production is around 45–50 Mt per year [18]. Most of it derives from natural gas (mainly methane for 48%) by steam reforming. The rest is principally obtained from oil (30%) and coal (18%) by the partial oxidation processes. Only 4% of the hydrogen worldwide is generated by electrolysis, invariably only when high purity is required or in the presence of special reasons that make this route economic. For instance, where a surplus of cheap hydroelectricity is available or when the hydrogen is a by-product of the chlor-alkali process.

By far the great part of the hydrogen produced in petrochemical complexes is consumed on site. For example, ammonia or methanol plants are likely located next to a hydrocarbon cracker or steam reformer. Distribution by pipeline is therefore confined to the same or to adjacent plants. Very little hydrogen is supplied to other consumers, for example, for use in the electronics or nuclear industries, and it is termed "merchant hydrogen." As the quantities tend to be relatively small, merchant hydrogen is generally distributed in the form of compressed gas in steel cylinders. A small proportion of hydrogen is liquefied and transported in cryogenic vessels, and it is mostly used in rather specialized applications (e.g., fuel for space rockets). Sometimes, hydrogen is liquefied for transportation purposes, particularly where substantial quantities are required and the distances between production and end-users are considerable.

The introduction of hydrogen in transport sector raises a fundamental question: how and where to produce it. The answer is not straightforward. Several options (centralized and/or decentralized facilities) have been proposed and critically analyzed. Hydrogen can be produced in large central plants several hundred kilometers from the point of end-use; in smaller, semicentral plants within 30–150 km of the end-use point or in small generation facilities located near or at the point of end-use.

Perhaps the easiest pathway to envision in the short term is one where users continue to fuel their cars with gasoline. An onboard reformer then generates hydrogen from the gasoline for use in the fuel cell. Another possibility is to produce hydrogen at small, decentralized locations, such as the household level or fueling stations. Finally, hydrogen could be produced at large, centralized facilities, and then transported to fueling stations by truck or pipeline, much like natural gas and gasoline are distributed today.

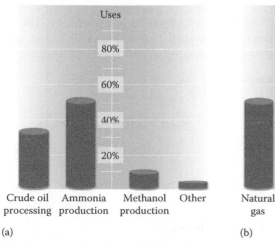
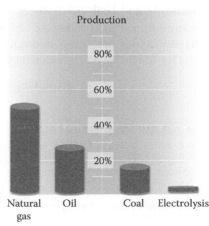

FIGURE 10.1   (a) Hydrogen uses. (b) Hydrogen sources.

Producing hydrogen onboard as needed would solve the infrastructure problem, but there would be also some drawbacks. The reformer would significantly increase the cost of the vehicle and take up additional room. Furthermore, the additional weight would generate a fuel penalty. A reformer placed at a centralized location might operate 80%–95% of the time, but an onboard version would operate a fraction of that time (only during driving) and be an underutilized capital asset. In addition, a hydrogen storage system is needed to guarantee hydrogen supply during start-up process or when high and fast power is required by the driver (i.e., instantaneous acceleration step). Most importantly, if the original fuel is gasoline, this option would not solve the problems associated with the current hydrocarbon energy system. For this reason, some suggest it only as a transitional solution to a more fully developed hydrogen economy.

Hydrogen could be produced at large centralized facilities and then transported to fueling stations. Several options for centralized production exist, including steam reforming of natural gas, coal gasification, electrolysis, noncatalytic partial oxidation of petroleum, and thermochemical splitting of water using high temperatures from either nuclear or solar sources. Once produced, the hydrogen would be delivered by truck, rail, ship, or pipeline to the fueling station and transported in either gaseous or liquid form or stored in the lattices of certain metals. Depending on the actual method of transportation, the hydrogen might have to be stored at the production plant and/or the fueling station. The end user would then directly refuel the vehicle with hydrogen.

Perhaps the most compelling case for centralized production is that it would allow for greater diversification in terms of energy feedstock. Of course, if there is a clear economic winner, such as coal gasification, then diversification may not happen.

The single largest drawback to the centralized approach is that the hydrogen must then be transported to the end user. Unfortunately, the low volumetric density of hydrogen relative to other fuel options makes it difficult to transport. A more decentralized approach to hydrogen production could eliminate, or at least greatly reduce, delivery and storage costs. Decentralized options include use of small-scale reformers or electrolyzers at either the fueling station or possibly even at the individual household level.

Hydrogen generation can be achieved by a variety of chemical, electrochemical, biological, and other methods. The production technologies can be fallen into three general categories: (1) thermal processes, (2) electrolytic processes, and (3) photolytic processes. Each approach has merits and limitations, and the choice of the best route depends on the economics of the location involved.

10.2.1 THERMAL PROCESSES

Thermal processes include reforming of hydrocarbons, gasification, renewable liquid fuel reforming, and high-temperature water splitting.

10.2.1.1 Reforming of Hydrocarbons

There are many different approaches and technologies for hydrocarbon reforming. The three most widely utilized methods are steam reforming (SR), partial oxidation (POX), and autothermal reforming (ATR), which can be represented by the following generic chemical equation:

$$C_nH_m + [Ox] \rightarrow xH_2 + yCO + zCO_2 \tag{10.1}$$

where
   C_nH_m is a hydrocarbon ($n \geq 1$ and $m \geq n$)
   $[Ox]$ is an oxidant such as O_2, H_2O, and CO_2

10.2.1.1.1 Steam Reforming
About 40% of the world hydrogen production is obtained by steam reforming of methane (SMR), the main component of natural gas [18,19]:

$$CH_4 + H_2O \rightarrow 3H_2 + CO \tag{10.2}$$

The technology is well developed and commercially available at a wide capacity range [20].

The reaction is highly endothermic and favored by low pressures. However, in most industrial application areas, hydrogen is required at the pressure of at least 20 bar; therefore, the reformers work at elevated pressures (usually, 20–30 bar) and at temperatures of about 850°C–900°C. High pressures allow for a more compact reactor design, thus increasing the reactor throughput and reducing the cost of materials. Industrial steam reformers typically use direct combustion of a fraction of methane (although other heat sources could be used) to provide the heat required for the process. In this respect, under investigation, there is also the possibility to use solar energy [21–26] or nuclear power [27,28].

Catalysts (e.g., nickel or nickel on aluminum oxide, cobalt, alkali, and rare earth mixtures) are widely used to accomplish the process at the practical range of temperature since the direct thermal (i.e., noncatalytic) interaction of methane with steam would require extremely high temperatures (in excess of 1000°C). The catalyst adsorbs methane molecules and step-by-step strips them of hydrogen atoms, ready for the subsequent conversion.

According to the reaction stoichiometry, the molar ratio of steam to methane is $H_2O:CH_4 = 1:1$; however, in practice, an excess of steam is used to prevent carbon (coke) deposition on the catalyst surface. The reactions responsible for carbon deposition on the reforming catalyst surface are the methane decomposition to H_2 and carbon equation (10.3) and the CO disproportionation equation (10.4). Addition of the supplemental steam shifts the reforming reaction equilibrium away from carbon formation:

$$CH_4 \rightarrow C + 2H_2 \tag{10.3}$$

$$2CO \Leftrightarrow C + CO_2 \tag{10.4}$$

Carbon may appear as filaments on the catalyst surface, encapsulating nickel metal particles and largely blocking the steam reforming reaction [29]. As mentioned earlier, excess steam is effective in preventing cracking or at least diminishing the negative effects by cleaning the catalyst surfaces.

The resulting mixture, H_2 and CO, is known as "synthesis gas" (or "syngas"). It may be used for the synthesis of a range of commodities including several organic chemicals such as methanol, formaldehyde, oxo-alcohols, and polycarbonates. This gaseous mixture, containing besides H_2 and CO, also steam and usually about 4% of unconverted methane, is rapidly cooled at 350°C and added to the reformer at high temperature. It is fed to Water Gas Shift (WGS) reactors, where CO reacts with steam over catalyst beds producing CO_2 and extra H_2:

$$CO + H_2O \Leftrightarrow CO_2 + H_2 \tag{10.5}$$

Therefore, the WGSR serves two important purposes: the reduction of the CO content of the primary reformate (up to 12% for steam reforming) and the increase of the hydrogen content. A number of extensive reviews of the WGS reaction were published [30–33]. The reaction is moderately exothermic. The increase in the catalyst temperature, while favorably increasing the rate of reaction, tends to shift the thermodynamic equilibrium to the left. In the practical system, a compromise is made between a high rate and unfavorable equilibrium and a lower rate and more favorable equilibrium [34]. To achieve high CO conversion, two reactors are commonly used in series: high temperature (HT)- and low temperature (LT)-WGS reactors, respectively. The HT-WGS reactor operates at the inlet temperatures of 340°C–360°C and uses an iron–chromium-based catalyst (90%–95% magnetite iron oxide, stabilized with 5%–10% of chromia, Cr_2O_3) [31]. A typical LT-WGS catalyst consists of CuO: 15%–30%, ZnO: 30%–60%, and Al_2O_3: balance, and it promotes a favorable reaction rate between 200°C and 300°C. The combination of HT- and LT-WGS reactors converts 92% of the CO in the reformate gas into H_2 [35] and lowers the CO content in the gas to about 0.1 vol%.

Finally, in the preparation of hydrogen, it is necessary to separate the gas production from the carbon dioxide and any other gas that may be present. Older SMR process used solvents to remove the acid gas (CO_2) from the gaseous stream after WGS reactors. Solvents used commercially for CO_2 removal in the gas separation unit include monoethanolamine (the most preferred and widely used solvent), water, ammonia solutions, potassium carbonate solutions, and methanol. This operation allows the reduction of CO_2 concentration in the process gas to about 100 ppm. The residual CO_2 and CO are removed in a methanation reactor where in the presence of H_2 they are converted to CH_4 (320°C, catalyst: Ni or Ru on oxide support).

Modern SMR plants incorporate a Pressure Swing Adsorption (PSA) unit for purifying hydrogen from CO_2, CO, and CH_4 impurities (moisture is preliminarily removed from the process gas).

The PSA unit consists of multiple (parallel) adsorption beds, most commonly filled with molecular sieves of suitable pore size operating at the pressure of about 20 atm. Impurities are selectively adsorbed, while pure hydrogen is withdrawn at high pressure. To allow continuous operation, multibeds connected in parallel are often used. Once a bed is saturated with impurities, the feed is switched automatically to another fresh bed to maintain a continuous flow of hydrogen. The reduction of the pressure in a number of discrete steps releases the adsorbed gases, and it regenerates spent adsorbent. Usually, for existing hydrogen plants, all the desorbed carbon dioxide is vented to the atmosphere, though in principle it is possible to separate it. Generally, SMR plants with PSA need only a HT-WGS stage, which may somewhat simplify the process.

Pressure swing adsorption yields high-purity hydrogen (up to 99.999%), and recovery of up to 90% is routinely achieved in industrial plants. The process is highly reliable, flexible, and easy to operate with a modest energy input. On the other hand, PSA suffers from limited capacity, and it requires to cool the gases to effect separation. After this operation, hydrogen is therefore at low pressure and low temperature. Such conditions make a nonideal feedstock for a gas turbine or a high-temperature fuel cell.

Because of a great practical importance of SMR as a major industrial process in H_2 manufacturing, the development of efficient steam reforming catalysts is a very active area of research. Nickel and noble metals are known to be catalytically active metals in the SMR process. The relative catalytic activity of metals in the SMR reaction is as follows [36]:

$$Ru > Rh > Ir > Ni > Pt > Pd$$

Although Ni is less active than some noble metals and more prone to deactivation (e.g., by coking), it is the most widely used catalyst for the SMR process due to its relatively low cost. The catalyst most commonly used in the reforming reaction is the high-content Ni catalyst (~12%–20% Ni as NiO) supported on a refractory material (e.g., α-Al_2O_3) containing a variety of promoters [31]. Key promoters include alkali ions to minimize carbon deposition on the catalyst surface. The Ni catalyst is manufactured in a variety of shapes to ensure high surface to volume ratio, optimal heat, and mass transfer, low pressure drop, high strength, etc. There are several stringent requirements to the reforming catalyst performance, which include long-term stability and high tolerance of the extreme operating conditions (e.g., very high temperature); robustness to withstand the stress of start-up and transient operational conditions; nonuniformity of the feedstock, which may expose the catalyst to poisons (e.g., sulfur); and excessive coke deposition. The industrial reforming catalysts are supposed to perform in excess of 50,000 h (or 5 years) of continuous operation before replacement [31].

The role of the support (or carrier) is to provide support for the catalytically active metal to achieve a stable and high surface area. The influence of the support on the activity of catalysts in the SMR reaction is often underestimated. It not only determines the dispersion of the catalytically active metal particles, but it also affects the catalyst's reactivity, resistance to sintering, coke deposition, and may even participate in the catalytic action itself. From this viewpoint, the support is an integral part of the catalyst and cannot be considered separately. Among the most common supports for SMR catalysts are α- and γ-Al_2O_3, $MgAl_2O_4$, SiO_2, ZrO_2, and TiO_2. These supports have relatively high surface area and porosity and suitable pore structure and surface morphology, which are conducive to better contact between the reactants and the catalyst. Furthermore, due to the nature of the chemical bonding between the support and the metal particles, the electronic properties of the metal, and hence, its catalytic activity is affected. For example, the supports with pronounced acidic properties are known to facilitate decomposition of methane. Generally, a strong interaction between a metal and a support makes the catalyst more resistant to sintering and coking, thus resulting in an enhanced long-term stability of catalysts [37].

Although SMR is a well-developed technology, there is room for further technological improvements, in particular, with regard to energy efficiency, gas separation, and H_2 purification stages. For instance, in the Sorption-Enhanced Reforming (SER) process, one of the gaseous reaction products (CO_2) of the catalytic reforming reaction is separated by sorption. As a result, the equilibrium of the reaction is shifted toward products. The advantages are fourfold: (1) fewer processing steps, (2) improved energy efficiency, (3) elimination of the need for shift catalysts, and (4) reduction in the temperature of the primary reforming reactor by 150°C–200°C. Efforts are being made to combine the WGS reaction and the removal of CO_2 into a single step by using a membrane reactor where hydrogen produced in the reforming reaction selectively permeates through a membrane leaving the reaction zone [38].

Typically, the several microns thick membranes are made of Pd or Pd/Ag or other Pd-based alloys. The main advantages of Hydrogen Membrane Reactor (HMR) are as follows: (1) the H_2 producing reactions are free from the limitations of chemical equilibrium (i.e., equilibrium is shifted toward products), (2) high methane conversions are reached at lower temperatures (compared to a conventional reactor), (3) the process produces separate H_2 and CO_2 flows, (4) there is no need for additional CO-shift converters, (5) the reactor has a more simple and compact configuration, and (6) overall efficiency is higher. Owing to the relatively low-temperature resistance of the Pd-based membranes, HMRs operate at 400°C–600°C (compared to 800°C–950°C typical of conventional reformers). As a result, HMR requires a low-temperature very active catalyst. A particular goal is the high quality of membranes operating at high temperatures, as this would obviate the need to cool and reheat large volumes of gases. Similarly, high-pressure making membranes are desirable in order to improve the efficiency of the process and to deliver the separated gases under pressure. The latter feature would save the considerable energy (and cost) involved in recompressing the gases for storage and transport.

10.2.1.1.2   Partial Oxidation

Partial oxidation (POX) is an alternative route for hydrogen production on a commercial scale.

$$C_xH_y + x/2O_2 = xCO + y/2H_2 \qquad (10.6)$$

In POX process, fuel and oxygen (or air) are combined in proportions, such that the fuel is converted into a mixture of H_2 and CO. There are several modifications of the POX process, depending on the composition of feed and on the type of the reactor used. The overall process is exothermic, and it can be carried out catalytically or noncatalytically. The noncatalytic one operates at high temperatures (1100°C–1500°C), and it utilizes any possible carbonaceous feedstock including heavy residual oils (HROs) and/or coal. The catalytic process is carried out at a significantly lower range of temperatures (600°C–900°C) and, generally, uses light hydrocarbon fuels as a feedstock.

If pure oxygen is used in the process, it significantly increases the cost of the system due to its production and storage. In contrast, if the POX process uses air as an oxidant, the effluent gas is diluted with nitrogen resulting in larger WGS reactors and gas purification units.

Although POX of hydrocarbons with air as an oxidant seems to be economically advantageous over the process using pure oxygen, the downstream process requirements cancel the benefits of using air. This is due to the fact that in air-blown POX, syngas is rather diluted requiring a more complex and expensive gas separation unit. In addition, the cost of compression of diluted syngas to >20 bar (which is the pressure typically required in the majority of downstream industrial processes) is high.

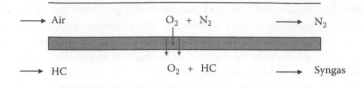

FIGURE 10.2 Oxygen permeable membrane scheme.

Recently, important advances were made in syngas production technology using oxygen-permeable membranes (OPMs) (Figure 10.2). The use of OPM integrating the oxygen separation allows POX processes in one single reactor (thus avoiding a costly oxygen production plant and reducing the cost of syngas production by 25%–40% [39]. Viable membrane material for the partial oxidation process should meet several stringent requirements; in particular, the membrane must be chemically and mechanically stable at elevated temperatures (1000°C and higher).

10.2.1.1.3 Autothermal Reforming

Autothermal reformers combine some of the best features of steam reforming and partial oxidation systems. A hydrocarbons feedstock (methane or liquid fuel) reacts with both steam and air (or oxygen) to produce a hydrogen-rich gas. The reaction takes place at high temperature (950°C–1100°C) and at pressures up to 100 bar. During this process, both the steam reforming and partial oxidation reactions concurrently run. Using the right mixture of input fuel, air/oxygen, and steam, the latter process supplies the heat required to drive the former. The autothermal reformer requires no external heat source and no indirect heat-exchangers. Consequently, the technology is at the same time simpler and more compact, and it will probably have a lower capital cost. Moreover, since the heat generated by partial oxidation is fully utilized to drive steam reforming, autothermal reformers typically offer higher efficiency (up to 80%–90% is possible) than partial oxidation, in which the excess of heat cannot easily recovered. Combustion conditions have to be carefully controlled to avoid carbon formation, but, overall, the unit has the advantage of being able to operate at low steam-to-carbon molar ratios. ATR is the preferred technology for the large-scale production of synthesis gas for methanol or synthetic diesel plants based on Fisher–Tropsch synthesis. For hydrogen plant production, using oxygen, ATR is economical only at very large-scale production because of the capital cost of the oxygen plant.

Nonoxidative conversion of hydrocarbon feedstocks to hydrogen generally occurs by the splitting of C–H bonds in the hydrocarbons in response to an energy input (heat, plasma, radiation, etc.), and it does not require the presence of oxidizing agents. The nonoxidative hydrocarbon-to-H_2 conversion processes are thermal, catalytic, and plasma hydrocarbon decomposition processes. The following general chemical reaction describes the nonoxidative transformation of hydrocarbons to hydrogen Equation 10.7:

$$C_nH_m + [\text{energy}] \rightarrow xH_2 + yC + zC_pH_q \qquad (10.7)$$

where
C_nH_m is the original hydrocarbon feedstock ($n \geq 1$, $m \geq n$)
C_pH_q represents relatively stable products of the feedstock cracking ($z \geq 0$, $p \geq 1$, $q \geq p$; in most cases, C_pH_q is CH_4 or C_2H_2)
[energy] is the energy input

10.2.1.2 Gasification

In our discussion, gasification is defined as the transformation of solid (fossil or renewable feedstock) into combustible gases in the presence of an oxygen carrier (e.g., air, oxygen, steam, and carbon dioxide) at high temperatures (e.g., 4700°C). The energy required for thermochemical conversion is provided by partial combustion of either the solid or the resulting syngas. Almost any fossil fuel can be treated in this way to produce hydrogen. The most widely distributed fossil fuels are the various types of coal, but other possibilities exist such as tar sands (now renamed "oil sands"), asphalts, heavy oils extracted from shale or refinery residues, and coke.

As supplies of natural gas that is mainly used in electricity production become depleted, gasification of coal is seen as the preferred technology for future electricity production, along with the renewable forms of energy. Coal is in fact plentiful and wide available, the gasification process is more efficient than conventional coal combustion, and the carbon dioxide can be extracted from the syngas before combustion and stored underground. All these benefits, however, have yet to be realized practically on a commercial scale. Major developments are needed to improve the performance and reduce the cost of the technologies.

Coal gasification facilities are being constructed around the world, particularly in China— primarily for the manufacture of chemicals and fertilizers, but also for more efficient generation of electricity. It is only now that the production of pure hydrogen from solid fuels, for use as an energy vector, is seen as a long-term goal. Here, the driving force is to utilize the world's extensive coal deposits in an environmentally friendly fashion to meet future energy demands.

Biomass, which includes agricultural waste and forestry wastes, energy crops, and municipal waste, may also be processed by gasification technology to produce hydrogen. In principle, hydrogen may be produced by gasification of any type of biomass. However, the most suitable materials are those with water contents in the range 5–30 wt.%. The gasification is generally conducted at lower pressures and temperatures than those used for coal and is limited to mid-size operation due to the heterogeneity of biomass, its localized production, and the relatively high costs incurred in its gathering and transportation.

The largest of such plants may process a few hundred tons of material per day. Properties such as moisture content, mineral fraction, density, and chemical composition (e.g., chlorine and nitrogen contents) differ widely depending on the source of the biomass. This may necessitate the development of a specific design of gasifier to accommodate a particular type of biomass. The wide diversity in the characteristics of biomass is a challenge for gasification processes, and in practice a range of technical designs of gasifier will be employed to convert different feedstocks into biogas and into hydrogen, as dictated by the operation scale and by the market. For localized production of hydrogen, the gasifier needs to be coupled to a gas cleanser and shift reactor. One of the major attractions of preparing hydrogen from biomass is that this fuel is rated as a renewable (since it is generally regarded as carbon neutral), and it provides the only direct route for bulk hydrogen production from a renewable source.

In addition to the nature and composition of the biomass, the feasibility of gasification is influenced by the entire supply system. Depending on the choice of gasification technology, various preparatory steps such as sizing (shredding, crushing, and chipping) and drying have to be included to meet process requirements. Storage, processing (e.g., densification) before transport and pretreatment affect moisture content, ash content, particle size, and sometimes even the degree of contamination (e.g., biomass obtained from salt marshes must be washed thoroughly).

Most biomass has been used for the heating of buildings, for cooking in rural areas and, more recently, for electricity production, either in conventional plant or in gasifiers. Where, in future, there hydrogen in modest amounts is locally requested and if a suitable source of biomass is available, gasification may be an acceptable means for disposing of waste with the added benefit of obtaining a useful product. It should be noted, however, that the overall energy efficiency of converting solar energy to hydrogen via the growing and harvesting of energy crops, gasification, and hydrogen separation is very poor, namely, less than 1%.

Moreover, the cost of the resulting hydrogen is several times that of hydrogen produced by the large-scale steam reforming of natural gas. Account must also be taken of the environmental impact associated with the growing, production, and transport of biomass, together with any potential degradation in land quality that might arise from the intensive farming that is required. Also, energy inputs in the form of fertilizers and transport fuel are not inconsiderable.

In a world that is both food-limited and carbon-constrained, there is no conflict that agricultural land is better utilized in growing food, rather than in raising energy crops. However, this leaves open space for second and third generation of biomass.

10.2.1.3 Reforming of Renewable Liquid Fuels

Renewable liquid fuels, like ethanol or bio-oils made from biomass resources, can be reformed to produce hydrogen by means of steam reforming, partial oxidation, and autothermal reforming like natural gas [40]. Besides these processes, aqueous phase reforming (APR) is another remarkably flexible process (see Chapter 3), which allows the production of hydrogen from biomass-derived oxygenated compounds such as sugars, glycerol, and alcohols using a variety of heterogeneous catalysts based on supported metals or metal alloys [41,42]. The APR process is a unique method that generates hydrogen from aqueous solutions of these oxygenated compounds in a single-step reactor process compared to more reaction steps required for hydrogen generation via conventional processes that utilize nonrenewable fossil fuels. The key breakthrough of the APR process is that the reforming is done in the liquid phase. Moreover, hydrogen is produced without volatilizing water, which represents a major energy saving. APR process occurs at temperatures (typically from 150°C to 270°C) and pressures (typically from 15 to 50 bar) where the WGSR is favorable, making it possible to generate hydrogen with low amounts of CO in a single chemical reactor. The latter aspect makes it particularly suited for fuel cells applications as well as for chemical process applications. The hydrogen-rich effluent can be effectively purified using pressure-swing adsorption or membrane technologies, and the carbon dioxide can also be effectively separated either by sequestration or used as a chemical. By taking place at low temperatures, APR also minimizes undesirable decomposition reactions typically encountered when carbohydrates are heated to elevated temperatures. Various competing reaction pathways are also involved in the reforming process.

The leaching of catalyst components into the aqueous phase during the reaction represents a possible disadvantage of the process. Therefore, the choice of catalyst-support materials has to be limited to those that exhibit long-term hydrothermal stability. Feed concentration is another important parameter that has to be taken into account.

Interestingly, some energy companies are currently commercializing the APR process. The Virent Energy Systems INC has recently developed APR systems for two applications: on-demand hydrogen generation for proton exchange membrane (PEM) and solid oxide fuel cell (SOFC) fuel cells and on-demand generation of hydrogen-rich fuel gas from biomass-derived glycerol and sorbitol to fuel a stationary internal combustion engine driven generator [43].

10.2.1.4 High-Temperature Water Splitting

High-temperature water splitting using solar concentrators is a technology under active development. A solar concentrator uses mirrors and a reflective or refractive lens to capture and focus sunlight to produce temperatures up to 2000°C. Researchers have identified more than 150 chemical cycles that, in principle, could be used with heat from a solar concentrator to produce hydrogen [44].

One of such pathways is the zinc/zinc oxide cycle, in which zinc oxide powder passes through a reactor that is heated by a solar concentrator operating at about 1900°C. At this temperature, the zinc oxide dissociates to zinc and oxygen gas. The zinc is cooled, separated, and reacted with water to form hydrogen gas and solid zinc oxide. The zinc oxide can be recycled and reused to produce hydrogen [45]. Other materials include reducible oxides such as CeO_2 and CeO_2-ZrO_2 mixed oxides [46].

10.2.2 Electrolytic Processes

Only a very small proportion of the world's hydrogen is produced directly by water electrolysis (~0.1%). Moreover, in the last years, even this small quantity has declined due to the increase of energy prices. On the other hand, the largest source of electrolytic hydrogen is the chlor-alkali industry where the main products are chlorine and sodium or potassium hydroxide, while hydrogen, although pure, is usually regarded only as a by-product of low value. This hydrogen may be pumped to a nearby oil refinery or burnt to raise steam.

Therefore, the electrolysis of water to generate hydrogen or oxygen is practiced when the cost of electricity is not a prime consideration. Large-scale plants have been built for instance in Brazil, Canada, Egypt, and Norway, which are countries that have surplus hydroelectric capacity or in France, Belgium, and Switzerland that have a large nuclear component.

The decomposition of water by electrolysis involves two partial reactions that take place at the two electrodes. The electron-conducting electrodes are separated by an electrolyte. Hydrogen is produced at the negative electrode (cathode) and oxygen at the positive electrode (anode). Energy to split the water molecules is supplied as electricity to the two electrodes, and the necessary exchange of charge occurs as ions flow through the electrolyte.

Conventional water electrolysis uses an alkaline aqueous electrolyte containing up to 30–40 wt.% of potassium hydroxide in order to minimize electricity consumption. The electrolyte must be prepared from very pure water with the aim to prevent fouling in the system. The gas sides are divided by a diaphragm that is made from polysulfone polymers or nickel oxide. Typical operating temperatures are between 80°C and 100°C. The electrode material is in most cases Raney-nickel. At output pressures of 2–5 bar, the process can reach efficiencies of around 65%. It is possible to improve the efficiency of the process designing high-pressure or high-temperature electrolyzers. However, during the process, it is necessary to maintain the concentration of the alkaline solution at a constant level through a process control system that adds as much water to the solution as is removed through water decomposition. In addition to this, the produced gases drain some electrolyte from the stack, a lye management unit controls the concentration of the alkaline electrolyte within the electrolyzer system. As in most cases, energy is available in the form of an alternate current power supply, a transformation station is necessary to supply direct current at the voltage level needed.

Several attempts have been made to develop solid electrolytes to replace alkaline liquid. Besides the higher gas quality, the advantages of this electrolyte type are a reduced corrosion, a constant electrolyte concentration (and therefore no need for concentration control), and the ability to use the electrolyte simultaneously as a diaphragm. Possible electrolyte materials for this are ion exchange membranes. Unfortunately, compared to alkaline electrolysis, material costs (Pt and Ir are used as catalyst) for solid electrolyzers are higher. To reduce costs, carbon-supported platinum is used to catalyze the hydrogen evolution reaction, but this is not possible on the anode side where the evolving oxygen under working potential would corrode the carbon material in a short time.

Among challenges for cost-effective electrolytic production of hydrogen are the reduction of the capital cost of the electrolyzer, lower-cost electrodes, and improving the energy efficiency.

A new and promising field for hydrogen production by electrolysis is offered by the idea of a renewable energy supply like wind, solar, geothermal, or hydroelectric power. In that way, hydrogen can be used as energy storage where intermittent power supply from renewables has to be collected. Hydrogen production via renewable energy electrolysis offers opportunities for synergy with nonconstant power generation. For example, though the cost of wind power continues to drop, the inherent variability of wind prevents its widespread use. Co-production of hydrogen and electric power could be integrated in a wind farm, allowing flexibility and the ability to shift production to match the availability of resources with the system's operational needs and market factors. In this respect, Iceland with its large reserves of hydropower and geothermal energy was the first country to tackle seriously the electrolytic production of hydrogen in bulk quantity to serve as a fuel and/or energy vector. However, the initiative has been motivated primarily by geopolitical considerations rather than by energy savings.

It was recently reported in literature that the electrolysis of aqueous solution of renewable alcohols in alkaline environment can produce simultaneously pure hydrogen and selectively added-value chemicals (e.g., carboxyl and polycarboxyl compounds of alkaline metals). This process shows great advantages over plain water electrolysis, which requires higher working potentials. The reduction

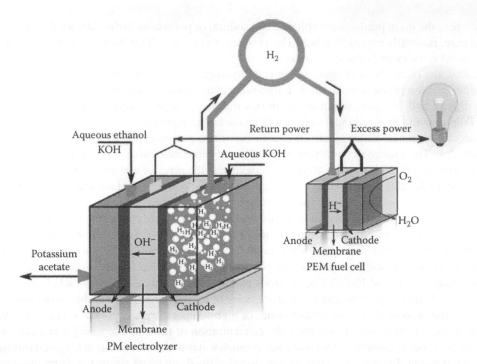

FIGURE 10.3 Polymer membrane electrolyzer coupled with a hydrogen-fuelled fuel cell. (Reproduced from Bambagioni, V. et al., *ChemSusChem*, 3, 851, 2010. With permission.)

of the potential applied to the electrolytic device lowers the amount of energy consumed decreasing significantly the production cost of electrolytic H_2 from 5–6 to 2 Euro/kg. From a technological point of view, the absence of O_2 evolution allows the design of a new type of electrolyzer operating at high pressure and without gas separator in which electrodes are soaked in the same electrolyte solution. Moreover, it is possible to couple this device with a hydrogen-fueled fuel cell as depicted in Figure 10.3. The power produced by this fuel cell is such to allow the design of a self-sustainable system, since part of the energy supplied by the fuel cell can be used to power the electrolyzer, whereas the remaining energy supplied by the fuel cell can be used to run a suitable load [47].

Finally, another type of electrolytic process, which is still under development by the U.S. Department of Energy, is the so-called nuclear high temperature electrolysis. In this case, the heat produced by a nuclear reactor generates high temperature (800°C–1000°C) steam that is electrolyzed to produce hydrogen and oxygen. The higher temperature reduces the amount of electricity required to split water molecules and also speed up the reaction kinetics, thus increasing the energy efficiency.

10.2.3 Photolytic Processes

Photolytic processes use light energy to split water into hydrogen and oxygen. These processes offer long-term potential for sustainable hydrogen production with low environmental impact. Photolytic processes include (1) photobiological water splitting, (2) photocatalytic water splitting, and (3) photoelectrochemical water splitting.

Photobiological hydrogen production is based on photosynthesis in bacteria and green algae. The organism's photosynthetic apparatus captures light, and the resulting energy is used to couple water splitting to the generation of a reducing agent, which is used to reduce a hydrogenase enzyme within the organism. Thus, photosynthesis uses solar energy to convert water to oxygen and hydrogen.

Hydrogen production in photobiological systems is presently limited by the low-energy conversion efficiencies. Solar conversion efficiencies are normally less than 1% in cultures of micro-algae, and a 10-fold increase is required before hydrogen production from micro-algae can become practical.

Another difficulty is the fact that hydrogenase enzymes are inhibited by oxygen at concentrations above 0.1%. This was a key problem during 30 years of research on hydrogen-producing microorganisms [48], but strains of algae that can overcome oxygen intolerance are now being developed [49].

Another serious barrier is the "light saturation effect," in which cells near the outside of the culture medium absorb all the available sunlight. In large-scale plants, it seriously reduces the yield of hydrogen. Photosynthetic bacteria also tend to absorb much more light than that they can effectively use to produce hydrogen, resulting in waste of energy.

Research in these areas is aimed at distributing the available light more evenly—by stirring the culture medium or by using reflecting devices to transfer light into the depths of the culture—and developing algal strains that absorb light less efficiently [48].

In photofermentation, photosynthetic bacteria produce hydrogen from organic acids, food processing and agricultural wastes, or high-starch biomass produced in turn by micro-algae grown in open ponds or photobioreactors.

Even if there are some reports of high hydrogen yields, these systems have several drawbacks: the nitrogenase enzymes used have high energy demands, solar energy conversion efficiencies are low, and huge areas of photobioreactors are needed [48].

The photocatalytic hydrogen production from aqueous solutions of renewable oxygenates compounds could represent an important alternative [50–54] to the more complex water splitting [55,56]. Noble and base metals, including Pt, Au, Pd, Ni, Cu, and Ag, have been reported to be very effective for enhancement of hydrogen production in TiO_2 photocatalysis [57].

In the photoelectrochemical water splitting, reactions take place at light-sensitive electrodes. Under irradiation with sunlight, the photoactive component of the electrode (e.g., pure or doped TiO_2) converts the energy of the photons into chemical energy. In particular, on the absorption of a photon by a n-type semiconductor like TiO_2, one electron is excited from the valence band and promoted to the conduction band to leave a positive charge, a hole, in the valence band. The electron in the conduction band is transferred to the counter electrode. At the electrode/electrolyte interface, the electrons react with protons producing hydrogen, while simultaneously the holes in the valence band of the semiconductor generate oxygen by oxidizing water.

Although the fundamental principles of the photoelectrochemical device for light-driven electrolysis of water are well known, its future as a device for the production of hydrogen from water rests solely on the ability to find the correct semiconducting material for the photoactive electrode.

Although this approach achieved encouraging results, there are still technical and cost issues to be addressed before it could become commercially viable. To date, a material must still be developed that is both efficient and durable to produce hydrogen for widespread use.

The known materials that are active in water splitting are not responsive to a wide portion of the solar radiation spectrum; they best perform in the ultraviolet region and thus with relatively low efficiencies. TiO_2 is a good example of a semiconductor subject to this limitation. Thus, one of the primary aims of research in this field is to find new photoelectrodes that can operate, either individually or in combination, across the solar spectrum and thereby they will allow more efficient conversion of sunlight to chemical fuels. To this end, extensive effort is being directed toward new nanostructured materials that are both satisfactory in performance and cheap to fabricate.

10.2.4 HYDROGEN PURIFICATION

10.2.4.1 Chemical Methods

As described earlier, there are many technological ways to produce hydrogen, each of them resulting in hydrogen with a different degree of purity. The impact of hydrogen purity on the system performance can be very different depending on the type and concentration of the impurity species. In order to power, for example, PEM fuel cell systems (see Section 10.5.2), hydrogen of excellent quality is a major demand.

Water Gas Shift reaction is commonly used in industrial scale production processes to purify hydrogen. The advantage of this process is that additional hydrogen is generated as CO is converted to CO_2. However, the additional hydrogen is obtained only by expending considerable energy for vaporizing additional water to drive the equilibrium to lower CO levels. Since the reaction is relatively slow, the size and weight of the reactor must be quite large to achieve significant conversions. Increasing the temperature to increase the rate of reaction unfortunately also causes the equilibrium CO level to increase because the reaction is exothermic.

After the WGSR step, the CO content of the reformate is on the order of 1%. CO must be reduced, as it is a poison for the electrodes of PEMFCs. For onboard application, there are a number of approaches that can, in principle, be considered to lower the final CO content of the stream to levels compatible for use in a fuel cell. These include preferential oxidation of CO (PrOx), CO methanation, Pd membrane separation, and the development of electrodes with greater CO tolerance [58]. PrOx (also called selective oxidation) involves preferentially or selectively oxidizing the CO in the presence of the large excess of H_2 and is the most studied method, as it is a relatively advanced technology. The oxygen needed to oxidize the CO is added after the shift step. For onboard application, the catalyst should ideally operate between the temperature of the shift reactor and the operating temperature of the PEMFC, usually between ca. 200°C and 80°C.

10.2.4.2 Membrane Technology

In recent years, incorporation of membrane technology into hydrogen production processes has become an area of interest. To date, separation membranes, usually based on Pd-alloys, have been used most extensively [59]. Separation membranes can be incorporated into various steps of the H_2 production process. At the reforming step, oxygen-selective membranes can be used to deliver highly active oxygen to the catalyst and eliminate the need for gas-phase oxygen. At the WGS stage, hydrogen-selective membranes can be used to remove hydrogen from the stream and therefore overcome thermodynamic limitations. There have also been reports of CO or CO_2—selective membranes—which can be used to purify hydrogen and eliminate the need for PrOx [60].

10.3   HYDROGEN STORAGE OPTIONS

The development of safe, reliable, compact, and cost-effective hydrogen storage technologies is one of the most technically challenging barriers to the widespread use of hydrogen as energy vector and the subsequent transition to the so-called hydrogen economy. This is however a tricky goal because hydrogen has physical characteristics that make it difficult to store in large quantities without taking up a significant amount of space.

It is important to highlight that hydrogen storage will be required onboard of the vehicles, at hydrogen production sites and at hydrogen-refueling stations. Storage of hydrogen is needed even considering the option of on-board hydrogen production, as the energy vector must be adequately supplied during critical driving steps such as engine start up or acceleration.

On a weight basis, hydrogen has nearly three times the energy content of gasoline but on a volume basis the situation is reversed. In practice, this means that the attainment of a driving range of 500 km for a conventional vehicle with today's Diesel technology requires a tank system weighing approximately 43 kg with a volume of 4 L. A zero-emission vehicle driven by a fuel cell with hydrogen will need a 700 bar high-pressure tank system of about 125 kg and 260 L to achieve the same driving range [61,62].

In general, the storage of hydrogen for stationary applications (e.g., for heating and air conditioning of buildings and industrial processes) is simpler than that needed for the propulsion of various transportation forms due to the existence of more severe constraints in terms of acceptable mass and volume, speed of charge–discharge, and, for some storage systems, heat dissipation and supply. This contributes to make hydrogen storage in transport sector a critical barrier.

TABLE 10.1
DOE Hydrogen Storage Targets

	Targets	
Storage Parameter	2010	2015
System gravimetric capacity wt.% H_2	4.5	5.5
System volumetric capacity g H_2/L	28	40
Durability/operability		
Min/max delivery T °C	−40/85	−40/85
Min/max operating p bar	5/12	5/12
Charging/discharging rates		
System fill time in min (for 5 kg)	3	2.5
Min full flow rate (g/s)/kW	0.02	0.02

Source: Adapted from http://www1.eere.energy.gov/hydrogenandfuelcells/
storage/pdfs/targets_onboard_hydro_storage.pdf

By translating currently vehicle performance requirements into storage system needs, the U.S. Department of Energy (DOE), in consultation with the U.S. Council for Automotive Research (USCAR), established an evolving set of technical targets for onboard hydrogen storage systems for 2010 and 2015 [63,64]. These targets, listed in Table 10.1, are not based on a particular method or technology for storing hydrogen but on equivalency to modern gasoline storage systems in terms of weight, volume, cost, and other operating parameters. For commercially acceptable system performance, they must be attained simultaneously. An important aspect of these targets is the fact that they refer to the properties of the whole storage system, and therefore take into account the mass, the volume, and the cost of any auxiliary components associated with the system, such as tank, valves, regulators, piping, mounting brackets, insulation, added cooling capacity, thermal management, and any other balance-of-plant components, in addition to any storage media and a full charge of hydrogen.

At this point, before describing hydrogen storage options, a number of general considerations have to be made.

One of the most widely quoted DOE targets is hydrogen storage system capacity, primarily in weight percent (wt.%). This target defines the net usable specific energy from the standpoint of the total onboard storage system, not just the storage medium. The term "net usable energy" is used to account both for unusable fuel and for hydrogen-derived energy used to extract the hydrogen from the storage medium (e.g., fuel used to heat a hydride to initiate or sustain hydrogen release).

Although system weight percent is the most widely quoted parameter in the literature, volumetric capacity (or system energy density) is another important target. As mentioned earlier, the onboard hydrogen storage system includes every component required to safely accept hydrogen from the delivery infrastructure, to store it onboard, and to release it to the power plant. Due to vehicle constraints and customer requirements, the storage system cannot take up too much volume. The so-called shape factor of the storage system also impacts the volume required to accommodate it, and, as mentioned earlier, any unusable fuel must be taken into account. Moreover, there is another factor considered in calculating the volumetric capacity target: the present gasoline tanks are considered conformable (or conformal). Conformability requires a tank to take irregular shapes and to fill the space available in the vehicle.

The storage system, and hence any material that stores hydrogen, must dependably store and deliver hydrogen at all conditions experienced by the vehicle. These include, for example, temperatures ranging from freezing to hot conditions under solar irradiation. Notably, less severe ambient conditions are selected for early demonstration fleets, and, therefore, interim targets are less stringent; but ultimately, consumers will expect vehicles all weather operative without any limiting factor. Besides operating requirements under external conditions, there are also requirements internal to the vehicle in terms of temperature and pressure (e.g., operative conditions of fuel cells). Pressure limits are also fixed for hydrogen delivered from the storage system to the power plant. These targets are different for fuel cells and for internal combustion engines (ICEs) due to the different hydrogen conversion technology.

Cycle life, or the ability to maintain constant performance over many refueling cycles, is clearly another requirement for the storage system. Customers expect the fuel system to be comparable with a gasoline tank (which requires little if any maintenance in current automobiles) and to last the life of the vehicle, typically, 250,000 km assuming a 500 km driving range between fill-ups. Taking into account that drivers generally refuel when the tank is partially rather than completely empty, a higher target for the number of fill cycles is required.

Finally, there are two other essential parameters for storage systems. They must supply hydrogen to the fuel cell at rates sufficient to meet the power demands of the driver and, for solid storage materials, to absorb/adsorb hydrogen fast enough to limit the time needed to refuel the vehicle.

Hydrogen can be stored as a gas, a liquid, or a solid. None of the existing methods for storing hydrogen are efficient in terms of energy density, neither on a volume nor a mass basis, and release rate at the same time. Accomplish the defined goals, extensive research studies are still required to optimize and improve the safety and accessibility for all, lower the production cost, increase the storage capacity, and enhance the sorption–desorption kinetics.

The following sections will describe the state of the art of these storage technologies.

10.3.1 COMPRESSED HYDROGEN GAS TANKS

Storing gaseous hydrogen under high pressure is an already common technique. High pressure tanks, which can be pressurized up to 300 bar, are available in the market [65]. These tanks are usually made of steel, and their capacities are not large enough for fuel cell applications in cars [66].

Some calculations clearly indicated the poor efficiency of distributing energy as compressed hydrogen at 200 bar in standard steel cylinders, compared with petroleum in tankers. For example, taking, in each case, a truck with a gross weight of 40 ton, for every petrol tanker on the road that is transporting fuel, there would need to be 22 hydrogen trucks. This makes the application of this option rather doubtful.

To make compressed hydrogen applicable to vehicular storage, the energy density of gaseous hydrogen can be increased by storing hydrogen at high pressures. However, this requires material and design improvements in order to ensure tank integrity. Advances in compression technologies are also required to get better efficiencies and reduce the cost of producing high-pressure hydrogen.

Intensive development taking place at present includes efforts to create lighter storage cylinders based on fiber-reinforced composites. The Canadian company Dynetek Industries Ltd., for example, claims weight savings of 20%–50% for its composite cylinders operating at 200–350 bar. Dynetek has also developed hydrogen cylinders that can operate at 825 bar for stationary use and 700 bar in transport applications [67]. Carbon fiber-reinforced 350 and 700 bar compressed hydrogen gas tanks are also under development by Quantum Technologies for its TriShield™ all-composite cylinder [68]. These tanks are already in use in some prototype hydrogen-powered vehicles. In the 700–1000 bar range, they can reach a gravimetric hydrogen density of up to 10 wt.%.

In this new family of cylinders, the inner liner of the tank is a high-molecular weight polymer that serves as a hydrogen gas permeation barrier. A carbon fiber-epoxy resin composite shell is placed over the liner and constitutes the gas pressure load-bearing component of the tank. Finally, an outer shell is placed on the tank for impact and damage resistance. The pressure regulator is located in the interior of the tank. There is also an in-tank gas temperature sensor to monitor the tank temperature during the gas-filling process when heating of the tank occurs.

The cost of these high-pressure compressed gas tanks is essentially dictated by the cost of the carbon fiber that must be used for light-weight structural reinforcement. Efforts are underway to identify lower-cost carbon fiber that can meet the required high pressure and safety specifications for hydrogen tanks. Lower-cost carbon fibers must still be capable of meeting tank thickness constraints in order to help to meet the volumetric capacity targets. Thus, lowering cost without compromising weight and volume is a key challenge. Furthermore, other intrinsic problems are the energy penalty of compressing gas to very high pressures together with the associated problem of cooling the compressed gas and the safety issues such as rapid loss of hydrogen in the event of an accident. Notably, in Japan, these high-pressure containers are prohibited on the roads for ordinary cars due to the considerable related risks.

Another issue in vehicular hydrogen storage is the shape of the tank. Recent gasoline tanks have the form close to a cylinder, so cylindrical design for hydrogen storage tanks is predictable. However, the ideal structure for a given amount of material is to construct conformal tanks [69]. It is reported that these tanks have 20% bigger volumetric capacities than of cylindrical tanks with the same amount of material and can utilize more than 80% of its envelope volume [70]. Nevertheless, cylindrical shapes are considered more appropriate for applications due to manufacturing ease.

Considering both storage and refueling technologies, compressed gas storage is probably the most promising short-term option.

10.3.1.1 Underground Storage of Gaseous Hydrogen

A special case of gaseous hydrogen storage is the possibility to use of large underground cavities similar to those now employed to store natural gas. In both cases, the quantities of energy involved have the potential to meet the needs of large communities for extended periods, such as might be needed to ensure security of supply or to meet seasonal variations in energy production and requirements. This gives underground storage a special importance and especially for storing large quantities of hydrogen for long periods.

Various underground hydrogen storage schemes have been proposed. One option is to store gaseous hydrogen in geological formations including depleted gas field or aquifers, caverns, and so on. Another possibility is underground storage in buried tanks, either in compressed gas form or in liquid form. Geological storage is generally close to the hydrogen production site, whereas buried tanks are close to the point of use, such as refueling stations.

In Tees Valley, United Kingdom, for example, a salt dome beneath an urban area is used to store 1000 ton of hydrogen for industrial use. Associated with the storage cavity is 30 km of hydrogen distribution pipeline [71].

However, further research is needed to evaluate the suitability of geological storage for hydrogen and to ensure proper engineering of the storage site and hydrogen containment.

10.3.2 LIQUID HYDROGEN

Liquid hydrogen can exist only at temperatures below −240°C that is the "critical temperature" of hydrogen. In practice, liquid hydrogen is usually stored at −253°C in cryogenic tanks because at this temperature it can be stored at atmospheric pressure, whereas liquid hydrogen at −240°C requires pressurized storage at 13 bar.

Liquefied hydrogen is denser than gaseous hydrogen and therefore has higher energy content on a per-unit-volume basis. Therefore, a far greater quantity of hydrogen can be carried compared with compressed gas, and fewer delivery trucks are therefore needed. However, liquefaction is energy-intensive, and the process consumes 30%–40% of the total energy content of the hydrogen. Furthermore, liquid hydrogen must be stored in insulated pressurized vessels to minimize hydrogen loss through evaporation or boil-off. In fact, hydrogen vapor that results from boil-off must be released from the tank through safety valves and vented into the atmosphere.

Besides issues with hydrogen boil-off and liquefaction process, weight, volume, and tank cost need to be addressed for economical use of liquid hydrogen. Safety concerns are also more of an issue with hydrogen in a liquid state. Liquid hydrogen fuel has a high-flammability range requiring thus special handling. Methods of storage and transport must be strictly adhered in order to avoid critical accidents. Sparks from electrical equipment, static electricity, open flames, or extremely hot objects must be excluded to insure one's safety. These safety measures for dealing with liquid hydrogen will not only have to take place inside the inner workings of the hydrogen vehicle, but in the fueling stations, transport vehicles, and other parts of manufacturing and distribution as well.

Liquid storage requires highly sophisticated tank systems with respect hydrogen gas. Heat transfer into the tank through conduction, convection, and radiation has to be minimized. Typically, the insulated vessels consist of an inner tank and an outer container with an insulating vacuum layer between them. The evacuated space between the nested containers is filled with multilayer insulation having several layers of aluminum foils alternating with glass fiber to avoid heat radiation. Nevertheless, due to the inevitable inward heat leakage, hydrogen evaporates in the container leading to an increase in pressure. Liquid hydrogen containers must therefore always be equipped with a suitable pressure relief system and safety valve. Liquid storage thus takes place in an open system in which released hydrogen has to be dealt with by means of catalytic combustion, dilution, or alternative consumption.

From an automotive engineering standpoint, there are practical difficulties in terms of designing, at acceptable cost, appropriate, and safe liquid hydrogen cryostats for onboard vehicular use.

Nevertheless, some progress has been made in Germany. The industrial gas supplier Linde has recently developed a modern liquid hydrogen storage for a hydrogen-powered bus [72]. The cylindrical tank has an outside diameter of 500 mm and an overall length of 5.5 m. Its capacity is 540 L of liquid hydrogen. It is designed to work at pressures from full vacuum up to 8 bar and at temperatures in the range $250°C–80°C$ [73]. It has to be noted that hydrogen vehicles such as buses require longer range capabilities and longer time of operation than other vehicles and liquid hydrogen-fueled vehicles meet these requirements. However, there are also aspects of vehicle refueling to be addressed, especially the design of transfer lines for liquid hydrogen that should be operated safely by the general public and without much loss through evaporation. Boil-off on standing would be a problem with the relatively small tanks that would typically be used in cars where their higher energy storage capacity translates into longer vehicle range between refueling.

10.3.3 Solid-State Hydrogen Storage Materials

Hydrogen can be stored within the structure or on the surface of certain materials as well as in the form of chemical compounds that undergo a reaction to release hydrogen [65,74–76].

For transportation use, a suitable solid-state storage material must be able to store a high weight percent and a high volume density of hydrogen. Furthermore, it must rapidly absorb and desorb hydrogen at or close to room temperature and pressure. Ideally, such a material should (1) be made from cheap constituents, using a low-energy preparation method, (2) be resistant to poisoning by trace impurities, (3) have good thermal conductivity in charged and uncharged conditions, (4) be safe and reusable on exposure to air, (5) have the ability to be regenerated, and (6) be readily recycled. One of the greatest challenges is the discovery and development of materials and compounds capable of storing enough hydrogen on-board to enable a 500 km range without adding significant weight or volume to today's conventional automobile. A minimum of 5 kg of hydrogen would need to be stored onboard to drive even the most fuel-efficient vehicle. At present, no single material is able to meet such a requirement.

Three generic mechanisms were up to now propose for storing hydrogen in materials: (1) absorption, (2) adsorption, and (3) chemical reaction. In absorptive hydrogen storage, hydrogen is adsorbed directly into the bulk of the material. In crystalline metal hydrides, this adsorption occurs by the incorporation of atomic hydrogen into interstitial sites in the crystallographic lattice structure.

Adsorption may be divided into physisorption and chemisorption, based on the energetics of the process. Physisorption has the advantages of higher energy efficiency and faster adsorption/desorption cycles, whereas chemisorption results in the adsorption of larger amounts of gas, but it could be irreversible, and a higher temperature is required to release the adsorbed gas. In both cases, the adsorption process typically requires highly porous materials to maximize the surface area available for hydrogen adsorption to occur.

Finally, the chemical reaction route for hydrogen storage involves chemical reactions for both hydrogen release and hydrogen storage. For reversible reactions onboard a vehicle, hydrogen release or hydrogen storage will take place by a simple reversal of the chemical reaction as a result of modest changes in the temperature and/or pressure. Sodium alanate-based complex metal hydrides are an example. In many cases, hydrogen generation reaction is not reversible under modest temperature/pressure changes. Therefore, although hydrogen can be generated onboard the vehicle, remaking the starting material by the reverse reaction must be done off-board. Sodium borohydride is an example.

The most studied hydrogen solid-state storage materials including carbon-based materials, metal hydrides, metal-organic frameworks (MOF), hollow glass microspheres and capillary arrays, clathrate hydrates, metal nitrides and imides, doped polymer, and zeolites will be briefly discussed.

10.3.3.1 Carbon-Based Materials

Carbon-based materials have been extensively studied as hydrogen storage media because of an ensemble of positive properties like high-specific surface area, microporosity, low mass, and good adsorption ability [77–83].

In general, their storage capacity is linearly proportional to the BET surface area [84–86]. The nature of hydrogen adsorption on carbon porous materials at moderate temperature is a result of molecular physisorption [87,88]. Since the molecular interaction in a physisorption process is very weak, only a relatively small amount of hydrogen can be adsorbed [89] even at a pressure of 90 bar [90]. Temperature also seems to be a key factor for hydrogen storage in pure carbon materials. The high temperature necessary for hydrogen release and the high degree of irreversibility involved in chemical hydrogenation of carbon materials are severe drawbacks for technical storage application.

It has to be highlighted that the data concerning hydrogen adsorption in carbon materials at room temperature are scattered over a wide range, and they do not seem reproducible. The reasons for these discrepancies can be attributed to the difficulty in measuring the hydrogen uptake and to the large differences in the sample quality. Unfortunately, all the reproducible results concerning maximum storage capacities of approximately 1 wt.% at RT are far less than required for technical applications.

Various types of nanostructured carbon materials (e.g., activated carbon, nanotubes, and graphite nanofibers) have been investigated.

Activated carbon (AC) is a high porosity, modified synthetic carbon containing crystallized graphite and amorphous carbon with a high-specific surface area, which can be prepared using either thermal or chemical procedures [91]. AC has been considered as a potential commercial candidate for hydrogen storage purposes because of its relative low cost and its accessibility. The hydrogen storage rate and capacity of the activated carbon can be influenced by its morphology and shape, that is, powder, fiber, and granular. It is shown that the hydrogen adsorption rate in fiber form is 2–50 times faster than in the granular form [92].

Hydrogen adsorption on various commercial and modified activated carbon products has been extensively studied [93]. Experimental results on conventional AC show that products with micropore volumes greater than 1 mL/g are able to store ca. 2.2 wt.% of hydrogen due to physisorption, and it is expected that optimization of the adsorbent and sorption conditions could lead to a storage capacity of 4.5–5.3 wt.%.

Jin et al. [85] prepared AC with different porosities using chemically activated coconut shell. They reported a maximum hydrogen adsorption capacity of 0.85 wt.% at 100 bar and RT. Sharon et al. [94] produced activated carbon fibers (ACF) using soybean and bagasse. The authors measured hydrogen storage capacities of 1.09–2.05 wt.% at a pressure of 0.11 mbar and room temperature. Another form of AC, the advanced AC monoliths, with good mechanical strength (maximum compression strength of 220 bar), high volume of micropores, and high density were shown to adsorb 29.7 g/L of hydrogen at −196°C and 40 bar [95,96].

Studies have shown that loading of precious metals, for example, Pt, onto AC increases the adsorption capacity [97]. The merging of the two adsorption phenomena, that is, chemisorption (on the Pt surface) and physisorption (on the carbon surface), gives rise to a significant amount of spillover hydrogen.

Dillon et al. [98], who were the first group that discovered the hydrogen storage capacity of Carbon NanoTubes (CNTs), presented promising results, which have triggered a worldwide tide of research in the area. There are two main different species of CNTs characterized by the structure of their walls, that is, the single-walled nanotubes (SWNT) and the multiwalled nanotubes (MWNT) [99]. Very different hydrogen storage values between 0.25 and 56 wt.% were reported in various experimental conditions for these materials.

In recent years, a new type of carbon nanostructure has been synthesized from catalytic decomposition of hydrocarbons. These fibrous materials are graphitic nanofibers (GNFs). GNFs consist of graphite platelets stacked together in various orientations to the fiber axis with an interlayer distance similar to that of graphite. Depending on the orientation angle, three distinct structures can exist: tubular, platelet, and herringbone.

Extremely high hydrogen uptake of herringbone GNFs, up to 67 wt.%, was reported in 1998 by Chambers et al. [100,101]. Ahn et al. [102] found however for GNFs with a significant fraction of herringbone structure a hydrogen storage capacity that does not exceed values of 0.2 wt.%. Recently, Lueking et al. [103] measured a maximum adsorption of only 3.3–3.8 wt.% for herringbone GNFs after hydrogen pretreatment at 700°C.

In the recent years, graphene, carbon formed into sheets a single atom thick, has attracted great attention since it appears to be a promising base material for capturing hydrogen, according to recent research at the National Institute of Standards and Technology (NIST) [104]. The easy synthesis, low cost, and nontoxicity of graphene make this material a promising candidate for gas storage applications. The material does not store hydrogen well in its original form, but if oxidized graphene sheets are stacked atop one another like the decks of a multilevel parking lot, connected by molecules that link the layers to one another and maintain space between them, the resulting graphene-oxide framework (GOF) can accumulate hydrogen in greater quantities. The GOFs can retain 1% of their weight in hydrogen at a temperature of −196°C and 1 bar. GOFs exhibit an unusual relationship between temperature and hydrogen absorption. In most storage materials, the lower the temperature, the higher the hydrogen uptake. GOFs, however, as reported by NIST group, behave quite differently. Although a GOF can absorb hydrogen, it does not take in significant amounts at below −223°C. Moreover, it does not release any hydrogen below this "blocking temperature," suggesting that GOFs might be used both to store hydrogen and to release it when it is needed, a fundamental requirement in fuel-cell applications. Some of the GOFs' capabilities are due to the linking molecules themselves. By keeping several angstroms of space between the graphene layers, they also increase the available surface area of each layer, giving it more spots for the hydrogen to latch on [104].

The hydrogen storage capacity in nanostructured carbon materials can be increased by atomic hydrogen spillover from a supported catalyst as described by Lachawiec et al. [105]. In this work, a supported catalyst (Pd-C) was used as the source of hydrogen atoms via dissociation and primary spillover, while an activated carbon (AC) or SWNT was secondary spillover receptors. In this way, the hydrogen adsorption amount increased by a factor of 2.9 for the AC receptor and 1.6 for the SWNT receptor at RT and 1 bar. Similar results were also obtained at 100 bar.

10.3.3.2 Zeolites

Zeolites are hydrated microporous crystalline aluminosilicates having infinite, open, and rigid 3D structures (pores and channels) with high-internal surface area (up to $1000 \, m^2/g$).

They have been intensively investigated for hydrogen adsorption capacities [106,107]. The presence of strong electrostatic forces inside the channels and pores should enable hydrogen retention in the free volumes of the zeolite. The electric field is produced by the additional metal ions, and it increases with increasing their charge and decreasing their size.

In spite of these exceptional characteristics, like high-specific surface area, high porosity, and the possibility to modulate the electric field inside the channels through ion exchange, very low hydrogen storage capacity has been reported for different types of zeolites [108–112].

All the results reported on zeolites are very congruent with each other and can be summarized by saying that the storage capacity for this inorganic material is less than 2 wt.% at liquid nitrogen temperature and less than 1 wt.% at room temperature.

Recently, a method that demonstrates the potential of hydrogen storage at high pressures in small pellets coated with a thin zeolite layer has been reported by Yu et al. [113]. The pellets are an inorganic material that adsorbs hydrogen and which is stable to the pH conditions used for zeolite membrane synthesis and to the temperature required to calcine the zeolite layer. Hydrogen can be adsorbed at high pressure in the pellets, which can then be sealed by reversible adsorption of a molecule in the thin zeolite layer.

10.3.3.3   Metal-Organic Frameworks

MOFs represent a new category of synthetic nanoporous material, which consists of metal ions linked together through organic ligands. The main characteristics of MOF structures are their low density, high surface area, and large porous volume. In recent times, the design and synthesis of porous MOFs with nanometer scale pores has expanded, and the number of available MOFs has increased dramatically [114].

The choice of rigid ligands during the synthesis is essential to reduce their degree of freedom, and, therefore, the number of possible framework geometries that can be formed. Moreover, the bridging anion is very important because of its coordinating abilities and its capacity to accommodate in the crystal structure [115].

The use of MOF materials as a medium for hydrogen storage was first reported in 1999 by Li et al. [116]. The authors indicated that MOF-5 shows a high-hydrogen storage capacity of 4.5 wt.% at −196°C and 0.8 bar, whereas its capacity at room temperature and at a pressure of 20 bar was almost 1 wt.% [116]. However, in 2004, it was observed that the maximum hydrogen uptake of MOF-5 was 1.32 wt.% at 1 bar and −196°C [117]. A hydrogen storage value of 1.6 wt.% was reported for MOF-5 at pressures above 10 bar, but the adsorption capacity of MOF-5 was very low at room temperature, with values of less than 0.2 wt.% at pressures up to 67 bar [118].

A hydrogen storage of 7.5 wt.% in MOF-177 material was reported by Wong-Foy et al. [119] correlated to surface area.

An interesting novel porous Mn-based MOF was recently synthesized, which showed a total hydrogen uptake of 6.9 wt.% at −196°C and 90 bar [120].

An improved synthesis of a high-quality MOF-5 material has been developed by Cheng et al. [121]. The hydrogen storage capacity of this MOF-5 synthesized in the presence of H_2O_2 was shown to be higher than a sample synthesized without H_2O_2. The synthesized MOF-5 exhibited a high hydrogen storage capacity of 2.96 wt.% at −196°C and 14.5 bar. However, the most promising results regarding MOFs have been reported by Yan et al. for the NOTT-112 structure [122]. The authors achieved 10 wt.% hydrogen storage capacity for NOTT-112 at 77 bar and −196°C. They also developed a new metal-organic polyhedral framework that shows an excess hydrogen uptake of 7.07 wt.% between 35 and 40 bar at −196°C.

The combination of high surface area and pore volume seems to be necessary to achieve the high level of hydrogen storage capacity found in these materials [122].

Latroche et al. [123] synthesized several giant-pore MOFs and further confirmed that small pores plays a major role, as well as specific surface area, in determining hydrogen storage capacity.

Yang and coworkers extensively studied the hydrogen spillover effect of hydrogen storage in MOFs [124,125] as well as for carbon materials [105,126] and zeolites [127]. They observed that the hydrogen storage capacity of porous materials could be remarkably improved by modification with metal-supported catalysts. The metal-doping process is believed to modify the chemical nature of the porous materials in order to strengthen subsequent hydrogen adsorption [128]. A good dispersion of the supported metal catalyst on the MOFs helps to effectively diffuse the material, which plays a key role in the final hydrogen storage capacity of doped and modified MOFs [129].

Ferey et al. [130] measured hydrogen adsorption in nanoporous metal-benzenedi-carboxylates, where the metal is trivalent chromium or aluminum. The maximum storage capacity obtained for the chromium compound is 3.1 and 3.8 wt.% for the aluminum compound at 16 bar and −196°C.

Changing the organic linkers and the metal gives the possibility to investigate an infinite number of new structures with tunable pore volume and surface area.

Pan et al. developed a microporous metal-organic material with pore dimension comparable to the length scale of the molecular diameter of hydrogen. In this case, two copper atoms share four carboxylate groups of the ligand. Each of these building units is connected to four others to give a 2D network. Two layers of two-dimensional structure are bound to give a 3D network with open channels. At room temperature, the reported storage capacity is 1 wt.% at 48 bar [131].

10.3.3.4 Clathrate Hydrates

Clathrate hydrates are a class of solid inclusion compounds in which guest molecules occupy cages formed from a hydrogen-bonded water molecule network. The empty cages are unstable and often collapse into conventional ice crystal structures. However, inclusion of appropriately sized molecules helps to stabilize them [132,133]. Clathrate hydrates of hydrogen often possess two different cages to meet the necessary storage requirements [134]. However, higher pressures of ca. 2000 bar are required to produce the material, which make them impractical.

10.3.3.5 Metal Hydrides

Metal hydrides are a category of promising materials that exhibit the potential to fulfill the defined targets for hydrogen storage. To date, simple metal hydrides, which contain magnesium, for example, MgH_2, and transition metals as well as more complex metal hydrides containing nontransition metals such as lithium, sodium, or calcium and boron or aluminum, for example, $NaAlH_4$ and $LaNi_5H_6$, have been considered for this purpose. These materials have a good energy density by volume, although their energy density by weight is often worse than that of hydrocarbons. Some of the liquid hydrides are easy-to-fuel at room temperature and pressure, whereas solid hydrides, which could be formed as pellets and granules, require temperatures of ca. 120°C to release their hydrogen content.

Metal hydrides show promising results in hydrogen storage, for example, MgH_2 stores up to 7.60 wt.% hydrogen [135], since they have higher hydrogen storage density as compared to compressed hydrogen gas or liquid hydrogen. Hydrogen can easily bond to metals under moderate pressure and temperature, and hence, metal hydride storage is much safer and volume-efficient for on-board applications [136]. Two possible routes for the production of metal hydrides consist of direct dissociative chemisorption and the electrochemical splitting of water. The addition of some components has been proved to improve the capacity and the reaction kinetics of the magnesium hydrides. For instance, Pd has been used to dope the magnesium surface due to its known characteristics as good catalyst to dissociate hydrogen [135]. Palladium is easy to recover from the oxide during exposure to hydrogen [137] and it enhances the adsorption rates even at room temperature. However, the cost of using such an expensive material is still the main disadvantage of this route [138].

Light metals, for example, lithium, beryllium, sodium, magnesium, boron, and aluminum-based storage materials, are some of the popular substances used for storage purposes.

Before 1980, several compounds including complex borohydrides, or aluminohydrides, and ammonium salts were also studied as potential hydrogen storage media. The maximum theoretical hydrogen yield of these hydrides is limited to ca. 8.5 wt.%. In 1953, Schlesinger et al. [139] published a paper about the generation of hydrogen from sodium borohydride, $NaBH_4$. Aiello et al. [140] has performed a feasibility study on $NaBH_4$ and $LiBH_4$ and concludes that both have a great potential as hydrogen storage materials. In fact, $NaBH_4$ has already been commercialized by the Millennium

Cells company. The waste product is borax, $NaBO_2$, which can be regenerated by just pumping hydrogen onto it or by hydrolysis using coke or methane.

Despite the fact that $LiBH_4$ has a gravimetric hydrogen density of 18 wt.%, which is a very high storage capacity, the desorption at low temperatures releases only a small amount of hydrogen (0.3 wt.%) [139]. In addition to this, $LiBH_4$ is an expensive compound [138]. In comparison to other nonboron-containing materials, the hydrolysis of boron-containing materials appears to have an obvious advantage in terms of capacity [141].

The boron nitride (BN) nanostructured in lightweight nanoballs and nanoforms is another candidate for hydrogen storage [142]. BN can store up to 2.6 wt.% H_2 after a milling process for 80 h and is able to desorb at 300°C, which is a relatively high temperature compared to other nanostructured materials [143]. BN fullerenes have also demonstrated a hydrogen storage capacity of 3 wt.% [144].

Various studies have shown that the alkali-metal amidoboranes, $LiNH_2BH_3$ and $NaNH_2BH_3$, fulfill some of the main criteria requests for hydrogen storage materials. $LiNH_2BH_3$ and $NaNH_2BH_3$ exhibit high storage capacities of 10.9 and 7.5 wt.%, respectively, with an easily accessible dehydrogenation temperature of ca. 90°C without the formation of the undesired borazine by-product. These materials are environmentally friendly compounds, nonflammable, nonexplosive, and stable solids at room temperature and pressure. To date, the only disadvantage of these materials is the lack of easy reversibility. Nevertheless, thermal neutral dehydrogenation of alkali-metal amidoboranes should facilitate on-board regeneration [145].

Several B-N-H compounds such as amine boranes, boron hydride ammoniates, hydrazine-borane complexes, and ammonium octahydrotriborates or tetrahydroborates have been extensively investigated as hydrogen carriers [136].

A process providing for the first time exceptionally high hydrogen yield values from ammonia borane (AB) at temperatures close to that of operating fuel cell and without use of catalyst has been very recently described by Diwan et al. [146]. AB contains 19.6% of hydrogen, resulting that relatively small quantity and volume of the material is needed to store a large amount of hydrogen making AB a very promising storage material. The process combines hydrolysis and thermolysis, two hydrogen-generating processes, which are not practical by themselves for vehicle applications. The maximum hydrogen storage capacity, obtained at 1 and 14 bar was 11.6 and 14.3 wt.%, respectively. These values are sufficiently higher to meet DOE requirements making AB hydrothermolysis methods promising for hydrogen storage in fuel cell-based vehicle applications.

10.3.3.6 Metal Nitrides, Amides, and Imides

Metal nitrides, amides, and imides show high hydrogen storage capacity at low operating temperatures compared to other competitive chemisorption approaches. Light metal-nitride compounds have also shown potential as a solid-state hydrogen storage medium. Lithium nitride that is usually used as a starting material for the synthesis of binary or ternary nitrides reversibly uptakes large amounts of hydrogen. However, the temperature required to release the hydrogen at usable pressures is too high for practical application of the present material [147].

Researchers from the National University of Singapore have made initial studies of Li-alkali earth metal imides as hydrogen storage materials. They have shown that lithium nitride is capable of storing 11.4% of its own weight in hydrogen, which is 50% higher than magnesium hydride. Other metal hydrides generally store only 2%–4% of their weight. Due to the high temperatures required to release the hydrogen, this novel material is not ready for practical applications, but it points the way to a practical hydrogen storage material. Metal imides have a maximum storage capacity of 7.0 wt.%. However, these materials require relatively high operating temperatures, which could limit their application [148]. Chen and coworkers have shown that the addition of alkali earth metals dramatically decreases the hydrogen storage temperatures, increases the desorption pressures, and allows high storage capacities to be achieved. This result takes the metal–N–H system a big step forward toward practical targets.

Furthermore, by reacting different metal amides with hydrides, the possibility of developing a broad range of metal N H systems such as ternary imides, for example, Mg–Na–N–H and

Li–Mg–N–H systems [149,150], and quaternary imides, for example, the Mg–Ca–Li–N–H system [151] for hydrogen storage, has been developed.

Luo and coworkers investigated the hydrogen storage properties and mechanisms of a novel Li–Al–N ternary system [152]. It is seen that ca. 5.2 wt.% of hydrogen is reversibly stored in a Li_3N–AlN (1:1) system, and the hydrogenated product is composed of $LiNH_2$, LiH, and AlN. A stepwise reaction is detailed for the dehydrogenation of the hydrogenated Li_3N–AlN sample. Further investigations have shown that the presence of AlN in the $LiNH_2$–2LiH system enhanced the kinetics of the first dehydrogenation step with a 10% reduction in the activation energy. This was mainly due to the higher diffusivity of lithium and hydrogen within AlN [152].

10.3.3.7 Other Hydrogen-Rich Compounds

Similarly to storing hydrogen in metal hydrides, it is possible to use hydrogen-rich compounds as hydrogen carrier. Potential carriers are, for example, hydrocarbons, ammonia (NH_3), methanol, or renewable alcohols like ethanol, ethylene glycol, and glycerol. All these substances are already produced and handled in large quantities, so they have the considerable advantage that a suitable infrastructure is already in place.

Their direct use as fuel in fuel cells makes them very attractive.

10.3.3.8 Doped Polymers

Incorporating metal nanoparticles (NPs) in polymer hosts has gained potential significance in technological applications, especially as advanced functional materials including optical sensors, hydrogen storage systems, and microwave absorbers [153]. Recently, a group of Korean researchers proposed functionalized organic molecules doped with titanium atoms as high-capacity hydrogen storage materials. The authors have studied six types of functional groups that form complexes with

Ti atoms found that each complex is capable of binding up to six hydrogen molecules. Among such complexes, Ti-decorated ethane-1,2-diol can store hydrogen with a maximum gravimetric density of 13 wt.% and, under ambient conditions, a practically usable capacity of 5.5 wt.%. They have also presented different forms of storage materials, which are obtained by modifying some well-known nanomaterials using Ti-functional group complexes [154].

In this area, the DOE has also planned to investigate possible utilization of carbon doped with organo-metallic compounds that exhibit rigid planar configuration and are rich in electronegative nitrogen atoms, [CNH]. From the results of Cabasso and Yuan [155], it is hoped that this class of materials will enable to meet the DOE targeted temperature and moderate pressure hydrogen storage system could be realized.

10.3.3.9 Hollow Glass Microspheres

In 1981, Teitel was the first person to propose the use of hollow glass microspheres (HGMs) as hydrogen storage media [156,157]. The characteristics of so-called super-high-strength micro-balloons toward hydrogen storage and extraction in comparison to other known methods were investigated [158]. HGM with approximate diameters of 1–200 μm and wall thicknesses of 1.5 μm and below are considered as viable containers for transporting and storing hydrogen. However, one of the main concerns with regard to the use of HGM for hydrogen storage purposes is their strength. If the HGMs were made of engineered glass, which can be up to 50 times stronger than normal glasses, the amount of hydrogen capable of being stored in the microspheres can be increased remarkably. This would lead to an increase in the energy density per unit volume by over an order of magnitude resulting in the advancement of hydrogen energy technology and in the improvement of the hydrogen market as a renewable energy source via cheaper and safer hydrogen storage [65,159].

Recently, researchers at the Savannah River National Laboratory (SRNL) reported a novel class of materials, that is, the so-called porous walled-hollow glass microspheres (PW-HGMs), for a variety of new applications. They consist of tiny micron-sized glass balloons where the interconnected porosity of their outer walls can be produced on a scale of 100–3000 Å. The porosity results in some unique properties. These open channels can be used to fill the micro-balloons with absorbents as well as other materials, thus providing a contained environment and a new type of glass-absorbent composite. Gaseous molecules such as hydrogen can enter the microspheres through the pores and be stored or cycled on absorbents inside, resulting in solid-state and contained storage. PW-HGMs, which are fluid like, recyclable and made from readily available resources, may be the ideal solution to hydrogen storage as well as for gas purification and targeted drug delivery [160].

In order to make HGMs more attractive for hydrogen storage purposes, researchers have been attempting to improve their fabrication procedures (i.e., using infrared [IR] irradiation).

A simple calculation for commercially available HGM reveals that the HGMs can endure higher pressures up to three times the pressure currently contained in typical metal composite cylinders that are in use today [161]. The filled microspheres can be reheated (at lower temperature than that required for filling) in a low-pressure vessel to outgas the hydrogen. Unfortunately, the inherently poor thermal conductivity of inorganic glasses has limited the further development and implementation of this hydrogen storage method. This poor thermal conductivity results in slow release rates of encapsulated hydrogen.

Studies have shown that the slow outgassing at low temperature can be accelerated with photoenhanced diffusion. This possible solution can be realized by the discovery of photoinduced outgassing, in which a high-intensity IR irradiation is used to facilitate hydrogen release of selectively doped glasses. The process results in much faster response times for the release of hydrogen in glasses in comparison to the normal heating procedure and may provide a path to superior performance for hydrogen storage in HGMs.

Another disadvantage of HGMs is the difficultly in producing an ideal spherical shape of uniform diameter and wall thickness. Usually, eccentricity and diameter spread of HGMs lead to the collapse of the spheres under the high external pressure applied during the hydrogen-filling process [165].

10.3.3.10 Capillary Arrays

Capillary arrays have also been developed as a new technology for hydrogen storage. The technology can be used effectively for safe transportation and storage of highly pressurized hydrogen in mobile systems, ranging from domestic electronic devices to ground and sea vehicles. Two types of capillary arrays for hydrogen storage systems are under consideration: (1) a bundle of cylinder capillaries sealed at both ends with semispherical caps and packed in a 2D hexagonal lattice and (2) an array of fused hexagonal capillaries sealed with a thin layer of a suitable substance.

The amount of stored hydrogen in each individual capillary is very small, which significantly reduces the possibility of explosions from improper handling or during accidents. Compared to HGMs, normal heating or IR radiation of the capillary arrays can be applied more easily to facilitate hydrogen release. The capillary arrays are very resistant to the external pressure, have a better packing ratio than the hollow spheres, and can be produced with uniform diameter and wall thickness [162].

10.4 TRANSPORT/DISTRIBUTION OPTIONS FOR HYDROGEN

The cost and the method of hydrogen delivery are greatly affected by the choice of the hydrogen production strategy. For example, larger, centralized facilities can produce hydrogen at relatively low costs due to economies of scale, but the delivery costs are high because the point of use is usually farther away. On the other hand, distributed production at the point of end-use, such as refueling stations or stationary power sites, eliminates the transportation costs, but results in higher production costs.

Although hydrogen has been used in industrial applications for decades, the current delivery infrastructure and technology cannot be sufficient to support a widespread use of hydrogen.

A wide variety of options exist for transporting hydrogen, ranging from gaseous or liquefied truck transport to large-scale pipelines. Getting the hydrogen to the end-use site may require multiple modes of transport, including large regional pipelines that connect to smaller, local pipelines, or to trucks, ships, or rail cars.

Hydrogen transportation or distribution is closely linked to its storage. Hydrogen can be distributed continuously in pipelines or batch by ships, trucks, railway, or airplanes. All transportation options require a storage system but also pipelines can be used as pressure storage system. Moreover, technologies for storing hydrogen on-board vehicles also affect the design and selection of a hydrogen delivery system and infrastructure.

The cost of hydrogen distribution is dependent on two primary factors: transport rate and distance. Some transportation options make sense for short distances and low transport rates, such as truck transport. Others, such as pipelines, are more feasible in the case of a fully developed hydrogen economy. Hydrogen's low volumetric density means that a truck can only transport roughly 180 kg of compressed hydrogen per trip or the energy equivalent of approximately 820 L of gasoline. Large-scale transport of hydrogen in this method, even if it made economic sense, would greatly increase truck traffic on the highways. This simply is not workable on a large scale.

It is evident that building a national hydrogen delivery infrastructure is still a big challenge. Currently, none of the options in the market satisfy the needs of the end users, which explains the growing interest and investment in hydrogen energy related research and development. Key challenges to hydrogen delivery include reducing delivery cost, increasing energy efficiency, maintaining hydrogen purity, and minimizing hydrogen leakage.

Delivery infrastructure needs and resources will vary by region and type of market (e.g., urban and rural). Infrastructure options will also evolve as the demand for hydrogen grows and as delivery technologies develop and improve.

10.4.1 Pipelines

Today, approximately 720 and 1500 km of hydrogen pipelines already exist in the USA and Europe, respectively. They are used to transport hydrogen as chemical commodity from one to another production site. The energy required to move the gas is irrelevant in this context, because energy costs are part of the production costs. This is not so for energy transport through pipelines.

Transporting gaseous hydrogen via existing pipelines is currently considered the lowest-cost option for delivering large volumes of hydrogen, but the initial capital cost of new pipeline construction constitutes a major barrier to expanding hydrogen pipeline delivery infrastructure. Research is focused on overcoming technical concerns related to pipeline transmission, including the potential for hydrogen to embrittle the pipeline steel and welds; the need to control hydrogen permeation and leaks; and the need for lower cost, more reliable, and more durable hydrogen compression technology.

One possibility for rapidly expanding the hydrogen delivery infrastructure is to adapt part of the natural gas delivery infrastructure. Converting natural gas pipelines to carry a blend of natural gas and hydrogen (up to about 20% hydrogen) may require only modest modifications to the pipeline; converting existing natural gas pipelines to deliver pure hydrogen may require more substantial modifications.

Another possible delivery process involves producing a liquid hydrogen "carrier," such as ethanol, at a central location, pumping it through pipelines to distributed refueling stations, and processing it at the station to produce hydrogen for dispensing. Liquid hydrogen carriers offer the potential of using existing pipeline and truck infrastructure technology for hydrogen transport.

It is difficult to give a generalized estimation for construction cost of new pipelines because it is very dependent on the location. A pipeline through a rural area without special environmental concerns can cost five times less than a pipeline of the same length and diameter through a dense urban area. The realization costs will include materials costs, labor, right of way, and miscellaneous costs (i.e., surveying, engineering, supervision, contingencies, allowances, overhead, and filing fees).

Hydrogen pipelines add an extra level of complexity due to a relative lack of experience in installing such a type of pipelines.

10.4.2 TRUCKS, RAILCARS, SHIPS, AND BARGES

Trucks, railcars, ships, and barges can be used to deliver compressed hydrogen gas, cryogenic liquid hydrogen, or novel hydrogen liquid or solid carriers.

Today, compressed hydrogen can be shipped in tube trailers at pressures up to about 200 bar. This method is expensive, and it becomes cost-prohibitive for distances greater than about 150 km. Researchers are investigating technology that might permit tube trailers to operate at higher pressures (up to 700 bar), which would reduce costs and extend the utility of this delivery option. However, it has been estimated that for typical delivery distances of 100–400 km, the diesel fuel consumed by the truck would equate to 20%–80% of the energy contained in the hydrogen. Apart from the energy efficiency and cost considerations, the larger number of heavy vehicles would add to congestion and emissions on the highway.

Currently, for longer distances, hydrogen is transported as a liquid in superinsulated, cryogenic tank trucks. Over long distances, trucking liquid hydrogen is more economical than trucking gaseous hydrogen because a liquid tanker truck can hold a much larger mass of hydrogen than a gaseous tube trailer. However, it takes energy to liquefy hydrogen as mentioned in Section 10.3.2. Research to improve liquefaction technology, as well as improved economies of scale, could help lower costs. Larger liquefaction plants located near hydrogen production facilities would also help reduce the cost of the process.

In the future, hydrogen could be transported as a cryogenic liquid in rail cars, barges, or ships that can carry large tanks and the larger the tank, the less hydrogen lost to evaporation. Marine vessels carrying hydrogen would be similar to tankers that currently carry liquefied natural gas, although better insulation would be required to keep the hydrogen liquefied (and to minimize evaporation losses) over long transport distances.

10.4.3 HYDROGEN-FUELING STATIONS

Hydrogen-fueling stations are an essential component to the success of future hydrogen vehicles running upon the nation's hydrogen highways. Indeed, in order for hydrogen to become attractive to millions of vehicle users, there will need to be an extensive network of these filling stations.

Refueling a vehicle with hydrogen is very similar to refueling a vehicle with compressed natural gas. A fueling station includes a hydrogen source (gaseous or liquid), a compressor, a storage unit, and a dispenser. Most of the current stations deliver gaseous hydrogen compressed at 350 or 700 bar in order to fill vehicle tanks quickly (refueling time usually less than 4 min). Composite tanks consisting of a metallic or polymeric liner in a fiber-reinforced composite structure have been developed to store hydrogen at high pressure and to increase the autonomy of vehicles (see Section 10.3). However, fast filling at high pressure results in a temperature increase in the composite vessel, which could damage it. As a consequence, there are a number refueling issues when combining high pressure, fast filling, composite cylinder use, and safety.

The dispensing nozzle must lock on to the vehicle receptacle before any hydrogen can flow. Hydrogen dispensers are also equipped with safety devices including breakaway hoses, leak detection sensors, and grounding mechanisms. These controls provide additional safety measures in the case of human error.

There are many issues that need to be worked out with hydrogen-fueling stations as they relate to hydrogen vehicle technology. For instance, currently, there are two kinds of hydrogen cars. One type uses hydrogen fuel cells and the other uses an ICE to burn hydrogen. A hybrid car also exists that switches back and forth between gasoline and hydrogen ICE's. Some hydrogen cars currently use compressed H_2 while others use liquid hydrogen. Issues of storage, containment, delivery, and safety all need to be addressed before hydrogen fuel stations become commonplace in the consumer market.

As an interim option, conventional filling stations could install reformers to manufacture hydrogen from, for example, natural gas, which is already widely distributed in many countries.

However, manufacturing hydrogen locally in this way would involve some CO_2 emissions, since capture and sequestration would probably be impractical on such a small scale.

A complete list of all known worldwide hydrogen-fueling stations can be found in Ref. [163]. The categories are broken down by location, type of hydrogen fuel dispensed, project, partners in building the maintaining the station, dates that it was opened or upgraded or decommissioned, H_2 production technique including electrolysis of water or reforming of natural gas, specifics, and comments plus a few even include a photograph of the station.

Road maps presenting the shorter-term aims and strategies for introduction of hydrogen for transport generally agree on the steps toward an extended hydrogen infrastructure. First, hydrogen-filling stations will be built in large cities to supply bus fleets and other vehicles in regular operation within a limited area. A single filling station in a densely populated area can serve a large number of users and maximize the environmental and health benefits of hydrogen. This is an optimal way to increase public awareness and acceptance of hydrogen technology. Step two is to build filling stations on major roads linking cities and large towns that already have their own hydrogen-filling stations. This allows existing fleets of hydrogen vehicles to increase their range and will help to increase the size of the hydrogen vehicle market.

10.5   HYDROGEN CONVERSION TECHNOLOGIES IN MOBILE APPLICATIONS

Hydrogen can be used as a vehicle fuel in two ways:

1. For ICEs
2. For FCs

10.5.1   DIRECT USE OF HYDROGEN

A hydrogen internal combustion engine (HICE) vehicle uses a traditional ICE that has been modified to use hydrogen fuel. One of the benefits of hydrogen-powered ICEs is that they can run on pure hydrogen or a blend of hydrogen and compressed natural gas. This fuel flexibility is very attractive as a means of addressing the widespread lack of hydrogen-fueling infrastructure in the near term. Gaseous hydrogen is injected into the engine, which then burns the hydrogen fuel much as it would burn gasoline for combustion. The engine is fueled by compressed hydrogen gas and stored in high-pressure tanks carried by the vehicle. To refuel, the vehicle must be driven to a specialized hydrogen-refueling station, capable of dispensing high-pressure hydrogen into the vehicle's tanks.

By using hydrogen as a fuel, GHGs (carbon monoxide and carbon dioxide) are almost completely eliminated in HICE vehicles. Future improvements may also reduce the small amounts of oxides of nitrogen (NO_x) produced by HICE engines to near-zero levels. The HICE system also eliminates particulate matter, which is a common dangerous byproduct of both diesel and natural gas engines. HICE therefore provides 99% of the emission reduction benefits of a fuel cell system at a fraction of the cost [164].

It is often claimed that HICEs perform well under most weather conditions because they require virtually no warm-up, have no cold-start issues even at sub-zero temperatures and achieve 25% better fuel efficiency than conventional engines. HICE vehicles operate at lower efficiency in comparison to fuel cell vehicles that are fueled with hydrogen. This means that a fuel-cell vehicle can travel much further on a given amount of hydrogen than a HICE vehicle. However, since the components and design of HICEs are similar to that of conventional gasoline engines, they are often regarded as an exciting and lower cost alternative to fuel cell vehicles. HICE could act as a bridging technology toward a widespread hydrogen infrastructure, since HICE vehicles can initially be designed for bi-fuel applications [165].

There are two types of hydrogen internal combustion engine vehicles: conventionally driven HICE vehicles and hybrid HICE vehicles. In conventionally driven HICE vehicles, the hydrogen-burning engine mechanically drives the vehicle's wheels, just as engines using gasoline or other fuels operate in conventional vehicles. In hybrid HICE vehicles, the hydrogen engine is used to run an electric generator, in a similar manner to series hybrid drive systems operating on other fuels. Power from the electric generator is then used to drive the vehicle's wheels and is generally augmented by power from a battery or ultracapacitor pack.

While no HICE vehicles are yet in mass production, several models have been demonstrated, and a select few, specially modified HICE vehicles, are commercially available. Ford and BMW are among current manufacturers of hydrogen-burning engines.

10.5.2 Fuel Cells

Although fuel cells have been around since 1839, it took 120 years until NASA demonstrated some of their potential applications in providing power during space flights. As a result of these successes, in the 1960s, industry began to recognize the commercial potential of fuel cells, but encountered technical barriers and high investment costs. Indeed, fuel cells were not economically competitive with existing energy technologies. After 50 years, despite technological progresses, the situation seems to be similar in many respects.

Today, a lot of companies around the world are working toward making fuel cell technology pay off. These companies are being driven by technical, economic, and social forces such as high-performance characteristics, reliability, durability, low cost, and environmental benefits. These attributes have provided the incentive for much of the research that has taken place on fuel cells in recent years.

The potential uses for fuel cells are extremely diverse and vary greatly in their power demand, from watts to megawatts. In particular, in transportation, they can be used to power electric road vehicles (cars, trucks, and buses), railway locomotives, submarines, and sub-sea vehicles.

Like combustion engine, a fuel cell utilizes a fuel and an oxidizer as reactant. Both systems derive the desired output of useful work from the chemical bond energy released via the oxidation of the fuel. However, while a combustion engine converts the chemical energy of the fuel and oxidizer into mechanical work, a fuel-cell engine converts the same initial chemical energy directly into electrical work. In other words, in the conventional engine, the fuel and oxidizer react via combustion to generate heat, which is then converted to useful work via some mechanical processes. An internal combustion engine in a car is a good example. Combustion expands the gas in the combustion chamber, which moves the pistons and is converted to rotational motion in the drive train. This turns the wheels and propels the vehicle. On the other hand, in an electrochemical engine, electricity is directly produced via an electrochemical oxidation process. This direct conversion of energy has a profound impact on the efficiency of electrochemical device. Indeed, because the process does not involve conversion of heat to mechanical energy, a fuel cell is intrinsically more efficient than engines based on combustion process.

Comprehensive descriptions of various fuel cells are given in the literature [166,167].

The core of a fuel cell consists of an electrolyte and two electrodes (anode and cathode). At the negative anode, a fuel such as hydrogen is oxidized, while at the positive cathode, oxygen is reduced. Ions are transported through the electrolyte from one side to the other. If hydrogen gas and oxygen are used as a fuel, in addition to electricity, the reaction by-products include heat and water. If methane or methanol is used directly as a fuel, a fuel cell also produces carbon dioxide as a by-product.

The type of electrolyte determines the temperature window of operation. This window of operation in its turn determines the catalyst that can be used, and the purity of the fuel to be used.

Table 10.2 summarizes the types of fuel cells showing their operating characteristics. Additional subclassification (not shown in Table 10.2) beyond the basic nomenclature can be assigned in terms of fuel used or the operating temperature range. Each fuel cell type has advantages that engender use for particular applications. In particular, for transportation applications, compact size, rapid start-up, robustness, and high efficiency are the primary technical goals.

TABLE 10.2

Main Types of Fuel Cells

	AFC	PEMFC	PAFC	MCFC	SOFC
Electrolyte	OH^- ions	H^+ ions	H^+ ions	CO_3^{2-} ions	O^{2-} ions
Operating T (°C)	80–200	60–90	180–220	600–700	800–1000
Efficiency (%)	45–60	40–60	35–40	45–60	50–65
Electrical power	Up to 20 kW	Up to 250 kW	>50 kW	>1 MW	>200 kW
Fuels	Pure H_2	H_2, natural gas, methanol	Natural gas, methanol, naphtha	Natural gas, H_2, CO	Natural gas, cola, methanol, petroleum
Advantages	Quick start-up; high performance	Long lifetime; quick start-up; high power density	High efficiency; can use impure hydrogen as fuel	High efficiency; impure H_2 as fuel	High stability; can use a variety of catalysts
Disadvantages	Requires pure H_2 and O_2	Sensitive to fuel impurities; expensive catalysts	Low current and power; large size/weight	Slow start-up; high T enhances breakdown of cell components	Slow start-up; high T enhances breakdown of cell components
Possible applications	Military and space vehicles	Vehicles	Power stations	Power stations	Power stations

In alkaline fuel cells (AFCs), the electrolyte consists of potassium hydroxide solution. The operating temperature is around 80°C but can be as high as 200°C. The AFC is currently being used for power generation on spacecrafts. The use of AFC's is limited because practically only pure hydrogen can be used as fuel. Air needs to be cleaned from CO_2, which limits the application for terrestrial applications considerably. The power density of the AFC is in the range of 0.1–0.3 W cm^{-2}. AFCs are especially available in the kilowatt range.

The electrolyte of the PEMFC consists of a cation-exchange membrane. The most commonly membrane currently used is Nafion® resin (Du Pont), which consists of a fluorocarbon polymer functionalized with sulfonic acid groups. This is an excellent proton conductor with good resistance to gas cross-over and does not suffer from electrolyte loss as the acid groups are fixed to the polymer backbone [168,169]. The operating temperature is around 80°C. Cold start, below 0°C, is possible. For transport applications, the PEMFC is the fuel cell of choice. For stationary applications, PEM fuel cells are developed as well. The PEMFC is rather sensitive toward impurities in the fuel. The power density of the PEMFC is in the range of 0.35–0.7 W cm^{-2}. PEM fuel cells are in development in the 1–250 kW range.

Phosphoric acid fuel cell (PAFC). Phosphoric acid is the electrolyte of the PAFC. The operating temperature is around 200°C. The PAFC can use reformate with CO concentrations up to 1%–2%. Commercially, it is the most successful fuel cell at the moment; in 2003, 245 samples of the 200 kW system have already been installed [170]. The power density of the PAFC is in the range of 0.14 W cm^{-2}.

In molten carbonate fuel cells (MCFC), a molten mixture of lithium, sodium, and potassium carbonate is used as the electrolyte in the MCFC. The operating temperature is between 600°C and 700°C. Due to the high operating temperature, internal reforming of hydrocarbon fuels is possible. The power density of the MCFC is in the range of 0.1–0.12 W cm^{-2}. The power of MCFC systems is in the 50 kW–5 MW range.

In SOFC, yttrium oxide stabilized zirconia is generally used as the solid electrolyte in the SOFC. Depending on the electrolyte and the material composition of the electrodes, the SOFC can be operated between 600°C and 1000°C. Fuels ranging from hydrogen to natural gas and higher hydrocarbons can be used. The SOFC is mainly in development for stationary power generation for systems in the 1 kW–5 MW range. However, it is also considered an important option for auxiliary power units on board of vehicles in the 5 kW range. The power density of the SOFC is in the range of $0.15–0.7\,W\,cm^{-2}$.

As mentioned earlier, while the higher temperature systems are well-suited for stationary applications, the low-temperature PEMFC are attractive candidates for applications in transport sector. However, PEMFCs have still many complex technical issues that have no simple solution.

The membrane used as electrolyte must be kept hydrated at all times. Thus, the PEMFC must operate in conditions where the water produced does not evaporate at a faster rate than it is produced. This necessity determines the normal operating temperature of 70°C–90°C, with an upper temperature limit of 120°C. Only noble metal-based (e.g., Pt) catalysts have sufficient activity in the low temperature range to meet power density targets for mobile applications. Another problem relates to the extreme sensitivity of the electrodes to CO and SO_2. These poison the electrodes by blocking the active surface. In the case of CO, the poisoning is somewhat reversible. In practice, it results that CO levels must be less than 10 ppm, and the stream must be sulfur free. With CO-tolerant electrodes, the upper limit may be 50 ppm, with some loss of performance, but even this level represents a significant barrier and illustrates how efficient the CO removal method needs to be.

Methanol and some other liquid fuels can be fed to a PEMFC directly without being reformed, thus forming a direct methanol fuel cell (DMFC), direct ethanol fuel cell (DEFC), direct formic acid fuel cell (DFAFC), and so on.

Most fuel cells designed for use in vehicles produce electricity far from enough to power a vehicle. Therefore, multiple cells must be assembled into a fuel-cell stack. The potential power generated by a fuel-cell stack depends on the number and size of the individual fuel cells that comprise the stack and the surface area of the PEM.

10.5.2.1 Fuel Processor Technology

The process of conversion of a carbon-based fuel to H_2 for use in fuel cells is commonly called fuel processing. One option to supply H_2 for automotive applications is that of onboard reforming, in which the H_2 is obtained from a hydrocarbon or oxygenate fuel by reforming reactions onboard the vehicle. Because of the absence of a H_2 infrastructure, this was initially considered to be the most accessible option, and much early work was devoted to this approach, with a view to utilizing the existing gasoline network or another liquid fuel that could be adapted (usually methanol).

The H_2 production methods discussed in Section 10.3 can be applied to onboard reforming, with suitable modification.

The tolerance toward carbon monoxide produced in reforming processes strongly depends on the temperature level of the fuel cell. The MCFC and SOFC can be fed with carbon monoxide, while the concentration of CO that can be tolerated by the PEMFC is in the range between 10 and 50 ppm. Other impurities with a negative impact on fuel-cell performance and durability have to be removed as well. Hydrogen purification methods are described in Section 10.2.4.

10.5.3 Commercial Prototypes of Hydrogen-Based Vehicles

Cars, buses, trains, bicycles, canal boats, cargo bikes, golf carts, motorcycles, wheelchairs, ships, airplanes, submarines, and rockets can already run on hydrogen, in various forms. NASA uses hydrogen to launch Space Shuttles into space. There is even a working toy model car that runs on solar power, using a regenerative fuel cell to store energy in the form of hydrogen and oxygen gas. It can then convert the fuel back into water to release the solar energy.

Here, we report briefly some recent examples of hydrogen-based vehicles.

Cars. Vehicle manufacturers have taken different approaches to commercializing fuel cell cars. Most car companies have at least one prototype fuel cell car [including Honda (Honda FCX), Ford (Ford Focus FCV and Ford P2000), DaimlerChrysler (NECAR and Mercedes F-Cell), Toyota (FCHV), and General Motors (HydroGen3, Sequel, and HyWire)].

Each car company has slightly different ideas on fuel-cell car design. Ford's strategy for instance is to modify existing models. Ford's rationale is that consumers are more likely to adopt a new technology if it provides comparable service and is not radically different in appearance from their current vehicles. In other words, consumers are comfortable driving existing models and may not be willing to purchase vehicles that look radically different.

General Motor's design track differs decidedly, from the use of proprietary fuel cells to the overall design. In 2002, GM introduced the Autonomy Concept Car, which is the first of this kind using fuel cells and by-wire technology. In 2003, GM expanded on this idea and introduced the first drivable concept car that utilizes both technologies: the HyWire, which resembles a futuristic vehicle. Whereas Ford's strategy assumes consumers will drive fuel cell cars only if they look and feel similar to existing vehicles, GM's philosophy is to change the way consumers look at their cars.

Most hydrogen cars are currently only available as demonstration models for lease in limited numbers and are not yet ready for general public use mainly due to the manufacture costs which are still too high.

In 2008, Hyundai announced its intention to produce 500 FC vehicles by 2010 and to start mass production of its FC vehicles in 2012. In early 2009, Daimler announced plans to begin its FC vehicle production in 2009 with the aim of 100,000 vehicles in 2012–2013. In 2009, Nissan started testing a new FC vehicle in Japan.

In September 2009, Daimler, Ford, General Motors, Honda, Hyundai, Kia, Renault, Nissan, and Toyota issued a joint statement about their undertaking to further develop and launch fuel-cell electric vehicles as early as 2015.

In February 2010, Lotus Cars announced that it was developing a fleet of hydrogen taxis in London, with the hope of them being ready to trial by the 2012 Olympic Games. London's deputy mayor, Kit Malthouse, said he hoped six filling stations would be available and that around 20–50 taxis would be in operation by then as well as 150 hydrogen-powered buses.

In March 2010, General Motors said that it had not abandoned the fuel-cell technology and still targeted to introduce hydrogen vehicles to retail customers by 2015. Charles Freese, GM's executive director of global powertrain engineering, stated that the company believes that both fuel-cell vehicles and battery electric vehicles are needed for reduction of GHG and reliance on oil, and the USA should follow Germany and Japan in adopting a more uniform strategy on advanced technology options. Both have announced plans to open 1000 hydrogen fuel stations.

Buses. Hydrogen buses were introduced in many cities for their low pollution as well as for social reasons, such as raising public hydrogen awareness and promotion of further research. One important hydrogen bus demonstration project worth mentioning in Europe is the HyFLEET:CUTE initiative that is comparing the advantages and disadvantages of hydrogen internal combustion engine (ICE) buses with fuel cell buses. CUTE stands for Clean Urban Transport for Europe and the goal of the project is to test and demonstrate hydrogen buses in 10 different cities in Europe, Asia, and Australia to reduce CO_2 emissions and move away from fossil fuels. The 10 cities currently involved in the HyFLEET: CUTE project include: Amsterdam, Barcelona, Beijing, Hamburg, London, Luxembourg, Madrid, Perth, Reykjavik, and Berlin [171]. Berlin, Germany, has been designated by the HyFLEET: CUTE project as a location of the largest supporting infrastructure for the hydrogen bus fleet. The filling station in Berlin will be equipped to supply both compressed gaseous hydrogen and liquid hydrogen and will be able to refuel 20 hydrogen buses or 200 hydrogen cars. The Berlin hydrogen-fueling station will also use new ionic liquid technology instead of pistons to compress the hydrogen gas to achieve greater psi than previous compressors.

Hydrogen produced for the HyFLEET: CUTE will come from many different sources both renewable and nonrenewable. The renewable generation of hydrogen will include hydrogen produced by solar, wind, and geothermal. The nonrenewable sources will include hydrogen produced from natural gas, LPG (Liquefied Petroleum Gas) and Bio-DME (Bio Dimethyl Ether).

One of the major goals of HyFLEET: CUTE is to demonstrate that both fuel cell and ICE buses can be cost effective alternative currently in operation. Comparisons will be drawn between the hydrogen buses and those that run on either diesel or CNG (compressed natural gas). Fuel cell buses in the HyFLEET: CUTE fleet are to remain in operation for 12 h a day with a minimum operation of 4000 h/bus.

Three of the fuel cell buses currently being tested in the HyFLEET: CUTE project include the Mercedes Benz Citaro Fuel Cell Bus, the Man Lion's City H 150 kW model, and the Man Lion's City H 200 kW model. A few other bus models also exist such as the liquid hydrogen hybrid bus running in Berlin.

Bicycles. The Chinese company Pear Hydrogen has developed hydrogen fuel cell bicycles. The 20″ wheel prototype weighs 32 kg and is powered by a PEM fuel cell and brushless electric motor. The top speed is 25 km/h and with the 600 L twin cylinders, fuel cells have a maximum range of 100 km. Of course, there are still technical challenges to overcome, like where people will refill the fuel cells. At present there is no hydrogen-refueling infrastructures in China, so customers will have to purchase refills from local suppliers. Horizon Fuel Cell Technologies made a hydrogen powered bicycle capable of doing more than 300 km at 25 km/h on a single fuel run. The fuel cell system in these light electric vehicle applications is much smaller than for automobiles or motorcycles, requiring less hydrogen, with readily available hydrogen storage technologies—making the proposition easy and attractive. With many fuel cell vehicles on the road, visibility is increased, meaning that the investment in public outreach and education is more efficient. Also, while providing mobility, the system on the bicycles is also small portable power system able to run radios, computers, lights, power tools, medical equipment, even generate heat. The possibilities are endless and the start of a critical mass can spark wider deployment of higher power applications including fuel cell powered automobiles.

Motorcycles and scooters. ENV develops electric motorcycles powered by a hydrogen fuel cell. Other manufacturers as Vectrix are working on hydrogen scooters.

Tractors. New Holland Agriculture has developed the NH_2 fuel cell tractor, the world's first hydrogen-powered tractor. The NH_2 was developed as part of New Holland Agriculture's Energy Independent Farm concept, a framework for future agriculture in which farmers produce their own compressed hydrogen from water using electricity produced by wind farms, solar panels, or biomass and biogas processes situated on the farm.

It runs on hydrogen and oxygen and produces no emissions at all. The zero-emissions tractor gets its power from a fuel cell that sends electrons to a 106-hp electric motor powering all four wheels. Problems include a major lack of range and an awfully high price, so New Holland is not expecting to put the model into production until 2013. The hydrogen tank at 300 bar holds enough fuel to power the tractor for 1.5–2 h.

More than just an idea, the NH_2 tractor is a 106-hp, working prototype able to perform all the tasks of a New Holland's T6000 tractor, only with no emissions and in near silence. The clean operation of the tractor brings added health benefits when working in confined areas, such as animal sheds or greenhouses.

Airplanes. Companies such as Boeing, Lange Aviation, and the German Aerospace Center pursue hydrogen as fuel for manned and unmanned airplanes. In February 2008 Boeing tested a manned flight of a small aircraft powered by a hydrogen fuel cell. Unmanned hydrogen planes have also been tested. For large passenger airplanes, however, hydrogen fuel cells are unlikely to power the engines of large passenger jet airplanes but could be used as backup or auxiliary power units onboard.

In July 2010 Boeing unveiled its hydrogen powered Phantom Eye UAV, powered by two Ford internal combustion engines that have been converted to run on hydrogen.

In Europe, the Reaction Engines A2 has been proposed to use the thermodynamic properties of liquid hydrogen to achieve very high speed, long distance antipodal flight by burning it in a pre-cooled jet engine, a concept for high speed jet engines that features a cryogenic fuel-cooled heat exchanger immediately after the air intake, to precool the air entering the engine. After gaining heat and vaporizing in the heat exchanger system, the fuel, for example, H_2 is burnt in the combustor. Precooled jet engines have never flown, but are predicted to have much higher thrust and efficiency.

Trucks. A HICE forklift or HICE lift truck is a hydrogen fueled, internal combustion engine powered industrial forklift truck used for lifting and transporting materials. The first production HICE forklift truck based on the Linde X39 Diesel was presented at an exposition in Hannover on May 27, 2008. It used a 2.0 L, 43 kW diesel internal combustion engine converted to use hydrogen as a fuel with the use of a compressor and direct injection. The hydrogen tank is filled with 26 L of hydrogen at 350 bar pressure.

Boats. In 2009, the world's first hydrogen fuel cell canal trip boat has been baptized in Amsterdam. The capacity of the fuel cell boat is about 87 passengers.

This very innovative ship is the first ship in the world solely designed and constructed for this type of production and use of energy as well as propulsion. The electricity for this propulsion system and all other on board functions will be produced by two large fuel cells. These two fuel cells will use high purity hydrogen. The hydrogen will be produced by a landside electrolyzing system; the electricity for the unit will in near future be produced at a North Sea wind farm.

10.5.4 HYDROGEN VERSUS COMPETING TECHNOLOGIES

What ICE, batteries, and fuel cells have in common is their purpose: all are devices that convert energy from one form to another.

ICEs run on noisy, high temperature explosions resulting from the release of chemical energy by burning fuel with oxygen form the air. ICEs convert chemical energy of fuel to thermal energy to generate mechanical energy.

Fuel cells and batteries are electrochemical devices and, by their nature, have a more efficient conversion process: chemical energy is converted directly to electrical energy. ICEs are less efficient because they include the conversion of thermal to mechanical energy, which is limited by Carnot's cycle.

If vehicles were powered by electricity generated for instance from direct hydrogen fuel cells, there would be no combustion involved. In an automotive fuel cell, hydrogen and oxygen undergo a relatively cool, electrochemical reaction that directly produces electrical energy. This electricity would be used by motors, including one or more connected to axles used to power the wheels of the vehicle.

The direct hydrogen fuel cell vehicle will have no emissions even during idling, and this is especially important during city rush hours. There are some similarities to an internal combustion engine, however. There is still a need for a fuel tank and oxygen is still supplied from the air.

An electric vehicle (EV) is a vehicle with an electric drive train powered by either an on-board battery or fuel cell. Batteries and fuel cells are similar in that they both convert chemical energy into electricity very efficiently and they both require minimal maintenance because neither has any moving parts. However, unlike a fuel cell, the reactants in a battery are stored internally and, when used up, the battery must be either recharged or replaced. In a battery-powered EV, rechargeable batteries are used.

With a fuel-cell powered EV, the fuel is stored externally in the vehicle's fuel tank, and air is obtained from the atmosphere. As long as the vehicle's tank contains fuel, the fuel cell will produce energy in the form of electricity and heat. The choice of electrochemical device, battery, or fuel cell depends upon use. For larger-scale applications, fuel cells have several advantages over batteries including smaller size, lighter weight, quick refueling, and longer range.

10.6 HYDROGEN INFRASTRUCTURE

The cost of building a hydrogen-fueling infrastructure is an often debated subject. Many economic actors will need to take coordinated action if hydrogen is to take a significant share of the market for transport fuel [172,173]. However, carmakers will not build more cars until an infrastructure is in place in which to refuel their cars. Energy companies, on the other hand, without large numbers of fuel-cell cars available at reasonable prices, will have little interest in building a costly new fueling infrastructure. Finally, customers will not purchase fuel cell vehicles unless adequate fuelling is available. This classic chicken-or-egg dilemma has long hobbled the development of most alternative fuels and has assured the supremacy of oil [174]. Thanks to low prices and abundant reserves just a few years ago, energy providers and automakers simply had little incentive to end the petroleum age. But, faced with the environmental issues (e.g., global warming) and soaring prices, automakers and oil companies have begun a hasty search for alternatives and have been working together to break the hydrogen logjam.

One of the key issues for the widespread introduction of hydrogen is whether the gas will be manufactured in a limited number of large central plants or in a multitude of strategically located, small units. Each approach has its merits as determined by the primary energy source employed (fossil or renewable), the use envisaged for the hydrogen and the amounts required. Centralized production is most appropriate for large fossil-fuel steam reforming or coal gasification operations. In principle, it allows the possibility of sequestrating the carbon dioxide that is formed as a by-product. Tonnage quantities of hydrogen will then have to be supplied to customers, either as a cryogenic liquid by tanker delivery or as a gas through a newly installed national grid of hydrogen pipelines—both of which are expensive procedures. A simpler transitional option, as mentioned in Section 10.4, may be to blend hydrogen produced centrally (with sequestration) into natural gas in existing pipelines, thereby reducing the overall carbon-to-hydrogen ratio of the transported fuel.

Localized, small-scale production of hydrogen is best suited to using renewable energy to generate electricity that is then employed to electrolyze water. Initially, this would be a limited option, but in the longer term, as renewables assume a greater share of the market, it may become more important. These forms of energy are themselves often well scattered (wind turbines, solar panels, etc.) and give rise to little or no carbon dioxide. The disadvantage of this route is that the hydrogen is likely to cost significantly more than that produced in bulk from fossil fuels. It is possible to construct small natural gas reformer units, appropriate for the distributed supply of hydrogen, but again the economies of scale would be lost.

Hydrogen infrastructure can evolve gradually with conversion and production technologies, since most of the infrastructure developed for fossil-fuel-based hydrogen will also work with hydrogen from renewable and nuclear sources. Infrastructure development will begin with pilot projects and expand to local, regional, and ultimately national and international projects.

10.6.1 FUTURE SUPPLY OPTIONS

The relevant hydrogen production options to supply the network of filling stations much depend on the local or regional conditions. In an initial stage, the hydrogen production and supply may take advantage of the existing infrastructure for electricity and natural gas.

Via electrolytic production, hydrogen may be generated and supplied at almost any location. The modularity of the electrolysis technologies means that both large and small-scale plants may be relevant. The individual hydrogen-filling stations may be equipped with electrolysis units, for example, for high-pressure electrolysis, compressor and storage facilities, and dispensers for high-pressure hydrogen. Even very small-scale residential dispenser units may be a future option based on electrolysis.

Where natural gas is available, reforming is another supply option for hydrogen. Reformer plants could be situated near, or in, hydrogen-filling stations. Economies of scale, however, may favor larger reformer plants, which, in turn, would require local pipelines to supply the filling stations.

Beyond the very initial stages of a hydrogen supply infrastructure build-up another option may be to distribute hydrogen mixed into the NG-grid in quantities up to about 15% volume, depending on pipeline material, pipeline pressure, and end-user technology.

Hydrogen supplied from numerous sources may be distributed via mixed gas grid lines.

Hydrogen-filling stations would be able to separate the hydrogen from the natural gas, purify, and pressurize it for use in vehicles. Most other users would simply be able to burn the natural gas/hydrogen mixture without problems.

The only difficulties would be with internal combustion engines running on natural gas.

Various other factors also limit the proportion of hydrogen than can be added to natural gas pipelines. The need to preserve the combustion quality of the gas (Wobbe index) would limit the hydrogen content to 25% by volume, while restrictions on density would allow a maximum of 17% hydrogen. The most stringent limit is set by leakage and material compatibility. Above about 15% hydrogen, problems of leakage, corrosion, degradation of polyethylene pipe, and embrittlement of steels are expected.

The energy content of hydrogen is only around one third that of the same volume of natural gas. As a consequence, pipelines and underground storage caverns for hydrogen would have to be three times the size of their natural gas equivalents, at the same pressure, to cover the same energy demand. For safety reasons, hydrogen pipelines will also operate at lower pressures than natural gas pipelines. Hydrogen is more demanding than natural gas in terms of materials of construction, and it is clear that pipelines for pure hydrogen will be more expensive than for natural gas also because of the great cost of the pipeline material.

Storage of gaseous hydrogen at high pressures is well-known technology. Hydrogen-filling stations would require pressurized storage tanks holding hydrogen equivalent to roughly one day sales. On a much larger scale, hydrogen can be stored underground in aquifers and cavities, in the same way as natural gas.

10.6.2 Long-Term Vision for a Hydrogen Infrastructure

The long-term vision is an energy system highly diversified, robust, environmentally benign, and affordable.

Fuel cells and electrolyzers or reversible fuels cells, plus hydrogen grids and storage systems, are likely to be key technologies for balancing electricity grids.

Hydrogen has the potential for being a link between the transport sector and the heat and power sectors. The flexibility of such system configurations can compensate for fluctuating power inputs, such as from wind power and solar cells, and thus promote the use of renewable energy for transport, heating, and electricity.

Barriers to this vision include the need to improve the performance of hydrogen and hydrogen-related technologies to the point where they can compete in economic terms with existing energy technologies. In particular, this means improving the performance of fuel cells and mobile hydrogen storage technologies. The vision for getting started would involve the transport sector in particular and the build-up of a limited number of filling stations primarily for serving car fleets like taxies and buses in urban areas. In following stages of building up a hydrogen infrastructure, pipelines with hydrogen or natural gas mixed with hydrogen may be made available, for example, to selected blocks with flats and public buildings like schools and hospitals. The technology to produce hydrogen from on site steam methane reforming is ready. The same applies for electrolytic hydrogen production.

The vision for the next 20–40 years is to make hydrogen fully competitive with conventional energy sources for both vehicles and individual dwellings. This will require extensive hydrogen distribution grids and a dense network of urban filling stations.

10.7   CONCLUSIONS AND PERSPECTIVES

The transportation sector is one of the main reasons for the air pollution emissions. This latter issue will become even more important in the near future, due to the globalization of economies and the exponential accelerated development of India and China and other developing and transitional countries. Hence, the only feasible way of reducing emissions produced by the transportation sector will be to introduce new technological approaches primarily with respect to the fuels utilized in vehicles as well as the traction systems applied. While biofuels or bioadditives, such as biodiesel or bioethanol, are an important reality in some countries, their use in conventional combustion engine leads to air pollution emissions even in the presence of state of the art end-of-pipe technologies (oxidation catalysts, particulate filters, etc.). To overcome this limitation, fuel cells are commonly considered as possible final solution. Fuel cells offer a number of advantages compared to internal combustion engines, the first of which is their theoretical higher conversion efficiency. In fact, the electrochemical conversion of the fuel into electricity will avoid generation of pollutants. In addition, the associated electric engine does not produce any noise. However, there are still a number of issues that must be addressed before considering a wide application of fuel cell technology. One of the most important questions is the choice of fuel. Hydrogen would the best option for fuel cells in transportation applications, as its clean conversion produces only water. However, the lack of a ready hydrogen infrastructure as well as the open question on the sustainable production of large quantities of hydrogen, dramatically limits a wide application of this energy vector. In addition, stability problems and cost of fuel cells must be still solved. In this context, other fuels are suggested and currently used, both as transition from a fossil based economy into a hydrogen-based economy, and for ultimate use.

The automotive industry is about to start the transition from ICE to batteries and fuel cells. However, at the moment, it is not clear whether there will be a significant market presence before 2020.

Before hydrogen can really compete for a prominent position in the top list of future energy vectors to be used in transportation the following technical limitations must be fully overcome.

- Identification and eventually realization of sustainable centralized large scale and/or delocalized small-scale hydrogen production plants. Careful evaluations of the specific regional socioeconomic situation, in accordance with state of the art Life Cycles Assessment methodology, must be used to identify the more convenient hydrogen source and production technology.
- Identification and realization of valuable technological options for hydrogen distribution and delivery. It must be considered that, once defined, the construction of these new infrastructures would require immense investments.
- Identification and realization of hydrogen storage technology must be significantly advanced.
- New materials and construction methods must be developed to reduce fuel-cell system cost to be competitive with the automotive combustion engine. The cost of the catalyst no longer dominates the price of most fuel-cell systems, although it is still significant. Manufacturing and mass production technology are also a key component to the commercial viability of fuel cell systems.
- Suitable reliability and durability of hydrogen fuel cells must be achieved. The performance of every fuel cell gradually degrades with time due to a variety of phenomena.
- Suitable system power density and specific power must be achieved according to the DOE 2010 targets.
- Desired performance and longevity of ancillary system components must be achieved. New hardware (e.g., efficient transformers and high-volume blowers) must be developed to suit the needs of fuel-cell power systems.
- More efficient hydrogen sensors and online control systems for fuel-cell systems are needed, especially for transient operation, where performance instability can become a major issue.

The lack of striking results in technologies and the high costs are dampening the enthusiasm by reducing the investment in R&D. Taking into account these aspects, and the increasing attention to second and third-generation biofuels (ethanol and biodiesel), a redefinition of future perspective of hydrogen in transportation is needed. In fact, liquid biofuels offer the great advantage to be more easily transported and stored and can take advantage of existing technologies and infrastructure. While large quantities of hydrogen will be used as a chemical feedstock and industrial gas for many years, its future as energy vector is less bright than it appeared in the past.

ACKNOWLEDGMENTS

Professor Mauro Graziani (University of Trieste) is warmly acknowledged for helpful discussions. University of Trieste, ICCOM-CNR, Regione Friuli Venezia Giulia, and Fondo Trieste are acknowledged for financial support.

REFERENCES

1. Mac Kay D.J.C. 2009. *Sustainable Energy—Without the Hot Air*, Cambridge, U.K.: UIT Cambridge, available at http://www.withouthotair.com/download.html
2. Hanjalic K., R. Van de Krol, and A. Lekic. 2008. *Sustainable Energy Technologies, Options and Prospects*, Dordrecht, the Netherlands: Springer.
3. USGS World energy Assessment Team, *World undiscovered assessment results summary*, U.S. Geological Survey Digital Data Series 60, available at http://greenwood.cr.usgs.gov/energy/WorldEnergy/DDS-60/sum1.html
4. Campbell C.J. and J.H. Laherrere. 1998. Preventing the next oil crunch—The end of cheap oil, *Sci. Am.* 278:77–83.
5. Crowley T.J. 2000. Causes of climate change over the past 1000 years, *Science* 289:270–277.
6. Hansen J.E. and A.A. Lacis. 1990. Sun and dust versus greenhouse gases—An assessment of their relative roles in global climate change, *Nature* 346:713–719.
7. Bray D. 2010. The scientific consensus of climate change revisited, *Environ. Sci. Policy* 13:340–350.
8. Ramanathan V. and Y. Feng. 2009. Air pollution, greenhouse gases and climate change: Global and regional perspectives, *Atmos. Environ.* 43:37–50.
9. International Energy Outlook 2010, available at http://www.eia.doe.gov/oiaf/ieo/ (accessed on July 2011)
10. Uherek E., T. Halenka, J. Borken-Kleefeld, Y. Balkanski, T. Berntsen, C. Borrego, M. Gauss et al. 2010. Transport impacts on atmosphere and climate: Land transport, *Atmos. Environ.* 44:4772–4816.
11. 2010 Production Statistics, http://www.oica.net/category/production-statistics/ (accessed on August 2011)
12. Heck R.M., R.J. Farrauto, and S.T. Gulati. 2009. *Catalytic Air Pollution Control*, Hoboken, NJ: John Wiley & Sons.
13. Searles R.A. 2003. Contribution of automotive catalytic converters, in *Materials Aspects in Automotive Catalytic Converters,* H. Bode (ed.), Weinheim, Germany: Wiley-VCH Verlag GmbH & Co. KGaA.
14. Kaspar J., P. Fornasiero, and M. Graziani. 1999. Use of CeO_2-based oxides in the three-way catalysis, *Catal. Today* 50:285–298.
15. Kaspar J., P. Fornasiero, and N. Hickey. 2003. Automotive catalytic converters: Current status and some perspectives, *Catal. Today* 77:419–449.
16. Farrauto R.J. and K.E. Voss. 1996. Monolithic diesel oxidation catalysts, *Appl. Catal B Environ.* 10:29–51.
17. van Setten B.A.A.L., M. Makkee, and J.A. Moulijn. 2001. Science and technology of catalytic diesel particulate filters, *Catal. Rev.* 43:489–564.
18. Armaroli N. and V. Balzani. 2011. The hydrogen issue, *ChemSusChem* 4:21–36.
19. Haussinger P., R. Lohmuller, and A. Watson. 2002. *Ullmann's Encyclopedia of Industrial Chemistry*, 7th edn., Weinheim, Germany: Wiley-VCH Verlag GmbH & Co.
20. Ibsen K. 2006. Equipment design and cost estimation for small modular biomass systems, synthesis gas cleanup, and oxygen separation equipment, *Contract Report NREL/SR-510-39943, NREL Technical Monitor*, Section 2, available at http://www.nrel.gov/docs/fy06osti/39943.pdf
21. Hirsch D. and A. Steinfeld. 2004. Solar hydrogen production by thermal decomposition of natural gas using a vortex-flow reactor, *Int. J. Hydrogen Energ.* 29:47–55.

22. De Maria G., C.A. Tiberio, L. D'Alessio, M. Piccirilli, E. Coffari, and M. Paolucci. 1986. Thermochemical conversion of solar energy by steam reforming of methane, *Energy* 11:805–810.
23. Tamme R., R. Buck, M. Epstein, U. Fisher, and C. Sugarmen. 2001. Solar upgrading of fuels for generation of electricity, *J. Solar Energy Eng.* 123:160–163.
24. Piemonte V., M. De Falco, A. Giaconia, P. Tarquini, and G. Iaquaniello. 2010. Life cycle assessment of a concentrated solar power plant for the production of enriched methane by steam reforming process, *Chem. Eng. Trans.* 21:25–30.
25. Xing Z., W. Hua, W. Yonggang, L. Kongzhai, and Y. Dongxia. 2010. Hydrogen production by two-step water-splitting thermochemical cycle based on metal oxide redox system, *Prog. Chem.* 22:1010–1020.
26. Abbas H.F. and W.M.A. Wan Daud. 2010. Hydrogen production by methane decomposition: A review, *Int. J. Hydrogen Energ.* 35:1160–1190.
27. Forsberg C.W. 2009. Is hydrogen the future of nuclear energy? *Nucl. Technol.* 166:3–10.
28. Elder R. and R. Allen. 2009. Nuclear heat for hydrogen production: Coupling a very high/high temperature reactor to a hydrogen production plant, *Prog. Nucl. Energ.* 51:500–525.
29. Clarke S., A. Dicks, K. Pointon, T. Smith, and A. Swann. 1997. Catalytic aspects of the steam reforming of hydrocarbons in internal reforming fuel cells, *Catal. Today* 38:411–423.
30. Liu Q., Q. Zhang, W. Ma, R. He, L. Kou, and Z. Mou. 2005. Progress in water-gas-shift catalysts, *Prog. Chem.* 17:389–398.
31. Armor J.N. 1999. The multiple roles for catalysis in the production of H_2, *Appl. Catal. A Gen.* 176:159–176.
32. Newsome D.S. 1980. The water-gas shift reaction, *Catal. Rev.* 21:275–281.
33. Smith R.J.B., M. Loganathan, and M.S. Shantha. 2010. A review of the water gas shift reaction kinetics, *Int. J. Chem. React. Eng.* 8:R4.
34. Kermode R. 1977. Hydrogen from fossil fuels, in *Hydrogen: Its Technology and Applications*, K. Cox and K. Williamson (eds.), Chapter 3, Boca Raton, FL: CRC Press.
35. Spath P. and M. Mann. 2001. Life cycle assessment of hydrogen production via natural gas steam reforming, Technical report, NREL, NREL/TP-570-27637, available at http://www.nrel.gov/docs/fy01osti/27637.pdf
36. Rostrup-Nielsen J.R. and J.H.B. Hansen. 1993. CO_2-reforming of methane over transition metals, *J. Catal.* 144:38–49.
37. Wang S. and G. Lu. 1998. CO_2 reforming of methane on Ni catalysts: Effect of the support phase and preparation technique, *Appl. Catal. B Environ.* 16:269–277.
38. Choudhary V., S. Banerjee, and A. Rajput. 2002. Hydrogen from step-wise steam reforming of methane over Ni/ZrO_2: Factors affecting catalytic methane decomposition and gasification by steam of carbon formed on the catalyst, *Appl. Catal. A Gen.* 234:259–270.
39. Hu Y. and E. Ruckenstein. 2004. Catalytic conversion of methane to synthesis gas by partial oxidation and CO_2 reforming, *Adv. Catal.* 48:298–345.
40. De Rogatis L. and P. Fornasiero. 2009. Catalyst design for reforming of oxygenates, in *Catalysis for Sustainable Energy Production*, P. Barbaro and C. Bianchini (eds.), pp. 173–233, Weinheim, Germany: Wiley-VCH Verlag GmbH & Co. KGaA.
41. Alonso D.M., J.Q. Bond, and J.A. Dumesic. 2010. Catalytic conversion of biomass to biofuels, *Green Chem.* 12:1493–1513.
42. Simonetti D.A. and J.A. Dumesic. 2009. Catalytic production of liquid fuels from biomass-derived oxygenated hydrocarbons: Catalytic coupling at multiple length scales, *Cat. Rev.* 51:441–484.
43. www.virent.com, (accessed on August 2011)
44. Kodama T. and N. Gokon. 2007. Thermochemical cycles for high-temperature solar hydrogen production, *Chem. Rev.* 107:4048–4077.
45. Steinfeld A., A. Frei, P. Kuhn, and D. Wuillemin. 1995. Solar thermal production of zinc and syngas via combined ZnO-reduction and CH_4-reforming processes, *Int. J. Hydrogen Energ.* 20:793–804.
46. Abanades S., A. Legal, A. Cordier, G. Peraudeau, G. Flamant, and A. Julbe. 2010. Investigation of reactive cerium-based oxides for H_2 production by thermochemical two-step water-splitting, *J. Mater. Sci.* 45:4163–4173.
47. Bambagioni V., M. Bevilacqua, C. Bianchini, J. Filippi, A. Lavacchi, A. Marchionni, F. Vizza, and P.K. Shen. 2010. Self-sustainable production of hydrogen, chemicals, and energy from renewable alcohols by electrocatalysis, *ChemSusChem* 3:851–855.
48. Hallenbeck P.C. and J.R. Benemannn. 2002. Biological hydrogen production: Fundamentals and limiting processes, *Int. J. Hydrogen Energ.* 27:1185–1193.
49. Elam C.C., C.E.G. Padro, G. Sandrock, A. Luzzi, A. P. Lindblad, and E.F. Hagen. 2003. Realizing the hydrogen future: The International Energy Agency's effort to advance hydrogen energy technologies, *Int. J. Hydrogen Energ.* 28:601–607.

50. Kawai T. and T. Sakata. 1980. Conversion of carbohydrate into hydrogen fuel by a photocatalytic process, *Nature* 286:474–476.

51. Kondarides D.I., V.M. Daskalaki, A. Patsoura, and X.E. Verykios. 2008. Hydrogen production by photo-induced reforming of biomass components and derivatives at ambient conditions, *Catal. Lett.* 122:26–32.

52. Gombac V., L. Sordelli, T. Montini, J.J. Delgado, A. Adamski, G. Adami, M. Cargnello, S. Bernal, and P. Fornasiero. 2010. CuO_x–TiO_2 photocatalysts for H_2 production from ethanol and glycerol solutions, *J. Phys. Chem. A* 114:3916–3925.

53. Yang Y.Z., C.-H. Chang, and H. Idriss. 2006. Photo-catalytic production of hydrogen form ethanol over M/TiO_2 catalysts (M = Pd, Pt or Rh), *Appl. Catal. B- Environ.* 67:217–222.

54. Daskalaki V.M. and D.I. Kondarides. 2009. Efficient production of hydrogen by photo-induced reforming of glycerol at ambient conditions, *Catal. Today* 144:75–80.

55. Fujishima A. and K. Honda. 1972. Electrochemical photolysis of water at a semiconductor electrode, *Nature* 238:37–38.

56. Khan S.U.M., M. Al-Shahry, and W.B. Jr. Ingler. 2002. Efficient photochemical water splitting by a chemically modified n-TiO_2, *Science* 297:2243–2245.

57. Korzhak A.V., N.I. Ermokhina, A.L. Stroyuk, V.K. Bukhtiyarov, A.E. Raevskaya, V.I. Litvin, S.Y. Kuchmiy, V.G. Ilyin, and P.A. Manorik. 2008. Photocatalytic hydrogen evolution over mesoporous TiO_2/metal nanocomposites, *J. Photochem. Photobiol. A Chem.* 198:126–134.

58. Farrauto R.J. 2005. Introduction to solid polymer membrane fuel cells and reforming natural gas for production of hydrogen, *Appl. Catal. B-Environ.* 56:3–7.

59. Rostrup-Nielsen J.R. and T. Rostrup-Nielsen. 2002. Large-scale hydrogen production, *CATTECH* 6:150–159.

60. Farrauto R., S. Hwang, L. Shore, W. Ruettinger, J. Lampert, T. Giroux, Y. Liu, and O. Ilinich. 2003. New material needs for hydrocarbon fuel processing: Generating hydrogen for the PEM fuel cell, *Ann. Rev. Mater. Res.* 33:1–27.

61. von Helmolt R. and U. Eberle. 2007. Fuel cell vehicles: Status 2007, *J. Power Sources* 165:833–843.

62. Felderhoff M., C. Weidenthaler, R. Von Helmolt, and U. Eberle. 2007. Hydrogen storage: The remaining scientific and technological challenges, *Phys. Chem. Chem. Phys.* 9:2643–2653.

63. DOE Targets for Onboard Hydrogen Storage Systems for Light-Duty Vehicles, http://www1.eere.energy.gov/hydrogenandfuelcells/storage/pdfs/targets_onboard_hydro_storage.pdf (accessed on August 2011)

64. Satyapal S., J. Petrovic, C. Read, G. Thomas, and G. Ordaz. 2007. The U.S. Department of energy's national hydrogen storage project: Progress towards meeting hydrogen-powered vehicle requirements, *Catal. Today* 120:246–256.

65. Schlapbach L. and A. Züttel. 2001. Hydrogen-storage materials for mobile applications, *Nature* 414:353–358.

66. Aceves S.M. and G.D. Berry. 1998. Onboard storage alternatives for hydrogen vehicles, *Energ. Fuel.* 12:49–55.

67. http://www.dynetek.com/ (accessed on August 2011).

68. http://www.qtww.com/tanks (accessed on August 2011).

69. Kohno T. *Conformal Field Theory and Topology*, *Translation of Mathematical Monographs*, Vol. 210, Providence, RI: American Mathematical Society, 2002.

70. Colozza A.J. 2002. Hydrogen storage for aircraft applications overview, *NASA/CR-211867*. available at http://gltrs.grc.nasa.gov/reports/2002/CR-2002-211867.pdf

71. Hydrogen case studies http://www.dti.gov.uk/renewable/hydrogen_casestudies.html (accessed on August 2011).

72. http://www.the-linde-group.com/en/index.html (accessed on August 2011).

73. Hydrogen Energy, http://www.linde-gas.com/en/innovations/hydrogen_energy/index.html/ (accessed on August 2011).

74. Yang J., A. Sudik, C. Wolverton, and D.J. Siegel. 2010. High capacity hydrogen storage materials: Attributes for automotive applications and techniques for materials discovery, *Chem. Soc. Rev.* 39:656–675.

75. Züttel A. 2007. Hydrogen storage and distribution systems, *Mitig. Adapt. Strat. Glob. Change* 12:343–365.

76. Hirscher M. 2010. *Handbook of Hydrogen Storage*, Weinheim, Germany: Wiley-VCH Verlag GmbH & Co. KGaA.

77. Wang L.F. and R.T. Yang. 2010. Hydrogen storage on carbon-based adsorbents and storage at ambient temperature by hydrogen spillover, *Catal. Rev.* 52:411–461.

78. Yurum Y., A. Taralp, and T.N. Veziroglu. 2009. Storage of hydrogen in nanostructured carbon materials, *Int. J. Hydrogen Energ.* 34:3784–3798.

79. Strobel R., J. Garche, P.T. Moseley, L. Jorissen, and G. Wolf. 2006. Hydrogen storage by carbon materials, *J. Power Sources* 159:781–801.
80. Zhou Y.P., K. Feng, Y. Sun, and L. Zhou. 2003. A brief review on the study of hydrogen storage in terms of carbon nanotubes, *Prog. Chem.* 15:345–350.
81. Yang Z.Q., Z.L. Xie, and Y.S. He. 2003. Hydrogen storage in new nanostructured carbon materials, *New Carbon Mater.* 18:75–79.
82. Hirscher M. and M. Becher. 2003. Hydrogen storage in carbon nanotubes, *J. Nanosci. Nanotechno.* 3:3–17.
83. Cheng H.M., Q.H. Yang, and C. Liu. 2001. Hydrogen storage in carbon nanotubes, *Carbon* 39:1447–1454.
84. Hirscher M. and B. Panella. 2005. Nanostructures with high surface area for hydrogen storage, *J. Alloys Compd.* 404:399–401.
85. Jin H., Y.S. Lee, and I. Hong. 2007. Hydrogen adsorption characteristics of activated carbon, *Catal. Today* 120:399–406.
86. Panella B., M. Hirscher, and S. Roth. 2005. Hydrogen adsorption in different carbon nanostructures, *Carbon* 43:2209–2214.
87. Schimmel H.G., G. Nijkamp, G.J. Kearley, A. Rivera, K.P. de Jong, and F.M. Mulder. 2004. Hydrogen adsorption in carbon nanostructures compared, *Mater. Sci. Eng., B* 108:124–129.
88. Züttel A., P. Sudan, P. Mauron, and P. Wenger. 2004. Model for the hydrogen adsorption on carbon nanostructures, *Appl. Phys. A Mater. Sci. Process.* 78:941–946.
89. Rzepka M., P. Lamp, and M.A. de la Casa-Lillo. 1998. Physisorption of hydrogen on microporous carbon and carbon nanotubes, *J. Phys. Chem. B* 102:10894–10898.
90. Zubizarreta L., A. Arenillas, and J.J. Pis. 2008. Carbon materials for H_2 storage, *Int. J. Hydrogen Energ.* 34:4575–4581.
91. Marsh H., H.E. Heintz, and F. Rodriguez-Reinoso. 1997. *Introduction to Carbon Technologies*, Alicante, Spain: University of Alicante Press.
92. Phan N.H., S. Rio, C. Faur, L. Le Coq, P. Le Cloirec, and T.H. Nguyen. 2006. Production of fibrous activated carbons from natural cellulose (jute, coconut) fibers for water treatment applications, *Carbon* 44:2569–2577.
93. Vasiliev L.L., L.E. Kanonchik, A.G. Kulakov, D.A. Mishkinis, A.M. Safonova, and N.K. Luneva. 2007. New sorbent materials for the hydrogen storage and transportation, *Int. J. Hydrogen Energ.* 32:5015–5025.
94. Sharon M., T. Soga, R. Afre, D. Sathiyamoorthy, K. Dasgupta, S. Bhardwaj, M. Sharon, and S. Jaybhaye. 2007. Hydrogen storage by carbon materials synthesized from oil seeds and fibrous plant materials, *Int. J. Hydrogen Energ.* 32:4238–4349.
95. Jorda-Beneyto M., D. Lozano-Castello, F. Suarez-Garcia, D. Cazorla-Amoros, and A. Linares-Solano. 2008. Advanced activated carbon monoliths and activated carbons for hydrogen storage, *Micropor. Mesopor. Mat.* 112:235–242.
96. Jorda-Beneyto M., F. Suarez-Garcia, D. Lozano-Castello, D. Cazorla-Amoros, and A. Linares-Solano. 2007. Hydrogen storage on chemically activated carbons and carbon nanomaterials at high pressures, *Carbon* 45:293–303.
97. Takagi H., H. Hatori, and Y. Yamada. 2004. Hydrogen adsorption/desorption property of activated carbon loaded with platinum, *Chem. Lett.* 33:1220–1221.
98. Dillon A.C., K.M. Jones, T.A. Bekkedahl, C.H. Kiang, D.S. Bethune, and M.J. Heben. 1997. Storage of hydrogen in single-walled carbon nanotubes, *Nature* 386:377–379.
99. Darkrim F., A. Aoufi, P. Malbrunot, and D. Levesque. 2000. Hydrogen adsorption in the NaA zeolite: A comparison between numerical simulations and experiments, *J. Chem. Phys.* 112:5991–5999.
100. Chambers A., C. Park, R.T.K. Baker, and N.M. Rodriguez. 1998. Hydrogen storage in graphite nanofibers, *J. Phys. Chem. B* 102:4253–4256.
101. Park C., P.E. Anderson, A. Chambers, C.D. Tan, R. Hidalgo, and N.M. Rodriguez. 1999. Further studies of the interaction of hydrogen with graphite nanofibers, *J. Phys. Chem. B* 103:10572–10581.
102. Ahn C.C., Y. Ye, B.V. Ratnakumar, C. Witham, R.C. Bowman, and B. Fultz. 1998. Hydrogen desorption and adsorption measurements on graphite nanofibers, *Appl. Phys. Lett.* 73:3378–3380.
103. Lueking A.D., R.T. Yang, N.M. Rodriguez, and R.T.K. Baker. 2004. Hydrogen storage in graphite nanofibers: Effect of synthesis catalyst and pretreatment conditions, *Langmuir* 20:714–721.
104. Burress J.W., S. Gadipelli, J. Ford, J.M. Simmons, W. Zhou, and T. Yildirim. 2010. Graphene oxide framework materials: Theoretical predictions and experimental results, *Angew. Chem. Int. Ed.* 49:8902–8904.
105. Lachawiec A.J., G. Qi, and R.T. Yang. 2005. Hydrogen storage in nanostructured carbons by spillover: Bridge-building enhancement, *Langmuir* 21:11418–11424.
106. Fraenkel D. and J. Shabtai. 1977. Encapsulation of hydrogen in molecular sieve zeolites, *J. Am. Chem. Soc.* 99:7074–7076.

107. Weitkamp J., M. Fritz, and S. Ernst. 1995. Zeolites as media for hydrogen storage, *Int. J. Hydrogen Energ.* 20:967–970.
108. Chahine R. and T. Bose. 1994. Low-pressure adsorption storage of hydrogen, *Int. J. Hydrogen Energ.* 19:161–164.
109. Nijkamp M.G., J.E.M.J. Raaymakers, A.J. van Dillen, and K.P. de Jong. 2001. Hydrogen storage using physisorption—Materials demands, *Appl. Phys. A*, 72:619–623.
110. Langmi H.W., A. Walton, M.M. Al-Mamouri, S.R. Johnson, D. Book, J.D. Speight, P.P. Edwards, I. Gameson, P.A. Anderson, and I.R. Harris. 2003. Hydrogen adsorption in zeolites A, X, Y and RHO, *J. Alloys Compd.* 356:710–715.
111. Arean C.O., O.V. Manoilova, B. Bonelli, M.R. Delgado, G.T. Palomino, and E. Garrone. 2003. Thermodynamics of hydrogen adsorption on the zeolite Li-ZSM-5, *Chem. Phys. Lett.* 370:631–635.
112. Kazansky V.B., V.Y. Borovkov, A. Serich, and H.G. Karge. 1998. Low temperature hydrogen adsorption on sodium forms of faujasites: Barometric measurements and drift spectra, *Micropor. Mesopor. Mat.* 22:251–259.
113. Yu M., S. Li, J.L. Falconer, and R.D. Noble. 2008. Reversible H_2 storage using a SAPO-34 zeolite layer, *Micropor. Mesopor. Mat.* 110:579–582.
114. Meek S.T., J.A. Greathouse, and M.D. Allendorf. 2011. Metal-organic frameworks: A rapidly growing class of versatile nanoporous materials, *Adv. Mater.* 23:249–267.
115. James, S.L. 2003. Metal-organic frameworks, *Chem. Soc. Rev.* 32:276–288.
116. Li H., M. Eddaoudi, M. O'Keeffe, and O.M. Yaghi. 1999. Design and synthesis of an exceptionally stable and highly porous metal-organic framework, *Nature* 402:276–279.
117. Rowsell J.L.C., A.R. Millward, K.S. Park, and O.M. Yaghi. 2004. Hydrogen sorption in functionalized metal-organic frameworks, *J. Am. Chem. Soc.* 126:5666–5667.
118. Panella B. and M. Hirscher. 2005. Hydrogen physisorption in metal-organic porous crystals, *Adv. Mater.* 17:538–538.
119. Wong-Foy G., A. J. Matzger, and O.M. Yaghi. 2006. Exceptional H_2 saturation uptake in microporous metal-organic frameworks, *J. Am. Chem. Soc.* 128:3494–3495.
120. Dinc M., A. Dailly, Y. Liu, C.M. Brown, D.A. Neumann, and J.R. Long. 2006. Hydrogen storage in a microporous metal-organic framework with exposed Mn^{2+} coordination sites, *J. Am. Chem. Soc.* 128:16876–16883.
121. Cheng S., S. Liu, Q. Zhao, and J.P. Li. 2009. Improved synthesis and hydrogen storage of a microporous metal-organic framework material, *Energ. Convers. Manage.* 50:1314–1317.
122. Yan Y., X. Lin, S.H. Yang, A.J. Blake, A. Dailly, N.R. Champness, P. Hubberstey, and M. Schroder. 2009. Exceptionally high H_2 storage by a metal-organic polyhedral framework, *Chem. Commun.* 9:1025–1027.
123. Latroche M., S. Surblé, C. Serre, C. Mellot-Draznieks, P.L. Llewellyn, J.H. Lee, J.S. Chang, S. Hwa Jhung, and G. Férey. 2006. Hydrogen storage in the giant-pore metal-organic frameworks MIL-100 and MIL-101, *Angew. Chem. Int. Ed.* 45:8227–8231.
124. Li Y.W. and R.T. Yang. 2006. Hydrogen storage in metal-organic frameworks by bridged hydrogen spillover, *J. Am. Chem. Soc.* 128:8136–8137.
125. Li Y.W. and R.T. Yang. 2006. Significantly enhanced hydrogen storage in metal-organic frameworks via spillover, *J. Am. Chem. Soc.* 128:726–727.
126. Lueking A.D. and R.T. Yang. 2004. Hydrogen spillover to enhance hydrogen storage—Study of the effect of carbon physicochemical properties, *Appl. Catal. A* 265:259–268.
127. Li Y.W. and R.T. Yang. 2006. Hydrogen storage in low silica type X zeolites, *J. Phys. Chem. B* 110:17175–17181.
128. Roland U., T. Braunschweig, and F. Roessner. 1997. On the nature of spilt-over hydrogen, *J. Mol. Catal. A: Chem.* 127:61–84.
129. Liu Y.Y., Z. Ju-Lan, Z. Jian, F. Xu, and L.X. Sun. 2007. Improved hydrogen storage in the modified metal-organic frameworks by hydrogen spillover effect, *Int. J. Hydrogen Energ.* 32:4005–4010.
130. Ferey G., M. Latroche, C. Serre, F. Millange, T. Loiseau, and A. Percheron-Guègan. 2003. Hydrogen adsorption in the nanoporous metal-benzenedicarboxylate M(OH)(O$_2$C–C$_6$H$_4$–CO$_2$) (M = Al^{3+}, Cr^{3+}) MIL-53, *Chem. Comm.* 24:2976–2977.
131. Pan L., M.B. Sandler, X. Huang, J. Li, M. Smith, E. Bitther, B. Bockrath, and J.K. Johnson. 2004. Microporous metal organic materials: Promising candidates as sorbents for hydrogen storage, *J. Am. Chem. Soc.* 126:1308–1309.
132. Rovetto L.J., T.A. Strobel, K.C. Hester, S.F. Dec, C.A. Koh, K.T. Miller, and E.D. Sloan. 2006. Molecular hydrogen storage in novel binary clathrate hydrates at near-ambient temperatures and pressures, in *Basic Energy Sciences*, Washington, DC: U.S. Department of Energy, available at http://www.hydrogen.energy gov/pdfs/progress06/iv_i_11_sloan.pdf

133. Sloan E.D. and C. Koh. 2008. *Clathrate Hydrates of Natural Gases*, 3rd edn., Boca Raton, FL: CRC Press.
134. Struzhkin V.V., B. Militzer, W.L. Mao, H.K. Mao, and R.J. Hemley. 2007. Hydrogen storage in molecular clathrates, *Chem. Rev.* 107:4133–4151.
135. Zaluska A., L. Zaluski, and J.O. Ström-Olsen. 1999. Nanocrystalline magnesium for hydrogen storage, *J. Alloys Compd.* 288:217–225.
136. Baitalow F., J. Baumann, G. Wolf, K. Jaenicke-Rossler, and G. Leitner. 2002. Thermal decomposition of B-N-H compounds investigated by using combined thermoanalytical methods, *Thermochim. Acta* 391:159–168.
137. Zaluski L., A. Zaluska, P. Tessier, J.O. Stromolsen, and R. Schulz. 1995. Catalytic effect of Pd on hydrogen absorption in mechanically alloyed Mg_2Ni, $LaNi_5$ and FeTi, *J. Alloys Compd.* 217:295–300.
138. Grochala W. and P.P. Edwards. 2004. Thermal decomposition of the non-interstitial hydrides for the storage and production of hydrogen, *Chem. Rev.* 104:1283–1315.
139. Schlesinger H.I., H.C. Brown, A.E. Finholt, J.R. Gilbreath, H.R. Hoekstra, and E.K. Hyde. 1953. Sodium borohydride, its hydrolysis and its use as a reducing agent and in the generation of hydrogen, *J. Am. Chem. Soc.* 75:215–217.
140. Aiello R., M.A. Matthews, D.L. Reger, and J.E. Collins. 1998. Production of hydrogen gas from novel chemical hydrides, *Int. J. Hydrogen Energ.* 23:1103–1108.
141. Fakioglu E., Y. Yürüm, and T.N. Veziroglu. 2004. A review of hydrogen storage systems based on boron and its compounds, *Int. J. Hydrogen Energ.* 29:1371–1376.
142. Oku T., M. Kuno, H. Kitahara, and I. Narita. 2001. Formation, atomic structures and properties of boron nitride and carbon nanocage fullerene materials, *Int. J. Inorg. Mater.* 3:597–612.
143. Wang P., S. Orimo, T. Matsushima, H. Fujii, and G. Majer. 2002. Hydrogen in mechanically prepared nanostructured h-BN: A critical comparison with that in nanostructured graphite, *Appl. Phys. Lett.* 80:318–320.
144. Oertel M., W. Weirich, B. Kügler, L. Lücke, M. Pietsch, and U. Winkelmann. 1987. The lithium-lithium hydride process for the production of hydrogen: Comparison of two concepts for 950°C and 1300°C HTR helium outlet temperature, *Int. J. Hydrogen Energ.* 12:211–217.
145. Xiong Z., C.K. Yong, G.T. Wu, P. Chen, W. Shaw, A. Karkamkar, T. Autrey, M.O. Jones, S.R. Johnson, P.P. Edwards, and W.I.F. David. 2008. High-capacity hydrogen storage in lithium and sodium amidoboranes, *Nat. Mater.* 7:138–141.
146. Diwan M., H.T. Hwang, A. Al-Kukhun, and A. Varma. 2011. Hydrogen generation from noncatalytic hydrothermolysis of ammonia borane for vehicle applications, *AIChE J.* 57:259–264.
147. Chen P., Z.T. Xiong, J.Z. Luo, J.Y. Lin, and K.L. Tan. 2002. Interaction of hydrogen with metal nitrides and imides, *Nature* 420:302–304.
148. Xiong Z.T., G.T. Wu, H.J. Hu, and P. Chen. 2004. Ternary imides for hydrogen storage, *Adv. Mater.* 16:1522–1524.
149. Xiong Z., J. Hu, G. Wu, and P. Chen. 2005. Hydrogen absorption and desorption in Mg-Na-N-H system, *J. Alloys Compd.* 395:209–212.
150. Xiong Z.T., J.J. Hu, G.T. Wu, P. Chen, W.F. Luo, K. Gross, and J. Wang. 2005. Thermodynamic and kinetic investigations of the hydrogen storage in the Li-Mg-N-H system, *J. Alloys Compd.* 398:235–239.
151. Liu Y.F., Z.T. Xiong, J.J. Hu, G.T. Wu, P. Chen, K. Murata, and K. Sakata. 2006. Hydrogen absorption/desorption behaviors over a quaternary Mg-Ca-Li-N-H system, *J. Power Sources.* 159:135–138.
152. Luo K., Y.F. Liu, F.H. Wang, M.X. Gao, and H.G. Pan. 2009. Hydrogen storage in a Li-Al-N ternary system, *Int. J. Hydrogen Energ.* 34:8101–8107.
153. Karthikeyan B. 2006. Novel synthesis and optical properties of Sm^{3+} doped Au- polyvinyl alcohol nanocomposite films, *Chem. Phys. Lett.* 432:513–517.
154. Lee H., M.C. Nguyen, and J. Ihm. 2008. Titanium-functional group complexes for high-capacity hydrogen storage materials, *Solid State Commun.* 146:431–434.
155. Cabasso I. and Y. Yuan. 2008. Annual progress report, Washington DC: U.S.DOE, available at: http://www.hydrogen.energy.gov/pdfs/progress08/iv_c_3_cabasso.pdf.
156. Teitel R. 1981. Hydrogen storage in glass microspheres, in *Report BNL 51439*, Suffolk County, NY: Brookhaven National Laboratories, U.S. DOE.
157. Teitel R. 1981. U.S. Patent 4 302 217.
158. Akunets A.A., N.G. Basov, V.S. Bushuev, V.M. Dorogotovtsev, A.I. Gromov, A.I. Isakov, V.N. Kovylnikov, Y.A. Merkulev, A.I. Nikitenko, and S.M. Tolokonnokov. 1994. Super-high-strength microballoons for hydrogen storage, *Int. J. Hydrogen Energ.* 9:697–700.
159. Rapp D.B. and J.E. Shelby. 2004. Photo-induced hydrogen outgassing of glass, *J. Non-Cryst. Solids* 349:254–259.

160. Wicks G.G., L.K. Heung, and R.F. Schumacher. 2008. SRNL's porous, hollow glass balls open new opportunities for hydrogen storage, drug delivery and national defense, *Am. Ceram. Soc. Bull.* 87:23–27.
161. Tsugawa R.T., I. Moen, P.E. Roberts, and P.C. Souers. 1976. Permeation of helium and hydrogen from glass-microsphere laser targets, *J. Appl. Phys.* 47:1987–1994.
162. Zhevago N.K. and V.I. Glebov. 2007. Hydrogen storage in capillary arrays, *Energ. Convers. Manage.* 48:1554–1559.
163. http://www.fuelcells.org/, and specifically http://www.fuelcells.org/info/databasefront.html. Welcome to Fuel Cells 2000's Worldwide Stationary Fuel Cell Installation database! (accessed August 2011) and http://www.fuelcells.org/info/charts/h2fuelingstations-US.pdf, U.S. Hydrogen Fuelling Stations (accessed August 2011).
164. Cho, A. 2004. Fire and ICE: Revving up for H_2, *Science* 305:964–965.
165. Verhelst S. and T. Wallner. 2009. Hydrogen-fueled internal combustion engines, *Prog. Energ. Combust.* 35:490–527.
166. Stolten D. 2010. *Hydrogen and Fuel Cells. Fundamentals, Technologies and Applications*, Weinheim, Germany: Wiley-VCH Verlag GmbH & Co. KGaA.
167. Song C.S. 2002. Fuel processing for low-temperature and high-temperature fuel cells—Challenges, and opportunities for sustainable development in the 21st century, *Catal. Today* 77:17–49.
168. Kordesch K. and G. Simader. 1996. Fuel cell powered electric vehicles, in *Fuel Cells and Their Applications*, Weinheim, Germany: Wiley-VCH Verlag GmbH & Co. KGaA.
169. Mauritz K.A. and R.B. Moore. 2004. State of understanding of Nafion, *Chem. Rev.* 104:4535–4585.
170. Breault R.D. 2003. Stack materials and stack design, in *Handbook of Fuel Cells, Fundamentals, Technology and Applications*, W. Vielstich, A. Lamm, and H.A. Gasteiger (eds.), pp. 797–810, Chichester, West Sussex, U.K.: Wiley-VCH Verlag GmbH & Co. KGaA.
171. http://www.global-hydrogen-bus-platform.com/, What is HyFLEET:CUTE?, (accessed August 2011) and http://hyfleetcute.com/data/HyFLEETCUTE_Brochure_Web.pdf, Hydrogen Transport Bus Technology & Fuel for TODAY and for a Sustainable Future (accessed August 2011).
172. Agnolucci P. 2007. Hydrogen infrastructure for the transport sector, *Int. J. Hydrogen Energ.* 32:3526–3544.
173. Köhler J., M. Wietschel, L. Whitmarsh, D. Keles, and W. Schade. 2010. Infrastructure investment for a transition to hydrogen automobiles, *Technol. Forecast. Soc.* 77:1237–1248.
174. Melaina M.W. 2003. Initiating hydrogen infrastructures: Preliminary analysis of a sufficient number of initial hydrogen stations in the US, *Int. J. Hydrogen Energ.* 28:743–755.

11 Technologies for Second-Generation Ethanol Based on Biochemical Platform

Francesco Cherchi, Tommaso Di Felice, Piero Ottonello,
Paolo Torre, Renzo Di Felice, and Marco Merlo

CONTENTS

11.1 INTRODUCTION

Energy consumption worldwide is characterized by a strong dependence on fossil fuels, oil, and natural gas; these are mostly produced by a limited number of countries that are not homogeneously distributed throughout the world. A representative case for this kind of scenario is the Italian situation, where the availability of fossil fuels is so irrelevant that practically all the energy requirements are satisfied by import (85.6% in 2008, compared to an EU average of about 56%). Energy demand is generally associated with industrial and civil sectors, but a more thorough examination reveals that energy consumption is also very important in other areas, transportation being the main example. In fact, energy consumption for transporting people and goods is larger than that directly related to the industrial, commercial, and residential sectors, and almost equal to the need for the production of electricity; this is clearly shown in Figure 11.1, which presents data on the final energy consumption, in 2007, divided by sector, in Italy, as elaborated by ENEA in their "Energy and Environment Report 2007–2008."

The problem, however, is not only limited to the Italian situation: energy demand for transportation is particularly critical as it accounts for over 30% of total energy demand in developed countries. Furthermore, energy used in transportation, unlike other sectors, originates almost entirely (97%–98%) from fossil fuels; therefore, it is responsible for more than half of the world's oil consumption [1].

The search for alternative and sustainable energy sources has become increasingly important because of possible short-term shortage of petroleum-derived fuels and the environmental threats posed by nonrenewable sources, particularly in terms of carbon dioxide (CO_2) emissions. Indeed, a diversification of primary energy will be required in order to produce fuel, for supply problems and environmental safety [2,3].

One of the pathways being explored is the progressive implementation of biofuels. The term biofuel refers to a gas or liquid fuel for use in the transportation sector, derived from biomasses. The advantages of biofuels compared to fossil fuels are potentially huge and include sustainability, reduction of greenhouse gas emissions, economic development and employment opportunities at regional level, increase in agricultural production, and security of supply.

The U.S. government has recently pledged to triple the production of bioenergy in the next 10 years. The European Union aims to replace 10% of diesel and gasoline fuels by 2020 with biomass originated alternatives.

The most promising biofuels are biodiesel and bioethanol [4–11]. Biodiesel, whose nomenclature indicates that it can be used in diesel engines instead of or in addition to fossil-originated fuel,

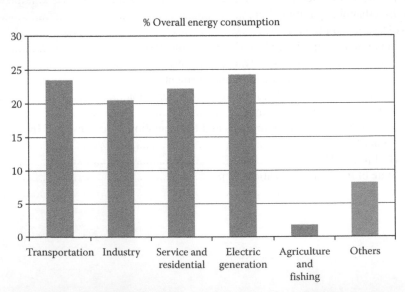

FIGURE 11.1   Final energy consumption dived by sector in Italy.

is most commonly made through routes that involve fatty acids (i.e., vegetable oils and animal fats) transesterification and has the potential to be economically competitive with conventional diesel if made from waste vegetable oil [12]. Bioethanol is the most promising and exciting biofuel as a potential substitute for gasoline in Otto cycle engines, since, in principle, it can be made from any material containing simple or complex sugars through fermentation to alcohol.

At industrial level, renewable ethanol is nowadays obtained from sugarcane and other starch-containing materials (corn, wheat, and potatoes), the so-called first-generation technology. The technology for bioethanol production from corn is well established, and many industrial plants exist and are in operation in the United States and other countries while production from sugarcane is a thriving reality in Brazil (United States and Brazil combined produce more than 90% of the 74 billion liters of bioethanol produced in the world in 2009). However, the corn to ethanol route [13] has been casting more than a question mark in the future development of the process, related, for example, with the fact that corn is a "food crop" and with quick price increases due to market fluctuations that often cannot be foreseen (e.g., the fall in the Russian production occurred in 2009).

It is not surprising therefore that alternative route for the bioethanol production is active sought and processes using lignocellulosic biomass as primary material, also known as "second-generation bioethanol production routes," are currently investigated. The most promising raw material is represented by lignocellulosic biomass from agricultural or forestry origin because of its abundance and low cost.

As said earlier, ethanol is made through the fermentation of sugars such as glucose, the most common simple sugar, that can be obtained from the cellulose present in the lignocellulosic biomass. This type of biomass, as evident from its name "lignocellulosic," contains cellulose, lignin, and hemicellulose. In second-generation processes, bioethanol is produced by converting cellulose into glucose and subsequent fermented to ethanol. The interest in this process is supported by the notion that cellulose is the most common biopolymer on Earth accounting for approximately 30%–40% of agricultural biomass on a dry basis. In addition, overall bioethanol productivity could be enhanced, as the hemicellulosic fraction of the lignocellulosic material, which consists primarily of sugars such as xylose, arabinose, mannose, and galactose can be as well converted into simple sugars and fermented to ethanol [14].

Agricultural and forestry resources are often abundant and locally available, and their use for the production of biofuels can play an important role in reducing greenhouse emissions.

Moreover, the net energy yield of corn, 100–130 GJ/ha-crop, is significantly lower than those of energy crops and agricultural (or forestry) residues (200–300 GJ/ha-crop) and sugarcane (~400 GJ/ha-crop) [15]. Also, the environmental costs of first-generation crops are very high: they cause more soil erosion (up to 100-fold) and require 7–10 times more pesticides and more fertilizers than grasses or wood [16]. For this reason, woody substrates, energy crops (e.g. *Arundo donax*, *Panicum virgatum*, and *Miscanthus giganteus*), corn stover, wheat straw, and sugarcane bagasse are among the biomasses that have attracted the greater interest of the research community.

Nonetheless, technological aspects to be addressed associated with the realization of a second-generation bioethanol process are quite challenging, both for each single step and for the overall configuration. It is perhaps not surprising that this process was chosen by the most authoritative academic journal in chemical engineering, the *American Institute of Chemical Engineers Journal*, as the prime example to describe a case study of "process design," with all the inherent challenges outlined, in a recent cover issue, as depicted in Figure 11.2.

11.2   FEEDSTOCK

Lignocellulosic materials are mainly composed of a mixture of cellulose, hemicellulose, and lignin.

Cellulose, a homopolysaccharide consisting of anhydrous glucose units, is by far the most abundant macromolecule on earth. The length of the linear cellulose chain varies between 2,000 and 20,000 linked glucose units with the disaccharide cellobiose as the basic repeating unit (see Figure 11.3). Cellulose chains are completely linear and have a strong tendency to form intra- and intermolecular hydrogen bonds. The presence of hydrogen bonds in cellulose makes it rigid and difficult to degrade [17,18].

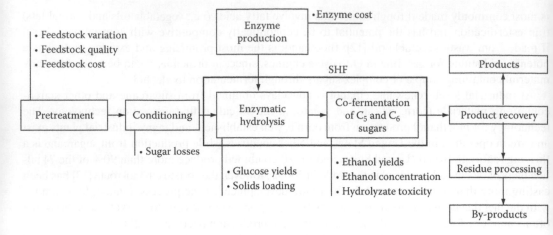

FIGURE 11.2 Process design study case: biochemical ethanol production.

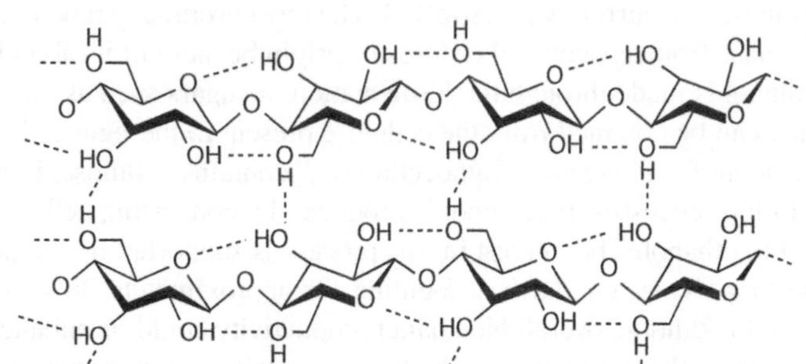

FIGURE 11.3 Cellulose chemical structure.

Hemicellulose comprises a group of highly branched heterogeneous polysaccharides [19,20], present in lignocellulosic materials (see Figure 11.4). Hemicellulose is more hydrophilic and is also easier to be degradated by acids into its monomeric components than cellulose. This possibility is under investigation in several proposed pretreatment methods to increase the accessible surface area of the substrate and make it available to enzymatic attack. Hemicellulose links covalently to lignin and through hydrogen bonds to cellulose, and its composition differs depending on the biomass origin.

Lignin is a highly complex compound, which is closely linked to cellulose and hemicellulose. Together with cellulose, it gives the plants their remarkable strength [21,22].

The chemical structure of lignin, Figure 11.5, is based on many complex carbon–carbon bonds and makes it very resistant to enzymatic or chemical degradation. The most common functional groups in lignin are methoxyl, aliphatic hydroxyl, phenolic hydroxyl, and carbonyl groups.

About 5%–10% of the lignocellulosic feedstock is ash (ash is defined as the inorganic residue remaining after dry oxidation at 600°C). Ash is composed of minerals such as silicon, aluminum,

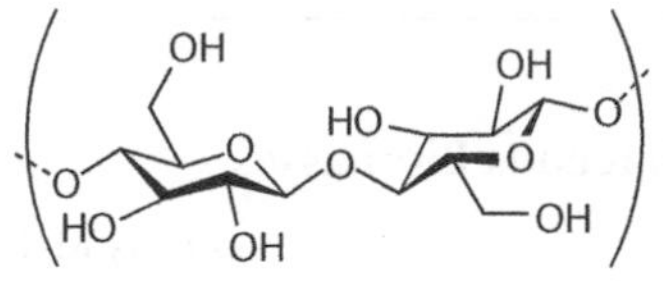

FIGURE 11.4   Hemicellulose chemical structure.

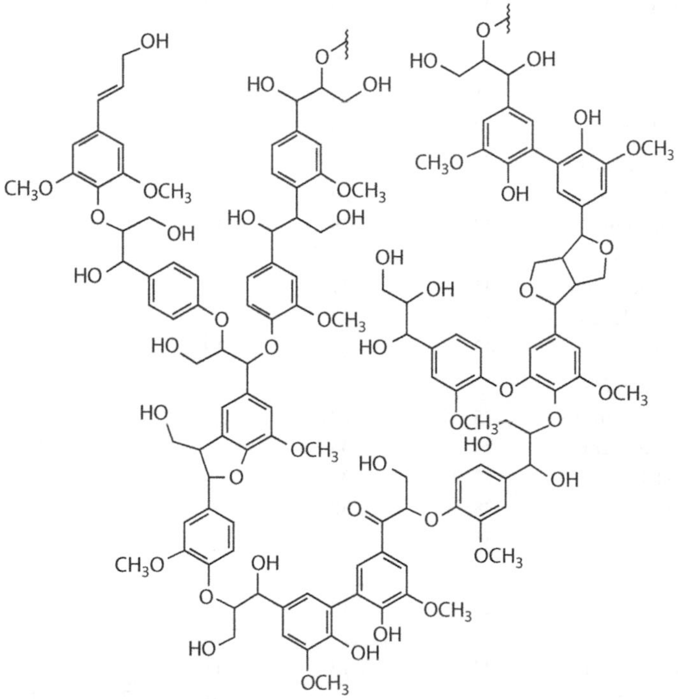

FIGURE 11.5   Lignin chemical structure.

calcium, magnesium potassium, and sodium derivatives. Other compounds present are known as extractives, as they can be extracted in organic solvents [23]. These include mainly fats and fatty acids, phenols, and salts.

Lignocellulosic material gross composition can differ significantly due to environmental (region, weather, and soil type) and genetic variability. However, cellulose is always the most significant fraction, up to 50% by weight in poplar wood, for example, whereas hemicelluloses and lignin weight fractions are generally in the region of 20%–30%. Lignocellulosic materials are also sometimes classified as softwood, hardwood, grasses, or agricultural residues, not only due to their visual appearance but also, more importantly, due to the average composition differences among these groups. Softwood generally has a higher lignin content, and the composition and the distribution of lignin within the wood cells differ compared to the other groups. In all materials, the most common carbohydrate is glucose, which is the constituent of cellulose and may also be present in hemicellulose. The hemicellulose in softwood is rich in mannan, whereas xylan dominates the hemicellulosic fraction of hardwood and agricultural residues [24,25].

Data summarizing composition of the most common lignocellulosic materials is now widely available on the web: the following Table 11.1, relative to herbaceous biomasses [26], is only one of the many possible examples.

It should be stressed, however, that gross chemical composition will only minimally influence the choice of the "best" biomass to be used in a second-generation bioethanol plant whereas other

TABLE 11.1

Composition of Herbaceous Biomasses

	% of Dry Matter					
	Cellulose	Hemicellulose	Lignin	ADL	Protein	Ash
Crop residues						
Corn stover	38	26	19	4	5	6
Soy bean	33	14	—	14	5	6
Wheat straw	38	29	15	9	4	6
Rice straw	31	25	—	3	3	6
Barley straw	42	28	—	7	7	11
Warm-season grasses						
Switchgrass	37	29	19	6	3	6
Big bluestem	37	28	18	6	6	6
Indian grass	39	29	—	6	3	8
Little bluestem	35	31	—	—	—	7
Praire cordgrass	41	33	—	6	3	6
Mischantus	43	24	19	—	3	2
Cool-season grasses						
Inter. wheatgrass	35	29	—	6	3	6
Reed canarygrass	24	36	—	2	10	8
Smooth bromeg	32	36	—	6	14	8
Timothyb	28	30	—	5	7	6
Tall fescue	25	25	14	—	13	11
Other crops						
Alfalfa	27	12	—	8	17	9
Forage sorghum	34	17	16	—	—	5
Sweet sorghum	23	14	11	—	—	5
Pearl millet	25	35	—	3	10	9
Sudan grass	33	27	—	8	12	12

factors, often overlooked, come heavily into play. As an example, some results of a comprehensive study recently carried out at CHEMTEX ITALIA Srl—Gruppo M&G evaluating the suitability of four different biomasses of interest for the specific project being developed (the four biomasses chosen after a preliminary screening were Sorghum, Panicum, Arundo, and Miscanthus) will be shown here to illustrate this point.

The four species have practically the same chemical composition, as illustrated in Figure 11.6:

Factors that must be taken into account in the choice of the "best" biomass include the following:

- Minimization of the production costs: This is an obvious parameter and its evaluation must consider, amongst others, how to maximize yield (in terms of tons of dry material produced per hectare), how to increase productivity in order to reduce the use of land, minimization of chemicals (fertilizers and herbicides), and water used, guarantee of a satisfactory economic return for the farmers.
- Exploitation of marginal land: One conflict that certainly has to be avoided is the use of land for fuel, in competition against food. Marginal land means land not in use because of low quality that cannot guarantee a satisfactory income to the farmers or also lands situated in hills but accessible to agricultural machinery or abandoned communal lands.

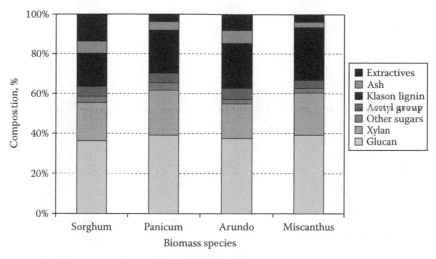

FIGURE 11.6   Chemical composition of the biomass species studied by Chemtex.

- Minimization of the distance from the bioethanol production plant: This parameter is of obvious relevance and in an ideal scenario the land is situated in an area as closest as possible to the production site. In fact, in this case, transportation to the plant can be carried out by simply using a tractor directly from the field, rather than trucks.
- Simplification of logistics with a minimization of losses and spaces: The ideal biomass can be produced throughout the year, so that storage costs can be knocked down, together with material losses associated to handling and undesired fermentation occurring during preliminary operations (haying and storage).
- Harvesting method: A biomass can be harvested, chopped, or baled. Chopping and consumption of the biomass without storage seem to be the preferred methods, as it is cheaper, the biomass is cleaner, and issues related to product formation during the conservation phase, typical of ensilage, can be avoided.

The results for the specific crop investigated at Chemtex are shown in Figure 11.7, in terms of relative production cost:

The average productivity (dry ton per hectare) evaluated both from literature and in-house agronomic experimentation, appears as in Table 11.2.

Figure 11.8 was obtained depicting relative biomass production costs expressed in unit cost per hectare (transportation costs are not included):

As a result, Arundo donax, which is a polyannual species, has the lowest relative production cost due to its very high productivity. Sorghum is an annual crop, but needs frequent irrigation and significant amount of fertilizers to get a reasonable productivity that increase the costs. Miscanthus and Switchgrass are perennial like Arundo, and they do not need significant operations on field, but productivity is low compared to Arundo, which increases the costs.

11.3   PROCESS OVERVIEW

Second-generation ethanol plants are directly derived from existing production units and their configuration must take advantage of the vast experience accumulated in the past decade of successful first-generation plant operation. Before a detailed description of the possible second-generation ethanol plant schemes is presented in detail, a short review of existing ethanol processes is given.

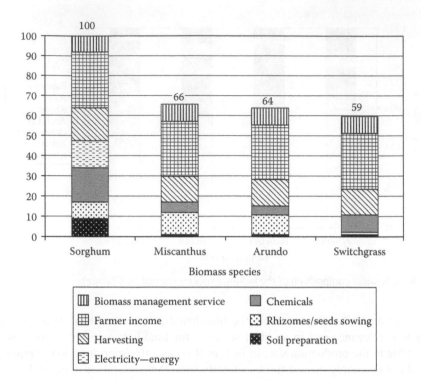

FIGURE 11.7   Relative production cost for biomasses considered.

TABLE 11.2
Evaluated Biomass Productivity (Northern Italy Case)

Biomass Species	Average Productivity (dry ton/ha)
Sorghum	25
Mischantus	20
Arundo	40
Switchgrass	18

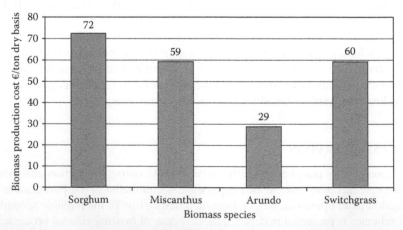

FIGURE 11.8   Relative production cost for biomasses considered.

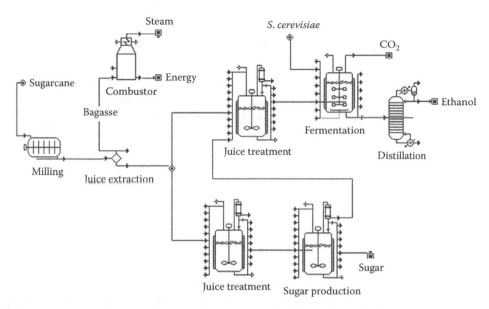

FIGURE 11.9 Simplified process for an ethanol or a sugar plant using sugarcane.

As already stated, global first-generation ethanol production is obtained from two different kinds of feedstock, with the process requirements varying accordingly: from sugary biomasses (sugarcane, sweet sorghum) or from starchy materials (corn, wheat, barley) [27–29].

Sugarcane contains about 55% (on dry basis) of sugars of which 90% is saccharose and 10% are glucose and fructose. The process outlined in Figure 11.9 refers to an integrated ethanol and sugar production. A sugar solution is extracted though a high-efficiency milling process (sugar recovery higher than 95%) with a solid bagasse byproduct sent to a combustor to increase overall process energy efficiency. A fraction of the juice is utilized to produce sugar, the remaining to produce ethanol. Ethanol is obtained from the cane juice through a fermentation step. Sugar content of the solution fed to the fermentation unit is on the order of 14%–18% that can be easily converted into ethanol by using a common strain of *Saccharomyces cerevisiae*. Fermentation generally takes place in batch reactors, at a temperature of 35°C with the final ethanol concentration of around 8%–10% w/w, maximum concentration allowed before the yeast is deactivated due to toxicity of the product. The liquor is then sent to a distillation unit where azeotropic aqueous ethanol (95% w/w) is produced. The remaining water is eventually eliminated and anhydrous ethanol produced, either by dehydration on molecular sieves or by distillation involving benzene or cyclohexane (azeotropic distillation).

The process used in producing ethanol from starch is, with a few minor adjustments, the standard procedure long since established by the food-starch industry.

The process of corn wet milling produces not just ethanol but several other products: corn oil, gluten feed, gluten meal, and corn steep liquor (CSL), each of which has a market value, see Figure 11.10.

Corn is first steeped in a weak acid SO_2 water solution at 50°C. Once the corn grains are soft enough, after up to 48 h, oil, protein, and a starch-rich fraction are separated out of the resulting meal. NaOH is added to control the pH at 5.5–6.2 prior to the addition of α-amylase, which breaks down the starch into smaller dextrin molecules (i.e., liquefaction step). After liquefaction, the paste moves on to the saccharification phase, during which the action of glucoamylase hydrolyzes dextrin into glucose with a reactor residence time on the order of 70 h. From this point, the process is similar to the one described earlier: at 35°C, *S. cerevisiae* yeast is added, with the chemical reaction converting the glucose into ethanol and carbon dioxide, followed by azeotropic distillation and further water removal.

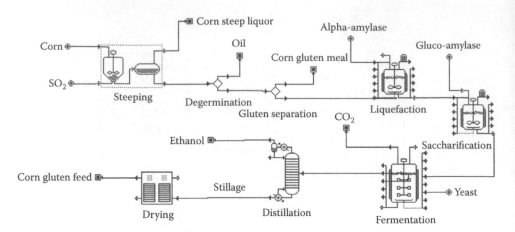

FIGURE 11.10   Simplified wet milling process for ethanol production from corn.

Current studies aimed at improving the present process are directed to two main aspects: the possibility of optimizing glucose extraction, and the subsequent fermentation, by carrying out the steps simultaneously in a single reactor (the SSF process), and optimization of the energy intensive ethanol–water separation [30,31].

In a different version of this process, gluten is not separated and remains in the final aqueous by-product stream. This stream can be subjected to further fractionation, centrifuged into a solid part rich in proteins, fiber and fats, and a liquid part rich in yeast, sugars, and tiny corn particles. Two products are eventually obtained, distillers dried grain (DDG) and distillers dried soluble (DDS), which can be typically employed as animal feed.

In the dry milling process, Figure 11.11, corn is simply milled, mixed with water, and heated before being hydrolyzed. Subsequently, starch is converted into ethanol, and the ethanol–water solution is separated with steps identical to the previous case.

Given that dry milling does not break down the cereal into its various components, the water-rich by-product stream is very rich in valuable nutrients (proteins, fiber, fats, and sugars) and always processed at the end of the distillation phase [32].

Numerous routes have been tried in order to produce bioethanol from lignocellulosic feedstock. In this review, attention will be limited to the so-called biochemical route for ethanol production, although we should remind that a "chemical" route also exists, which was quite popular after the Second World War. Some old plants are still running in the former Soviet Union, and some new

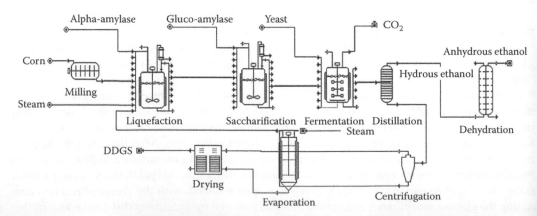

FIGURE 11.11   Simplified dry milling process for ethanol production from corn.

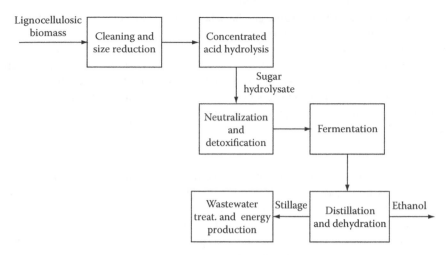

FIGURE 11.12 An example of chemical route for bioethanol production.

ones have been recently built in Sweden [33]. The chemical route is obviously very different from the biochemical route, the main difference concerning the hydrolysis step from biomass to simple sugars, an example is shown in Figure 11.12.

A chemically driven hydrolysis, based on concentrated acid, such as hydrochloric or sulfuric, is used. Hydrolysis is characterized in this case by high yields and short residence time. However, this approach has numerous disadvantages: a very high consumption of chemicals, high energy demand for acid recovery, and, obviously, several corrosion problems that affect the choice of the construction material. Moreover, undesirable by-products can also be formed, which are detrimental for the successive fermentation and require a costly detoxification step. All these factors do not play in favor of the chemical route and indicate that a biochemical route should be sought for instead.

The biochemical route can be summarized into four basic phases [34–40], which will be analyzed in details later in the chapter: pretreatment, hydrolysis, fermentation, and separation (Figure 11.13). These steps can be combined in different ways depending on the process configuration.

Pretreatment is needed because lignocellulosic materials are naturally very difficult to break down. It reduces cellulose fiber crystallinity and increases porosity and enzymatic access. It can be achieved through physical, chemical, and biological treatments. Once the pretreatment is completed, the next step consists in hydrolyzing the sugar polymer chains into monomers. Typically, this reaction is catalyzed by enzymes or acid. After hydrolysis, the glucose and sugar monomers are fermented into ethanol. Simultaneous saccharification and fermentation (normally identified as "SSF") may be employed to combine the two process steps of hydrolysis and fermentation into one, by using enzymes to convert cellulose and hemicellulose into reducing sugars, parallel to a fermentation process occurring in the same reactor. The "beer" (4%–8% v/w. of ethanol content) is then sent to a purification section, where ethanol content is increased up to the required value, thanks to a series of subsequent concentration steps (i.e. evaporation, distillation, and molecular sieves dehydration). In parallel, lignin and residual solid by-products, characterized by an interesting energy content, in most cases, are exploited through a dedicated cogeneration plant section, where combined heat and power are produced and sent back to other process sections so as to optimize energy integration.

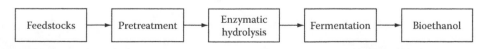

FIGURE 11.13 Overall view of biochemical ethanol production.

In case of excess electric power, the same can be sent to the local electrical grid, generally consisting of additional revenue for the technology.

Given the fact that a second-generation ethanol process differs from a first-generation version due to the starting material used, the major differences between the two processes are to be expected in the raw material treatment. Once the sugar has been extracted from the lignocellulosic material and the solution obtained, fermentation and separation can be similar, although overall process integration may again lead to quite different process designs.

11.3.1 Pretreatment Step

Goal of the pretreatment step is a modification of the lignocellulosic material structure in order to make sugar fractions more easily hydrolysable. It is well known that, in lignocellulosic material, lignin acts as a strong and protective layer around the cellulose structure with the consequence of making its solubilization rather difficult. Means to facilitate sugar solubilization are to reduce cellulose fiber crystallinity and to increase material porosity and access to hydrolytic agents [41]. Figure 11.14 is a good visual representation of the pretreatment process.

At the same time, however, care should be taken during a pretreatment step, so that degradation or loss of carbohydrate is reduced, and the formation of byproducts inhibitory to the subsequent hydrolysis and fermentation processes, such as formic acid or furfural, is avoided or at least minimized [42]. Pretreatment can be a physical, chemical, or biological transformation, although in the majority of cases, a combination of these is actually used. Table 11.3 [43] provides a good overview of some pretreatment technologies and their main qualitative effects on the lignocellulosic material characteristics.

Not all possible pretreatments will be reviewed here (a very comprehensive summary has been published by Sun et al. and Mosier et al. [42,43]), but only the ones with the most relevant potential industrial impact will be described and analyzed more in detail.

11.3.1.1 Mechanical Comminution

Mechanical comminution is the least sophisticated pretreatment a biomass can be subjected to: chipping, grinding, and/or milling can be used to reduce the size of the material typically to 10–30 mm (chipping) or to 0.1–1 mm (milling or grinding). The reduction in size corresponds to a beneficial increase of biomass specific area; moreover, it has been observed that the comminution process also induces a reduction in cellulose crystallinity. The fundamental parameter to be considered here is the energy consumption to carry out material size reduction, and this parameter depends both on

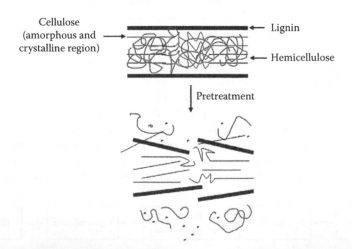

FIGURE 11.14 Schematic of the role of pretreatment in the conversion of biomass to fuel.

TABLE 11.3

Qualitative Effect of Various Pretreatments on the Biomass Structure

	Increases Surface Area	Decrystallizes Cellulose	Removes Hemicellulose	Removes Lignin	Alters Lignin Structure
Uncatalyzed steam explosion	L		L		S
Liquid hot water	L	ND	L		S
pH controlled hot water	L	ND	L		ND
Flow through liquid hot water	L	ND	L	S	S
Dilute acid	L		L		L
Flow through acid	L		L	S	L
AFEX	L	L	S	L	L
ARP	L	L	S	L	L
Lime	L	ND	S	L	L

L, large effect; S, small effect; ND, not determined.

TABLE 11.4

Energy Requirement for Mechanical Communition for Selected Biomass

		Energy Consumption (kJ/kg)	
Lignocellulosic Material	Final Size (mm)	Knife Mill	Hammer Mill
Hardwood	1.60	470	470
	2.54	290	430
	3.20	180	420
	6.35	90	340
Straw	1.60	30	150
	2.54	25	100
Corn stover	1.60	—	50
	3.20	70	35
	6.35	55	—
	9.35	10	—

the biomass characteristics and on the final particle size. Table 11.4 reports some data on energy consumption for three different biomasses and two different processes as a function of final material size [44].

Table 11.4 seems to indicate that in the worst case scenario, the energy needed for physical communition is comparable with the energy content of the biomass itself. This observation, coupled with contradictory experimental results on the effectiveness of such a simple treatment, makes physical communition not a popular choice for biomass processing.

11.3.1.2 Chemical Attack with Dilute Acids or Bases

A concentrated acid, for example, sulfuric or hydrochloric, can in principle be used to treat lignocellulosic materials. Understandably, concentrated acids are powerful agents, and they can break down the lignin and the hemicellulose structure, but, at the same time, they are toxic, corrosive, and hazardous; and therefore, their use is not recommended. Various investigators have suggested the use of dilute acids instead. Dilute sulfuric acid pretreatment has shown to lead to high reaction rates and to significantly improve cellulose hydrolysis. For example, Esteghlalian et al. [45] investigated the effect of acid concentration and temperature on sugar recovery from three different biomasses. They found out that temperature, not acid concentration, was the more important parameter in determining process yield. Dilute acid pretreatment can be carried out both at high (over than 150°C) or low temperature (less than 150°C). A main drawback of dilute-acid hydrolysis processes is the degradation of the sugars in hydrolysis reactions and formation of undesirable by-products. This not only lowers the yield of sugars, but also several of the by-products generated strongly inhibit the formation of ethanol during the fermentation process. Recently, research on dilute acid hydrolysis processes has being aimed at using milder process conditions to minimize inhibitors formation. Although dilute acid pretreatment can significantly improve the cellulose hydrolysis, its cost is usually higher than some physical–chemical pretreatment processes such as steam explosion or AFEX and a neutralization of pH is always necessary before entering the downstream enzymatic hydrolysis or fermentation processes.

Similarly, some bases can also be used for pretreatment of lignocellulosic materials [46]. Alkaline hydrolysis apparently is effective by breaking the bounds between lignin and hemicelluloses. Dilute NaOH treatment of lignocellulosic materials causes swelling, leading to an increase in internal surface area, a decrease in the degree of polymerization, a decrease in crystallinity, separation of structural linkages between lignin and carbohydrates, and basically a disruption of the lignin structure [47]. However, the overall effect of dilute base pretreatment is still not very well known with contradicting results reported in the literature (the digestibility of NaOH-treated hardwood increased from 14% to 55% with the decrease of lignin content from 24%–55% to 20% [48]), but no effect of dilute NaOH pretreatment was observed for softwoods with lignin content greater than 26% [49], indicating that this kind of pretreatment is characterized by low flexibility toward different feedstock, and there is still a need of further studies and improvements.

11.3.1.3 Ammonia Recycle Percolation

Another type of process utilizing a chemically driven attack is the ammonia recycle percolation (ARP) method. The primary factors influencing the reactions occurring in the ARP are reaction time, temperature, ammonia concentration, and the amount of liquid throughput [50]. When choosing this process, aqueous ammonia (10–15 wt.%) passes through biomass at elevated temperatures (150°C–170°C) with residence time from 10 to 20 min, after which the ammonia is then recovered, separated, and recycled. Under these conditions, aqueous ammonia reacts primarily with lignin, causing its depolymerization and cleavage of lignin–carbohydrate linkages. Lignin liberated during pretreatment is known to inhibit enzymatic hydrolysis and microbial activity, thus limiting the conversion efficiency of ethanol production by fermentation. In addition to the effect on lignin fraction, the ARP process solubilizes 40%–60% of the hemicellulose but retains more than 92% of the cellulose content [50].

The cost of ammonia, and especially of ammonia recovery, is a very important factor to keep in mind in this process.

11.3.1.4 Hot Water Treatment (Auto-Hydrolysis)

Water treatment at elevated temperatures (180°C–230°C) and pressures has shown the potential to increase the biomass surface area and to solubilize hemicellulose. Next to the thermal action, a chemical action also takes place due to the acetic acid release, which is capable of acting as a catalyst for the hydrolysis reaction.

Three types of reactors are used for hot-water pretreatment including co-current (biomass and water are heated together for a certain residence time), countercurrent (water and biomass move in opposite directions), and flow through (hot-water passes over a stationary bed of biomass).

The advantage of autohydrolysis is that acid addition and size reduction are not needed. A disadvantage of these methods is that they form sugar degradation products (furfural from pentoses and HMF from glucose). The degradation products can be minimized by controlling the pH of the reaction mixture by addition of bases such as potassium hydroxide.

Uncatalyzed hydrothermal hydrolysis and mechanical comminution of lignocellulosic biomass are regarded as environmentally benign processes because no chemicals are used. Hydrothermal treatment using steam or hot compressed water (HCW) is an effective process for the enzymatic hydrolysis of hardwoods and agricultural residues, although it has a relatively small effect on softwoods. Under hot-water treatment conditions, the hydronium ion initially causes xylan depolymerization and cleavage of the acetyl group. The autohydrolysis reaction then follows, in which the acetyl group catalyzes the hydrolysis of the hemicelluloses [51–53].

The autohydrolysis reaction involves the formation of acids from the solubilization of acidic components in hemicellulose, such as acetic acid, formic acid, and glucuronic acid [54,55].

11.3.1.5 Steam Explosion

Steam explosion is one of the reference pretreatment processes due to the fact that it was largely studied in the past decades and that it is well consolidated at industrial scale being utilized for different applications treating lignocellulosic materials. In this process, the biomass is first contacted with a water stream at high temperature and pressure, and then the pressure is suddenly reduced causing an extremely rapid vaporization of the water (simulating an *explosion*), which breaks the physical and chemical bonds between the various components of the original biomass [46]. The effect of this process on the biomass structure can be easily explained by the fact that the explosive decompression causes the "flashing off" of the liquid water from the cellular structure of the substrate. The actual phenomena connected to steam explosion are still not completely known; however, it seems clear that a combined mechanical and hydrolytic effect produces defiberization and cell disruption. In particular, it seems that a physical separation of the biomass residues into its principal components (hemicellulose, cellulose, and lignin) is actually obtained, separation that seems to depend on the *severity* of the adopted treatment [56].

Steam explosion has already proven successful on a wide range of substrates: wood chips, macroalgae, sugarcane bagasse, oil palm cake, wheat straw, hemp, corn silage, corn stover, switchgrass, and bananagrass [57] both at laboratory and pilot plant scale. Biomass residence time is on the order of minutes, the pressure usually ranges from 1.5 to 4 MPa while the temperatures are in the 180°C–230°C range.

As expected, the effectiveness of steam explosion pretreatment depends on the severity of the pretreatment regime adopted, although the results are not fully quantified yet. It is known that under too severe pretreatment conditions degradation products that inhibit the fermentation process are likely to be formed. On the other hand, excessively mild conditions might not be sufficient to break the lignocellulosic material structure.

A semiempirical parameter, R_0, function of temperature and treatment time, has been suggested in order to quantify the severity of the operating conditions [58]:

$$R_0 = t \cdot \exp\left(\frac{T - 100}{14.75}\right) \tag{3.1}$$

11.3.1.6 AFEX

AFEX stand for "ammonia fiber explosion." The idea behind AFEX pretreatment is similar to steam explosion, with the obvious difference that ammonia rather than steam is used. Typical process operating parameters are ammonia to dry biomass weight ratio on the order of one, with temperature below 100°C and residence time of 30 min. The use of ammonia considerably reduces component solubilization, with material composition reported essentially the same as the original one. One of the positive consequences is that this pretreatment therefore does not produce inhibitors for the further biological hydrolysis and fermentation processes, so water washing may not be necessary. There is some indication, however, that hydrolysis carried out after AFEX process pretreatment is not very effective for biomasses with high lignin content (higher than 20%).

The use of ammonia imposes its recycling, for obvious economic and environmental reasons. Various recovery processes have been suggested, but all of them introduce a further complication to the overall system, which must be taken into account [59–62].

11.3.2   Selection Criteria for a Pretreatment Process: The Industrial Point of View

Giving the wide range of alternative pretreatment processes described in the literature, there is a need for a guideline defining the criteria to be adopted in order to design a stable and efficient second-generation ethanol process.

It is our aim here to outline a number of criteria that play an important role in the pretreatment process selection.

Of course, the main drivers in the choice of a pretreatment are the conservation of the sugars during the phase of pretreatment and the accessibility at the enzymatic hydrolysis of the material that has passed through the pretreatment step. An ideal process is one that is able to guarantee complete access to the sugar during the subsequent enzymatic hydrolysis step without degradation of the feedstock and inhibitory compound formation. In addition to the loss of sugar yield, inhibitory compounds, even in small quantity, can represent an obstacle to the fermentation of the sugars by microorganisms, and it is thus necessary to orient the choice toward processes that do not produce high amounts of these compounds. In fact, even if the elimination of these substances in the downstream of the pretreatment is usually possible, the cost associated to the removal can be high.

A balance among the two effects (accessibility and degradation) has to be found for each of the processes, but, in general, it can be said that acid pretreatments are generally associated to a high accessibility and a high inhibitory compound generation, while alkaline processes are generally associated to the use of milder conditions and lower inhibitor generation, although, in this case, the hemicellulose fraction tends to remain untouched during the process and will then require more complex enzymatic cocktails set for its further depolymerization during the hydrolysis step.

Furthermore, it is necessary to take into consideration that the use of chemicals has a strong impact over operation cost. Beside the direct cost of chemicals that usually is not very high, process neutralization cost has to be taken into consideration for acid and alkaline pretreatment, together with the impact associated to the disposal of the product formed during neutralization.

The use of chemicals has also an important effect on the investment cost. Pretreatments that use acid, such as concentrated acid hydrolysis or dilute acid hydrolysis, require special construction material for the machinery, such as titanium or Hastelloy. In the case of an alkaline process, in particular those that make use of ammonia, expensive catalyst recovery sections are generally required, having again an important impact on the cost and complexity of the plant.

Other important aspects to be considered are the simplicity of the fiber preparation setup and the flexibility toward different feedstock. Processes that do not require any further size reduction, a part from that occurring during collection, are to be preferred, because resizing processed are energy intensive. Although a very fine powder such as the one that can be obtained by a milling process represents a highly accessible material, the cost associated to this type of process makes it completely uneconomical.

Because of the high variety of climates and available feedstock in the world, and considering the fact that typically not all raw materials are available all year long in a single climatic area, it is highly desirable that the pretreatment is able to accept a wide variety of raw materials (hardwood, softwood, agricultural residues, and dedicated energy crops). Of course, each raw material will require an adjustment of the process conditions, depending on its nature and structure. For this reason, a pretreatment characterized by a wide process operating window (temperature, residence time, and pH) is to be preferred.

Another desired feature is to the possibility of efficiently separating the feedstock, during the pretreatment process, into its constituent fractions (cellulose, hemicellulose, and lignin) in a way that they are split into different streams, each one rich in one of the constituents. This biorefinery approach represents an intrinsic advantage for pretreatments because allowing this fractionation, it is offering the possibility of alternative process configurations downstream and allowing a better exploitation of the constituents, not only for ethanol production but also for productions routes toward different chemicals.

From an industrial point view, a last key factor to be taken into account is the scalability of the process. Literature offers a wide range of attractive processes that are promising from the point of view of the previously described aspects, but up to now only limited to the laboratory scale. The production of bioethanol in a continuous plant, which is exploiting the economy of scale, requires the scale up of the pretreatment equipment to a commercial scale. Giving the complexity of the system in term of number of component and its multiphase nature, attention has to be paid to ensure the continuity of the process, especially for those technologies and those steps in which a sudden pressure change is required, such as feeding and discharging in a steam explosion or the AFEX pretreatment process. An approach that may be beneficial in the selection of the pretreatment is to adopt a process that can be performed with machinery taken from already existing industrial processes that deal with similar material. Although design modifications will always be necessary, in this way, scale-up risk is minimized, and a shorter time-to-market for the technology is obtained.

Even if a wide range of possible pretreatments is available, there is not a single solution able to satisfy all the aforementioned criteria. For this reason, it is necessary to establish what are the most critical points that have to be absolutely satisfied and what are of secondary importance. After a series of different considerations (technical and economical), the most relevant factor seems to be the necessity to operate without adding any chemicals (acid, alkaline, and oxidants); therefore, the preferred class of processes is the one that employs only steam or hot water (physical pretreatments). Moreover, this kind of processes has the added advantage that, being known since the late-1970s, it is relatively easy to scale up and implement at the industrial scale. However, the issues traditionally associated with them, that is, high degradation for the steam explosion and low sugar accessibility with high-liquid-to-solid ratio for the hot-water pretreatment, still have to be overcome.

A clever approach that has been followed by some of the leading technologies for second-generation ethanol production is to increase the flexibility and the number of process options. In this way, it is possible to capture the inherent benefits and to balance the disadvantages, by combining two or more different processes in a single pretreatment line.

11.3.3 THE HYDROLYSIS STEP

In this section, attention will be given exclusively to the hydrolysis carried out through an enzymatic route. For enzymatic hydrolysis, we mean the process in which the sugar fractions of the cellulose and hemicelluloses are solubilized thanks to the action of a cocktail of specific enzymes. It poses far more challenging problems, since the starting lignocellulosic material is an interconnected matrix of cellulose, hemicelluloses, and lignin, which make the whole process more difficult to carry out due to, for example, lignin blockage, low surface area, unreactive crystalline material, and substrate inhibition.

Figure 11.15 depicts, in a very simplified way, the main step of an enzymatic hydrolysis that occurs on the cellulosic fraction, although the real mechanism is actually quite complex and still under investigation. Cellulose is first converted into a linear, more reactive, form which is then broken down into soluble oligosaccharides and finally transformed into glucose.

Efficient hemicellulose hydrolysis is important not only for recovery of sugars from residual hemicellulose, but also because hemicellulose appears to hinder the access of cellulases to cellulose fibers. In some feedstocks, similar considerations may be applied to residual pectin [63].

Like any other chemical reaction, enzymatic hydrolysis is strongly dependent on the temperature. Experimental evidence indicates that an increase in temperature of 20°C could lead to a three- to fivefold increase in overall reaction rate. However, it is well known at the same time that high temperature can cause undesirable denaturation with catastrophic reduction in enzymatic activity.

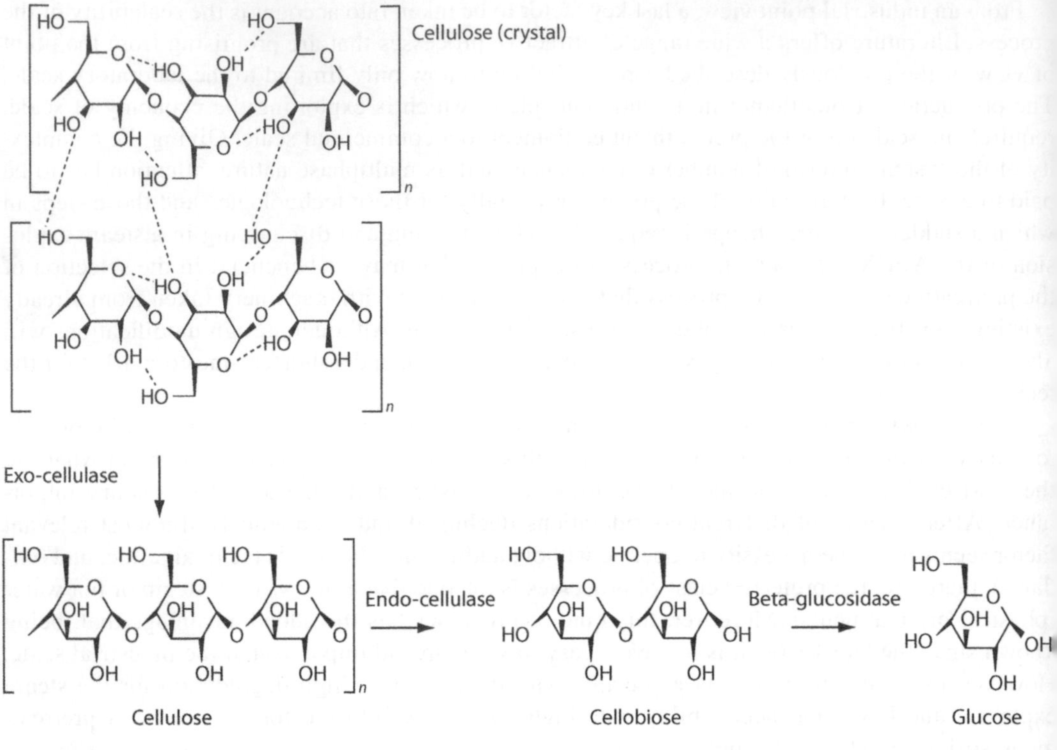

FIGURE 11.15 Simplified hydrolysis mechanism.

The rate of hydrolysis by a specific enzyme cocktail differs also depending on the substrate. In lignocellulosic materials, the cellulose is highly crystalline and surrounded by hemicellulose and lignin, thus making it recalcitrant to the enzymatic attack. One goal of the pretreatment step is to enhance the rate of hydrolysis by increasing the accessibility to the enzyme action. Initially, enzymatic hydrolysis is fast, but the rate slows down as the amorphous areas of cellulose decrease and the number of free chain-ends decreases. Additionally, the rate of hydrolysis is affected by the thermal and mechanical deactivation of the enzymes [17,64]. To improve the yield and rate of the enzymatic hydrolysis, research focuses both on enhancing enzyme activity in distinctive hydrolysis and fermentation steps [65] as well as combining the different steps in fewer reactors.

Due to the insoluble nature of native cellulose, cellulases primarily work in a two-dimensional environment involving the unidirectional movement of cellobiohydrolases along the cellulose chain. Thus, with the lack of parameters normally used for evaluating enzyme kinetics (substrate concentration, freely diffusing enzymes), it is not surprising that the synergistic degradation of lignocellulose does not follow classic Michaelis–Menten kinetics. The understanding of mechanisms involved in the interfacial solid–liquid hydrolysis system is further complicated by a number of factors that include the heterogeneous nature of lignocellulose. Even when working with model substrates (e.g., filter paper), the interaction between enzyme system and substrate creates innumerable factors, which influence the rate and extent of the hydrolysis [66,67]. These factors often cannot be fully isolated, even under lab conditions. Much research has gone into cellulase kinetics, resulting in various models. Unfortunately, few have been able to extensively predict limiting factors or process design optima through kinetic models.

As an example, Table 11.5 summarizes glucose and xylose yields measured after 24 h of hydrolysis for different substrates and with enzyme of different origin [68,69].

For industrialization and process scale-up, there are also a variety of factors of great importance that need to be considered. These include, among others, enzyme efficiency and cost, possibility of enzyme recycling, the solid content in the stream, and product inhibition.

All these aspects will be discussed in the next sections.

TABLE 11.5

Glucose and Xylose Yield after 24 h of Hydrolysis

	Substrate						
	Glucose Yield (%)				Xylose Yield (%)		
Enzyme	Solka Floc	Corn	Spruce	Willow	Solka Floc	Corn	Willow
SF	36	55	23	28	26	33	10
CO	46	59	33	53	25	28	19
SP	31	30	20	25	49	13	6
WI	40	34	21	48	24	13	15
CE	52	41	20	37	32	21	7
EC	47	34	20	44	18	18	17

11.3.3.1 Enzymes for Pretreated Lignocellulosic Hydrolysis

In contrast to enzymatic hydrolysis of starch, as used in first-generation ethanol production, which requires a single family of amylases, the challenge here is much more significant as the actions needed are much more varied and they strongly depend on the substrate utilized. A large variety of microorganisms naturally produce enzymes that degrade lignocellulosic materials in order to provide substrate for their own survival. Cellulose and hemicellulose degrading enzymes are often grouped into cellulases and hemicellulases, respectively. Endoglucanases cut the cellulose chain, preferably at the amorphous regions, while cellobiohydrolase attack the end of the cellulose chain. Cellobiohydrolases and endoglucanases together depolymerize cellulose to cellobiose. The cellobiose is then hydrolyzed to monomeric glucose units by β-glucosidases.

Furthermore, also the hemicellulosic fraction needs to be depolymerized to allow its conversion into simple sugars (xylose, mannose, galattose, arabinose, etc.) that have the potential to be fermented to ethanol. Due to the complex nature of the hemicellulose and the fact that its structure can be quite different from species to species, in addition to the three mayor groups of cellulases, a number of ancillary enzymes that attack the hemicellulose backbone, such as glucuronidase, acetylesterase, xylanase, β-xylosidase, galactomannanase, and glucomannanase, are required [70].

It is of paramount importance to have defined as precisely as possible the activity needed from the enzymes, so that the "best" mixture (or cocktail) can be prepared and used in the enzymatic hydrolysis step [71]. Enzymes must be able to break the various chemical bonds; therefore, the ideal enzyme cocktail should certainly contain cellulases in order to break glucane and glucoligomer chains, but also hemicellulases and deacetylases in order to attack the hemicellulose structure with the results of producing xylose and, at the same time, create access to the remaining cellulose and should maybe contain laccases too, in order to attack and partially degrade the lignin fraction.

Both bacteria and fungi can produce cellulases for the hydrolysis of lignocellulosic materials [72,73]. These microorganisms can be aerobic or anaerobic, mesophilic, or thermophilic.

The most commonly used and commercialized enzymatic cocktails for lignocellulosic degradation are obtained from the fungus *Trichoderma reesei*. *Trichoderma reesei* secretes mainly endoglucanases and cellobiohydrolases but is deficient in β-glucosidase, which should be supplemented to avoid the accumulation of cellobiose, causing end-product inhibition. Genetic modification of *Trichoderma reesei* to supplement the cocktail activity is a practice widely used in the energy industry.

11.3.3.2 Solid Content Effects

Solid content in the stream at the start of the process is a parameter of fundamental importance. The justification is quite simple: water is an inert compound as far as the transformations involved are concerned, so high solid content means low volumes being treated for a given amount of final product, with all the relative savings in equipment size and in pumping energy, associated with a

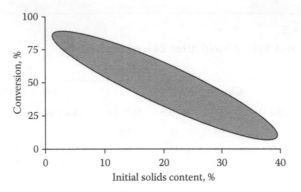

FIGURE 11.16   The effect of initial solid content on the cellulose conversion.

substantial reduction in cost when the final ethanol–water mixture needs to be separated through distillation. Increasing the solid content is likely however to produce some side effects, which may negatively influence the overall process efficiency. For example, for some technologies, the energy to produce a satisfactory mixing in the original pretreated material-water mixture is a strong function of the solid fraction, and the mixing itself is important in order to distribute the added enzyme as homogeneously as possible. Diffusion of the enzyme in the suspension is also going to be negatively affected by the presence of a solid phase, with the considerable size of the enzyme itself complicating the whole process. Moreover, the solid content will directly influence the final sugar concentration in the main stream to be subsequently fermented: as we will see later, there is an optimum value for this parameter in order to maximize sugar to alcohol transformation.

Quantitative information in the open literature about these parameters is scarce, and, generally, overall effects are reported. Figure 11.16, gathered from a good number of different experimental works, broadly shows that as the initial solid content increases, the overall enzymatic process conversion decreases [74–78].

11.3.3.3 Enzyme Dosage

How much enzyme should we use? Obviously, the more enzyme we use, the faster the reaction is, as demonstrated in Figure 11.17, which depicts experimental data obtained at Chemtex laboratories expressed in arbitrary units. However, with the increase of enzyme dosage, the cost rises a lot, and so it is expected that a certain enzyme concentration would result in an overall minimum cost for this step. This optimum is going to be dependent on the enzyme unit cost and on the equipment cost. Needless to say, much uncertainty is due to the first factor.

At this moment, R&D efforts of all the enzyme suppliers are focused on both increasing enzyme performance and reducing production costs. There has been quite a lot of research, and some results

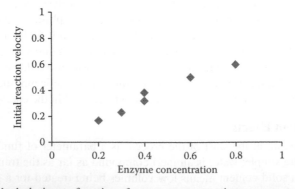

FIGURE 11.17   Initial hydrolysis rate function of enzyme concentration.

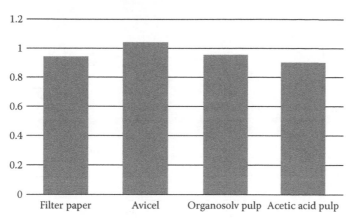

FIGURE 11.18 The ratio of glucose concentration for the case of free enzyme over immobilized enzyme.

claim to have reduced the cost by one order of magnitude. One of the problems is that each solution is tailor-made, and so it cannot be generalized [79].

One way to reduce enzyme cost could be by recycling: enzymes, like all other catalysts, are unchanged at the end of the reaction, and so, in theory, they could be recycled by treating multiple batches of feedstock using the same batch of enzymes. Unfortunately, the problem has no simple technological solution: only a fraction of the enzyme is dispersed in the liquid solution, as the remaining is bound to the lignocellulosic solid material and, once attached, is practically impossible to recover, and it is carried through the rest of the process. Moreover, even for the fraction, which is dispersed in the liquid, a recovery system to separate the enzyme from the fluid and the other dissolved components should be envisaged. A possible solution would be to attach the enzyme to a nonreactive substance prior to the hydrolysis step, so that recovery and recycling are facilitated. This technique has been shown to produce encouraging results [80], where an epoxy-activated support was used to immobilize a commercial β-glucosidase. As the result presented in the following clearly shows (Figure 11.18), sugar conversion rates remained nearly unchanged when the immobilized enzyme was used, compared to the free enzyme for a variety of substrates.

11.3.3.4 Enzyme Inhibition

The correct choice of the enzyme dosage is made even more uncertain by a further important process characteristic of the enzymatic hydrolysis, which has not been considered so far: enzymatic activity inhibition. It has been well known for a long time that cellulase activity is inhibited by the end products of the process: cellobiose and glucose. Moreover, further inhibition is caused by the specific process that is used upstream to the enzymatic hydrolysis. Enough evidence has been gathered, which indicate lignin as a cause of enzymatic activity loss. Berlin et al. [81] reported a drop in enzymatic activity by a factor of up to 5 due to the presence of lignin during enzymatic hydrolysis of cellulose, whereas Pan et al. [82] measured a marked drop in cellulose to glucose yield when the process was carried out in the presence of a large amount of lignin (41%) compared to the case when the lignin concentration was below 10%, see Figure 11.19.

There is also some evidence that pretreatment by-products, such as furfural or formic acid or levulinic acid, may play a negative role in enzyme activity. For example, Tenborg et al. [83] carried out enzymatic hydrolysis of lignocellolusic material in two different ways: in the first one, only the solid fraction of the pretreated material was used whereas in the second type, both the solid and liquid result of the pretreated material was subjected to enzymatic attack. The result, summarized in Figure 11.20, clearly indicates a superior cellulose conversion when a washed solid was used compared to when the original mixture was hydrolyzed.

Given the uncertainty surrounding enzyme inhibition phenomena, it is not surprising that a clear way to minimize its negative effect has still not been established.

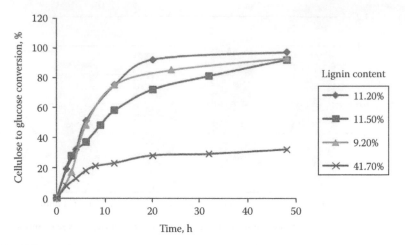

FIGURE 11.19 Effect of lignin content on cellulose to glucose conversion.

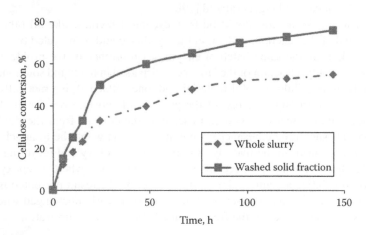

FIGURE 11.20 Influence of by-products on enzymatic hydrolysis.

11.3.4 FERMENTATION STEP

The word "fermentation" has been traditionally used to identify a chemical process in which carbohydrates are converted in various substances by a means of microorganism without the need of oxygen. In a later stage, the meaning has evolved to include any kind of transformation, in aerobic or anaerobic conditions and in the presence of proper nutrients, which make use of a microorganism. The most known fermentation is still the conversion of glucose into ethanol.

$$C_6H_{12}O_6 \rightarrow 2C_2H_5OH + 2CO_2$$

It is clear that if 1 mol of glucose produces 2 mol of ethanol, the maximum yield is 51% by weight.

Although one may expect the fermentation process in second-generation ethanol to be very similar to the one in first-generation plants, there are still some differences that require careful analysis in the design step. Basically, there are two factors that must be taken into consideration and that do not allow a simple replica of existing plants: the first one is due to the fact that the stream to be fermented does not contain exclusively 6-carbon sugar like glucose, but 5-carbon sugars, such as xylose, are also present in not negligible amounts. Second, a number of inhibitor compounds are also present in the process stream, their composition, and concentration depending on the hydrolysis and pretreatment process used upstream [84].

The most widely employed microorganism in glucose fermentation is the common yeast *S. cervisiae* since, in contrast to most other species, it is able to carry out alcoholic fermentation under complete anaerobic conditions. Other microorganisms require a minimal supply of oxygen to trigger the alcoholic fermentation. Given that it is very difficult to achieve an optimal oxygen level in large-scale fermentations, non-*Saccharomyces* yeasts are therefore less practical for large-scale alcohol production [85].

Although ethanol is the major product of *S. cerevisiae* when it is grown anaerobically, it can also exercise an inhibiting effect on the fermentation process itself. In general, ethanol concentrations of around 10% seem to be sufficient to inhibit both yeast growth and ethanol production in most yeast strains [86,87] to an extent, which depends on the nutritional status of the *medium* as well as osmotic strength and temperature [88]. The identification of a microorganism that is also effective at relatively high ethanol concentrations is one of the goals to be reached in this field in order to improve the overall efficiency of the process.

Xylose fermentation to ethanol can be represented by the following reaction:

$$3C_5H_{10}O_5 \rightarrow 5C_2H_5OH + 5CO_2$$

Again, theoretical xylose to ethanol yield calculation by weight is 51%, identical to that computed for the corresponding glucose to ethanol fermentation.

Unfortunately, naturally occurring *S. cerevisiae* yeasts are unable to ferment 5-carbon sugars such as xylose into ethanol. While the technology to convert hexoses to ethanol is well known and established, the fermentation of pentoses is much more problematic.

In recent years, various yeasts and fungi that can convert xylose into ethanol have been identified. However, the use of their cultures in the presence of pentose to obtain rapid and efficient ethanol production is somewhat more complex than the alcoholic fermentation of glucose with *S. cerevisiae*. Among the factors strongly influencing ethanol yields in xylose fermentation, aeration is probably the most crucial. For example, Abbi et al. [89] reported that *Candida shehatae* could ferment 100 g/L of xylose solution to ethanol at 45°C and Converti et al. [90] starting from a 40 g/L of xylose solution obtained 12 g/L of ethanol after 60 h using *Pachysolen tannophilus*.

Some thermophilic and mesophilic bacteria have also been identified, which are able to ferment xylose to ethanol. Among these, the most thoroughly investigated are *Clostridium thermohydrosulfuric* and *Clostridium ethanolicus* [91]. The use of these bacteria is not advantageous for various reasons: the bacteria react poorly to certain levels of ethanol in the fermentation medium; the fermentation process is not selective, and, in addition to ethanol, which is the most abundant constituent, some organic acids such as acetic acid are also produced.

Since 1980, numerous researchers have investigated the strategies of genetically modifying microorganisms capable of fermenting both xylose and glucose simultaneously. The recombinant bacterium *Zymomonas mobilis* CP4 (pZB5) has been proven to be effective in co-fermenting both xylose and glucose [92–94]. Krishnan et al. found that using a laboratory scale fluidized-bed bioreactor (FBR), xylose conversion to ethanol was 91.5% for a feed containing 50 g/L of glucose and 13 g/L of xylose. Ho et al. [95] developed strains of genetically engineered *S. cereviasiae* yeasts. The engineered strains contain the following xylose-metabolizing genes: a xylose reductase gene, a xylitol dehydrogenase gene (both from *Pichia stipitis*), and a xylulokinase gene (from *S. cerevisiae*). As a result, these recombinant yeasts can effectively ferment both glucose and xylose to ethanol, but, more importantly, can also effectively coferment both glucose and xylose when they are present at the same time in the process stream.

The fermentation process using microorganisms requires, along with the sugar compounds, metabolizable nitrogen, essential fatty acids and sterol, vitamins (growth factors), and inorganic ions. In order for the yeast to grow, a known amount of nitrogen is needed, which increases the rate of fermentation. Yeast can assimilate only low-molecular-weight nitrogen compounds such as ammonium ions, urea, and amino acids. The most popular addition of nitrogen comes as urea added directly to fermentors.

Sterols and unsaturated fatty acids are necessary in order to maintain yeast membrane integrity and help in the resistance of cells to stress, whereas vitamins are needed in small quantity for enzyme structure and functionality and minerals for enzyme stability and metabolism.

It is not clear what the overall quantitative need of these supplements is; generally, each microorganism type requires its own personal level, which, for nitrogen, can be on the order of milligram per liter, whereas, for the metallic ions and vitamins, on the order of parts per million. In spite of such relatively small quantities, additions to the fermenting stream still represent a non-negligible cost, which needs to be taken into account. Ideally, it is expected to minimize this contribution by choosing a microorganism whose nutrient request is coming directly from the original biomass material and therefore is already present in the stream to be fermented. This is not generally the case although, whenever possible, it is preferable to limit addition to nitrogen-based low cost compounds such as urea.

Just like any other chemical reaction, temperature and concentration in the reactor are basic parameters to optimize. The best temperature is basically determined by the microorganism utilized in the fermentation step, as each microorganism is characterized by a very narrow range of temperature at which it can carry out the reaction at an acceptable rate. For *S. cerevisiae*, for example, the best temperature window is between 28°C and 35°C. At lower temperature, the reaction rate is unacceptably low, and, at higher temperature, the microorganism cannot survive. The second limit is given by the product concentration. Ethanol in high concentration is not tolerated by many microorganisms; in fact, for the *S. cerevisiae*, the threshold limit is about 120 g/L, so that this number should be kept in mind when the initial sugar concentration in the fermentation stream is designed.

Moreover, the performances of both yeasts and bacteria are affected by the unwanted presence of compounds in the fermentation medium, compounds which are the results of the upstream lignocellulosic material processing. These inhibitors are divided in three main groups based on their origin: weak acid, furan derivatives, and phenolic compounds. In order to eliminate their negative influence, a detoxification step may be necessary, a step that can be based on biological, physical, or chemical methods.

The biological step may use peroxidase and laccase enzymes to completely remove phenolic monomers and *Trichoderma reesei* fungus to eliminate acetic acid, furfural, and benzoic acid derivatives. Biological detoxification has shown to be capable to increase ethanol productivity by up to a factor of four [96,97]. Physical methods take advantage of well-known physical processes well established in industry such as selective volatilization, selective adsorption on solid material, or liquid–liquid extraction. Palmqvist and Hahn-Hägerdal [98] reported that the most volatile fraction of a willow hemicellulose hydrosylate suffered much less inhibition compared to the less volatile fraction, and a more easy fermentation was achieved after extracting the original hydrolysate stream with diethyl ether. A typical chemical detoxification consists in increasing the pH to 9–10 by adding $Ca(OH)_2$ (so that a large precipitate is formed) and then readjusting it to 5.5 with H_2SO_4. Again, ethanol productivity and yield have been shown to improve sharply following this operation.

Detoxification step, when needed, should selectively remove inhibitors, and be cheap and easy to integrate into the overall process. Integration is a characteristic that is often overlooked but is, instead, of extreme importance. For example, enzymatic detoxification can be carried out by simply introducing the active compounds into the fermentation vessel without the need of introducing a further unit, although enzyme production should be taken into account at the same time. Selective evaporation requires the introduction of an evaporator with consequent increase of the overall energy consumption. The base-acid pH adjustment is fairly easy and cheap to carry out, but the overall cost is bound to increase due to the separation unit, which needs to be introduced in order to remove the precipitate from the main process stream [99].

Fermentation itself can be carried out in a batch, fed-batch, or continuous mode. In batch mode, the microorganisms operate at high substrate concentration during the initial phase and high product concentration during the final phase of the process. Generally, batch systems are characterized by low-ethanol production rates, and, moreover, they are labor intensive. Nevertheless, the majority of plant installed today operate in batch mode, because contamination is feared when the process is operating in continuous mode. However, batch mode does not seem to be appropriate when inhibitor concentrations are high in the hydrosylate as the yeast is quickly deactivated, and ethanol generation comes to a halt. Fed-batch mode offers some advantages in the presence of inhibitors as the rate at which hydrosylate can be fed into the reactor is adjustable, and, in this way, inhibitor concentration can be kept under control. Therefore, an optimum feeding rate exists for every specific hydrosylate. If inhibitor concentration is low, then feeding rate can be high; if inhibitor concentration is high, then feeding rate can be low. The precise knowledge of these concentrations is therefore of great impor tance. Continuous fermentation is also possible. In this case, wash-out has to be prevented by feeding the reacting mixture at low rates, and, by doing so, reducing productivity. Cell recirculation, after immobilization, is a possible solution that would allow the use of a continuous process at acceptable rates without incurring in wash-out. Typically, the process configurations chosen for a big scale plant can allow both operation in batch and continuous modes, with the first being normally preferred to reduce the risk of losing much of the intermediate product in case of contamination.

11.3.5 BIOETHANOL SEPARATION STEP

Once bioethanol has been finally produced in the fermenter, it must be separated from the water and the other by-products. Understandably, this step will not be any, if at all, different from the first-generation plant: a classical distillation unit, producing an azeotropic ethanol–water mixture, followed by a further separation, for example, by selective adsorption to concentrate the ethanol to its final grade. Nevertheless, attention should still be devoted to this step as it is one of the most energy intensive. The large amount of energy required for recovery of the ethanol from the fermentation broth is a significant economic limiting step: while distillation is effective in separating the ethanol from water, it is not energy efficient. It has been claimed that the largest energy input in producing ethanol from biomass is the fermentation/distillation step [100]. For this reason, operation conditions that can lead to a reduction in the energy demand related to this step are highly desirable.

An improvement of this specific aspect would greatly benefit the overall process energy demand. The suggested improvements are therefore not just limited to second-generation plants. They will not be treated here but they are trying to take advantages, for example, of membrane technology by using a pervaporation unit or solid adsorption.

Pervaporation is a separation process in which a liquid feed containing two or more components comes into contact with one side of a membrane while vacuum (or purged gas) is applied on the other side to produce a permeate vapor. Generally, this process is more energy efficient than a classical distillation process.

Jones et al. [101] put forward an even more revolutionary idea by combining fermentation and separation in a single unit: they suggested to remove a small stream from the fermentation broth and selectively extract ethanol from the broth by passing it through an adsorption column. In this way, as the ethanol is being produced, fermentation is allowed to continue while suppressing inhibition. Furthermore, the ethanol adsorbed can be desorbed, and the process can be made cyclic by adding parallel columns, and alternating between adsorption and desorption. If the separation is highly selective in its preference for ethanol, the product stream desorbed could be expected to be highly concentrated.

Unfortunately, none of these futurist suggestions has gone any further than preliminary laboratory testing.

11.4 PROCESS DESIGN AND INTEGRATION

In the previous sections, various possible alternatives have been described for the different process steps: pretreatment, enzymatic hydrolysis, fermentation, and final products separation. For each step, possible pathways can be followed, and the choice of the best one must be made not only by taking into consideration the step itself but mainly by considering how each step integrates with the others. Second-generation ethanol production is a classic example that shows how process integration is a

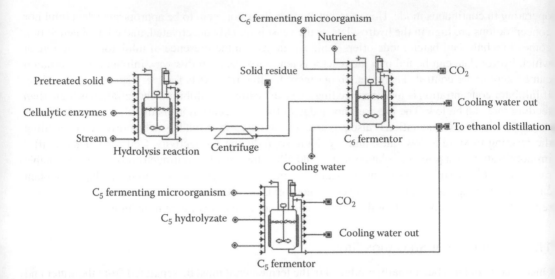

FIGURE 11.21　Schematic representation of the SHF process arrangement.

fundamental aspect and how the choice of the various pathways is determined by considering not only the single operation, but the whole package. This subject is far too vast and complicated to give a complete treatment here; nevertheless, some possible arrangements are presented and discussed, referencing mainly the work published by Taherzahed and Karimi [102].

11.4.1　SEPARATE ENZYMATIC HYDROLYSIS AND FERMENTATION

The basic idea behind this strategy is that each operation is carried out in a separate unit in order to optimize the operating conditions for each operation. A simple representation of the separate enzymatic hydrolysis and fermentation (SHF) process arrangement is depicted in Figure 11.21. The assumption is to pretreat the lignocellulosic material, producing two main streams: a liquid solution where the majority of pentose components of the original material has been solubilized and a solid stream containing the cellulose and all the original lignin. This solid stream is then sent to the hydrolysis reactor where specific enzymes are added, and the process is carried out at the optimum temperature of 45°C–50°C. The output constituted by a liquid solution containing suspended solids can be centrifuged, so that a lignin-rich solid stream is separated and the liquid solution, containing mainly hexose simple sugars, is sent to an ethanol producing fermenter, working at a temperature between 30°C and 35°C. In parallel to this fermenter, a second unit is also utilized to convert the pentose rich solution into ethanol. Needless to say, in each fermenter, specific microorganisms are added. Both streams coming out from the two fermenters are then sent to the final product separation.

As said, the main advantages of this strategy are to have the hydrolyzers and the fermenters working at the optimum operational conditions (temperature, pH, and agitation) using the best possible enzyme and microorganisms needed for the specific process (i.e., pentoses and hexoses fermentation). Main drawbacks are the inhibition of cellulase activity by cellobiose and glucose (with reported activity reduced up to 60% for 6 g/L of cellobiose, e.g.) and the fact that by operating separately on the two streams the investment cost for the reactors will be higher, and the final concentration of ethanol in the two streams will be low.

11.4.2　SEPARATE ENZYMATIC HYDROLYSIS AND COFERMENTATION

This configuration is really close to the one described in the previous section, with the only difference that pentoses and hexoses fermentations are carried out in the same reactor and not separately (Figure 11.22). The main advantage related to this process is that there is no need for a

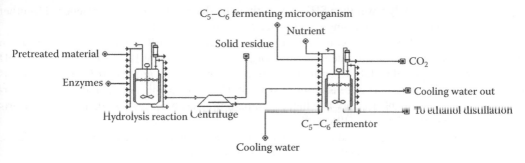

FIGURE 11.22 Schematic representation of the SHCF process arrangement.

previous separation of the two different sugar fractions present in the feedstock. This can simplify the upstream configuration, and in particular the pretreatment step. Indeed, as for SHF case, other positive returns are related with working in each step at the optimal process conditions and with the possibility to separate the insoluble lignin before fermenting, avoiding possible interaction phenomena between microorganisms and solids. The main issue regards the co-fermentation step, due to the fact that there is the need for a microorganism able to simultaneously transform C_5 and C_6 is needed as described in Section 11.3.4. Also, for separate enzymatic hydrolysis and cofermentation (SHCF), product inhibition phenomena in enzymatic hydrolysis step can occur.

11.4.3 SIMULTANEOUS SACCHARIFICATION AND FERMENTATION

In this process, Figure 11.23, the glucose produced by the enzymatic hydrolysis reaction is consumed, as soon as it is made available, by the fermenting microorganism also present in the same reactor. In theory, there is a great advantage in using the simultaneous saccharification and fermentation (SSF) strategy, since the inhibition effect of cellobiose and glucose is minimized by keeping these sugars at a low concentration during the process. Laboratory tests [103–105] indicate higher ethanol yields and lower enzyme use when SSF is used compared to SHF. Moreover, the risk of

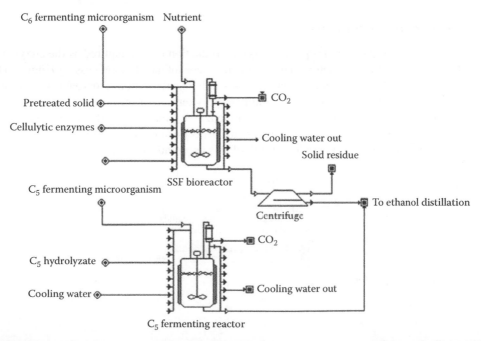

FIGURE 11.23 Schematic representation of the SSF process arrangement.

contamination is probably lower than SHF, since ethanol reduces contaminants presence. The other obvious advantage is the use of a single unit instead of two separate ones.

On the other hand, operating conditions in a SSF reactor are still a major open point, given that hydrolyzing enzymes and fermenting microorganisms have quite different ideal operating pH and temperature. Typical cellulases optimum working temperature is around 40°C–45°C, whereas *S. cerevisiae* optimum temperature is around 30°C–35°C. To tackle this problem, the use of different microorganism, capable to work at a higher temperature, has been suggested, and many efforts are being made in that direction [106–110].

A further problem is represented by the ethanol inhibition of cellulases, as some studies indicate a drop in enzyme activity for an ethanol concentration as low as 30 g/L.

In the layout depicted in Figure 11.23, a pentose fermenter may also be present, and, again, the two streams coming out from the reactor are merged and sent to the ethanol–water separation section. A solid–liquid separation step must also be envisaged before the distillation to recover the solid residue (mainly lignin) at the exit of the SSF reactor. In this case, there is no advantage related with the absence of insoluble solids during the fermentation.

11.4.4 SIMULTANEOUS SACCHARIFICATION AND COFERMENTATION

Another obvious set up, which is currently being explored, sees the simultaneous saccharification and fermentation of all the sugars, both pentose and hexose, in a single unit operation (Figure 11.24). In this case, product streams from the pretreatment are not separated but sent to the hydrolyzer/fermenter reactor [111]. The challenge consists in fermenting both hexoses and pentoses in a bioreactor using a single microorganism. Separation of the unreacted solid is carried out at the exit of the reactor. Lawford and Rousseau [112] showed that this process can be possible using a metabolically engineered strain of *Zymomonas mobilis*. The integrated system converted more than 50% of all sugar potentially available in the biomass fed to the bioreactor. Even better conversion was reported by Kim et al. [113]. The economic sustainability of this approach has been positively evaluated by Aden et al. [114] in a comprehensive study, which also took into consideration feedstock handling and storage, product purification, wastewater treatment, and by-product utilization.

11.4.5 CONSOLIDATED BIOPROCESSING

In all the cases presented so far, a separate enzyme production unit is required as the enzymes are provided externally to the bioreactors. The idea behind consolidated bioprocessing (Figure 11.25) is that all the required enzymes are produced in a single reactor by a microorganism community or by a single microorganism also able to simultaneously ferment pentose and hexose sugars. The potential of this approach is enormous as the cost for enzyme would be greatly diminished. It is still unclear how far advanced the state of the art is in this aspect, but it seems certain that there is still a long way to go before it will be possible to think of this as an industrial application.

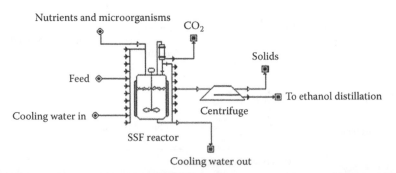

FIGURE 11.24   Schematic representation of the SSCF process arrangement.

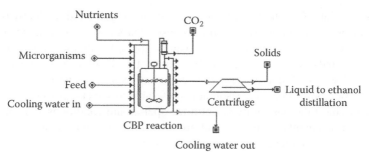

FIGURE 11.25 Schematic representation of the CBP process arrangement.

11.4.6 SELECTION CRITERIA FOR HYDROLYSIS AND FERMENTATION PROCESSES: THE INDUSTRIAL POINT OF VIEW

As for pretreatment, also for enzymatic hydrolysis and fermentation sections, there are many different options, in terms of process configuration and operational conditions. Each of these solutions gives a series of advantages, and so the choice of an efficient design is strictly related with the process criteria adopted for the other sections and with improvements in enzyme and microorganism development and cost reduction.

It is necessary to define a series of key elements that can lead to the optimal configuration both by a process and a mechanical point of view.

Process conditions (e.g., temperature, pH, and stirring) are very dependent on the working windows delineated for biological catalysts, due to the fact that each enzyme and microorganism has specific optimal activity conditions. Cellulases operate optimally at temperatures (>40°C) higher than those tolerated by ethanologenic organisms, and so these two processes currently cannot be up to now consolidated into a single process step. Enzyme cocktails that can effectively release the cellulose and hemicellulose present in pretreated solids are also important for achieving the high yields needed for large-scale competitiveness.

The hydrolysis of crystalline cellulose is the rate-limiting step in biomass conversion to ethanol because aqueous enzyme solutions have difficulty acting on this insoluble, highly ordered structure. Cellulose molecules in their crystalline form are packed so tightly that enzymes and even small molecules such as water are unable to permeate the structure.

Enzyme solution is the major cost factor in second-generation ethanol. Hydrolysis and fermentation sections must be optimized considering a minimal enzyme dosage that can guarantee satisfactory cellulose and hemicellulose solubilization with a limited impact on final cost of ethanol. Another requirement is to work at high-dry matter content in enzymatic hydrolysis to produce a final ethanol concentration above 4 w/w% to achieve a significant reduction of energy required in the distillation and purification steps.

Enzymatic reactor selection criteria are a key point in the overall process feasibility. In the enzymatic hydrolysis at high-dry content matter, the mixing system in the reactor has to be selected in order to guarantee close contact between the enzyme and the cellulose-containing biomass, avoiding diffusion problems, preventing local accumulations of high cellobiose concentration that can inhibit the enzyme, and assuring low energy requirements. Conventional stirred reactors are not able to handle these types of slurries, and gravity-based reactors such as a drum reactor can guarantee a good degree of mixing but still require high energy consumption.

A continuous set up of this section would be a preferred alternative, especially if coupled with the possibility of operating at high dry content, thus limiting energy needed for mixing and the possibility of diffusion issues in the process.

An ideal process is the one that is able to guarantee full access to the sugar in the enzymatic hydrolysis step without degradation of the feedstock and inhibitory compound formation and the consequent fermentation of all the sugar to ethanol with good yield and productivity. For the feasibility of the whole process, cellulose and hemicellulose sugars have to be fermented to ethanol by a fermenting organism. Alternative configurations can be applied to this section, each of them giving different advantages and drawbacks. The search for an "optimal" set up can be tricky, due to the fact that a compromise has to be found between the best solutions for each issue related with hydrolysis and fermentation operations. For this reason, it is preferable to select a process that is as flexible as possible, in order to have the possibility to optimize operational conditions and configurations and be ready for the improvements in enzyme and microorganism behavior that are expected to occur in the near future as a consequence of biotech R&D efforts.

11.4.7 BY-PRODUCT UTILIZATION

The final separation of the ethanol from the fermentation broth is usually carried out by means of a distillation section. The stream coming from the top will consist in an ethanol rich mixture containing also water and traces of other volatile compounds. The stillage current contains water, microbial biomass and nonvolatile compounds. It may also contain lignin, the main process by-product, if it has not been separated before the fermentation step.

Lignin, the main component of the lignocellulosic material together with cellulose and the hemicelluloses is only minimally solubilized during the pretreatment and hydrolysis steps, and, therefore, most of this fraction entering a biomass-to-ethanol process (10%–30% in weight depending on the material used) will be present in the final solid residue.

This residual solid could simply be used, once the water content has been brought down to an acceptable level by a further separation step, in a boiler to generate steam and electric energy for the energy demand of the plant. This would definitely improve the overall energy efficiency of the process.

Another more sophisticated possibility would be to gasify the lignin in order to produce a valuable gas stream. Biomass gasification processes have been extensively studied during the past decades and important steps forward have been made. Nevertheless, full industrialization of the process is still somewhat distant due to the cost and problems related, for example, to cleaning of the gas stream.

These solutions, acceptable nowadays, are not optimal for the future, as they do not try to exploit the rich variety of chemical compounds making up the lignin materials: capturing the value of the lignin residue will significantly enhance the economic competitiveness of biomass-to-ethanol conversion. If it would be possible to recover the basic lignin constituent, such as phenols, a whole new series of very interesting by-products could be obtained [115–118]. Some effort has been devoted to this direction, one example being a new process to convert lignin into a fuel addictive studied at NREL. In this process, lignin is depolymerized with an alkali-catalyzed step, obtaining phenolic intermediates, which are then hydroprocessed into the final product.

Another suggestion [119] indicates the use of phenol obtained from lignin, for phenol–formaldehyde resins.

11.4.8 STATE OF THE INDUSTRIALIZATION STEP

Attempts to industrialize the second-generation bioethanol process date back to 1970s. Because of the high cost for cellulose degrading enzymes, the processes being preferred in the past were those based on a chemical route. An example of this kind of process is still under investigation by Bluefire Ethanol, a company based in California, whose project has the aim of converting solid waste by means of a patented concentrated acid hydrolysis. However, the drawbacks associated to the chemical route, together with historical circumstances, did not favor a full scale development of this ethanol technology.

With the increase of R&D efforts of biotech companies to deliver more efficient enzymes and reduce the cost of their production, the exploitation of more efficient and less capital intensive biochemical routes has been enabled.

Leading industrial biotech companies like Novozymes, Genencor-Danisco, and Verenium have been involved in the last years in ambitious development programs for producing more efficient biocatalysts with the final aim of making the second-generation ethanol economically sustainable.

In the past decades, with the increase in oil price and at the same time the raise of concerns about global warming and climate change, the government policies in United States, Europe, and China have played an important role in terms of direct support to R&D and demonstrative projects and in terms of mandatory policies for blending of renewable fuel into existing fuel.

This overall scenario has created a framework that has given birth to several projects having the final purpose of scaling up the processes tested and developed at laboratory and pilot scale to a demonstrative and commercial level.

The landscape of active players now consists mainly in relatively small technology players (Sekab, Biogasol, Lignol, etc.) but also in an increasing number of large companies that are eyeing the "science" companies for strategic alliance purposes (mergers, acquisitions, JVs, etc.) and are willing to invest in sustainable energy (Iogen-Shell, Verenium-BP, KL energy-Petrobras). Nonetheless, a number of technology clusters, networks, and partnerships are developing, often composed of companies with complementary expertise along the biomass value chain, particularly with regard to second-generation technologies (DDCE: a JV Dupont—Danisco; Sunopta—Mascoma; Linde—Südchemie, etc.).

Some of the companies active in these sectors are coming from first generation ethanol industry (POET, Abengoa, Pacific ethanol) or are already involved in energy production from biomass (DONG Energy-Inbicon), further stressing the fact that the knowledge of the supply chain will play an important role in these developments. Interestingly, because there are a number of technological challenges concerning the design of an efficient second-generation ethanol process, the engineering companies who decide to invest in this new frontier undoubtedly will have a competitive advantage. So far, there have been only few engineering companies that entered the competition for lignocellulosic ethanol development. Chemtex is leading the way among them.

Technologies chosen are quite varied in each of the process steps. They range from an approach that is trying to remain close to the corn to ethanol process, thereby using a pretreatment followed by enzymatic hydrolysis and fermentation (Chemtex, Sekab, DONG, Abengoa, POET) and others that are directed toward an integrated consolidated bioprocessing (CBP) approach (Mascoma, Verenium, Qteros). Others are using mixed versions that include a thermochemical step, like gasification, followed by fermentation (Zeachem, Coskata, Ineos BIO, etc.).

Based on the announced plans of companies that are developing second-generation biofuel facilities, the first commercial-scale operations should be seen as early as 2011–2012.

Based on Chemtex experience, in order to reach this goal without relying on subsidies and be able to provide a technology that can be suitable worldwide, some general guidelines have to be followed:

1. Process needs to be flexible in order to accept different biomasses as raw material, without stringent size requirements (agricultural wastes or energy crops), because there is no unique solution for feedstock in all geographic locations and all during the year.
2. Integration of the different steps along the value chain (feedstock collection and handling, pretreatment, hydrolysis and fermentation, distillation) is of high importance. In fact, optimizing conditions in one step of the process can influence performance in the other steps. The challenge is to find the right combination of trade-offs that optimizes the integrated process.
3. Preference will be given to those technologies that are not using chemicals (acid, alkali, or solvents) in the pretreatment step, since besides direct impact on the operative cost, chemicals often force into expensive recycling and neutralization and often require the use of special construction material, affecting capital expenditure significantly.

4. Capability of maximizing the biomass to ethanol conversion yield, while operating at high solids content with a low amount of enzymes will be important aspects to be considered to maximize the production efficiency.
5. Process should be self-sustained from an energy point of view even at small industrial scale. Integrated production of thermal and electric energy, able to satisfy plant requirements and preferably to provide a surplus in order to improve the revenues, is a key point for a winning technology.
6. The use of simple machinery coupled with the possibility of operating on single line even at very high production scale is another key point for implementing the process in plants characterized by different capacities, maintaining a low investment requirement.

Market will be open to technologies able to produce low-cost ethanol, competitive with fossil fuel without the need of incentives.

REFERENCES

1. Brown, L.R. 2006. *Plan B 2.0: Rescuing a Planet under Stress and a Civilization in Trouble*, New York: W.W. Norton & Company.
2. International Energy Agency (IEA). Biofuels for transport: An international perspective. <http://www.iea.org/textbase/nppdf/free/2004/biofuels2004.pdf>
3. Lynd, L.R., C.E. Wyman, and T.U. Gerngross. 1999. Biocommodity engineering. *Biotechnology Progress*, 15: 777–793.
4. Morrow, W.R., W.M. Griffin, and H.S. Matthews. 2006. Modeling switchgrass derived cellulosic ethanol distribution in the United States. *Environmental Science and Technology*, 40(9): 2877–2886.
5. Sims, R., M. Taylor, J. Saddler, and W. Mabee. 2008. From 1st to 2nd generation biofuels technologies. An overview of current industry and RD&D activities. IAE Bioenergy Report. November 2008.
6. Japan for Sustainability, Japan. Commercial-scale wood-based ethanol production begins. http://www.japanfs.org/db/1674-e
7. Farrell, A.E., R.J. Plevin, B.T. Turner, A.D. Jones, M.O'Hare, and D.M. Kammen. 2006. Ethanol can contribute to energy and environmental goals. *Science*, 311: 506–508.
8. Kszos, L.A., S.B. McLaughlin, and M. Walsh. 2001. *Bioenergy from Switchgrass: Reducing Production Costs by Improving Yield and Optimizing Crop Management*, Oak Ridge, TN: ORNL. http://www.ornl.gov/wwebworks/cppr/y2001/pres/114121.pdf (accessed on June 2010).
9. Kim, S. and B.E. Dale. 2005. Environmental aspects of ethanol derived from no-tilled corn grain: Nonrenewable energy consumption and greenhouse gas emissions. *Biomass and Bioenergy*, 28: 475–489.
10. Kim, S. and B.E. Dale. 2006. Life cycle assessment of various cropping systems utilized for producing biofuels: Bioethanol and biodiesel. *Biomass and Bioenergy*, 29: 426–439.
11. Adler, P.R., S.J.D. Grosso, and W.J. Parton. 2007. Life-cycle assessment of net greenhouse-gas flux for bioenergy cropping systems. *Ecological Applications*, 17(3): 675–691.
12. Zhang, Y., M.A. Dube, D.D. McLean, and M. Kates. 2003. Biodiesel production from waste cooking oil: 1. Process design and technological assessment. *Bioresource Technology*, 89: 1–16.
13. Patzek, T.W. 2004. Thermodynamics of the corn-ethanol biofuel cycle: 1. Introduction. *Critical Reviews in Plant Sciences*, 23(6): 519–567.
14. Brown, R.M. 2004. Cellulose structure and biosynthesis: What is in store for the 21st century? *Journal of Polymer Science*, 42: 487–495.
15. Rogner, H.H. 2000. Energy resources. In: Goldemberg, J. (Ed.), *World Energy Assessment*, pp. 135–171, New York: United Nations Development Programme.
16. Berndes, G., M. Hoogwijk, and R. Van den Broek. 2003. The contribution of biomass in the future global energy supply: A review of 17 studies. *Biomass and Bioenergy*, 25: 1–28.
17. Zhang, Y.H.P. and L.R. Lynd. 2004. Toward an aggregated understanding of enzymatic hydrolysis of cellulose: Noncomplexed cellulase systems. *Biotechnology Bioengineering*, 88: 797–824.
18. Fengel, D. and G. Wegener. 1989. *Wood: Chemistry, Ultrastructure, Reactions*, Berlin, Germany: Walter de Gruyter & Co.
19. Saha, B.C. 2003. Hemicellulose bioconversion. *Journal of Industrial Microbiology and Biotechnology*, 30: 279–291.

20. Shimizu, K. 1991. Chemistry of hemicellulose, Chapter 5. In: Hon, D.N.S. and Shiraishi, N. (Eds.), *Wood and Cellulosic Chemistry*, New York and Basel, Switzerland: Marcel Dekker, Inc.

21. Fan, L.T., Y.H. Lee, and M.M. Gharpuray. 1982. The nature of lignocellulosics and their pretreatments for enzymatic hydrolysis. *Advanced Biochemical Engineering*, 23: 158–187.

22. Lee, Y.L. and J.M. Sanchez. 1997. Theoretical study of thermodynamics relevant to tetramethylsilane pyrolysis. *Journal of Crystal Growth*, 178: 513–517.

23. NERL technical report NREL/TP 510-42622, January 2008.

24. Grethlein, H.E., D.C. Allen, and A.O. Converse. 1984. A comparative study of the enzymatic hydrolysis of the acid-pretreated white pine and mixed hardwood. *Biotechnology Bioengineering*, 26: 1498–1505.

25. Ramos, L.P., C. Breuil, and J.N. Saddler. 1992. Comparison of steam pretreatment of eucalyptus, aspen, and spruce wood chips and their enzymatic hydrolysis. *Applied Biochemical and Biotechnology*, 34: 37–48.

26. Lee, D.K., V.N. Owens, A. Boe, and P. Jeranyama, Composition of Herbaceous Biomass Feedstokes. http://ncsungrant.sdstate.org/uploads/publications/SGINC1-07.pdf (accessed on June 2010).

27. Van der Laaka, S.R.P., J.M. Raven, and G.P.J. Verbong. 2007. Strategic niche management for biofuels: Analysing past experiments for developing new biofuel policies. *Energy Policy*, 35: 3213–3225.

28. McCormick-Brennan, K., C. Bomb, E. Deurwaarder, and T. Kåberger. 2007. Biofuels for transport in Europe: Lessons from Germany and the U.K. *Energy Policy*, 35: 2256–2267.

29. Tibelius, C. 1996. Coproducts and near coproducts of fuel ethanol fermentation from grain. Final report, Contract No. 01531-5-7157. Agriculture and Agri-food, Ottawa, Canada.

30. Jacquet, F., L. Bamiere, J.C. Bureau, and L. Guinde. 2007. Recent developments and prospects for the production of biofuels in the EU: Can they really be "part of solution"? In: *Proceedings of the Conference on "Biofuels, Food and Feed Tradeoffs,"* St. Louis, MO, April 12–13 2007.

31. Elander, T.E. and V.L. Putsche. 1996. Ethanol from corn: Technology and economics. In: Wyaman, C.E. (Ed.), *Handbook on Bioethanol: Production and Utilization*, pp. 351–347, Washington, DC: Taylor & Francis.

32. Maiorella, B., H. Blanch, and C. Wilke. 1983. Distillery effluent treatment and by-product recovery. *Process Biochemistry*, 18: 5–8.

33. Taherzadeh, M.J. and K. Karimi. 2007. Process for ethanol from lignocelluloses. I. Acid base hydrolysis processes. *BioResources*, 2: 472–499.

34. Wooley, R., M. Ruth, J. Sheehan, K. Ibsen, H. Majdeski, and A. Galvez. 1999. Lignocellulosic biomass to ethanol process design and economics utilizing co-current acid prehydrolysis and enzymatic hydrolysis. Current and futuristic scenarios. Report No. NREL/TP-508-26157. National renewable Energy Laboratory, Golden, CO.

35. Wooley, R., M. Ruth, D. Glassner, and J. Sheejan. 1999. Process design and costing of bioethanol technology: A tool for determining the status and direction of research and development. *Biotechnology Progress*, 15: 794–803.

36. Lynd, L.R. 1996. Overview and evaluation of fuel ethanol from cellulosic biomass: Technology, economics, the environment, and policy. *Annual Reviews, Energy Environment*, 21: 403–465.

37. McAloon, A., F. Taylor, W. Yee, K. Ibsen, and R. Wooley. 2000. Determining the cost of producing ethanol from corn starch and lignocellulosic feedstocks. Technical Report NREL/TP-580-28893. National Renewable Energy Laboratory, Golden, CO, 35pp.

38. Hahn-Hägerdal, B., M. Galbe, M.F. Gorwa-Grauslund, and G. Zacchi. 2006. Bioethanol—The fuel of tomorrow from the residues of today. *Trends Biotechnology*, 24: 549–556.

39. Cardona, C.A. and O.J. Sànchez. 2006. Energy consumption analysis of integrated flowsheet for production of fuel ethanol from lignocellulosic biomass. *Energy*, 31: 2447–2459.

40. Hamelinck, C.N., G. van Hooijdonk, and A. Faaij. 2005. Ethanol from lignocellulosic biomass: Techno-economic performance in short-, middle- and long-term. *Biomass and Bioenergy*, 28: 384–410.

41. Hsu, T.-A. 1996. Pretreatment of biomass. In: Wyman, C.E. (Ed.), *Handbook on Bioethanol, Production and Utilization*, Washington, DC: Taylor & Francis.

42. Sun, Y. and J. Cheng. 2002. Hydrolysis of lignocellulosic materials for ethanol production: A review. *Bioresource Technology*, 83: 1–11.

43. Mosier, N., C. Wyman, B. Dale, R. Elander, Y.Y. Lee, M. Holtzapple, and M. Ladisch. 2005. Features of promising technologies for pretreatment of lignocellulosic biomass. *Bioresource Technology*, 96: 673–686.

44. Cadoche, L. and G.D. Lopez. 1989. Assessment of size reduction as a preliminary step in the production of ethanol from lignocellulosic wastes. *Biological Wastes*, 30: 153–157.

45. Esteghlalian, A., A.G. Hashimoto, J.J. Fenske, and M.H. Penner. 1997. Modeling and optimization of the dilute-sulfuric-acid pretreatment of corn stover, poplar and switchgrass. *Bioresources Technology*, 59: 129–136.

46. McMillan, J.D. 1994. Pretreatment of lignocellulosic biomass. In: Himmel, M.E., Baker, J.O., and Overend, R.P. (Eds.), *Enzymatic Conversion of Biomass for Fuels Production*, pp. 292–324, Washington, DC: American Chemical Society.

47. Fan, L.T., Gharpuray, M.M., and Y.H. Lee. 1987. *Cellulose Hydrolysis Biotechnology Monographs*, p. 57, Berlin, Germany: Springer.

48. Bjerre, A.B., A.B. Olesen, and T. Fernqvist. 1996. Pretreatment of wheat straw using combined wet oxidation and alkaline hydrolysis resulting in convertible cellulose and hemicellulose. *Biotechnology Bioengineering*, 49: 568–577.

49. Millet, M.A., A.J. Baker, and L.D. Scatter. 1976. Physical and chemical pretreatment for enhancing cellulose saccharification. *Biotechnology Bioengineering Symposium*, 6: 125–153.

50. Kim, T.H., J.S. Kim, C. Sunwoo et al. 2003. Pretreatment of corn stover by aqueous ammonia. *Bioresources Technology*, 90: 39–47.

51. Casebier, R.L., J.K. Hamilton, and H.L. Hergert. 1969. Chemistry and mechanism of water prehydrolysis of southern pine wood. *TAPPI*, 52: 2369–2377.

52. Fernandez-Bolanos, J., B. Felizon, A. Heredia, and A. Jimenez. 1999. Characterization of the lignin obtained by alkaline delignification and of the cellulose residue from steam-exploded olive stones. *Bioresources Technology*, 68: 121–132.

53. Lora, J.H. and M. Wayman. 1978. Delignification of hardwoods by autohydrolysis and extraction. *TAPPI*, 61: 47–50.

54. McGinnis, G.D., W.W. Wilson, and C.E. Mullen. 1983. Biomass pretreatment with water and high pressure oxygen. The wet-oxidation process. *Industrial Engineering Chemistry Product Research Design*, 22: 352–357.

55. Timell, T.E. 1967. Recent progress in the chemistry of wood hemicelluloses. *Wood Science and Technology*, 1: 45–70.

56. Ibrahim, M. and G.W. Glasser. 1999. Steam assisted biomass fractionation, Part III: A quantitative evaluation of the "clean fractionation" concept. *Bioresource Technology*, 70: 181–192.

57. Turn, S.Q., C.M. Kinoshita, W.E. Kaar, and D.M. Ishimura. 1998. Measurement of gas phase on steam explosion of biomass. *Bioresource Technology*, 64: 71–75.

58. Abatzoglou, N., E. Chornet, and K. Belkacemi. 1992. Phenomenological kinetics of complex systems: The development of a generalized severity parameter and its application to lignocellulosics fractionation. *Chemical Engineering Science*, 47: 1109–1122.

59. Mes-Hartree, M., B.E. Dale, and W.K. Craig. 1988. Comparison of steam and ammonia pretreatment for enzymatic hydrolysis of cellulose. *Applied Microbiology and Biotechnology*, 29: 462–468.

60. Holtzapple, M.T., J.E. Lundeen, and R. Sturgis. 1992. Pretreatment of lignocellulosic municipal solid waste by ammonia fiber explosion (AFEX). *Applied Biochemical and Biotechnology*, 34/35: 5–21.

61. Tengerdy, R.P. and J.G. Nagy. 1988. Increasing the feed value of forestry waste by ammonia freeze explosion treatment. *Biological Wastes*, 25: 149–153.

62. Vlasenko, E.Y., H. Ding, J.M. Labavitch, and S.P. Shoemaker. 1997. Enzymatic hydrolysis of pretreated rice straw. *Bioresources Technology*, 59: 109–119.

63. Mabee, W.E., D.J. Gregg, C. Arato et al. 2006. Updates on softwood-to-ethanol process development. *Applied Biochemistry and Biotechnology*, 129: 55–70.

64. Gregg, D.J. and J.N. Saddler. 1996. Factors affecting cellulose hydrolysis and the potential of enzyme recycle to enhance the efficiency of an integrated wood to ethanol process. *Biotechnology Bioengineering*, 51: 375–383.

65. Gregg, D.J., A. Boussaid, and J.N. Saddler. 1998. Techno-economic evaluations of a generic wood-to-ethanol process: Effect of increased cellulose yields and enzyme recycle. *Bioresource Technology*, 63: 7–12.

66. Lee, Y.H. and L.T. Fan. 1982. Kinetic studies of enzymatic hydrolysis of insoluble cellulose: Analysis of the initial rates. *Biotechnology Bioengineering*, 24: 2383–2406.

67. Lee, Y.H. and L.T. Fan. 1982. Kinetic studies of enzymatic hydrolysis of insoluble cellulose: Analysis of extended hydrolysis time. *Biotechnology Bioengineering*, 25: 939–966.

68. Bezzerra, R.M.F. and A.A. Dias. 2005. Enzymatic kinetic of cellulose hydrolysis. *Applied Biochemistry and Biotechnology*, 126: 49–60.

69. Malherbe, S. and T.E. Cloete. 2002. Lignocellulose biodegradation: Fundamentals and applications. *Reviews in Environmental Science and Biotechnology*, 1: 105–114.

70. Duff, S.J.B. and W.D. Murray. 1996. Bioconversion of forest products industry waste cellulosics to fuel ethanol: A review. *Bioresources Technology*, 55: 1–33.

71. Beguin, P. and J.P. Aubert. 1994. The biological degradation of cellulose. *Microbiological Reviews*, 13: 25–58.

72. Sternberg, D. 1976. Production of cellulase by trichoderma. *Biotechnology and Bioengineering Symposium*, 10: 35–53.
73. Coughlan, M.P. and L.G. Ljungdahl. 1988. Comparative biochemistry of fungal and bacterial cellulolytic enzyme system. In: Aubert, J.-P., Beguin, P., and Millet, J. (Eds.), *Biochemistry and Genetics of Cellulose Degradation*, pp. 11–30, New York: Academic Press.
74. Jørgensen, H., J.B. Kristensen, and C. Felby. 2007. Enzymatic conversion of lignocellulose into fermentable sugars: Challenges and opportunities. *Biofuels, Bioproducts and Biorefining*, 1(2): 119–134.
75. Kristensen, J.B., L.G. Thygesen, C. Felby, H. Jørgensen, and T. Elder. 2008. Cell-wall structural changes in wheat straw pretreated for bioethanol production. *Biotechnology for Biofuels*, 1: 5.
76. Kristensen, J.B., J. Börjesson, M.H. Bruun, F. Tjerneld, and H. Jørgensen. 2007. Use of surface active additives in enzymatic hydrolysis of wheat straw lignocelluloses. *Enzyme and Microbial Technology*, 40: 888–895.
77. Kristensen, J.B., C. Felby, and H. Jørgensen. 2009. Determining yields in high solids enzymatic hydrolysis of biomass. *Applied Biochemistry and Biotechnology*, 156: 557–562.
78. Kristensen, J.B., C. Felby, and H. Jørgensen. 2009. Yield determining factors in high solids enzymatic hydrolysis of lignocelluloses production. *Biotechnology for Biofuels*, 2: 11.
79. Huanga, H.J., S. Ramaswamya, W. Al-Dajania, U. Tschirnera, and R.A. Cairncross. 2009. Effect of biomass species and plant size on cellulosic ethanol: A comparative process and economic analysis. *Biomass and Bioenergy*, 3: 234–246.
80. Tu, M., X. Zhang, A. Kurabi, N. Gilkes, W. Mabee, and J. Saddler. 2006. Immobilization of *b*-glucosidase on Eupergit C for lignocellulose hydrolysis. *Biotechnology Letters*, 28: 151–156.
81. Berlin, A., N. Gilkes, D. Kilburn et al. 2006. Evaluation of cellulase preparations for hydrolysis of hardwood substrates. *Applied Biochemistry and Biotechnology*, 130: 528–545.
82. Pan, X.J., N. Gilkes, J. Kadla et al. 2006. Bioconversion of hybrid poplar to ethanol and co-products using an organosolv fractionation process: Optimization of process yields. *Biotechnology Bioengineering*, 94: 851–861.
83. Tengborg, C., M. Galbe, and G. Zacchi. 2001. Reduced inhibition of enzymatic hydrolysis of steam-pretreated softwood. *Enzyme and Microbial Technology*, 28: 835–844.
84. Lin, Y. and S. Tanaka. 2006. Ethanol fermentation from biomass resources: Current state and prospect. *Applied Microbiology and Biotechnology*, 69: 627–642.
85. van Dijken, J.P., R.A. Weusthuis, and J.T. Pronk. 1993. Kinetics of growth and sugar consumption in yeasts. *Antonie Van Leeuwenhoek*, 63: 343–352.
86. Aiba, S., M. Shoda, and M. Nagatani. 2000. Kinetics of product inhibition in alcohol fermentation. *Biotechnology Bioengineering*, 67: 671–690.
87. Oliveira, S.C., T.C. Paiva, A.E. Visconti, and R. Giudici. 1998. Discrimination between ethanol inhibition models in a continuous alcoholic fermentation process using flocculating yeast. *Applied Biochemistry and Biotechnology*, 74: 161–172.
88. Neway, J.O. 1989. *Fermentation Process Development of Industrial Organisms*, New York: Marcel Dekker.
89. Abbi, M., R.C. Kuhad, and A. Singh. 1996. Fermentation of xylose and rice straw hydrolysate to ethanol by Candida shchatac NCL-3501. *Journal of Industrial Microbiology*, 17: 20–23.
90. Converti, A. and M. Del Borghi. 1998. Inhibition of the fermentation of oak hemicelluloses acid-hydrolysate by minor sugars. *Journal of Biotechnology*, 64: 211–218.
91. Ogier, J.-C., D. Ballerini, J.P. Leygue, L. Rigal, and J. Pourquiè. 1999. Production d'ethanol à partir de biomasse lignocellulosique. *Oil & Gas Science and Technology-Revue de l'IFP*, 54: 67–94.
92. Krishnan, M.S., M. Blanco, C.K. Shattuck, N.P. Nghiem, and B.H. Davison. 2000. Ethanol production from glucose and xylose by immobilized *Zymomonas mobilis* CP4(pZB5). *Applied Biochemistry and Biotechnology*, 84–86: 525–541.
93. Joachimsthal, E.L. and P.L. Rogers. 2000. Characterization of a high-productivity recombinant strain of *Zymomonas mobilis* for ethanol production from glucose/xylose mixtures. *Applied Biochemistry and Biotechnology*, 84–86: 343–356.
94. Lawford, H.G. and J.D. Rousseau. 2000. Comparative energetics of glucose and xylose metabolism in recombinant *Zymomonas mobilis*. *Applied Biochemistry and Biotechnology*, 84–86: 277–293.
95. Ho, N.W., Z. Chen, and A.P. Brainard. 1998. Genetically engineered *Saccharomyces* yeast capable of effective cofermentation of glucose and xylose. *Applied Environmental Microbiology*, 64: 1852–1859.
96. Palmqvist, E., B. Hahn-Hägerdal, Z. Szengyel, G. Zacchi, and K. Reczey. 1997. Simultaneous detoxification and enzyme production of hemicellulose hydrolysates obtained after steam pretreatment. *Enzyme and Microbial Technology*, 20: 286–293.

97. Jonsson, L.J., E. Palmqvist, N.O. Nilvebrant, and B. Hahn-Hägerdal. 1998. Detoxification of wood hydrolysates with laccase and peroxidase from the white-rot fungus *Trametes versicolor*. *Applied Microbiology and Biotechnology*, 49: 691–697.

98. Palmqvist, E. and B. Hahn-Hägerdal. 2000. Fermentation of lignocellulosic hydrolysates. II. Inhibitors and mechanism of inhibition. *Bioresources Technology*, 74: 25–33.

99. Palmqvist, E. and B. Hahn-Hägerdal. 2000. Fermentation of lignocellulosic hydrolysates. I. Inhibition and detoxification. *Bioresources Technology*, 74: 17–24.

100. Pimentel, D., R. Doughty, C. Carothers et al. 2001. Energy inputs in crop production in developing and developed countries. *Workshop on Food Security and Environmental Quality in the Developing World*, Ohio State University, Columbus, OH.

101. Jones, H.L., A. Margaritis, and R.J. Stewart. 2007. The combined effects of oxygen supply strategy, inoculum size and temperature profile on very-high-gravity beer fermentation by *Saccharomyces cerevisiae*. *Journal of the Institute of Brewing*, 113: 168–184.

102. Taherzadeh, M.J. and K. Karimi. 2007. Enzyme based hydrolysis processes for ethanol from lignocelluloses: A review. *BioResources*, 2: 707–738.

103. Eklund, R. and G. Zacchi. 1995. Simultaneous saccharification and fermentation of steam-pretreated willow. *Enzyme and Microbial Technology*, 17: 55–259.

104. Karimi, K., G. Emtiazi, and M.J. Taherzadeh. 2006. Ethanol production from dilute-acid pretreated rice straw by simultaneous saccharification and fermentation with *Mucor indicus*, *Rhizopus oryzae*, and *Saccharomyces cerevisiae*. *Enzyme and Microbial Technology*, 40: 138–144.

105. McMillan, J.D., M.M. Newman, D.W. Templeton et al. 1999. Simultaneous saccharification and cofermentation of dilute-acid pretreated yellow poplar hardwood to ethanol using xylose-fermenting *Zymomonas mobilis*. *Applied Biochemistry and Biotechnology*, 77: 649–665.

106. Tenborg, C., M. Galbe, and G. Zacchi. 2001. Influence of enzyme loading and physical parameters on the enzymatic hydrolysis of steam-pretreated softwood. *Biotechnology Progress*, 17: 110–117.

107. Ballesteros, M., J.M. Oliva, M.J. Negro et al. 2004. Ethanol from lignocellulosic materials by a simultaneous saccharification and fermentation process (SFS) with *Kluyveromyces marxianus* CECT 10875. *Process Biochemistry*, 39: 1843–1848.

108. Golias, H., G.J. Dumsday, G.A. Stanley et al. 2002. Evaluation of a recombinant *Klebsiella oxytoca* strain for ethanol production from cellulose by simultaneous saccharification and fermentation: Comparison with native cellobiose-utilising yeast strains and performance in co-culture with thermotolerant yeast and *Zymomonas mobilis*. *Journal of Biotechnology*, 96: 155–168.

109. Hong, J., Y. Wang, H. Kumagai et al. 2007. Construction of thermotolerant yeast expressing thermostable cellulase genes. *Journal of Biotechnology*, 130: 114–123.

110. Kadam, K.L. and S.L. Schmidt. 1997. Evaluation of *Candida acidothermophilum* in ethanol production from lignocellulosic biomass. *Applied Microbiology and Biotechnology*, 48: 709–713.

111. Teixeira, L.C., J.C. Linden, and H.A. Schroeder. 2000. Simultaneous saccharification and cofermentation of peracetic acid-pretreated biomass. *Applied Biochemistry and Biotechnology*, 84: 111–127.

112. Lawford, H.G. and J.D. Rousseau. 1998. Improving fermentation performance of recombinant Zymomonas in acetic acid-containing media. *Applied Biochemistry and Biotechnology*, 70: 161–172.

113. Kim, S.B., H.J. Kim, and C.J. Kim. 2006. Enhancement of the enzymatic digestibility of waste newspaper using Tween. *Applied Biochemistry and Biotechnology*, 130: 486–495.

114. Aden, A., M. Ruth, K. Ibsen et al. 2002. Lignocellulosic biomass to ethanol process design and economics utilizing co-current dilute acid prehydrolysis and enzymatic hydrolysis for corn stover. Technical report NREL/TP-510-32438, Golden, CO.

115. Johnson, D.K., E. Chornet, W. Zmierczak, and J. Shabtai. 2002. Conversion of lignin into a hydrocarbon product for blending with gasoline. *Fuel Chemistry Division Preprints*, 47: 380–383.

116. Aden, A., R. Wooley, and M. Yancey. 2000. Oregon biomass to ethanol project: Pre-feasibility study and modeling results. NREL report http://www.oregon.gov/ENERGY/RENEW/Biomass/docs/OCES/OCES_A.PDF (accessed on June 2010).

117. Miller, J.E., L.R. Evans, J.E. Mudd, and K.A. Brown. 2002. Batch microreactor studies of lignin depolymerization by bases. II. Aqueous Solvents SAND REPORT SAND 2002-1318.

118. Yamamoto, K., S. Ohara, K. Magara et al. 2008. Bioethanol Production and Lignin Utilization in the Biomass Town System of Kita-akita City. http://www.biomass-asia-workshop.jp/biomassws/05workshop/poster/P-14.pdf (accessed on June 2010).

119. Martin, C., G.J.D.M. Rocha, M. Perez et al. 2007. Acid prehydrolysis, alkaline delignification and enzymatic hydrolysis of rice hulls. *Cellulose Chemistry and Technology*, 41: 129–135.

12 Efficient Distributed Power Supply with Molten Carbonate Fuel Cells

Peter Heidebrecht and Kai Sundmacher

CONTENTS

12.1 INTRODUCTION

Fuel cells are devices for electric power production. Similar to their classical counterparts which are based on a cycle process, they convert chemical energy (usually the enthalpy of combustion of a combustible substance) into electric energy. The substantial difference between these two classes of processes is that cycle processes usually use three energy transformation steps—from chemical enthalpy to thermal, then kinetic, and finally electric energy—whereas fuel cells directly convert chemical into electrical energy and thereby offer a chance to obtain higher degrees of efficiency [1–4].

The environmental impact of fuel cells strongly depends on the fuel the cell is fed with. Fuel production based on fossil resources neither is free of greenhouse gases, because carbon dioxide is emitted, nor is it sustainable. Fuel based on renewable resources has overall zero carbon dioxide emissions, but the combination of it with fuel cells is not compulsory. Thus, fuel cells are not a sustainable technique by themselves, but they promise a more efficient use of available fuels, be they based on fossil or renewable resources.

Among the known fuel cell types, the two high temperature fuel cells, namely, the molten carbonate fuel cell (MCFC) and the solid oxide fuel cell (SOFC), do not require hydrogen as their primary fuel gas, but they can be fed with any fuel gases containing short-chained hydrocarbons, carbon monoxide, and hydrogen. While a dominant part of low temperature fuel cell systems is occupied by the reforming process, which transforms fuel gas into hydrogen, this process can simply be integrated into high temperature fuel cells. This so-called internal reforming concept not only offers a simpler system design compared to that of low temperature fuel cells with their external reforming units, but it also increases the overall electric system efficiency significantly. In addition, their insensitivity with respect to carbon monoxide allows a wide spectrum of fuels to be used in high temperature fuel cells.

But they not only provide electric power. Due to their operating temperature, these concepts combine high electric efficiency with a wide spectrum of heat utilization, for example, steam production, cold chillers, or a downstream cycle process. Compared to classical concepts, they offer a higher electricity/heat ratio, which is preferable in many stationary applications. Due to their combined heat and power production together with their fuel flexibility, high temperature fuel cells are attractive for several areas of application. They can be used in small power plants based on natural gas, where they accommodate residential areas or a single larger building, for example, a hospital or an office building.

Beyond the replacement of classical units in today's stationary applications, where they offer superior efficiency, high temperature fuel cells are also strong candidates for sustainable energy systems. The most prominent example is the use of bio gases from fermentation processes (e.g., from wastewater treatment or from cattle-breeding farms). Further applications are the consumption of lean waste gases from processes in food or chemical industries, landfill gases or mining gases. All these applications use fuels for on-site production of electricity in combination with heat or steam, depending on the specific demands of each application. High temperature fuel cells are more suitable for these applications than a cycle process with low efficiency in combination with a boiler. They offer a chance to alter today's centralized energy supply system toward a distributed system where numerous opportunities for renewable energy supply can be exploited.

Although these properties apply to both known high temperature fuel cell types, namely, the MCFC and the SOFC, we will focus on the first mentioned in the following sections. This is mainly because the MCFC is technically better developed with most vital questions about material stability, system reliability and production procedures already solved and with the first commercial systems available on the market.

In the following, after a technical introduction into MCFC, their general advantages and drawbacks are discussed. Afterward, several existing technical realizations are described and compared against each other. Several actual examples of application of MCFCs are discussed and finally some future development trends are indicated.

12.2 TECHNICAL BACKGROUND

12.2.1 WORKING PRINCIPLE

Like any other fuel cell, the MCFC consists of several layers (Figure 12.1). The electrolyte layer of the MCFC consists of a eutectic carbonate melt (38% K_2CO_3, 62% Li_2CO_3), which is immobilized in a porous aluminum oxide structure (γ-$LiAlO_2$/α-Al_2O_3). It serves as a semi-permeable layer that only allows carbonate ions (CO_3^{2-}) to pass through the layer. Other substances, especially dissolved non-ionic gases, may not pass through.

On each side of the electrolyte layer, a porous catalyst layer, an electrode, is placed. These layers consist of an electron-conducting solid material, which also serves as a catalytic promoter for the reactions occurring at the respective electrode. Alternatively, the functionality of electron conduction and reaction promotion may be separated by using a carrier material which mainly serves as the conductor, and placing the catalyst in a thin layer upon the surface of the carrier material. In the case of MCFC, nickel and nickel oxide are preferred materials for the anode and the cathode electrode, respectively. A part of the molten carbonate is also located in the electrode pores, held in place by capillary forces. The other part of the pores is filled with gas through which the educts and products of the reactions inside the electrodes are transported. The electrochemical reactions (see below)

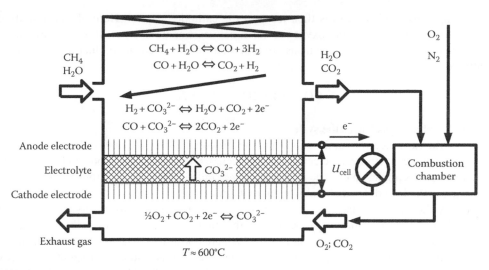

FIGURE 12.1 Working principle of an MCFC.

basically happen at the three-phase boundary between gas, liquid and catalyst, so a large interfacial area is required in the pore structure.

At each electrode, on the opposite side of the electrolyte layer, a gas channel is located. The anode channel is fed with a mixture of steam and fuel gas, for example, methane. Prior to the reaction at the electrode, this gas has to be converted to hydrogen in the reforming process. The major chemical reactions in this process are the steam-reforming and the water–gas shift reaction:

$$CH_4 + H_2O \leftrightarrow CO + 3H_2$$

$$CO + H_2O \leftrightarrow CO_2 + H_2$$

This process requires two things: heat, because it is endothermic, and high temperatures, because its conversion is severely limited by its chemical equilibrium at low temperatures. Both are available in the MCFC. The catalyst commonly used to promote these reactions is based on nickel.

The reforming products, mainly hydrogen and carbon monoxide, diffuse into the anode electrode and dissolve in the electrolyte inside the electrode pores. There they react at the electrode catalyst surface, thereby consuming carbonate ions from the electrolyte and producing free electrons, which are located on the electron conducting solid phase of the electrode after the reaction. In addition, carbon dioxide and water are produced. Generally, the hydrogen oxidation is intrinsically faster than the carbon monoxide oxidation. But especially with carbon monoxide rich fuel gases, this reaction becomes important:

$$H_2 + CO_3^{2-} \leftrightarrow H_2O + CO_2 + 2e^-$$

$$CO + CO_3^{2-} \leftrightarrow 2CO_2 + 2e^-$$

Because full conversion of hydrogen and carbon monoxide is not possible for thermodynamic and energetic reasons, the anode exhaust gas contains significant amounts of hydrogen and carbon monoxide, as well as a small portion of unreformed fuel gas. This gas has to be oxidized completely, so it is mixed with air and fed into a combustion unit. Along with the heat releasing electrochemical reactions, this combustion is the main heat source within the MCFC system and it is used to heat up the fresh air to the process temperature.

The completely oxidized gas is then fed into the cathode channel. Here, the carbon dioxide is consumed together with some oxygen to form new carbonate ions, closing the carbonate ion loop. In the same reaction, two electrons are taken out of the electron conducting solid phase of the cathode:

$$CO_3^{2-} \leftrightarrow 1/2O_2 + CO_2 + 2e^-$$

The cathode exhaust gas leaves the system.

Between the extra electrons at the anode and the "missing" electrons, that is, the positive electron holes at the cathode electrode, an electric voltage occurs. Connecting the anode and cathode via an electric load, for example, an electric engine or a light bulb, allows the electrons to move from the anode to the cathode and do electrical work at that device.

Obviously the MCFC exhausts carbon dioxide. This is in contrary to the common saying that fuel cells do not emit greenhouse gases. In fact, this fuel cell even requires carbon dioxide in its cathode reaction; otherwise, the carbonate ions which are consumed at the anode reaction could not be replaced and the cell would quickly run out of electrolyte. It is also not sufficient to exclusively recycle the carbon dioxide which is produced at the anode electrode. This would require that every molecule of carbon dioxide fed into the cathode channel is converted to carbonate ions, which is not possible for thermodynamic reasons. Thus, a continuous feed of carbon to the system is necessary causing a continuous exhaust of carbon dioxide. A global "zero emission" operation can only be achieved by using biofuels.

12.2.2　Features of the MCFC

One of the major aspects in MCFC is temperature. With a typical operating temperature of about 550°C–650°C, relatively inexpensive metal materials can be used—in contrast to the SOFC, which operates at 800°C–1000°C and consequently needs expensive ceramic materials. On the other hand the MCFC temperature is high enough to obtain sufficiently high reaction rates with inexpensive and less active catalysts like nickel. For the reforming process, a temperature of 700°C–800°C would be preferable. At this temperature, not only the reaction rate is high but also the chemical equilibrium which limits the conversion in this process is significantly more favorable than at lower temperatures. In the internal reforming concept shown in Figure 12.1, the continuous removal of the reforming products, that is, hydrogen and carbon monoxide by the oxidation process helps to obtain high degrees of reforming conversion, although the MCFC temperature is relatively low.

Because of the absence of highly precious metals like platinum, high temperature fuel cells are tolerant with respect to carbon monoxide. This is what makes the MCFC suitable for a wide range of different fuels. In principle, the MCFC would even operate on carbon monoxide only. Whereas sulfur is a catalyst poison in MCFC, so it must be removed from the feed gas.

A further advantage of the MCFC is its potential in the combined production of heat and electric power. Even if the exhaust gas is used to pre-heat the feed gas of the system it still has a temperature of about 400°C, which is sufficient to generate pressurized steam in an industrial application or hot water for a residential building. Such coproductions are also possible with classical apparatuses, but often times the electric power demand is equal or even higher than the demand for heat, a ratio which cannot be satisfied by engines or turbines alone. Due to their high efficiency, high temperature fuel cells can meet these requirements. A hybrid system consisting of an MCFC and a downstream turbine can further increase the portion of electric energy produced by the system. Because of the low efficiency and high costs of very small turbines, this is economically useful for systems above 1 MW.

Like most other fuel cells, the MCFC principle promises low maintenance costs. Except for the blowers, which move the gases through the channels with comparably low pressure drop, there are no moving parts in the system. The major part of maintenance effort is the replacement of the cell stack, which has to take place after a certain degradation of the electrode catalysts. Today's stack lifetime expectance is about 2–4 years, during which the system can continuously deliver heat and power.

12.3 TECHNICAL REALIZATIONS

12.3.1 CELL STACK

A single fuel cell, as depicted in Figure 12.2, is capable of delivering a voltage of about 0.7–0.9 V under operating conditions. According to today's standards, a typical current density of an MCFC is at about 150 mA cm^{-2}, so a cell of 1 m^2 size delivers approximately 1 kW electric power. To obtain systems with higher power, several cells are combined in a cell stack (Figure 12.2). From the chemical engineering point of view, these are parallel reactors; from the electric point of view, this is a series of current sources.

Channel structures between the cells distribute the gases across the electrode area and simultaneously collect the produced charges at the electrodes and transfer them to the neighboring cell. Because they connect the anode of a cell to the cathode of the next one, they are referred to as bipolar plates. Cell stacks can contain up to several hundred cells.

One practical issue with the stacking of fuel cells is the sealing. Due to temperature gradients in the system, and because of different expansion coefficients of the materials, the layered structure of the stack has to be pressed with a pressure of several bars.

The plates at either end of the stack are often times massive structures. This is not only because they have to take up the pressure the stack is under, but also because here the electric current is collected in one point. While in bipolar plates the electric current flows mainly directly through the plate in perpendicular direction to the cell plane, the current in the end plate has to flow along the plane. To minimize resistance, a high cross-sectional area is required; thus, the end plates are comparably thick.

12.3.2 PERIPHERAL DEVICES

In addition to the fuel cell itself, a fuel cell system requires several additional peripheral process units. These are mainly as follows: a desulfurization unit for the feed gas, an evaporator for the steam necessary for the reforming process, an upstream pre-reformer, a combustion unit between the anode exhaust and the cathode inlet, and eventually one or more heat exchangers to recover the thermal energy of the exhaust stream.

Especially for fuel gases from fermentation processes, sulfur plays an important role. For an MCFC—as for any other fuel cell, too—sulfur is a strong catalyst poison and has to be removed from the feed gas. Depending on the fuel gas in the respective application, this unit has to fulfill different requirements. Processed natural gas contains a well-defined, constant portion of sulfuric

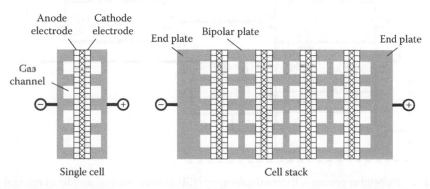

FIGURE 12.2 Fuel cell stack, bipolar plates, and end plates.

components. In this case, active coal filters are applied, but this choice is not optimal due to the high replacement frequency and its costs. As an alternative an adsorption bed with zinc oxide under cyclic operation is being discussed. It transforms the sulfur containing components to hydrosulfide and then adsorbs it during the first cycle period. Once the catalyst bed is loaded to a certain extent, the feed gas is switched to a second identical adsorption bed in parallel and air is fed into the first one. With this, the sulfur is oxidized and removed from the bed as sulfur dioxide, leaving a cleaned adsorption bed behind.

Gases containing larger portions of sulfur require additional measures. Bio gas, for example, often contains significant amounts of sulfur with frequently changing concentrations and varying molecular composition. Here, biological anaerobic processes have been proposed, which produce elementary sulfur. The advantage of this concept is that the sulfur is not emitted as sulfur oxide, but it is available in a pure and less toxic form. On the other hand, these processes cannot clean the gas sufficiently for application in MCFC. Consequently, a combination of biological and adsorptive desulfurization methods seems advisable here [5–7].

In addition to hydrocarbons or carbon monoxide, steam is an important component in the feed gas. The amount of water in the feed gas is usually described by the so-called steam to carbon ratio, S/C. In most applications, this ratio is between 2 and 3. Mixtures with significantly less steam tend to reversibly deposit carbon on the surfaces of the pipes and electrodes. This leads to the blocking of the anode electrode and the reforming catalyst in the fuel cell and quickly decreases the performance of the system [8–12]. On the other hand, too high steam dosage only dilutes the feed gas and leads to decreased system efficiency, especially because of the energy costs of the water evaporation. Prior to usage in the MCFC, the water has to be cleaned. Reverse osmosis is a preferred principle here to remove most of the undesired ions from the water, which could possibly act aggressively toward the piping, the catalysts and the electrolyte at the high temperature inside the MCFC. Furthermore, an evaporator is obviously required in state-of-the-art MCFC systems.

While in principle the reforming process can take place exclusively inside the anode channel—as depicted in Figure 12.1—a part of it usually is relocated to an external, separate reactor outside the cell. This concept, which is common use with low temperature fuel cells, is known as external reforming (ER) or pre-reforming (PR). Another concept is the indirect internal reforming (IIR), in which the reforming takes place inside a stack or thermally directly attached to it, but not within the cells themselves. Consequently, the concept of performing the reforming process inside the anode channel is called direct internal reforming (DIR). All three concepts are shown schematically in Figure 12.3 [13–21].

The advantage of the ER is that the design of the reactor and its operating parameters are largely independent from the fuel cell itself, giving several degrees of freedom with respect to geometry,

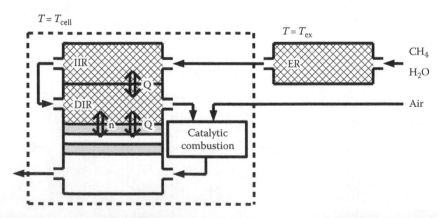

FIGURE 12.3 Reforming concepts. External reforming (ER), indirect internal and direct internal reforming (IIR and DIR).

heat management, fuel flexibility and control. On the other hand, as already mentioned, the reforming process requires heat and high temperature, which in the case of ER have to be provided from the outside of the reactor. This can either be done by using the heat of the exhaust gas, accepting a limit process temperature equal to that of the cells exhaust gas, or by direct oxidation of a portion of the fuel with air inside or outside of the reforming reactor. The choice of direct oxidation not only contradicts the fuel cell principle of turning reaction enthalpy into electric energy, but it also tends to lower the overall electric efficiency of the system. Furthermore, the ER is subject to the chemical equilibrium of the reforming process, which at an operating temperature of 400°C is at about 50% conversion, and for 800°C approaches full conversion, depending on the feed gas composition and pressure. The application of the ER is still advisable in MCFC systems for two reasons. First, it increases the fuel flexibility. If the feed gas contains short-chained hydrocarbons like ethane or propane, these have to be converted to methane before being reformed to hydrogen. This cracking of carbon–carbon bonds is significantly slower than the steam reforming process and thus requires larger amounts of catalysts. These can be provided by a larger external reforming reactor. Thus the external reformer allows adopting the MCFC system to the specific fuel of each individual application without the need to manipulate the design parameters of the fuel cell itself. Secondly, the external reformer serves as a kind of warning system in applications with strongly alternating sulfur content. In the case of a sulfur breakthrough, the external reformer is the first unit to be poisoned by sulfur. This can easily be detected by a set of thermocouples which show the position of the reaction zone moving through the catalyst bed. In this case, a shut down or the fix of the desulfurization unit can save the fuel cell from harm. Even in the worst case, that is, if the external reforming unit is irreversibly damaged, it is cheaper to replace than an irreversibly damaged fuel cell stack. Thus, application of the ER concept is advisable especially in systems with an unreliable desulfurization unit or with strongly alternating sulfur content.

The IIR concept is based on the idea to provide the required heat for the reforming process by direct coupling to the heat releasing electrochemical reactions in the fuel cells. This can be realized either by a separate reactor inside the hot housing of the cell stack or by inserting a reforming reactor between several fuel cells. This promotes the heat exchange toward the reforming process, so it takes place at about cell temperature.

The highest degree of process integration is the DIR concept, in which the reforming process is located directly inside the anode channel. This not only leads to an intense energetic exchange between heat releasing electrochemical reactions and the endothermic reforming process, but it also includes the mass coupling of the production of hydrogen and carbon monoxide and their consumption at the anode electrode. This direct consumption of the reforming products accelerates the reforming process and shifts its chemical equilibrium toward an extent of conversion to almost 100% even at lower temperature of about 600°C.

As already mentioned in the technical introduction, the anode exhaust and the cathode inlet are coupled via a combustion device. In the MCFC, a catalytic unit is preferred, because it allows oxidizing diluted gas/air mixture over the complete cell stack without a pilot light. In addition, it helps to avoid extreme temperatures and thus suppresses the formation of nitrogen oxides, NO_x. In fact, nitrogen oxide concentration in the exhaust gas is below the detection limit.

12.3.3 ACTUAL SYSTEM DESIGN

MCFCs are developed by different companies around the world. Although their products are working on the same basic principle, they differ in system design, size, and intended applications. In the following, the technical solutions of the most prominent developers are discussed.

12.3.3.1 MTU Onsite Energy

In Germany, the MTU Onsite Energy company has developed the so-called Hotmodule MCFC. MTU and Fuel Cell Energy have been partners to a technology and supply exchange contract.

Development started in 1990, and the functionality of the concept was first publicly demonstrated in 1997 at a prototype plant in Dorsten, Germany [22]. This was followed by a series of more than 30 demonstration plants, which where installed in various applications in Germany, Europe and other parts of the world [23,24]. Exemplary applications are power supply in telecommunications, combined heat and power supply in hospitals and a university and combined steam and power supply at a tire manufacturer. Currently, preparations for a series production of the Hotmodule are ongoing.

The Hotmodule is a complete MCFC system with a nominal power of 250 kW. The stack consists of 343 cells, in which the anode and cathode gas channels are arranged in a cross-flow design. The stack is in horizontal position so that the cells are standing upright, with the anode channels running from the bottom to the top of the stack and the cathode channels going from one side to the other. The stack is located in a cylindrical vessel in which all the hot compartments of the system are located, thus the name Hotmodule (Figure 12.4).

As shown in the flow scheme (Figure 12.5), the Hotmodule includes all three reforming concepts. The IIR is realized by inserting a flat reforming reactor after each package of eight

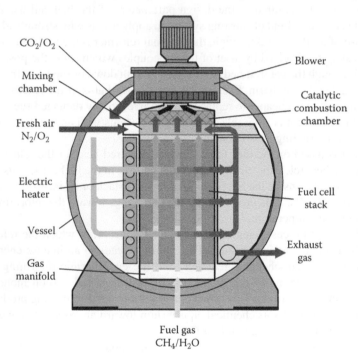

FIGURE 12.4   Cross section of the Hotmodule.

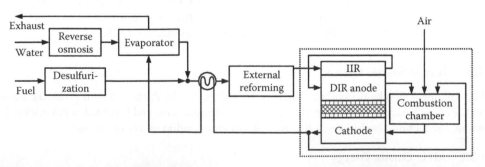

FIGURE 12.5   Flow scheme of the Hotmodule. The dotted line indicates the cylindrical vessel inside which the hot compartments are located.

fuel cells. Before the external reformer, the feed gas is heated up and mixed with steam in a combined heat exchanger and humidifier.

The Hotmodule has all the general advantages of the MCFC concept and combines them with a simple, efficient design. The cylindrical housing of the rectangular cell stack automatically segments the surrounding space into four compartments. One of them contains the gas manifold, which distributes the pre-reformed feed gas to the anode channels, and another one contains the combustion chamber. The other two volumes connect the combustion chamber with the cathode inlet, respectively realize the splitting of the cathode exhaust gas into the cathode recycle stream and the exhaust gas stream. The fact that the cell stack is completely surrounded by hot gases reduces the mechanical stress due to thermal gradients. As the stack is mounted on rails, it can be replaced by simply opening the vessel front cover, pulling the old stack out and moving the new one in. A complete replacement procedure takes about 3–4 days. This is mostly due to the long time required to heat up the new stack and cool down the old one.

The latest development by MTU is the HM400. It applies improved electrodes and contains more fuel cells than the first generation of Hotmodules. Due to this, it has a nominal electrical power output of 345 kW.

12.3.3.2 Ansaldo Fuel Cells

In Trieste, Italy, Ansaldo Fuel Cells has successfully demonstrated the feasibility of a 100 kW MCFC system and is working on their "2TW" model, a 500 kW system including four stacks [25–27]. In contrast to the Hotmodule by MTU, the 2TW system is operated at elevated pressure of several 3–5 bars. The compressor for the feed gas is coupled with a turbine in the exhaust gas stream, and the energy balance of these two is positive, meaning that the 2TW provides electrical power not only from the cell stack but also from the combined turbine/compressor.

Another remarkable characteristic trait of the 2TW is the focus on a single reforming unit (Figure 12.6). This unit is combined with the combustion chamber to provide the energy required by the endothermic reforming process and is located inside the containment vessel of the hot system parts. The second advantage of this combination is that high operating temperatures and consequently high degrees of conversion can be obtained in the reformer. Although this reforming unit is located in the hot cell housing, it must be considered as an ER, because its temperature is dominated by the combustion temperature instead of the cell temperature. The air is inserted at the cathode inlet, so it is heated up by the hot combustion exhaust gas and a high oxygen concentration is obtained at the cathode electrode. In the Hotmodule, the cool air is fed into the combustion chamber, which reduces the temperatures inside the chamber, but in the 2TW system, a high temperature in the combustion and consequently in the reforming unit is favorable.

The MCFC systems by Ansaldo are also realized in a cylindrical vessel, which is advantageous for pressurized operating conditions.

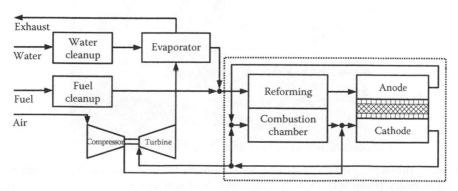

FIGURE 12.6   Flow scheme of the MCFC system by Ansaldo.

12.3.3.3 Ishikawajima-Harima Heavy Industries

In Japan, MCFC development is conducted by a consortium of companies led by Ishikawajima-Harima Heavy Industries [28,29]. The MCFC system they are focusing on is a 300 kW stack with similar features as the 2TW by Ansaldo. It is operated under pressurized conditions and only contains a single ER unit combined with the combustion chamber. An important addition in this system is the water recovery system, which by condensation of the steam in the exhaust gas provides the water required for the reforming process, so under regular operating conditions, no external water supply and cleanup is required. A first plant has been demonstrated at the Kawagoe test site and plans also include the development and demonstration of a plant of the size of several MW.

12.3.4 FUEL CELL ENERGY

Fuel Cell Energy, located in Danbury, Connecticut, has been one of the first suppliers of MCFC components and complete systems [30–32]. Their system applies direct internal reforming and an elevated pressure level. During the last few years, FCE has installed a large number of systems in a wide variety of applications worldwide. They offer MCFC systems with 300 kW, 1.2 and 2.4 MW electrical power output. The larger systems are created by combination of several cell stacks in a larger housing. In addition, a combined system with a fuel cell stack and turbine is available [33].

12.3.5 ACTUAL APPLICATIONS

MCFC fuel cell systems have proven their potential in a large number of trial plants in quite different applications. They demonstrate the full functionality of the fuel cell and its superior electric system efficiency in comparison with classical units of the same size. The efficiency reaches up to 48%, depending on the system size and the individual application.

One typical application is the implementation of an MCFC system in a power plant delivering electricity and heat to a larger building complex, as is the case at the university hospital in Magdeburg, Germany. The specific Hotmodule system in Magdeburg is fed with natural gas and it is combined with a tube-bundle heat exchanger utilizing the exhaust gas for the heat generators and cold chillers of the plant (Figure 12.7). While the system has an overall electric efficiency of 47%, its combined electric and thermal efficiency is at about 70%.

FIGURE 12.7 Hotmodule MCFC system made by MTU in the IPF power plant at the university hospital in Magdeburg, Germany. (From MTU Onsite Energy Gmbh. With permission.)

A similar MCFC system operated with natural gas is installed at a technical center of a telecommunication provider in Munich, Germany. The highly complex electronic devices in telecommunication require direct current, which is produced by the fuel cell, so AC/DC converters can be omitted and their energetic losses can be avoided. Secondly, the MCFC serves as an on-site uninterruptible power supply, which helps to avoid costly power failures in critical electronic systems. Finally, the heat produced by the Hotmodule is used to operate absorption chillers for the air-conditioning of the offices and technical rooms of the facility. In this application, the full range of the potential products of the MCFC is unfolded.

During the last few years, numerous MCFC systems have been installed based on renewable fuel sources. One example is the Hotmodule plant at a waste treatment plant in Leonberg near Stuttgart, Germany. Domestic organic waste is converted to a mixture of methane and carbon monoxide in a digester tower. This gas is then used in a Hotmodule to produce electricity, which is supplied to the public grid. The off-heat from the system is sufficient to heat the digester tower and to dry the digestate. Another example is the application of a system from FCE at a wastewater treatment plant in Southern California. Biological components in the wastewater are decomposed by microorganisms in an anaerobic digester, producing a gas mixture rich of methane. This gas is supplied to an FCE's system to produce electricity and heat. Experiences with these systems regarding reliability and performance are very good.

12.4 FUTURE CHALLENGES AND DEVELOPMENTS

Application of MCFC systems with conventional fuels such as natural gas and diverse renewable sources of fuel gas have been successfully demonstrated during the last years. It has opened new opportunities of utilization of renewable energy sources, which could not be exploited with classical plants so far.

Three main issues have been identified in the actual development. The life-time of the fuel cell stacks is still not sufficient. Although MCFC systems have reached 30,000 h of operating time before shut down, this lifespan needs to be increased further. This basically demands better control of the temperature distribution inside the fuel cell stacks, which can be achieved by optimization of system design and operating conditions.

Simultaneously, production costs need to be further decreased. It is expected that relative investment costs (price related to installed nominal power) will remain slightly higher than those of classical units. However, this will be compensated by the higher system efficiency which leads to lower fuel consumption and therefore lower operating costs. Especially with increasing fuel prices, fuel efficiency and fuel costs will become increasingly important. Investment costs will certainly decrease with increasing production rates of MCFC systems, as this makes fully or semi-automated production economically attractive. However, this is a classical "hen and egg" dilemma: Production rates will increase once the prices will go down.

A third important aspect is the scale-up of systems to the megawatt class, which will open new opportunities on the market. Currently, FCE creates systems in this class by combination of several fuel stacks ("numbering up"). MTU is increasing the power output of their system by applying more cells in a stack, which effectively is also a kind of "numbering up." This allows increasing the system's power by a certain factor, but this factor is limited, and it does not make full use of the economy of scale. In the long term, fuel cell designs with considerably larger electrode areas are going to be needed in order to provide cheap systems with a power output of several megawatts.

It is reasonable to assume that these issues can be solved within the next few years. With that, MCFC will not only be applicable in niche markets as is the situation today, but they will be serious competitors to traditional energy conversion apparatuses, offering higher efficiency and lower emissions at affordable prices.

REFERENCES

1. Kordesch K. and G. Simader. 1996. *Fuel Cells and Their Applications*, 1st edn., Weinheim, Germany: Wiley VCH.
2. Larminie J. and A. Dicks. 2003. *Fuel Cell Systems Explained*, 2nd edn., Chichester, West Sussex, U.K.: John Wiley & Sons.
3. Vielstich W., A. Lamm, and H. Gasteiger. 2003. *Handbook of Fuel Cells: Fundamentals, Technology, Applications*, Chichester, West Sussex, U.K.: John Wiley & Sons.
4. Carrette L., K.A. Friedrich, and U. Stimming. 2001. Fuel cells—Fundamentals and applications, *Fuel Cells* 1(1):5.
5. Shennan J.L. 1996. Microbial attack on sulphur-containing hydrocarbons: Implications for the biodesulphurisation of oils and coals, *J. Chem. Tech. Biotechnol.* 67:109–123.
6. Maxwell S. and J. Yu. 2000. Selective desulphurisation of dibenzothiophene by a soil bacterium: Microbial DBT desulphurisation, *Proc. Biochem.* 35:551–556.
7. Bagreev A., S. Katikaneni, S. Parab, and T.J. Bandosz. 2005. Desulfurization of digester gas: Prediction of activated carbon bed performance at low concentration of hydrogen sulfide, *Catal. Today* 99:329–337.
8. Trimm D.L. 1999. Catalysts for the control of coking during steam reforming, *Catal. Today* 49:3–10.
9. Tomishige K., Y.G. Chen, and K. Fujimoto. 1999. Studies on carbon deposition in CO_2 reforming of CH_4 over nickel-magnesia solid solution catalysts, *J. Catal.* 181:91–103.
10. Swaan H.M., V.C.H. Kroll, G.A. Martin, and C. Mirodatos. 1994. Deactivation of supported nickel-catalysts during the reforming of methane by carbon-dioxide, *Catal. Today* 21:571–578.
11. Sperle T., D. Chen, R. Lodeng, and A. Holmen. 2005. Pre-reforming of natural gas on a Ni catalyst—Criteria for carbon free operation, *Appl. Cat. A* 282:195–204.
12. Shamsi A. 2004. Carbon formation on Ni-MgO catalyst during reaction of methane in the presence of CO_2 and CO, *Appl. Cat A* 277:23–30.
13. Dicks, A.L. 1996. Hydrogen generation from natural gas for the fuel cell systems of tomorrow, *J. Power Sources* 61:113–124.
14. Rostrup-Nielsen J.R. and L.J. Christiansen. 1995. Internal steam reforming in fuel-cells and alkali poisoning, *Appl. Cat A* 126:381–390.
15. Clarke S.H., A.L. Dicks, K. Pointon, T.A. Smith, and A. Swann. 1997. Catalytic aspects of the steam reforming of hydrocarbons in internal reforming fuel cells, *Catal. Today* 38:411–423.
16. Freni S. 2001. Rh based catalysts for indirect internal reforming ethanol applications in molten carbonate fuel cells, *J. Power Sources* 94:14–19.
17. Berger R.J., E.B.M. Doesburg, J.G. van Ommen, and J.R.H. Ross. 1996. Nickel catalysts for internal reforming in molten carbonate fuel cells, *Appl. Cat. A* 143:343–365.
18. Bradford M.C.J. and M.A. Vannice. 1996. Catalytic reforming of methane with carbon dioxide over nickel catalysts. 1: Catalyst characterization and activity, *Appl. Cat. A* 142:73–96.
19. Effendi A., Z.G. Zhang, K. Hellgardt, K. Honda, and T. Yoshida. 2002. Steam reforming of a clean model biogas over Ni/Al_2O_3 in fluidised- and fixed-bed reactors, *Catal. Today* 77:181–189.
20. Marquevich M., R. Coll, and D. Montane. 2000. Steam reforming of sunflower oil for hydrogen production, *Ind. Eng. Chem. Res.* 39:2140–2147.
21. Comas J., F. Marino, M. Laborde, and N. Amadeo. 2004. Bio-ethanol steam reforming on Ni/Al_2O_3 catalyst, *Chem. Eng. J.* 98:61–68.
22. Bischoff M. and G. Huppman. 2002. Operating experience with a 250 kW(el) molten carbonate fuel cell (MCFC) power plant, *J. Power Sources* 105:216–221.
23. Bischoff M. 2006. Molten carbonate fuel cells: A high temperature fuel cell on the edge of commercialization, *J. Power Sources* 160:842–845.
24. http://www.mtu-online.de (accessed on September 2010).
25. De Simon G., F. Parodi, M. Fermeglia, and R. Taccani. 2003. Simulation of process for electrical energy production based on molten carbonate fuel cells, *J. Power Sources* 115:210–218.
26. Bosio B., P. Costamagna, F. Parodi, and B. Passalacqua. 1998. Industrial experience on the development of the molten carbonate fuel cell technology, *J. Power Sources* 74:175–187.
27. http://www.ansaldofuelcells.com (accessed on September 2010).
28. Morita H., M. Komoda, Y. Mugikura, Y. Izaki, T. Watanabe, Y. Masuda, and T. Matsuyama. 2002. Performance analysis of molten carbonate fuel cell using a Li/Na electrolyte, *J. Power Sources* 112:509–518.

29. http//www.ihi.co.jp/index-e.html
30. Doyon J., M. Farooque, and H. Maru. 2003, The direct fuel cell™ stack engineering, *J. Power Sources* 118:8–13.
31. Lukas M.D., K.Y. Lee, and H. Ghezel-Ayagh. 2002. Modelling and cycling control of carbonate fuel cell power plants, *Control Eng. Pract.* 10:197–206.
32. http://www.fce.com (accessed on September 2010).
33. Ghezel-Ayagh H., J. Walzak, D. Patel, J. Daly, H. Maru, R. Sanderson, and W. Livingood, 2005. State of direct fuel cell/turbine systems development, *J. Power Sources* 152:219–225.

13 Solid Oxide Fuel Cells and Electrolyzers for Renewable Energy

Michael D. Gross and Raymond J. Gorte

CONTENTS

13.1 INTRODUCTION

Fuel cells and electrolyzers are simply devices for converting between chemical and electrical energy. As such, they are not an essential part of renewable energy and can be used even when the primary energy source is a conventional fossil fuel, such as petroleum or coal. Fuel cells are often included with renewable energy technologies because they exhibit high efficiencies, having the potential to greatly reduce the amount of fuel that must be consumed in order to generate a given amount of electrical power. Their high efficiency relative to heat engines comes from the fact that they produce electricity directly, with no moving parts and no Carnot-cycle limitations. They are also intrinsically "clean" in that even high temperature fuel cells operate well below the conditions where NO_x is generated. In addition, the exhaust stream can be scrubbed of CO_2 easily because it is not diluted with large quantities of N_2. Finally, fuel cells allow flexible conversion of chemicals to energy and back. The same device can operate as a fuel cell or an electrolyzer by simply reversing the direction of the current.

There are a number of different fuel-cell types and they are usually classified by the type of electrolyte that is employed [1]. The types that are most discussed in the popular literature (e.g., phosphoric acid fuel cells [PAFC], proton-exchange membrane fuel cells [PEMFC]) are based on proton-conducting electrolytes that operate at low temperatures (below 473 K). However, this chapter will focus on solid oxide fuel cells (SOFC) and solid oxide electrolyzers (SOE), which are all solid-state devices that operate at high temperatures (e.g., 773–1273 K) and use an electrolyte that is a ceramic material.

The high operating temperatures and solid-state nature offer several important advantages: (1) Electrode "overpotentials" (i.e., the energy losses in the electrodes, which are primarily caused by sluggishness in the electrochemical reactions) are much lower at high temperatures; indeed, it is common to operate SOFC just a few hundred millivolts below the Nernst potential. (2) High temperatures significantly decrease the Nernst potentials for SOE [2], decreasing the electrical energy required for electrolysis. (3) Because SOE are all solid state, it is possible to have a pressure difference across the electrolyte and to produce H_2 at that higher pressure. We will elaborate on these points later.

While there are serious materials issues that must be solved with SOFC/SOE because of the high operating temperatures, the high operating temperatures do not affect the system efficiency. With all fuel cells, the chemical energy of the fuel that does not go into electrical power is given up as heat. Removal of this heat can be a problem for low temperature fuel cells. With SOFC, it can be controlled by flowing excess air at the cathode side.

Although proton-conducting ceramics exist [3,4], SOFC and SOE are usually based on ceramic electrolytes that are oxygen-ion conductors, most commonly yttria-stabilized zirconia (YSZ) or Gd- or Sm-doped ceria (SDC and GDC). In addition to the fact that these materials have high ionic conductivities and reasonable mechanical stability, oxygen-ion conductors have a number of important advantages as electrolytes. First and foremost, SOFC based on oxygen-ion conductors are able, in principle, to operate on any combustible fuel [5]; and their SOE counterparts are capable of converting CO_2 to CO almost as easily as they convert H_2O to H_2 [6–8]. The primary disadvantage of SOFC based on oxygen-ion conductors is that the products of fuel conversion (e.g., H_2O and CO_2) are produced in the fuel stream, causing dilution of the fuel and making it difficult to achieve fuel utilization above 90%. With SOFC based on protonic conductors and operating on H_2, the H_2O is produced at the air electrode.

Here, we will review the basics behind SOFC and SOE operation and their applications to renewable energy. While thorough reviews of the SOFC literature can be found elsewhere and such a review is beyond the scope of this chapter [1,5,9,10], we will focus on those aspects of SOFC that are most applicable to the use of biofuels.

13.2　BASIC FUEL CELL PRINCIPLES

13.2.1　Principles

A schematic of an SOFC operating on H_2 is shown in Figure 13.1. As with any fuel cell, the three main components of an SOFC are the cathode, the electrolyte, and the anode. O_2 from the air supplied to the cathode is catalytically reduced to oxygen anions, O^{2-}, via the half-cell reaction, Equation 13.1:

$$\tfrac{1}{2}O_2 + 2e^- \rightarrow O^{2-}. \tag{13.1}$$

The cathode must maintain good electronic conductivity, in air at high temperatures, in order to provide electrons from the external circuit for O_2 reduction. The oxygen ions are then transported through the electrolyte membrane, to the anode, where they react with H_2 to produce H_2O and electrons according to the half-cell reaction, Equation 13.2:

$$O^{2-} + H_2 \rightarrow H_2O + 2e^-. \tag{13.2}$$

The electrons produced at the anode are at a higher potential than those consumed at the cathode and are therefore able to do work in the external circuit.

Anode reaction:

$$H_2 + O^{2-} \rightarrow H_2O + 2e^-$$

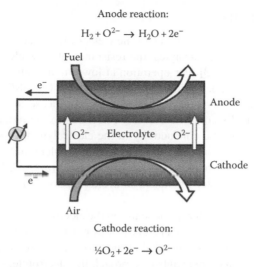

FIGURE 13.1 Schematic of SOFC operating on H_2 fuel.

Cathode reaction:

$$\tfrac{1}{2}O_2 + 2e^- \rightarrow O^{2-}$$

The driving force for migration of oxygen ions across the electrolyte membrane is the difference in oxygen fugacities at the cathode and anode. At the cathode, the oxygen fugacity is equal to the partial pressure of O_2; at the anode, the fugacity is established by the O_2 partial pressure that would be in equilibrium with the H_2 and H_2O present in the anode compartment. The reversible work for an ion migrating across the electrolyte is equal to ΔG_{rxn}° (=RT ln{$P(O_{2,anode})$/$P(O_{2,cathode})$}), which becomes the Nernst Equation, Equation 13.3, when converted to electrical units:

$$V_{Nernst} = V^\circ + \frac{RT}{nF} \ln\left(\frac{P_{H_2,anode} \cdot P_{O_2,cathode}{}^{1/2}}{P_{H_2O,anode}} \right). \tag{13.3}$$

The anode oxygen fugacity is established by the ratio of $P(H_2, anode)$ and $P(H_2O, anode)$ as shown in Equation 13.3.

In this equation, V° is the equilibrium potential at standard pressure and some specified operating temperature and is related to ΔG_{rxn}° through Equation 13.4:

$$V^\circ = -\frac{\Delta G_{rxn}^\circ}{nF}. \tag{13.4}$$

where
   n is the number of electrons involved in each half-cell reaction
   F is Faraday's constant, the number of coulombs in a mole of electrons

The equations describing operation of a fuel cell on hydrocarbon fuels are similar to those mentioned earlier, except that oxygen fugacity would be established by oxidation of hydrocarbons in Equation 13.2, rather than H_2. The appropriate equations have been presented elsewhere [11].

Equations 13.1 through 13.4 apply equally well to electrolyzers as to fuel cells, with the only difference being the direction of the arrows in the half-cell reactions. Also, because the definition of a cathode is the electrode where reduction takes place and the anode is where oxidation takes place, the "air" electrode where O_2 is produced becomes the anode and the "fuel" electrode where H_2 is produced becomes the cathode in an SOE. Therefore, when working with reversible cells that cycle between SOFC and SOE operation, it is sometimes easier to refer to the electrodes as the "air" and "fuel" electrodes.

13.2.2 Standard Materials

Although there have been extensive searches for new electrolyte materials, YSZ remains the standard. When using thin (<15 μm) electrolytes, the resistance of YSZ electrolytes gives rise to negligible losses down to 973 K [12,13]. For operation at lower temperatures, the doped cerias (GDC and SDC), Sc-doped ZrO_2 (SDZ), and $La_{0.8}Sr_{0.2}Ga_{0.8}Mg_{0.2}O_{3-\delta}$ (LSGM) are preferred due to their higher ionic conductivities. However, each of these materials has serious limitations as well. GDC and SDC become electronically conductive at low $P(O_2)$ due to the reducibility of ceria. This is a serious problem because electronic conductivity in the electrolyte allows consumption of the fuel without generating power. SDZ does not suffer from this problem and has mechanical properties even better than that of YSZ, but scandium is rare and expensive. LSGM is being considered by at least one developer but requires steps be taken to avoid reaction with some of the more common electrode materials [14,15].

While the choice of electrode materials depends on the electrolyte, the basic requirements are the same in all cases and are as follows: (1) The electrodes must be electronically conductive in their environment so that electrons can be provided at the cathode and removed at the anode. (2) They must be porous to allow gas-phase reactants to approach the electrolyte. (3) They must catalyze the two half-cell reactions. While not essential, it is also helpful if the electrode has ionic conductivity. Ionic conductivity increases the electrochemically active region in the electrode by allowing transport of ions some depth into and out of the electrode from the electrolyte interface.

The traditional fuel electrode is a porous composite of Ni and the electrolyte (e.g., YSZ). Ni has good electronic conductivity and is an excellent oxidation catalyst. Ni is also a good reforming catalyst, allowing H_2 to be generated within the anode compartment of an SOFC by co-feeding of steam and natural gas [16,17]. The electrolyte phase within the composite electrode provides ionic conductivity, prevents the Ni from sintering into a dense layer at high temperatures, and also helps modify the coefficient of thermal expansion (CTE) so that there is a better match with the electrolyte phase. The main problems with Ni-based electrodes are that they cannot tolerate re-oxidation [18] and they are severely poisoned by small amounts of sulfur in the fuel [19]. When using hydrocarbon fuels, caution must be exercised to prevent formation of carbon fibers that lead to loss of Ni by metal dusting [20,21] and to mechanical failure [22,23].

For the air electrode, the primary challenge is in maintaining electronic conductivity under strongly oxidizing conditions at high temperatures. Precious metals are too expensive and coinage metals are readily oxidized. Therefore, the air electrodes are usually made from conducting oxides, with the perovskite Sr-doped $LaMnO_3$ (LSM) being the most common. Because LSM has very low ionic conductivity [24], it is again common to form a composite of LSM with the electrolyte material. More recently, other perovskites, especially Sr-doped $LaCo_{0.2}Fe_{0.8}O_3$ (LSCF), have been gaining in popularity because they have better electrochemical performance [25–27]. However, because many of these alternative perovskites are more reactive with YSZ than is LSM, it is necessary to incorporate barrier layers between the conducting oxide and the YSZ electrolyte in order to take advantage of their improved properties [28–32].

It should be recognized that there are important materials issues associated with the system surrounding the cells as well. Since real fuel-cell and electrolyzer systems require that cells be connected in series to produce a stack, interconnect materials are required to make electrical connection between the anode of one cell and the cathode of the next. What makes this a very difficult problem is that the interconnect material must provide electronic conductivity under both oxidizing *and* reducing conditions. The traditional material for this application is doped $LaCrO_3$, which is expensive and has relatively low conductivity. With lower operating temperatures, it has more recently been possible to use ferritic stainless steel [13]. The other major materials issue involves seals. Because the questions concerning materials for the seals and the interconnects are coupled together with stack design, the interested reader is referred to other reviews for more information on these subjects [33–36].

13.2.3 THREE-PHASE BOUNDARY

While we have hinted at the concept of a three-phase boundary (TPB) line in the previous section, it is so important for understanding electrochemical reactions in SOFC and SOE electrodes that it merits further discussion. Considering the SOFC cathode reaction in Equation 13.1, there are three distinct species that must be present in the same proximity: gas-phase O_2, electrons, and oxygen anions. The sites where these three species can coexist are referred to as TPB sites. In order for gas-phase O_2 to be present, the pores of the electrode must maintain a continuous path to allow diffusion of O_2 from the main cathode compartment to these sites. In order that electrons can be transferred to these sites from the external circuit, there must be a continuous path of the electronic conductor to these sites. Finally, in order that ions can be transferred to the electrolyte, there must be a continuous path in the ion conducting component of the electrode from these sites to the electrolyte interface. In the absence of mixed conductivity in the electrode material, reaction can only occur at the sites where all three phases are in contact. This contact requires that all three phases "percolate" the structure. Because the percolation threshold for random structures is approximately 30 vol%, developing the correct electrode structure is an important challenge.

In principle, the perovskites used in SOFC cathodes can have mixed electronic and ionic conductivity; however, this is not the case for the standard material, LSM. While LSM has good electronic conductivity at 800°C and atmospheric conditions, ~200$S \cdot cm^{-1}$ (1 S = 1 Ohm^{-1}), its ionic conductivity has been reported to be less than 4×10^{-8} $S \cdot cm^{-1}$ [24]. (For comparison, YSZ has an ionic conductivity of 0.043$S \cdot cm^{-1}$ and negligible electronic conductivity at 800°C [37].) The standard material for SOFC anodes, Ni, is also not a mixed conductor. Therefore much of the effort in electrode development has gone into fabricating structures with a long TPB.

13.2.4 PERFORMANCE CHARACTERISTICS

In ideal fuel cells, the electrons would be produced at the Nernst potential, given in Equation 13.3; and electrolyzers would only require the electrons to have the potential given by the Nernst Equation. In real fuel cells and electrolyzers, there are irreversible voltage losses associated with processes that occur in the electrolyte and electrodes. These losses generate heat, some of which may be recovered and used for other purposes; but it is clearly preferable to minimize these losses.

Since the electrolyte can be treated as a simple resistor, with area specific resistance equal to R_E, the potential energy losses in the electrolyte are simply equal to iR_E. R_E is inversely proportional to the electrolyte thickness, so that losses in the electrolyte can be reduced by fabricating thinner electrolyte membranes. By using one of the electrodes for mechanical support, electrolyte layers are routinely less than 10 μm in thickness. It is straightforward to calculate R_E and the electrolyte losses because the ionic conductivities of most common materials are well documented [12].

Voltage losses in the electrodes are linked to the same factors that were discussed in Section 13.2.2. Specifically, losses occur due to (1) transport of electrons, (2) mass transport of the gaseous reactants, (3) reaction kinetics, and (4) ionic conduction within the electrode. In general, the materials used for the electrodes have very high electronic conductivity, so that the contribution from electronic resistance is negligible. Losses due to mass transport of gaseous species, also referred to as concentration polarization, occur when the concentration in the main electrode compartment differs from that which exists at the electrolyte interface due to diffusion through the pores. The effective Nernst potential, obtained from the concentrations at the electrolyte interface, will then be lower than the real Nernst potential, calculated from the concentrations in the electrode compartments. This problem is most serious when the concentration of the reactant is low. For example, at high conversions of H_2 in an SOFC, the H_2 is diluted in steam. A gradient in the H_2 concentration must exist through the porous electrode in order to generate a mass flux equivalent to that required to generate a given current density. This gradient can in turn cause the H_2:H_2O ratio at the electrolyte interface to be quite different from that used to calculate the Nernst potential in Equation 13.3.

Limitations due to reaction kinetics, also called activation polarization, are coupled together with ionic conductivity and are less well understood. These losses are typically modeled using Butler–Volmer kinetics, a simple exponential relationship between current density and voltage. The Butler–Volmer expression is derived by assuming that potential gradients at the reaction site lower the energy barrier for reaction. Although this exponential dependence between voltage and current density is routinely observed in low temperature fuel cells, the relationship between current density and voltage is often observed to be linear in SOFC over a wide range of current densities. Furthermore, while large potential gradients, on the order of $1\,V \cdot nm^{-1}$, are observed in the Helmholtz layer in solution-phase electrodes [38], it is not clear that such large gradients exist in SOFC or SOE electrodes, particularly when using mixed-conducting materials.

For those cases where the electrochemical oxidation and reduction reactions depend primarily on temperature, the electrode losses may increase linearly with current density and be described by impedances, R_A and R_C. In this case, the cell potential will be given by Equation 13.5:

$$V = V_{Nernst} - i(R_E + R_A + R_C). \tag{13.5}$$

Such a linear relationship is observed in Figure 13.2a, which is a plot of the cell potential as a function of current density for operation under both fuel-cell and electrolyzer conditions. In this plot, positive currents indicate operation as a fuel cell, and negative currents indicate operation as an electrolyzer. This example shows that the V–i relationship has a similar slope under both fuel-cell and electrolysis conditions, a result that is difficult to rationalize in terms of the Butler–Volmer picture.

Although R_A and R_C have been written as if they were resistances, they can be separated from R_E using transient-response techniques, such as impedance spectroscopy, because the rate processes that give rise to these impedances are time dependent. An example where impedance spectroscopy has been used to determine electrode and electrolyte losses is shown in Figure 13.2b. This Cole–Cole plot of the impedance spectrum, taken at open circuit, corresponds to the V–i curve in Figure 13.2a. The high-frequency intercept with the real axis at $0.42\,\Omega \cdot cm^2$ is equal to the ohmic losses in the cell and is close to the value expected for R_E based on the thickness and known conductivity of the electrolyte [37]. The length of the arc separating the zero-frequency and high-frequency intercepts on the real axis is equal to $R_A + R_C$, approximately $0.3\,\Omega \cdot cm^2$ in this example.

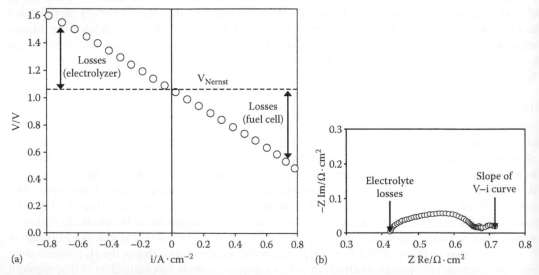

FIGURE 13.2 (a) V–i polarization curve of an SOFC operating at 973 K under both fuel cell and electrolyzer conditions and (b) corresponding Cole–Cole plots taken at open circuit voltage.

13.3 REVERSIBLE FUEL CELLS

A significant problem for many sources of renewable energy, such as solar and wind power, is the fact that these sources are not continuous. Energy storage is required in order to take full advantage of the opportunities that they present. An interesting approach to energy storage involves the use of reversible fuel cells/electrolyzers that would generate H_2 through electrolysis at times of peak power, then reverse the direction of the current to generate electricity when it is needed.

While presently available electrolyzer systems based on alkaline or polymer membranes could certainly be used for this application, solid-oxide systems would have a number of important advantages. First, unlike SOFC and SOE, fuel cells and electrolyzers based on alkaline and polymer electrolytes typically require high loadings of precious metals in order to reduce electrode losses. Second, stability and contamination are serious problems with both alkaline and polymer electrolytes. For example, electrolyzers based on alkaline electrolytes readily absorb CO_2 from the air to form carbonates. For fuel-cell operation, this CO_2 sensitivity implies that the air must be purified. Polymer membrane systems must use very pure de-ionized water or they will accumulate cations that displace protons and increase the cell resistance over time. Third, because of the higher operating temperatures, the theoretical potentials required for electrolysis in SOE are much lower. This results from the fact that the Nernst potential is temperature dependent due to the $T\Delta S$ term in ΔG_{rxn}. At typical SOE operating temperatures, V_{Nernst} is approximately 0.2 V lower than it would be at the normal boiling point of water. Fourth, again because of the higher operating temperatures, the actual electrode overpotentials are typically much smaller than that found with low-temperature fuel cells. Finally, technologies based on alkaline or polymer membranes are also not easily amenable to high pressure operation in which there is a differential pressure across the membrane, because of the fluid-like nature of the membrane. By comparison, it is possible to electrochemically pressurize H_2 in an SOE because the electrolyte is a solid. Increasing the overpotential by 0.1 V in an SOE is theoretically sufficient to increase the pressure of the H_2 that is produced by a factor of 10.

A nontrivial challenge in the use of reversible H_2–H_2O fuel cells is the need for H_2 storage. In this regard, it is important to notice that SOFC can, in principle, operate on any combustible fuel and that electrolysis in an SOE is not limited to the reduction of H_2O to H_2. For example, an SOFC stack showing stable performance while operating on CO has been demonstrated with nearly identical performance characteristics to operation in H_2 [8]. Others have shown that electrolysis of CO_2 can be accomplished with overpotentials similar to that found in H_2O electrolysis [6,7]. Data from one of these studies are given in Figure 13.3, which shows V–i polarization curves for a cell operating in CO–CO_2 mixtures having a range of compositions. The data show that high current densities, above $1 \text{ A} \cdot \text{cm}^{-2}$ (~400 sccm $CO_2 \cdot \text{cm}^{-2} \cdot \text{h}^{-1}$), are possible in a cell operating at 1.5 V. For comparison, typical operating voltages in low temperature electrolyzers would be greater than 2 V, even for H_2O [39]. Obviously, safety concerns would need to be addressed in the storage of CO for reversible fuel-cells operating in CO–CO_2 mixtures; however, the technical challenges of finding adsorbents for CO and CO_2 would be trivial compared to that for H_2.

It may also be possible to use redox cycles involving molten metals for reversible solid-oxide systems. For example, a recent study indicated that the overpotentials associated with oxidation of molten Sb to molten Sb_2O_3 at YSZ surfaces at 973 K were very low [40]. Therefore, it should be possible to flow molten Sb through an SOFC stack to produce power, converting the Sb to Sb_2O_3. During periods of excess electricity, the molten Sb_2O_3 would be reduced in the same stack. The advantage here would be that the volumes of the metallic fuel and the oxidized product would be very low, so that the overall systems could be very simple. The equivalent number of electrons in 1 m^3 of H_2 at STP would have a volume of Sb equal to 0.00027 m^3. Although the use of molten metals in fuel-cell systems has only recently received significant attention [41–45], their potential seems to justify additional work.

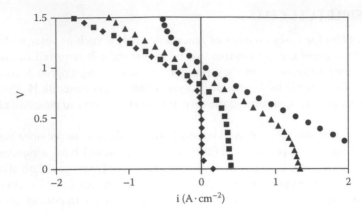

FIGURE 13.3 V–i polarization curves for mixtures of CO_2 and CO at 1073 K. Negative currents correspond to electrolysis of CO_2: 100% CO_2—0% CO (♦); 90% CO_2—10% CO (■); 50% CO_2—50% CO (▲); 10% CO_2—90% CO (●). The cell composition was as follows: 40 wt% LSF in YSZ|YSZ(65 μm)| 0.5 wt% Pd, 5 wt% $Ce_{0.48}Zr_{0.48}Y_{0.04}O_2$, and 45 wt% LSCM in YSZ. (Reproduced from Bidrawn, F. et al., *Electrochem. Solid-State Lett.*, 11, B167, 2008. With permission of The Electrochemical Society.)

13.4   OPERATION OF SOFC ON BIOFUELS

As discussed earlier, the primary advantage of using SOFC compared to other types of fuel cells is intrinsic fuel flexibility that results from high temperature operation and the fact that the mobile species are oxygen anions, rather than protons. Theoretically, one can fuel an SOFC with anything that is combustible. However, in practice, there can be serious problems with using biofuels.

There are obviously a range of different types of biofuels, and their application to powering SOFC will depend on the particular fuel that is being considered. First, conventional SOFC require that the fuel be vaporized, since the fuel must gain access to the TPB sites. (Later in this chapter, we will discuss some recent ideas for Direct Carbon Fuel Cells (DCFC) that may avoid this limitation.) Although any hydrocarbon can be gasified to syngas (a mixture of CO and H_2) by partial oxidation, this would not be the preferred route for extracting the energy of biomass. SOFC would have an advantage over low temperature fuel cells due to their high tolerance for CO, but syngas could be used by any type of fuel cell. Since conventional use of fossil fuels in SOFC involves reforming and operation of the fuel cell on syngas, we refer the interested reader to other reviews for this mode of operation [16,17]. In this section, we will focus on the use of biofuels that could be vaporized without reforming.

13.4.1   CARBON FOULING

A complication that is common for application of many biofuels in SOFC is carbon fouling, a problem that can occur with any hydrocarbon fuel [11,46], not just fuels derived from biomass. In general, there are two distinct mechanisms responsible for carbon fouling: (1) tar formation due to homogeneous reactions and (2) carbon formation catalyzed by the electrode material itself [47]. Tar formation occurs when hydrocarbons decompose and polymerize at high temperatures. For example, gas-phase alkanes larger than methane undergo free-radical reactions that lead to nonvolatile products beginning at approximately 923 K [48]. More complicated molecules, such as pyrolysis oil derived from the high temperature reaction of cellulosic materials [49], are much more reactive. Preventing polymerization of these molecules into solids is a major challenge in the use of biofuels, even in conventional applications. In the case of SOFC, the formation of nonvolatile products would fill the pores of the electrode [47], so that no additional reactions could occur. In principle, if the electrode were stable toward oxidation, these deposits could be removed by periodic oxidation cycles.

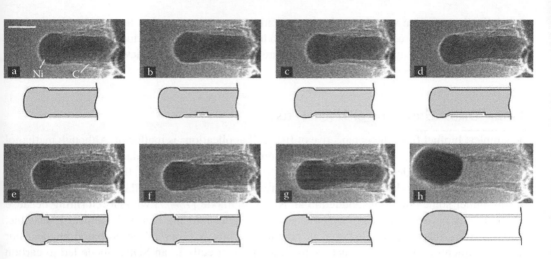

FIGURE 13.4 Image sequence of a growing carbon nanofiber. Images (a–h) illustrate the elongation/contraction process. Drawings are included to guide the eye in locating the positions of mono-atomic Ni step edges at the C–Ni interface. The images are acquired *in situ* with CH_4:H_2 = 1:1 at a total pressure of 2.1 mbar with the sample heated to 809 K. Scale bar, 5 nm. (Reprinted by permission from Macmillan Publishers Ltd: *Nature*, Helveg, S., Lopez-Cartes, C., Sehested, J., Hansen, P.L., Clausen, B.S., Rostrup-Nielsen, J.R., Abild-Pedersen, F., and Norskov, J.K., Atomic-scale imaging of carbon nanofibre growth, 427, 426–429, Copyright 2004.)

Carbon formation catalyzed by conventional Ni anodes is a far more serious problem. This reaction has been studied extensively because of its importance in steam-reforming catalysis [50–55] and in dry corrosion, also called metal dusting [20,21]. The mechanism for carbon fiber formation involves deposition of carbon on the Ni surface, dissolution of the carbon into the bulk of the Ni, and precipitation of carbon as a fiber at some nucleation point. Because these reactions involve more than the Ni surface, carbon fiber formation can lead to loss of Ni from the surface (hence the term "metal dusting"), a process that occurs when Ni is physically lifted from the sample by its attachment to the growing carbon fiber. Figure 13.4 shows a micrograph of an Ni catalyst that was exposed to a CH_4/H_2 mixture at 809 K [56]. The Ni particle can be seen at the tip of the carbon filament. In addition to the dusting of Ni, carbon fibers can also cause the electrodes to delaminate from the electrolyte or even to fracture because of the mechanical stresses induced by the growth of the fibers [22].

Carbon fiber formation with Ni can be avoided if there is sufficient steam so that carbon is removed faster than it deposits. While thermodynamics are often used to predict the H_2O:C ratios required to avoid carbon [57,58], fibers can and do form at H_2O:C ratios that are much higher than would be predicted from equilibrium calculations [59]. Therefore, it is the relative rates of carbon deposition and carbon removal that determine stability of Ni anodes. Because the rate of carbon deposition with CH_4 is relatively low, the required H_2O:C ratio for avoiding carbon is approximately one, making direct utilization of CH_4 with added steam (i.e., internal reforming) feasible. However, carbon deposition rates on Ni are much higher with hydrocarbons larger than methane, and operation on these fuels requires higher H_2O:C ratios to prevent carbon fiber formation [60]. Stable operation with higher hydrocarbons has been demonstrated on Ni-based anodes under special conditions [61]; however, the intrinsic instability of the system and the catastrophic consequences of fiber formation make application of this approach impractical.

Although Ni is not unique in catalyzing carbon formation and similar chemistry has been reported with Fe, Co, Ru, and Pd [62–64], there are conductors that will be stable in the presence of hydrocarbons at high temperatures. For example, Pt and Cu do not catalyze carbon formation in the presence of methane at high temperatures. Based on the mechanism for carbon formation in metals, it would seem unlikely that any oxide would catalyze carbon formation, since it is difficult to visualize how carbon would dissolve in an oxide. Therefore, there has been a significant effort aimed at developing direct-utilization SOFC anodes based on Cu or conducting ceramics. Again, this work is not primarily focused on biofuels and a number of good reviews are available for the interested reader [11,65–69].

13.4.2 DIRECT UTILIZATION OF LIQUID FUELS

The simplest biofuel from the perspective of fuel-cell applications is methanol. Although methanol is typically made from natural gas [70,71], it is sometimes called "wood alcohol" because it can in principle be formed by destructive distillation of wood. It is a simple fuel in that it is easily reformed back to CO and H_2 at relatively mild temperatures [72]. In SOFC, it has been shown that methanol can, under some conditions, be fed to an Ni-based anode [73–77] without reforming because it theoretically carries sufficient oxygen to avoid entering into the thermodynamically unstable, carbon-forming regime [78]. However, in practice, carbon is observed on Ni-based anodes. For example, Cimenti et al. reported that feeding methanol directly to an SOFC anode led to carbon formation that caused the delamination of some Ni–YSZ anodes [79]. This same group reported that the addition of a ceria–zirconia mixed oxide to their Ni–YSZ anodes stabilized the electrodes.

However, there is an additional interesting aspect associated with using methanol in an SOFC. Methanol will readily decompose to syngas on many metals, such as Cu, via Equation 13.6:

$$CH_3OH \rightarrow CO + 2H_2. \tag{13.6}$$

This reaction results in a significant loss in free energy due to the increase in entropy associated with forming three moles of product from each mole of methanol. At 973 K, the effect of this lost free energy is a decrease in the Nernst potential of 0.2 V, independent of fuel conversion [80]. If it were possible to electrochemically oxidize methanol, reversibly, a cell could achieve significantly higher voltages than are possible with CO or H_2. This increased potential for methanol compared to H_2 has indeed been observed using a carbon–ceria anode, in which the carbon provided conductivity and the ceria was the oxidation catalyst [80]. The measured open-circuit voltage (OCV) with the carbon–ceria anode was 100 mV higher than that observed with a Cu-based anode, presumably because the reaction in Equation 13.6 was rapid in the presence of Cu. The high Nernst potential associated with methanol can also affect the materials in the anode because of the very low $P(O_2)$ that it represents. At 1% fuel utilization and 973 K, the calculated $P(O_2)$ are 3.3×10^{-30} atm for methanol and 1.5×10^{-25} atm for H_2. The equilibrium properties for CeO_2 would predict the composition to be $CeO_{1.97}$ (~6% of Ce atoms in the +3 oxidation state) for the $P(O_2)$ established by equilibrium with humidified (3% H_2O) H_2 but significantly more reduced for methanol [81]. This appears to cause severe deactivation of ceria-based anodes.

The next simplest biofuel, ethanol, has little similarity to methanol when used in an SOFC. It forms tar via gas-phase pyrolysis in a manner similar to that found with simple alkanes [48]. Because scission of the carbon–carbon bond results in adsorbed carbon, it tends to form carbon fibers readily on Ni-based anodes [73]. Just as it is possible to use alternative anode materials, such as Cu, to avoid carbon formation and utilize small alkanes directly, it should be possible to do the same with ethanol.

In order of complexity for reforming, the next interesting biofuel is glycerol. Although glycerol is difficult to vaporize and carbon formation can still be a problem [82], the molecule carries a significant amount of oxygen with it, so that very little steam is required to stabilize the anode against carbon formation. Still, carbon formation over Ni is an issue that must be dealt with.

As discussed earlier, other biofuels tend to polymerize at the high temperatures where SOFC operate, so that direct utilization is not practical, even with on-anode reforming. For use in traditional SOFC, it will likely be necessary to first reform these fuels to syngas.

13.4.3 Operation on Solid Biofuels

There has recently been significant interest in development of fuel cells that can generate power from carbonaceous fuels, including coal and biomass [83–86]. In principle, such fuel cells could provide electricity at much higher efficiency than is possible using combustion processes. Furthermore, with fuel cells based on electrolytes that are oxygen-ion conductors, the CO_2 would be produced in a concentrated form at the anode effluent, making it much easier to employ sequestration strategies for when CO_2 emissions become regulated. While most approaches to using solid fuels in fuel cells involve some kind of gasification followed by steam reforming to produce syngas that would then be used by the fuel cell [84], a process that could utilize the carbon directly would have the advantage of process simplicity.

The first Direct Carbon Fuel Cell (DCFC) dates back to the nineteenth century and was based on a molten NaOH electrolyte [87]. This system had many problems, not least of which was the fact that the NaOH electrolyte was consumed when it reacted with CO_2 to form Na_2CO_3. A second approach to DCFC involves carbon gasification by CO_2 to form CO, which can be used by a traditional SOFC [88]. Again, this approach is probably not useful with biofuels since it requires that the fuel be heated to very high temperatures (~1000°C) in the presence of CO_2. In the case of most biofuels, the high temperatures would result in many products besides CO_2 and these products would cause rapid deactivation of a traditional fuel-cell anode.

An alternative approach has been to use molten anodes, with the majority of this work using a molten carbonate mixture (e.g., $Li_2CO_3 + K_2CO_3 + Na_2CO_3$) as the liquid anode. The molten carbonate mixture can then be used with either a molten carbonate electrolyte [89] (for which CO_3^{2-} ions are the mobile species) or with a ceramic oxygen-ion conductor as the electrolyte. Molten carbonates have been shown to be good oxidizers of carbonaceous fuels that are immersed in them; however, they are not electronically conductive. Therefore, a metallic current collector must be incorporated into the anode structure. State-of-the-art MCFC use Ni alloys, typically Ni–Cr or Ni–Al, for this purpose. Dissolution of Ni into the molten carbonate is a problem with MCFC and will likely be even more of a problem when the Ni is completely immersed in the solution.

The low electronic conductivity of molten carbonates represents a much bigger problem than Ni dissolution for the anode reaction, Equation 13.7:

$$C + 2O^{2-} \rightarrow CO_2 + 4e^-. \tag{13.7}$$

Removal of electrons is essential for the oxidation reaction to occur, and oxidation of carbon is likely limited by the inability of electrons to leave the site of the reaction in molten carbonate solutions. The situation is completely analogous to what happens in normal SOFC electrodes, where reaction can only occur at TPB sites. In order to allow electron transfer from within the carbonate solution to the metal current collector and in order to increase the concentration of sites where the reaction in Equation 13.7 can occur, developers have used a conductive form of carbon as the fuel and maintained a high concentration of that carbon within the carbonate solution. While this approach has resulted in impressive performance [90], it obviously limits what fuels can be used, since the fuel itself is part of the anode. Most biofuels, such as cellulose, are not conductive and could not be used in this application.

One of the more interesting schemes for the development of DCFC involves the use of SOFC with molten-metal anodes [43–45]. In this approach, which is shown diagrammatically in Figure 13.5, oxygen from the electrolyte is transferred to the molten metal to form the metal oxide via the reaction in Equation 13.8:

$$M + nO^{2-} \rightarrow MO_n + 2ne^-. \tag{13.8}$$

The metal oxide is then reduced by the fuel, either in the anode compartment itself or by removing the oxygen-saturated metal and reducing it in a separate reactor. Implementation of this strategy is under way with molten Sn anodes [41,86]. However, there are very little published data available

FIGURE 13.5　Schematic of DCFC based on an SOFC with a molten-metal anode.

on the factors that limit the performance of SOFC with molten Sn electrodes, and the data that are available suggest that the cells exhibit relatively high impedances, requiring that they be operated at temperatures above 1273 K [45]. Finally, the solubility of oxygen in Sn is very low and SnO_2 has a very high melting point, 1400 K.

One of our groups has studied fuel cells with molten Sn anodes more closely [43]. The cells were studied in the "battery" mode, operating them with oxidation of the molten metal in the absence of fuel in order to focus on the reaction in Equation 13.8. With Sn, the Nernst potential for Sn oxidation is 0.93 V at 973 K. This potential, which is roughly 0.1 V lower than the equilibrium value for carbon oxidation to CO_2, is nearly ideal for a DCFC because it implies reduction of the oxide by carbon should be thermodynamically spontaneous, while still allowing the fuel cell to produce electrons at a high enough potential for a high system efficiency. The problem with Sn is that oxygen solubility in the metal is very low. Therefore, oxidation at the electrolyte surface causes the formation of an insulating SnO_2 layer at the electrolyte interface, leading to large increases in the cell impedance after transfer of just a few $C \cdot cm^{-2}$ of charge.

The presence of SnO_2 and its effect on cell performance is shown in Figures 13.6 and 13.7. Figure 13.6 shows the V–i characteristics of the cell with the molten Sn anode at 973 K. In this experiment, the anode compartment was reduced in H_2 while holding the cell at open circuit. After flushing the H_2 from the anode compartment with an inert gas, the initial OCV of the freshly reduced cell was approximately 0.93 V, very close to Nernst potential [43]. Next, the current generated by the cell was measured while decreasing the potential at $10 \, mV \cdot s^{-1}$. After the cell potential reached zero, the potential was ramped up at $10 \, mV \cdot s^{-1}$. The data show that the current reached a maximum as the potential was lowered and then began to decrease. The decrease in current was not reversed by increasing the potential. Impedance data, measured before and after the current cycle, showed that the cell ohmic resistance remained unchanged, while the non-ohmic impedance increased dramatically following this V–i cycle. This result is consistent with formation of a SnO_2 layer on the electrolyte interface. SnO_2 is an electronic semiconductor, so the thin layer over the electrolyte does not affect the ohmic resistance. Unfortunately, it is insulating to oxygen ions, which in turn leads to the high non-ohmic losses.

Confirmation that cell deactivation was due to formation of an oxide layer at the YSZ interface was obtained using SEM and EDX. After performing a V–i cycle like the one shown in Figure 13.6, the cell was cooled to room temperature, broken, and then examined. The total charge transferred in this experiment was $8.2 \, C \cdot cm^{-2}$, which corresponds to the formation of a SnO_2 layer that is 4.6 μm thick. As shown in Figure 13.7, this oxide layer is readily apparent in the cross-sectional SEM image. The SEM and EDX data indicate that there is indeed a SnO_2 layer, approximately 10 μm thick, at the YSZ interface. The somewhat larger SEM thickness is likely due to the SnO_2 layer having some porosity. Also, EDX results indicated some spherical SnO_2 particles farther from the electrolyte interface. It is unclear if this results from the process of fracturing the Sn–YSZ interface or if this is an indication that some SnO_2 may extend some distance from the YSZ electrolyte interface.

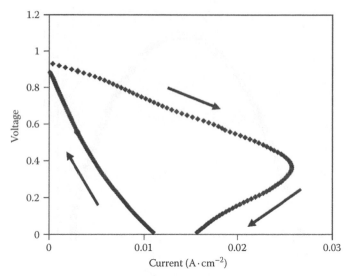

FIGURE 13.6 V–i polarization curve for a cell with molten Sn anode at 973 K. After reduction of the Sn in humidified H_2, the anode compartment was exposed to dry, flowing He while ramping the voltage from open circuit and back at 10 mV · s⁻¹. (Reproduced from Jayakumar, A. et al., *J. Electrochem. Soc.*, 157, B365, 2010. With permission of the Electrochemical Society.)

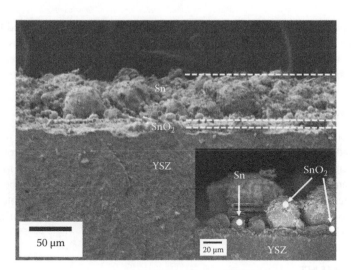

FIGURE 13.7 SEM and EDX results obtained at the molten Sn/YSZ electrolyte interface. The micrograph was obtained after passing 8.2 C · cm⁻² of charge through the electrolyte at 1073 K, then quenching to room temperature, and shows the formation of a SnO_2 layer at the YSZ interface. (Reproduced from Jayakumar, A. et al., *J. Electrochem. Soc.*, 157, B365, 2010. With permission of The Electrochemical Society.)

In order to obtain a process like that shown in Figure 13.5, it would be necessary that both the metal and its oxide be molten at the operating temperature. The metal that has nearly ideal properties for this is Sb, for which the melting temperature is 904 K, with Sb_2O_3 melting at 929 K. (Sb was discussed earlier in connection with Energy Storage). Figure 13.8 shows a V–i polarization curve for a cell with a molten-Sb anode, operated in the battery mode at 973 K. This cell had a 100 μm thick SDZ electrolyte and an infiltrated $La_{0.8}Sr_{0.2}FeO_3$ cathode, so that the sum of the electrolyte and cathode losses are expected to be 0.4 Ω · cm² [37,91]. The results in Figure 13.8 show that the OCV is 0.75 V, equal to the Nernst potential for Sb. Furthermore, the voltage decreases linearly with current

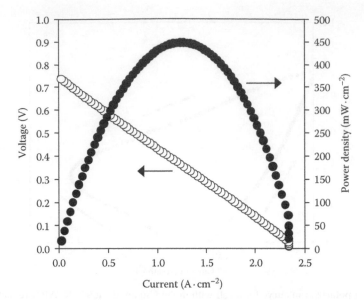

FIGURE 13.8 V–i and P–i curves for a cell with a molten Sb anode, operated at 973 K. The cathode was LSF–YSZ composite formed by infiltration of the LSF in porous YSZ. The electrolyte was 100 μm, Sc-doped ZrO_2.

density, with a slope of $0.4\,\Omega\cdot cm^2$. Since this slope is equal to the losses expected from the cathode and electrolyte, the losses in the anode must be negligible. Finally, there were no changes in the V–i curve with current until a significant fraction of the Sb had been oxidized. To complete the cycle, it is obviously necessary that the Sb_2O_3 be reduced by biofuels. Work on this is presently under way.

13.5 SUMMARY

Solid oxide fuel cells and solid oxide electrolyzers have important roles to play in our sustainable-energy future. In addition to their intrinsically high efficiencies for interconversion between chemical and electrical energy, they offer fuel flexibility that is unmatched by other fuel-cell technologies. Recent developments have demonstrated that high fuel-cell performance can be achieved with a wide range of materials, indicating that more research breakthroughs are possible. Finally, reports of direct utilization of carbon fuel with molten-metal anodes offer the exciting possibility of converting cellulosic biomass into useful energy at high efficiency.

ACKNOWLEDGMENT

RJG expresses his gratitude to the U.S. Department of Energy's Hydrogen Fuel Initiative (grant DE-FG02-05ER15721).

REFERENCES

1. Stambouli, A. B. and E. Traversa. 2002. Solid oxide fuel cells (SOFCs): A review of an environmentally clean and efficient source of energy. *Renew. Sust. Energy Rev.* 6:433–455.
2. Ni, M., M. K. H. Leung, and D. Y. C. Leung. 2008. Technological development of hydrogen production by solid oxide electrolyzer cell (SOEC). *Int. J. Hydrogen Energy* 33:2337–2354.
3. Norby, T. 1999. Solid-state protonic conductors: Principles, properties, progress and prospects. *Solid State Ionics* 125:1–11.
4. Kreuer, K. D. 2003. Proton-conducting oxides. *Annu. Rev. Mater. Res.* 33:333–359.
5. Ormerod, R. M. 2003. Solid oxide fuel cells. *Chem. Soc. Rev.* 32:17–28.

6. Ebbesen, S. D. and M. Mogensen. 2009. Electrolysis of carbon dioxide in solid oxide electrolysis cells. *J. Power Sources* 193:349–358.

7. Bidrawn, F., G. Kim, G. Corre, J. T. S. Irvine, J. M. Vohs, and R. J. Gorte. 2008. Efficient reduction of CO_2 in a solid oxide electrolyzer. *Electrochem. Solid State Lett.* 11:B167–B170.

8. Homel, M., T. M. Gur, J. H. Koh, and A. V. Virkar. 2010. Carbon monoxide-fueled solid oxide fuel cell. *J. Power Sources* 195:6367–6372.

9. Minh, N. Q. 1993. Ceramic fuel cells. *J. Am. Ceram. Soc.* 76:563–588.

10. Minh, N. Q. 2004. Solid oxide fuel cell technology—Features and applications. *Solid State Ionics* 174:271–277.

11. McIntosh, S. and R. J. Gorte. 2004. Direct hydrocarbon solid oxide fuel cells. *Chem. Rev.* 104:4845–4865.

12. Steele, B. C. H. 1996. Ceramic ion conducting membranes. *Curr. Opin. Solid State Mater. Sci.* 1:684–691.

13. Steele, B. C. H. and A. Heinzel. 2001. Materials for fuel-cell technologies. *Nature* 414:345–352.

14. Zhang, X., S. Ohara, H. Okawa, R. Maric, and T. Fukui. 2001. Interactions of a $La_{0.9}Sr_{0.1}Ga_{0.8}Mg_{0.2}O_{3-\delta}$ electrolyte with Fe_2O_3, Co_2O_3 and NiO anode materials. *Solid State Ionics* 139:145–152.

15. Maffei, N. and G. de Silveira. 2003. Interfacial layers in tape cast anode-supported doped lanthanum gallate SOFC elements. *Solid State Ionics* 159:209–216.

16. Ahmed, K. and K. Foger. 2000. Kinetics of internal steam reforming of methane on Ni/YSZ-based anodes for solid oxide fuel cells. *Catal. Today* 63:479–487.

17. Peters, R., R. Dahl, U. Kluttgen, C. Palm, and D. Stolten. 2002. Internal reforming of methane in solid oxide fuel cell systems. *J. Power Sources* 106:238–244.

18. Sarantaridis, D. and A. Atkinson. 2007. Redox cycling of Ni-based solid oxide fuel cell anodes: A review. *Fuel Cells* 3:46–58.

19. Matsuzaki, Y. and I. Yasuda. 2000. The poisoning effect of sulfur-containing impurity gas on a SOFC anode: Part I. Dependence on temperature, time, and impurity concentration. *Solid State Ionics* 132:261–269.

20. Toh, C. H., P. R. Munroe, D. J. Young, and K. Foger. 2003. High temperature carbon corrosion in solid oxide fuel cells. *Mater. High Temp.* 20:129–136.

21. Chun, C. M. and T. A. Ramanarayanan. 2005. Metal dusting corrosion of austenitic 304 stainless steel. *J. Electrochem. Soc.* 152:B169–B177.

22. Kim, H., C. Lu, W. L. Worrell, J. M. Vohs, and R. J. Gorte. 2002. Cu-Ni cermet anodes for direct oxidation of methane in solid-oxide fuel cells. *J. Electrochem. Soc.* 149:A247–A250.

23. He, H. P. and J. M. Hill. 2007. Carbon deposition on Ni/YSZ composites exposed to humidified methane. *Appl. Catal. A* 317:284–292.

24. Ji, Y., J. A. Kilner, M. F. Carolan. 2005. Electrical properties and oxygen diffusion in yttria-stabilised zirconia (YSZ)-$La_{0.8}Sr_{0.2}MnO_{3\pm\delta}$ (LSM) composites. *Solid State Ionics* 176:937–943.

25. Simner, S. P., M. D. Anderson, M. H. Engelhard, and J. W. Stevenson. 2006. Degradation mechanisms of La-Sr-Co-Fe-O_3 SOFC cathodes. *Electrochem. Solid State Lett.* 9:A478–A481.

26. Esquirol, A., N. P. Brandon, J. A. Kilner, and M. Mogensen. 2004. Electrochemical characterization of $La_{0.6}Sr_{0.4}Co_{0.2}Fe_{0.8}O_3$ cathodes for intermediate-temperature SOFCs. *J. Electrochem. Soc.* 151:A1847–A1855.

27. Serra, J. M. and H.-P. Buchkremer. 2007. On the nanostructuring and catalytic promotion of intermediate temperature solid oxide fuel cell (IT-SOFC) cathodes. *J. Power Sources* 172:768–774.

28. Tsoga, A., A. Gupta, A. Naoumidis, and P. Nikolopoulos. 2000. Gadolinia-doped ceria and yttria stabilized zirconia interfaces: Regarding their application for SOFC technology. *Acta Mater.* 48:4709–4714.

29. Simner, S. P., J. P. Shelton, M. D. Anderson, and J. W. Stevenson. 2003. Interaction between La(Sr)FeO$_3$ SOFC cathode and YSZ electrolyte. *Solid State Ionics* 161:11–18.

30. Anderson, M. D., J. W. Stevenson, and S. P. Simner. 2004. Reactivity of lanthanide ferrite SOFC cathodes with YSZ electrolyte. *J. Power Sources* 129:188–192.

31. Shiono, M., K. Kobayashi, T. L. Nguyen, K. Hosoda, T. Kato, K. Ota, and M. Dokiya. 2004. Effect of CeO_2 interlayer on ZrO_2 electrolyte/La(Sr)CoO_3 cathode for low-temperature SOFCs. *Solid State Ionics* 170:1–7.

32. Rossignol, C., J. M. Ralph, J.-M. Bae, and J. T. Vaughey. 2004. $Ln_{1-x}Sr_xCoO_3$ (Ln=Gd, Pr) as a cathode for intermediate-temperature solid oxide fuel cells. *Solid State Ionics* 175:59–61.

33. Zhu, W. Z. and S. C. Deevi. 2003. Development of interconnect materials for solid oxide fuel cells. *Mater. Sci. Eng. A* 348:227–243.

34. Fergus, J. W. 2004. Lanthanum chromite-based materials for solid oxide fuel cell interconnects. *Solid State Ionics* 171:1–15.

35. Mahapatra, M. K. and K. Lu. 2010. Glass-based seals for solid oxide fuel and electrolyzer cells—A review. *Mater. Sci. Eng. R* 67:65–85.

36. Singh, R. N. 2007. Sealing technology for solid oxide fuel cells (SOFC). *Int. J. Appl. Ceram. Technol.* 4:134–144.

37. Sasaki, K. and J. Maier. 2000. Re-analysis of defect equilibria and transport parameters in Y_2O_3-stabilized ZrO_2 using EPR and optical relaxation. *Solid State Ionics* 134:303–321.

38. Paunovic, M. and M. Schlesinger. 2006. *Fundamentals of Electrochemical Deposition*. New York: John Wiley & Sons, p. 48.

39. Marcelo, D. and A. Dell'Era. 2008. Economical electrolyser solution. *Int. J. Hydrogen Energy* 33:3041–3044.

40. Jayakumar, A., J. M. Vohs, and R. J. Gorte. 2010. Molten-metal electrodes for solid oxide fuel cells. *Ind. Eng. Chem. Res.* 49:10237–10241.

41. Tao, T., M. Slaney, L. Bateman, and J. Bentley. 2007. Anode polarization in liquid tin anode solid oxide fuel cell. *ECS Trans.* 7:1389–1397.

42. Program on Technology Innovation: Systems Assessment of Direct Carbon Fuel Cells Technology. 2008. Electric Power Research Report Number 1016170. EPRI, Alto, CA.

43. Jayakumar, A., S. Lee, A. Hornes, J. M. Vohs, and R. J. Gorte. 2010. A comparison of molten Sn and Bi for solid oxide fuel cell anodes. *J. Electrochem. Soc.* 157:B365–B369.

44. Pati, S., K. J. Yoon, S. Gopalan, and U. B. Pal. 2009. Hydrogen production using solid oxide membrane electrolyzer with solid carbon reductant in liquid metal anode. *J. Electrochem. Soc.* 156:B1067–B1077.

45. McPhee, W. A. G., M. Boucher, J. Stuart, R. S. Parnas, M. Koslowske, T. Tao, and B. A. Wilhite. 2009. Demonstration of a liquid-tin anode solid-oxide fuel cell (LTA-SOFC) operating from biodiesel fuel. *Energy Fuels* 23:5036–5041.

46. Cimenti, M. and J. M. Hill. 2009. Direct utilization of liquid fuels in SOFC for portable applications: Challenges for the selection of alternative anodes. *Energies* 2:377–410.

47. Kim, T., G. Liu, M. Boaro, S.-I. Lee, J. M. Vohs, R. J. Gorte, O. H. Al-Madhi, and B. O. Dabbousi. 2006. A study of carbon formation and prevention in hydrocarbon-fueled SOFC. *J. Power Sources* 155:231–238.

48. Gupta, G. K., A. M. Dean, K. Ahn, and R. J. Gorte. 2006. Comparison of conversion and deposit formation of ethanol and butane under SOFC conditions. *J. Power Sources* 158:497–503.

49. Lin, Y.-C., J. Cho, G. A. Tompsett, P. R. Westmoreland, and G. W. Huber. 2009. Kinetics and mechanism of cellulose pyrolysis. *J. Phys. Chem. C* 113:20097–20107.

50. Baker, R. T. K., P. S. Harris, J. Henderson, and R. B. Thomas. 1975. Formation of carbonaceous deposits from the reaction of methane over nickel. *Carbon* 13:17–22.

51. Baker, R. T. K., P. S. Harris, and S. Terry. 1975. Unique form of filamentous carbon. *Nature* 253:37–39.

52. Bartholomew, C. H. 1982. Carbon deposition in steam reforming and methanation. *Catal. Rev. Sci. Eng.* 1:67–112.

53. Zhang, T. and M. D. Amiridis. 1998. Hydrogen production via the direct cracking of methane over silica-supported nickel catalysts. *App. Catal. A* 167:161–172.

54. Monnerat, B., L. Kiwi-Minsker, and A. Renken. 2001. Hydrogen production by catalytic cracking of methane over nickel gauze under periodic reactor operation. *Chem. Eng. Sci.* 56:633–639.

55. Sehested, J. 2006. Four challenges for nickel steam-reforming catalysts. *Catal. Today* 111:103–110.

56. Helveg, S., C. Lopez-Cartes, J. Sehested, P. L. Hansen, B. S. Clausen, J. R. Rostrup-Nielsen, F. Abild-Pedersen, and J. K. Norskov. 2004. Atomic-scale imaging of carbon nanofibre growth. *Nature* 427:426–429.

57. Eguchi, K., H. Kojo, T. Takeguchi, R. Kikuchi, and K. Sasaki. 2002. Fuel flexibility in power generation by solid oxide fuel cells. *Solid State Ionics* 152–153:411–416.

58. Sasaki, K. and Y. Teraoka. 2003. Equilibria in fuel cell gases. *J. Electrochem. Soc.* 150:A878–A884.

59. Sperle, T., D. Chen, R. Lodeng, and A. Holmen. 2005. Pre-reforming of natural gas on a Ni catalyst: Criteria for carbon free operation. *App. Catal. A* 282:195–204.

60. Farrauto, R. J. and C. H. Bartholomew. 1997. *Fundamentals of Industrial Catalytic Processes*, 1st edn. London, U.K.: Blackie Academic and Professional, pp. 341–57.

61. Murray, E. P., T. Tsai, and S. A. Barnett. 1999. A direct-methane fuel cell with a ceria-based anode. *Nature* 400:649–651.

62. Toebes, M. L., J. H. Bitter, A. J. van Dillen, and K. P. de Jong. 2002. Impact of the structure and reactivity of nickel particles on the catalytic growth of carbon nanofibers. *Catal. Today* 76:33–42.

63. Baker, R. T. K. and J. J. Chludzinski. 1986. In-situ electron microscopy studies of the behavior of supported ruthenium particles. 2. Carbon deposition from catalyzed decomposition of acetylene. *J. Phys. Chem.* 90:4730–4734.
64. Atwater, M. A., J. Phillips, and Z. C. Leseman. 2010. Formation of carbon nanofibers and thin films catalyzed by palladium in ethylene-hydrogen mixtures. *J. Phys. Chem. C* 114:5804–5810.
65. Tao, S. and J. T. S. Irvine. 2004. Discovery and characterization of novel oxide anodes for solid oxide fuel cells. *Chem. Rec.* 4:83–95.
66. Sun, C. and U. Stimming. 2007. Recent anode advances in solid oxide fuel cells. *J. Power Sources* 171:247–260.
67. Gorte, R. J. and J. M. Vohs. 2003. Novel SOFC anodes for the direct electrochemical oxidation of hydrocarbons. *J. Catal.* 216:477–486.
68. Atkinson, A., S. Barnett, R. J. Gorte, J. T. S Irvine, A. J. McEvoy, M. Mogensen, S. C. Singhal, and J. Vohs. 2004. Advanced anodes for high-temperature fuel cells. *Nat. Mater.* 3:17–27.
69. Gross, M. D., J. M. Vohs, and R. J. Gorte. 2007. Recent progress in SOFC anodes for direct utilization of hydrocarbons. *J. Mater. Chem.* 17:3071–3077.
70. Cybulski, A. 1994. Liquid-phase methanol synthesis: Catalysts, mechanism, kinetics, chemical equilibria, vapor-liquid equilibria, and modeling—A review. *Catal. Rev. Sci. Eng.* 36:557–615.
71. *Kirk-Othmer Encyclopedia of Chemical Technology.* 2006. Hoboken, NJ: John Wiley & Sons, Inc. v. 16, 5th edn., p. 299.
72. de Wild, P. J. and M. J. F. M. Verhaak. 2000. Catalytic production of hydrogen from methanol. *Catal. Today* 60:3–10.
73. Jiang, Y. and A. V. Virkar. 2001. A high performance, anode-supported solid oxide fuel cell operating on direct alcohol. *J. Electrochem. Soc.* 148:A706–A709.
74. Saunders, G. J., J. Preece, and K. Kendall. 2004. Formulating liquid hydrocarbon fuels for SOFCs. *J. Power Sources* 131:23–26.
75. Sahibzada, M., B. C. H. Steele, K. Hellgardt, D. Barth, A. Effendi, D. Mantzavinos, and I. S. Metcalfe. 2000. Intermediate temperature solid oxide fuel cells operated with methanol fuels. *Chem. Eng. Sci.* 55:3077–3083.
76. Feng, B., C. Y. Wang, and B. Zhu. 2006. Catalysts and performances for direct methanol low-temperature (300 to 600°C) solid oxide fuel cells. *Electrochem. Solid State Lett.* 9:A80–A81.
77. Brett, D. J. L., A. Atkinson, D. Cumming, E. Ramirez-Cabrera, R. Rudkin, and N. P. Brandon. 2005. Methanol as a direct fuel in intermediate temperature (500°C–600°C) solid oxide fuel cells with copper based anodes. *Chem. Eng. Sci.* 60:5649–5662.
78. Sasaki, K., H. Kojo, Y. Hori, R. Kikuchi, and K. Eguchi. 2002. Direct-alcohol/hydrocarbon SOFCs: Comparison of power generation characteristics for various fuels. *Electrochemistry* 70:18–22.
79. Cimenti, M., V. Alzate-Restrepo, and J. M. Hill. 2010. Direct utilization of methanol on impregnated Ni/YSZ and Ni-$Zr_{0.35}Ce_{0.65}O_2$/YSZ anodes for solid oxide fuel cells. *J. Power Sources* 195.4002–4012.
80. Kim, T., K. Ahn, J. M. Vohs, and R. J. Gorte. 2007. Deactivation of ceria-based SOFC anodes in methanol. *J. Power Sources* 164:42–48.
81. Kim, T., J. M. Vohs, and R. J. Gorte. 2006. Thermodynamic investigation of the redox properties of ceria-zirconia solid solutions. *Ind. Eng. Chem. Res.* 45:5561–5565.
82. da Silva, A. L. and I. L. Muller. 2010. Operation of solid oxide fuel cells on glycerol fuel: A thermodynamic analysis using the Gibbs free energy minimization approach. *J. Power Sources* 195:5637–5644.
83. Cao, D., Y. Sun, and G. Wang. 2007. Direct carbon fuel cell: Fundamentals and recent developments. *J. Power Sources* 167:250–257.
84. Li, S., A. C. Lee, R. E. Mitchell, and T. M. Gur. 2008. Direct carbon conversion in a helium fluidized bed fuel cell. *Solid State Ionics* 179:1549–1552.
85. Jain, S. L., B. Lakeman, K. D. Pointon, and J. T. S. Irvine. 2007. Carbon-air fuel cell development to satisfy our energy demands. *Ionics* 13:413–416.
86. Tao, T., L. Bateman, J. Bentley, and M. Slaney. 2007. Liquid tin anode solid oxide fuel cell for direct carbonaceous fuel conversion. *ECS Trans.* 5:463–472.
87. Jacques, W. W. 1896. Method of converting potential energy of carbon into electrical energy. US Patent 555,511.
88. Gur, T. M. 2010. Mechanistic modes for solid carbon conversion in high temperature fuel cells. *J. Electrochem. Soc.* 157:B751–B759.
89. Nabae, Y., K. D. Pointon, and J. T. S. Irvine. 2008. Electrochemical oxidation of solid carbon in hybrid DCFC with solid oxide and molten carbonate binary electrolyte. *Eng. Environ. Sci.* 1:148–155.
90. Heydorn, B. and S. Crouch-Baker. 2006. *Direct Carbon Conversion: Progressions of Power.* New York: IOP.
91. Huang, Y., J. M. Vohs, and R. J. Gorte. 2004. Fabrication of Sr-doped $LaFeO_3$ YSZ composite cathodes. *J. Electrochem. Soc.* 151:A646–A651.

Part IV

Trends, Needs, and Opportunities in Selected Biomass-Rich Countries

14 Research and Prospective of Next Generation Biofuels in India

Arvind Lali

CONTENTS

14.1 INTRODUCTION

Development of an alternative and renewable energy is today as relevant and acute, and probably more, to India as to any other country in the world. Intense research and development activity in this area is witnessed in many parts of the world, notably the developed countries. Mounting pressure on developing economies to cut down their greenhouse gas emissions and find renewable energy sources for their needs instead of depending on imports arises more from the dual need of energy security and cleaner fuels than mere global policies. It can be expected that energy demand in India will rank among the top 3 in the world by the year 2030. It is important that at this time in our history it is realized by both developed and developing economies that the problems faced in this energy crisis must be shared responsibility of all nations of the planet and that joint efforts should be made to solve local as well as global energy problems.

All renewable energy sources on the planet depend on sunlight as the sole source of energy. One of the promising sources, though not the most efficient in terms of solar energy capture, is biomass majorly available in the form of terrestrial plants. This biomass can be converted to one or the other energy forms, including liquid biofuels. Plants therefore can be designed and grown specifically for energy generation. Rising population, especially in Asia, however, requires that any land that can be used for any kind of agriculture may not be used for needs other than food. With increasing population, the requirement of land for growing food is most certain to become increasingly critical over next two decades. Also noteworthy is the fact that India has the largest livestock population in the world and this population almost wholly feeds on agricultural residues than grains like elsewhere in the developed world. India with about 150 million hectares of cultivable land (out of a total land mass of about 300 million hectare) needs to grow as much food and fodder as possible for food security. The use of any fraction of this land for growing energy crops is out of question. However, agriculture at massive scale as done in India also produces considerable amount of nonfood and nonfodder class of lignocellulosic biomass waste, a significant fraction of which is today inefficiently burnt in open air with damaging effects on the environment.

Several reports (e.g., prepared for/by National Thermal Power Corporation, India; and TIFAC under Department of Science and Technology, India) suggest that the quantum of "available" lignocellulosic waste is in the range of 150–200 million metric tons a year. All of this biomass if converted to, say, bioethanol or hydrocarbon fuel can provide a minimum of 50–70 million tons of biofuel that can potentially fully replace gasoline and diesel consumption today in the country if made available at competitive price.

India is one of the largest producers of cane sugar and concomitantly produces 25 million tons of bagasse and another 40 million tons of cane trash. While all of the trash is currently burnt at source, bagasse is increasingly being diverted to power cogeneration plants. A total of at least 50 million tons of bagasse is actually available for conversion to ethanol, especially since most sugar refineries also produce ethanol from molasses produced [1]. This much bagasse has the potential to provide ethanol for much more than 20% blending Indian government has targeted by the year 2020.

Mention must be made here on the potential of biodiesel (e.g., as fatty acid methyl esters) or green diesel (obtained after hydrotreating vegetable oil followed by cracking to desired hydrocarbons fuels) derived from vegetable oils as a biofuel option for India. Huge efforts both from government and industry have been invested in developing production of nonedible oils (vegetable oils and more particularly tree-bound oils) like Jatropha (*Jatropha curcas*) and Karanja (*Pongamia pinnata*) oils. Department of Biotechnology (DBT), under Ministry of Science and Technology of Government of India, launched a mission program on *Jatropha curcas* (http://www.dbtjatropha.gov.in). Under the mission, research programs have been funded across the length and breadth of the country. These programs focus on collection, selection, and screening of available *Jatropha curcas* germplasm in the country, develop elite planting material, and demonstrate cultivation of the selected material at different agro climatic regions of India. Besides efforts from the central government, most state governments have launched vigorous Jatropha cultivation programs. Several large industries, both public and private sector, have undertaken large plantations of *Jatropha curcas* across the country. Similar attempts are also being made by several other organizations like Centre for Jatropha Promotion and Biodiesel (CJP), which is the global authority for scientific commercialization of Jatropha and other nonfood biofuel crops.

The liquid fuel consumption pattern in India is heavily skewed toward diesel than gasoline. As a result, there has been an understandable thrust to find blending options for diesel than gasoline. Thus, the interest in bioethanol as liquid fuel substitute was slower to take off than biodiesel. The other reason for large interest in biodiesel was ready availability of imported as well as indigenously developed technologies. However, the initial explosion of interest in vegetable or tree-bound oil-derived biodiesel has begun to subside in 2010. There are several reasons for the slowdown on biodiesel activity.

At 50 million tons/year, consumption of petro-diesel in the country today, a 5% blending, demands annual production of 2.5 million tons of biodiesel produced from roughly 2.5 million tons of vegetable oil per year. The total annual vegetable or tree borne oil production in India stands at about 10 million tons, of which about 8 million tons is edible oil against a current demand of about 12 million tons a year with the gap of 4 million tons being met through imports. Against this backdrop, indigenous production of vegetable oils for biodiesel production to meet the current 5% blending target appears difficult. Higher blending target is an even more unattainable target unless the country decides to exercise the option of diverting a considerable portion of agricultural or potentially agricultural land to growing vegetable or tree borne oil as feedstock for biodiesel. The government has indicated possible use of about 40 million hectares of marginal or nonarable land for Jatropha cultivation to meet its 20% blending target by 2015. However, by the end of 2009, only about a million hectares was under Jatropha cultivation. This figure is unlikely to grow since the oil yields have less than expected yield and the cost of cultivation has been higher than expected. Thus, while quite a few state governments and industrial enterprises are still committed to expanding the area under the Jatropha crop, most plans have been put on hold owing to the poor success of the crop worldwide.

Other reasons of waning interest of petroleum refining companies in biodiesel are related to instability of the biodiesel produced from unsaturated fatty acids present in all triglyceride plant oils. Despite addition of antioxidants the shelf life of biodiesel can be at best extended to 6 months though at a concomitant cost [2]. Such is the complex nature of plant oil-based biodiesel that as many as 21 quality assurance tests, including oxidation stability at 110°C, are necessary for a blending company to feel safe in blending with petro-diesel (see website http://www.siamindia.com/scripts/bio-diesel.aspx). The problem of instability has, however, been solved to an extent by development of "green diesel" from plant oils. Green diesel is a result of hydrogen treating of the fatty oils/acids followed by cracking to result in a product that is almost indistinguishable from petro-diesel. Green diesel technology, however, is as yet not commercially viable but should replace the concept of biodiesel almost completely in the near future.

Under the circumstances, unless a far more viable and sustaining solution to cost-effective availability of triglyceride oils, such as algal oil, is made available in future, potential availability of lignocellulosic biomass for making biofuels assumes greater importance in the Indian context. The two major roadblocks to successful emergence of cellulosic ethanol as sufficient and sustainable liquid fuel in India are (a) the actual "collectable" availability of the biomass and (b) a technology that can produce biofuel in a commercially viable manner.

14.2 LIGNOCELLULOSIC BIOMASS TO BIOFUELS

There are two possible biomass–based liquid fuel options that will emerge in the near future. These are

1. Lignocellulosic ethanol
2. Lignocellulose to hydrocarbon fuels

14.2.1 BIOMASS TO LIQUID BIOFUELS

There is intense activity all around the world on development of technologies for conversion of lignocellulosic biomass (LBM) to liquid or other forms of fuel. Several BTL (biomass-to-liquid) and BTG (biomass-to-gas) technologies are being explored for converting biomass to hydrocarbon fuels, much like those derived from petroleum crude or coal. Several catalytic and thermal technology options have been under examination for over a decade, the major among these being fast pyrolysis, thermal gasification, thermochemical liquefaction, and supercritical water gasification. Many commercial liquid fuel plants, today, are able to combine a small percentage of LBM with coal for gasification (CTL + BTL). However, large-scale plants that gasify biomass alone are yet to take shape. One of the major bottlenecks in these technologies is the high capital investment required and low yields of the final fuel. Maximum BTL fuel yield reported is about 180 L per ton dry biomass. It is estimated that these technologies become competitive to petroleum-derived fuels only when the crude oil price is above USD 100 per barrel. Another major issue with these chemical routes is that these are viable only in mega scale like petro-refineries. In the Indian context, small farmland holdings, diverse crop patterns, biomass collection logistics at the scale required, and transportation of the low bulk density biomass to a single large plant site, all make feedstock availability not only difficult but also expensive for viability of a mega scale biomass to biofuel refinery.

Two options are available for conversion of LBM to liquid fuels. One is conversion to alcohols (ethanol or butanol), and the other is conversion to liquid hydrocarbons, that is, gasoline or diesel, respectively, called biomass-to-alcohol (BTA) and biomass-to-liquid (BTL) technologies. Both BTA and BTL routes to renewable liquid biofuels, today, face technological roadblocks in terms of economical and ecological viability. Both routes as of today produce biofuels at more than 1.50 USD/L and entail high capital cost for setting up plants. Unless breakthroughs are made in biomass pretreatment and/or catalytic technologies, these technologies may not see meaningful commercialization. BTA and BTL routes to liquid biofuels have their respective advantages and disadvantages, and like every other country India will probably see a combination of these providing partial solutions to its liquid fuel problems. Concerted efforts are the need of the hour, and the Indian government has decided to put R&D and pilot plant technology tests on a fast track on both these technology fronts. Both public and private agencies from across the country are today engaged in trying to find innovative and time-bound solutions.

In the context of small land holdings and farming patterns in Asian and African countries, biomass collection logistics at the scale required and transportation of the low bulk density biomass to the factory site make the feedstock availability not only difficult but also expensive. Thus, capital intensive and essentially large-scale technologies that are based on thermochemical routes will find difficulties in successful implementation in Asia and Africa, unlike in countries like the United States and elsewhere where super scale farms are a norm. One way out of this problem for developing world is conversion of low density biomass into high density forms like high density pellets and thermochemically produced bio-oil followed by transportation to factory sites for further processing. Both these options have been, partly or in full, attempted in different parts of the world, and technology has been tested at pilot scale and may find successful commercialization by the end of 2012. Lignocellulose biomass derived bio-oils can be shipped, stored, and utilized like conventional liquid fuels. Though currently easily usable for primary energy purposes like in boilers and furnaces, there are quite a few challenges that will need to be overcome before bio-oil derived fuels become a viable option in terms of both capital and operating cost.

Alternative technologies to chemical routes are biological routes that aim at producing alcohols via production of fermentable sugars. Iogen in Canada, NREL in the United States, and many other industries and institutes have worked for decades on development of a sustainable technology for biological conversion of LBM to ethanol or other alcohols. The overall technology comprising multiple steps is still far from being economically and ecologically sustainable and attractive. Much work needs to be done in areas of biomass feedstock improvement, technologies for pretreatment, fractionation of LBM to cellulose, lignin, and hemicellulose, and their conversion to sugars and finally to alcohol and other value-added products. Any resultant technology will need to be complete, cost-effective, as well as eco-friendly.

Table 14.1 gives an idea of the costs involved in production of energy from different renewable feedstock. The data in Table 14.1 has been estimated and compiled on the basis of information, direct or indirect, from different and varied sources.

Attempts are being made to bring down the cost of production of lignocellulosic ethanol or hydrocarbon fuels through the two routes, viz., chemical and biochemical. As on today, chemical routes for biomass conversion to ethanol do not appear possible in India for various reasons, the simplest being that not much developmental work is going on in this direction. In any case, given the complexity associated with large biomass collection and availability at a single location, and given the complex nature of the technologies involved, bioethanol seems a preferred biofuel for India, considering that it, indeed, was the first country that began large-scale production of the product from cane molasses in the early 1960s.

Production of ethanol (or other alcohols) from LBM is also a complex technology as biomass is a complex mixture that comprises different components that potentially can be converted to different products as depicted in Figure 14.1.

From publicly available information, 40–50 pilot plants have been set up all over the world in sizes that have handled biomass from 100 kg/day (NREL, USA) to about 1 ton/day (NREL, Denver, USA; Gulf Oil Chemicals, Pittsburgh; Procter & Gamble, Pennsylvania). Apparently, the largest lignocellulosic ethanol plant started operating in Denmark in November 2009 at a capacity of 100 ton biomass/day designed to produce about 18,000 L ethanol/day. The plant cost incurred was about 60 million USD which comes to about USD 2.5 million per MW (compare with an average figure of USD 1.2 million per MW for coal-fired, thermal power plants). POET in South Dakota is putting up a large cellulosic ethanol plant scheduled to go into operation by 2011 to produce 100 million liters ethanol per year. Department of Energy, USA, has recently funded about a dozen demonstration

TABLE 14.1

Production Cost Comparison of Different Energy Sources

Source	Etd. Equipment Cost per MW[a] or Equivalent (in Million USD)	Etd. Production Cost per kWh or kWh Equivalent (in USD)	Etd. Cost[b] USD/L
Biomass to bioethanol	1.20	0.05	1.0
Corn to bioethanol	0.70	0.06	0.60
Biomass to liquid hydrocarbon fuel (BTL)	1.60[c] 0.6[d]	0.07	2.1
Biomass to biomethane[e] (at scale of >1000 m³/h)	3.50	0.25	—
Coal to electricity	1.20	0.01	—
Gas	0.70	0.02	—
Atomic energy	1.60	0.005	—

[a] 1 MW ~ 1 million liters EtOH/year.
[b] Incl. cost of capital.
[c] If set up all new (Biomass-syngas-FT + HCs-cracking-fractionation to liquid fuel).
[d] Only gasification/pyrolysis + cleanup + FT part.
[e] When cleaned for use as CNG for transportation.

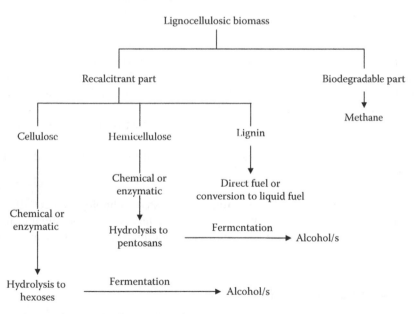

FIGURE 14.1 Typical product chart for lignocellulosic ethanol.

projects with different companies for cellulosic ethanol. These plants using different technologies and funded at a cost of about US 450 million dollars are expected to be operational by 2012–2013.

Many different technologies are being attempted in different plants and the overall process scheme used can vary greatly. Conceptually, the process comprises four main steps as depicted in Figure 14.2. The figure also indicates where the production costs today are prohibitive for a viable technology. Thus, while the pretreatment step 1 is capital intensive, the saccharification step 2 incurs high operating expenses. Innovations are also required at the fermentation step 3 and recovery and purification step 4 in order to make the technology truly competitive.

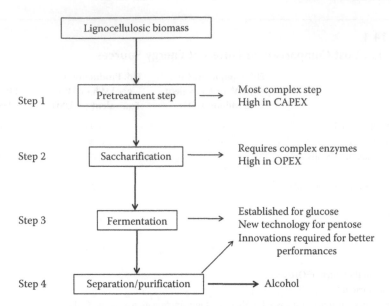

FIGURE 14.2 Typical schematic process for lignocellulosic alcohol.

In order that a lignocellulosic alcohol technology is sustainable and finds widespread use across the world in all economies small or big, it must meet some essential criteria given as follows:

1. The technology must be implementable at a scale suitable to be located near biomass production locations, for example, in agricultural heartland is Asian countries.
2. The technology must be robust enough to be operated by local semiskilled operators.
3. The technology must not produce any waste or by-product with limited use (compared to biofuel) in order to be ecologically sustainable.
4. The technology must result in a manufacturing plant that is compact and low in capital investment.
5. The biofuel (alcohol) must be produced at a cost equal to or less than production cost of petroleum fuel.

It is generally perceived that none of the available and tested technologies meet all of the aforementioned criteria and, not surprisingly therefore, none has been found suitable for scale-up. Significant scientific and engineering innovations are required if a technology must succeed and be sustainable.

14.2.2 Biomass to Bioethanol Technologies

Different process philosophies have emerged for LBM ethanol. Most technologies deployed at pilot scale across the world involve pretreatment and enzymatic hydrolysis of celluloses without removing lignin from the biomass. The pretreated and hydrolyzed products is a mixture of hexoses and pentoses in presence of lignin, all of which is cofermented to ethanol and lignin recovered as residue from the fermentation broth. The other concept involves pretreatment to delignify the biomass and "soften" celluloses before subjecting to enzymatic hydrolysis. Separation of hexose (glucose) yielding cellulose and pentose (predominantly xylose) yielding hemicellulose is also carried out where either a xylose to ethanol fermenting strain is not available, or pentose is intended for alternative use. A large amount of work has been focused on developing strains that will ferment both xylose and glucose to ethanol separately or together. Figure 14.3 gives a more

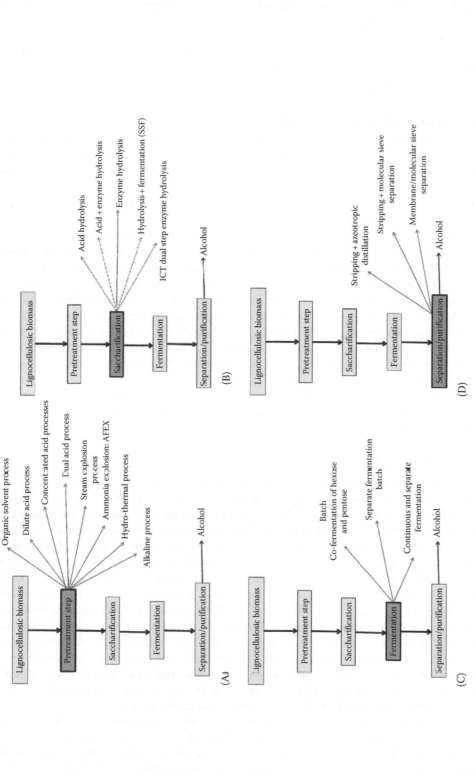

FIGURE 14.3 Technology options in the four major steps in lignocellulosic ethanol technology: (A) pretreatment step; (B) saccharification step; (C) fermentation step; (D) ethanol separation step.

detailed flowchart for lignocellulose to ethanol conversion with the several technology options attempted and reported at each step.

There are many reports that give detailed discussion on the merits and demerits of each of the option in each of the steps shown in Figure 14.3 [3–6]. While a detailed discussion would be too long, a brief discussion on each of the steps is given in the following especially as a platform to discuss Indian situation later.

14.2.2.1 Biomass Selection and Pretreatment

LBM can be derived in many forms, from grasses to crop residues to woods from forests. Availability of biomass varies from region to region and would also depend on competing uses. For example, sugarcane bagasse, though available easily with the sugar mills, finds use as direct fuel for steam and power generation. Other crop residues find important use as cattle feed. In any case, it is amply proven that each agricultural region does have some or the other surplus biomass available provided the logistics of its collection and transport to the plant are worked out.

Pretreatment is aimed at loosening the bonds between cellulose, hemicellulose, and lignin. As shown in Figure 14.3 there are various options possible. Acid pretreatment is the most widely used method, but suffers from being non-ecofriendly and requires expensive material of construction besides giving lower yields of monosugars and forming fermentation toxic side products. Hydrothermal or steam explosion is the next popular choice but suffers from scalability problems. Alkali hydrolysis is expensive but produces high quality cellulosic residues and gives higher yields of fermentable sugars [5]. Other options like AFEX and solvent processes are not likely to find acceptance on account of the costs involved. The choice of a method is decided by many factors, the most important being the type of biomass. Biomass is graded on the basis of severity of pretreatment required to obtain enzyme hydrolysable biomass. Thus, while bagasse is a "low-severity" biomass; wood-chips and cotton or jatropha plant waste are "high-severity" biomass. Most agricultural residues in India would classify as low-to-medium severity biomass with some exceptions like cotton stalk.

14.2.2.2 Saccharification of Pretreated Biomass

Most pretreatment technologies would leave residual solid mass that is "de-lignified" and "softened" for hydrolysis to fermentable sugars. Options are between sending the whole biomass through saccharification step without fractionation into separated components, and sending the biomass into saccharification after separation of the fractionated components namely cellulose, hemicellulose, and lignin. Saccharification can be either chemical or enzymatic. Combined or simultaneous saccharification and fermentation (SSF) to ethanol using engineered microbes has been shown possible but not found acceptance at large scale. Chemical saccharification or hydrolysis is known to give lower sugar yields than enzymatic methods. Besides, they require acidic conditions under which reaction mass handling problems are severe. Enzymatic methods have been generally regarded as more acceptable. Treatment of pretreated biomass without fractionation has been shown to be successful whereby the resulting sugar syrup containing lignin is charged as such into the fermenter with suitable strains that either convert hexose or both hexose and pentose to ethanol. Consolidated Bioprocessing (CBP) is a further advanced concept whereby saccharification and fermentation of biomass are done in a single step and has advantages of minimal processing steps, thereby reducing operating and capital costs [7].

Composite or whole biomass processing (WBP), i.e., whereby no component of the biomass is separated from pretreatment to fermentation, and CBP though have assumed centre stage, there is a growing interest in technologies that permit separation of cellulose and hemicellulose fractions and follow their separate conversions to respective sugars which are then fermented to alcohol. This results from the two major aspects:

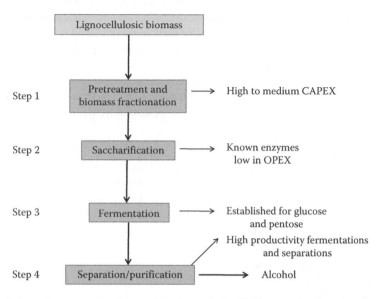

FIGURE 14.4 Schematic process for lignocellulosic alcohol with biomass fractionation.

1. CBP or WBP are based on pretreated biomass and thus their performance greatly depends upon the biomass and the pretreatment technology employed. This is so since the success of the enzymes and/or microorganisms in conversion processes are strongly related to biomass structure and the products (especially the by-products) produced in the pretreatment step.
2. Second, the overall processes are rather slow (may take 48–150 h) and, thus, present severe scale-up challenges.

Clean separation of biomass components permits better design and control of both enzymatic and microbial processes. Most enzymes and microbes under development today are attempting to address the cost and rate of the processes. Extensive efforts and large funds are being put into design of "super enzymes" that will reduce the enzyme cost to one-tenth of current costs. While this may take some time to happen, a shift from the case of low-cost pretreatment technologies and high enzyme cost to medium cost pretreatment + fractionation and low cost enzyme process may emerge as a useful option. Figure 14.2 indicates the current situation which can change to the situation depicted in Figure 14.4.

14.2.2.3 Fermentation and Ethanol Purification/Dehydration

Fermentative conversion of hexose, that is, glucose, is a well-established technology. However, pentose fermentation is not straightforward and many different attempts have been made to convert pentose like xylose to ethanol. Recombinant or mutated strains of *Saccharomyces sp.*, *Pichia stipitis*, *Zymomonas mobilis*, etc., have been developed and successfully used. While hexose to ethanol volumetric productivities have touched 5–10 g/L/h, xylose fermentations are slower at 1–3 g/L/h. Reported best ethanol yields on glucose and xylose achieved are 0.5 and 0.4 g/g, respectively [8,9].

Higher productivities, and hence smaller fermenters, have been shown possible through the use of continuous fermenters with or without cell recycle and with extractive fermentation techniques. Extractive fermentations, whereby the ethanol formed is continuously "extracted" through evaporation, pervaporation, membrane separation, or L-L extraction, have resulted into increase in productivities up to 50 g/L/h [10]. However, this requires a very clean and defined feed which can seldom come from known lignocellulosic processes.

Recovery of ethanol and purification is also a major cost contributor being an energy intensive step. Heat integration in ethanol distillation plants and the use of heat generated from lignin residue obtained in the process can result in an acceptable overall economy. Advent of molecular sieves for drying of 95% ethanol to anhydrous ethanol has proven successful and is used widely. New technologies like those based on membrane separation are under development and may further ease the energy intensiveness of ethanol recovery step.

Ethanol forms a low boiling azeotrope with water and hence ethanol concentration finally obtained in fermentation has finite but not grave impact on cost of recovery. Attempts are underway to increase ethanol tolerance of microbes more from productivity viewpoint than recovery viewpoint.

14.2.3 Overall Economics of LBM Ethanol Production in India

India, though fresh on the road to development of lignocellulosic ethanol, is likely to see emergence of large-scale plants by the end of 2012 as a result of a number of initiatives of all kinds both on public and private fronts. However, the country needs to address some major issues as follows:

1. Biomass type, availability, and logistics challenges
2. Production technology challenges
3. Distribution technology challenges
4. Usage technology challenges
5. Sociopolitical challenges

As mentioned in the beginning here, several surveys have indicated surplus availability of biomass for conversion to liquid biofuels. Lignocellulosic ethanol makes, at present, better sense in Indian context given the size and nature of its largely agrarian economy. Availability of the low density biomass and its transportation cost necessitates that the bioethanol plants will need to be based in the agricultural heartland. Figure 14.5 indicates how much biomass can be available in India per day, from low fertility land to high fertility land, as a function of the area that an ethanol plant is expected to cover. Also indicated is the cost of biomass transportation with an average cost of transportation considered as 0.67 USD/dry ton/km.

Thus, a 100 sq km area would provide biomass from 40 to 200 ton/day depending upon the soil conditions and/or competing uses of the biomass. This indicates that the average LBM ethanol plant size will not be too small even if the production plants were to be located in a decentralized fashion and a minimum plant would require a capital investment in excess of 5–25 million dollars.

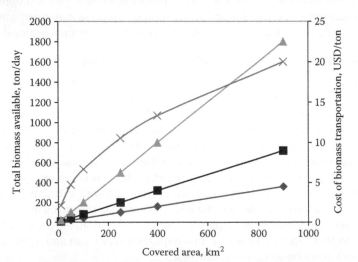

FIGURE 14.5 Biomass availability in ton/day and biomass transportation cost as a function of area covered by a lignocellulosic ethanol plant and average surplus biomass production per hectare: ◆ 1.0 ton/year; ■ 2.0 ton/year; ▲ 5.0 ton/year; × cost of biomass transportation.

One of the important factors in the supply chain of biomass for conversion to biofuels is the price that a farmer would get for his biomass "waste." Rural India is resource poor, and general practice of the people is to put each bit of agricultural produce to some or the other use. With the largest cattle or livestock population in the world, India struggles to feed its animals. Besides, biomass is used as direct fuel for cooking and heating. Indeed, 40% of India's primary energy need is met directly through not-so-efficient burning of wood and agri-waste. In addition, the agri-waste is used to build houses, fences, houseroofs, etc. It is in this context that the terms "surplus" and "available" biomass assume importance and it is indeed creditable that the reports made by different agencies have considered these aspects in making an estimate of the available biomass at about 150 million tons/year. It can be however argued that if priced rightly, the biomass availability for biofuel will be able to compete well with low-value applications. A useful example can be given here on how biomass for biofuel may find more takers than hitherto imagined. India burns its bagasse (about 25 million tons/year) for direct heat. There is increasing interest in using this heat for power generation. One ton of bagasse at maximum generates 800 kWh which would fetch about USD 84 if added to the grid at the tariff recently decided by the government. On the other hand, 1 ton of bagasse can yield 300 L of ethanol (1972 kwh equivalent) which will earn USD 180 at the federal price of USD 0.6/L ethanol. It is thus possible to imagine that alcohol producer will be able to give farmers a price of up to USD 50/ton of biomass. This compares almost at par with the price that sugar refineries give for the cane produce in the country.

It therefore appears that agricultural waste may not present serious problems should the technology for cellulosic ethanol production become viable. This brings us to the question of which cellulosic biofuel technologies are relevant to India. Of the technologies known today, the best can produce ethanol at USD 2.50/gal. This just about borders on being an acceptable cost and efforts are on to reduce this figure to USD 2 or lower. From the indications available a viable ethanol from biomass may be a reality within the next 2–3 years, that is, by the end of 2013. What is, however, not known is how efficient or effective any given technology would be with respect to generating and handling the waste streams. Also unknown is the capital cost associated with the technology which often appears to be alarmingly high in most reported or known cases. Thus, the overall viability of the technology will become visible only when plants higher than 100 ton/day capacity go into full operation over extended periods. The Danish plant put up by Inbicon is a case that merits close watch but only uses the hexose fraction of the biomass and generates C5 sugar stream as side product whose use at the scale produced is yet to be analyzed.

14.3 BIOREFINERY CONCEPT

LBM is a multicomponent resource that can lead to more than one product. Even if the sheer scale at which relatively minor components like arabinose, acetic acid, or polyhydroxyalkanoids (PHA) may be produced is much larger than their market demand and is produced alongside large production of alcohol, it may be worthwhile to examine the concept of a *biorefinery* using biomass as feedstock. Design of a bioalcohol facility as a multiproduct biorefinery would perhaps be the key to emergence of an ecologically and economically sustainable biomass-to-bioenergy conversion technology at least in the initial years when, for example, ethanol may be produced at high cost of more than USD 0.5/L. Thus, besides working on bioconversion processes for biomass to liquid fuel alternatives, technologies also need to be developed for conversion of biomass components to other value-added materials like arabinose, xylose, PHA; conversion of lignin to fuel and chemicals; acetate to acetic acid; and the use of silica for material science applications.

In this scenario, the lignocellulosic ethanol technology that proposes fractionation of biomass to its components namely, cellulose, hemicellulose, and lignin, seems a preferred alternative to ones that are CBP- or WBP-based technologies. Part of the higher production cost would have to come from minor more valuable chemicals produced as by-products or coproducts. Diversion of 10%–30% of the glucose and xylose produced may have to be used for production of chemicals' values higher than fuel ethanol. It is in this sense that India will be well advised to choose the concept of biomass biorefineries working alongside sugar biorefineries and agriculture industry.

India presents a very different kind of challenge with respect to the lignocellulosic ethanol technology that is both sustainable and suitable in the rural backdrop where the bulk of biomass is available. Plants using technologies that are sophisticated enough to be robustly simple and viable at small and medium scale are required. The biggest advantage offered by decentralized production of lignocellulosic ethanol would be the concept of "local produce-to-local consumption," thus negating the disadvantages of mega sized plants that must incur transportation and distribution costs associated with both raw material and finished goods.

In addition to the technological challenges, there are other challenges associated with possible emergence of bioethanol as fuel option in India. Ethanol is a hygroscopic chemical and tends to corrode the normal mild steel pipe lines and storage tanks. It is also lower in calorific value and is required in 1.3–1 ratio when compared to petro gasoline. Increasing the blending to higher than 20% ethanol in gasoline would require changes to be made in engine designs in the vehicles. These problems have indeed been solved by Brazil and the United States at the associated cost of the required changes. India will need to follow their example. While engine manufacturers will need to come up with flexi-fuel designs, it will be necessary that the gasoline distributing companies do not seek to have all the required ethanol for blending transported to their refinery sites. Oil companies will need to plan to put up blending stations across the country much like what was done by Indian Oil Corporation, the largest petroleum fuel producer and distributor in India, for biodiesel, and is currently practiced in Brazil. Localized blending would not only reduce the cost of the distributed ethanol but also ease the pressure on the technology in terms of the net cost of manufacture.

Issues, however, arise on handling of large ethanol volumes across the country that treats ethanol as a complicatedly taxable and tightly controlled commodity. This forms a part of the sociopolitical challenge packet.

14.3.1 INDIAN INITIATIVES

A number of initiatives have been taken over the last one decade in India by different ministries through various scientific promotion agencies in promoting development of components of a potentially viable lignocellulosic ethanol technology. There have been programs funded by the Department of Science and Technology (under NMITLI program or otherwise), the DBT, and CSIR. However, the overall investment has been not more than USD 20 million which is small compared to what has been spent by the United States and European countries. The largest single effort has come from the Department of Biotechnology in setting up a dedicated DBT-ICT Centre for Energy Biosciences with Institute of Chemical Technology in Mumbai. This centre has been functional since mid-2008 and has developed a lignocellulosic ethanol technology that is being scaled up to a 10 ton biomass/day pilot plant at the India Glycols Ltd. at Kashipur in Uttrakhand. The LBM fractionation technology developed by NCL, Pune, has been scaled up by Godavari Sugar Mills which aimed at producing good quality celluloses for various. Another centre for biofuels has been started by the Department of Science and Technology with NIIST, Trivandrum to work specifically on cellulosic ethanol.

The number of private initiatives in this area has been very small. Major efforts are being made by Praj Industries, Reliance Industries, and Mission NewEnergy (an Australia-based company). Tata Chemicals and Indian Oil Corporation and other oil companies like Hindustan Petroleum Corporation Ltd. (HPCL) and Bharat Petroleum Corporation Ltd. (BPCL) are actively exploring the technology options for putting up a lignocellulosic ethanol pilot or demonstration plants.

14.4 BIOMASS BEYOND LIGNOCELLULOSE

It can be easily imagined that LBM availability in countries like India with limited land resources will sooner or later hit stagnancy. With rapidly rising food and fuel demand in these countries, there is a need to find more sustainable biofuel solutions. One of the promising next generation fuel options is algae derived biofuels. Technologies for production of algal biofuel, however, are in a nascent stage worldwide despite increasingly intense attempts being taken up across the world including India. Unlike existing agriculture derived biomass, dedicated technologies must be developed for large-scale generation of algal biomass and other algae derived components like lipids and carbohydrates. Design or development of robust algal species capable of giving fast and dense growths with minimal inputs of energy and nutrients is required. The next step of design of suitable photo-bioreactors and downstream processing stages that operate at super scale is also a tough challenge that will need to be overcome in time. The task of finding solutions to these problems is seemingly more complex than problems faced by BTA and BTL technologies. Even with the innumerable efforts going on around the globe, it is difficult at present to make a safe guess as to what will finally be the shape of a successful algal biofuel technology.

India with its large coastline and availability of brackish/marine water has begun to focus on capacity building in algal biotechnology. This is important since, as mentioned earlier, India cannot divert its limited fresh water resources and cultivable land for fuel production purposes. Near constant ocean temperatures round the year, plenty of sunshine and nutrient rich coastal waters are ideally suited to systematic and large-scale algal farming. Under such favorable conditions microalgae can grow several folds faster than multicellular lignocellulosic plants, and thus not only provide an effective carbon sink but also can act as source of rapid generation of biomass for conversion to biofuels. Many algal species generate large amounts of vegetable oils and/or starchy biomass. For example, oil generating algal species can give more than 50 ton fatty oil per hectare compared to best yields of 5 ton/ha for palm oil [11]. The dry algal biomass generation can very well touch 200 ton/ha/year compared to the best possible yield of 100 ton/ha for sugarcane.

Considerable work in the area of strain screening, taxonomy, and characterization has been already done at more than a dozen institutes around the country in India. Focused work has been undertaken on specific strategies for algal strain improvement and their culturing conditions to impart the desired robustness so that the selected algae would grow in large-scale industrial environments. Design of photo-bioreactors involving suitable fluid dynamics, light distribution systems design, technology for supply of enriched carbon dioxide and minor nutrients, and temperature control are some aspects that need attention and will involve expertise from chemical engineers, mechanical and electrical engineers, and optical physicists, in addition to algal biotechnologists.

REFERENCES

1. Ghosh P. and T.K. Ghose. 2003. Bioethanol in India: Recent past and emerging future. *Adv. Biochem. Eng. Biotechnol.* 85: 1–27.
2. Schober S. and M. Mittelbach. 2004. The impact of antioxidants on biodiesel oxidation stability. *Eur. J. Lipid. Sci. Technol.* 106: 382–389.
3. Tereck C.D. and P.W. Madson. 2004. Lignocellulosic feedstocks for ethanol production: The ultimate renewable energy ethanol as transportation fuel—Production technology developments, *AIChE Annual Meeting*, Austin, TX.
4. Cardona C.A., J.A. Quintero, and I.C. Paz. 2010. Production of bioethanol from sugarcane bagasse: Status and perspectives. *Bioresour. Technol.* 101: 4754–4766.
5. Sanchez O.J. and C.A. Cardona. 2008. Trends in biotechnological production of fuel ethanol from different feedstocks. *Bioresour. Technol.* 99: 5270–5295.
6. Balat M., H. Balat, and O. Cahide. 2008. Progress in bioethanol processing. *Prog. Energ. Combust.* 34: 551–573.
7. Vertes A.A., N. Qureshi, H.P. Blascheck, and H. Yukawa. 2010. *Biomass to Biofuels: Strategies for Global Industries*. John Wiley & Sons, New York.

8. Rogers P.L., Y.J. Jeon, K.J. Lee, and H.G. Lawford. 2007. *Zymomonas mobilis* for fuel ethanol and higher value products. *Adv. Biochem. Eng. Biotechnol.* 108: 263–288.
9. Hahn-Hägerdal B., K. Karhumaa, M. Jeppsson, and M.F. Gorwa-Grauslund. 2007. Metabolic engineering for pentose utilization in *Saccharomyces cerevisiae*. *Adv. Biochem. Eng. Biotechnol.* 108: 147–177.
10. Maiorella B., C.R. Wilke, and H.W. Blanch. 1981. Alcohol production and recovery. *Adv. Biochem. Eng. Biotechnol.* 20: 43–92.
11. Sivakumar G., D.R. Vail, J. Xu, D.M. Burner, Jr. O.L. Jackson, X. Ge, and P.J. Weathers. 2010. Bioethanol and biodiesel: Alternative liquid fuels for future generations. *Eng. Life Sci.* 1: 8–18.

15 Catalytic Technologies for Sustainable Development in Argentina

Carlos R. Apesteguía

CONTENTS

15.1 INTRODUCTION

Sustainable industrial development requires a balance between economic, technological, and environmental aspects. In particular, the development of cleaner and renewable technologies as well as fundamental to process optimization, waste reduction, and pollution prevention are becoming imperative for developing countries, such as Latin American transition-economies countries. The use of renewable resources in chemical industry represents also an opportunity of the growth for agricultural productions, while reducing the impact of oil-based processes on the environment. However, Latin American countries encounter significant difficulties in the transition toward market economy, which are in part related to the presence of old and environmentally non-friendly technologies.

Catalysis is a powerful tool for the development of clean catalytic technologies in chemical manufacturing and the use of renewable resources, particularly crop-derived raw materials, and thereby meets the economic and social interest of Latin American countries. Expected benefits of catalytic routes are higher process efficiency, decreased waste production, and elimination of hazardous reagents and by-products. Thus, increasing research work has been lately focused on the development of novel catalyst formulations and catalytic processes in sectors including polymer, agricultural, petrochemical, pharmaceutical and fine chemical industries. However, the industrial development of catalysts is an expensive and labor-intensive activity because research in catalysis is still dominated largely by experimental studies. The development and transfer of highly advanced and sophisticated catalytic technologies and processes and their industrial application and adoption is therefore crucial for the industrialization programs of Latin American countries.

Argentina is a developing country requiring sustainable economy growth based on industrial development for improving the population income. But industrial development may not be conceived without sustainable chemistry, and protection of the environment and quality of life. Advanced research in catalysis is needed for developing novel clean technologies into chemical design and achieving these economic and environmental goals. In this work, three examples of innovative catalytic processes developed by the Argentinean academic community in the field of renewable resources, clean technologies, and biomass derivatives valorization are presented. Two examples deal with the use of solid catalysts for the synthesis of fine chemicals. Traditionally, fine and specialty chemicals have been produced predominantly by non-catalytic or homogeneously catalyzed synthesis. But these processes have been characterized by the coproduction of large amounts of unwanted products, the use of toxic or corrosive reagents and solvents, and harmful liquid catalysts. In particular, the use of strong liquid acids (HF, H_2SO_4, HCl) and Friedel–Crafts ($AlCl_3$, $TiCl_4$, $FeCl_3$) catalysts in homogeneous commercial processes poses problems of high toxicity, corrosion, and spent acid disposal.

New industrial strategies for fine chemical synthesis demand the use of renewable raw materials and the replacement of liquid acids or bases by solid catalysts. Precisely, the first example describes the use of acid zeolites for replacing $AlCl_3$ Friedel–Crafts catalyst in the gas-phase synthesis of aromatic ketones from acylation of phenol. The second example deals with the development of bifunctional solid catalysts for producing menthols from citral, a renewable raw material, in a one-step process. Finally, the third example illustrates the employment of catalytic technologies for promoting a larger use of renewable low-value raw materials. Specifically, the use of solid catalysts for efficiently producing monoglycerides from glycerol, a low-value by-product in the biodiesel production, is explored.

15.2   GREEN CATALYTIC PROCESSES FOR FINE CHEMICAL SYNTHESIS

15.2.1   Introduction

Aromatic ketones are valuable intermediate compounds in the synthesis of important fragrances and pharmaceuticals. In many cases, these compounds are currently obtained in homogeneous processes via Friedel–Crafts acylations. However, Friedel–Crafts acylations are real and alarming examples of very widely used acid-catalyzed reactions that are based on 100 years old chemistry and are extremely wasteful. Replacement of $AlCl_3$ Friedel–Crafts catalysts by solid acids for acylation reactions would drastically reduce the aqueous discharge and solid waste.

In particular, hydroxyacetophenones are useful intermediate compounds for the synthesis of pharmaceuticals. For example, para-hydroxyacetophenone (p-HAP) is used for the synthesis of paracetamol, a well-known anti-pyretic drug, and ortho-hydroxyacetophenone (o-HAP) is a key intermediate for producing 4-hydroxycoumarin and warfarin, which are both used as anticoagulant drugs in the therapy of thrombotic disease [1,2]. In the classical commercial process, p-HAP is obtained via the Fries rearrangement of phenyl acetate in a liquid-phase process involving the use of homogeneous catalysts, such as $AlCl_3$, $TiCl_4$, $FeCl_3$, HF, which pose problems of high toxicity, corrosion, and spent acid disposal [3]. In an attempt to develop a suitable and environmentally benign process for producing p-HAP, strong solid acids such as ion exchange resins, zeolites, Nafion, and heteropoly acids have been tested in liquid phase for the Fries rearrangement of phenyl acetate. However, solid acids form significant amounts of phenol together with p-HAP, and are in general rapidly deactivated [4,5].

Hydroxyacetophenones may be also obtained by the acylation of phenol in liquid or gas phases by employing different acylating agents. In liquid phase, the reaction produces mainly p-HAP using either Friedel–Crafts or solid acid catalysts, but the process is hampered because of environmental constraints and catalyst activity decay [6,7]. In gas phase, the phenol acylation on solid acids forms predominantly o-HAP, but the reported experimental o-HAP yields are still moderate, particularly because of significant formation of phenyl acetate [8,9]. The previous analysis of bibliography shows that the potential use of solid acids for obtaining hydroxyace-tophenones in gas or liquid phase via either phenol acylation or phenyl acetate rearrangement reactions is limited because of relatively low yields and rapid activity decay. The development of more selective and stable catalysts is therefore required to efficiently promote the synthesis of o-HAP. Taking this into account, our research group decided to perform a detailed study of the gas-phase acylation of phenol with acetic acid over different solid acids. The goal was to relate the structure properties and the surface acid site density and strength of the solids with their ability for the efficient catalysis of the phenol acylation reaction to yield o-HAP. Our studies also focused on catalyst activity decay.

15.2.2 Sample Preparation and Characterization

The o-HAP synthesis was studied on HPA(30%)/MCM-41, Al-MCM-41 (Si/Al = 18), zeo-lites HY (UOP-Y54) and HZSM-5 (ZeoCat PZ-2/54), and SiO_2–Al_2O_3 (Ketjen LA-LPV) cata-lysts. Sample preparation and characterization are detailed in Padró and Apesteguía [10]. Table 15.1 shows the physicochemical characteristics (surface area, pore diameter, chemical composition) and the acidity of the samples. Sample acid properties were probed by temperature-programmed desorption (TPD) of NH_3 preadsorbed at 373 K and by infrared spectroscopy (IR) of preadsorbed pyridine. The NH_3 surface densities for acid sites in Table 15.1 were obtained by deconvolution and integration of TPD traces (not shown here). Sample HPA/MCM-41 showed a sharp NH_3 desorption peak at about 910 K which accounts for the strong Brønsted acid sites present on this material. The evolved NH_3 from HY, HZSM-5, and SiO_2–Al_2O_3 gave rise to a peak at 483–493 K and a broad band between 573 and 773 K. In contrast, Al-MCM-41 did not exhibit the high-temperature NH_3 band but a single asymmetric broad band with a maximum around 482–496 K. Zeolites HY and HZM-5 exhibited the highest surface acid densities per m^2 (about 2.2 μmol/m^2).

The density and nature of surface acid sites were determined from the IR spectra of adsorbed pyridine. The relative contributions of Lewis and Brønsted acid sites were obtained by decon-volution and integration of pyridine absorption bands appearing at around 1450 and 1540 cm^{-1}, respectively (Table 15.1). In agreement with the results obtained by TPD of NH_3, the amount of pyridine adsorbed on Al-MCM-41 after evacuation at 423 K, in particular on Brønsted sites, was clearly lower as compared to acid zeolites or SiO_2–Al_2O_3, reflecting the moderate acidic character of mesoporous Al-MCM-41 sample. The areal peak relationship between Lewis (L) and Brønsted (B) sites on Al-MCM-41 was L/B = 4.2, higher than on SiO_2–Al_2O_3 (L/B = 3). The L/B ratio on HY was 1.5 while zeolite HZSM-5 contained a similar concentration of Brønsted and Lewis acid sites.

TABLE 15.1
Sample Physical Properties and Acidity

Catalyst	Surface Area S_g (m²/g)	Pore Diameter $\bar{d}_p$ (Å)	Si/Al	TPD of NH_3		IR of Pyridine	
				μmol/g	μmol/m²	B Area/g	L Area/g
HY	660	7.4	2.4	1380	2.1	310	465
HZSM-5	350	5.5	20	770	2.2	337	341
Al-MCM-41	925	30	18	340	0.4	32	135
HPA/MCM-41	505	29	—	352	0.7	—	—
SiO_2–Al_2O_3	560	45	11.3	1005	1.8	68	204

15.2.3 Catalytic Results

The gas-phase acylation of phenol (P) with acetic acid (AA) was carried out in a fixed bed, continuous-flow reactor at 553 K and 101.3 kPa. Standard catalytic tests were conducted at a contact time (W/F_P^0) of 146 g h/mol. Main products of phenol acylation with acetic acid were phenyl acetate (PA), o-HAP, and p-HAP; *para*-acetoxyacetophenone (p-AXAP) was detected in trace amounts. Phenol conversion (X_P, mol of phenol reacted/mol of phenol fed) was calculated as: $X_P = \Sigma Y_i /$ ($\Sigma Y_i + Y_P$), where ΣY_i is the molar fraction of products formed from phenol, and Y_P is the outlet molar fraction of phenol. The selectivity to product i (S_i, mol of product i/mol of phenol reacted) was determined as $S_i(\%) = [Y_i/\Sigma Y_i]100$. Product yields ($\eta_i$, mol of product i/mol of phenol fed) were calculated as $\eta_i = S_i X_P$.

On all the samples, o-HAP and PA were the predominant products. At similar phenol conversion levels, the initial o-HAP selectivity was between 67.1% (HZSM-5) and 39.1% (SiO_2–Al_2O_3) while the $S_{p\text{-HAP}}^0$ values were always lower than 8%. Figure 15.1 shows the evolution of the formation rate of o-HAP as a function of time on stream. By comparing the experimental data in Figure 15.1 at the beginning of the reaction, it is inferred that zeolites HZSM-5 and HY are clearly more active than HPA/MCM-41 and SiO_2–Al_2O_3 samples for producing o-HAP from phenol, while Al-MCM-41 shows intermediate $r_{o\text{-HAP}}$ values.

By determining the effect of contact time on the product distribution we identified the primary and secondary reaction pathways involved in the synthesis of o-HAP from phenol and acetic acid [10]. Specifically, we proposed (Figure 15.2) that o-HAP is formed from phenol and AA via two parallel pathways: (1) The direct C-acylation of phenol; (2) the O-acylation of phenol forming the PA intermediate which is consecutively transformed to o-HAP via intramolecular Fries rearrangement or intermolecular phenol/PA C-acylation. The relative rate of the different pathways involved in Figure 15.2 greatly depends on the solid acid employed. In fact, as shown in Figure 15.1, the o-HAP formation rate was higher on acid zeolites containing strong Brønsted and Lewis acid sites (zeolites HY and HZSM-5) as compared to samples containing only Brønsted acid sites (HPA/MCM-41), or exhibiting moderate acidity (SiO_2–Al_2O_3 and Al-MCM-41). This result suggested that both strong Brønsted and Lewis sites are required to produce efficiently o-HAP via both the direct C-acylation of phenol and the acylation of phenyl acetate intermediate formed from O-acylation of phenol. In contrast, the Fries rearrangement of phenyl acetate to o-HAP occurs only on samples containing very strong Brønsted, such as HPA/MCM-41 [11].

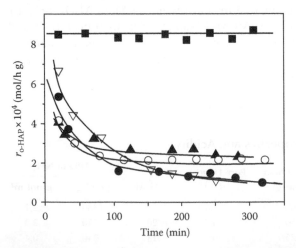

FIGURE 15.1   Formation rate of o-HAP as a function of time on stream on HZSM-5 (■), HY (▽), HPA/MCM-41 (●), SiO_2–Al_2O_3 (○), Al-MCM-41 (▲) (553 K, 101.3 kPa total pressure, $W/F_P^0 = 146$ g h/mol, P/AA = 1, N_2/(P + AA) = 45).

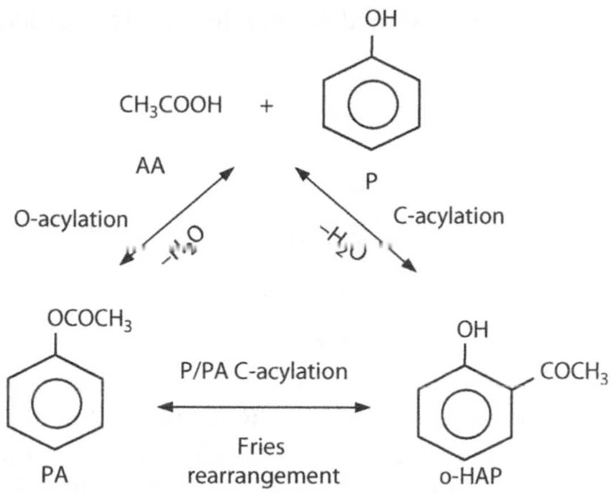

FIGURE 15.2 Synthesis of o-HAP by acylation of phenol with acetic acid.

Figure 15.1 also shows that the o-HAP formation rate does not change with time on HZSM-5, but rapidly decreases on the other samples, particularly on HPA/MCM-41 and HY. Coke formation was determined by analyzing the samples after the catalytic tests by temperature-programmed oxidation. The amount of carbon on the samples ranged from 18.4%C on HY to 2.9%C on ZSM-5. The %C formed on Al-MCM-41, SiO_2–Al_2O_3, and HY increased with the sample acidity, that is, %C was in the order Al-MCM-41 < SiO_2–Al_2O_3 < HY. On the other hand, it was observed that catalyst deactivation increased with the amount of carbon on the sample, thereby suggesting that the activity decay for the formation of o-HAP is caused by coke formation. Additional studies were performed in order to determine the catalyst deactivation mechanism, and to ascertain the causes for the superior stability of zeolite HZSM-5 on stream. Previous work [12], reported that formation of coke via the irreversible polymerization of highly reactive ketene is the main reason of the rapid deactivation observed during the liquid-phase Fries rearrangement of PA on solid acids. Ketenes may be formed via the conversion of PA to P according to Reaction 15.1, and are extremely reactive and unstable compounds that dimerize to diketenes and polymerize very quickly. However, we determined that ketenes are not present during the gas-phase acylation of phenol with AA because ketenes rapidly react with water formed in reaction (see Figure 15.2) to produce acetic acid [10].

In an attempt to ascertain the nature of the species responsible for coke formation in the

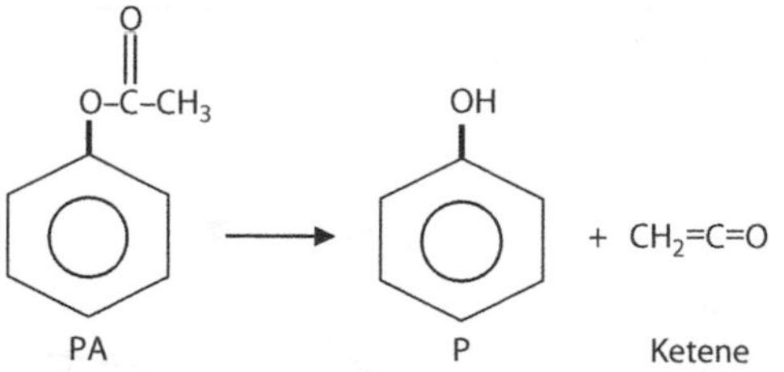

$$\text{(15.1)}$$

synthesis of o-HAP from acylation of P with AA, we studied the conversion of o-HAP with AA on zeolites HY and ZSM-5. The coinjection of o-HAP with AA on HY formed P, PA, and o-acetoxyacetophenone (o-AXAP), and rapid activity decay was observed. In contrast, HZSM-5 did not produce o-AXAP and did not deactivate during o-HAP/AA conversion reactions. The observed HY deactivation was therefore related with the formation of o-AXAP. Neves et al. [13] reported that coke formed on MFI zeolites during the acylation of phenol with AA is mainly constituted by methylnaphtols, 2-methylchromone and 4-methylcoumarine. Formation of 2m-chromone and

4m-coumarine may take place from o-HAP and AA via the initial formation of o-AXAP as depicted in Reaction 15.2:

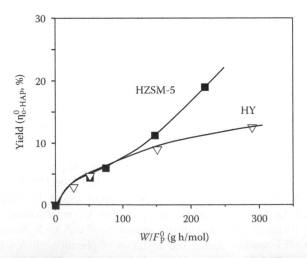

$$\text{o-HAP} + \text{AA} \longrightarrow \text{o-AXAP} \longrightarrow \begin{array}{l} 2m\text{-Chromone} \\ 4m\text{-Coumarine} \end{array} \qquad (15.2)$$

The assumption that coke is formed essentially via Reaction 15.2 is also consistent with results showing that zeolite HZSM-5 that does not produce o-AXAP when cofeeding o-HAP and AA, does not deactivate. It seems therefore that the superior stability of zeolite HZSM-5 is due to a shape-selectivity effect that avoids formation of the coke precursor specie. In other words, zeolite HZSM-5 does not deactivate because its narrow pore size structure hinders the formation of o-AXAP that is the key coke precursor in the gas-phase acylation of phenol with acetic acid.

The o-HAP yield on HZSM-5 may be improved by selecting proper reaction conditions. Figure 15.3 shows the evolution of initial o-HAP yield ($\eta^0_{\text{o-HAP}}$) as a function of contact time over zeolites HZSM-5 and HY. $\eta^0_{\text{o-HAP}}$ increases with W/F^0_P on both zeolites, but formation of o-HAP is clearly favored on HZSM-5 at high W/F^0_P values. On the other hand, in Figure 15.4 is plotted the evolution of $\eta^0_{\text{o-HAP}}$ as a function of P_{AA} over HZSM-5 and Al-MCM-41 samples. By changing the reactant AA/P ratio from 0.5 to 4, the $\eta^0_{\text{o-HAP}}$ is increased from 6.0% to 38.6% on HZSM-5. Figure 15.4 also confirms the superior activity of HZSM-5 to produce o-HAP as compared to Al-MCM-41.

In summary, our studies show that zeolite HZSM-5 is an active and stable catalyst for efficiently promoting the synthesis of o-HAP from phenol acylation in gas phase. Zeolite HZSM-5 contains strong Lewis and Brønsted acid sites and effectively catalyzes the two main reaction pathways leading from phenol to o-HAP, that is, the direct C-acylation of phenol and the O-acylation of phenol forming the PA intermediate which is consecutively transformed via intermolecular phenol/PA C-acylation. Besides, on zeolite HZSM-5 the o-hydroxycetophenone yield remains stable on stream and formation of coke is drastically suppressed. The superior stability of zeolite HZSM-5 is due to the fact that the microporous structure of this zeolite avoids the formation of bulky o-acetoxyacetophenone, which is the key intermediate for coke formation.

FIGURE 15.3 Initial o-HAP yield as a function of contact time (553 K, 101.3 kPa total pressure, P/AA = 1).

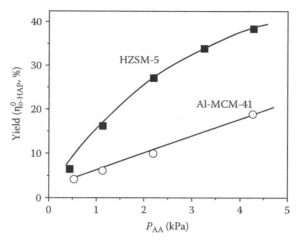

FIGURE 15.4 Initial o-HAP yield as a function of P_{AA} (553 K, 101.3 kPa total pressure, $P_P = 1.10$ kPa, $W/F_P^0 =$ 146 g h/mol).

15.3 SUSTAINABLE CATALYTIC PROCESSES FOR FINE CHEMICAL SYNTHESIS

15.3.1 INTRODUCTION

Menthol is a fine chemical of significant industrial interest because it is widely employed in pharmaceuticals, cosmetics, toothpastes, and chewing gum as well as in cigarettes. The menthol molecule is a terpenoid containing three quiral centers which account for the four resulting enantiomers: menthol, isomenthol, neomenthol, and neo-isomenthol. Because each individual enantiomer forms two optical isomers (the dextro(+) and levo(−) isomers) menthol comprises eight optically active isomers, but only (−)-menthol possesses the characteristic peppermint odor and exerts a unique cooling sensation on the skin and mucous membranes. Most of (−)-menthol is obtained from natural essential oils, but its production by synthesis has increased lately. In 1998, the production of synthetic menthol was 2500 metric tons which represented about 20% of the world production of menthol [14]. Synthetic menthol is currently produced in the world by two companies, Symrise (ex-Haarmann & Reimer) and Takasago, respectively. The Haarmann & Reimer process [15,16] uses thymol as raw material for obtaining racemic (±)-menthol, which is then resolved into pure (−)-menthol by a crystallization process. Takasago developed in the early 1980s an asymmetric synthesis technology for producing (−)-menthol from myrcene [17]. The key of the Takasago process was the use of a chiral Rh BINAP catalyst for transforming diethylgeranylamine obtained from myrcene to the chiral 3R-citronellal enamine with more than 95% enantiomeric excess.

Considerable effort has been devoted to the production of (−)-menthol by synthetic or semisynthetic means from other more readily reliable raw materials. Our research group recently reported [18,19] for the first time the selective synthesis of menthols from citral in a one-step process which involves the initial hydrogenation of citral to citronellal, followed by the isomerization of citronellal to isopulegols, and the final hydrogenation of isopulegols to menthols. Menthols from citral is an attractive synthetic route because citral is a renewable raw material that is mainly obtained by distillation of essential oils, such as lemongrass oil that contains ca. 70%–80% citral. However, the citral conversion reaction network potentially involves a complex combination of series and parallel reactions, as is depicted in Figure 15.5. The direct synthesis of menthols from citral requires, therefore, the development of highly selective bifunctional metal-acid catalysts. The main results obtained by our research group with the aim of developing novel bifunctional catalysts to efficiently promote the liquid-phase synthesis of menthols from citral are described here.

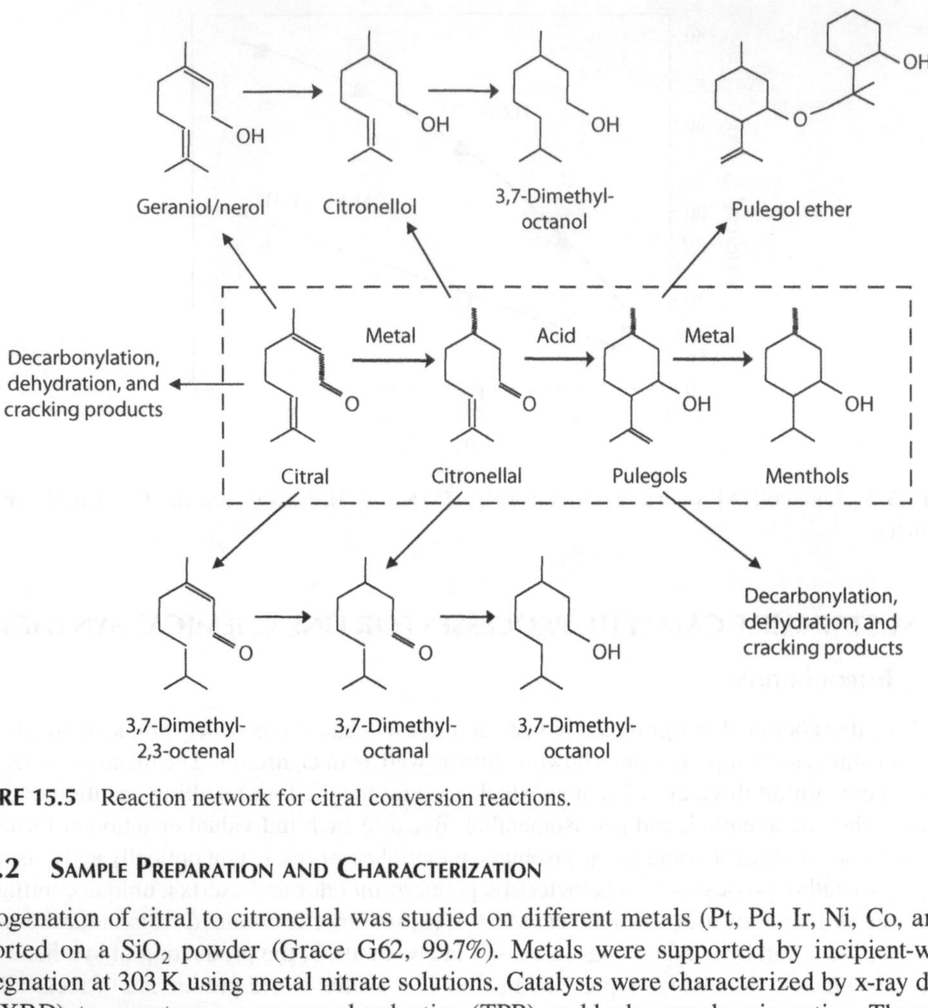

FIGURE 15.5 Reaction network for citral conversion reactions.

15.3.2 SAMPLE PREPARATION AND CHARACTERIZATION

Hydrogenation of citral to citronellal was studied on different metals (Pt, Pd, Ir, Ni, Co, and Cu) supported on a SiO_2 powder (Grace G62, 99.7%). Metals were supported by incipient-wetness impregnation at 303 K using metal nitrate solutions. Catalysts were characterized by x-ray diffraction (XRD), temperature-programmed reduction (TPR), and hydrogen chemisorption. The catalyst metal loadings and surface areas together with characterization results are shown in Table 15.2.

TABLE 15.2
Catalyst Characterization and Catalytic Data for Citral Hydrogenation
($T = 393$ K, $P = 1013$ kPa, $W_{cat} = 1$ g)

Catalyst	XRD	TPR T_{Max} (K)	S_g (m²/g)	H₂ Chemisorption (cm³/mol Metal)	$S_{Cit}^{max\ a}$ (%)	S_i^b (%) Citronellal	G–N[c]	Others
Ni(12%)/SiO₂	NiO	665	250	0.53	99	97	0	3
Cu(12%)/SiO₂	CuO	523	218	0.08	53	53	41	6
Co(12%)/SiO₂	CoO	656	240	0.16	14	11	81	8
Pt(0.3%)/SiO₂	—	395	280	1.49	59	59	25	16
Pd(0.7%)/SiO₂	—	393	250	1.98	91	71	0	29
Ir(1%)/SiO₂	—	359	230	1.95	50	32	60	8

[a] Maxima selectivities to citronellal.
[b] Selectivities at 60 min.
[c] Selectivity to geraniol and nerol isomers.

TABLE 15.3

Characterization of Sample Acidity

Catalyst	TPD of NH$_3$ (μmol/g)	IR of Pyridine		
		B (μmol/g)	L (μmol/g)	L/(L + B)
H-BEA	496	100	120	0.55
Al-MCM-41	110	20	36	0.64
ZnO/SiO$_2$	2200	0	50	1
Cs-HPA	37	—	—	—

Isomerization of citronellal to pulegols was carried out on Al-MCM-41 (Si/Al = 10, S_g = 780 m^2/g), zeolite Beta (Zeocat PB, Si/Al = 25, S_g = 630 m^2/g), ZnO(25%)/SiO$_2$, and Cs-HPA (Cs$_{0.5}$H$_{2.5}$PW$_{12}$O$_{40}$, S_g = 130 m^2/g). Sample preparation details have been reported elsewhere [18]. Acid site densities were determined by deconvolution and integration of TPD of NH$_3$ and the results are given in Table 15.3. On a weight basis, ZnO/SiO$_2$ exhibited the highest density of acid sites, probably reflecting the presence of chloride ions on the surface (sample was prepared using an aqueous solution of Cl$_2$Zn), but the TPD peak maximum appeared at relatively low temperature (about 523 K). The evolved NH$_3$ from zeolite Beta (HBEA) and Al-MCM-41 gave rise to a single asymmetric broad band with a maximum around 482–496 K. Cs-HPA desorbed NH$_3$ in a single peak centered at about 893 K, reflecting superior acid site strength compared to the other samples. Table 15.3 also shows the nature of the surface acid sites as determined from IR spectra obtained after adsorption of pyridine at room temperature and evacuation at 423 K. As it is well known, Cs-HPA contains only Brønsted sites [20]. Sample ZnO/SiO$_2$ contained only Lewis sites while the relative concentration of Brønsted and Lewis acid sites (L/B) on Al-MCM-41 and H-BEA were 1.8 and 1.2, respectively. Results of Tables 15.2 and 15.3 reveal that zeolite H-BEA contains a higher density of stronger acid sites compared with Al-MCM-41.

15.3.3 CATALYTIC RESULTS

Figure 15.5 shows that the selective formation of menthol from citral requires bifunctional metal/acid catalysts with the ability of not only promoting coupled hydrogenation/isomerization reactions of the citral-to-menthol pathway but also minimizing the parallel hydrogenation reactions of citral to nerol/geraniol or 3,7-dimethyl-2,3-octenal, and of citronellal to citronellol or 3,7-dimethyloctanal. In other words, from a kinetic point of view it is required that (1) the formation rate of citronellal from citral was much higher than the hydrogenations rates of citral to nerol/geraniol and 3,7-dimethyl-2,3-octenal, and (2) the formation rate of pulegols from citronellal was much higher than the hydrogenations rates of citronellal to citronellol and 3,7-dimethyloctanal. Thus, the individual steps involved in the reaction pathway leading to menthols from citral were studied separately in order to select the metallic and acid functions of the bifunctional catalyst.

Monometallic catalysts of Table 15.2 were tested for the liquid-phase hydrogenation of citral (T = 393 K, P = 1013 kPa, W_{cat} = 1 g) using solvent isopropanol. Pd/SiO$_2$ and Ni/SiO$_2$ selectively hydrogenated the conjugated C=C bond of the citral molecule giving initially more than 90% selectivity to citronellal (Table 15.2). This result showed that on both catalysts the citral hydrogenation to citronellal is clearly favored compared to parallel hydrogenations leading to nerol/geraniol and 3,7-dimethyl-2,3-octenal. The citronellal selectivity then decreased with reaction time because citronellal is in turn hydrogenated to citronellol or 3,7-dimethyloctanal. In contrast, the maximum selectivity to citronellal was never higher than 60% on the other catalysts which formed significant amounts of nerol/geraniol isomers. Overall, results of Table 15.2 are consistent with previous works on citral hydrogenation showing that Ni and Pd favor C=C bond hydrogenation [21,22] while Co and Ir are more selective for C=O bond hydrogenation [22,23].

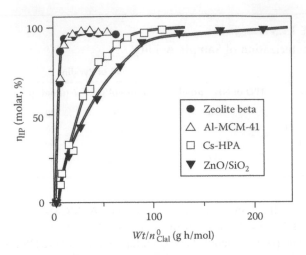

FIGURE 15.6   Cyclization of citronellal to isopulegols: Isopulegol yields as a function of parameter Wt/n_{Clal}^0 (343 K, 506.5 kPa nitrogen, $W = 0.200$ g, citronellal:toluene = 2:150 [mL]).

Solid acids of Table 15.3 were tested in the liquid-phase isomerization of citronellal to isopulegols ($T = 343$ K, $P_{N2} = 506.5$ kPa, $W = 0.200$ g, citronellal:toluene [mL] = 2:150) using as solvent toluene. Isopulegol yields are shown in Figure 15.6 as a function of Wt/n_{Clal}^0, where W is the catalyst weight, t the reaction time, and n_{Clal}^0 the initial moles of citronellal. The local slope of each product in Figure 15.6 gives its rate of formation at a specific value of reactant conversion and contact time. In all the cases, pulegol isomers were the only products detected.

Figure 15.6 shows that the citronellal cyclization rate was clearly higher on zeolite Beta and Al-MCM-41 as compared to both ZnO/SiO₂ and Cs-HPA. The exact nature of the surface active sites required for efficiently catalyzing the cyclization of citronellal to isopulegols is still debated. While several authors [24,25] reported that the reaction is readily catalyzed on Lewis acids, others [26] correlated the cyclization activity on acid zeolites with accessible Brønsted acid sites. Chuah et al. [27] found that catalytic materials containing strong Lewis and weak Brønsted acidity show good activity and selectivity for cyclization of citronellal to isopulegol. Solid acids of Table 15.3 contain either Lewis (ZnO/SiO₂), Brønsted (Cs-HPA), or both Lewis and Brønsted acid sites (Al-MCM-41 and H-BEA). The superior activity showed by zeolite Beta and Al-MCM-41 samples for the formation of isopulegols are consistent with the assumption that Lewis/weak Brønsted dual sites are required to efficiently catalyze the citronellal cyclization.

Based on the aforementioned results, three bifunctional catalysts containing one of the metals most selective for hydrogenating citral to citronellal (Pd or Ni) and one of the solid acids more active for converting citronellal to isopolugelos (zeolite Beta or Al-MCM-41) were prepared: Pd(1%)/H-BEA, Ni(3%)/H-BEA, and Ni(3%)/Al-MCM-41. These bifunctional catalysts were tested for the conversion of citral to menthols ($T = 343$ K, $P_T = 506.5$ kPa, $W = 1$ g) using toluene as solvent. In all the cases, citral and citronellal were totally converted after 5 h of reaction. Figure 15.7 shows the evolution of the yield of total menthols as a function of reaction time. It is observed that the menthols yield was only about 20% on Pd/H-BEA catalyst at the end of the catalytic test. This poor selectivity for menthol synthesis reflected the high activity of Pd/H-BEA to hydrogenate the C=C bond of citronellal, thereby forming considerable amounts of 3,7-dimethyl-octanal. Pd/H-BEA also formed significant amounts of undesirable products (35%) via secondary decarbonylation and hydrogenolysis reactions. In contrast, Figure 15.7 shows that the menthols selectivity on Ni/H-BEA was 81% at the end of the catalytic test. None of the byproducts formed from hydrogenation of citral or citronellal was detected on Ni/H-BEA suggesting that this bifunctional catalyst satisfactory combines the hydrogenation and isomerization functions needed to selectively promote the reaction pathway leading from citral to menthols. However, formation of secondary compounds formed probably via decarboxylation and cracking reactions on the strong

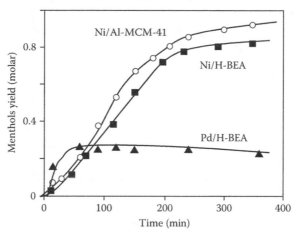

FIGURE 15.7 Menthol synthesis from citral: total menthol yields as a function of time (343 K, 506.5 kPa total pressure, $W = 1$ g, citral:toluene = 2:150 [mL]).

acid sites of zeolite Beta was significant. The best catalyst was Ni/Al-MCM-41 that yielded ca. 90% of menthols. The observed menthol yield improvement on Ni/Al-MCM-41 is explained by considering that the moderate acid sites of Al-MCM-41 do not promote the formation of byproducts via side cracking reactions. Table 15.4 shows the distribution of menthol isomers obtained at the end of catalytic runs. (±)-Neo-isomenthol was never detected in the products. On Ni-based catalysts, the menthol mixture was composed of 70%–73% of (±)-menthols, 15%–20% of (±)-neomenthol, and 5%–10% of (±)-isomenthol. On Pd/Beta the racemic (±)-menthol mixture represented only about 50% of total menthols.

Finally, an additional test was performed on Ni/Al-MCM-41 by increasing the hydrogen pressure to 2026 kPa. Results are presented in Figure 15.8 and Table 15.4. Figure 15.8 shows the evolution of product yields and citral conversion as a function of time. Citral was totally converted to citronellal on metallic Ni crystallites, but the concentration of citronellal remained very low because it was readily converted to pulegols on acid sites of mesoporous Al-MCM-41 support. Pulegols were then totally hydrogenated to menthols on metal Ni surface sites. The menthols yield reached 94% at the end of the test showing the beneficial effect of increasing P_{H2}, probably because it diminishes the formation of undesirable products via secondary reactions. In contrast, menthol isomer distribution was not changed by increasing the hydrogen pressure (Table 15.4).

In summary, results presented here shows that the liquid-phase synthesis of menthols from citral was successfully achieved using proper bifunctional catalysts. In fact, at P_{H2} = 2026 kPa Ni(3%)/Al-MCM-41 yields 94% menthols directly from citral and produces about 72% of racemic (±)-menthol into the menthol mixture.

TABLE 15.4
Menthol Synthesis from Citral: Menthol Isomer Distribution
(343 K, $W = 1$ g, Citral:Toluene = 2:150 [mL])

		Menthol Isomer Distribution (%)		
Catalyst	Pressure (kPa)	(±)-Menthols	(±)-Neomenthol	(±)-Isomenthol
Pd/H-BEA	506.5	47.2	15.6	37.2
Ni/H-BEA	506.5	72.0	21.3	6.7
Ni/Al-MCM-41	506.5	72.3	20.2	7.5
Ni/Al-MCM-41	2026.0	71.1	20.0	8.9

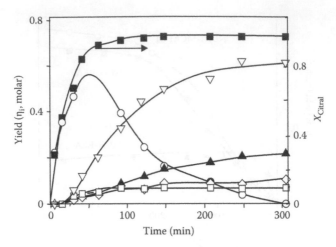

FIGURE 15.8 Menthol synthesis from citral on Ni(3%)/Al-MCM-41. Product yields and citral conversion as a function of time. Isopulegols (O), (±)-menthol (□), (±)-neomenthol (▲), (±)-isomenthol (◇), others (□) (343 K, 2026 kPa total pressure, W = 1 g, citral:toluene = 2:150 [mL]).

15.4 VALORIZATION OF RENEWABLE LOW-VALUE RAW MATERIALS

15.4.1 INTRODUCTION

Argentina is a top supplier of soybean oil and the fourth largest producer of biodiesel after the EU, USA, and Brazil. The biodiesel production capacity in Argentina stands in 2010 at 2.5 million tons per year. In 2009, Argentina's biodiesel exports reached 1.15 million tons. Regarding the local market, the government law requiring all diesels to be mixed with 7% of the plant-based fuel represents production of about 1.1 million tons of biodiesel in 2010.

Commercial processes of transesterification of vegetable oils with methanol to biodiesel produce about 10% of glycerol (Gly) as byproduct. The increasing production of low-cost glycerol due to current and future biodiesel demands causes environmental and economical concerns since the drop of the Gly price has forced the producers to burn or sell Gly without even refining it [28]. There is a need, therefore, of developing novel catalytic processes to convert Gly to valuable chemicals. In this regard, the synthesis of monoglycerides (MG) via the transesterification of Gly with fatty acid methyl esters (FAME) is an interesting alternative technology for obtaining fine chemicals from bio-resources [29]. Monoglycerides are widely used as emulsifiers in food, pharmaceutical, cosmetic, and detergent industries [30]. Although MG can be also formed from glycerol using triglycerides or fatty acids, the process using FAME has several advantages, for example, FAME is less corrosive than fatty acids, has lower hydrophobic character than triglycerides and exhibits higher miscibility with glycerol.

The current technology for producing MG involves the use of corrosive liquid bases and therefore, neutralization and elimination of the resulting salts from the reaction media is necessary [31]. Moreover, the process is not selective and forms a mixture of mono-, di- (DG) and triglycerides (TG), as shown in Reaction 15.3. The use of heterogeneous catalysis may certainly improve this technology, since solid catalysts can be easily separated from the reaction media and are often reusable. Previous research on this reaction using solid catalysts includes the use of metal oxides such as ZnO, La_2O_3, MgO and CeO_2 [32], Al-Mg mixed oxides [29], and silica-immobilized bases [33]. Results have shown that the reaction activity and selectivity greatly depend on the catalyst basicity. Here, we present our results on the glycerolysis of FAME using solid catalysts with different surface acid–base properties.

$$\underset{\text{FAME}}{\overset{R}{\underset{O}{\bigvee}}\overset{O}{\underset{CH_3}{\bigvee}}} + \underset{\text{Glycerol}}{\overset{OH}{\underset{OH}{\bigvee}}OH} \longrightarrow \underset{\text{Monoglyceride}}{\overset{CH_2\text{-}OCO\text{-}R}{\underset{CH_2\text{-}OH}{\overset{|}{\underset{|}{CH\text{-}OH}}}}} + CH_3OH \overset{+\text{ FAME}}{\longrightarrow} \underset{\text{Diglycerides}}{\overset{CH_2\text{-}OCO\text{-}R}{\underset{CH_2\text{-}OH}{\overset{|}{\underset{|}{CH\text{-}OCO\text{-}R}}}}} + \underset{\text{Triglycerides}}{\overset{CH_2\text{-}OCO\text{-}R}{\underset{CH_2\text{-}OCO\text{-}R}{\overset{|}{\underset{|}{CH\text{-}OCO\text{-}R}}}}} + CH_3OH \quad (15.3)$$

15.4.2 Sample Preparation and Characterization

Y_2O_3, ZnO, and CeO_2 single oxides were prepared as detailed in [34]. Magnesium oxide was prepared by hydration with distilled water of low-surface-area commercial MgO (Carlo Erba 99%; $27\,m^2/g$). Then the resulting $Mg(OH)_2$ was decomposed and stabilized in a N_2 flow for 1 h at 373 K, then for 1 h at 623 K, and finally for 18 h at 773 K to obtain high-surface-area MgO. $Zr(OH)_4$ precursor was obtained by precipitation of zirconium oxychloride with ammonium hydroxide; then, it was decomposed in air at 723–773 K overnight in order to obtain ZrO_2. Nb_2O_5 was prepared by decomposition at 773 K of commercial hydrated niobium pentoxide HY340 from Companhia Brasileira de Metalurgia e Mineracao (CBMM). Alumina was a commercial sample of γ-Al_2O_3 Cyanamid Ketjen CK 300.

The BET surface area, electronegativity and acid and base site densities of the different oxides used in this work are reported in Table 15.5. Oxide electronegativities (χ_{oxide}) were calculated as the geometric mean of the atomic electronegativities using the Pauling's electronegativity scale. For an oxide with the formula $MnOx$, the bulk oxide electronegativity is [35]

$$\chi_{oxide} = [(\chi_M)^n (\chi_O)^x]^{1/(n+x)}$$

The catalyst acidic properties were determined by TPD of NH_3. Acid site densities (n_a) calculated by integration of the NH_3 TPD profiles are reported in Table 15.5. As expected, the n_a values increased with increasing the catalyst electronegativity. The catalyst basic properties were determined by TPD of CO_2. The base site density, n_b, was calculated by integration of TPD curves as the total amount of CO_2 desorbed from the catalysts, and the resulting values are reported in Table 15.5. It is observed that the total base site density decreased with increasing oxide electronegativity. Thus, the n_b value was 655 μmol/g on the most basic single oxide (MgO) while the density of basic sites was only 5 μmol/g on the most electronegative catalyst (Nb_2O_5).

15.4.3 Catalytic Results

The transesterification of methyl oleate (FAME) with glycerol was carried out at 473–523 K in a seven necked cylindrical glass reactor with mechanical stirring equipped with a condenser to remove the methanol generated during reaction. Details of the four-phase reactor, operation conditions,

TABLE 15.5
Electronegativity, Surface Area, and Catalyst Base–Acid Properties

Catalyst	Electronegativity χ_{oxide} (Pauling Unit)	Surface Area (m²/g)	Base Site Density n_b (µmol/g)	Acid Site Density n_a (µmol/g)
MgO	2.12	192	655	15
Y_2O_3	2.27	54	280	12
CeO_2	2.37	75	1457	21
ZnO	2.38	18	3	3
ZrO_2	2.51	83	58	22
Al_2O_3	2.54	230	14	101
Nb_2O_5	2.76	84	5	125

and sample analysis and quantification are detailed elsewhere [36]. Conversion (X_{FAME}, referred to the total content of esters in the reactant), selectivity (S) and yield (Y) were calculated through the following equations (n_j, mol of product j; MG = monoglycerides (both isomers), DG = diglycerides (both isomers), TG = triglyceride):

$$X_{FAME}(\%) = \frac{n_{MG} + 2n_{DG} + 3n_{TG}}{n_{MG} + 2n_{DG} + 3n_{TG} + n_{FAME}} \cdot 100$$

$$S_{MG}(\%) = \frac{n_{MG}}{n_{MG} + 2n_{DG} + 3n_{TG}} \cdot 100$$

$$S_{DG}(\%) = \frac{2n_{DG}}{n_{MG} + 2n_{DG} + 3n_{TG}} \cdot 100$$

$$S_{TG}(\%) = \frac{3n_{TG}}{n_{MG} + 2n_{DG} + 3n_{TG}} \cdot 100$$

$$Y_j(\%) = X_{FAME}\, S_j \cdot 100$$

Figure 15.9 shows FAME conversion and glyceride yields obtained at 493 K on CeO_2 and typically illustrates the time-on-stream behavior of the catalysts during the reaction. Initially, only MG formation was observed, but as the reaction proceeded DG was also formed, according to the consecutive reaction pathway described in Reaction 15.3. At the end of the run, the yields for MG and DG on CeO_2 were 63% and 28%, respectively. No TG formation was observed.

All the samples of Table 15.5 were tested for the glycerolysis of FAME. Similarly to the results shown in Figure 15.9 for CeO_2, the products detected were only MG and DG in all the cases. In Figure 15.10 we have plotted the MG yields as a function of reaction time. Basic oxides such as MgO, Y_2O_3, and CeO_2 efficiently catalyzed the reaction and converted more than 90% of FAME in less than 8 h, yielding more than 60% of monoglycerides. In contrast, acidic oxides such as Al_2O_3, ZrO_2, and Nb_2O_5 were clearly less active for MG synthesis than basic oxides producing less than 20% MG at the end of the runs.

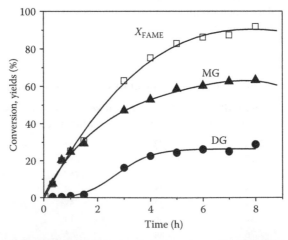

FIGURE 15.9 FAME conversion and glyceride yields on CeO_2 (Gly/FAME = 4.5 (molar), T = 493 K, W/n_{FAME}^0 = 30 g/mol).

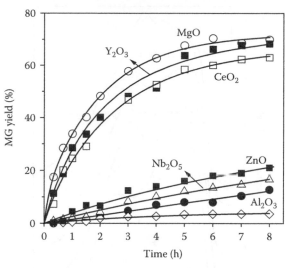

FIGURE 15.10 Monoglyceride yields on different single oxides. Reaction conditions as in Figure 15.9.

From the plots of Figure 15.10, we determined the initial MG formation rate (r_{MG}^0, mmol/gh) by calculating the initial slopes of the MG yield curves according to

$$r_{MG}^0 = \frac{n_{FAME}^0}{W_{cat}} \left[\frac{dY_{MG}}{dt} \right]_{t=0}$$

where
   W_{cat} is the catalyst weight
   n_{FAME}^0 is the molar amount of FAME initially loaded in the reactor

The obtained r_{MG}^0 values were plotted as a function of the density of basic sites (Table 15.5) in Figure 15.11. The initial MG formation rate increased monotonically with the sample basicity, thereby confirming that the glycerolysis of FAME is promoted on basic sites.

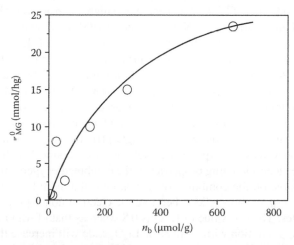

FIGURE 15.11 Initial formation rate of the MG as a function of base site density. Reaction conditions as in Figure 15.9.

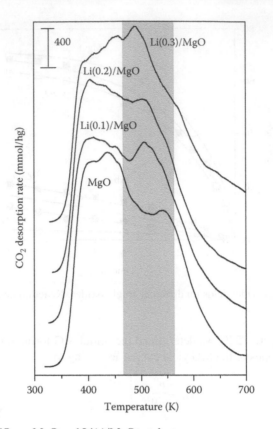

FIGURE 15.12   TPD of CO_2 on MgO and Li(x)/MgO catalysts.

Figure 15.11 shows that the highest MG yield at 493 K was obtained on MgO that converted FAME totally in 6 h and produced 70% MG. This MG yield is higher than those typically reported using homogeneous catalysts (40%–60%) [29,30].

Magnesium oxide was doped with increasing amounts of lithium with the aim of improving the MgO activity for the glycerolysis of FAME. Promotion of metal oxides with alkaline cations has been widely employed to catalyze base-promoted reactions, such as alcohols coupling [37], aldol condensations [38,39], methane coupling [40], and alcohol decompositions [41]. In particular, we have noted [30] that the addition of Li to MgO generates a higher density of high-strength basic site in comparison to other alkaline cations (Na, K, Cs). Three Li(x)/MgO samples were prepared by incipient wetness impregnation; x represents the wt% of lithium. Details of the catalyst preparation and characterization are reported elsewhere [42]. The surface base properties of the Li(x)/MgO samples were investigated by TPD of CO_2 preadsorbed at room temperature. Figure 15.12 shows the CO_2 desorption rate as a function of desorption temperature for MgO and Li(x)/MgO catalysts. The CO_2 desorption peak at high temperature (500–550 K) increased with the Li content on the sample. This high-temperature peak is attributed to the release of unidentate carbonates from low coordination oxygen anions, the stronger surface basic sites on MgO [41]. The increase of the base site strength with increasing Li content may be explained by taking into account that the basicity of an oxide surface is related to the electrodonating properties of the combined oxygen anions, so that the higher the partial negative charge on the combined oxygen anions, the more basic the oxide. The oxygen partial negative charge ($-q_0$) would reflect therefore the electron donor properties of the oxygen in single-component oxides. The $-q_0$ value of Li_2O is 0.8 whereas that of MgO is 0.5, and therefore it is expected that surface promotion with more basic Li_2O oxide will increase the basicity of MgO.

Besides, the ionic radius of Li^+ is similar to that of Mg^{2+} ion, and replacement of Mg^{2+} by Li^+ in the MgO lattice may take place. Substitution of a divalent ion by a monovalent one in the MgO matrix

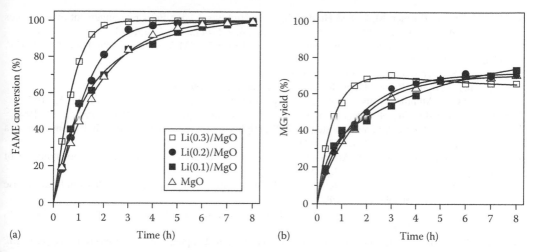

FIGURE 15.13   (a) FAME conversions and (b) MG yields on MgO and Li(x)/MgO catalysts. Reaction conditions as in Figure 15.9.

requires the formation of O^- anions in order to maintain electroneutrality, resulting in strained Mg–O bonds and formation of $[Li^+O^-]$ species, which causes the generation of strong basic sites.

Figure 15.13 shows the evolution of FAME conversion and MG yields with reaction time on Li(x)/MgO samples. The catalyst activity increased with Li content on the sample. For example, Li(0.3)/MgO converted 100% FAME in 3 h while on MgO total FAME conversion was reached in about 7 h. This result confirmed that the transesterification of Gly with FAME is particularly promoted by strong base sites. Regarding the production of MG, the Li(0.3)/MgO sample yielded 70% MG after 3 h reaction, when X_{FAME} reached 100%. Then, Y_{MG} decreased slowly, probably because of the formation of secondary products.

In summary, the liquid-phase monoglyceride synthesis by glycerolysis of methyl oleate is efficiently carried out on solid base catalysts such as MgO, Y_2O_3, and CeO_2. Therefore, solid bases offer an alternative technology to the current commercial processes that use environmentally unfriendly liquid bases.

REFERENCES

1. Commarieu, A. et al., Fries rearrangement in methane sulfonic acid, an environmental friendly acid, *J. Mol. Catal. A Chemical*, 182, 137, 2002.
2. Uwaydah, I. et al., U.S. Patent 5, 696, 274, 1997.
3. Fritch, J., Fruchey, O., and Horlenko, T., U.S. Patent 4, 954, 652, 1990.
4. Vogt, A., Kouwenhoven, H., and Prins, R., Fries rearrangement over zeolitic catalysts, *Appl. Catal.*, 123, 37, 1995.
5. Jayat, F., Sabater Picot, M.J., and Guisnet, M., Solvent effects in liquid phase Fries rearrangement of phenyl acetate over a HBEA zeolite, *Catal. Lett.*, 41, 181, 1996.
6. Mueller, J. et al., U.S. Patent 4508924, 1985.
7. Freese, U., Hinrich, F., and Roessner, F., Acylation of aromatic compounds on H-Beta zeolites, *Catal. Today*, 49, 237, 1999.
8. Subba Rao, Y.V. et al., An improved acylation of phenol over modified ZSM-5 catalysts, *Appl. Catal. A General*, 133, L1, 1995.
9. Jayat, F. et al., Acylation of phenol with acetic acid. Effect of density and strength of acid sites on the properties of MFI metallosilicates, in *Studies Surface Science Catalysis*, vol. 105, Chon, H., Ihm, S.K., and Uh, S., Eds., Elsevier, Amsterdam, the Netherlands, 1997, p. 1149.
10. Padró, C.L. and Apesteguía, C.R., Gas-phase synthesis of hydroxyacetophenones by acylation of phenol with acetic, *J. Catal.*, 226, 308, 2004.
11. Padró, C.L., Sad, M.E., and Apesteguía, C.R., Acid site requirements for the synthesis of o-hydroxyacetophenone by acylation of phenol with acetic acid, *Catal. Today*, 116, 184, 2006.

12. Heidekum, A., Harmer, M.A., and Hölderich, W.F., Highly selective Fries rearrangement over zeolites and Nation in silica composite catalysts: A comparison, *J. Catal.*, 176, 260, 1998.
13. Neves, I. et al., Acylation of phenol with acetic acid over HZSM-5 zeolite, reaction scheme, *J. Mol. Catal.*, 93, 169, 1994.
14. Clark, G.S., Menthol, *Perfum. Flavor.*, 23, 33, 1998.
15. Fleischer, J., Bauer, K., and Hopp, R., German Patent, DE 2, 109, 456, 1971.
16. Davis, J.C., L-Menthol synthesis employs cheap, available feedstocks, *Chem. Eng.*, 11–12, 62, 1978.
17. Misono, M. and Nojiri, N., Recent progress in catalytic technology in Japan, *Appl. Catal.*, 64, 1, 1990.
18. Trasarti, A.F., Marchi, A.J., and Apesteguía, C.R., Highly selective synthesis of menthols from citral in a one-step process, *J. Catal.*, 224, 484, 2004.
19. Trasarti, A.F., Marchi, A.J., and Apesteguía, C.R., Design of catalyst systems for the one-pot synthesis of menthols from citral, *J. Catal.*, 247, 155, 2007.
20. Kozhevnikov, I.V., Catalysis by heteropoly acids and multicomponent polyoxometalates in liquid-phase reactions, *Chem. Rev.*, 98, 171, 1998.
21. Aramendía, M.A. et al., Selective liquid-phase hydrogenation of citral over supported palladium, *J. Catal.*, 172, 46, 1997.
22. Maki-Arvela, P. et al., Liquid phase hydrogenation of citral: Suppression of side reactions, *Appl. Catal. A General*, 237, 181, 2002.
23. Singh, U.K. and Vannice, M.A., Liquid-phase citral hydrogenation over SiO_2-supported group VIII metals, *J. Catal.*, 199, 73, 2001.
24. Ravasio, N. et al., Intramolecular ene reactions promoted by mixed cogels, in *Studies in Surface Science Catalysis*, vol. 108, Blaser, H.U., Baiker, A., and Prins, R., Eds., Elsevier, Amsterdam, the Netherlands, 1997, p. 625.
25. Milone, C. et al., Isomerisation of (+)citronellal over Zn(II) supported catalysts, *Appl. Catal. A General*, 233, 151, 2002.
26. Fuentes, M. et al., Cyclization of citronellal to isopulegol by zeolite catalysis, *Appl. Catal.*, 47, 367, 1989.
27. Chuah, G.K. et al., Cyclisation of citronellal to isopulegol catalysed by hydrous zirconia and other solid acids, *J. Catal.*, 200, 352, 2001.
28. Jérome, F., Pouilloux, Y., and Barrault, J., Rational design of solid catalysts for the selective use of glycerol as a natural organic building block, *ChemSusChem*, 1, 586, 2008.
29. Corma, A. et al., Lewis and Brönsted basic active sites on solid catalysts and their role in the synthesis of monoglycerides, *J. Catal.*, 234, 340, 2005.
30. Zheng, Y., Chen, X., and Shen, Y., Commodity chemicals derived from glycerol, an important biorefinery feedstock, *Chem. Rev.*, 108, 5253, 2008.
31. Corma, A. et al., Catalysts for the production of fine chemicals: Production of food emulsifiers, monoglycerides, by glycerolysis of fats with solid base catalysts, *J. Catal.*, 173, 315, 1998.
32. Bancquart, S. et al., Glycerol transesterification with methyl stearate over solid basic catalysts I. Relationship between activity and basicity, *Appl. Catal. A General*, 218, 1, 2001.
33. Kharchafi, G. et al., Design of well balanced hydrophilic-lipophilic catalytic surfaces for the direct and selective monoesterification of various polyols, *New J. Chem.*, 29, 928, 2005.
34. Ferretti, C. et al., Monoglyceride synthesis by glycerolysis of methyl oleate on solid-base catalysts, *Chem. Eng. J.*, 161, 346, 2010.
35. Sanderson, R.T., *Chemical Bonds and Bond Energy*, 2nd edn., Academic Press, New York, 1976.
36. Ferretti, C.A. et al., Heterogeneously-catalyzed glycerolysis of fatty acid methyl esters: Reaction parameter optimization, *Ind. Eng. Chem. Res.*, 48, 10387, 2009.
37. Xu, M. et al., Isobutanol and methanol synthesis on copper catalysts supported on modified magnesium oxide, *J. Catal.*, 171, 130, 1997.
38. Tanabe, K., Zhang, G., and Hattori, H., Addition of metal cations to magnesium oxide catalyst for the aldol condensation of acetone, *Appl. Catal.*, 48, 63, 1989.
39. Di Cosimo, J.I., Díez, V.K., and Apesteguía, C.R., Basic catalysis for the synthesis of α,β-unsaturated ketones from the vapor-phase aldol condensation of acetone, *Appl. Catal.*, 137, 149, 1996.
40. Lunsford, J.H., The catalytic conversion of methane to higher hydrocarbons, *Catal. Today*, 6, 235, 1990.
41. Díez, V.K., Apesteguía, C.R., and Di Cosimo, J.I., Acid-base properties and active site requirements for elimination reactions on alkali-promoted MgO catalysts, *Catal. Today*, 63, 53, 2000.
42. Díez, V.K., Apesteguía, C.R., and Di Cosimo, J.I., Aldol condensation of citral with acetone on MgO and alkali-promoted MgO catalysts, *J. Catal.*, 240, 235, 2006.

16 Biofuels and Biochemicals in Brazil

Eduardo Falabella Sousa-Aguiar, Nei Pereira, Jr.,
Donato Alexandre Gomes Aranda, and
Adelaide Maria de Souza Antunes

CONTENTS

16.1 INTRODUCTION

Nowadays, refining industry faces a number of daunting challenges. There is increasingly stringent environmental regulation resulting from the growing demand for cleaner fuels. The quality of today's crude oil is inferior to the crude oil of 30 years ago; today's is very heavy and acidic, resulting in a more laborious refining process [1–3].

These issues in addition to the growing pressure to reduce emissions, globalization, and customer behavior need to be balanced with the maintenance of profits. The requirements for clean burning fuels are changing the traditional goals of petroleum refineries, frequently imposing a conundrum. In order to reduce the emission of sulfur oxides and to avoid the formation of acid rains, sulfur compounds present in the fuels must be reduced. However, desulfurization in diesel fuel production requires the production of hydrogen through steam reforming processes. These operations require energy, have low thermal efficiency, and emit CO_2. So, a cleaner diesel fuel could come at the expense of a higher polluting production process.

An intelligent alternative solution must be sought to meet all of these needs. The use of natural gas and biomass as feedstock is playing an important role as well as synthetic fuels. It is necessary to examine both the desired fuel as well as the production process. In this situation, the role of catalysis is paramount. Rather than changing entire production systems to generate a new or different fuel, change only the catalyst.

Global climate change has been considered the most hazardous environmental problem of humankind and may become the biggest challenge ever to be confronted by the oil and gas industry. Recently [4–6], the divulgation of the summary for decision makers of fourth report of IPCC (Intergovernmental Panel on Climate Change) along with the fact that IPCC and Albert Arnold (Al) Gore Jr. were awarded of the Nobel Peace Prize "for their efforts to build up and disseminate greater knowledge about manmade climate change, and to lay the foundations for the measures that are needed to counteract such change," has caused a great impact on the public opinion. Such actions will probably bring about the adoption of additional measures to those already existing in the Kyoto protocol [7].

The combination of regulations to limit emissions coming from both the utilization of oil-derived fuels and their production in the refineries, which try to control emissions of sulfur oxides (SO_x), nitrogen oxides (NO_x), carbon monoxide (CO), particulates, and greenhouse gases (GHGs) such as carbon dioxide (CO_2), methane (CH_4), and others, has made the oil refining industry one of the most regulated industries in the world.

Eventually, consumers are responsible for the oil derivative demand and require fuels that are safer and less pollutant, presenting high performance and lower prices. Furthermore, consumers' choices regarding next-generation vehicles will certainly have a great influence on the refining sector of the oil industry. Indeed, next-generation vehicles, a rather encompassing term, generally refer to hybrid electric vehicles (HEVs), which combine an internal combustion engine and one or more electric motors. This concept may include various technological options such as hybrid electric-petroleum vehicles, which use internal combustion engines, generally gasoline or diesel engines powered by a variety of fuels and electric batteries to power electric motors, flexible-fuel vehicles that can use a mixture of input fuels mixed in one tank—typically gasoline and ethanol, or methanol, or biobutanol—and plug-in hybrid electrical vehicle (PHEV) or fuel cell hybrid, not to mention bi-fuel vehicle using liquefied petroleum gas (LPG) and natural gas.

It must be borne in mind that global competition has conducted to an overwhelming restructuring of the refining industry. The number of refineries has drastically diminished since 1980, having the remaining ones larger capacities and efficiency. In fact, refineries have dealt with several economic impacts caused by changes in the oil price, oscillations in its quality, and periods of low profit margin; and they are also obligated to meet a growing demand of refined products with increasingly stringent specifications. In the future, the refining industry will have to cope with this balance between higher amounts of increasing quality products and profitability.

In summary, the main challenges of refining industry are as follows:

- Increasing stringent environmental regulation
- Growing demand for cleaner fuels
- Globalization
- Increase in the production of derivatives from declining quality oil
- Uncertainty about the consumer choice
- Growing pressure of several segments of the society aiming at the reduction of GHG
- Maintenance of its profitability
- Search for alternative raw materials such as biomass and coal

16.1.1 SCENARIOS FOR THE REFINING INDUSTRY

The future refining schemes may display several configurations, ranging from the traditional ones, based on hydroprocessing, to more innovative schemes, using natural gas, biomass, and residues, comprising processes of pyrolysis, gasification, and synthesis, as Fischer–Tropsch (FT), or even fermentation and biorefining.

The U.S. Department of Energy (DOE) has promoted a partnership among oil companies aiming at identifying the needs of this industry in terms of research and development (R&D). This partnership, called "Petroleum Refining Industry of the Future," was finalized after generating two documents concerning an overview of the American refining industry about the future. According to both documents, in 2020, the refining industry will evolve by carrying out continuous improvements related to a more efficient usage of raw materials, a better environmental performance of the refineries, and the products thereof. Refineries will be safer and simpler, using more comprehensive processes as far as fundaments are concerned. In order to reach such goals, it will be necessary to take actions resulting from three strategic vectors:

- Energy efficiency and process improvement
- Environmental performance
- Inspection technologies and materials science

To improve energy efficiency, refinery will incorporate advanced technologies of low energy intensity and economically viable. The result will be a highly flexible refinery, more efficient and capable of producing a large variety of derivatives from crude oils of oscillating quality and nonconventional feedstock. Refineries will take advantage of opportunities in terms of generation and cogeneration of energy, which will be sold, thereby increasing their profitability. There will be an increasing utilization of biological processes, for instance, oil bioprocessing, biotreatment of residual waters, and soil bioremediation.

On the other hand, to improve its environmental performance, the refining industry will try to reduce emissions. All the steps of the process (production, storage, and transportation) will be under control by means of sensors, thereby detecting, correcting, and avoiding pollutant emissions. Vehicle emissions will be reduced via a combination of new regulation and improvement in the vehicle design and fuel formulation. A well-to-wheel life cycle assessment (LCA) will be employed to minimize pollution in the whole process.

New technologies of inspection and materials science will reduce the cost of maintenance and will increase industrial safety and the useful life of equipment. Inspection technologies will be on line, noninvasive, and, in some cases, remotely controlled and operated. Equipment will be highly instrumented, so its structural integrity may be monitored.

DOE also sponsors a program called "Vision 21" [8], aiming at developing a modular plant capable of producing electric energy, heat, fuels, and chemicals from distinct raw materials such as coal, heavy oils, natural gas, biomass, and residues, without local pollutants emissions. Such plant will also utilize carbon sequestration to reduce GHG emissions [9]. Although this program had been conceived to make viable U.S. coal reserves, its main objectives have evolved and, currently, are very ambitious. Should its main objectives be attained, this program may bring about a revolution in the fuels and energy industry in the next years. Specifically, this program aims at the development of a group of modular technologies which may be interconnected in various ways, generating the expected products from different sources. The expected efficiencies are as follows:

- Sixty percent for the conversion of coal into electricity
- Seventy-five percent for the conversion of gas into electricity
- Seventy-five percent for the production of H_2 from coal

Another interesting program sponsored by U.S. DOE is called "Biomass Programme," whose main focus is the development of technologies for the production of biofuels, bioproducts, and bioenergy. This program's main goal is the production of lignocellulosic ethanol; nevertheless, also establishes as one of its objectives the implementation of biorefineries, or rather a facility that integrates biomass conversion processes and equipment to produce fuels, power, heat, and value-added chemicals from biomass [10]. Besides biochemical processes (enzymatic hydrolysis, fermentation, etc.), biorefineries may also use thermochemical processes, as pyrolysis to produce bio-oil or gasification followed by FT synthesis to yield biofuels and chemicals [11–13]. Moreover, the improvement of these processes creates the possibility of coprocessing in a conventional refinery, which may help the faster implementation of biofuels [14].

A recent study performed by the Instituto Mexicano del Petróleo to guide its activities of R&D for the next 20 years has developed three scenarios for the refining industry. These scenarios, hereinafter called inertial, incremental, and innovative, are depicted in Table 16.1.

The inertial scenario takes into consideration traditional pretreatment of petroleum (desalting and hydrotreating); however, both atmospheric distillation and vacuum distillation are to be replaced by reactive distillation. Fluidized bed catalytic cracking (FCC) will operate with higher conversion to olefin. Delayed coking and hydroconversion processes are also included, being hydrogen produced via steam reforming. Such scenario also proposes the minimization of environmental liabilities.

The incremental scenario is a vision shared among various stakeholders. It is attractive in part due to its use of existing technologies and infrastructure and integration with petrochemistry. Smaller equipment and improved technologies will be introduced gradually within this scenario. The refinery will no longer house all of the refining operations as pretreatment of oil will happen in the fields before arriving at the refinery, using new technologies such as ultrasound or microwaves. The C_1–C_2 stream is converted to alcohols by means of selective oxidation. LPG is converted to gasoline via alkylation and polymerization. The possibility of using nanocatalysis for distillate hydrotreating is proposed. Interestingly, bottoms are the feedstock for hydrogen production via gasification. As in the other scenario, the minimization of environmental liabilities is also considered. Nevertheless, alternatives for CO_2 capture and sequestration are disregarded.

The innovative scenario deserves special attention. It involves using any carbonaceous compound as feedstock for fuel production. It requires brand new automotive technologies such as the use of dimethylether (DME) which can replace diesel and LPG. Radical changes in current production operations, including the gasification of crude oil to produce synthesis gas (syngas), a mixture containing carbon monoxide and hydrogen, will take place. This syngas will undergo FT synthesis to produce

---

TABLE 16.1
Scenarios for the Refining Industry

Scenario	Raw Material	Market	Process	Focus
Inertial (predominant vision among refiners)	Increasing proportion of heavy crude oil	Traditional fuels with more stringent regulation Growing market	Traditional ones	Higher profitability
Incremental (vision shared among various stakeholders such as refiners, contractors, catalyst industry, automobile industry, and government agencies)	Heavy oils Natural gas	Potential use of hydrogen as fuel Growing market	Use of proven technologies More compact equipment Oil pretreatment in the production fields	Integration with petrochemistry
Innovative	Heavy oils Natural gas Coal Biomass Residues	Brand new automotive technologies	Radical change in the technological paradigms Renewable energy Gasification of crudeoil	Minimum environmental impact

TABLE 16.2
Brazilian Refining Scenarios for 2020

Scenario	Raw Material	Market	Process	Focus
	Increasing utilization of heavy crudes, however, with necessary modifications to process heavy oils and acidic oils without pretreatment in the production fields	Traditional fuels and growing market, however, taking into account problems related to demand and new specifications	Use of proven technologies, with special attention to hydrotreating and hydroconversion	Integration with petrochemistry and minimum environmental impact
	Inertial	Inertial	Incremental	Incremental/innovative

hydrocarbon products that may be upgradable to either fuels or lubricants and food grade paraffin. Furthermore, the refinery produces electricity through cogeneration to increase its profitability. Also, CO_2 sequestration is forecast. This scenario includes routes that would allow the utilization of alternative raw materials. The ultimate goal of the innovative scenario is environmental sustainability.

16.1.2 Scenarios' Evaluation

The choice of the best scenario, that is to say, the one which has the best possibilities of becoming a reality, is influenced by three factors:

- The oil share in the future energy matrix
- The main characteristics of the oil to be processed
- The demand of the consumer market

According to the DOE scenario for the world energy consumption, in 2030 fossil fuels will continue supplying the major part of world commercialized energy. Liquid fuels will represent the largest share (34% in 2030), equivalent to 118 million bpd. Out of this amount, ca. 11 million bpd will be generated by nonconventional sources such as GTL and biofuels. In spite of the significant growth of liquid fuels from nonconventional sources, petroleum will continue to be the main source of world primary energy.

As far as Brazil is concerned, despite the great contribution of biomass (and biofuels) in the energy matrix, petroleum is still the main source of primary energy. In 2006, petroleum represented 37% of the Brazilian energy matrix, equivalent to 76 million tons [15].

According to Perissé, the refining scheme to be adopted in Brazil will be influenced by all three scenarios previously described, as depicted in Table 16.2. Due to the characteristics of the Brazilian fuel market, which has a gasoline surplus and a need to import diesel, a change in the current refining scheme will occur. Hydrocracking units shall be introduced in the existing refineries; hence, the Brazilian refining industry will become similar to the European model, which is more dedicated to diesel production.

The refinery of the future is a sophisticated technological concept requiring innovative technologies. These technologies will be implemented gradually. With each new piece, refineries will become more dynamic, being able to use different raw materials based on availability and logistic characteristics.

16.2 BIOFUELS' CLASSIFICATION

Undoubtedly, there is growing interest in biofuels in many developing countries. The use of biofuels will certainly provide such countries with better solution for abundant biomass sources, providing greater access to clean liquid fuels while helping to address energy costs, energy security, and global warming concerns associated with petroleum fuels.

Generally, biofuels are classified as "first-generation biofuels" and "second-generation biofuels," although recently the term "next-generation biofuels" has been proposed after much controversy about technologies of ethanol production has arisen. Also, the term "third-generation biofuels" has been attributed to those fuels produced from algal materials. However, there are no strict technical definitions for these terms, being the main distinction between them the feedstock used.

"First-generation biofuels" [16] are biofuels made from sugar, starch, vegetable oil, or animal fats using conventional technology. The basic feedstock for the production of first-generation biofuels is often sugar-containing plants, such as sugarcane, or seeds and grains such as wheat, which yields starch that is fermented into bioethanol, or oily seeds, which are pressed to yield vegetable oil that can be transformed into biodiesel via transesterification with methanol or even ethanol. Therefore, first-generation biofuels use only a specific (often edible) portion of the aboveground biomass produced by a plant, which could instead enter the animal or human food chain.

The most common first-generation biofuels are as follows:

- Bioalcohols
- H-BIO (green diesel)
- Biodiesel
- Vegetable oil
- Bioethers
- Biogas
- Synthesis gas
- Solid biofuels

According to a UN report on biofuels [17], "second-generation fuels are made from ligno-cellulosic biomass feedstock using advanced technical processes." Lignocellulosic sources include "woody," "carbonous" materials that do not compete with food production, such as leaves, tree bark, straw, or wood chips. Nevertheless, in the longer term, biofuels could also be produced from materials that are not even dependent on arable land, such as algae, which grow in water. While first-generation biofuels are already being produced in significant commercial quantities in several countries, to the best of our knowledge, second-generation biofuels are not yet being produced commercially in any country.

Among the second-generation biofuels, the following may be highlighted:

- Biohydrogen
- Biomethanol
- DMF
- Bio-DME
- FT diesel
- Mixed alcohols

Figure 16.1 displays the main biofuels and the petroleum derivatives they may replace. First-generation biofuels such as ethanol produced via fermentation or even fatty acid methyl ester (FAME), produced via esterification of vegetable oils, may replace only one petroleum refined derivative, whereas more sophisticated second-generation biofuels produced by means of FT synthesis may replace several derivatives.

Another interesting taxonomy of biofuels is based on the processes used to convert biomass into liquid fuels. Figure 16.2 presents an overview of the main processes that can be employed to generate products that may replace traditional petroleum derivatives. Generally, such processes are divided into three main areas, or rather biochemical processes, which use fermentation to yield products, the thermochemical processes, in which solid/liquid biomass undergoes gasification to produce syngas $(CO + H_2)$, and oil chemistry, related to esterification/transesterification processes to generate biodiesel. Recently, a new process to produce H-BIO or green diesel via hydrodeoxygenation of vegetable oil has been proposed. For the sake of clarity, however, since thermal transformations take place, this process will be included in the thermochemical area.

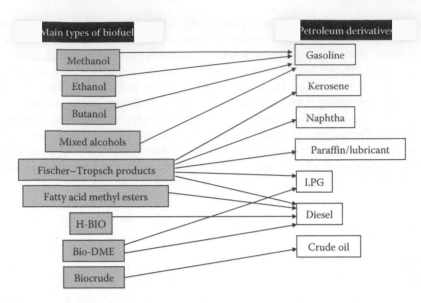

FIGURE 16.1 Main types of biofuels and the petroleum derivatives they may replace.

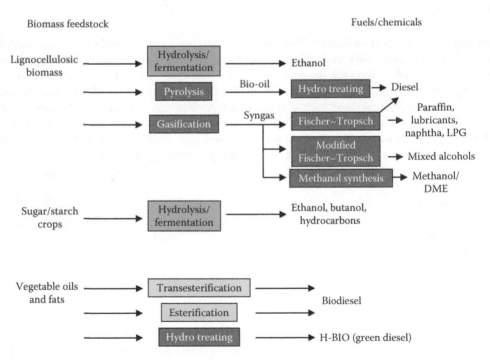

FIGURE 16.2 Main processes to generate products that may replace traditional petroleum derivatives from various biomass feedstocks.

Concerning Brazil, first-generation ethanol produced from sugarcane has an enormous tradition. Also, first-generation biodiesel production is being commercialized. However, second-generation biofuel systems require more sophisticated processing equipment, more investment per unit of production, and, in some cases, larger-scale facilities than first-generation biofuels. Many efforts are ongoing worldwide to commercialize second-generation biofuels, and Brazil is also investing considerable amounts in R&D to make commercial plants a reality.

16.3 BIOCHEMICAL PROCESSES

16.3.1 INTRODUCTION

The growth of human population has originated a rising and associated energy demand which is mostly supplied by nonrenewable sources, especially by petroleum. However, the progressive worldwide exhaustion of this fossil carbon source is each time a less denied reality, in the way that the most pessimistic previsions confirm its total depletion in about 41 years [18]. Due to extraction difficulty, consumption increment, and scarcity, its prices will continue to increase [19].

The environmental protection is another factor that has brought great concern to modern society. The burning of fossil fuels and the deforestation emit great quantities of gases in the atmosphere, especially CO_2 (anthropic emissions). The rising emissions of this gas and others as methane (CH_4), nitrous oxide (N_2O), hydrofluorinecarbonates (HFCs), perfluorcarbonates (PFCs), and sulfur hexafluoride (SF_6) in the atmosphere have caused serious environmental problems, accentuating the global warming. Due to the quantity in which it is emitted, the CO_2 is a gas that most contributes to the global warming. Its persistence in the atmosphere can take decades. This means that the emissions nowadays have long-term effects, resulting in climate system impacts along the centuries, which already affects in present time, as follows: increase of the global temperature, rise of the ocean level, reduction of the arctic ice layer, adverse impacts on agriculture, and loss of biodiversity. These environmental impacts are a reality with which society has to live, to control, and to adapt itself, needing to be conscious of the importance of this problem, and requiring changes in various consumption habits.

Fortunately, great progresses could be seen in the manner of looking for alternative energy sources, particularly biomasses, and for the implementation of actions as for reducing CO_2 concentration in the atmosphere, such as preservation of native forests, implementation of agroforestal systems, and recuperation of degraded areas. The participating nations of the convention on climate changes, which took place in June 1992 in Rio de Janeiro, compromised themselves by proposing actions to reduce GHG emissions. The convention came into force in 1994, and from that time onward, the countries have met to discuss the topic and have tried to look for solutions for this serious problem which can compromise life in our planet.

In 1997 in Kyoto was established an important agreement known as the Kyoto protocol, in which were defined mechanisms and limits for the reduction of GHG emissions to 5.2% between 2008 and 2012, in relation to the verified levels of 1990. The negotiations of the protocol were extended until 2004 when Russia ratified the document. With its ratification, becoming a treatise, measures to aim the reduction of these gases come into force and, together with other inconveniences of petroleum, create a great opportunity for the use of alternative energy sources, produced from renewable materials, collectively denominated biomasses.

With this treatise, a worldwide carbon market has already been created. The countries that cannot reduce their GHG emissions can buy credits from other countries which contribute to reduce these gases from the atmosphere in a larger quantity as they emit. This is one of the principal aspects of the *Kyoto* treatise, which transforms environmental concerns into economic worries.

Brazil finds itself in a sufficient privileged position to assume the leadership in the integral utilization of biomass, since it is one of the major planet potentials in renewable feedstocks with the great availability of agricultural cultures in large extension, with prominence for the cane industry, with intense sun radiation, plenty of water, climate diversification. It is the pioneer in the production of a biofuel in large scale, the ethanol. Moreover, the country unifies conditions to be the principal receiver of investments, arising from the carbon market in the segment of production and use of bioenergy, having in its environment its greatest richness and owning enormous atmospheric absorption and regeneration capacity.

In Brazil, the fuel ethanol is produced from sugarcane juice, a material directly fermentable by possessing its main substrate (sucrose) in a form of direct utilization from the producing biological agent, the yeast *Saccharomyces cerevisiae*, therefore eliminating previous hydrolysis procedures. Even when sugar and alcohol, in lots of cases, are produced concomitantly, it does not seem to be rational that considerable fertile soil extensions should be used for the production of fuel in detriment to the food production, of major social importance. On the other side, it would not be strategically secure, for a long time, to depend only on sugarcane for the ethanol production when Brazil has, as alternative sources, starchy feedstocks and abundant residues generated by the agricultural and forestall sectors.

Additionally, studies indicate that the projected demand for ethanol for 2013 in Brazil will be 32 billion liters, which corresponds to almost twice that of the 2005 production [20]. Several factors contribute to this increase, including the growth of bifueled car selling (*flex fuel*), the worldwide increase of ethanol demand and consequently of the Brazilian exportations, the increment on gasoline demand, which is mixed with 20%–25% of anhydrous ethanol, and the implementation of the biodiesel mixture, produced with ethanol, in the diesel oil.

However, it can be foreseen that the harvested sugarcane in the same period will not be sufficient to supply such demand, ratifying the need of producing ethanol based on other sources.

In this context, R&D has been intensified in a more diversified way for the utilization of renewable feedstocks in substitution of fossil sources. Emphasis has been placed on the utilization of abundant agricultural (those produced in the field, resulted from the harvesting activity) and agro-industrial (those generated in the industrial processing units) residues, both of lignocellulosic composition. The utilization of these residues, named residual biomasses, is of great interest and relevance, constituting one of the most important topics of modern biotechnology, since there is no additional demand for extension area for agricultural activities. The aim is to transfer them from the position of solid residues to valuable feedstocks for the production of fuels and a variety of chemical substances, within the context of what has been called by biorefinery. Progresses in this area indicate that the utilization of renewable feedstocks, including their residues, should revert the worldwide dependence on fossil sources.

16.3.2 AGRICULTURAL AND AGRO-INDUSTRIAL RESIDUES GENERATED IN BRAZIL

Analyzed for a long term, the economical viability of ethanol produced from sucrose and starch is conditioned to the availability of feedstocks and to the problem of land occupation. In accordance to some authors, the use of annual cultures for energetic purposes (grains and cereals and sugarcane) reduces the soil fertility and its availability for food production. There are competitive demands for soil resources for the production of biomasses and for their alternative uses. In reality, biomass is a precious resource, is unlimited, and has diverse uses. Therefore, the expansion of these cultures gives reason to great controversies, because such activities can become a socioeconomical problem if there will not be a concomitant establishment for an appropriated agricultural policy. Besides, with the fast growth of the current ethanol market, the agricultural surplus can become exhausted, causing an ascendant pressure on feedstock prices. Therefore, a lot needs to be done that the ethanol from sugarcane and/or cereals, or even the biodiesel from vegetable oils, might become, in short terms, real substitutes of petroleum derivative fuels [21–23]. Besides the great search for ethanol, another factor which should be considered is the commercial sugar market, for which worldwide demand grows 3.4 million tons/year. Brazil, in spite of detaining 36% of the world market, can only currently amplify its sugar exportation to 1.7 million tons without compromising the alcohol supplies on the Brazilian market [20].

A great feedstock source for bioconversion into ethanol is concentrated in lignocellulosic materials, normally in the form of residues, and even though some of them have been used as solid fuel, for animal feeding, and as recycling material, there are enormous surpluses. They are found in abundance and in great use availability. Approximately, 350 million tons of agricultural and agro-industrial residues were produced in 2006 in Brazil, being originating mainly from sugarcane, followed by soya, which present the major generated amounts (Figure 16.3).

Because of a vast biodiversity found in its territory, Brazil owns a great variety of residual biomasses of which utilization will be of great economical and social interest, since they possess a very low economical value and they are seen as additional cost inside the production process, due to the need of its final disposal [22–24].

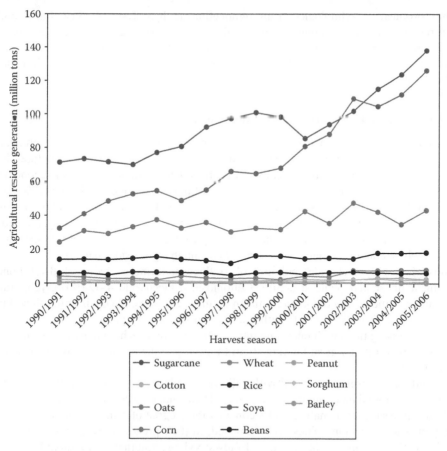

FIGURE 16.3   Production of residues from the main national cultures.

As an example, the generation of cane bagasse in 2004 was approximately 101.8 million tons [19]. This agro-industrial residue is mostly burned for energy generation in sugar mills. However, the viability of new technologies for residual biomass utilization could deviate part of this bagasse for the production of fuels and other chemicals of major economic value. Therefore, it is important to encourage investments for the development of new technologies to obtain energetic gains from the renewable resources, which are produced in great quantity in the country [23].

The contained energy in 1 ton of bagasse, with 50% of humidity, corresponds to 2.85 GJ [25]. The bagasse is the macerated stem, not including the straw and the quills, which present 55% of the accumulated energy in the cane plantation. This fabulous potential is rarely used, and, in the majority of the cases, is burned on the field. Bagasse can offer energetic use outside of the sugar mills and can be used as fodder. Also, it has application in the production of cellulose/paper, structural elements, and, more recently, its utilization has been contemplated as feedstock for the production of ethanol, through hydrolysis and fermentation technology.

Not all produced residues can or should be used for the production of bioenergy. The indiscriminate removal of residues can cause decline in soil quality with lasting adverse environment impacts. The return of agricultural residues to the soil improves its quality through its impact on reducing erosion risk, recycling of nutrients, improvement of soil structure and stabilization, water retention, energy supply for soil microbiological processes, and increase of agronomical productivity. In view of these factors, in general only 40%–50% of the agricultural residues can be used for different kinds of activities [22,24].

Ethanol production technologies from lignocellulosic hydrolysis of agricultural and agro-industrial residues are in development, being realized in developed countries, portentous research investments for this aim, and will reach commercial period of probation in a few years. It is estimated that in 2020 about 30 billion liters of alcohol can be obtained from lignocellulosic source only in the United States [19,26].

16.3.3 BIOMASS OF LIGNOCELLULOSIC COMPOSITION

The lignocellulosic materials are the most abundant organic compounds in the biosphere, participating in approximately 50% of the terrestrial biomass. The term "lignocellulose structure" is related to the part of the plant that forms the cell wall (half lamella, primary wall, and secondary wall), composed of fibrous structures, basically constituted of polysaccharides (cellulose (40%–60%) and hemicellulose [20%–40%]). These components are associated to a macromolecular structure containing aromatic substances, denominated lignin (15%–25%) [27]. In a general way, it can be stated that those materials possess in their compositions approximately, 50%–70% of polysaccharides (in a dry basis), which contain in their monomeric units valuable glycosides (sugars).

Cellulose is a polysaccharide, polymer of D-glucose, forming chains of β-1,4 bonds, and maintaining a linear and plane structure. Cellobiose, disaccharide 4-O-(β-D-glycopyranosil-D-glucopyranose), is the repeated polymer unit. In natural celluloses, the chains are aligned in a way of forming complex organized fibrils, either in crystalline or amorphous structures. These fibrils are established among them with inter- and intra-hydrogen bonds, which individually are weak, but collectively, they result in a great binding strength, giving to the cellulose a high resistance to the hydrolysis attack.

Hemicellulose is closely associated with cellulose in plant tissues and together with cellulose they are the most abundant components in plants. These macromolecules, contrarily to cellulose, present heteropolysaccharide nature and a considerable degree of ramification, consequently not presenting crystalline regions. They are composed, in their great majority, of a mixture of polysaccharides with a low molecular mass, as follows: xylans, arabinans, arabinoxylans, mannans, and galactomannans. The fundamental units (monomers) are, basically, molecules of D-xylose, D-mannose, D-galactose, D-glucose, L-arabinose, D-glucuronic acid, D-galacturonic acid, α-D-4-O-methylglucuronic acid, and also some oxidation products, as, for example, acetates.

Different to cellulose, the hemicellulose structure does not present a high crystallinity, therefore being more susceptible to the chemical hydrolysis under milder conditions. The varieties of bindings and ramifications, just as the presence of different monomeric units, contribute to the hemicellulose structure complexity and its different conformations [28,29].

Table 16.3 summarizes the main differences between the polysaccharides, components of lignocellulosic materials. The understanding of these characteristics is of fundamental importance, in order to define strategies for the use of these biomasses as feedstocks for second-generation ethanol production and other chemicals.

TABLE 16.3
Differences between Cellulose and Hemicellulose

Cellulose	Hemicellulose
Consists of glucose units	Consists of various units of pentoses and hexoses
High degree of polymerization (2,000–18,000)	Low degree of polymerization (50–300)
Forms fibrous arrangement	Does not form fibrous arrangement
Presents crystalline and amorphous regions	Presents only amorphous regions
Slowly attacked by diluted inorganic acid in hot conditions	Rapidly attacked by inorganic acid diluted in hot conditions
Insoluble in alkalis	Soluble in alkalis

Lignin is a natural macromolecule composed by p-propylphenolic units with methoxyl substituents on the aromatic ring and, between these units, exist principally ether-type bounds. It presents a highly complex structure, formed by polymerization of three different monomers: coumaric alcohol (I), coniferyl alcohol (II), and synapyl alcohol (III), which differ from one another by possessing different substituents in their aromatic ring. This structure is also responsible for the hardness of the cell wall, constituting in a binding material ("glue-like substance"), which holds the cellulosic fibers. Lignin possesses high molecular mass and presents about 25% of the photosynthesis biomass produced yearly on earth, retaining 50% more carbon than cellulose.

The lignin biodegradation is an oxidation process that involves a complex extracellular enzymatic system, produced principally by fungus species which live in wood; among them, the most studied is the filamentous fungus of the white rottenness *Phanerochaete chrysosporium*. Some bacteria, especially the actinomycetes, are also able to degrade lignin. The main involved enzymes are lignin peroxidase, manganese peroxidase, and laccase. These enzymes present great utilization potential for the pulp and paper industry, especially for the treatment of their recalcitrant effluent [30,31].

Besides these three main components, there are others in minor proportions in lignocellulosic biomass, such as resins, tannin, and fat acids. Nitrogen compounds are found in small quantities, in general in the form of proteins. Among the mineral salts, the salts of calcium, potassium, and magnesium are the most frequent [32,33].

Such as form and size of the cell wall of lignocellulosic materials vary from species to species, their chemical composition is distinctly in function of their origin. Table 16.4 presents the composition of some lignocellulosic materials, expressed in their three main components. In a general way,

TABLE 16.4
Composition of Lignocellulosic Residues

Material	Composition (%)				Reference
	Cellulose	Hemicellulose	Lignin	Other	
Cane bagasse	36	28	20	NR	2
Cane straw	36	21	16	27	1
Maize straw	36	28	29	NR	2
Corncob	36	28	NR	NR	2
Corn straw	39	36	10	NR	3
Barley straw	44	27	7	NR	3
Rice straw	33	26	7	NR	3
Oat straw	41	16	11	NR	3
Cotton straw	42	12	15	NR	4
Peanut shell	38	36	16	NR	4
Rice shell	36.1	19.7	19.4	20.1	5
Barley bran	23	32.7	21.4	NR	5
Pine tree	44	26	29	NR	2
MSW	33	9	17	41	1
Willow	37	23	21	NR	2
Grass	32	20	9	39	1
Paper	43	13	6	NR	2
Cardboard	47	25	12	NR	2
Newspaper	62	16	21	1	1

MSW, municipal solid wastes; NR, not reported values; (1) Shleser [167]; (2) Olsson and Hahn-Hägerdal [168]; (3) Awafo et al. [61]; (4) Ghosh and Singh [170]; (5) Couto and Sanromán [169].

cellulose can be found in greater proportions, followed by hemicellulose and lignin, finally. Even presented in minor quantities than the cellulose fraction, lignin offers sufficient limitation to delay or even to hinder completely the microbial attack on the material.

Among the lignocellulosic residues of greater importance are cane bagasse and straw, corn straw and stover, rice straw, wheat straw, processed wooden wastes, and municipal residues from paper. In the Brazilian context, it can be estimated that only the sugar-alcohol sector generates, approximately, 6.6 million tons of exceeded cane bagasse (surplus) and 76 million tons of straw [19,34]. An estimation of the potential of these residues in the sugar-alcohol sector for ethanol production, for example, let us conclude that, with the technological knowledge we have today, we could double the Brazilian production of this fuel, without needing to expand the agricultural areas. Considering even the food biomass residues, the total generated quantity in our country reaches the value of approximately 350 million tons/year (Figure 16.3). This is an enormous potential, which should not be neglected and point out that these materials should have a more rational use.

16.3.4 RESIDUAL BIOMASS AND THE BIOREFINERY-ASSOCIATED CONCEPT

The chemical industry has been discovering the value of the molecules contained in the lignocellulosic materials, and due to the abundance of their generation in the form of residues, the market starts to focus on their use. A promising area has been developed, which has been named as "biorefinery."

Biorefinery is a relatively new term, referring to the use of renewable materials (*biomass*) and their residues, in a most integral and diversified way for the production of fuels, chemicals, and energy, with minimal generation of wastes and emissions.

As previously discussed, the biorefinery concept is analogous to today's petroleum refineries, which produce multiple fuels and products from petroleum, and where an industrial segment works as a generating pole of raw materials to others. Industrial biorefineries have been identified as the most promising route to the creation of a new domestic bio-based industry. Saying it in a simple way, it means that products, secondary products, and agricultural and agro-industrial residues sustain different types of processes.

The utilization of lignocellulosic biomass within the context of biorefinery is based on two different platforms. Both platforms aim at providing building blocks to obtain a variety of valuable products. The biochemical platform is based on biochemical conversion processes of glucosides (sugars) extracted from the biomass by hydrolytic (chemical and/or enzymatic) processes. Yet, the thermochemical platform, as the name implies, is based on thermochemical conversion processes by reacting the raw material at high temperatures with a controlled amount of oxygen (gasification) to produce syngas ($CO + H_2$) or in the absence of oxygen (pyrolysis) to produce a bio-oil, which after hydrodeoxygenation process produces a liquid mixture of hydrocarbons similar to those of petroleum crude oil. Figure 16.4 illustrates the two routes of using lignocellulosic feedstocks for the production of fuels and other chemicals of industrial interest.

Figures 16.5 through 16.7 describe schematically the biorefinery concept, following the biochemical platform and its application around the agro-industry, having its lignocellulosic biomass as a center of the production processes of a great variety of molecules. Such concept has been a target of heavy investments from the North Americans, aiming at restructuring their industry, especially the alcohol industry. The idea of creating a "broad belt" of processes presents enormous logistical advantages, principally for the transportation of feedstock, products, and services. In Brazil, this new industrial structure is still on seed stage, being increasingly studied.

The total hydrolysis of cellulose generates only glucose, which can be converted into a series of chemical and biochemical substances (Figure 16.5). We can say that the glucose, because of the existence of an exclusive and common metabolic pathway for the great majority of living beings, can be biologically converted into a wide range of substances such as ethanol, organic acids, glycerol, sorbitol, mannitol, fructose, enzymes, and biopolymers, among others. It can still be chemically

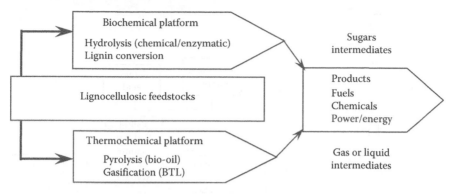

FIGURE 16.4 Utilization of lignocellulosic feedstock within the context of biorefinery (BTL: biomass to liquids).

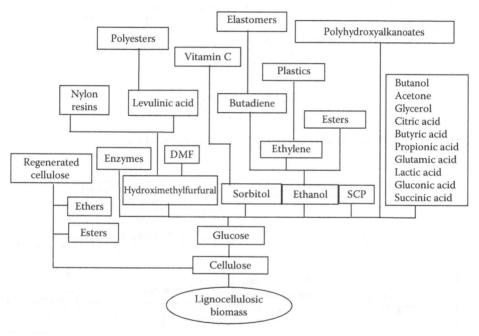

FIGURE 16.5 Lignocellulosic feedstock biorefinery: cellulose products.

converted to hydroximethylfurfural, which is an important intermediate for the production of dimethylfuran (DMF) or furan-based polymers.

The hydrolysis product of the hemicellulose fraction is a mixture of sugars, as described before, with predominance, in most cases, of xylose. The traditional market for this sugar has been the production of furfural, a selective solvent, very reactive, being used in large scale in the purification of mineral, vegetable and animal oils, as well as in the vitamin A concentration of fish liver. Alternatively, xylose can be hydrogenated to produce xylitol, which presents applications as a non-carcinogenic sweetener, with the same sweetening power of sucrose and with metabolization in the humans independent of insulin.

Nevertheless, xylose can be biologically converted to single-cell protein (SCP) and to a variety of fuels and solvents, such as ethanol by yeasts with the ability to ferment this pentose (*Pichia stipitis, Candida sheratae,* or *Pachysolen tannophilus*); xylitol by microorganisms with exclusively NADPH-dependent reductase activities on xylose, as, for example, *Candida guilliermondii, Debaromyces hanseni,* and *Candida tropicalis* [35,36]; biopolymers (polyhydroxyalkanoates, PHAs; polylactate; etc.);

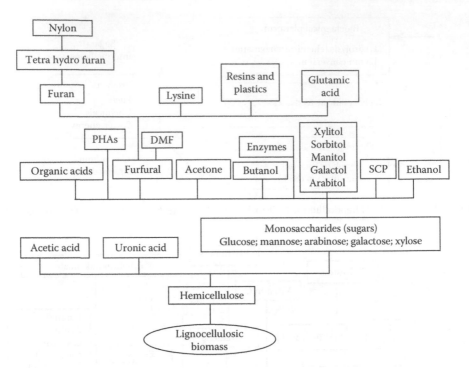

FIGURE 16.6 Lignocellulosic feedstock biorefinery: hemicellulose products.

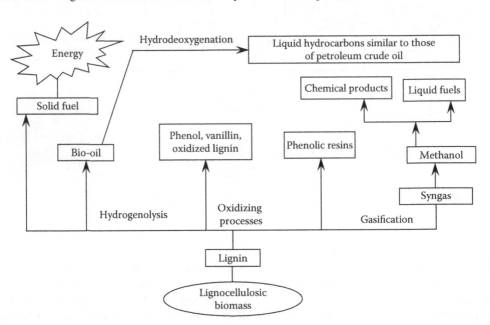

FIGURE 16.7 Lignocellulosic feedstock biorefinery: lignin products.

a series of organic acids (succinic, propionic, acetic, lactic, and butyric acids); solvents (butanol and acetone); and other fuels/fuel additives (DMF, butanol, 2,3-butanediol).

As said before, the lignocellulosic biomass is the most abundant organic material on earth. Approximately, 50 million tons of lignin are generated worldwide per year, as residues from the production processes of the cellulose pulp and paper industry [37]. Most of the residual lignin is burned to generate energy in this industrial segment. However, in view of its interesting functional properties, lignin offers useful perspectives to obtain high-value products, such as carbon fibers, emulsifiers, dispersants, sequestrants, surfactants, binders, and aromatics [38].

The physical and chemical properties of lignin differ depending on the extraction technology (sulfite, Kraft, alkaline, and organosolv processes). For example, the lignosulfates are hydrophilic and the *Kraft lignins* are hydrophobic [39].

The industry started first using lignin in the 1880s, when lignosulfonates were used in leather tanning and dye baths. From that time onward, lignin has even found applications in food products, serving as emulsifiers in animal feed and as raw material in the production of vanillin, which is extensively used as flavoring in food, as component in the pharmaceutics product formulation and also as fragrance in the perfume industry. The derivative product applications of lignin will expand literally, creating impacts in a lot of industrial segments [38].

Although hundreds of applications for lignin can be pointed out, its main use in the pulp and paper industry is as biofuel to replace fossil fuels in heat or power generation, and the lignin-depleted black liquor can be reused in the cooking operation. All those well-established industrial knowledge of the pulp and paper industry can be incorporated in the biorefinery concept using other sources of lignocellulosic feedstocks.

By producing multiple chemicals, the biorefinery takes advantage of the various components in biomass and their intermediates, therefore maximizing the value derived from the biomass feedstock. A biorefinery could, for example, produce one or several low-volume/high-value chemical or biochemical products and a low-value/high-volume liquid transportation fuel such as bioethanol or DMF, and at the same time generating electricity and process heat, through combined heat and power technology for its own use, and perhaps enough for sale of electricity to the local utility. The high-value products increase profitability, the high-volume fuel helps to meet energy demands, and the power production helps to lower energy costs and reduce GHG emissions from traditional power plant facilities. Although some of these facilities already exist, the biorefinery has yet to be fully realized.

16.3.5 Fractionation of Lignocellulosic Biomass Components

To make possible the use of lignocellulosic materials as feedstocks for the production of ethanol and other chemicals following the biochemical platform, it is necessary to separate their main components. For this separation, a pretreatment stage is essential, which aims at disorganizing the lignocellulosic complex. The pretreatment can be carried out through physical, physical–chemical, chemical, or biological processes, and can be either associated or followed by hydrolysis procedures of the polysaccharides (hemicellulose and cellulose) in their respective monomeric units (pentoses and hexoses).

The result of the pretreatment stage is the partial depolymerization and dissolution of the hemicelluloses. From the remaining material (cellulose + lignin), the cellulose can be separated, through lignin dissolution with alkalis or organic solvents (delignification), remaining the cellulose component with its increased digestibility for the enzymatic hydrolysis, or lignin can be separated through the cellulose hydrolysis with strong mineral acids (concentrated or diluted) in high temperatures [40]. The latter has been abandoned, since toxic substances can be generated in the hydrolysis process.

Figure 16.8 shows a simplified scheme for the fractionation of the main components of the lignocellulosic materials.

16.3.5.1 Lignocellulosic Biomass Pretreatments

In the lignocellulosic biorefinery context, pretreatment is understood as a process through which the cellulose molecule becomes more susceptible to enzymatic hydrolysis by cellulases. In the literature, frequently the terms "prehydrolysis" and "autohydrolysis" are used as synonyms of pretreatment. The increasing enzyme accessibility to the cellulose molecule is due to the removal of the hemicellulose fraction, as well as to the partial lignin removal (acid-soluble lignin), promoting a sort of "opening up" of the lignocellulose matrix. Additionally, as described further on, the usual pretreatment techniques involve a synergism between the heat action, the medium pH and the time

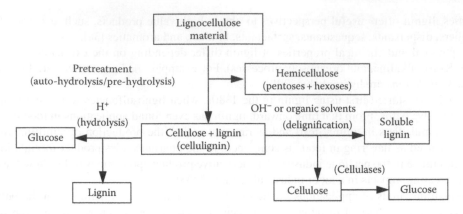

FIGURE 16.8   Fractionation of lignocellulosic components.

of exposition under process conditions. This results in a decrease in cellulose crystallinity, and, consequently, making it more susceptible to the action of cellulases [23,27,41–45].

The pretreatments can be divided into four types: physical (comminution of the material through fragmentation or grinding); physical–chemical (steam-explosion, catalyzed or not); chemical (acid hydrolysis in mild conditions, ozonolysis, or oxidizing delignification); and biological (microbial or enzymatic), according to the agent that acts in the structural alteration [27,43].

Because of the heterogeneity of the lignocellulose, it is not possible to choose only one pretreatment as being considered the best. The choice will depend, basically, on the nature/source of the material which needs to be treated, as well as on the use of the hydrolysate material. Various pretreatment processes have been developed aiming at increasing the efficiency in the removal of the hemicellulose fraction [41]. Thus, pretreatment processing conditions must be tailored to the specific chemical and structural composition of several sources of lignocellulosic biomass. As follows, the main available pretreatment technologies for utilizing lignocellulosic materials are described with more details.

16.3.5.1.1   Thermal Pretreatments

A quite efficient alternative for the extraction and partial hydrolysis of the hemicelulose is the technology of compression and fast decompression, carried out by the so-called steam-explosion, also denominated "autohydrolysis." Its operation works through the impregnation of the lignocellulosic material in water, in a system under high pressure (7–50 atm) and temperature (160°C–190°C) [27]. After that, the pressure is alleviated instantaneously. This change provokes a violent explosion, resulting in the rupture of the structural bindings of the lignocellulosic complex [46]. A wet solid material is obtained with the lignocellulosic complex disorganized (cellulignin); and a liquid phase, separated by filtration, composed of xylose, xylooligosaccharides, and uronic and acetic acids, is also obtained. The partial hydrolysis of hemicellulose, in special of the highly acetylated xylans, results fundamentally from their acid characteristics, therefore the term "autohydrolysis."

The structures of the hemicelluloses differ significantly as a function of their origins [27,47]. The hemicelluloses of *hardwood* are composed in their greater part of highly acetylated heteroxylans, generally classificated as 4-*O*-methyl glucuronoxylans. Hexosans, in the form of glucomanans, are also present, but in much lower quantity. Due to the acid characteristics and the chemical properties, the xylans of *hardwood* are relatively labile to the acid hydrolysis and suffer autohydrolysis in relatively moderate conditions. In contrast, the hemicelluloses of *softwood* have a higher proportion of glucomanans and galactoglucomanans in part acetylated, and the xylans only correspond to a small fraction of their total structure. In consequence, the hemicelluloses of *softwood* (in its major part composed of hexosans) are more resistant to the hydrolysis processes than the hemicelluloses of *hardwood* (in its major part composed of pentosans) [23].

In the last decades, various studies have been published involving the use of chemical agents, aiming at increasing the process efficiency of the steam-explosion. In this case, the denomination "catalyzed steam-explosion" has been used. The main chemical agents used are sulfuric acid, with concentration varying between 0.1% and 5% v/v, and sulfurous anhydride. When using sulfuric acid, previously to the steam-explosion, the material is soaked in the acid solution. After this phase, the steam-explosion process is operated. In the case of the use of sulfurous anhydride, the pretreatment is carried out by introducing in the vapor phase a rich stream with this gas. In both cases, the temperature range and the exposition time are not different from the simple steam-explosion [40,44,46].

Other chemical substances can be used, as carbon dioxide, which in solution forms carbonic acid [48] or ammonium, a known process as *ammonia fiber expansion* (AFEX), which principle is based on the high solubility of the hemicellulose in alkaline environments [49]. However, it needs to be considered that, similarly, lignin presents high solubility in these alkaline environments, and it might be necessary a detoxification step of the medium generated from this process. The retention of the cellulose in the solid fraction, in all cases, is very high (superior of 90% of the original structure).

16.3.5.1.2 Chemical Pretreatments

Various chemical pretreatments were studied, aiming at the removal of the hemicellulosic fraction, the cleavage of the bindings between the lignin and the polysaccharides, and the reduction of the cellulose crystallinity degree before the enzymatic hydrolysis of cellulose. Even though a lot of these processes reach high efficiency, there exists the disadvantage of them requiring plants constructed with materials which have great resistance for drastic reaction conditions, especially concerning the aspect of environment corrosivity. The mainly used chemical agents are acids, alkalis, gases, oxidant agents, solvents, etc. [26,44,47].

The alkaline pretreatment is frequently used to increase the digestibility of lignocellulosic materials. This process was originally developed in the paper and cellulose industry in the pulping processes to attain paper with long fiber, being indicated, especially when working with straws, due to their lignin content. The normally used conditions in this pretreatment are concentration of NaOH between 8% and 12% of the dry biomass to be treated, time of exposition between 30 and 60 min, and temperature between 80°C and 120°C [44]. The disadvantage of this process is related to the caustic soda price and the difficulty of its recuperation, which still involves prohibitive costs [27,47].

An alternative to the alkaline pretreatment is the simultaneous use of peroxide ("alkaline peroxide medium"). The delignification of the lignocellulosic materials with peroxide of hydrogen depends strongly on the pH, since its dissociation occurs in pH values around 11.5. Such dissociation results in the formation of highly reactive radicals, which act with the lignin molecule leading to its solubilization and oxidation. Some variations of this process involve two phases: the first using caustic soda and the second soda and peroxide. The oxidative delignification with peroxide occurs in low temperatures (25°C–40°C), and as a general rule, the generated residues have a low pollutant load. Another agent that is being recognized by its high oxidant power and selectivity in breaking the structure of lignin is the peracetic acid. This acid promotes the opening of the aromatic rings of the lignin, generating dicarboxylic acids and their lactones [50]. Similarly to the previous case, the process also can be operated in two phases, trying to minimize the costs with peracetic acid, since its price is high [27,45,47].

Notoriously, the advantage of the alkaline or the alkaline-oxidative pretreatment is the low energetic demand. Nevertheless, these processes present some potential disadvantages. Strongly alkaline environments can degrade the hemicellulose in saccharinic acids, which are not substrates for fermentation processes and the oxidized degradation of lignin generates an accumulation of phenolic monomers and oligomers, which are inhibitors of the biological transformation processes. The process named *organosolv*, involving the use of diluted alkalis together with solvents (e.g., ethanol) has been considered as a promising alternative for the delignification [27]. Nevertheless, such technology is still being studied and recent experiences of its application in Brazil shows that the problem of the generation of toxic inhibitors was still not solved.

As already described, the acidity of the medium is one of the fundamental aspects for increasing the efficiency of the pretreatment. Based on this, the acid pretreatment processes—in special those which employ diluted sulfuric acid—are more and more becoming a target of worldwide studies. The high reaction rates, the reduced consumption of acid and its low cost, when compared with alkalis, constitute in advantages of these processes. As mentioned previously, the disadvantages reside in the question of the corrosivity and, also, depending on the imposed operational conditions, in the formation of the inhibitors. The acid concentration ranges from 0.1% to 5%, the temperature between 110°C and 220°C, and the exposition time from 10 to 180 min. Several studies indicate that the pretreatment performed with more than one stage can reach high efficiency, causing a lower consumption of cellulases during the enzymatic hydrolysis phase [41,44,45].

AFEX is another promising method for pretreating agricultural material for bioenergy production. During this process, liquid ammonia is added to the biomass under moderate pressure (100–400 psi) and temperature (70°C–200°C) before rapidly releasing the pressure. Major process parameters are the temperature of the reaction, residence time, ammonia loading, and water loading. This process decrystallizes the cellulose, hydrolyzes hemicellulose, removes and depolymerizes lignin, and increases the size and number of micropores in the cell wall, thereby significantly increasing the rate of enzymatic hydrolysis [44].

16.3.5.1.3 Biological Pretreatments

The biological pretreatments consist in the use of a "pool" of enzymes, aiming at the hydrolysis of the hemicellulose and the delignification. In the case of the hydrolysis of the hemicelluloses, in spite of the specificity of xylanases, where the action is carried out through the synergy of the β-xylosidase, endo 1,4-β-xylanases, acetyl-xylanaesterase, α-glucoronidase, and L-arabinofuranosidase enzymes, there are problems related to the costs of those enzymes, which still consist in impediment for the implementation of these on an industrial scale. In this way, studies have been developed with the objective of producing enzymes from the xylanase complex [51,52]. However, the principal focus has been to the paper and cellulose sector, in which a crescent interest can be observed to use such enzymes in the stage of pulping, in substitution of chlorine and chlorine derivatives substances. This can be seen, principally, due to the irreversible tendency in favor of the total chlorine-free bleaching (TCF systems) and elemental chlorine free (ECF systems) [53].

The commercial development of hemicellulases is not so advanced than that of the cellulases; therefore, the current commercial preparations of cellulase have been developed for the hydrolysis of pretreated biomass with diluted acid, where the hemicellulose is removed before the cellulose saccharification. Nevertheless, with the development of nonacid pretreatments, in which the hemicellulose fraction stays intact or partially hydrolyzed, the hemicellulases will be compulsory required.

The present commercial cellulases, as of the *Trichoderma reesei* strains, tend to posses low hemicellulase activity and are not adequate for the complete conversion to the hemicellulose monomeric sugars. It has been expected that the development of low-cost hemicellulases production, working in synergism with cellulases, will be intensively focused in the next future. Figure 16.9 shows the different enzymes of the xylanolytic complex.

Lignin imposes challenges for the enzymatic cellulose hydrolysis, due to its nonproductive binding with the cellulases, which results in a decrease in their catalytic power and inactivation. Berlin et al. [54] proposed a new approach to improve the activity of the cellulases during the hydrolysis of lignocellulosic materials using enzymes which bind themselves weakly to the lignin. The authors show that cellulases of naturally occurring microorganism, with similar catalytic activity, differ significantly in relation to their affinity for lignin and, therefore affecting the performance of the enzymes on the native substrates. Palonen [55] showed that the localization and the structure of lignin affect more the enzymatic hydrolysis than the absolute lignin quantity in the lignocellulosic complex. The study revealed, furthermore, that modifications of the lignin surface by oxidants treatments with laccase led to a rise of the cellulose hydrolysis. However, there still not exist

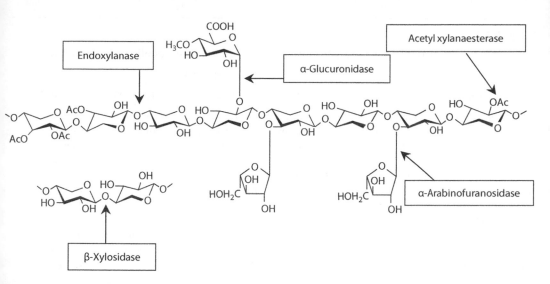

FIGURE 16.9 Enzymes involved in the hydrolysis of xylans.

sufficient studies about the transposition of a laboratorial scale to pilot plant which can show their technical and economical viability for substituting chemical pretreatments.

Due to the importance of cellulose enzymatic hydrolysis as well as its connection with the fermentative process, its main features will be described in details further on.

16.3.5.1.4 Other Technologies

Other pretreatment processes are being studied as well. Some studies propose the use of liquid hot water (LHW). This technology, called thermohydrolysis, involves the washing of the material with preheated water in high pressure, with temperatures of circa 220°C and times around 2 min, but the efficiencies are still low, when compared with the steam-explosion or with the acid prehydrolysis [44].

The use of irradiation with microwaves has been a target of some researches [56]. Commonly used conditions are irradiation of 240 W/10 min. However, contrary to all described pretreatment technologies, irradiation is still studied on bench scale, and it remains uncertain its application on an industrial scale, in view of the inherent energetic demand for the process.

Relevant pretreatments are steam-explosion process and diluted acid hydrolysis [27,41,44,57]. It should be pointed out that, when the intention is to hydrolyze the cellulose fraction with enzymes, the steam-explosion pretreatment is one of the most adopted technological tendencies. In this case, the hydrolytic process is carried out in several stages, as follows: comminution of the lignocellulosic material, steam-explosion, followed by the removal of the hemicellulosic fraction (liquid phase). The liqueur from the prehydrolysis step contains some soluble pentoses and hexoses, lignin, and oligomers of pentoses. The oligomers are then hydrolyzed to monomers, or at least to dimers or trimers, by any conventional technique. Alternatively, the steam-explosion process can be integrated (catalyzed steam-explosion), by using auxiliary inputs, as acidic gases (being SO_2 mostly used), which will result in a monomer-containing liqueur. Lastly, cellulases are used in the remaining solid (cellulignin) for the attainment of a glucose-rich medium [42]. A delignification step might be necessary to improve the enzymatic efficiency. Some technological "bottlenecks" can be identified in the pretreatment using acid medium and high temperatures as follows: formation of unwanted toxic compounds, derived from the sugar and lignin degradation, which can cause inhibition to the biological conversion processes, and problems related to equipment corrosion when working in acidic medium with high temperatures.

Table 16.5 summarizes the principal characteristics of the more employed and more current pretreatment technologies.

TABLE 16.5
Characteristics of the Main Pretreatment Technologies for Lignocellulosic Feedstocks

	Pretreatment Technology				
Characteristics	SE	CSE	DAH	TH (LHW)	AFEX
Typical operational conditions	Batch or continuous 2–10 min	Batch or continuous 2–10 min	Batch or continuous 5–30 min	Batch 5–60 min	Batch 5%–15% ammonium 10–30 min
Temperature (°C)	190–270	160–200	150–180	170–230	100–180
Consumption of chemical inputs	No	Yes	Yes	No	Yes
Xylose yield	10% xylose; 90% xylosaccharides	70%–90%	85%–95%	50% hemicellulose-derived sugars	60% hemicellulose-derived sugars
Removal of lignin (effect)	Minor	Moderate	Moderate	Minor	Major
Formation of inhibitors	Yes, under severe conditions	Yes, under severe conditions	Yes, under severe conditions	Few	Yes, under severe conditions
Reduction of the required particle size	Medium	Medium	High	Medium	High
Cellulose enzymatic hydrolysis efficiency (%)	80–90	80–85	80–85	80–90	50–90
Waste generation	Less significant	Moderate	Significant	Less significant	Significant
Corrosivity of the medium	Low	Low to moderate	Moderate to high	Low	Low to moderate

Sources: Adapted from Lynd, L.R., *Annu. Rev. Energy Environ.*, 21, 403, 1996; Ogier, J.C. et al., *Oil Gas Sci. Technol.*, 54, 67, 1999; Mosier, N. et al., *Bioresour. Technol.*, 96, 673, 2005.

Finally, it should not be neglected that the efficiency of the pretreatment processes is a result of the synergism between temperature, time, and medium acidity/alkalinity. The combination of these factors defines the parameter "degree of severity," which is intrinsically associated with the toxicity and fermentability of the hydrolysates. Generally, there exists an optimum degree of severity, under which the hydrolysis efficiency will be lower, and above which there will be degradation of sugars and formation of other derivative inhibitors from lignin [43].

16.3.6 DETOXIFICATION PROCESSES OF LIGNOCELLULOSIC HYDROLYSATES

Depending on the application of the hydrolysates and on the adopted pretreatment technology, their detoxification can be necessary. Preferably, the generation of inhibitors during the pretreatment stage should be minimized, since, in a lot of cases, the detoxification technology can lead to a partial loss of sugars originating from the hemicellulose hydrolysis [58,59]. Table 16.6 shows some usual detoxification procedures.

Recent studies developed in our laboratories on the prehydrolysis of cane bagasse showed that the tendency is to minimize, or even, to abolish the use of techniques of detoxification. Through the progressive acclimatization of yeasts in non-detoxified hydrolysates, Fogel et al. [35] and Betancur and Pereira [60] reached good results in the production of xylitol and ethanol, respectively. The absence of further treatments after prehydrolysis potentially makes the lignocellulosic biomass utilization economically more competitive.

TABLE 16.6

Procedures for the Detoxification of Hemicellulose Hydrolysates

Procedures	Effects
Treatment with fluent steam	Removal of volatiles (furfural, phenols, and acetic acid)
Neutralization with CaO, NaOH, KOH; treatment with active coal; filtration	Reduction in acetic acid concentration
Neutralization (pH=6.5) or alkalinization (pH=10) with $Ca(OH)_2$, CaO or KOH; removal of the precipitated; addition of H_2SO_4 (pH=6.5)	Precipitation of acetate, heavy metals, furfural, tannins, terpenes, phenolic compounds
Ionic exclusion chromatography	Aromatics removal
Neutralization (pH=6.5) with $CaCO_3$; removal of the precipitated; treatment with active coal; filtration	Clarification; removal of SO_4^- and phenolic compounds
Extraction with ether	Removal of furfural
Vacuum evaporation	Removal of acetic acid
Extraction with ethyl acetate	Removal of derivative compounds from the lignin degradation

16.3.7 ENZYMATIC HYDROLYSIS OF CELLULOSE

The cellulose hydrolysis processes can be chemical or enzymatic. The first, of greater knowledge, is run under established conditions of temperature (pressure), exposition time, type and concentration of acid, as well as solid: liquid ratio, similarly to the prehydrolysis, which was described previously.

The option for the enzymatic hydrolysis of the cellulose comes from the absence of severe conditions, typically of the chemical hydrolysis. This technological strategy differs from the conception of old processes in which the chemical hydrolysis of cellulose and hemicellulose (polysaccharides with different susceptibilities to the hydrolytic attack) was taking in one step. These processes generated hydrolysates with high toxicity, which hindered the metabolism of the microorganism agent of the fermentative process. Nevertheless, the cellulose chemical hydrolysis has been left behind and substituted by the enzymatic hydrolysis. Therefore, the chemical hydrolysis of cellulose will not be emphasized here.

Cellulases have in nature a fundamental role, through the degradation of cellulose present in plant biomass to establish a basic link in the development of the carbon cycle. To face the challenge of degradating the cellulose, cellulolytic microorganisms produce a complex mixture of enzymes: *the cellulases*. These enzymes, which collectively present specificity for the glycosidic linkages β-1,4, are all necessary for the complete solubilization and hydrolysis of cellulose (amorphous and crystalline), existing a synergism in their way of acting. The soil surface is the principal habitat of the cellulolytic aerobic microorganisms.

The enzymes of the cellulolytic complex are classified into three groups: endoglucanases, which cleave the internal bindings of the cellulose fiber producing cellodextrins; exoglucanases, which act in the external region of the cellulose producing cellobiose; and β-glucosidases, which hydrolyze soluble oligosaccharides to glucose [42]. Figure 16.10 outlines the action of these enzymes on cellulose.

The group of exoglucanases is constituted in its majority of 1,4-β-D-glucan-glucanohydrolases enzymes (EC 3.2.1.74), also known as cellodextrinases and 1,4-β-D-glucan-cellobiohydrolases enzymes (EC 3.2.1.91). Among both types, certainly the most referred in literature is the cellobiohydrolases (CBHs).

The CBHs are distinguished in two types: The enzymes of type I (CBH I) hydrolyze reducing terminals, while the ones of type II (CBH II) hydrolyze nonreducing terminals. These enzymes generally suffer inhibition by their hydrolysis product (cellobiose) [61].

The structure of CBHs presents a region in the form of a "hook," whose function is to bind the cellulosic fiber, facilitating its access to the catalytic site. Additionally, it is referred that the CBH I possesses ten active subsites under catalytic domain, whose function is to bind itself physically to the cellulose and initiate the chemical reactions that hydrolyze the chain to cellobiose.

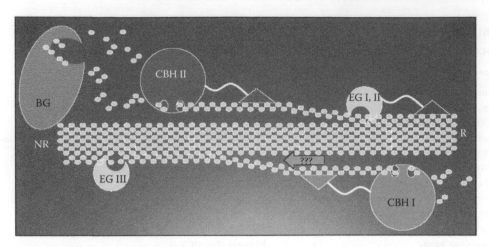

FIGURE 16.10   Enzymes involved in the hydrolysis of the cellulose. (1) EG, endoglucanase; (2) CBH, cellobiohydrolase (exoglucanase); and (3) BG, β-glucosidase (*Note:* NR: cellulose nonreducing terminal; R: cellulose reducing terminal).

The third and last great group of enzymes of the cellulolytic complex include the β-glucosidases, or β-glucohydrolases enzymes (EC 3.2.1.21). The β-glucosidases have the property to hydrolyze soluble cellobiose and oligosaccharides (with less than seven monomeric units) to glucose. Equal to the CBHs, they are also reported to suffer inhibition of their hydrolysis product [61].

When acting together, the cellulases present a better yield than the sum of the individual yields when acting isolated. Such effect is known as synergy. There are at least three forms of synergy [42]:

- *Endo-exo synergy*: The endoglucanases, acting in the amorphous regions of the fiber, provide reducing and nonreducing terminals for the action of the CBH I and CBH II, respectively.
- *Exo-exo synergy*: The CBH I and CBH II acting simultaneously in the hydrolysis of the reducing and nonreducing terminals released by the action of the endoglucanases.
- *Exo-BG synergy*: The CBHs liberate cellobiose, which is substrate for the β-glucosidases.

The hydrolysis of the cellulose polymer by cellulases involves basically two phases: The adsorption of the cellulases on the surfaces of the cellulosic substrate (fiber) and the hydrolysis of the cellulose in fermentable sugars. For this, the following steps happen [61]:

1. Diffusion of the cellulolytic complex from the bulk of the fluid to the location region of the cellulose substrate. In the case of insoluble substrate, the diffusion happens in direction to the film immediately surrounding the solid substrate.
2. Adsorption of the cellulolytic complex to the available sites in the cellulosic substrate.
3. Formation of an active cellulase-substrate complex.
4. Hydrolysis of the glycosidic linkages of the cellulose polymer.
5. Diffusion of the hydrolysis products from the active sites "cellulases-substrates" to the bulk of the medium.
6. Desorption of the cellulase complex from the hydrolyzed substrate.

16.3.8   Engineering of Cellulases

Cellulases are produced, among other microorganisms, by different filamentous fungi, being the greatest producers, belonging to the species *Trichoderma*, *Penicillium*, and *Aspergillus*. Although, generally, the levels of this enzymatic complex, being secreted from the fungus, fulfill in the nature to the necessity of the lignocellulose decomposition and the availability of sugars for their metabolism, the industrial use of the cellulases requires the attainment of enzymatic preparations with high activity and stability levels, being necessary to modify strains of naturally occurring filamentous fungi in hypersecretors, using techniques of classic genetic or molecular biology. Studies in this direction have been developed by different national and international laboratories of universities and enterprises, deserving distinction for the hyperproducing strains of *Trichoderma reesei*.

Several approaches have been used to improve the performance of cellulases and to decrease the amount of the necessary enzymes for an efficient hydrolysis of lignocellulosic materials. The first goal for cellulases engineering has been the CBHs, since they tend to constitute 60%–80% of the natural cellulolytic complex [42]. Teter et al. [62] demonstrated that the utilization of combined techniques of genetic engineering (*site-directed mutagenesis, site-saturation mutagenesis, error-prone PCR, and DNA shuffling*) generated highly productive strains of CBH (*Trichoderma reesei* Cel7a), which surpassed the wild strain in the hydrolysis of pretreated agricultural residue of the corn processing.

Another approach that has been utilized is the insertion of heterologous genes that codify for the production of cellulases in already existing systems, so that the overall performance of the recombinant strain is improved. Bower et al. [63] introduced various bacterial genes that codify for the endoglucanase in *Trichoderma reesei*. One of them, the GH5A of *Acidothermus cellulolyticus*, was fusionated with one of the CBHs (CBH1) of *T. reesei*. The fusion product was expressed in *T. reesei* and demonstrated being more effective in the cellulose saccharification of the corn-processing residue than that originating from the parental strain (time reduction of the cellulose hydrolysis from 10 to 6 h).

These findings show that further developments should occur in the next future in order to make viable the production of cellulases, particularly in the same industrial plant (*dedicated cellulase production*) where the producing processes are in operation, simply because great quantities of enzymes will be required for the efficient hydrolysis of the abundant lignocellulosic material, be it for the production of ethanol or other chemicals. Process integration of the production of cellulases with other producing processes in the same industrial installation is also within the context of biorefinery.

16.3.9 Hexose and Pentose Fermentation

The ethanol production technologies of sugary (sugarcane juice) and starchy (corn) feedstocks are commercially established and carried out by the yeast *Saccharomyces cerevisiae*. However, as described before, the hydrolysates of the polysaccharides of the lignocellulosic materials possess a mixture of hexoses (mainly glucose) and pentoses (mainly xylose), being the naturally occurring strains of *S. cerevisiae* unable to metabolize xylose.

The fermentation of glucose occurs primarily when the glucose concentration is high or when oxygen is not available. The cells attain a maximum specific growth rate of about $0.45\,h^{-1}$ with a low biomass yield of $0.15\,g$ dry mass per gram glucose consumed and a high respiratory quotient (the ratio of CO_2 production rate to O_2 consumption rate), resulting in a low energy yield of only about 2 ATP molecules per mol of glucose metabolized. The stoichiometry of this reaction is as follows:

$$C_6H_{12}O_6 \rightarrow 2C_2H_5OH + 2CO_2 + \varepsilon$$

The term "glucose fermentation" refers to the whole of the sugar's breakdown from the glucose molecule itself to ethanol and carbon dioxide. In other words, the glycolysis is a catabolic metabolism (Embden–Meyerhof–Parnas pathway), which represents the conversion of glucose to pyruvate followed by the conversion of pyruvate to ethanol.

Xylose-fermenting microorganisms use the pentose-phosphate shunt and the *Embden–Meyerhof–Parnas* pathway. Pyruvate is finally converted to different end products, such as ethanol, organic acids, ketones, and volatile products, depending on the microorganism and on the regulation of carbon flow through available metabolic routes, which can be controlled by the process variables.

Whereas in most bacteria the metabolism of D-xylose proceeds via direct isomerization to D-xylulose catalyzed by xylose isomerase, in the great majority of yeasts and filamentous fungi the formation of xylulose occurs via a two-step reaction in which xylose is reduced to xylitol by NADPH-dependent xylose reductase, with subsequent oxidation of xylitol to xylulose by NAD+-dependent xylitol dehydrogenase.

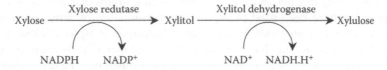

Yeasts show different abilities in the utilization of xylose, this being closely related to their requirements for oxygen; some can consume xylose only aerobically, others are able to fermenting it *quasi*-anaerobically or under oxygen restriction conditions. The latter capability is closely related to the dual specificity for the cofactors involved in the two initial steps of xylose catabolism displayed by some xylose-fermenting yeasts. Xylulose is then incorporated into the pentoses-phosphate pathway, originating glyceraldehyde 3-P and fructose 6-P. Both are converted into pyruvate through the glycolytic pathway, which gives origin to ethanol through two sequential reactions (decarboxylation and reduction).

Even though certain naturally occurring yeasts, as, for example, *Pichia stipitis*, *Pichia segobiensis*, *Candida tenius*, *Candida shehatae*, and *Pacchysolen tannophilus* [64] are capable to ferment xylose to ethanol, the production rates are more reduced, compared to those of the alcoholic fermentation of glucose.

Aiming at the integration of these two processes (xylose and glucose fermentations), researches have basically been developed with two approaches. In the first, it is searched for the construction of a recombinant with additional hability to ferment xylose [65], inserting genes that codify for xylose transport and metabolism and in the second, it is aimed at increasing the ethanol yield through the genetic engineering in microorganisms which already possess the hability to ferment hexoses and pentoses [66].

As the majority of the sugars in the lignocellulosic hydrolysates are constituted of glucose and xylose (with smaller quantities of arabinose, galactose, and mannose), the initial efforts for the construction of ethanologenic strains have focalized the cofermentation of glucose and xylose. In this approach, genes that codify for the xylose catabolism have been inserted into wild strains of the yeast *Saccharomyces cerevisiae* and of the bacterium *Zymomonas mobilis* [65,66]. Recombinant strains of *S. cerevisiae* with the ability of coferment glucose and xylose have been constructed through the addition of genes of *Pichia stipitis* (*XYL1* and *XYL2*), which codifies for NADPH-dependent xylose reductase and NAD+-dependent xylitol dehydrogenase, as well as through the improvement of the xylulokinase expression [65]. In this way, xylose is converted to xylulose-5-phosphate, which is a central metabolite of the pentoses-phosphate pathway.

Even though these mutants have been constructed with success and demonstrated satisfactory performance on a laboratorial scale, the anaerobic cofermentation of glucose and xylose has not yet reached the requirements for the industrial production. This is because the metabolism of xylose with these recombinants presents an imbalance of the cell redox potential in relation to the cofactors, in particular concerning the ratio NAD+/NADH.H+, which leads the cell to require oxygen, even in low tensions. With the aim at circumventing this problem, other strategies have been utilized, with the insertion of an NADP+-dependent glyceraldehyde-3-phosphate dehydrogenase which aids for the regeneration of the NADPH [67] or through the construction of a mutant which expresses a xylose reductase with greater affinity for the NADH and, therefore, decreases its consumption for NADPH [68].

Additionally, in contrast to the promptly fermentable streams of sucrose and of the starchy hydrolysates, the lignocellulosic hydrolysates tend to have fermentation inhibitors, arisen from the pretreatment (acetic acid, furfurals, and aromatics), which need to be removed when their concentration is very high or their direct utilization will require the development of robustness strains which can be resistant to these inhibitors.

Developments in this area have advanced speedily and in the next future recombinant yeasts, which ferment efficiently glucose and xylose from the hydrolylates of lignocellulosic biomass with production rates compatible to the industrial requirements, will be available.

16.3.10 STRATEGIES FOR THE ETHANOL PRODUCTION FROM LIGNOCELLULOSIC MATERIALS

The transformation of lignocelulosic materials for the production of ethanol has been studied under different strategies of processing. Due to the presence of different sugars, very often the multiprocessing is made necessary, in other words, the use of enzymes simultaneously to the action of microorganisms. Or even the use of different microorganisms in successive stages, or of recombinant microorganisms as for utilizing the utmost of the available sugars (substrates). In this sense, four strategies are conceived, each one in a different development stage, which will be described as follows.

Separated hydrolysis and fermentation (SHF): It is the most ancient conception, in which the cellulose hydrolysis, after the pretreatment of the feedstock for hemicellulose solubilization and hydrolysis, occurs in a separated fermentation stage. The fluxogram shown in Figure 16.11 depicts a process that utilizes diluted acid for hemicellulose hydrolysis. Next, the cellulose is hydrolyzed enzymatically, before the alcoholic fermentation stage. This strategy has been left behind, due to the low efficiency of the enzymatic hydrolysis of the cellulose, when it occurs separately from the glucose fermentation.

Simultaneous saccharification and fermentation (SSF): As the name implies, the cellulose enzymatic hydrolysis and the fermentation occur in the same stage. The hemicellulose fraction is hydrolyzed and fermented in a separate stage, as well as the enzyme production (Figure 16.12). Contrary to what occurs with the hemicellulose, from which sugars can be obtained through its hydrolysis, when it is aimed at hydrolyzing cellulose enzymatically, this should be associated to a transformation process. This comes from the fact that (even they present high catalytic activities) the enzymes of the cellulolytic complex are inhibited by their own hydrolysis final products, particularly glucose. Thus, the alternative to sort out the inhibition problems consists in moving the equilibrium of the hydrolysis reaction, through the "glucose removal" from the reactive medium. To achieve this goal, the adopted strategy is to couple the enzymatic reaction to a fermentative process, which should

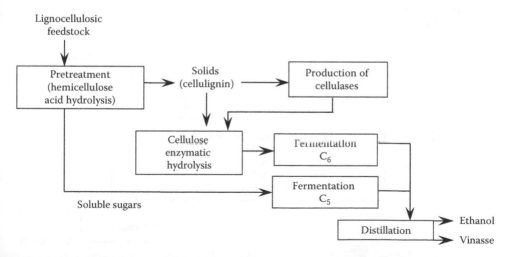

FIGURE 16.11   Diagram of the separated hydrolysis and fermentation (SHF) process.

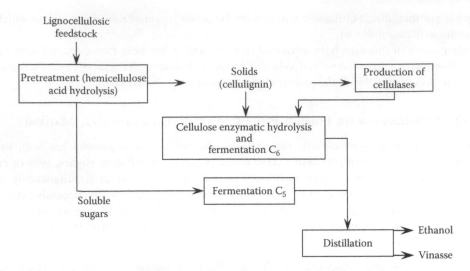

FIGURE 16.12   Diagram of the simultaneous saccharification and fermentation (SSF) process.

occur simultaneously, while glucose is being formed. This process is called in the literature as "simultaneous saccharification and fermentation."

On one hand, this process offers the advantage of minimizing the inhibition problems, on the other, the optimum operational conditions for an efficient enzymatic hydrolysis are not necessarily the same as those from the fermentation. In relation to this aspect, efforts have been made in the sense of producing enzymes which act in temperatures and pH values close to the optimum of the fermentation process.

Simultaneous saccharification and cofermentation (SSCF): This process involves three stages, of which the hydrolysis of the hemicellulose and the production of cellulases take place separately, as illustrated in Figure 16.13. In accordance to this conception, the liquid stream, rich in pentoses, obtained after the pretreatment remains in the bioreactor to which cellulases are added, followed by inoculation with a recombinant strain (capable of fermenting pentoses and hexoses). The main advantage of this strategy resides in the fact that only one reactor is used for ethanol production.

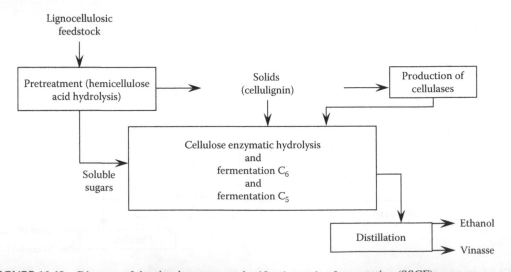

FIGURE 16.13   Diagram of the simultaneous saccharification and cofermentation (SSCF) process.

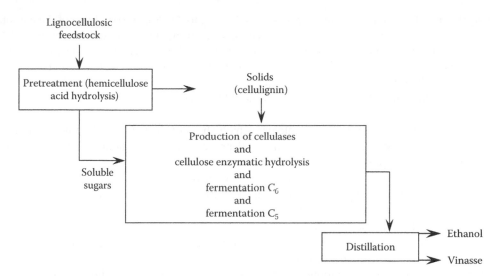

FIGURE 16.14   Diagram of the CBP.

Consolidated bioprocess (CBP): It is the most advanced process conception, in which all, or at least three of the stages, can be carried out in the same equipment. In the CBP, ethanol and all required enzymes are produced in the same bioreactor. With the modern tools of the molecular biology, there exist possibilities to get expressed several activities in only one microorganism, be it associated to the producing capacity of the enzymes of the xylanase and cellulase complexes, as well as to the efficient fermentation capacity of pentoses or hexoses. Figure 16.14 displays this conception of the process, which seems to be the final logical point of the evolution of the lignocellulosic biomass conversion technologies, being a middle-/long-term perspective, where molecular biology plays a fundamental role.

16.3.11   CONCLUDING REMARKS

The dependence of petroleum remains as the most important factor which affects the worldwide distribution of wealth, global conflicts, and the quality of the environment. The population's growth and the associated demand for fuel and goods have intensified R&D for the utilization of renewable feedstocks in substitution to the fossil sources. The advances in this area point out that the utilization of renewable raw materials, including their residues, will revert this dependence.

The lignocellulosic materials, especially the residues of the agro-industry, have been the object of intensive researches all over the world because they are renewable feedstocks of carbon and energy available in great quantities. The integral and rational utilization of these abundant feedstocks can revolutionize a series of industrial segments, such as the liquid fuels, the food/fodder, and the chemical supplies, bringing immeasurable benefits for countries with great territorial extensions and with high productivity of biomass, among them, Brazil occupies a distinguished position. The sugarcane bagasse is the main Brazilian agro-industrial residue, being produced in approximately 250 kg/ton of sugarcane. In spite of the great potential of this residual biomass of lignocellulosic composition (50%–70% carbohydrates) for the production of fuels and chemicals, the majority of it is burned in sugar mills and alcohol distilleries for energy generation, and a smaller fraction is used for animal feeding, yet there still have surpluses [69].

The effective utilization of the lignocellulosic materials in biological/fermentative processes faces us two principal challenges: the crystalline structure of the cellulose, highly resistant to the hydrolysis and the lignin–cellulose association, which forms a physical barrier that hinders the enzymatic access to the cellulose fibers. Additionally, the cellulose acid hydrolysis presents the inconvenience of requiring the use of high temperatures and pressures, leading to the destruction of part of the carbohydrates (sugars) and the generation of toxic substances, derived from lignin partial degradation [28]. On the other hand, the enzymatic saccharification requires the use of physical (grinding, heating, and irradiation) or chemical (sulfuric acid, phosphoric acid, and alkalis) pretreatments, to reach viable yields.

To make possible that the ethanol production technology from lignocellulosic biomass can be implemented industrially, the following aspects should be focused:

1. Development of pretreatment technologies which should be efficient and do not generate toxic substances that can hinder biochemical processes, neither should require onerous high pressure equipment.
2. Combination of cellulose enzymatic conversion with alcoholic fermentation (and other fermentations) to maintain low levels of sugars, resulting in improvements of the enzymatic conversion rates due to the minimization of enzymes inhibition by their final hydrolysis products (cellobiose and glucose).
3. Construction of "optimum" microorganisms, through molecular biology, for an efficient fermentation of pentoses and hexoses.
4. Development of cellulase production processes by submerged and solid-state fermentations (with natural-occurring or recombinant microorganisms), as well as to develop a deep knowledge about their structures and properties in order to formulate an enzymatic preparation (*product engineering*) for an efficient hydrolysis of cellulose.
5. The industrial production of cellulases should be *in plant* as for reducing the inherent costs with enzymes in the process.
6. Incorporation of reduced temperatures for ethanol separation to allow the enzyme recycling without thermal denaturation.
7. Realization of a detailed study of process integration (mass and energy), including all the streams, be they of the process or utilities, in order to favor the *input/output* ratio of energy.
8. Realization of a detailed technical-economic evaluation of the process viability for the utilization of agricultural and agro-industrial residues, including the logistic issues.

For countries like Brazil, with a strong agriculture tradition, the fermentation industry of lignocellulosic feedstocks is of great importance to the creation of technologies for the production of a range of useful compounds, within the context of biorefinery. This concept offers innovative possibilities, since it can bring solutions to supplant technologies which pollute the biosphere or contribute to the depletion of finite sources. Nevertheless, the industry, the scientific community, and the government need to work together in order to allow Brazil to reach its industrial/economical/environmental sustainability and to follow its natural vocation for the biomasses. In developed countries, integrated researches and the development of chemical and biological processes from lignocellulosic residues have advanced speedily, and commercial plants for the utilization of such materials are becoming reality.

16.4　OLEOCHEMISTRY

Oleochemistry refers to the transformation of fats and vegetable oils through different processes such as hydrogenation, esterification, hydrolysis, among others, generating products of considerable value and multiple applications [70].

The oleochemical industry acts as a "raw material center" of oleochemical derivatives for various industries such as food, cosmetics and toiletries, paints and varnishes, rubber, metals, plastics, cellulose, biotechnology, soap and detergents, lubricants, textile and leather, pharmaceuticals, and chemical industries. The main basic products of the oleochemical complex are fatty acids, fatty esters, fatty alcohols, and glycerin.

16.4.1 INDUSTRIAL OLEOCHEMICAL PLANT

An oleochemical plant can produce the following:

- Fatty acids C_8–C_{18} (saturated and unsaturated) and their fractions
- Fatty esters C_8–C_{18} (saturated and unsaturated) and their fractions
- Fatty alcohols C_8–C_{18} (saturated and unsaturated) and their fractions
- Glycerin 85% and 99.5%
- Biodiesel

Like esters, fatty acids, fatty alcohols, and glycerin are the most important chemical products for the worldwide oleochemical industry development.

Raw materials generate a multitude of products, serving different segments and having high added value (Figure 16.15).

16.4.1.1 Fatty Esters

Esters are considered one of the most important classes of organic compounds and can be obtained by using different methods: (1) reaction between alcohols and carboxylic acids (esterification) by eliminating water; (2) interesterification reactions, where acyl groups exchanges occur between esters and carboxylic acids (acidolysis), between esters and alcohols (alcoholysis) or glycerol (glycerolysis), and between esters (transesterification); (3) through natural sources either by distillation and extraction with adequate solvents or by chemical processes, and recently, by biocatalysis.

Esters resulting from acids and short-chain alcohols (2–8 carbon atoms) are important components of aromas and flavorings used in food, beverages, cosmetics, and pharmaceutical industries. The long-chain carboxylic acids esterification (12–20 carbon atoms) products with long-chain alcohols are used as lubricants and plasticizers in high-precision machines. The esters resulting from the reactions between long-chain acids with short-chain alcohols (2–8 carbon atoms) are used in industries as additives in foods, detergents, cosmetics, and medicines. Sucrose esters, for example, are known as good emulsifiers in the food, pharmaceutical, and cosmetic industries; isopropyl palmitate is used in medicinal preparations for cosmetics, in which good absorption of the product by the skin is necessary. Long-chain fatty acid esters such as oleate, palmitate, and linolenate are the major components of biodiesel. Ethyl oleate is used as a biological additive, plasticizer of polyvinyl chloride (PVC), waterproof agent, and hydraulic fluid.

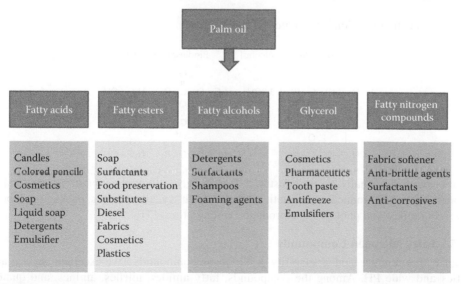

FIGURE 16.15 Products obtained by oleochemical route.

16.4.1.2 Fatty Acids (Saturated and Unsaturated) and Their Fractions

Fatty acids, fundamental units of most lipids, are organic acids, having 4–24 carbon atoms. They may have short chains (4–6 carbons), medium chains (8–12 atoms), and long chains (over 12). Besides the size of the carbon chain, fatty acids differ in the number and double bond position. Nearly all have even number of carbon atoms. The most abundant are the ones with 16 or 18 carbon atoms, the palmitic and stearic acids, respectively. The medium-chain ones are more abundant in butter and coconut oil.

Fatty acids are not found free in cells and tissues, but covalently linked to different types of lipids. The type and configuration of fatty acids in fats are responsible for differences in flavor, texture, melting point, and absorption.

16.4.1.3 Saturated Fatty Alcohols and Their Fractions

These are produced either from vegetable oils or petrochemically synthesized. Fatty alcohol is a raw material largely used in the manufacture of specialty chemicals derived from ethylene oxide, which are widely used as an input of personal hygiene products, also having several applications in household cleaning products (cleaners, powder and liquid detergents, and fabric softeners), agrochemicals, and textiles, among others.

16.4.1.4 Glycerol

Glycerol is produced by different routes. The routes of synthesis are as follows:

- Saponification: Saponification or alkaline hydrolysis is a reaction between alkalis and oils under pressure and moderate heat, which originates, in addition to glycerol, fatty acids in the form of soaps (alkali salts).
- Hydrolysis: Hydrolysis is a chemical reaction between fat (or oil) with water, generating glycerol and fatty acids:

$$\text{Triglycerides} + 3H_2O \leftrightarrow 3 \text{ fatty acids} + \text{glycerol}$$

- Transesterification: In the process of transesterification, oils or fats react with short-chain alcohols producing esters (either methyl or ethyl) and glycerol using homogeneous and heterogeneous catalysts [71,72].

Transesterification reaction is presented as follows:

Glycerin is used in virtually every industry. The main glycerin application areas are food, pharmaceutical, beverages production, cosmetics, tobacco industry, alkyd resins, packaging, lubricants, adhesives, ceramics, and photographic products, among others [71].

16.4.1.5 Fatty Nitrogen Compounds

Although with unknown biodegradability grade, these compounds have excellent surface activity properties and wide PH. Among the compounds, fatty amides, nitriles, amines, and quaternary ammonium compounds—quats—may be mentioned.

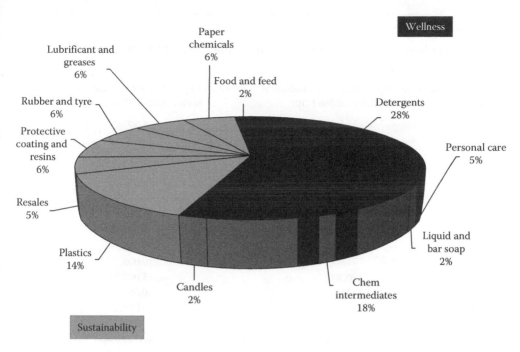

FIGURE 16.16 Esters and fatty acids market obtained in the oleochemical industry.

16.4.2 Oleochemical Sector: An Overview

Figure 16.16 shows the several applications of esters and fatty acids, noticing that they serve a large number of markets without excessive concentration in a given segment, which is very important.

The worldwide production capacity of fatty acids is distributed as follows:

- Asia with 60% of world's production
- The United States with 15.7% of world's production
- European Community with 22.8% of world's production
- Only 1.4% of world's production distributed to other parts of the world

The production of oleochemical derivatives managed to change the panorama. More than 100 years old, this industry had lost prestige with the onset of petrochemistry, supported by relatively lower prices, at least until the 1970s.

Besides the price, petrol and natural gas derivatives were seen as modern, offering unlimited molecule variant possibilities. However, the ecological appeal brought back the charm of products obtained from natural and renewable raw materials, even if requiring more complex processes and suffering from seasonality and crop failures of the agricultural products.

In the current view of modernity, fatty acid and alcohol derivatives displace synthetic ingredients of the personal care products and cosmetics formulations, besides showing strength even in some domisanitaries.

National oleochemistry is still limited to the basic distillation and hydrogenation operations, being restricted to the offer of mixed fatty acids, with honorable exceptions. The most significant advances are in the field of food and food additives, with the recent offer of hydrogenated fats and oils, free of trans isomers, controvertibly dangerous. Yet in the area of industrial inputs, the situation is more conservative.

According to experts, the oleochemical industry does not compete with food supply such as edible oils and margarine, but uses by-products of these lines. They also say that the problem is that

TABLE 16.7

Brazilian Import in the Industrial Sector in the 1996–2009 Period, Total and Chemicals

Year	Brazilian Import in the Total Industrial Sector (US$ million FOB)	Brazilian Import in the Chemicals Industrial Sector (US$ million FOB)
1996	54.500	8.500
1997	59.890	8.600
1998	58.600	8.600
1999	49.780	8.430
2000	56.345	8.600
2001	56.825	8.600
2002	47.600	8.600
2003	47.968	8.700
2004	61.000	10.000
2005	71.900	10.000
2006	90.300	13.000
2007	120.000	20.000
2008	172.000	30.000
2009	124.000	20.000

throughout Latin America, there are only three or four industries with portfolio, scale, and world-class technology in the sector.

In Brazil, the import of those chemicals has increased (Table 16.7); altogether, imports in the industrial sector had an 76% increase over a 13 year period (1996–2008); in 2009, there was a $8.847 billion decline in imports due to the financial crisis impact on the Brazilian trade balance. That effect has evidenced a variable behavior of the Brazilian imports, which relates positively to the country's economic growth and negatively with the trade balance. By the import projection period—January–June 2010—a gradual stabilization of the trade balance can be observed.

In Brazil, the historical local deficiency in fractionated acids is justified by the lack of adequate raw material, since the most abundant oil, the soybean oil, is limited to carbon chains ranging C_{18}.

Experts emphasize that the ideal would be to have a large and stable supply of coconut oil, palm kernel, or babassu oil, with a more diverse composition, which allows obtaining nobler fractions. In this framework, Asian competitors take advantage, operating facilities for more than 100,000 ton/year of acid, while the national average oscillates between 5,000 and 10,000 ton/year.

In 2005, the worldwide capacity of basic oleochemicals was of approximately 6.5 million tons/year. The long-term trend is favorable for oleochemicals and the current production capacity of 10.8 million tons/year is expected to rise to 12 Mt in 2010.

The leading regions, Europe, Asia, and the United States maintain occupancy rates exceeding 85%; Latin America and the others oscillate from 50% to 60% on average; on the other hand, not counting on the large-scale production, Latin America imports large volumes of oleochemicals.

In Brazil, the first oleochemical plant (Oleoquímica Indústria e Comércio de Produtos Químicos Ltda.) (Oleochemical Industry and Trade of Chemicals) has been working since 2008, with capacity to produce about 100 ton of fatty alcohols. Organized by Oxiteno S.A. with an investment of US$120 million, it is located in the Industrial Pole of Camaçari/BA. It uses renewable raw material (coconut and palm kernel oils, the former being extracted from the almond of the palm) to produce fatty acids and alcohols as well as an excellent quality USP/Kosher glycerin, highly pure, thermally stable, with applications in the personal care, food, and pharmaceutical markets, as well as in synthesis intermediates. The fatty alcohol line comprises lauryl alcohol, keto-stearyl alcohol and its fractions, cetyl alcohol, and stearyl alcohol. Another product of the plant will be the caprylic-capric acid, with applications in food, agrochemicals, animal feed, and pharmaceuticals.

In Brazil and Latin America, altogether, there is room for growth, since they are rich in raw materials of plant and animal origin; the use of regional products simply has to be promoted.

16.4.3 OLEOCHEMICAL PROCESSING ROUTES

The technological routes of an oleochemical plant that transform vegetable oils into high value products are summarized in Figure 16.17.

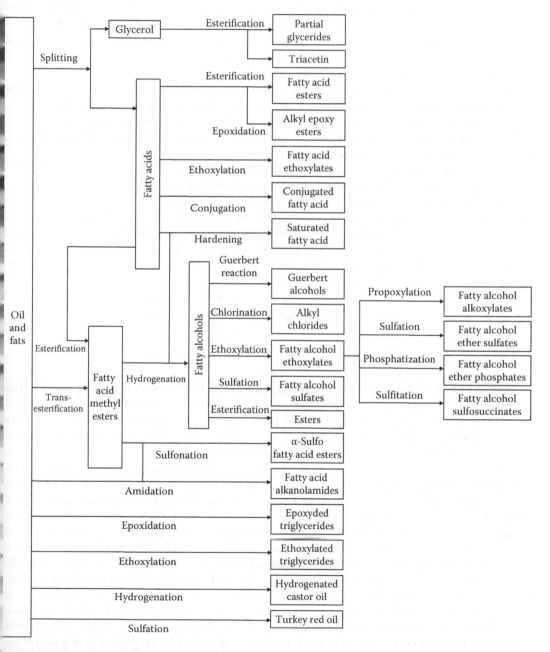

FIGURE 16.17 Oleochemical transformation routes.

16.4.3.1   Hydroesterification

The hydroesterification process is the ultimate alternative for biodiesel production. It allows the use of any raw fat (animal fat, vegetable oil, used frying oil, refining sludge acid of vegetable oils, among others). These raw materials are thoroughly transformed into biodiesel, regardless of their acidity and moisture [72].

This is a big difference when compared to the conventional transesterification process. Industrial transesterification occurs by alkaline catalysis, inevitably generating soaps, invariably requiring semi-refined—more expensive—raw materials [70]. This problem affects the yield of those plants and biodiesel/glycerin separation is difficult as well. To solve this problem, transesterification makes use of large amounts of acids in order to break the emulsion, which generates a high operating cost [73].

It is estimated that a transesterification biodiesel plant has an operational cost of \$70/ton of biodiesel (electric power, thermal power, chemicals, and labor). As hydroesterification does not need homogeneous catalysts or acid and base washing, it has a \$35/ton operating cost. In a 100,000 ton/year plant, this represents a US\$3.5 million savings per year.

Besides, benefits are even greater because of the use of acid raw materials such as crude palm tree oils (palm, macaw palm, and babassu), raw castor oil, or high acidity animals' fat viscera (pork, beef, and chicken). Transesterification is infeasible for these cheaper raw materials. As about 80% of the biodiesel production cost derives from the cost of raw material, hydroesterification allows a significant feasibility leap in a biodiesel project.

Hydroesterification is a process that involves a hydrolysis stage followed by esterification [75].

Hydrolysis is a chemical reaction between fat (or oil) with water, generating glycerol and fatty acids. Certain substances cleave into two or more pieces and these new molecules complement their chemical bonds with the H^+ and OH^- groups, a consequence of the chemical bond break that occurs in several water molecules. Hydrolysis can be divided into acid hydrolysis, basic hydrolysis, and neutral hydrolysis [74,75]. Thus, varying according to the type of catalyst used, the split of triacylglycerides allows the obtainment of fatty acids and glycerol waters, according to the following typical reaction:

$$
\begin{array}{cccc}
\text{RCOOCH}_2 & & \text{RCOOH} & \text{CH}_2\text{OH} \\
| & & & | \\
\text{R'COOCH} + 3\text{H}_2\text{O} \rightleftharpoons & \text{R'COOH} + & \text{CHOH} \\
| & & & | \\
\text{R''COOCH}_2 & & \text{R''COOH} & \text{CH}_2\text{OH} \\
\text{Triglyceride} \quad \text{Water} & & \text{Fatty acid} & \text{Glycerine}
\end{array}
$$

In this reaction, the existence of three reaction stages can be observed. In the first one, the triacylglyceride is transformed into diacylglyceride; in the second, this is transformed into monoacylglyceride (monoacylglycerol), which eventually, in the third stage, is hydrolyzed to give rise to fatty acid, generating glycerol as a by-product, just as diacylglyceride and monoacylglyceride revert to triacylglyceride and diacylglyceride by adding up a water molecule.

This is a process known worldwide, even in Brazil, where there are currently three plants in operation. In these plants, conversions above 99% are reached. Regardless of the acidity and moisture (which is the reagent of the process) of the raw material, the hydrolysis end product has over 99% acidity. Therefore, instead of reducing the acidity through refining, hydrolysis increases, on purpose, the acidity of the raw material. In addition, much more pure glycerin than the glycerin coming from the transesterification is obtained. Food grade raw material generates food grade glycerin from hydroesterification. This never occurs in transesterification, where a significant content of salts, alcohols, and other impurities are present in the glycerin. The obtained glycerin can be submitted to a purification process either by evaporation or distillation to provide distilled commercial glycerin with 99% or more purity. About 75% of the glycerol used in Europe comes from this process. Glycerol is also used as antifreeze and wetting agent in the cosmetics and tobacco industry, as well as to obtain alkyd resins (used in the preparation of paints); it can be transformed into polyol by the process of polymerization and it can be used in plants spraying as a polymer adjuvant (a water retention polymer).

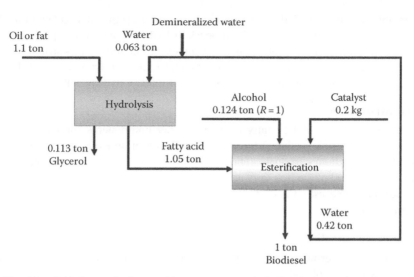

FIGURE 16.18 Material balance (hydroesterification per ton of biodiesel).

The fatty acids obtained from hydrolysis undergo an esterification process; the combination of these processes is called hydroesterification, which yields highly pure methyl ester. The esterification stage consists in obtaining esters from the reversible reaction of a carboxylic acid (fatty acid) with an alcohol (either methanol or ethanol), with water formation as a by-product. The reaction is equimolar, since each fatty acid molecule present in the oil will react with a used alcohol molecule [76]. There is no glycerin contact (already removed in the hydrolysis) with biodiesel (produced in the esterification).

$$\underset{\text{Acid}}{RCOOH} + \underset{\text{Alcohol}}{R'OH} \underset{\text{Catalyst}}{\overset{\text{Acid}}{\rightleftharpoons}} \underset{\text{Ester}}{RCOOR'} + \underset{\text{Water}}{H_2O}$$

Esterification reactions are facilitated by the increase of the reaction medium temperature and the presence of acid catalysts, such as sulfuric or niobic acid. The conversion rate of fatty acid esters depends directly on the way the reaction will be conducted as well as on the process conditions. Thus, the esterification course will be influenced by many factors that include raw material quality (free fatty acids content and the presence of water), reaction temperature, fatty acid molar ratio, alcohol, type, and catalyst concentration [75].

Currently, Agropalma's (Belém-PA) biodiesel plant already operates esterification [77–78]. Only water is generated as a by-product. This water returns to the hydrolysis process. This avoids biodiesel contamination problems with free or total glycerol residues (mono-, di-, and triglycerides). The result is a higher purity of biodiesel without the need for washing steps that generate effluents and high consumption of chemicals. Biobrax, a company based on Bahia state (Una-BA) is the first hydroesterification biodiesel plant with capacity of 60,000 million tons/year. Figure 16.18 shows the mass balance for this plant.

16.4.3.2 Ethoxylation

Process of obtaining raw materials where there is ethylene oxide molecule(s) insertion in compounds that have active hydrogen atoms.

16.4.3.3 Hydrogenation

The process of hydrogenation consists in transforming an unsaturated organic compound in another one, saturated by the addition of hydrogen. In the case of natural products, it is common to have vegetable oils hydrogenation carried out; these oils are composed of triglycerides, rich in unsaturated fatty acids or polyunsaturated. Hydrogenation increases the stability of vegetable oil, changes the texture, and increases its melting point. The process of hydrogenation takes place in the presence of hydrogen and catalysts, and the natural product undergoes changes in its structure, exchange of saturated into unsaturated bonds, therefore slight changes to its characteristic natural chemistry.

Vegetable margarines made from oils are typical examples of hydrogenated active cosmetics.

16.4.3.4 Neutralization

This is the removal process of free fatty acids and other components (proteins, fatty oxidation, and glyceride decomposition products) through the addition of aqueous solution of alkali such as sodium hydroxide or sodium carbonate. The amount of alkaline solution required for the process will depend on the content of the free fatty acids in the oil, mixing time, neutralization temperature, and process adopted. The use of sodium carbonate reduces the saponification of neutral oil to a minimum, but it also eliminates the phosphatides, pigments, and other impurities.

16.4.3.5 Transesterification

In the process of transesterification, oils or fats react with short-chain alcohols producing esters (methyl or ethyl) and glycerol [74].

The esters obtained in this process have multiple applications [72], among which the following may be mentioned:

- Sulfonated to produce sulfonated ∞-methyl esters
- Hydrogenated to produce fatty alcohols that later can be ethoxylated and sulfonated
- Used directly as biodiesel or as feedstock in the production of cosmetics, textiles, and plastics

Currently, there are 64 biodiesel industrial plants in Brazil running with transesterification processes. Total capacity of production is about 5 billion liters/year.

16.5 THERMOCHEMICAL ROUTES

As previously mentioned, the thermochemical routes are in principle related to second-generation biofuels, that is to say, they represent technologies capable of transforming biomass residues such as sugarcane bagasse or straw into liquid fuels. Since thermochemistry is the branch of chemistry that deals with the relationship of heat to chemical change, thermochemical routes may be defined as those in which biomass chemical transformations take place upon thermal treatment; hence, cracking and hydrocracking of vegetable oils may be considered thermochemical routes.

In Brazil, the main areas of R&D in the field of thermochemical routes are the following:

- BTL (comprising gasification, FT, and hydrotreating)
- H-BIO (also called *green diesel*)
- Bio DME/biomethanol
- Pyrolysis

16.5.1 Biomass to Liquids (BTL)

BTL is defined as "chemical transformation of biomass residues into liquid fuels and specialties with no contaminants and high performance." This definition takes into consideration the fact that BTL may be employed to produce not only fuels but also nonenergetic chemicals such as paraffin and lubricants.

There are four main steps to yield BTL products: (1) biomass pretreatment, (2) syngas generation with gas purification, (3) FT synthesis, and (4) product upgrading [79].

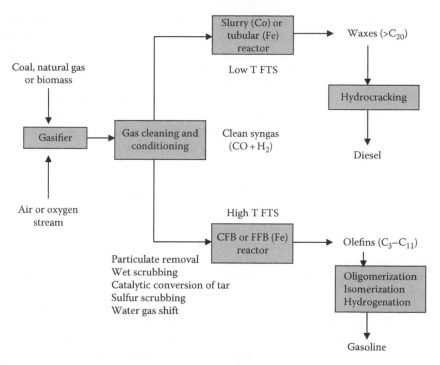

FIGURE 16.19 The BTL scheme for the production of middle distillates and gasoline.

1. Biomass pretreatment is often required before biomass can be admitted to step 2, the type of pretreatment depending on the source of biomass: wood, bagasse, straw, agricultural wastes, etc.
2. High-temperature gasification to obtain synthesis gas ($CO + xH_2$). Purification of the raw synthesis gas is a critical step.
3. FT synthesis to yield high-molecular-weight n-paraffin, diesel, naphtha, and LPG.
4. Hydrocracking of the paraffin fraction to yield more middle distillates or hydroisomerization to produce lubricants.

Figure 16.19 depicts a generic process flow diagram. When using natural gas as the feedstock, many authors [80] have recommended autothermal reforming or autothermal reforming in combination with steam reforming as the best option for syngas generation. This is primarily attributed to the resulting H_2/CO ratio and the fact that there is a more favorable economy of scale for air separation units than for tubular reactors (steam methane reforming—SMR).

Should the feedstock be biomass, the conversion to an $H_2 + CO$ containing feed gas suitable for FT synthesis takes place through gasification but, in this case, a pretreatment prior to gasification is required and generally consists of screening, size reduction, magnetic separation, "wet" storage, drying, and "dry" storage. Gasification can take place at different pressures, either directly or indirectly heated (lower temperatures) and with oxygen or air. Direct heating occurs by partial oxidation of the feedstock; while indirect heating occurs through a heat exchange mechanism [81]. A wide variety of biomass resources can be used as feedstock; however, in Brazil, bagasse and straw from the Brazilian ethanol production seem to be the most attractive ones to generate FT diesel. If the feedstock is coal, the syngas is produced via high-temperature gasification in the presence of oxygen and steam. Depending on the types and quantities of FT products desired, either low- (200°C–240°C) or high-temperature (300°C–350°C) synthesis is used with either an iron or cobalt catalyst. FTS temperatures are usually kept below 400°C to minimize CH_4 production.

Generally, cobalt-based catalysts are only used at low temperatures [82]. This is because at higher temperatures, a significant amount of methane is produced. Low temperatures yield high-molecular-weight linear waxes, while high temperatures produce naphtha and low-molecular-weight olefins. The FT reactors are operated at pressures ranging from 10 to 40 bar. Upgrading usually means a combination of hydrotreating, hydrocracking, and hydroisomerization in addition to product separation.

Compared to conventional fuels, FT fuels contain virtually no sulfur and low aromatics. These properties, along with a high cetane number, result in superior combustion characteristics. Tests performed on heavy-duty trucks showed decreases in vehicle emissions of HC, CO, NO_x and PM when using a FT fuel. FT diesel has been tested in a variety of light- and heavy-duty vehicles and engines. Alleman et al. [83] summarized FT diesel fuel property and emission information found in the literature. Also, several LCAs have been performed on a variety of transportation fuels including FT diesel and gasoline. Most studies have examined only GHG emissions and energy consumption with the exception of the one carried out by General Motors, in which the influence of five types of pollutants (VOCs, CO, NO_x, PM10, and SO_x) have also been examined. The results of the studies vary based on the feedstock procurement, technology conversion, and vehicle assumptions. However, in general, there is not a big advantage for FT liquids from fossil fuels in terms of energy consumption and green house gas emissions. This will not be the case for biomass systems. Because of the improved combustion characteristics of the FT liquids, a complete LCA will mostly likely show the overall benefits of FT liquids compared to conventional transportation fuels.

Brazil has a long time tradition in the use of renewable energy. The sugarcane sector in Brazil produces and processes more than 300 million metric tons of sugarcane. More than 50% of the sucrose is used in the production of ethanol, generating approximately 85 L of ethanol per ton of sugarcane. The sugarcane bagasse provides all energy required to process the sugarcane and several mills are already generating surplus power, selling it. This surplus power generation of the sugar/ethanol mills could be highly increased by the use of more efficient energy conversion systems, such as biomass gasification integrated with gas turbines and recovery of part of the sugarcane trash currently burned or wasted today, so as to supplement the bagasse as fuel. Both BIG-CC and trash recovery are emerging technologies that need development and demonstration in order to reach the market [84].

Under normal conditions, Brazil annually produces and processes a quarter of the 1300 million tons grown in more than 100 countries worldwide. The Brazilian sugarcane sector gross annual income of US$10 billion represents around 2% of the gross national product. Sugarcane production and processing are highly energy-intensive activities that require, under Brazilian conditions, for each ton of cane 190 MJ in agricultural area (in the form of fossil fuels, fertilizers, and other chemicals) and 1970 MJ in industry (in the form of chemicals and bagasse), the latter providing nearly 100% of the industry's energy requirement.

A life cycle analysis for ethanol production has indicated, however, that for each unit of fossil energy input to the agro-industrial system, follow approximately nine units of renewable energy output (ethanol and surplus bagasse) to be used outside the system. This situation has a huge potential for improvement if we bear in mind that ethanol represents only one-third of the energy available in sugarcane, being the other two-thirds represented by fiber in the cane stalks (bagasse) and in cane leaves (straw). It is worth mentioning that 93% of the bagasse is used as fuel in cane processing, in a very inefficient way and 85% of the straw is burned prior to cane harvesting to reduce the cost of this operation. This fact indicates that with some effort and investment, this potentially available fuel (sugarcane bagasse and straw) can be used to generate electric power [85]. The following three actions are required to accomplish this

- Improve process energy efficiency to generate more bagasse surplus.
- Harvest unburned cane and recover a reasonable fraction of the total trash.
- Use an efficient technology to generate power.

The introduction of BTL technology may be an excellent solution to improve energy efficiency in the Brazilian ethanol production facilities. A simplified scheme is presented in Figure 16.20.

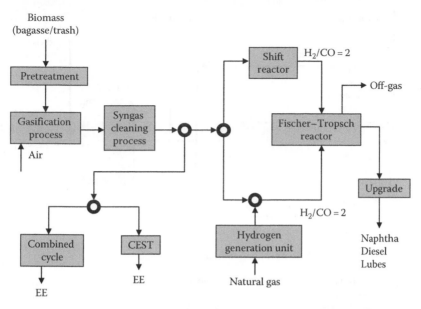

FIGURE 16.20 Simplified BTL scheme to improve energy efficiency in Brazilian ethanol production industry.

According to this scheme, a BTL industrial unit with a production capacity of 5000 bpd of liquid by-products will be able to generate around 47 MWe, using 7000 ton of biomass residues per day.

16.5.2 H-BIO

Contrary to what might be expected, H-BIO does not stand for a product. Rather than a derivative, H-BIO is a technology developed by Petrobras. Such technology allows the production of diesel from renewable feedstock such as vegetable oils by processing them in the existing refining scheme, which clearly represents an advantage. Interestingly, in the H-BIO technology, vegetable oils are coprocessed with petroleum in hydrotreating (HDT) units.

The H-BIO process was developed to introduce a renewable oil source in the diesel fuel production scheme taking advantage of existing plants as depicted in Figure 16.21. The vegetable oil

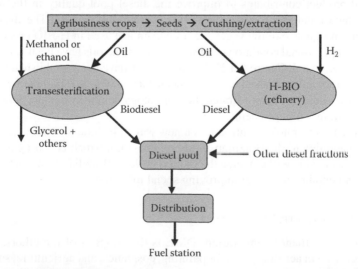

FIGURE 16.21 The logistics of H-BIO.

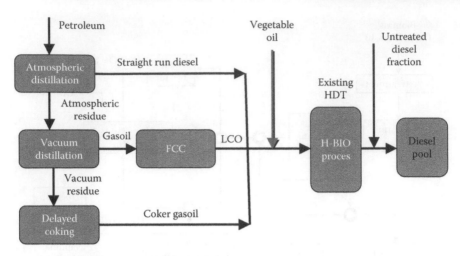

FIGURE 16.22   The scheme of H-BIO production in a typical refinery.

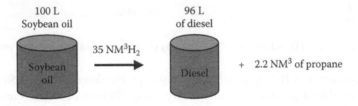

FIGURE 16.23   H-BIO process yields from soybean oil.

stream blended with mineral diesel fractions is hydroconverted in HDT units, which are mainly used for diesel sulfur content reduction and quality improvement in petroleum refineries.

This process involves a catalytic hydroconversion of the mixture of diesel fractions and vegetable oil in an HDT reactor under controlled conditions of high temperature and hydrogen pressure. The triglycerides from the vegetable oil are transformed into linear hydrocarbon chains, similar to that already existing in the diesel coming from petroleum (Figure 16.22). The most important aspect of H-BIO process is its very high conversion yield, at least 95% v/v to diesel, without residue generation and a small propane production as a by-product (Figure 16.23).

The converted product contributes to improve the diesel pool quality in the refinery, mainly increasing the cetane number, reducing the sulfur content and density. The diesel pool quality upgrade will be a consequence of the vegetable oil percentage used in H-BIO process.

A large range of operational conditions and several vegetable oils have been tested in pilot plants located at Petrobras Research Centre (CENPES) and yields have been established for various types of raw materials. Afterward, industrial tests have been carried out in HDT units for technical evaluation. Such industrial runs have demonstrated the technology flexibility and the potential use thereof in the existing Brazilian refining scheme.

The Petrobras H-BIO technology introduces a new way to include renewable feedstocks for biofuel production in addition to the Brazilian Biodiesel Program, which is getting ahead following a fast-track development. This will enhance the biomass role in Brazilian transportation fuel supply, generating environmental benefits and improving social inclusion.

16.5.3   BIO-DIMETHYLETHER/BIOMETHANOL

Usually obtained via methanol dehydration, DME is the simplest of the ethers. Currently, this substance has been used in aerosol sprays for painting, cosmetics, and agriculture, substituting chlorine- and fluorine-based compounds, which are harmful to the environment. Recently, DME has

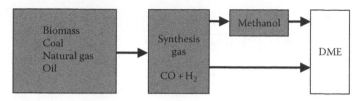

FIGURE 16.24 DME production routes from different potential raw materials.

come to the attention of various companies, research centers, and universities in leading countries, due to its potential for use as a fuel [86–88]. In fact, this substance can be used in diesel-powered engines, thermoelectric power plants, and fuel cells, as well as a substitute for LPG. The fact that DME is obtained from either natural gas or biomass residues should also be highlighted, which allows production costs to be independent of the swings in the price of oil, as well as the wide availability of the raw material, bearing in mind the current world reserves of natural gas and growing amounts of biomass trash. All of these facts together make DME known currently as the "fuel of the twenty-first century." Various recent studies undertaken in Japan, the United States, and Korea, among other countries [89–91], irrefutably showed the technical viability of the use of DME in diesel-powered engines. In fact, DME has a very high cetane index (60), it does not emit particles or sulfur oxides upon burning, making it one of the best fuels in respect to the pollution issue. However, recent information obtained from long duration engine tests showed that some development still needs to be carried out. In this context, the following is highlighted: the need for the development of low-cost elastomers, as DME may attack traditional polymers, additives seeking to alter the properties of lubricants and the viscosity of the DME and injection pumps specifically for DME [92].

DME is colorless; its boiling point is 25.1°C; and it has properties similar to propane and butane, the principal constituents of LPG. As it is easily liquefied, it can be distributed and stored employing the same technology used for LPG. These properties allow its use as a substitute for LPG, despite having a net heating value (6900 kcal/kg) slightly lower than the oil derivatives previously mentioned. Recently, disclosed technical–economical evaluation work indicates that DME production costs are competitive with those for LPG on the international market. It was also shown that modifications to ovens and containers are not very relevant [93]. As shown in Figure 16.24, DME can be produced in two distinct ways: the first, called the indirect route, produces methanol, then promoting its dehydration; the second, known as the direct route, produces DME in a single stage, using bifunctional catalysts.

Although synthesis from methanol is simpler, direct synthesis is more interesting as far as catalysis is concerned, and it may also become more attractive economically. The principal reactions involved in DME direct synthesis are as follows:

$$2CO + 4H_2 \rightarrow 2CH_3OH \tag{16.1}$$

$$2CH_3OH \rightarrow CH_3OCH_3 + H_2O \tag{16.2}$$

$$CO + H_2O \rightarrow CO_2 + H_2 \tag{16.3}$$

It is evident that methanol is the main intermediary in direct synthesis and that the water displacement reaction is a strong competitor to the principal reaction. The catalysts must be bifunctional, having methanol synthesis characteristic metallic sites, yet containing sufficient acidity for its dehydration reaction to occur. They must also be capable of promoting the CO_2 formation reaction, which becomes ever more significant as water is being generated by the dehydration stage.

CENPES, together with National Institute of Technology (INT), has started a platform to develop catalysts and process conditions for DME production. Some results are presented in Figure 16.25,

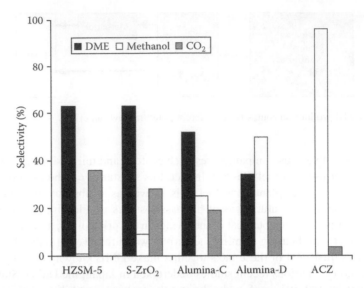

FIGURE 16.25   The importance of the catalyst composition (concentration of acid sites vs. metallic sites) in the formation of DME by direct synthesis. (From Ramos, F. et al., *Catal. Today*, 101, 39, 2005. With permission.)

in which data from bifunctional catalysts are shown [94]. These data elucidate the influence the acid and metallic sites have on the selectivity of the process. Greater acidity favors the formation of DME, whereas an increase in the concentration of metallic sites promotes conversion to methanol. The formation of CO_2 follows that of DME, as a clear indication that water generated by the dehydration continues to react with CO [95]. Therefore, direct synthesis has some interesting challenges in the development of catalysts. Among these, one can mention preparation methods that allow the introduction of classic metallic sites for methanol synthesis (Cu and Zn) without blocking acid sites, presuming that these may be in zeolite cavities, or the introduction of acid sites in a hydrotalcite Cu/Zn/Al, recognized as the best catalyst for methanol synthesis. Equally interesting is the search for a system that catalyzes CO_2 hydrogenation to methanol, thereby closing the cycle shown in reactions (16.1) through (16.3).

Furthermore, Brazil is now devoting considerable effort in the development of the so-called Bio-DME, that is to say, DME produced from syngas which has been generated from biomass residues via gasification. The gasification step would be the same carried out in the BTL process; therefore, joint programs of research have been discussed.

16.6   BIOCHEMICALS

16.6.1   INTRODUCTION

In Brazil, the use of ethanol as a fuel dates back to the founding in 1923 of a new government entity, Estação Experimental de Combustíveis e Minérios, the precursor of today's Instituto Nacional de Tecnologia. On February 20, 1931, a decree was passed (n° 19717) making it mandatory for ethanol to be added to gasoline at a proportion of 5%.

Biomass has been a traditional source of energy in Brazil for years. Indeed, in 1950, 55% of all primary energy consumption was derived from biomass [96].

As for the chemicals industry, at the beginning of the 1920s Brazilian ethanol was used in the production of ethyl chloride, ethyl ether, and acetic acid by Rhodia (Rhône-Poulenc) in São Paulo. In 1950, Usina Victor Sense in Rio de Janeiro started to produce butanol and acetone by the butanol-acetone fermentation of molasses. In the 1950s, a number of chemicals were produced from ethanol: butadiene was produced for rubber production, ethylene for polyethylene production, acetaldehyde and acetic acid for vinyl acetate production, and butyraldehyde for 2-ethylhexanol. The main use of ethyl alcohol was as a raw material for the Brazilian chemicals industry (44%), followed by its use as a fuel in an ethanol–gasoline mix (41%), while the remaining 15% went to other uses (beverages, pharmaceuticals, etc.).

In the early 1970s, interest in alcohol chemistry declined as the price of oil dropped to US$2/ barrel. However, after first oil shock in 1973, Brazil launched its pro-ethanol program in 1975, while other oil-dependent countries started to investigate the potential of different raw materials as fuels and inputs for the chemicals industry.

In Brazil, larger-scale sugarcane ethanol production started to be developed with the use of its by-products (bagasse). Large distilleries were built that were independent from sugar mills with a view to assuring the production of sufficient quantities of ethanol for its use as a fuel and in the chemicals industry. The bagasse was used as an energy input for the distilleries themselves and as a raw material in the production of fiberboard, paper, pulp, plastics, furfural, and levulinic acid.

The creation of the National Ethanol Programme in 1975 was an important political step forward in Brazil's long-term commitment to producing a substitute for imported oil. However, it must be borne in mind that biomass can also be the feedstock for large-scale plants to produce chemicals and intermediates.

A series of national alcohol chemistry conferences were held in Brazil in the early 1980s with the aim of presenting the latest research into fermentation processes for ethanol production and its use as a substitute for oil in the production of acetaldehyde, butanol, 2-ethylhexanol, ethylene, vinyl chloride, vinyl polychloride, polyethylene, and polystyrene [97].

The terms "biochemical," "biomass," and "biotechnology" have basically the same root and sometimes are indistinctly misused, since biomass is often the raw material for biochemical processes, which are a branch of biotechnology area.

In renewable energy systems, biomass is defined as the part of the organic matter that can be used as a source of energy and/or chemicals (i.e., biochemicals). Biotechnology is the use of organisms, enzymes, or biological processes to convert raw materials into products of higher added value, or biochemicals.

Biotechnology can be split into two main areas: the traditional area of fermentation for the production of beverages, foods, antibiotics, fuels, and biochemicals, plus the new area that involves the production and use of genetically modified organisms for several applications, including the large-scale production of proteins, vaccines, and monoclonal antibodies for use, diagnoses, and treatments.

Brazil is a world leader in some areas of biotechnology, such as agriculture and biofuels. To develop the sector, the country has introduced a number of incentives: the Biotechnology Sector Fund, coordinated by the Ministry of Science and Technology; the Forum for Competitiveness in Biotechnology, coordinated by the Ministry of Development, Industry and Foreign Trade's, which culminated in the creation of a new biotechnology policy for the country; and the introduction by the Ministry of Education of several undergraduate and postgraduate courses with an eye to training skilled workers and people with masters and doctorates in the area. Production has been fostered by a number of incubators, including BIORIO in Rio de Janeiro, BIOMINAS in Minas Gerais, and an incubator in Ribeirão Preto, São Paulo state.

The renewable chemicals market encompasses all chemicals obtained from renewable feedstock, such as agricultural raw materials, agricultural waste products or biomass, microorganisms, sucrose, starch, cellulose, lignin, oil, fats, proteins, and, in Brazil, sugarcane.

The global renewable chemicals industry is expected to be worth some US$60 billion by 2014, with the United States and Europe taking a 30% and 35% market share, respectively. Government support and initiatives to encourage the use of renewable chemicals are expected to provide the necessary boost to the markets in these economies. The rapidly developing economies of India, China, and Russia still largely consume chemicals from petrochemical feedstock (marketsandmarkets.com).

Biochemicals can be used in several sectors of the economy, including manufacturing, transportation, textiles, food safety, environment, communications, housing, recreation, health, and hygiene. Platform chemicals include C_1–C_6 building blocks.

The great variety of applications for biochemicals and biofuels produced directly from reusable sources, including waste materials, has led to the emergence of biorefineries, which are built exclusively for the production of biofuels and biochemicals.

The worldwide quest to find viable fuel alternatives to oil has encouraged ethanol producers to focus on increasing their agricultural and industrial output by increasing scale and minimizing costs. R&D efforts have been put into finding more efficient ways of harnessing the biomass present in raw material by-products. As a result, second-generation ethanol has started to be produced, which comes from sugarcane bagasse and sugarcane straw in Brazil, or maize and other agricultural sources in other countries.

Around the world there are several success cases of biorefineries:

Japanese company Idemitsu Kosan has partnered with the Research Institute of Innovative Technology for the Earth (RITE) to develop a bioethanol production method based on nonedible cellulosic biomass. The process makes use of corynebacteria in an anaerobic environment to metabolize C_6 and C_5 sugars into organic chemicals, including aromatics, butanol, carboxylic acid, ethanol, and propanol [98].

In the United States, the Alternative Energy Technology Center Inc. is at the completion phase of its program for a 20–100 ton/day vertically integrated biorefining system that produces ethanol, gasoline, diesel, and other lubricants as well as a number of intermediate compounds. The plant will consist of a cellulosic biomass reduction unit, which reduces and fractionates cellulosic biomass to nano- and microscale particles generating higher material surface area for simultaneous saccharification and fermentation of the cellulose and hemicelluloses by enzymatic conversion and microbic fermentation [99].

Also in the United States, GlycosBio has demonstrated the feasibility of industrial-scale production of chemicals such as lactates, propanediol, and succinic acid using biotechnology. It has developed microbial strains to process nonfood carbon sources such as glycerine [100]. GlycosBio is a spin-off from Houston-based Rice University and Myriant Technologies LLC in collaboration with the University of Florida's Institute of Food and Agricultural Sciences and Buckeye Technologies Inc., for the development of a first-of-its-kind biorefinery pilot project to be located in Florida in 2010. The 5 ton/day facility will produce high-value specialty chemicals and biofuels from cellulosic materials, demonstrating the path to maximize value in a renewable, sustainable manner. Scientists from the University of Florida are operating the ethanol research platform, while Myriant will operate the specialty chemicals component of the project. The project will be capable of producing in excess of 140,000 gallons of biofuels a year or 1000 ton a year of bio-based chemicals from a variety of feedstocks, including wood, sugarcane bagasse, and sweet sorghum [101].

In Great Britain, Green Biologics Ltd. is developing fermentation methods to produce both biofuels and biochemicals, while the North East Process Industry Cluster (NEPIC), combining Akzo Nobel, GrowHow, Scott Bader, Jacobs Engineering, Graphite Resources, and Link2Energy and three supporting academic bodies, has set up its Assessing Biomass to Chemicals (ABC) Project, which has selected chemicals for in-depth analysis, including methanol, ethanol, and butanol. The ABC Project has to date explored a wide range of possible biomass sources: wood chips, biomass recovered from municipal waste, cereals, and cereal residues [102].

Figure 16.26 shows the range of biochemicals that could be produced at future biorefineries. The biochemicals manufacturing trends in Brazil and the world are presented per chemical function and number of atoms.

16.6.2 ALCOHOLS

16.6.2.1   C₁ Alcohol

16.6.2.1.1   Biosyngas/Methanol

In green chemistry, biomass is generally considered to be agricultural and biological feedstocks, which are often made up of a variety of lignocellulosic substances. These can be excellent options for biochemical and biofuel production. Pyrolysis and gasification of biomass produces biosyngas and biomethanol.

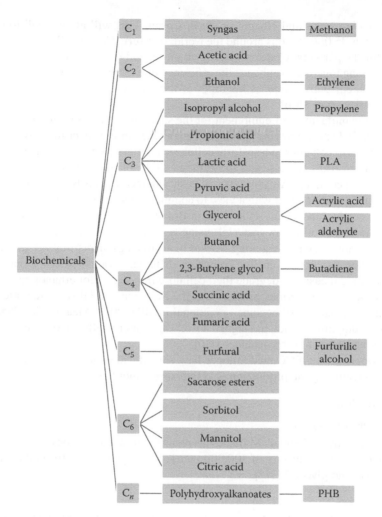

FIGURE 16.26 Biochemicals opportunities.

Several companies, including Dow, Coskata (an American biology-based renewable energy company), Range Fuels (an American company engaged in biomass conversion), Fulcrum Bioenergy, INEOS Bio (whose technology platform is an innovative biochemical process), and LanzaTech, have been conducting research into the potential of syngas as a source of higher-chain alcohols such as ethanol, butanol, and propanol [103].

Canada-based Syntec Biofuel has a thermochemical technology for gasifying different raw materials and converting them into methanol, ethanol, propanol, and butanol. The process utilizes Syntec's high-performance catalyst technology exclusively licensed from the Energy and Environmental Research Center Foundation [104].

The BioGold Fuels Corporation was formed as a result of a merger with Full Circle Industries in 2007 with the objective of reducing or eliminating refuse being dumped into landfills by processing the waste into usable by-products, producing many kinds of renewable energy and syngas transformation to liquid fuels [105].

16.6.2.2 C₂ Alcohol

16.6.2.2.1 Second-Generation Ethanol

Brazilian oil company Petrobras and France's Tereos have entered into a deal to develop and produce first- and second-generation bioethanol. In Brazil, Shell has also agreed to enter into a joint venture with Brazilian sugar and ethanol producer Cosan. This will allow Shell to sell Brazilian ethanol at some of its 45,000 stations around the world. In return, Cosan will gain access to research being done by Shell's partners, Codexis and Iogen, on second-generation ethanol. Codexis, Shell, and Iogen are working together to enhance the efficiency of enzymes. Iogen's demonstration facility in Ottawa, Canada, currently produces hundreds of thousands of liters of cellulosic ethanol from wheat, oat, and barley straw [106].

Brazil's largest manufacturer of equipment for the sugar and ethanol industry, Dedini, has partnered with Danish industrial enzyme specialist Novozymes in a project to produce ethanol from cellulose. They will use enzymes to transform cellulosic biomass from sugarcane bagasse and straw into ethanol. By using 100% bagasse and straw, they will double their ethanol production [107].

U.S.-based biotech company Amyris's Brazilian subsidiary, Amyris Brasil S.A., has developed a pilot plant and demonstration plant with a view to commercial-scale production. It is also setting up a joint venture to conduct industrial-scale trials in Brazil on second-generation bioethanol derived from sugarcane waste and other raw materials using genetically modified yeasts [108].

American company Agrivida is among several biotechnology firms that are aiming to surpass existing technologies. Its current pipeline comprises different varieties of crops that produce their own enzymes, making it easier to degrade their cellulose into sugars for ethanol manufacture [109].

In the United States, Novozymes predicts that the first commercial-scale facilities for the production of cellulosic ethanol will become operational in 2011–2012. Meanwhile, Canada's leading cellulosic ethanol company, Lignol, has signed a R&D agreement with Novozymes to make ethanol from wood chips and other forestry residues [110].

In China, Novozymes, COFCO, and Sinopec have signed a memorandum of understanding with the aim of commercializing the production of cellulosic ethanol from corn stover [111].

16.6.2.3 C$_3$ Alcohol

16.6.2.3.1 Propylene Glycol

In the United States, Archer Daniels Midland has plans to build a new propylene glycol plant using renewable sources. It has a capacity of 100,000 metric tons a year and will produce the propylene glycol from sorbitol and glycerol [112].

16.6.2.3.2 Glycerol

Glycerol is a raw material for methanol, ethanol, glycerol carbonate, and epichlorhydrin production [113].

The glycerin obtained from biodiesel production is used to produce methanol. The European Union Renewable Energy Directive states that carbon dioxide emissions from this process are negligible [114].

Around the world, there is an effort to develop oleochemical processes to enhance the profitability of biodiesel plants. Houston-based Glycos Biotechnologies has a high-efficiency process for the conversion of crude glycerin into ethanol.

In Ireland, Ceimici Novel has developed a smart catalyst, a new heterogeneous catalyst system that is simply recovered using filtration. The smart catalyst produces glycerol at 98% or over. The processing of glycerol to produce derivatives such as glycerol carbonate, which is used as a solvent in personal care and as a CO_2 absorption solvent, and glycidol, used as a stabilizer for natural oils and vinyl polymers and also in chemical synthesis, may add extra value [115].

In Tavaux, France, Solvay has opened a pilot epichlorohydrin production plant using glycerol as an input. Meanwhile, Czech company Spolchemie is producing epichlorohydrin, a key ingredient for epoxy resins, from glycerin [116].

16.6.2.3.3 1,3-Propanediol

The rapid development of industrial biotechnology is allowing more cost-effective production of monomers. DuPont and Tate & Lyle have developed 1,3-propanediol (bio-PDO) based on renewable sources. PDO is used in the production of several different materials, from polyesters to adhesives and paints [117].

16.6.2.4 C₄ Alcohol

16.6.2.4.1 Butanol

The global *n*-butanol market is expected to grow 3.2%/year until 2025. It can be used as a drop-in biofuel to be blended with gasoline, diesel, and ethanol, converted into jet fuel, brake oil, or plastics, or sold as a solvent for use in paints, cleaners, adhesives, and flavorings [118].

In Brazil, Oxiteno is using fuel oil, sugarcane juice, and fermented molasses to produce 10,000 ton a year of butanol at a multipurpose plant. Meanwhile, the sugar chemistry process involving molasses fermentation used by Usina Victor Sence, and the Elekeiroz alcohol chemistry process involving the aldol condensation of acetaldehyde into butanol were deactivated in Brazil in the 1990s [97].

In the United States, biobutanol is not yet produced on a commercial scale, but there are some firms that are keen to make the technology commercially feasible. For instance, U.S.-based Gevo Inc. started on a demonstration-scale basis in 2009. One of its technologies is a genetically engineered biocatalyst, a strain of yeast that has been biologically modified to produce isobutanol rather than ethanol. Another technology is the Gevo Integrated Fermentation Technology (GIFT), which is compatible with different feedstocks, including all conventional raw materials for ethanol and cellulosic biomass [119].

Also in the United States, biobutanol is one of the products that are set to be produced by a biorefinery owned by Patriarch Partners as of 2012 [120].

Other projects in the United States include a partnership between Cobalt and Colorado State University to produce biobutanol from forest waste and mill residue into *n*-butanol; Genencor Inc.'s conversion of cellulosic biomass to butanol using its Accellerase® 1500 enzyme complex; and Dyadic International's strategic collaboration in bioproducts, refocusing its efforts on growing a sustainable industrial enzyme business through a nonexclusive license agreement with Abengoa Bioenergy New Technologies Inc. for the right to use Dyadic's patent rights and know-how relating to the platform enzymes for use in manufacturing cellulosic ethanol and butanol.

Butamax Advanced Biofuels LLC is setting up a plant in the UK to demonstrate its biobutanol technology. It expects to have a commercial-scale plant by late 2012.

In Japan, cellulose and castor oil are being investigated by Mitsui Chemicals as feedstocks for the production of isopropanol and butanol [121,122].

16.6.2.4.2 2,3-Butanediol

In China, an investment plan for a 100 ton/year project for the production of 2,3-butanediol has already been completed, as well as an upgrade for a 1000 ton/year plant. More than ten new genes have been cloned that have an impact on *Serratia marcescens* in the synthesis of 2,3-butanediol by fermentation. A researcher from the East China University of Technology (ECUT) has applied for patents for six of the new genes [123].

16.6.2.5 C₅ Alcohol

16.6.2.5.1 Furfuryl Alcohol

Ferghana in Uzbekistan specializes in the production of furfuryl alcohol obtained from furfural (aldehyde), which is used in the pharmaceuticals, agrochemicals, and smelting industries [124].

16.6.3 Aldehyde

16.6.3.1 C₃ Aldehyde

16.6.3.1.1 Acrolein

Japan-based Nippon Shokubai Co. Ltd. is developing a high-performance catalyst for manufacturing acrolein from glycerin. Acrolein is a raw material for making acrylic acid. The new production platform involves the dehydration of glycerin. The acrolein is then oxidized via a gas-phase oxidation technology into acrylic acid and generates about 30% less carbon dioxide than production based on petroleum-derived feedstock [125].

16.6.3.1.2 Propanal

It is possible to use ethylene from ethanol that in the presence of a carbonylation catalyst facilitates the production of propanal, which can be further processed to produce propionic acid (by oxidation) or propanol (by hydrogenation).

16.6.3.2 C₅ Aldehyde

16.6.3.2.1 Furfural

Furfural is a renewable material derived mainly from sugarcane bagasse, corncobs, and other agricultural by-products. It is an important solvent and starting material for organic synthesis and is used as a solvent in the refining of crude oil. Production plants have been set up in Nebraska, US, and Piedmont, Italy, using rice waste and olive waste [126].

Biofuels America is setting up a lignocellulosic biorefinery using wood waste and municipal waste as the raw materials for the production of ethanol and furfural [127].

16.6.3.2.2 Valeric Aldehyde

In 2011, Sweden's Perstorp will produce 150,000 ton/year of valeric aldehyde. This is used as a raw material for the production of valeric acid, which is in turn used in agrochemicals [128].

16.6.4 ACIDS

16.6.4.1 C₂ Acids

16.6.4.1.1 Acetic Acid

In Brazil, acetic acid is produced by three companies using a route based on the ethanol production process from sugarcane: Butilamil, at a multipurpose facility with a 9,000 ton/year capacity; Cloroetil, which produces acetic acid from acetic aldehyde and has a production capacity of 13,000 ton/year; and Rhodia Poliamida, with a 40,000 ton/year installed capacity [129].

In the United States, ZeaChem has plans to start producing acetic acid and ethyl acetate by the fermentation of cellulosic ethanol in 2011 [130].

Acetic acid is a building block for over 500 products.

16.6.4.2 C₃ Acids

16.6.4.2.1 Lactic Acid

In Brazil, Purac Sínteses produces lactic acid from sugarcane sucrose.

In Denmark, a technology for producing lactic acid from biomass carbohydrates without fermentation using an inorganic, heterogeneous catalyst has been developed by researchers at Haldor Topsoe and the Technical University of Denmark [131].

In France, Mabiolac is developing lactic acid–based biodegradable composite materials. Textiles and rigid packaging are two potential applications for the material [132].

16.6.4.2.2 Acrylic Acid

Acrylic acid and its esters are probably the most versatile of all monomers.

Arkema, a French chemical company, has advanced its bio-based acrylic program with the installation of an acrylics pilot plant at its R&D center in Carling. The company is focused on developing a direct glycerin-to-acrylic acid process to be used in superabsorbent polymers or cosmetics. The process uses a tungstated zirconia catalyst [133].

The first commercial product of OPXBIO Biotechnologies, an American company that uses biotechnology to convert renewable raw materials into biochemicals and biofuels, will be bioacrylic acid, which is used in a range of industrial and consumer products including paints, adhesives, nappies, and detergents. The company is currently producing bioacrylic acid on a pilot scale in advance of opening its demonstration plant in 2011 and a full-scale commercial plant in 2013. OPXBIO has used its Efficiency Directed Genome Engineering technology to reduce bioacrylic acid production costs by 85% [134].

16.6.4.2.3 Propionic Acid

Swedish company, Perstorp, in partnership with Lund University has received government funding to develop and manufacture propionic acid and 3-hydroxypropionic acid from organic raw materials. Propionic acid is used in feedgrain preservation and is also included in the production of more than 40 products, including vitamin E [128].

16.6.4.2.4 Pyruvic Acid

Pyruvic acid is obtained by glycolysis (the degradation of glucose), and is mainly used as a precursor of lactic acid and a chemical input.

In China, Tianfu Biochemical Technology Co. Ltd. has a project to build a 1500 ton/year pyruvic acid production facility. It will use biological enzymolysis technology [135].

16.6.4.3 C₄ Acids

16.6.4.3.1 Succinic Acid

This acid is a building block for use in polymer chemistry, personal care, intermediates for the chemical industry, and other applications.

Various companies are currently developing a fermentation route for succinic acid. DSM (the Netherlands) and Roquette (France) are working together on a project, and two universities are also developing technologies. Rice University has licensed a genetically modified *Escherichia coli* strain that generates succinic acid from glucose, while the University of Georgia has licensed a process that allows a significant increase in production using the enzyme pyruvate carboxylase. The first test volumes of this renewable and versatile building block, used in the manufacture of polymers, resins, and many other products, have already been produced at a demonstration facility built in Lestrem (France) in 2009. DSM and Roquette will each have 50% stake in the new entity, Reversida V.o.f., which will be headquartered in the Netherlands.

In France, Bioamber (a joint venture between DNP Green Technology and Agro-industrie Recherches et Developpements) recently commissioned the world's first bio-based succinic acid production facility in France with a capacity of 2 million tons/year by fermentation of various renewable feedstocks [136].

In the United States, MBI has a process that involves the use of a novel bacterium, *Actinobacillus succinogenes*, to produce succinic acid, while Myriant has announced that it has begun to draw funds from a DOE fund for its bio-based succinic acid facility in the United States [101].

In Canada, DNP Green Technology and GreenField Ethanol have agreed to invest $50 million in a new plant that will produce bio-based succinic acid for deicing applications [137].

BASF Future Business, a subsidiary of BASF, has formed an alliance with a Purac subsidiary, CSM NV, for the commercial-scale production of bio-based succinic acid. Production is anticipated to commence in 2010 [138].

16.6.4.3.2 Fumaric Acid

The incorporation of fumaric acid to ruminant feeds improves growth rates by 10% and reduces methane emissions from cattle and sheep. This reduction in methane production is particularly significant to countries like New Zealand, where ruminant animals are responsible for large quantities of methane emissions [139].

Jiangsu Polytechnic University of China has patented a technology for the joint production of fumaric acid with DL-malic acid using green chemistry techniques. The total yield using this process is over 99.5% fumaric acid and malic acid. The product's quality meets U.S. Food and Drug Administration and British Pharmacopoeia (BP2000) standards [121].

16.6.4.4   C$_5$ Acids

16.6.4.4.1   Itaconic Acid

Itaconic acid is used in the production of styrene butadiene latex, synthetic resins, acrylic fibers, and adhesives. The world's biggest producers are the United States, Japan, and Russia. U.S.-based Itaconix, which is involved in environmentally friendly technology development, produces itaconic acid by corn glucose fermentation. The company also develops research into the application of lignocellulosic and waste biomass as a raw material [140].

16.6.4.5   C$_6$ Acids

16.6.4.5.1   Citric Acid

There are two companies that produce citric acid in Brazil: Cargill, which started production in 2,000 with 30,000 ton/year capacity, and Tate & Lyle, which acquired Bayer's citric acid manufacturing facility in 2002. Tate & Lyle is the world's largest citric acid producer and Cargill is the third largest. Cargill manufactures anhydrous citric acid by the submerged fermentation of glucose or sucrose [129].

Citric acid is a natural preservative, and is used in detergent production. It is starting to be used as an alternative to sodium tripolyphosphates (STTPs), which harm the environment. One way of producing citric acid is to use biotechnology that involves using *Aspergillus niger* and sucrose as a substrate.

16.6.4.5.2   Adipic Acid

American company Verdezyne is working to develop bioprocesses for adipic acid production. Their single-step fermentation process for producing the acid relies on combinatorial pathway engineering and reduces costs by at least 20%, the company claims. The process involves multiple gene-specific mutations and the use of combinations of enzymes to identify the optimal pathway. The two major targets in Verdezyne's pipeline include ethanol (from hexose and pentose sugars) and adipic acid. Its proprietary technology enhances the metabolic pathways for the production of these compounds [134].

16.6.5   ESTERS

16.6.5.1   Sucrose Esters

Sucrose octaacetate is produced by Italian Euticals and American Syntex Agribusiness as a bulk active ingredient for the pharmaceuticals industry [141].

16.6.5.2   Acrylic Esters

Acrylic esters are produced by reacting acrylic acid with an alcohol. The main markets for methyl, ethyl, and butyl alcohols are coatings, superabsorbents, water treatment, paper, and textiles. The main application of 2-ethylhexyl acrylate, which is growing fast in the world market, is in the manufacture of pressure-sensitive adhesives.

16.6.5.3   Esters from Fatty Acids

U.S. firm C-Tech is developing enhanced processes for the production of cosmetic esters from plant-based fatty acids [117].

16.6.5.4   DPHP

Sweden's Perstorp has plans to produce DPHP (di-(2-propylheptyl)phthalate) and the alcohol 2-PH (2-propylheptanol). There is growing demand for DPHP for the production of electrical cables, plastic flooring, and car interior features, which are expected to grow by 5% a year [142].

16.6.6 Olefins

16.6.6.1 C₂ Olefins

16.6.6.1.1 Ethylene

In Brazil, Braskem and Dow Chemical plan to produce sugarcane ethanol-based ethylene for polyethylene, while Solvay plans to produce it for PVC.

16.6.6.2 C₃ Olefins

16.6.6.2.1 Propylene

Existing techniques are to be modified to enable the production of propylene using a new bioprocess developed by Mitsui Chemicals.

16.6.6.3 C₄ Olefins

16.6.6.3.1 Butadiene

American company Arzeda and the University of Washington have received a grant to develop a process for producing bio-butadiene using specially designed enzymes [143].

16.6.7 Biopolymers/Bioplastics

The market for natural polymers that is set to grow the most in the coming years is packaging. Demand for cellulose, starch, and fermentation products is expected to grow by more than 10%, with larger-scale production and products with better properties. The main producers of biopolymers from potato, maize, and tapioca starch are Biotec, Novamont, Cereplast, Limagrain, Fkur, NatureWorks, Telles, Mirel, Purac, Sphere, and PSA. These companies make polylactic acid (PLA), polyhydroxyalcanoate (PHA), and polyhydroxybutyrate (PHB) [144].

Brazil is a nascent market for bioplastics when compared to its European and American counterparts.

Cargill's subsidiary in Brazil is contemplating producing bioplastics. The company has the infrastructure required, including the raw materials from its sugar and citric acid production unit in Minas Gerais [145]. Japanese cosmetics company Shiseido, in collaboration with Braskem (Brazil), has an R&D effort for polyethylene terephthalate (PET) and PP containers made from sugarcane bioethanol. The two firms are currently working together to produce polyethylene containers using bioethanol made from sugar-refining residue as a feedstock. Commercial-scale manufacturing is anticipated to commence in 2011. Shiseido will introduce haircare products for the Chinese market packaged in containers with around 60% PLA content in 2009 [146].

16.6.7.1 Polyethylene (PE)

Brazil-based company, Braskem, was granted a license in July 2010 to produce 200,000 ton/year of polyethylene (PE) from sugarcane ethanol.

Also in Brazil, the Dow Chemical Company should have the capacity to produce 350,000 ton/year of PE made from ethylene from sugarcane by 2011. However, the company is still looking for a partner to provide the ethanol feedstock for Crystalsev (50% owned by Dow) [129].

16.6.7.2 Polyvinyl Chloride

In Brazil, Solvay Indupa uses ethanol to make 360,000 ton/year of PVC. Globally, more than 750,000 ton of PVC filled with wood fibers was produced in 2006, of which 85% was from North America.

16.6.7.3 Polyethylene Terephthalate

In Brazil, Petrobras and Braskem make a partnership to develop green PET.

16.6.7.4 Polyitaconic Acid (PIA)

Itaconix is a producer of polyitaconic acid, a water-soluble polymer produced from itaconic acid. PIA has a broad range of applications, including superabsorbents, anti-scaling agents in water treatment, co-builders in detergents, and dispersants for minerals in coatings.

16.6.7.5 Polylactic Acid

Thanks to the greater availability and improved manufacturing processes of such polymers as PLA, the demand for natural polymers is anticipated to climb by 7.1%/year to four billion dollars in 2012 [147].

Cargill has recently taken over full control of NatureWorks, which has a facility in the United States that produces 140,000 ton/year of PLA from renewable feedstocks such as maize and sugarcane.

NatureWorks LLC is an independent company that is wholly owned by Cargill. It has developed a new PLA grade for various applications currently dominated by GPPS, HIPS, and ABS. IngeoTM, a natural plastic derived from 100% annually renewable resources, is the world's first polymer showing a significant reduction in GHG emissions. The material was particularly formulated for use in consumer goods sectors like electronics, cosmetics, and houseware. Potential applications include semidurable products that are either transparent or opaque, and those which require a level of impact resistance comparable to medium-impact polystyrene [148].

NatureWorks has plans for a second PLA production facility which is set to enter service by 2013 with a 140,000–150,000 ton/year capacity in one of several possible locations, including Thailand, Singapore, Malaysia, China, Europe, and Brazil. In 2011, it will have a pilot plant with a capacity of 5,000–15,000 ton/year to validate its technology for producing PLA [149].

In Italy, Novamont has the capacity to produce 70,000 ton/year of a plastic (PLA) from maize starch.

In Thailand, CSM is to build a lactide plant to produce PLA. The facility should be ready in 2011 and will produce up to 75,000 ton/year [150]. Also in Thailand, the National Innovation Agency should provide incentives for technology research ventures for bioplastics. Two major sugar producers intend to manufacture lactic acid, each facility with the capacity to produce 3000–5000 ton. The project will start by 2011–2012 [151].

Belgian company Futerro, a joint venture between Total Petrochemicals and Galactic, will set up a new pilot plant in 2010 with the capacity to produce 1500 ton of sugar beet–based PLA a year. Futerro envisages new products combining other plastics with PLA as well as patenting new applications for the automotive, agrofeed, and building industries.

In South Korea, LG Chem in collaboration with researchers at the KAIST University of Korea has developed a one-step fermentation process for the production of PLA and its copolymers. The process makes use of an engineered strain of *E. coli* and combines enzyme and metabolic engineering [152].

In Japan, Mitsubishi Chemical Holdings is making progress in its biopolymer business, with a commercial-scale manufacturing plant for its GS Pla® biodegradable plastic to come on stream in 2013. Full-scale manufacture, 20,000 ton/year, is set for 2015. Mitsubishi's GS Pla® contains a thermoplastic aliphatic polyester, polybutylene succinate (PBS), and is produced from succinic acid and 1,4-butanediol. The material offers excellent heat-sealing, gas permeability, printability, and water-resistance properties [153].

In 2005, BASF unveiled its first biodegradable plastic based on renewable raw materials. Called Ecovio, it is made from a mixture of corn-based PLA and BASF's Ecoflex biodegradable plastic [154].

A French player in the global auto industry, Faurecia, has organized a consortium of agricultural cooperatives, fiber producers, and biomaterial research centers to develop a material entirely based on renewable resources with a matrix of PLA [155].

The medical applications of poly(lactic acid) (PLA) fibers include suture thread for use in surgery and bone material.

16.6.7.6 Furfuryl Alcohol Resins

Bac2, a UK cleantech materials company, has expanded its materials portfolio to include solid CSR latent acid catalysts to control acid catalyzed polymerization processes, including furfuryl alcohol resins used in the manufacture of laminates, composites in glass-reinforced plastics, foam insulation, abrasives, and other products [156].

16.6.7.7 Biopolyester Resins

Japan U-Pica Co. Ltd. makes bioresins from glycolic acid and carboxylic acid obtained from nonedible plant materials and fumaric acid, which it has applied to sheet molding compounds and compounds for injection molding reinforced with natural kenaf, hemp, and bamboo fibers. It also produces biomass-based thermosetting unsaturated polyester resins with four new specialty grades with superior reactivity and flexibility [157].

16.6.7.8 Polyhydroxyalkanoates

These are polyesters produced from renewable raw materials. They have thermoplastic properties that can be applied to a wide variety of products.

American firms NatureWorks, Telles, Cereplast, and Mirel started PHA production from sugar in 2010. Mirel, a joint venture between Archer Daniel Midlands and Metabolix, is beginning 50,000 ton/year production from corn sugar-based PHA resin [158].

16.6.7.9 Polyhydroxybutyrates

BASF is channeling some of its product R&D efforts into biomass-based processes for producing PHB, which are biodegradable polymers.

In Brazil, PHB Industrial produces 50 ton/year of these polymers from sugar [129].

16.6.7.10 Polyvinyl Alcohol (PVA)

PVA has a wide range of applications, including medical grades PVA05-88, PVA17-88, and PVA-124 [159].

In Brazil, several companies produce PVA from vinyl acetate: BASF (21,265 ton/year); Denver, with a multipurpose plant that can produce up to 8,400 ton/year; DFM (8,000 ton/year capacity); EMZ Quimica (up to 7,500 ton/year at a multipurpose plant); IQT (up to 6,400 ton/year at a multipurpose plant); and Resinac (up to 6,000 ton/year at a multipurpose plant) [129].

16.6.7.11 Polybutylene Succinate

U.S.-based Showa Highpolymer is producing a biomass-based version of its Bionelle biodegradable plastic, made up of PBS produced from 1,4-butanediol and bio-succinic acid. The company is aiming to add other uses for Bionelle, including agricultural mulching film, compost bags, shopping bags and vertical drains [160].

Mitsubishi Motors in collaboration with Aichi Industrial Technology Institute has developed a material which uses a plant-based resin, PBS, combined with bamboo fiber for use in the company's new mini car. The PBS is made from 1,4-butanediol and succinic acid (obtained from fermented corn or sugar) [153].

The market for polymers with added vegetable fibers is estimated to be 40,000 ton/year, of which 10,000 ton is used in cars, amounting to 3–5 kg/vehicle. The vegetable fibers used include hemp, flax, and wood.

Japan's Mitsubishi Chemical Corp. and Thailand's PTT Public Company Ltd. have entered into an agreement to conduct a feasibility study into the manufacture of bio-PBS. Mitsubishi Chemical Corp. produces green sustainable plastic (GS Pla) made of PBS, a biodigradable polymer made from petro-based succinic acid, developing an original process to produce the succinic acid from biomass. PTT is Thailand's biggest public company and is heavily involved in the development of bio-related businesses such as biofuels and bio-based polymers [161].

16.6.7.12 Polycarbonate

Mitsubishi Chemical intends to start sample distribution of its biomass-based polycarbonate in 2010. Biopolycarbonate can be applied in light-emitting diode lighting and touch panels because it delivers excellent optical properties.

16.6.7.13 Final Considerations

Green chemistry-based innovations in the chemicals sector are growing apace across the globe due to increased demand and the finite reserves of petroleum and natural gas. Biomass is an attractive alternative to fossil-based feedstock and is used as the foundation for several profitable industries. The markets most likely to take advantage of bio-based platform chemicals as feedstocks include polymers, solvents, resins, and surfactants/detergents [117].

There are a number of biorefineries in development worldwide:

- Myriant Technologies in association with the University of Florida will set up a pilot scale cellulosic refinery to produce ethanol and bio-based chemicals. Its raw materials will be sugarcane bagasse, sweet sorghum, wood waste, and waste-based biomass [101].
- ZeaChem's biorefinery in Oregon produces a number of bio-based alternative products in the C_2 (ethylene and ethanol), C_3 (propylene), C_4 (butanol), and C_6 (hexane and hexanol) [130].
- Finland's Chempolis has started operations in its demonstration plant at Oulu, which will focus on paper and pulp fiber using vegetable matter to produce many biomass-based products and chemicals [114].

The oil industry is reacting by carrying out its own R&D initiatives, sponsoring academic R&D and forming alliances with creative companies doing white biotech:

- Shell has a pilot plant in Hawaii for growing marine algae.
- Shell has partnered with Codexis (the United States) and Logan Energy (Canada) to produce improved enzymes for making cellulosic ethanol, and with Choren Industries (Germany) for making biodiesel from biomass.
- ExxonMobil has an R&D program with Synthetic Genomics for making biofuels from algae (pilot plant in the United States).
- BP is working with DuPont on commercializing biobutanol (the United Kingdom). ENI and UOP are working on processes for making biodiesel [162].

The governments of developed countries are aware of the need to invest in renewable sources, as can be seen from the following examples:

In the United Kingdom: The Bioscience for Business knowledge transfer network, the Chemistry Innovation knowledge transfer network, the Royal Society of Chemistry, and IChemE have established the Renewable Platform Chemicals to Value-Added Products special interest group for renewable chemicals to assist industrialists and policy-makers in identifying present capabilities, technology gaps, and adoption barriers that hinder growth in the chemical sector.

In Japan: The RITE has collaborations with several companies, including Honda R&D, Dow Chemical, and Sumitomo Rubber Industries to enhance the production of biochemicals [163].

In the United States: The U.S. DOE has set the target to increase the percentage of chemicals derived from biomass to 20% by 2020 and 50% by 2050; 12 chemicals have been identified for producing biochemicals, glycerol being one of them. Glycerol's end products are ethylene glycol, 1,2-propylene glycol, and lactic acid. The use of catalytic methods is anticipated to develop in the coming years. Glycerol can also be used as a substrate for the microbial fermentation of various chemicals:

1,3-propylene glycol, butanol, ethanol, methanol, hydrogen, and propionic acid. Biofuel-Solution, a research organization, has developed gas-phase reactions of glycerol encompassing dehydration and hydrogenation reactions to obtain mono-alcohols, alkene monomers for polymer production, and energy gases [164].

In Sweden: The government has a long-term research partnership with Lund University aimed at developing strategically important biochemical processes from renewable raw materials to produce acrylic acid, methacrylic acid, and 1,3-propandiol from propionic acid and 3-hydroxypropionic acid [165].

In Brazil: Large-scale production is expected to boost this market in the region. Competitive production scale and increased demand will be crucial for making bioplastics a growing and profitable market in the region. Legislation and government incentives, which are currently scarce for bioplastics, are also important at this stage to support small, domestic producers. The Brazilian bioplastics market has promising growth projections for the next 5 years. Brazil is the leading producer of sugarcane in the world, delivering attractive production costs for this raw material. As sugarcane and ethanol production in Brazil is constantly increasing, this will be a competitive advantage for the country to expand its bioplastics production based on ethanol. Brazil's competitiveness in the bioplastics market will be strongly dependent on product demand, R&D, local infrastructure, and incentives. Investments from private investors and the government are expected to grow about 25% by 2013 and about 35% by 2015 [166].

There are some enterprises in Brazil with pilot plants to scale-up bioprocesses and biochemicals, like GCTbio and Alfa Rio Química.

While alcohols currently form the largest segment of the renewable chemicals market, the polymers segment has the greatest growth potential at an expected CAGR of 11% for the next 5 years.

Finally, the production of biorenewables is among the top ten biggest trends in the global chemicals industry, and coordination between private enterprise and universities will be the key to accelerating the R&D thrust, which in turn will depend on government incentives.

16.7   CONCLUSIONS

As clearly demonstrated in the previous sessions, both biomass and biomass derivatives are of primordial importance for the development of Latin America. Indeed, in Latin America and Caribbean, agriculture represents 10% of GDP whereas agricultural products account for 30% of exports. Also, rural people still make up 30% in these regions.

As far as Brazil is concerned, it is worth mentioning that the country has about 394 MM ha of total available surface for agriculture, of which only 66 MM ha are already occupied. These figures are quite impressive, if one compares them with the figures displayed by United States (269 vs. 188), Russia (220 vs. 132), China (138 vs. 96), and India (169 vs. 169). Hence, the potential use of biomass as raw material for the production of both biofuels and biochemicals is undisputable. For this reason, the country has invested considerable amounts of money in R&D, trying to overcome technology bottlenecks in the production of bio-based fuels and chemicals. Although the Brazilian Ethanol Programme is a reality, there are still plenty of technological developments to be carried out.

It must be borne in mind, also, that all biofuel routes must deal with the problem of cost reduction. Moreover, the problem of incentives has to be discussed, since biofuels are in principle cleaner fuels. Finally, questions regarding LCA for different raw materials are a key issue, since LCA must take into consideration cost efficiency and social impact of the different raw materials in distinct countries.

REFERENCES

1. Perissé, J.B. 2007. *Evolução do refino de petróleo no Brasil*. Master's thesis. Instituto de Química. Universidade do Estado do Rio de Janeiro. 158p.
2. American Petroleum Institute. 2000. *Technology Roadmap for the Petroleum Industry*. Washington, DC: API. 38p.

3. American Petroleum Institute. 1999. *Technology Vision 2020: A Report on Technology and the Future of the U.S. Petroleum Industry.* Washington, DC: API. 12p.

4. Intergovernmental Panel on Climate Change. 2007. Mudança do clima 2007: A base das ciências físicas. Sumário para os formuladores de políticas. Contribuição do Grupo de Trabalho I para o quarto relatório de avaliação do Painel Intergovernamental Sobre Mudança do Clima. Genebra: IPCC. 25p.

5. Intergovernmental Panel on Climate Change. 2007. Mudança do clima 2007: Impactos, adaptação e vulnerabilidade. Contribuição do Grupo de Trabalho II ao quarto relatório de avaliação do Painel Intergovernamental sobre Mudança do Clima. Sumário para os formuladores de políticas. Genebra: IPCC. 30p.

6. Intergovernmental Panel on Climate Change. 2007. Mudança do clima 2007: Mitigação para mudança do clima. Contribuição do Grupo de Trabalho III ao quarto relatório de avaliação do Painel Intergovernamental sobre Mudança do Clima. Sumário para os formuladores de políticas. Genebra: IPCC. 42p.

7. United Nations Framework Convention on Climate Change. *Kyoto Protocol to the United Nations Framework Convention on Climate Change.* 2007. http://unfccc.int/essential_background/kyoto_protocol/items/1678.php (accessed November 17, 2007).

8. *Vision 21—The Ultimate Power Plant Concept.* 2007. http://fossil.energy.gov/programs/powersystems/vision21 (accessed December 17, 2007).

9. Moure, G.T. 2003. *Perspectivas para a indústria do petróleo no futuro.* Lecture presented at Petrobras Research Center, Rio de Janeiro, Brazil.

10. Briens, C., J. Piskors, and F. Berruti. 2008. Biomass valorization for fuel and chemicals production: A review. *Int J Chem React Eng* 6, R2: 1–49.

11. Erickson, J.C. 2007. Overview of thermochemical biorefinery technologies. *Int Sugar J* 109: 163–173.

12. Huber, G.W., S. Iborra, and A. Corma. 2006. Synthesis of transportation fuels from biomass: Chemistry, catalysts, and engineering. *Chem Rev* 106: 4044–4098.

13. Hayes, D.J. 2007. *State of Play in the Biorefining Industry.* Available at http://www.lufpig.eu/documents/StateofPlayinTheBiorefiningIndustry-DanielJohnHayes.pdf (accessed December 18, 2007).

14. Szklo, A. and R. Schaeffer. 2006. Alternative energy sources or integrated alternative systems? Oil as a modern lance of Peleus for the energy transition. *Energy* 31: 2177–2186.

15. Brasil. Ministério de Minas e Energia. 2007. *Balanço Energético Nacional 2007: Ano base 2006.* Rio de Janeiro, Brazil: Empresa de Pesquisa Energética. 192p.

16. Farrell, A.E., R.J. Plevin, B.T. Turner, A.D. Jones, M. O'Hare, and D.M. Kammen. 2006. Ethanol can contribute to energy and environmental goals. *Science* 311: 506–508.

17. United Nations. 2006. *Sustainable Bioenergy: A Framework for Decision Makers.* http://esa.un.org/un-energy/pdf/susdev.Biofuels.FAO.pdf (accessed March 12, 2010).

18. Oil Depletion Analysis Center (ODAC). 2007. Available at http://www.odac-info.org (accessed February 2007).

19. Brasil. Ministério da Agricultura, Pecuária e Abastecimento. 2005. *Plano Nacional de Agroenergia: 2006–2011.* Available at http://www.agricultura.gov.br (accessed March 24, 2006).

20. Pereira, R.E. 2006. *Avaliação do potencial nacional de geração de resíduos agrícolas para a produção de etanol.* Master's thesis. Escola de Química. Universidade Federal do Rio de Janeiro.

21. Schuchardt, U., M.L. Ribeiro, and A.R. Gonçalves. 2001. A indústria petroquímica no próximo século: Como substituir o petróleo como matéria-prima? *Quim Nova* 24: 247–251.

22. Lal, R. 2005. World crop residue production and implications of its use as a biofuel. *Environ Int* 31: 575–586.

23. Ramos, L.P. 2003. The chemistry involved in the steam treatment of lignocellulosic materials. *Quim Nova* 6: 863–871.

24. Vieira, J.A. 2006. Álcool de mandioca atrai investimentos. *R Assoc bras Prod Amido de Mandioca* 4, no. 13 (janeiro-março). Available at http://www.abam.com.br/revista/revista13/alcooldemandioca.php (accessed January, 2007).

25. Resíduo da cana-de-açúcar: energia. *Biodieselbr.com.* 2006. Available at http://www.biodieselbr.com/energia/residuo/residuo-setor-sucroalooeiro.htm (accessed Fevereiro 2006).

26. Kim, S. and B.E. Dale. 2006. Ethanol fuels: E10 or E85—Life cycle perspectives. *Int J Life Cycle Ass* 11: 117–121.

27. Sun, Y. and J. Cheng. 2002. Hydrolysis of lignocellulosic materials for ethanol production: A review. *Bioresour Technol* 83: 1–11.

28. Jacobsen, S.E. and C.E. Wyman. 2000. Cellulose and hemicellulose hydrolysis models for application to current and novel pretreatment processes. *Appl Biochem Biotechnol* 84: 81–96.

29. Malburg Jr., L.M., J.M.T. Lee, and C.W. Forsberg. 1992. Degradation of cellulose and hemicelluloses by rumen microorganisms. In *Microbial Degradation of Natural Products*, G. Winkelmann (ed.), pp. 127–159. Weinheim, Germany: Wiley-VCH Verlag.

30. Odier, E. and I. Artaud. 1992. Degradation of cellulose and hemicelluloses by rumen microorganisms. In *Microbial Degradation of Natural Products*, G. Winkelmann (ed.), pp. 161–191. Weinheim, Germany: Wiley-VCH Verlag.

31. Breen, A. and F.L. Singleton. 1999. Fungi in lignocellulose breakdown and biopulping. *Curr Opin Biotechnol* 10: 252–258.

32. D'Almeida, M.O. 1988. Celulose e papel. Tecnologia de fabricação de pasta celulósica. 2nd edn. São Paulo: Instituto de Pesquisas Tecnológicas.

33. Wayman, M. and S. Parekh. 1990. Biotechnology of biomass conversion: Fuels and chemicals from renewable resources. Englewood Cliffs, NJ: Prentice Hall. (Biotechnology Series, p. 30).

34. Dedini Indústrias de Base. 2005. Available at http://www.dedini.com.br (accessed October 18, 2005).

35. Fogel, R., R.R. Garcia, R.S. Oliveira, D.N.M. Palacio, L.S. Madeira, and N. Pereira Jr. 2005. Optimization of acid hydrolysis of sugarcane bagasse and investigations on its fermentability for the production of xylitol by *Candida guilliermondii*. *Appl Biochem Biotechnol* 123: 741–752.

36. Vásquez, M.P., M.B. Souza Jr., and N. Pereira Jr. 2006. RSM analysis of the effects of the oxygen transfer coefficient and inoculum size on the xylitol production by *Candida guilliermondii*. *Appl Biochem Biotechnol* 129: 256–264.

37. Dille, O. 2008. Available at http://www.otto-dille.de/indexe.html (accessed 2008).

38. Lignin Institute. 2006. Available at http://www.lignin.org (accessed January 2007).

39. van Dam, J., R. Gosselink, and E. Jong. 2004. *Lignin Applications*. Wageningen, the Netherlands: Agrotechnology & Food Innovations. Available at http://www.biomassandbioenergy.nl/infoflyers/LigninApplications.pdf (accessed January 2007).

40. Harris, J.F. 1975. Acid hydrolysis and dehydration reactions for utilizing plant carbohydrates. *Appl Polym Symp* 28: 131–144.

41. Lynd, L.R. 1996. Overview and evaluation of fuel ethanol form cellulosic biomass: Technology, economics, the environment, and policy. *Annu Rev Energy Environ* 21: 403–465.

42. Lynd, L.R., P.J. Weimer, W.H. Van Zyl, and I.S. Pretorius. 2002. Microbial cellulose utilization: Fundamentals and biotechnology. *Microbiol Mol Biol Rev* 66: 506–577.

43. McMillan, J.D. 1994. Pretreatment of lignocellulosic biomass. In *Enzymatic Conversion of Biomass for Fuel Production*, M.E. Himmel, J.O. Baker, and R.P. Overend (eds.), pp. 292–324. Washington, DC: American Chemical Society.

44. Mosier, N., C. Wyman, B. Dale, R. Elander, Y.Y. Lee, M. Holtzapple, and M. Ladisch. 2005. Features of promising technologies for pretreatment of lignocellulosic biomass. *Bioresour Technol* 96: 673–686.

45. Ogier, J.C., D. Ballerini, J.P. Leygue, L. Rigal, and J. Pourquie. 1999. Production d'éthanol à partir de biomasse lignocellulosique. *Oil Gas Sci Technol* 54: 67–94.

46. Negro, M.J., P. Manzanares, J.M. Olivia, I. Ballesteros, and M. Ballesteros. 2003. Changes in various physical/chemical parameters of Pinus pinaster wood after steam explosion pretreatment. *Biomass Bioenergy* 25: 301–308.

47. Hamelinck, C.N., G.V. Hooijdonk, and A.P.C. Faaij. 2005. Ethanol from lignocellulosic biomass: Techno-economic performance in short-, middle-, and long-term. *Biomass Bioenerg* 28: 384–410.

48. Hohlberg, A.I., J.M. Aguilera, E. Agosín, and R. San Martín. 1989. Catalyzed flash pretreatments improve saccharification of pine sawdust. *Biomass* 18: 81–93.

49. Teymouri, F., L. Laureano-Perez, H. Alizadeh, and B.E. Dale. 2005. Optimization of the ammonia fiber explosion (AFEX) treatment parameters for enzymatic hydrolysis of corn stover. *Bioresour Technol* 96: 2014–2018.

50. Teixeira, L.C., J.C. Linden, and H.A. Schroeder. 2000. Simultaneous saccharification and cofermentation of peracetic-acid pretreated biomass. *Appl Biochem Biotechnol* 84–86: 111–127.

51. Ferreira, V., P.C. Nolasco, A.M. Castro, J.N.C. Silva, A.S. Souza, M.C.T. Damaso, and N. Pereira Jr. 2006. Evolution of cell recycle om *Thermomyces lanuginosus* Xylanase: A production by *Pichia pastoris* GS 115. *Appl Biochem Biotechnol* 129–132: 226–233.

52. Damaso, M.C.T., A.M. Castro, C.M.M.C. Andrade, and N. Pereira Jr. 2004. Application of xylanase from *Thermomyces lanuginosus* IOC-4145 for enzymatic hydrolysis of corncob and sugarcane bagasse. *Appl Biochem Biotechnol* 113–116: 1003–1112.

53. Viikari, L., A. Kantelinen, J.M. Sundquist, and M. Linko. 1994. Xylanases in bleaching: From an idea to the industry. *FEMS Microbiol Rev* 13: 335–350.

54. Berlin, A., N. Gilkes, A. Kurabi, R. Bura, M. Tu, D. Kilburm, and J. Saddler. 2005. Weak lignin-binding enzymes: A novel approach to improve activity of cellulases for hydrolysis of lignocellulosics. *Appl Biochem Biotechnol* 121–124: 163–170.

55. Palonen, H. 2004. Role of lignin in the enzymatic hydrolysis of lignocellulose. *VTT Publications* 520: 1–80.

56. Kitchaiya, P., P. Intanakul, and M. Krairiksh. 2003. Enhancement of enzymatic hydrolysis of ligno-cellulosic wastes by microwave pretreatment under atmospheric-pressure. *J Wood Chem Technol* 23: 217–225.

57. Glasser, W.G. and R.S. Wright. 1998. Steam-assisted biomass fractionation. II. Fractionation behavior of various biomass resources. *Bioresour Technol* 14: 219–235.

58. Brito, F.H.X. 2000. Bioprodução de etanol de hidrolisado de bagaço de cana utilizando diferentes formas de operação do bioprocesso. Master's thesis. Escola de Química. Universidade Federal do Rio de Janeiro.

59. Mussato, S.I. and I.C. Roberto. 2004. Alternatives for detoxification of diluted-acid lignocellulosic hydrolisates for ethanol production. *Bioresour Technol* 96: 1–10.

60. Betancur, G.V. and N. Pereira Jr. 2010. Sugar cane bagasse as feedstock for second generation ethanol production. Part I: Diluted acid pre-treatment optimization. *Electron J Biotechnol* 13(3): 1–9. Available at http://www.ejbiotechnology.cl/content/vol13/issue3/full/3/3.pdf

61. Awafo, V.A., D.S. Chahal, and B.K. Simpson. 1998. Optimization of ethanol production by *Saccharomyces cerevisiae* (ATCC 60868) and *Pichia stipitis* Y-7124: A response surface model for simultaneous hydrolysis and fermentation of wheat straw. *J Food Biotech* 22: 49–97.

62. Teter, S., J. Cherry, C. Ward, A. Jones, P. Harris, and J. Yi. 2005. *Variants of Glycoside Hydrolases*. US patent 2005048619 (A1).

63. Bower, B.S., E.A. Larenas, and C. Mitchinson. 2005. CBH1-E1 Fusion construct. T. reesei cbh1, linker (no CBD). Acidothermus cellulolythicus endoglucanase 1 core (E1). WO patent 2005093073 (A1).

64. Toivola, A., D. Yarrow, E. van den Bosch, J.P. van Dijken, and W.A. Scheffers. 1984. Alcoholic fermentation of D-Xylose by yeasts. *Appl Environ Microbiol* 47: 1221–1223.

65. Jeffries, T.W. and Y.S. Jin. 2004. Metabolic engineering for improved fermentation of pentoses by yeast. *Appl Microbiol Biotechnol* 63: 495–509.

66. Dien, B.S., M.A. Cotta, and T.W. Jeffries. 2003. Bacteria engineered for fuel ethanol production: Current status. *Appl Microbiol Biotechnol* 63: 258–266.

67. Verho, R., J. Londesborough, M. Penttilä, and P. Richard, 2003. Engineering redox cofactor regeneration for improved pentose fermentation in *Saccharomyces cerevisiae*. *Appl Environ Microbiol* 69: 5892–5897.

68. Jeppson, L.R., H.H. Keifer, and E.W. Baker. 2005. *Mites Injurious to Economic Plants*. Berkeley, CA: University of California. 614p.

69. Zanin, G.M., C.C. Santana, E.P.S. Bon et al. 2000. Brazilian bioethanol program. *Appl Biochem Biotechnol* 84–86: 1147–1161.

70. Dieckelman, G. and H.J. Heinz. 1988. *The Basics of Industrial Oleochemistry*. Columbia, MO: South Asia Books, 192p.

71. Aranda, D.A.G., S. Zhao, D.P. Tolle, R.R. Joao, R.T.P. Santos, and G.L.M. Souza. 2007. *Catalytic Process for the Transesterification of Vegetable Oils and Fats Using Basic Solid Catalizers*. BRPI patent 0504759 (A).

72. Tapanes, N.C.O., D.A.G. Aranda, J.W.M. Carneiro, and O.A.C. Antunes. 2008. Transesterification of *Jatropha curcas* oil glycerides: Theoretical and experimental studies of biodiesel reaction. *Fuel* 87: 2286–2295.

73. Gonçalves, J.A., A.L.D. Ramos, L.L.L. Rocha et al. 2011. Niobium oxide solid catalyst: Esterification of fatty acids and modeling for biodiesel production. *J Phys Org Chem* 24(1): 54–64.

74. Aranda, D.A.G. Transesterification and hydroesterification: Theoretical and experimental analysis of biodiesel production using heterogeneous catalysis. In *International Conference Catalysis for Renewable Sources: Fuel, Energy, Chemicals*. Paper presented at the Boreskov Institute of Catalysis SB RAS, St. Petersburg, Russia, June 28–July 2, 2010.

75. Aranda, D.A.G., J.G. Araújo, J.S. Perez et al. 2009. The use of acids, niobium oxide, and zeolite catalysts for esterification reactions. *J Phys Org Chem* 22: 709–716.

76. Cemici Unveils Biodiesel Catalyst. 2010. *Biofuel Int* 3, 10: 24–24.

77. Pereira, A.T., K.A. Oliveira, R.S. Monteiro, D.A.G. Aranda, R.T.P. Santos, and R.R. João. 2007. Production process of biodiesel from the esterification of free fatty acids. US patent 20070232817 (A1).

78. Aranda, D.A.G. and O.A.C. Antunes. 2004. Catalytic process to the esterification of fatty acids present in the acid grounds of the palm using acid solid catalysts. WO patent 2004096962 (A1).

79. Casanave, D., J.L. Duplan, and E. Freund. 2007. Diesel fuels from biomass. *Pure Appl Chem* 79: 2071–2081.
80. Schulz, H. 1999. Short history and present trends of FT synthesis. *Appl Catal A Gen* 186: 3–12.
81. Sie, S.T. and R. Krishna. 1999. Fundamentals and selection of advanced FT-reactors. *Appl Catal A Gen* 186: 55–70.
82. Espinoza, R.L. and A.P. Steynberg. 1999. Low-temperature Fischer–Tropsch synthesis from a Sasol perspective. *Appl Catal A Gen* 186: 13–26.
83. Alleman, T.L. and R.L. McCormick. 2003. *Fischer–Tropsch Diesel Fuels—Properties and Exhaust Emissions. A Literature Review.* Paper presented at the SAE 2003 World Congress, Detroit, MI, March 3–6, 2010. (SP-1737).
84. Hassuani, S.J., M.R.L.V. Leal, and I.C. Macedo. 2005. *Biomass Power Generation: Sugarcane Bagasse and Trash.* Piracicaba, Sao Paulo: Centro de Tecnologia Canavieira.
85. Faaij, A., R. van Ree, L. Waldheim et al. 1997. Gasification of biomass wastes and residues for electricity production. *Biomass Bioenerg* 12: 387–407.
86. Lunsford, J.H. 2000. Catalytic conversion of methane to more useful chemicals and fuels: A challenge for the 21st century. *Catal Today* 63: 165–174.
87. Adachi, Y., M. Komoto, I. Watanabe, Y. Ohno, and K. Fujimoto. 2000. Effective utilization of remote coal through dimethyl ether synthesis. *Fuel* 79: 229–234.
88. Sorenson, S.C. 2001. Dimethyl ether in diesel engines: Progress and perspectives. *J Eng Gas Turb Power* 123: 652–658.
89. McCandless, J. 2002. *DME: The Next Generation Diesel Fuel.* Paper presented at the Fifth DME Meeting, Rome, Italy, December 2002.
90. Boehman, A.L. 2002. *Development of a DME—Fueled Shuttle Bus.* Paper presented at the Fifth DME Meeting, Rome, Italy, December 2002.
91. Sato, Y. 2002. *Project of Heavy Duty DME Truck in Japan.* Paper presented at the Fifth DME Meeting, Rome, Italy, December 2002.
92. Lee, D. 2002. *DME Vehicle Research and a Forecast of Its Spread in Korea.* Paper presented at the Fifth DME Meeting, Rome, Italy, December 2002.
93. Marchionna, M., S. Dellagiovanna, and D. Romani. 2002. *DME as LPG Substitute: Economics and Markets Considerations.* Paper presented at the Fifth DME Meeting, Rome, Italy, December 2002.
94. Sousa-Aguiar, E.F., L.G. Appel, and C. Mota. 2005. Natural gas chemical transformations: The path to refining in the future. *Catal Today* 101: 3–7.
95. Ramos, F., A. Farias, L. Borges, J. Monteiro, M. Fraga, E.F. Sousa-Aguiar, and L. Appel. 2005. Role of dehydration catalyst acid properties on one-step DME synthesis over physical mixtures. *Catal Today* 101: 39–44.
96. Rosillo-Calle, F. 1986. The Brazilian ethanolchemistry industry: A review. *Biomass* 11: 19–38.
97. Souza, A.M.L. 1979. Alternativas para o uso industrial do álcool etílico no Brasil. Master's thesis. Programa de Engenharia Química. Coordenação dos Programas de Pós-Graduação de Engenharia. Universidade Federal do Rio de Janeiro. 91p.
98. Voegele, E. 2009. Retrofitting for alternatives. *Ethanol Producer Mag* (October). Available at http://www.ethanolproducer.com/article.jsp?article_id=5956
99. Mordekhay, D. 2008. AETE Technology Unveiled. *Press* (March 24). Available at http://www.nanotech-now.com/news.cgi?story_id=28633
100. GlycosBio announces first production. 2010. *Chimie Pharma Hebdo* 500: 10.
101. Green Car Congress. 2010. Myriant Technologies receiving funds under $50M DOE award for succinic acid biorefinery project (April 7). Available at http://www.greencarcongress.com/2010/04/myriant-20100407.html
102. North East Process Industries Cluster. 2010. ABC investigates eight chemicals for sustainable production (May 20). Available at http://www.icis.com/Articles/2010/05/20/9372565/abc-investigates-eight-chemicals-for-sustainable-production.html
103. Alcohols from syngas—Methanol from syngas is a well-established process. However, with the continuing interest in ethanol as a fuel, especially in the United States, a great deal of research money is now being poured into producing ethanol and other higher chain alcohols via a syngas step. 2009. Nitrogen+Syngas reports. *Nitrogen Syngas* 302: 22–27.
104. Business and people: Syntec biofuel ethanol gasification process. 2010. *Ethanol Producer Mag* (January). http://ethanolproducer.com/article.jsp?article_id=6200
105. Gamer, R. 2008. Renewable fuels expert forms organic chemical division. Available at http://www.bio-goldfuels.com/images/BGF-PR-5-29-08.doc

106. Boost for bioethanol. 2010. *Oils Fats Int* 26: 10–11.
107. Novozymes Brazilian partnership: Bioethanol production, Dagbladet Borsen. Available at http://www. borsen.dk (accessed July 16, 2010).
108. Amyris and Cosan creating joint venture for production and commercialization of cane based renewable chemicals. Available at http://www.amyrisbiotech.com (accessed June 22, 2010).
109. Voith, M. 2010. Growing chemicals: New developments enable fuel and polymer production in crops. *Chem Eng News* 88: 25–28.
110. Novozymes and Lignol sign deal to make biofuel from wood. http://www.novozymes.com (accessed June 15, 2010).
111. Commercial production of cellulosic biofuel on fast track in China: COFCO, Sinopec and Novozymes to construct biofuel demonstration facility by 2011 (May 27, 2010). http://www.novozymes.com/en/ MainStructure/PressAndPublications/PressRelease/2010/COFCO+and+Sinopec+MOU.htm
112. Chang, J. 2009. ADM delays opening renewable PG plant. *ICIS Chem Bus.* September 28.
113. Arkema plans acrolein and acrylic acid production using plant resources. 2009. *Chimie Pharma Hebdo* 459: 9.
114. A matter of refinement. 2009. *Biofuel Int* 3: 56–58.
115. A speedy option. 2010. *Biofuel Int* 3: 57–60.
116. Spolchemie's François Vleugels lays plans to lead producer to safety. 2010. *ICIS Chem Bus* 277(10).
117. Skibar, W. 2009. Who needs oil anyway? *TCE* 816: 38–39.
118. Cobalt Technologies is first to create renewable biobutanol fuel from beetle-killed pine (April 7). http:// www.cobalttech.com/news/news-item/renewable-biobutanol-fuel-from-beetle-killed-pine
119. Voegele, E. 2010. Biobutanol: Friend or foe? *Ethanol Producer Mag* (March). http://www.ethanolproducer. com/article.jsp?article_id=6337
120. Koepenick, M. 2010. From pulp and paper to helicopter fuel! *Pulp Paper Int* 52(4): 24, 26, 27, 29. http:// www.risiinfo.com/magazines/April/2010/PPI/From-pulp-and-paper-to-helicopter-fuel.html
121. Frost & Sullivan lauds Genencor for its innovative product line of enzymes that address key challenges in the biofuel industry (January 13, 2010). http://www.frost.com/prod/servlet/press-release-print. pag?docid=189889977
122. Dyadic International reports fiscal year 2009 financial results: Reports record revenues, income from operations and profit (March 25, 2010). http://www.dyadic.com/pdf/DyadicFinancialResults2010.pdf
123. ECUT achieved progress in making 2,3-butanediol biologically. 2010. *China Chem Report* 21: 21.
124. Uzbekistan: Tender for chemical plant extended. 2010. *Financial Times*. March 24.
125. Shokubai advances glycerine-to acrylic acid process. 2009. *Chem Week*. December 1.
126. Commercial derivatives of furans. 2009. *Chimica News (Supplement)* 51: 12.
127. Biofuels America proposes low-cost 30 Mgy cellulosic ethanol plant in Tennessee. http://www.biofuels- digest.com/blog2/2009/04/15/biofuels-america-proposes-low-cost-30-mgy-cellulosic-ethanol-plant-in- tennessee/ (accessed July 19, 2010).
128. Perstorp launches biotech project. 2008. *Chem Eng News* 86: 19–20.
129. *Guia da Indústria Química Brasileira.* 2010. São Paulo: Associação Brasileira de Indústria Química.
130. Marquart, J. 2010. ZeaChem celebrates groundbreaking of biorefinery in Boardman, Oregon. http:// www.zeachem.com/press/pdf/Groundbreaking060210.pdf (accessed June 2, 2010).
131. Haldor Topsoe develops novels lactic acid process. 2010. *Chem Week*. May 24.
132. Mabiolac: From beet to socks. 2006. *Circuits Cult* 401: 22.
133. Innovative catalysts promise to improve the economics of chemical production. 2009. *ICIS Chem Bus* 276: 28–29.
134. Challener, C. 2010. Flexible feedstock venture. *Chem Ind* 5: 11.
135. Tianfu to build pyruvic acid project in China. 2006. *China Chem Report* 17: 7.
136. Hartmann, M. 2010. Bioamber and Mitsui & Co., Ltd partner for biobased succinic acid distribution in Asia (April 15). http://www.dnpgreen.com/img/pdf/Bioamber_Mitsui_PR_final.pdf
137. Succinic acid plant planned for Canada. 2010. *Chem Eng News* 88: 12.
138. Partnerships for bio-based succinic acid. 2009. *Plast Rubber Asia* 24: 9.
139. Lawrence, C. 2006. Adding fumaric acid helps keep gas emissions low. *Farmers Wkly* 144: 37.
140. Itaconic acid: Qingdao Langyatai (Group) Co. Ltd. 2002. *China Chem Report* 13: 23.
141. Sucrose octaacetate. 1985. *Chem Mark Report* 228: 22.
142. Perstorp 2007: Annual report (May 7, 2008): 17–18. http://www.perstorp.com/upload/annual_report_2007.pdf
143. After isoprene, the world can expect butadiene: Start-up firm wins grant to develop sugar-derived butadiene. 2010. *Eur Rub J* 192: 25.
144. Green polymers. 2009. *Plast Caoutchoucs Mag* 865: 32–34,36,38.

145. De Biagio, F. 2009. Cargill mulls producing bioplastics locally: Brazil. *BN Americas*. July 20.
146. Shiseido eyes sugarcane-based PP, PET containers with Brazilian partner. 2009. *Jpn Chem Web*. Available at: http://www.japanchemicalweb.jp (accessed on July 1, 2010).
147. Natural polymer demand is rising. 2009. *ʲHAPPI* 46: 14.
148. PLA grade targets styrenics applications. 2009. *Mod Plast Worldw* 86: 23.
149. NatureWorks weighing six sites for its no 2 polylactic acid plant. 2010. *Jpn Chem Web*. Available at: http://www.japanchemicalweb.jp (accessed on July 1, 2010).
150. CSM's Purac builds lactides plant in Thailand. 2010. *Kemivärlden Biotech med Kemisk Tidskrift* 1–2: 6.
151. Wongsamuth, N. 2010. Industry pushes for research subsidies. *Bangkok Post*. Available at: http://www.bangkokpost.net (accessed on July 1, 2010).
152. Badina, J. 2010. Futerro begins European polylactic acid production. *Chimie Pharma Hebdo* 506: 4.
153. Mitsubishi Chemical HD making strides in biopolymer arena. *Jpn Chem Web*. Available at: http://www.japanchemicalweb.jp (accessed on July 1, 2010).
154. Alperowicz, N. 2006 . BASF ponders use of alternative feedstocks. *Chem Week* 168:28.
155. Markets and production of polymers with added vegetable fibres. 2010. *Plast Caoutchoucs Mag* 876: 27.
156. Bac2's latent acid catalyst for low-temperature control of pre-polymeric mixes now available in solid form (June 7, 2010). http://bac2.co.uk/news/pr&pr=bac2–035
157. Uses widening for U-Pica's bio-based unsaturated polyester resins. 2009. *Jpn Chem Web*. Available at: http://www.japanchemicalweb.jp (accessed on July 1, 2010).
158. Eposito, F. US companies stepping up biopolymer production. 2010. *Plast News*. Available at: http://www.plasticsnews.com/headlines2.html?id=18152 (accessed on July 1, 2010).
159. Bright market prospects for biodegradable medical fibres. 2009. *China Chem Report* 20: 20–21.
160. Showa Highpolymer testing biomass-based biodegradable plastic. 2010. *Jpn Chem Web*. Available at: http://www.japanchemicalweb.jp (accessed on July 1, 2010).
161. Plant polymer for Mitsubishi: interior car components. 2006. *Plast Rub Wkly* 1:2.
162. Comyns, Alan E., 2009. After petroleum, what next? *Focus Cat* 2009 (11): 1–2.
163. RITE and Idemitsu to advance process to make biochemicals. 2009. *Jpn Chem Web*. Available at: http://www.japanchemicalweb.jp (accessed on July 1, 2010).
164. Alternative uses for glycerine. 2009. *Biofuel Int* 3: 54–54.
165. Perstorp's long-term investment in industrial biotechnology (June 23, 2008). http://www.perstorp.com/News/PressReleases/Pressrelease_Archive_2008/2008–06–23%20Vinnova.aspx?pagelang=en
166. Abundance of raw materials and new production units drive bioplastics market in Brazil and Mexico, finds Frost & Sullivan (June 3, 2010). http://www.frost.com/prod/servlet/press-release.pag?docid=203238843
167. Shleser, R. Ethanol Production in Hawaii: Processes, Feedstocks, and Current Economic Feasibility of Fuel Grade Ethanol Production in Hawaii. (1994) Hawaii State Department of Business, Economic Development and Tourism. Available in http://www.hawaii.gov (accessed January 5, 2007).
168. Olsson, L. and B. Hahn-Hägerdal. 2000. Fermentation of lignocellulosic hydrolysates for ethanol production. *Enzyme Microb Technol* 18: 312–331
169. Couto, R. and M.A. Sanromán. 2005. Influence of redox mediators and metal ions on synthetic acid dye decolourization by crude laccase from *Trametes Hirsute*. *Chemosphere* 58: 417–422.
170. Ghosh, S. and R. Singh. 1993. Citrus in South Asia. FAO/RAPA Publication No. 1993/24. Bangkok, Thailand, p. 70.

17 Biofuels and Biochemicals in Africa

Dorsamy (Gansen) Pillay and
Ademola Olufolahan Olaniran

CONTENTS

17.1   INTRODUCTION

Interest around biofuels at the international level has exploded over the past few years. Biofuels are now high on the agenda of the regional banks (most noticeably, the Inter-American Development Bank), the UN institutions (including UNCTAD and the FAO), as well as the World Bank and the OECD. In spring 2007, Presidents George W. Bush from the United States and Luiz Inãcio da Silva from Brazil announced a pact to build toward international biofuel trade. Today, the European Union, South Africa, China, India, the United States, and Brazil have joined what is now referred to as the International Biofuels Forum (IBF).

Biofuels are generally divided into two groups—first- and second-generation biofuels. Those biofuels that have already been introduced into the fuel market, such as bioethanol and biodiesel, are defined as first-generation biofuels. The advantages of the first-generation biofuels are their availability and the possibility of utilizing them with the currently available technologies. The main disadvantage however is that only a certain range of crops can be utilized, for instance, bioethanol produced from sugar-bearing crops such as sugarcane and sugar beet. Second-generation biofuels on the other hand are produced from forest and crop residues as well as municipal waste and are defined as those that are still in developmental stages and that are not yet available on a large scale. The main goal of second-generation biofuels is to increase the amount of biofuels that can be produced and in the process reduce carbon emissions, increase energy efficiency, and reduce increased dependency on energy sources. Unfortunately, due to the high production costs and infrastructure developments required, the production of second-generation biofuels on a large enough scale has been hampered. Furthermore, additional developments of enzymes, pretreatment, and fermentation processes are needed (Antizar-Ladislao and Turrion-Gomez, 2008).

Although biofuels investment in Africa has started occurring at a quick pace and has major social and environmental implications, little is known about it still. This chapter is therefore aimed at providing an overview of the biofuels and biochemicals status in Africa.

17.1.1   Need for Biofuels and Biochemicals

17.1.1.1   Fossil Fuel and Other Nonrenewable Sources of Energy: Limitations, Challenges, and Associated Environmental Problems

According to an IEA World Energy Outlook (2009), the demand for energy will increase by 40% between now and 2030, reaching 16.8 billion tons of oil equivalent. One of the major challenges for society in the twenty-first century is to meet the increasing energy demands and to provide sustainable raw materials with a substantial reduction in green house gas emissions (Hahn-Hägerdal et al., 2006). Recent industrial and economic growth has led to major increase in the electricity usage with the energy demand being expected to drastically increase over the next 50 years, as both economic development and employment patterns change (Banks and Schäffler, 2006). Approximately two-thirds of known petroleum reserves are located in the Middle East, with the global reserves seriously declining while the demand has doubled (Campell, 1995). Although scarce, these resources may last for more than a century if used at current rates but of more concern are the larger power plants, as these will need to be replaced as the supplies run low (Banks and Schäffler, 2006). Another major concern is the production of carbon dioxide (CO_2) when coal is burned, which serves as a serious environmental problem contributing to climate change as well as local air pollution (Banks and Schäffler, 2006). Increase in human agricultural activities, waste disposal, and fossil fuel usage has resulted in increased concentrations of CO_2 and several other greenhouse gases, such as methane, nitrous oxide, and chlorofluorocarbons (CFCs), being released into the atmosphere (Lightfoot and Green 1992). This has raised concerns about increases in the earth's temperature. Currently, possible approaches to reduce the major greenhouse gas emissions include improving the conversion and utilization efficiency of energy, utilizing nonfossil energy sources, increasing biomass production, and sequestration of carbon from currently available fossil fuels (Steinberg, 1999).

There are three major environmental issues surrounding the use of fossil fuels—acid precipitation, stratospheric ozone depletion, and the greenhouse effect (Dincer and Rosen, 1999). The combustion of fossil fuels may lead to the production of acids, namely, SO_2 and NO_x, which can be transported great distances throughout the atmosphere and deposited via precipitation onto various ecosystems resulting in acidification of water bodies, damage to aquatic life, and agricultural crops as well as deterioration of surrounding buildings and metal structures (Dincer and Rosen, 1999). The distortion and depletion of the stratospheric ozone layer is another major concern as this may result in increased exposure levels to harsh ultraviolet radiation, causing considerable negative health effects such as skin cancer and eye damage (Dincer, 1998). Temperature increases due to increasing concentrations of greenhouse gases such as, CO_2, CH_4, CFCs, and N_2O in the atmosphere is another problem associated with energy utilization as these results in increased confinement of heat radiated from the earth's surface (Dincer and Rosen, 1999).

17.1.1.2 Inherent Advantages of Biofuels and Biochemicals

Irrespective of the fact that only certain parts of the plant biomass can be utilized for energy purposes, the continuous growth of plants on the earth's surface provides a source of energy that can be unlimitedly utilized for energy gain. The implementation of renewable sources of energy has many advantages associated with it such as increased job creation, biofuel manufacture, and transportation as well as distribution of feedstocks and resulting products (Kojima and Johnson, 2006). This was seen with the increased production of sugarcane in Brazil during 2004 resulting in the creation of 700,000 jobs and around 3.5 million indirect jobs (Coelho, 2005). Biofuel production could also lead to a major reduction in poverty levels, especially in rural areas due to increased use of unskilled labor, reduction in the time spent on basic survival activities such as gathering water and firewood such as applicable in rural areas of most African countries, increased use of educational media and reduction in pollution and deforestation. However, one of the major concerns associated with biofuel production is competition with food production and the associated price increases of major commodities such as corn and wheat. Industries that use the same raw materials required for biofuel production such as the food and forest-based industries are most likely to be affected. In addition, most of the industries using biomass also produce additional products such as dried distiller's grain that can be used as animal feed, thus causing the price of feed derived from biofuel production to decrease significantly (FAPRI, 2007).

17.2 OVERVIEW OF AVAILABLE BIOFUELS AND BIOCHEMICALS

17.2.1 Bioethanol

Bioethanol is one of the most common biofuels, accounting for more than 90% of the total biofuel production, and can be produced from any feedstock that contains considerable amounts of biomass that can be converted into sugars (IEA, 2007). There are many different abundant biomass sources (Table 17.1) that can be utilized for bioethanol production (Rutz and Janssen, 2007). The utilization of these feedstocks is beneficial due to the low fossil fuel inputs resulting in a high level of energy production with little CO_2 emissions (Wyman, 1996).

Conventional bioethanol production occurs via enzymatic conversion of starch-containing biomass into sugars, fermentation of six-carbon sugars followed by final distillation of ethanol to fuel

TABLE 17.1

Examples of Feedstocks Utilized for Ethanol Production

Feedstock	Crop Type	Examples
Sugar	Root crops	Sugar beets
	Stalk crops	Sugarcane
		Sweet sorghum
Starch	Cereals	Corn
		Barley
		Rye
		Wheat
		Sorghum
	Root crops	Potatoes
		Cassava
Cellulose	Energy crops	Willows
		Poplar
		Switch grass
	Agricultural waste	Straw
		Corn stover
		Bagasse

Source: Adapted from Rutz, D. and Janssen, R., *Biofuel Technology Handbook,* WIP Renewable Energies, Munich, Germany, 2007.

grade. Research is currently focusing on improving the current processes, so that all available lignocellulosic materials can be utilized for bioethanol production as opposed to using only the sugar or starch-containing components (Wuebbles and Jain, 2001). Furthermore, the large-scale production of feedstock requires substantial amounts of land with fertile soil and adequate water, hence making highly industrialized areas unsuitable (Rutz and Janssen, 2007).

One of the major advantages of bioethanol is that it is a clean fuel that can be utilized in combustible engines. The high octane number of these ethanol-blend fuels when compared to ordinary petrol (116:86) influences the antiknocking property of the fuel itself (Rutz and Janssen, 2007). Knocking refers to the uncontrolled combustion in spark-ignition engines, which ultimately puts a great deal of stress on the engine (IEA, 2007). These ethanol-blended fuels now offer considerable fuel savings as well as reduction in greenhouse gas emissions. Additional advantages include cleaner combustion, considerable benefits to energy security as it shifts the need for energy production from foreign-produced oil to locally produced renewable energy sources and ensuring that fuel spills can be cleaned quickly and more efficiently due to the fuel being more biodegradable.

The volume of ethanol as a fuel, replacing gasoline, is presently around 500,000 barrels of oil equivalent per day or 0.7% of the world's oil consumption of 86 million barrels per day and approximately 3% of the gasoline in use in the world today. Ethanol is produced mainly in the United States (from corn) and in Brazil (from sugarcane). In 2008, the United States produced 34 billion liters, Brazil 22.5 billion liters, and the European Union 2.7 billion liters (mainly from sugar beets) with a grand total of 65.6 billion liters/year (RFA, 2009a). However, utilizing these raw materials, in conjunction with the conversion process have proved to be very costly and hence inexpensive lignocellulosic biomass is now being utilized as a low cost alternative (RFA, 2009b). Bioethanol production

TABLE 17.2

Comparison of Bioethanol Production from Different Crops

Crops	Yield (Ton ha^{-1} Year^{-1})	Conversion Rate to Bioethanol (L Ton^{-1})	Bioethanol Yield (L ha^{-1} Year^{-1})
Sugarcane	70	70	4900
Cassava	45	150	6000
Sweet sorghum	30	80	2800
Maize	5	410	2050
Wheat	4	390	1560
Rice	5	450	2250

Source: Adapted from Jannson, C. et al., *Appl. Energy*, 86, 595, 2009.

from biomass involves process that is largely dependent on the feedstock utilized varying from sugar and starch to cellulose. Utilizing sugar feedstocks involves a simple method utilizing the yeast *Saccharomyces cerevisiae* to convert glucose to ethanol. The bioethanol production potential from different crops is shown in Table 17.2.

17.2.2 BIODIESEL

Biodiesel serves as a biodegradable and nontoxic alternative to the currently used diesel, and is produced from a range of biological sources through a process known as transesterification (Ma and Hanna, 1999). There are a range of production processes currently used, varying based on the initial raw material used. There are significant advantages when using vegetable oil as a means to produce biodiesel; these include renewability, availability, and improved heat content (Shay, 1993). In addition to this, it provides a market for the excess production of vegetable oils and animal fats, does not contribute to global warming, and results in lower emissions of carbon monoxide when added to regular diesel, thus improving the lubricating properties of the fuel (Van Gerpen, 2005). However, there are some major disadvantages associated with biodiesel production as well—such as freezing during cold weather, fuel degradation during prolonged periods of storage as well as clogging of fuel filters (Hill et al., 2006). Filter blockage is usually encountered when biodiesel fuel blends are initially introduced to equipment that has been operated with pure hydrocarbons for a prolonged period of time. This is due to the loosening of the hydrocarbon deposits that forms on the inside of the respective tanks. These problems however can be overcome with proper maintenance following the introduction of the biodiesel fuel (Bozbas, 2005).

17.2.3 BIOHYDROGEN

Biohydrogen production is one such promising alternative associated with sustainable development and efficient energy production. Hydrogen itself has enormous potential as a renewable energy source due to the fact that it has the highest energy content per unit weight when compared to the current known gaseous fuels as well as producing minimal carbon emissions upon combustion, thereby yielding mainly water as end products (Kotay and Das, 2008; Levin et al., 2004). Hence, the utilization of hydrogen as a fuel source has great potential as it does not contribute to greenhouse emissions, acid rain, or ozone depletion. Biohydrogen production is generally not as energy intensive as chemical processes because they are mainly carried out at ambient temperatures and pressures. The key enzyme during this process is the hydrogenase or nitrogenase enzyme that regulates the hydrogen-metabolism of microorganisms (Chen et al., 2008, Manish and Banerjee, 2007).

Biohydrogen production can divided into two broad groups—one which is light dependent (direct or indirect biophotolysis and photofermentation) (Yu and Takahashi, 2007), and the other, which is a light-independent process (dark fermentation) (Kotay and Das, 2008).

17.2.4 MICROBIAL FUEL CELL

The microbial fuel cell (MFC) concept has been explored for a long time since the 1970s, however, only recently have microbial fuel cells with enhanced power output been researched and developed, thus providing possible breakthroughs in their practical applications (Figure 17.1).

A MFC works much in the same way as a bacterial cell, harnessing, and storing energy in the form of adenosine triphosphate (ATP). The main components of a MFC are the anode, cathode, and the electrolyte. Bacteria catalyze the oxidation of reduced substrates, resulting in the release of electrons during respiration, which then flow toward the anode, through an external circuit to the cathode, thereby creating a current. This process results in water production at the cathode in the presence of a catalyst. Apart from oxygen, other chemicals such as ferricyanide can be used resulting in greater overall potentials (Logan and Regan, 2006). The bacteria of the MFC gain metabolic energy by transferring electrons from the electron donor to the acceptor, with a greater potential difference between the donor and acceptor resulting in a larger gain for the bacterium. Three main types of MFCs have been distinguished—photoautotrophic, heterotrophic, and sediment biofuel cells (Rabaey et al., 2005). The use of MFCs provides considerable advantages such as the direct conversion of substrate energy into electricity, efficient operation at ambient or low temperatures does not require additional energy input and can be utilized in a variety of locations lacking electrical infrastructures (Rabaey and Verstraete, 2005).

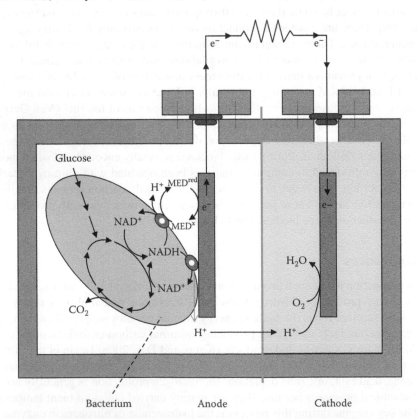

FIGURE 17.1   The functioning of a microbial fuel cell. (From Rabaey, K. and Verstraete, W., *Trends Biotechnol.*, 23, 291, 2005. With permission.)

17.3 AFRICA'S BIOFUEL AND BIOCHEMICAL PRODUCTION POTENTIAL

17.3.1 POTENTIAL FOR PLANT BIOMASS PRODUCTION: AVAILABILITY OF OLEAGINOUS PLANT SOURCES

17.3.1.1 *Jatropha curcas*

Jatropha curcas is a versatile tree belonging to the family of *Euphorbiaceae* that has been used to control erosion, reclaim land, and formulate a barrier for livestock (Figures 17.2 and 17.3). It is found in many parts of the world, namely, North America, many of the tropics as well as in Africa and Asia and can be successfully cultivated in both irrigated and rainfall conditions, growing quickly in a short period of time (Wood, 2005). The *Jatropha* plant has few insect or fungal pests and is not a host to many diseases that attack agricultural plants. The *J. curcas* plant has been view as a multipurpose plant due to certain important advantages, ranging from land reclamation, production of animal feeds after detoxification, provision of medicinal chemicals, applications in pharmaceutical and biopesticides as well as reductions in greenhouse gas emissions (Sustainable Design Update, 2008). The leaves and stems of the *Jatropha* plant itself are toxic to animals but after sufficient treatment, the seeds left behind can then be used as an animal feed. Apart from this, various parts of the plant offer medicinal value, the plant itself can be utilized for its honey production potential, and the wood and fruit can be utilized for fuel production (Openshaw, 2005). *Jatropha curcas* is valued for biomass production due to its rich oil (curcas oil), which is high in fats. This oil is obtained from the seeds and can be used in place of kerosene and diesel as well as to aid rural areas in terms of cooking and light availability (Augustus et al., 2002). *Jatropha*

FIGURE 17.2 Seeds and leaves of the *Jatropha curcas* plant.

FIGURE 17.3 *Jatropha curcas* plantation (From sustainabledesignupdate.com).

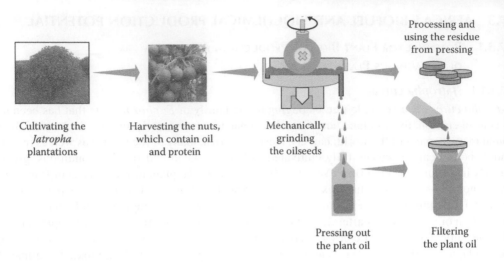

FIGURE 17.4 Process of bio-oil production from *Jatropha curcas*. (From biodieselprocessor.org/biodiesel %20 jatropha).

cultivation geared toward biofuel production has to take a range of factors into consideration, ranging from land area, plantation establishment, and management practices such as machines, infrastructure and energy requirements, to the seeds, husks, and GHG emissions. *Jatropha* can be cultivated in a wide range of soils ranging from sandy to gravelly and should never be planted on soils where potential water logging may occur (Achten et al., 2008). The *J. curcas* fruits should be harvested after ripening and turning from green to yellow when the lipid content is the highest. Maturation is reached approximately 90 days after flowering, but it should be noted that the fruits do not all mature simultaneously and thus should be harvested at regular intervals.

The next step involves the extraction of the oil from the seeds using one of two methods—mechanical extraction or chemical extraction (Figure 17.4). The seeds dried either in an oven or via the sun, following which, mechanical or chemical extraction takes place. Mechanical extraction may utilize either whole seeds or kernels or a mix of both while chemical extraction utilizes only ground kernels. The shells can then be used as a combustible by-product or feedstock for gasification (Achten et al. 2008). However, due to a vast amount of technical and economical drawbacks, poor marketing of *Jatropha*'s products, and poor information availability, the full potential of *Jatropha* has not yet been realized (Openshaw, 2000). The estimated contribution of ethanol derived from biomass to energy supply in South Africa is represented in Table 17.3.

17.3.2 BIOTECHNOLOGY ENTERPRISES

The demands for biofuel production have increased drastically over the past few years indicating that the sole cultivation of energy crops on noncultivated land will not be sufficient; hence, modern plant breeding techniques and biotechnology are being utilized to increase land productivity in the form of more biomass output per hectare, as well as increasing crop quality in the form of more fermentable carbohydrates and higher oil content. Biotechnology is today one of the most effective tools that can be used to improve biofuel production in a variety of ways such as the improved development of enzymes that can convert hemicelluloses more efficiently. Furthermore, biotechnology aids increased biomass yield per hectare while reducing the needs for production costs, improved crop quality, reduced land-use, higher productivity, and reduced losses from a range of biotic (insect and viruses) and abiotic (drought, wind, and salinity) stresses. It assists in the cultivation of energy crops in areas with marginal conditions as well as aiding in the development of efficient microorganisms and enzymes to convert hemicelluloses to sugars, which can then be fermented into biofuel (EuropaBio, 2007). Internationally, agricultural biotechnology has aided

TABLE 17.3

Estimated Contribution of Ethanol Derived from Biomass to Energy Supply in South Africa

Crop or Biomass Source	Energy Content of Potential Annual Ethanol Production PJ (TWh)
Cassava	72.3 (20.1)
Sugarcane	11.1 (3.1)
Bagasse	5.6 (1.6)
Molasses	2.3 (0.6)
Maize	22.5 (6.3)
Sorghum straw	5.1 (1.4)
Wheat straw	7.0 (1.9)
Forest	3.4 (0.9)
Sawmills	0.7 (0.2)
Total	130 (36.1)

Source: Adapted from Eberhard, A.A. and Williams, A., *Renewable Energy Resources and Technology Development in South Africa,* Elan Press, Cape Town, South Africa, 1988.

in the increase of crop yields by 8.34 billion pounds in 2005, with a greater than 33.1% increase in corn production since the introduction of biotech corn in 1996 (Sankula, 2006). Further advancements in agricultural biofuel production include improvements with regards to biodiesel production that has traditionally been produced from soy. This involves mustard that is the new expected feedstock. Mustard can be grown on cheaper land than that used to grow corn and soy and is an adaptable crop that can be genetically altered to meet specific needs (Sexton et al., 2006).

Biotechnology has proven to be effective in a range of areas, for example, the enzymes needed to convert cellulose feedstock's are both expensive and inefficient, thus providing an area for scientists to research and develop new enzymes that will make this process more efficient. In addition to this, new enzymes that can convert starch to sugar more quickly and efficiently are currently being researched. Additional biotechnological research has focused on working to replace yeast with bacteria that are less prone to infection and able to withstand extreme temperatures through genetic manipulation, thus making the bacteria more efficient. However, it should be noted that the growth and development of agricultural biotechnology is hampered by a range of regulations and bans, which thus reduce both productivity and possible opportunities to improve this technology. A large amount of pressure on agricultural biofuel production has been relieved due to new irrigation technology, better pest abatement tools, and crop breeding. The current transgenic crops have allowed staple food sources like corn and rice to be infused with the naturally occurring pesticide *Bacillus thuringiensis* resulting in approximately 80% increase in crop yields and reduced chemical pesticide applications by 70% (Sexton et al., 2006).

17.3.3 LAND AVAILABILITY IN AFRICA

Land availability is unlikely to be the limiting factor for biofuels development in Africa. For example, the available land for food and nonfood production in Tanzania was estimated at more than 40 million ha (Janssen et al., 2005). This view is supported by a study estimating the global bioenergy production potential in 2050 for several agricultural investment levels. Furthermore, Hoogwijk et al. (2009) reported that Africa has the potential to become an important producer and exporter of raw biomass produced on abandoned and rest land.

TABLE 17.4

Suitable and Available Areas for Bioenergy Crops in Sub-Sahara's Arid/ Semiarid Regions

Country	Senegal	Burkina Faso	Mali	Kenya	Tanzania	Zambia	Botswana	South Africa
Total area, km²	196,013	272,339	1,252,281	581,871	941,375	751,920	587,337	1,221,361
Arid and semiarid, km²	111,147	149,973	637,960	457,908	316,738	160,281	581,605	901,345
Arid and semiarid available, km²	15,783	22,756	192,438	379,698	147,252	67,383	291,860	722,874
Percentage arid and semiarid available and suitable	14	15	30	82	46	42	51	79

Source: Adapted from Watson, H.K., COMPETE Competence platform on energy crop and agroforestry systems for arid and semi-arid ecosystems—Africa, Second Task Report on WP1 Activities—Current land use patterns and impacts, Sixth Framework Programme, 2008; Watson, H.K., *Energy Policy J.—Sustainabili. Biofuels*, 2009.

However, even though land is generally available for bioenergy production in sub-Sahara Africa, the feasibility and sustainability of specific bioenergy projects need to be evaluated on a case-by-case basis, carefully taking into account local environmental constraints as well as potential competition over land and water resources. This will become increasingly important in the future in the light of natural resource limitation due to climate change and an expected population growth placing pressure on the supply of affordable and adequate food.

The identified available and suitable arid and semiarid land in the COMPETE study countries Senegal, Burkina Faso, Mali, Kenya, Tanzania, Zambia, Botswana, and South Africa is presented in Table 17.4 (Watson, 2008, 2009). The fraction of suitable and available arid and semiarid land varies between 15% for the Western African countries Senegal and Burkina Faso and more than 80% for Kenya.

17.4 SUSTAINABLE FEEDSTOCKS FOR BIOFUEL AND BIOCHEMICAL PRODUCTION IN AFRICA

17.4.1 STARCH FEEDSTOCKS

17.4.1.1 Cereal Grains: Corn, Wheat, Barley, Sorghum, Oat, and Rice

Cereal grain crops are most commonly grown for their edible grain or seeds, thus serving as an efficient food supply. These cereal crops are grown on a larger scale than any other and account for more than 80% of all grain production worldwide (Rutz and Jannsen, 2007). Corn is one of the main feed stocks utilized for ethanol production in both the United States and Canada, with more than 95% of the ethanol production in the United States coming solely from corn. Apart from the high starch content, the corn kernel also contains approximately 10% protein, 4.5% oil, and 10%–15% of other materials such as fiber and ash, which can be recovered during the fermentation process and sold as coproducts (Drapcho et al., 2008). Wheat, rye, oats, barley, and spelt are classified as cool-season cereals, which grow well in moderate temperatures but which cease to grow in extremely hot temperatures of 30°C and over. These cool-season cereal crops can be further divided into winter and spring-type crops based on their growth cycles, thus allowing for optimal water and land usage (Rutz and Janssen, 2007).

Barley is one such alternative grain feedstock that can be used for ethanol production; however, there are several problems associated with utilizing this feedstock. First, barley contains a much lower starch content than corn; hence, the ethanol yield per unit mass will be significantly lower. Furthermore, the barley hull is abrasive and can cause damage to both milling and grain-handling equipment. In addition to this, there are significantly high levels of β-glucan, which can cause inadequate mixing of the mash. Sweet sorghum is another sustainable crop that can be utilized to produce biofuels at a cost-effective level for both rural communities and industries due to its robustness with regards to low water and nitrogen availability as well as tolerance to salinity and drought. In many developing countries, grain sorghum serves as a staple food source for both humans and livestock and is now bred in four distinct groups—grain, fiber, multipurpose, and sweet sorghum. Grain sorghum, apart from serving as a staple food source, is also produced as a means of improving other production areas related to paper and starch production, while the sweeter varieties of sorghum is mainly utilized to maximize ethanol production (Woods, 2005).

Another important grain feedstock is rice, which is defined as being the third most important grain crop worldwide. Asia has been classified as a primary rice production region with a large area utilized for rice harvesting and accounting for over 90% of international rice production (Kim and Dale, 2004). Approximately 88% of global production is utilized to provide food for communities while about 2.6% of global production is used as animal feed. Rice has several characteristics that make it a good potential feedstock for bioethanol production, such as its high starch content and several recoverable high-value nutraceuticals. Despite these advantages, rice has not been seriously considered as a feedstock for commercial production of fuel ethanol because its cultivation requires very large quantities of water for irrigation and its planting is a very labor-intensive process. Furthermore, in many regions, virtually all the rice produced is used for food or exported.

17.4.1.2 Other Grains, for Example, Pearl Millet

Pearl millet (*Pennisetum glaucum*) is a widely grown food crop in Africa and in India and is the fifth most important cereal crop in Zimbabwe. It has shown great promise as a potential feedstock that can be utilized in biofuel production due to its ability to grow in arid conditions with very low rainfall (Chakauya and Tongoona, 2008). It has the potential to survive in areas where corn and sorghum cannot proliferate. Pearl millet grain contains approximately 27%–32% more protein, higher concentrations of essential amino acids, and approximately 70% starch, thereby yielding an approximate theoretical bioethanol yield of 0.43 L/kg (Drapcho et al., 2008). This crop shows a number of advantages (tolerance to drought and heat) that have made it a traditional staple crop in low-resource, hot semiarid regions. Apart from this, it has exhibited the ability to proliferate in short favorable periods (Sparks, 1992). Stalk rot, grain molds as well as a range of insects such as the European corn borer, and corn ear worm tend to reduce the yields obtained. Bird damage is another problem and is considerably greater in smaller fields than larger fields. These problems can be minimized by keeping pearl millet fields away from tree lines and efficient crop monitoring and subsequent timely harvest.

17.4.2 Tuber and Root Crops, for Example, Cassava Production in West Africa

Root and tuber crops can be grown in both humid and subhumid conditions and are regarded as potential bioethanol feedstocks due to their ability to concentrate and store sugars and starch at or below the soil surface (O'Hair et al., 1983). Cassava (*Manihot esculenta* spp. *esculenta*) is classified as one of the most important tropical starch-containing root crops with more than 70% of cassava production occurring in the subtropical and tropical regions of Africa, America, and Asia (Jannson et al., 2009). Within these areas, this crop is grown as a staple animal feed and food source and can survive drought, low nutrient conditions, and heat tolerance (McGraw-Hill). Cassava is known to be a high starch producer with levels of up to 90% of its total root dry weight. Furthermore, it gives fairly good yields in dry and poor soils and does not require high levels of maintenance or cost to manage. Starch is synthesized and deposited in the storage roots underground, which varies in size from 1 m in length and over 10 cm in diameter. In Africa, cassava results in production potential of twice as much as maize and three times as much millet and sorghum. With Nigeria being the biggest producer in Africa and cassava covering one-third of the dietary needs of the African population, cassava has taken over thousands of hectares and has become the staple food for over 200 million Africans. Cassava has a range of characteristics that make it suitable for biofuel production ranging from high drought and heat tolerance, small applications of agricultural fertilizers as well as its high starch content even in dry and poor soil conditions (Jannsen et al., 2009).

17.4.3   OTHER CROPS AND PLANT BIOMASS

Bagasse is classified as the fibrous plant material that is left behind once all of the juice has been squeezed from the plant itself. Generally, bagasse consists of water and dry plant matter such as stalk fibers as well as leaves and other biomass components. Soybeans have been regarded as being one of the dominant oilseed crops that are used to produce vegetable oil worldwide. When compared to other oilseed crops like rape and cottonseed, these crops yield a fairly low amount of biodiesel per hectare; however, their desirability lies in their ability to grow in both temperate and tropical climates and their ability to be grown in rotation with other crops such as corn and sugarcane. In addition to this, soybeans also have a nitrogen-fixing potential, thus requiring less fertilizer additions (Rutz and Jannsen, 2007). There are a range of other crops that can be used for bio-oil and biodiesel production. Peanuts account for approximately 8.7% of the international oilseed production with the major producers being China, India, and the United States. Field studies have indicated that a peanut can produce approximately 1138 kg/ha of peanut oil.

17.4.4   MICROALGAE FOR BIODIESEL PRODUCTION

Microalgae are defined as a diverse group of prokaryotic and eukaryotic photosynthetic microorganisms that grow rapidly due to their simple structure. These microorganisms can be utilized to produce large amounts of lipids that subsequently can be utilized for the production of a number of different biofuels such as biodiesel, bio-oil, bio-syngas, and bio-hydrogen. In the 1970s, algae were viewed as a possible replacement fuel, but high production costs and limitations discouraged the commercial development of algae-based fuel production. Microalgae have high growth rates and photosynthetic efficiencies due to their simple structures as well as a higher overall biomass productivity when compared with other plants. Additional advantages include less water requirements, high CO_2 tolerance, and reduction in nitrous oxide emissions (Li et al., 2008).

17.5   NEW CEREAL STRAINS WITH HIGHER BIOFUEL AND BIOCHEMICAL YIELDS

New cereal strains (GM) are appearing with much higher yields. If Africa used these strains successfully, there would be enough food as well as biomass over for biofuel production. South Africa currently produces far more maize than what it consumes, yet the production of biofuels from the excess are forbidden. Recently, the SA Minister of Agriculture suggested that a change might be on its way. Example of this is shown in the production of isopropanol, one of the secondary alcohols produced by microbes used in place of methanol to esterify various fats and oils, thereby reducing the tendency of biodiesel to crystallize at low temperatures. Isopropanol can also be dehydrated to yield propylene that is widely used for plastic production. Generally, isopropanol is produced in *Clostridium* from the acetone pathway, however, due to limited genetic tools in *Clostridium*, other hosts such as *E. coli* are now of interest. Acetone production in *E. coli* has been demonstrated by introducing four genes from *Clostridium acetobutylicum* ATCC 824 and resulting in 5.4 g/L acetone production with 0.5 g/L/h production rate indicating that *E. coli* could be a suitable host for both acetone and isopropanol production (Atsumi and Liao, 2008).

Thus far, large-scale fermentation of hexose sugars using baker's yeast, *Saccharomyces cerevisiae*, has been well established; however, a major challenge is now the conversion of pentose sugars to ethanol as there are no known naturally occurring yeast strains that can ferment pentose sugars from lignocellulosic biomass currently. In recent years, major advances toward this goal have been made through the metabolic engineering of *S. cerevisiae*. There are a range of existing recombinant strains that effectively convert pentose sugars to ethanol, namely, *rSaccharomyces*, *rE.coli*, and *rZymomonas. rKlebsiella, rPichia stipitis, rB. Stearothrermophilus*, and *rKluveromyces. Saccharomyces* yeast strains are one of the currently used strains for ethanol production, which have been used for brewing and fermentation of distilled products. In many of the production plants, sterilization of feed streams is not required as contamination is controlled by varying the pH, thereby saving both capital and operating costs. A certain fraction of the yeast is dried and sold as an additional product such as an animal feed additive. Genetic instability is avoided due to modified chromosomes, which results in increased stability with performance over multiple cycles (Hettenhaus, 1998).

Ethanol-fermenting organisms are selected based on a range of performance parameters such as temperature range, pH range, alcohol tolerance, growth rate, productivity, osmotic tolerance, specificity, yield, genetic stability, and inhibitor tolerance. In addition to this, other requirements such as compatibility with existing products, processes, and equipment also need to be considered (Hettenhaus, 1998). Yeast strains initially adapt to the substrate and after acclimatization increase their alcohol production. These strains ability to remain genetically stable after multiple generations is essential for steady performance (Hettenhaus, 1998).

17.6 PRETREATMENT OF BIOMASS TO OBTAIN A FERMENTABLE SUBSTRATE TO BIOETHANOL

The main aim of pretreatment is to remove lignin and hemicellulose, reduce the crystallinity of cellulose, and increase the porosity of the lignocellulosic materials. The presence of lignin in lignocelluloses leads to a protective barrier that prevents plant cell destruction by fungi and bacteria for conversion to fuel. In order for pretreatment to be successful, the following criteria must be met: the production of sugars via hydrolysis should be improved, the degradation or loss of carbohydrates as well as the formation of byproducts that are inhibitory to the hydrolysis and fermentation processes should be avoided, and, overall, these pretreatment processes should be cost-effective. A range of pretreatment techniques ranging from alkali treatment to ammonia explosions have been developed to make the cellulose accessible for conversion to fuels by changing the structure of the lignocellulosic biomass and improving the hydrolysis rates. Hemicellulose can be readily hydrolyzed by dilute acids, but much more extreme conditions are required for cellulose hydrolysis. Pretreatment methods can be divided into different categories: physical (milling and grinding), physicochemical (steam pretreatment, hydrolysis, and wet oxidation), chemical (alkali, dilute acid, oxidizing agents, and organic solvents), biological, electrical, or a combination of these.

17.6.1 PHYSICAL PRETREATMENT

Mechanical pretreatment involves a combination of chipping, grinding, and/or milling that can be applied to reduce the crystal nature of cellulose and results in materials with a size range of approximately 10–30 mm after chipping and 0.2–2 mm after milling or grinding. Pyrolysis has also been used for the pretreatment of lignocellulosic materials in which cellulose rapidly decomposes to gaseous products and residual char when treated at temperatures greater than 300°C.

17.6.2 Physicochemical Pretreatment

Steam explosion is the most commonly used pretreatment method whereby biomass is treated with high-pressure saturated steam. This process involves the sudden reduction in pressure, which results in an explosive decompression of the biomass. This process is initiated at a temperature of 160°C–260°C for several seconds to a few minutes before the material is exposed to atmospheric pressure. The next step involves the biomass-steam mixture being held for a period of time to promote hemicelluloses hydrolysis followed by termination with an explosive decompression. There are a range of factors that affect steam explosion ranging from residence time, temperature, chip size, and moisture content (Duff and Murray, 1996). Removal of hemicelluloses from the microfibrils results in exposure of the cellulose surface, thereby increasing enzyme accessibility to these microfibrils. The removal of hemicellulose and lignin increases the volume of the pretreated sample. In the dilute-acid process, the reaction is carried out at high temperature and pressure. The dilute-acid hydrolysis process uses high temperatures ranging from 160°C to 230°C and pressures of approximately 10 atm. The acid concentration in the dilute-acid hydrolysis process ranges between 2% and 5% while more concentrated acid-hydrolysis processes utilize between 10% and 30%. The concentrated-acid hydrolysis involves lower operating temperatures and atmospheric pressures as well as longer retention times and results in higher ethanol yields than that obtained during the dilute-acid hydrolysis process.

17.7 RECENT DEVELOPMENTS IN BIOFUEL AND BIOCHEMICAL PRODUCTION IN AFRICA

17.7.1 African Biofuel Market

The African biofuel market is currently in its developmental stage. Work is currently being undertaken by most oil-producing countries to develop strategies, formulate legislation, and implement support policies that will help the industry develop and grow. However, biofuel production is still limited and is mostly restricted to ethanol production in Malawi and Zimbabwe. However, there are many small-scale projects currently ongoing in Africa. As can be expected, the volume of fuel currently being produced in Africa is negligible in the overall fuel market; however, there are many opportunities within African biofuel markets. A recent research conducted into biofuel markets in seven African countries found that all of them will become involved in the production of biofuels before the end of 2013. If the mandatory blending levels envisaged by each country were to be achieved, the picture by 2010 would be the following (Table 17.5):

TABLE 17.5
The Sub-Saharan Biofuels Markets: Market Potential for Selected African Countries (2010)

Country	B Level	E Level	Volume of Biofuel (2010)
Zimbabwe	B10	E10	217 million liters
Mozambique	B10	E10	57 million liters
Tanzania	B10	E10	161 million liters
Kenya	B10	E10	207 million liters
Malawi	B10	E10	47 million liters
Zambia	B10	E10	23 million liters
Ghana	B10	E10	116 million liters
Total			853 million liters

Source: Frost & Sullivan, *Biofuels Developments in Key African Countries,* Frost & Sullivan Ltd, London, U.K., 2007.

This forecast is based on current fuel demand, the fuel growth in each country, and expected government blending rates. From the forecast, it would appear that Zimbabwe holds the most potential. However, the current economic and political instability in this country mean that it may not be the ideal entry point into the African biofuel market.

The second largest market is in Kenya, and the government as well as the private sector (and NGO's) here have already made some significant progress in the development of sufficient feedstock. Amongst others, the Vanilla Foundation has already ensured that over 3000 ha of Jatropha have been planted, and the seeds are being used to grow new saplings for further distribution. The country envisages using the Nucleus Farming Approach, where small-scale farmers cooperate with commercial farmers to increase production and also transfer skills and equipment.

Tanzania is also a significant consumer of fuel in the sub-Saharan African region. It is expected that the country will use sugarcane and sweet sorghum to produce ethanol and *Jatropha* to produce biodiesel. Fuel imports into Tanzania currently account for as much as 40% of all imports (in value), and hence, the government set up a biofuels task force to spearhead the development of the industry in 2006. It is envisaged that current sugarcane plantations will increase from 23,000 ha in 2006 to 39,000 ha in 2013. *Jatropha* is also being planted and over 100,000 ha of land could be used for this. This would make Tanzania one of the largest biofuels feedstock producers.

The Ghanaian government is taking a cautious approach to biofuels. The country has experienced hype and bust agro industries before, and the government is hesitant to overinvest in an industry that is only forming. However, Ghana has significant potential in terms of feedstock production in sugarcane and palm oil and has formed strong links with other countries, such as Brazil, to ensure that the biofuels industry develops smoothly.

Mozambique, Malawi, and Zambia all use less fossil fuel, and therefore, the production targets set by their governments are lower than other countries. However, all three of these countries have significant potential to develop and supply feedstock to the regional biofuels industry. In Mozambique and Malawi, the potential to increase sugarcane production is significant, and both governments are keen to have increased economic activity at grassroots level. In Zambia, there is good potential to grow both sugarcane and *Jatropha*, and a lot of attention is currently being focused on this country. Currently, only 15% of the 25 million hectares of arable land in Zambia is being used for food crop production. It is expected that the country will be a major feedstock provider for the global biofuel market over the next 10–15 years by making use of this potential.

At present, the only investments in biofuel production that are actively producing commercial biofuels are in Malawi and Zimbabwe. The Zimbabwe operation is however currently undergoing a much-needed upgrade, although funding is a problem. In Malawi, the volume of ethanol being produced is sufficient for local demand, and some ethanol is even exported to be used in the food industry in neighboring countries. Many feasibility studies are being carried out to determine the most appropriate location, feedstock production models, and distribution networks to establish biofuels on the continent. Legislative uncertainty is responsible for the delay observed.

17.7.2 Biofuel Activities in Africa and Bioeconomic Prosperity Potential

Like food security, energy security is crucial in sustaining development and technological progress in Africa. Some 39 least developed countries [LDCs] in the African continent face the never-ending problems of energy and food insecurities. Traditional biomass fuels available on an insecure and nonsustainable basis help meet the energy and food cooking requirements of virtually all African LDCs in Africa (Table 17.6). The various biofuel activities taking place in Africa and the corresponding socioeconomic benefits and opportunities are summarized in Table 17.7 (Pillay and Da Silva, 2009).

TABLE 17.6

Food Situation in African Regions

Region	Percentage Undernourished			
	1969–1971	1979–1981	1990–1992	1996–1998
A: *Percentage of Population Undernourished in African Regions*				
Sub-Saharan Africa	34	37	35	34
Near East and North Africa	25	9	8	10

B: *African Countries Obtaining 70% or More of the Diet from Cereals, Roots, and Tubers in 1996–1998*
Central Africa: Democratic Republic of Congo (75)
Eastern Africa: Eritrea (78); Ethiopia (79)
West Africa: Benin (74); Burkina Faso (75); Ghana (75); Mali (73); Niger (74); Togo (77)
Southern Africa: Lesotho (80); Madagascar (74); Malawi (74); Namibia (79); Zambia (79)

Source: FAO Corporate Repository—Assessment of The World Food Security Situation, 27th Session, Rome, May 28– June 1, 2001, http://www.fao.org/docrep/meeting/003/Y0147E/Y0147E00.htm

17.7.3   KEY BARRIERS PREVENTING THE DEVELOPMENT OF BIOFUEL PRODUCTION IN AFRICA

In many regions, traditional biomass resources are readily available for the local population; however, their production and use causes a variety of negative impacts—such as overuse of natural resources leading to deforestation and adverse health effects due to indoor air pollution (IAP) in the poorer areas of developing countries as summarized in Table 17.8.

17.7.4   BIOFUELS IN AFRICA: STATUS, POLICIES, AND PERSPECTIVES

Many African countries are currently engaging in the establishment and development of suitable policies and strategies to ensure efficient bioenergy applications that aid economic development. The Competence Platform on Energy Crop and Agroforestry Systems for Arid and Semi-arid Ecosystems—Africa (COMPETE) project is one such development that aims to stimulate sustainable bioenergy implementation in Africa by providing support for policy developments, aiding organization of policy workshops, and the development of policy recommendation documents, which highlight Africa's approach to implementing sustainable bioenergy. The COMPETE partnership comprises 20 European and 23 non-European partners, 11 of which are from 7 African countries.

The main aim of COMPETE is to carry out efficient assessment of current land use and energy demands as well as to identify pathways for the sustainable provision of bioenergy in Africa. This itself aims to improve the quality of life and create a source of income for rural populations in Africa as well as aid the dissemination of knowledge between the EU and other developing countries.

The COMPETE Conference and Policy Debate on "Biofuels Sustainability Schemes—An African Perspective" on June 16–18, 2008 in Arusha, Tanzania, brought together more than 60 high-level participants including decision makers from several African countries. The main aim of this conference was to elaborate recommendations addressing the opportunities and challenges of global bioenergy development from an African perspective. The *COMPETE Declaration on Sustainable Bioenergy for Africa* was elaborated in cooperation with high-level decision-makers from Kenya, Mozambique, Tanzania, Uganda, and Zambia (COMPETE Declaration on Sustainable Bioenergy, 2008).

TABLE 17.7
Biofuel Activities in Africa

Country	Activity/Goal	Socioeconomic Benefits and Opportunities
Angola	Joint Brazilian (Petrobras)—Italian (Eni Spa)—Angola initiative tapping vast potential of biofuel production	Institutes new source of income and job opportunities for Angolan farmers and researchers *via* South–North–South cooperation
	Three-partner cooperative benefits are as follows: reinforcement of South–South shared technical knowledge, North–South teamwork ensuring biofuel security via Euro-African-South American partnership, and elimination of Angolan rural poverty	Brazilian partner provides technical expertise; Italian partner provides financial resources for construction of biodiesel plant and Angolan partner provides investment openings for biofuel production and export to financial partner
	With Portugal a biodiesel plant near Ambriz in the Bengo province using oil from the African oil palm is scheduled to start in 2008	Opening of new labor markets and rural industries
Burkina Faso	Burkina Faso, landlocked nation in West Africa enters partnership with Taiwan, China for the production of sweet sorghum-based ethanol	Development of exchange program to train Burkinabese students in Taiwan to use sweet sorghum for ethanol production
Cameroon	Oil palm production from *Elaeis guinea* to develop a new biofuel foreign-exchange market	Development of oil palm plantations for generation of income and new labor markets
Democratic Republic of Congo	Biodiesel production from palm oil is envisaged through a Spanish enterprise *Aurantia* to build four oil palm mills	Production of biodiesel from palm oil in line with feasibility studies dealing with logistics, labor opportunities, and environment-friendly infrastructural inputs
Ethiopia	Ethiopia-U.K. collaboration from 2005 results in the first operational *Jatropha* nursery	Reduction in rural poverty through creation of employment opportunities
Equatorial Guinea	Development of methanol plant on Bioko island	Production at a gas-fired power station to supply high grade electricity
Ghana	Oil palm production—establishment of the Bio-fuel Implementation Committee in 2005	Focus on development of regulatory protocols for biofuel production, use of biodiesel and bio-alcohol as substitutes for fossil-fuel imports
Kenya	In 2006, half a million *Jatropha* seedlings planted in the Eastern, Rift Valley, and Nyanza Provinces	Focus on halting the onset of desertification and Aiding reforestation activities
	Fossil-fuel substitution in Kenya using biogas	Development of capacity building in fuel substitution through a microfinancing approach in Kenyan schools
Mali	Mali-FolkCenter Fuels from Agriculture in Communal Technology (FACT Foundation) in Eindhoven, Netherlands; and the Regional Economic Commission for West Asia (ECOWAS)	Women produce locally biodiesel to run *posho* (corn meal) mills and produce soaps, cosmetics, and biofertilizers to strengthen rural bioeconomic prosperity
Malawi	The Biodiesel Agricultural Association encourages farmers to strengthen the energy sector through the planting of *Jatropha curcas*	Benefits: savings in foreign exchange expenses, minimization of pollution of the environment, and instituting employment and income generation activities

(continued)

TABLE 17.7 (continued)
Biofuel Activities in Africa

Country	Activity/Goal	Socioeconomic Benefits and Opportunities
Mozambique	Joint Mozambique-Norway collaboration in 2006 on African Green Conference *Mozambique BioFuels* highlights Mozambique strategic location as gateway to the landlocked countries of Malawi, Swaziland, Zimbabwe and Zambia	Focus on production of bioethanol from sugarcane; of biodiesel from copra seed oil, cotton seed oil, sunflower seed oil, and *Jatropha curcas* to reduce foreign exchange expenses; minimize pollution of the environment; and expand labor and income generation activities
Namibia	A Bio-Oil Energy committee oversees plantation of about 63,000 ha of the *Jatropha* bush by 2013	Development of an environmental-friendly biofuel
Niger	Reclamation of desert land through innovative agricultural practices	Simple innovative village agricultural practices fuel rural energy and feed markets
South Africa	Development of a *Jatropha curcas* nursery to produce biodiesel in the North West Province	Reduce dependence on high cost fossil fuels in the transportation sector
	On-stream production by *Ethanol Africa* of biofuel at South Africa's first bioethanol plant in Bothaville in the Free State built by a German enterprise	
	Development of new technology to produce biodiesel	Use of algae as a feedstock for biofuel production
	Organization of *"African Biofuels"* conferences aimed at helping the African continent on the merits and demerits of biofuel use as an alternative to reliance on fossil-fuel imports	State-of-the art and awareness conferences on the beneficial uses of biofuels (http://www.africanbiofuels.co.za/Press.pdf)
	Promotion of liquid biofuels in the City of Cape Town	Reduction in dependence on fossil-fuels in the transportation sector
	Fiat will launch its Brazilian-built flex-fuel Uno in South Africa in late 2007 which runs on gasoline, ethanol or a combination of the two fuels as displayed at the Durban Auto Show (Automotive World: March 27, 2007)	
Sudan	*Jatropha*—widely encountered in the Bahr El Gazal, Bahr El Jebel, Kassala, Khartoum, and Kordofan States	Mainly used as a medicinal plant due to its molluscicidal properties
Tanzania	Transition to *Jatropha* biofuels is in a nascent stage	Reduction of dependence on fossil-fuel imports and developing ancillary rural industries
Uganda	*Jatropha* grown in rural areas of the Karamoja region is an ideal bioresource for production of biodiesel; Feasibility study shows molasses are an uneconomical feedstock which is used to produce local gin—*Uganda Waragi*	Reinforces and improves women's welfare and their participation in rural governance Technical collaboration with India and Mali is foreseen
	A flower farm in Mukono in Central Uganda has begun producing biodiesel using *Jatropha curcas*	Raw materials used for soap and lubricant Manufacture offer new employment opportunities in financing Mukono's district markets
	Provision of energy services in rural areas	Facilitating access through biofuel energy use derived from Multi Functional Platforms

TABLE 17.7 (continued)
Biofuel Activities in Africa

Country	Activity/Goal	Socioeconomic Benefits and Opportunities
Zambia	In the central province of Mkushi, *Jatropha* is planted to produce biodiesel. About 300,000 small-scale farmers of the National Association for Peasant and Small-Scale Farmers of Zambia are expected to grow some 150,000 ha of *Jatropha*	Sustainable agriculture of *Jatropha* crops helps in poverty alleviation
	Promotion of bio-energy use in Zambia and establishment of a bio-oil processing plant in Kabwe aids national economic progress. Currently, a total of 400,000 *Jatropha* plants are being grown by 5,000 farmers in the Chibombo and Kapiri Mposhi districts	Self- empowerment for farmers, provision of new employment opportunities, and conservation of valuable foreign-exchange reserves
Zimbabwe	*Jatropha curcas* (*Mujirimono*) is a cash crop for biodiesel production	Alternative to fossil-fuel used in the transportation sector
	A National Biodiesel Feedstock Production Programme focuses on achieving self-sufficiency in use of biofuels through *Jatropha curcas* plantations	Improvement of Zimbabwe's energy security through Annual contribution of 360 000 ton of *Jatropha curcas* seeds for processing into 110 million L of biodiesel by 2010
	The National Oil Company of Zimbabwe (NOCZIM) is set to grow 25,000 *Jatropha* plant seedlings to distribute to farmers for the 2006–2007 season	Create national employment opportunities for research, production, and processing of biodiesel
Regional initiatives		
Common Market for Eastern and Southern Africa (COMESA)	Fourth COMESA Business Forum recommends investment in biofuels [*Final Communiqué, May 18–19, 2007; Document CBF/IPPSD/1, May 2007; Original English*]	Reduction in the dependence on fossil-fuels through use of liquid biofuels in the transportation sector
Economic Community of West African States (ECOWAS)	Focus on poverty reduction and the development of national and regional economic infrastructure	Focus on health, rural development, and small-scale enterprises that contribute to national and regional bioeconomic prosperity
Economic Community of Central African States (ECCAS)	Focus on creation of an awareness of the potential benefits of using clean and green biofuels	Focus on rural development and small-scale reduction in dependence on the use of fossil-derived fuels
Desert Margins Programme (DMP)	A collaborative effort convened by International Crops Research Institute for the Semi-Arid Tropics (ICRISAT) that unites nine African countries with desert margins that ring the heart of Africa: Botswana, Burkina Faso, Kenya, Mali, Namibia, Niger, Senegal, South Africa, and Zimbabwe	An answer to eradication of the twin scourges—*poverty and environmental degradation*
		Reinforcement of rural energy, food and market enterprises in disadvantaged and poverty-prone nonurban and nomadic communities

(*continued*)

TABLE 17.7 (continued)
Biofuel Activities in Africa

Country	Activity/Goal	Socioeconomic Benefits and Opportunities
Southern African Development Community (SADC)	Assessment of economic benefits accruing from production of biofuels in the SADC region—Angola, Botswana, Democratic Republic of the Congo (since September 8, 1997), Lesotho, Madagascar (since August 18, 2005), Malawi, Mauritius (since August 28, 1995), Mozambique, Namibia (since March 31, 1990), Seychelles (joined SADC on September 8, 1997 and left on July 1, 2004), Swaziland, South Africa (since August 30, 1994), Tanzania, Zambia, and Zimbabwe	Feasibility study for the production and use of biofuels—*Straight* or *Recycled* vegetable oils (SVO or RVO) in the SADC region as an answer to the reduction of dependence on high-cost imported fossil fuels (http://www.nab.com.na/jdocs/ biiofuels_study_final_report.pdf)

North–south and south–south cooperation

North–South Cooperation	Development of the biodiesel industry in Lesotho, Tanzania and Zambia	Development of small- and medium-scale industries; job opportunities; and acquisition of financing through the African Sustainable Fuels Centre, Cape Town, RSA
	India helps West Africa develop biofuels	Provision of US $250 million from India to boost biofuel production in 15 West African countries
	Potato Value Chain Development in West Africa (Guinea-Senegal)	Supported by Common Fund for Commodities, the potato sector in West Africa is integrated
	Green OPEC member countries are: Bénin, Burkina Faso, the Democratic Republic of the Congo, Gambia; Ghana; Guinée; Guinea-Bissau; Madagascar; Mali; Morocco; Niger; Sénégal; Sierra Leone; Togo; Zambia	into competitive markets. Creation of mechanisms to (1) share financial burdens of high oil prices in solidarity system that protects ongoing national projects aimed at socio-poverty alleviation and (2) institute financial governance of large-scale use of biofuels in African continent that facilitates African oil producing countries to invest in biofuels in non-oil producing African countries
	The island states—Mauritius, Madagascar and Reunion (French territory) in the Indian Ocean team up in a collaborative effort wherein Mauritius, Malaysia, and China provide the required *savoir-faire* and technical expertise while Madagascar and Réunion provide land for plantations	Reduction of crippling dependency on rising fossil-fuels oil costs that erode the bio-economic prosperity of these island states

Globalized cooperation

International Sugarcane Biomass Utilization Consortium (ISBUC)	Established after meetings in Mount Edgecombe, RSA (July 2006), and Alagoas, Brazil (November 2006), under umbrella of the International Society of Sugarcane Technologists with representatives from institutions in eight sugar-producing countries: Australia, Brazil, India, Mauritius, South Africa, Swaziland, Thailand, and the United States	Promote use of sugarcane biomass residual products for production of bioenergy and biofuels

Source: Adapted from Pillay, D. and DaSilva, E.J., *Afri. J. Biotechnol.*, 8, 2397, 2009.

TABLE 17.8

Key Barriers Preventing the Development of Biofuel Production in Africa

Barrier	Short Term (1–2 Years)	Medium Term (3–4 Years)	Long Term (5–7 Years)
Sufficient feedstock production	High	High	Medium
Food versus fuel			
Hunger in Africa remains a problem			
Water shortage widely experienced			
Biofuel plant construction	High	High	High
Equipment remains expensive and skills are in short supply			
Government policies and legislation	High	Medium	Medium
No real commitment as yet (no mandatory blending as yet)			
At present no subsidies or support			
Supply chain management	High	High	Medium
Ensuring a continuous feedstock supply			
Fuel distribution (poor infrastructure)			
Production plants close to feedstocks very important			

The following visions were identified to provide the guiding principles for bioenergy policy development in African countries:

- Rural development and improved livelihoods for the rural population in African countries
- Increased energy access and income generation opportunities
- Sustainable large-scale production of biofuels involving communities, smallholders, cooperatives, local enterprises, and foreign investors
- Modernization of agricultural practices and sustainable soil and land management to exploit complementarities of food and bioenergy production

17.7.5 Biofuels: The New Scramble for Africa: Gebremedhine Birega, African Biodiversity Network

Biofuel production is growing in sub-Saharan Africa. The "Green OPEC" is a group of 15 African countries (including: Benin, Burkina Faso, The Democratic Republic of Congo, Gambia, Ghana, Guinee, Guinea-Bissau, Madagascar, Mali, Morocco, Niger, Senegal, Sierra Leone, Togo, and Zambia) that have identified a desire to expand in biofuel production. These biofuels are different than previous "bioenergy" production in Africa (i.e., energy produced from biomass such as wood or charcoal). Many foreign companies are coming into Africa to clear forests or take over crop-producing land for biodiesel production. Notably, more and more Asian companies (from India, China, Malaysia, and Japan) are investing in this area. Africa is considered a good environment for biofuel production because of available land space, favorable climate and "cheap" labor. Moreover, African governments have been providing foreign companies incentives to invest in the region.

Unfortunately, most African countries lack regulatory frameworks to monitor this new sector geared for exports. This vacuum allows for situations such as in Ghana, where a corporation illegally seized 38,000 ha of land for biofuel production and only backed down when civil society made it a major public issue, or in Ethiopia, where 10,000 ha were cleared, of which 86% were part of an elephant sanctuary (Frost & Sullivan, 2007).

17.7.6 POLICY SITUATION IN WEST AFRICA: BABACAR NDAO, ROPPA

In West Africa, governments are incorporating biofuels into their agriculture and energy policies. This is the case in Mali, where the recently adopted *Loi d'orientation agricole*, or National Agriculture Strategy, focuses on biofuel production to meet rural energy needs. In Niger, a coordination structure at the ministerial level has recently been put in place. Senegal has created a Ministry for Biofuels and Renewable Energy. The West African Economic and Monetary Union (UEMOA) have a regional agricultural policy that addresses bioenergy, but it ignores many of the key challenges Africa faces today. The Economic Community of West African States' (ECOWAS) agriculture policy ignores energy challenges altogether. West Africa is a particularly sensitive area with regard to competition for land and water. Investment in agriculture has been extremely low in the past three decades, mainly due to the economic policy conditions imposed by international financial institutions. While the demand for energy in the region is huge, the energy needs in urban areas are prioritized.

Biofuels could help to provide energy for rural Africa, but it will require significant investment and careful planning. The balance between energy security and food security is sensitive. ROPPA believes that food security must be prioritized and that food crops must be given most attention. While it is also true that there might be opportunities to explore nonfood crops that could be grown on marginal land, such as *Jatropha*, there is a need for more research in this area.

17.7.7 OTHER AFRICAN GOVERNMENT INITIATIVES AND BIOFUELS PROGRAMME

The Government of Mozambique has constantly been promoting the introduction of biofuels in order to save foreign currency, to reduce environmental problems, to reduce dependence on unpredictable and volatile market oil prices, and to contribute to rural development by creating employment opportunities. Furthermore, the Government of Mozambique supports a range of biofuel-based rural electrification projects and places high priority on increasing access of energy to a range of rural areas. In addition to this, a biofuel commercialization program (PCB) has been established to purchase ethanol and biodiesel for blending with fossil fuels. In terms of biofuels export, Mozambique will act as exporter of processed biofuels in order to enable local producers to add value to their current production.

The Government of Tanzania aims to contribute to the replacement of fossil transport fuels and to stimulate socioeconomic development through rural electrification projects, which are further enhanced due to the suitable climate as well as available arable land and water resources within Tanzania. Currently, the Government of Tanzania is engaged in the implementation of a biofuels action plan, which involves a thorough review of existing policies as well as legal and regulatory frameworks with the aim to develop new national bioenergy policies. The current range of projects include the *Jatropha* plantations for biofuel production, which is being managed by the German company Prokon in the Mpanda region as well as a range of rural electrification projects being run multifunctional platforms (MFP) coordinated by the Tanzanian NGO. The government in Zambia is also undertaking a range of bioenergy policies that aim to ensure the sustainable exploitation of biomass resources. Activities undertaken include the utilization of appropriate cost-effective equipment, which will stimulate biomass production as well as the implementation of public awareness campaigns.

17.8 FUTURE DIRECTION AND CONCLUDING REMARKS

The African biofuels industry holds significant potential as a result of the possibility of producing feedstocks as well as growing internal demand. However, participants have to be aware of the challenges in the market and develop mitigating strategies in order to increase their chances for success. Biofuels could help economic growth in Africa and contribute to poverty alleviation and rural development. For example, biofuel production resulting from *Jatropha* is projected to increase both production and reduce poverty due to increased use of unskilled labor as well as due to the constant increase of land out-sourcing on a small scale as opposed to large plantation owners (Smeets et al., 2007). However, the issue of food security needs to be addressed, and the use of nonfood crops for biofuel production should be encouraged. The large majority of the population in sub-Saharan Africa, especially in the rural areas, depends on traditional biomass for cooking and heating (Human Development Report, 2003). Furthermore, by 2030, over 700 million people will rely on traditional biomass such as charcoal, firewood, and agricultural wastes as well as animal and human wastes (IEA World Energy Outlook, 2006). Another concern is that the expansion of biofuels may have negative impacts on biodiversity and the use of natural resources through increasing competition over land and water resources, hence the need for proper implementation of policies and development plans to ensure socially, economically, and environmentally sustainable bioenergy production in Africa.

Governments have many roles to play and finding the right balance between development, support, and independence is essential. Legislation to develop the industry is critical and is the only way to assure investors, farming communities, and equipment suppliers that the market for biofuels in Africa is real, sustainable, and profitable. Without this legislation, it can be expected that African countries will become feedstock providers for other countries and that the true value of biofuels will be lost. Subsidy levels, potential participants, and mandatory blending levels have to be spelt out. Currently, several African governments are in the process of developing bioenergy policies and implementation strategies. Ideally, the legislation being developed should clearly stipulate the following: What level of government support will be provided within the full value chain (from farmer to end consumer); The procedures to qualify for a license and the responsibilities of producers; The level of local representation needed for foreign companies; To what extent profits can be repatriated; Who may produce feedstocks, where and the pricing models to be used to ensure prices remain stable (either removed from food production or else in a noncompetitive manner); and the mandatory blending levels and time frames for implementation (Frost & Sullivan, 2007).

Future research into the extensive water usage and expenditure during biofuel production is needed as certain feed stocks such as sugarcane, typically requires irrigation and therefore has implications on the much limited water resources. Further assessments with regard to increases in transport and infrastructure costs due to biofuel production are also needed (Arndt et al., 2008).

Government support in the form of incentives, tax relief, and subsidies has been identified as a major factor that will influence bioethanol production in Africa in a positive manner. In addition, mandatory blending targets and defined price and quality for bioethanol production would aid lowering the cost of biofuel production. Due to increased concerns over food security, there has been a significant interest in nonfood crops for biofuel production such as the *Jatropha* plant. These nonfood crops can be grown on marginal lands in unfavorable conditions, and their large production should be encouraged.

REFERENCES

Achten, W.M.J., L. Verchot, Y.J. Franken, E. Mathijs, V.P. Singh, R. Aerts, and B. Muys. 2008. *Jatropha* biodiesel production and use. *Biomass Bioenergy* 32:1063–1084.

Antizar-Ladislao, B. and J.L. Turrion-Gomez. 2008. Second generation biofuels and local bioenergy systems. *Biofuels, Bioproducts Biorefin.* 2:455–469.

Arndt, C., R. Benfica, F. Tarp, J. Thurlow, and R. Uaiene. 2008. Biofuels, poverty, and growth: A computable general equilibrium analysis of mozambique. Int. Food Policy Res. Inst. Discussion Paper: 00803.

Atsumi, S. and J.C. Liao. 2008. Metabolic engineering for advanced biofuels production from *Escherichia coli*. *Curr. Opin. Biotech.* 19:414–419.

Augustus, G.D.P.S., M. Jayabalan, and G.J. Seiler. 2002. Evaluation and bioinduction of energy components of *Jatropha curcas*. *Biomass Bioenergy* 23:161–164.

Banks, D. and J. Schäffler. 2006. The potential contribution of renewable energy in South Africa. Draft Update Report. Sustainable Energy and Climate Change Project (SECCP). RAPS Consulting Pty Ltd, Wellington, NZ.

BioDiesel Processor. http://biodieselprocessor.org/Biodiesel_System.htm (accessed on 15 October 2010).

Bozbas, K. 2005. Biodiesel as an alternative motor fuel: Production and policies in The European Union. *Renew. Sust. Energy Rev.* 12:542–552.

Campell, C.J. 1995. The coming crisis. *Sun World* 19:16–19.

Chakauya, E. and P. Tongoona. 2008. Analysis of genetic relationships of pearl millet (*Pennisetum glaucum* L.) landraces from Zimbabwe using microsatellites. *Acad. J.* 2:1–7.

Chen, C.Y., M.H. Yang, K.L. Yeh, C.H. Liu, and J.S. Chang. 2008. Biohydrogen production using sequential two-stage dark and photo fermentation processes. *Int. J. Hydrogen Energy* 33:4755–4762.

Coelho, S.T. 2005. Biofuels—advantages and trade barriers. *United Nations Conference on Trade and Development*, pp. 3–28 (accessed on 17 November 2010).

COMPETE. 2008. Competence platform on energy crop and agroforestry systems for arid and semi-arid eco-systems—Africa. *International Conference and Policy Debate on Bioenergy Sustainability Schemes—An African Perspective*. Declaration on Sustainable Bioenergy for Africa. Final Version.

Dincer, I. 1998. Energy and environmental impacts: Present and future perspectives. *Energy Sources* 20:427–453.

Dincer, I. and M.A. Rosen. 1999. Energy, environment and sustainable development. *Appl. Energy* 64:427–440.

Drapcho, C., J. Nghiem, and T. Walker. 2008. Biofuel feedstocks, in: *Biofuels Engineering Process Technology*, Chapter 4, pp. 69–78. New York: The McGraw Hill Companies.

Duff, S.J.B. and W.D. Murray. 1996. Bioconversion of forest products industry waste cellulosics to fuel ethanol: A review. *Biores. Techn.* 55:1–33.

Eberhard, A.A. and A. Williams. 1988. *Renewable Energy Resources and Technology Development in South Africa*. Cape Town, South Africa: Elan Press.

EuropaBio Factsheet—Biofuels and Food. 2007. The European Association for Bioindustries. Biofuels and Developing Countries.

FAO Corporate Repository—Assessment of the world food security situation, 27th Session, Rome, May 28–June 1, 2001, http://www.fao.org/docrep/meeting/003/Y0147E/Y0147E00.htm

FAPRI—Agricultural Outlook. 2007. The European Association for Biofuels. EuropaBio Fact Sheet: Biofuels and Food. www.fapri.org/outlook2007 (accessed on 25 November 2010).

Frost & Sullivan. 2007. *Biofuels Developments in Key African Countries*. London, U.K.: Frost & Sullivan Ltd.

Hahn-Hägerdal, B., M. Galbe, M.F. Gorwa-Grauslund, G. Lidén, and G. Zacchi. 2006. Bio-ethanol—The fuel of tomorrow from the residues of today. *Trends Biotech.* 24:549–556.

Hettenhaus, J.R. 1998. Ethanol fermentation strains present and future requirements for biomass to ethanol commercialization. United States Department of Energy (Office of Energy Efficiency & Renewable Energy Ethanol Program) and National Renewable Energy Laboratory, pp. 1–25.

Hill, J., E. Nelson, D. Tilman, S. Polasky, and D. Tiffany. 2006. Environmental, economic, and energetic costs and benefits of biodiesel and ethanol biofuels. *Proc. Natl. Acad. Sci.* 103:11206–11210.

Hoogwijk, M., A. Faaij, B. deVries, and W. Turkenburg. 2009. Exploration of regional and global cost—Supply curves of biomass energy from short-rotation crops at abandoned cropland and rest land under four IPCC SRES land-use scenarios. *Biomass Bioenergy* 33:26–43.

Human Development Report, 2003, UNDP, New York, United Nations Development Programme.

IEA (International Energy Agency) 2009. World Energy Outlook 2009, OECD/IEA, Paris, France.

IEA (International Energy Agency) 2003. World Energy Outlook 2003, OECD/IEA, Paris, France.

IEA Energy Technology Essentials—Biofuel Production. 2007. www.iea.org/Textbase/techno/essentials.htm (accessed on 25 November 2010).

Jannson, C., A. Westerbergh, J. Zhang, X. Hu, and C. Sun. 2009. Cassava, a potential biofuel crop in the People Republic of China. *Appl. Energy* 86:595–599.

Jannssen R., D. Rutz, P. Helm, J. Woods, and R.D. Chavez. 2005. *Bioenergy for Sustainable Development in Africa—Environmental and Social Aspects*. Munich, Germany: WIP Renewable Energies.

Kim, S. and B.E. Dale. 2004. Global potential of bioethanol production from wasted crops and crop residues. *Biomass Bioenergy* 26:361–375.

Kojima, M. and T. Johnson. 2006. Potential for biofuels for transport in developing countries. *Knowledge Exchange Ser.* 4:1–4.

Kotay, S.M. and D. Das. 2008. Biohydrogen as a renewable energy resource—Prospects and potentials. *Int. J. Hydrogen Energy* 33:258–263.

Levin, D.B., L. Pitt, and M. Love. 2004. Biohydrogen production: Prospects and limitations to practical application. *Int. J. Hydrogen Energy* 29:173–185.

Li, Y., M. Horseman, N. Wu, C.Q. Lan, and N. Dubois-Calero. 2008. Biofuels from microalgae. *Biotechnol. Prog.* 24:815–820.

Lightfoot, H.D. and C. Green. 1992. The dominance of fossil fuels: Technical and resource limitations to alternative energy sources. C^2GCR Report No. 92-6.

Logan, B.E. and J.M. Regan. 2006. Microbial challenges and fuel cell applications. *Environ. Sci. Technol.* 40:5172–5180

Ma, F. and M.A. Hanna. 1999. Biodiesel production: A review. *Biores. Technol.* 70:1–15.

Manish, S. and R. Banerjee. 2007. Comparison of biohydrogen production processes. *Int. J. Hydrogen Energy* 33:279–286.

O'Hair, S.K., R.B. Forbes, S.J. Locascio, J.R. Rich, and R.L. Stanley. 1983. Cassava root starch content and distribution varies with tissue age. *Hort. Sci.* 18:735–737.

Openshaw, K. 2005. A review of *Jatropha curcas*: An oil plant of unfulfilled promise. *Biomass Bioenergy* 19:1–15.

Pillay, D. and E.J. DaSilva. 2009. Sustainable development and bioeconomic prosperity in Africa: Biofuels and the South African gateway. *Afri. J. Biotechnol.* 8:2397–2408.

Rabaey, K., W. Ossieur, M. Verhaege, and W. Verstraete. 2005. Continuous microbial fuel cells converts carbohydrates to electricity. *Water Sci. Technol.* 52:515–523.

Rabaey, K. and W. Versraete. 2005. Microbial fuel cells: Novel biotechnology for energy generation. *Trends Biotechnol.* 23:291–298.

RFA (Reneweables Fuel Association). 2009a. Statistics—2008 world fuel ethanol production. Available at: http://www.ethanolrfa.org/industry/statistics/ (accessed on 5 December 2010).

RFA (Reneweables Fuel Association) 2009b. "Ethanol facts agriculture: feeding the world, fueling a nation." Available at: http://www.ethanolrfa.org/resource/facts/agriculture (accessed on 5 December 2010).

Rutz, D. and R. Janssen. 2007. *Biofuel Technology Handbook*. Munich, Germany: WIP Renewable Energies.

Sankula, S. 2006. Executive Summary: Quantification of the impacts on US agriculture of biotechnology-derived crops planted in 2005. National Center for Food and Agricultural Policy, Washington, DC. Available at: http://www.ncfap.org (accessed on 1 December 2010).

Sexton, S.E., L.A. Martin, and D. Zilberman. 2006. Biofuel and biotech: A sustainable energy solution. *Update Agric. Resour. Econ.* 9:1–4. University of California Giannini Foundation.

Shay, E.G. 1993. Diesel fuel from vegetable oils: Status and opportunities. *Biomass Bioenergy* 4:227–242.

Smeets, E.M.W., A.P.C. Faaij, I.M. Lewandowski, and W.C. Turkenburg. 2007. A bottom-up assessment and review of global bio-energy potentials to 2050. *Prog. Energy Combus. Sci.* 33:56–106.

Sparks, D.L. 1992. *Advances in Agronomy*, Vol. 48. Newark, CA: Academic Press.

Steinberg, M. 1999. Fossil fuel decarbonization technology for mitigating global warming. *Int. J. Hydrogen Energy* 24:771–777.

Sustainable Design Update. 2008. *Jatropha*: The biodiesel plant. http://sustainabledesignupdate.com /2008/09/jatropha-the-biodiesel-plant (accessed on 10 November 2010).

Van Gerpen, J. 2005. Biodiesel processing and production. *Fuel Process. Technol.* 86:1097–1107.

Watson H.K. 2008. COMPETE: Competence platform on energy crop and agroforestry systems for arid and semi-arid ecosystems—Africa. Second Task Report on WP1 Activities—Current land use patterns and impacts. Sixth Framework Programme.

Watson, H.K. 2009. Potential to expand sustainable bioenergy from sugarcane in southern Africa. *Energy Policy J.—Sustainabil. Biofuels.*

Wood, P. 2005. Out of Africa: Could *Jatropha* vegetable oil be Europe's biodiesel feedstock? *Biofuels* 6:40–44.

Wuebbles, D.J. and A.K. Jain. 2001. Concerns about climate change and the role of fossil fuel use. *Fuel Process. Technol.* 71:99–119.

Wyman, C.E. 1996. *Handbook on Bioethanol: Production and Utilization* Applied Energy Technology Series. Washington, DC: Taylor & Francis.

Yu, J. and P. Takahashi. 2007. Biophotolysis-based hydrogen production by cyanobacteria and green microalgae. In *Communicating Current Research and Educational Topics and Trends in Applied Microbiology*, ed. A. Méndez-Vilas, pp. 79–89. Badajoz, España: Formatex.

Index

Printed and bound by CPI Group (UK) Ltd, Croydon, CR0 4YY

21/10/2024

01777105-0016